Student's Solutions Manual

to accompany

Barnett • Ziegler • Byleen

College Algebra

Fred Safier

City College of San Francisco

Special thanks to Jeanne Wallace for the preparation of this manuscript.

Boston   Burr Ridge, IL   Dubuque, IA   Madison, WI   New York   San Francisco   St. Louis
Bangkok   Bogotá   Caracas   Lisbon   London   Madrid
Mexico City   Milan   New Delhi   Seoul   Singapore   Sydney   Taipei   Toronto

WCB/McGraw-Hill

A Division of The McGraw-Hill Companies

Student Solutions Manual to accompany
COLLEGE ALGEBRA

3 4 5 6 7 8 9 0 QPD/QPD 3 2 1 0 9

ISBN 0-07-365584-8

http://www.mhhe.com

TABLE OF CONTENTS

Suggestions for Success in Algebra

1. Go to class, participate in the class discussions, and don't hesitate to ask questions.

2. Do all assigned homework, as soon as possible after class. Try each assigned problem on your own. If you need help, first check the text for similar examples. Once you have found an answer, check it with the answer in the back of the text. If you have difficulty, study the solution given here, then try to work the problem again without looking at the solution.

3. Use the answers in the texts and the solutions in this manual as guides. However, do not always try to work toward the answer in the back of the text. You should be able to do problems without looking at the answers. On a test (and in real life) you are seldom provided with the answers, so you cannot work toward them.

4. At the end of each chapter, work the Chapter Review exercises. These exercises provide an excellent review for a test. In areas of weakness, return to an appropriate section in the chapter for a refresher.

5. If you are still having difficulty, be sure to talk to your instructor and to use any tutorial help that is available to you. However, do not let a tutor do the work for you.

6. Take all tests and quizzes that the instructor gives.

This manual contains solutions for all odd-numbered problems in the text exercises plus all problems in the Chapter Review exercises. The Chapter Review exercises are keyed to corresponding text sections.

CHAPTER 1

Exercise 1-1

Key Ideas and Formulas

Basic Properties of the Set of Real Numbers

Let R be the set of real numbers and x, y and z arbitrary elements of R.

ADDITION PROPERTIES

CLOSURE:	$x + y$ is a unique element in R.
ASSOCIATIVE:	$(x + y) + z = x + (y + z)$
COMMUTATIVE:	$x + y = y + x$
IDENTITY:	0 is the additive identity; that is, $0 + x = x + 0 = x$ for all x in R, and 0 is the only element in R with this property.
INVERSE:	For each x in R, $-x$ is its unique additive inverse; that is, $x + (-x) = (-x) + x = 0$, and $-x$ is the only element in R relative to x with this property.

MULTIPLICATION PROPERTIES

CLOSURE:	xy is a unique element in R.
ASSOCIATIVE:	$(xy)z = x(yz)$
COMMUTATIVE:	$xy = yx$
IDENTITY:	1 is the multiplicative identity; that is $1x = x1 = x$, and 1 is the only element in R with this property.
INVERSE:	For each x in R, $x \neq 0$, $\frac{1}{x}$ is its unique multiplicative inverse; that is, $x(1/x) = (1/x)x = 1$, and $\frac{1}{x}$ is the only element in R relative to x with this property.

COMBINED PROPERTY

DISTRIBUTIVE:

$x(y + z) = xy + xz$ $\qquad$ $(x + y)z = xz + yz$

Properties of Negatives:

For all real numbers a and b

1. $-(-a) = a$
2. $(-a)b = -(ab) = a(-b) = -ab$
3. $(-a)(-b) = ab$
4. $(-1)a = -a$
5. $\dfrac{-a}{b} = -\dfrac{a}{b} = \dfrac{a}{-b}$ $\qquad b \neq 0$
6. $\dfrac{-a}{-b} = -\dfrac{-a}{b} = -\dfrac{a}{-b} = \dfrac{a}{b}$ $\qquad b \neq 0$

Zero Properties:

For all real numbers a and b

1. $a \cdot 0 = 0$
2. $ab = 0$ if and only if $a = 0$ or $b = 0$ or both.

---

1. Since 4 is an element of {3,4,5}, the statement is true. T

3. Since 3 is an element of {3,4,5}, the statement is false. F

5. The statement is false, since not every element of {1,2} is an element of {1,3,5}. 2 is an element of {1,2}, but not an element of {1,3,5}. F

7. Since each element of the set {7,3,5} is also an element of the set {3,5,7}, the statement is true. T

9. The commutative property (+) states that, in general, $x + y = y + x$. Comparing this with $x + 7 = ?$, we see that 7 takes the place of y. Hence, $x + 7 = 7 + x$. $7 + x$

11. The associative property ($\cdot$) states that, in general, $(xy)z = x(yz)$. Comparing this with $x(yz) = ?$, we see that $x(yz) = (xy)z$. $(xy)z$

13. The identity property (+) states that, in general, $0 + x = x + 0 = x$. Comparing this with $0 + 9m = ?$ we see that $9m$ takes the place of x. Hence, $0 + 9m = 9m$. $9m$

15. Commutative ($\cdot$) $ym = my$ is a special case of $xy = yx$.

17. Distributive. $7u + 9u = (7 + 9)u$ is a special case of $xz + yz = (x + y)z$

19. Inverse($\cdot$). $(-2)(\frac{1}{-2}) = 1$ is a special case of $x(\frac{1}{x}) = 1$.

21. Inverse (+). $w + (-w) = 0$ is a special case of $x + (-x) = 0$.

23. Identity (+). $3(xy + z) + 0 = 3(xy + z)$ is a special case of $x + 0 = x$.

25. Negatives. $\frac{-x}{-y} = \frac{x}{y}$ is a special case of $\frac{-a}{-b} = \frac{a}{b}$. (Theorem 1, Part 6)

27. The even integers between -3 and 5 are -2, 0, 2, and 4. Hence, the set is written $\{-2,0,2,4\}$.

29. The letters in "status" are: s,t,a,u. Hence, the set is written $\{s,t,a,u\}$, or, equivalently, $\{a,s,t,u\}$.

31. Since there are no months starting with B, the set is empty. $\emptyset$

33. Commutative (+). $(3x + 5) + 7 = 7 + (3x + 5)$ is a special case of $x + y = y + x$.

35. Associative (+). $(3x + 2) + (x + 5) = 3x + [2 + (x + 5)]$ is a special case of $(x + y) + z = x + (y + z)$

37. Distributive. $x(x - y) + y(x - y) = (x + y)(x - y)$ is a special case of $xz + yz = (x + y)z$

39. Zero. $(2x - 3)(x + 5) = 0$ if and only if $2x - 3 = 0$ or $x + 5 = 0$ is a special case of $ab = 0$ if and only if $a = 0$ or $b = 0$. (Theorem 2, Part 2)

41. Yes. This restates Zero Property (2). (Theorem 2, Part 2)

43. (A) True.
(B) False, $\frac{2}{3}$ is an example of a real number that is not irrational.
(C) True.

45. $\frac{3}{5}$ and -1.43 are two examples of infinitely many.

47. (A) -3 is an integer. It belongs to Z, therefore also to Q and R.
(B) 3.14 is a rational number ($\frac{314}{100}$). It belongs to Q, therefore also to R.
(C) π is an irrational number. It belongs to R.
(D) $\frac{2}{3}$ is a rational number. It belongs to Q, therefore also to R.

49. (A) 0.88888888...   (B) 0.27272727...   (C) 2.23606797...   (D) 1.37500000...

51. (A) True; commutative property for addition.
(B) False; for example $3 - 5 \neq 5 - 3$.
(C) True; commutative property for multiplication.
(D) False; for example $9 \div 3 \neq 3 \div 9$.

53. (A) List each element of A. Follow these with each element of B that is not yet listed. $\{1,2,3,4,6\}$
(B) List each element of A that is also an element of B $\{2,4\}$.

55.   Let c = 0.090909…

Then $100c$ = 9.0909…

$100c - c$ = (9.0909…) - (0.090909…)

$99c$ = 9

$c = \dfrac{9}{99} = \dfrac{1}{11}$

57.

$$
\begin{array}{rl}
23 & \\
\underline{12} & \\
46 & 23 \cdot 2 \\
\underline{230} & 23 \cdot 10 \\
276 &
\end{array}
$$

$23 \cdot 12 = 23(2 + 10)$

$= 23 \cdot 2 + 23 \cdot 10$

$= 46 + 230$

$= 276$

Exercise 1-2

Key Ideas and Formulas

For n a natural number and a any real number, $a^n = \underbrace{a \cdot a \cdots a}_{n \text{ factors}}$

First Property of Exponents: For any natural numbers m and n, and a any real number: $a^m a^n = a^{m+n}$.

A polynomial in x is an algebraic expression of the form

$$a_n x^n + a_{n-1} x^{n-1} + \cdots + a_1 x + a_0$$

where the coefficients a_0, a_1, …, a_n are real numbers and n is a non-negative integer.

Two terms in a polynomial are called like terms if they have exactly the same variable factors to the same powers. Like terms in a polynomial are combined by adding their numerical coefficients.

$$a - b = a + (-b)$$

Distributive Properties:

$$a(b + c) = (b + c)a = ab + ac$$
$$a(b - c) = (b - c)a = ab - ac$$
$$a(b + c + \cdots + f) = ab + ac + \cdots + af$$

FOIL method:

	F First Product	**O** Outer Product	**I** Inner Product	**L** Last Product
$(a + b)(c + d) =$	ac +	ad +	bc +	bd

Special Products:

$$(a - b)(a + b) = a^2 - b^2$$
$$(a + b)^2 = a^2 + 2ab + b^2$$
$$(a - b)^2 = a^2 - 2ab + b^2$$

1. 3

3. $(2x^3 - 3x^2 + x + 5) + (2x^2 + x - 1)$

$= 2x^3 - 3x^2 + x + 5 + 2x^2 + x - 1$

$= 2x^3 - x^2 + 2x + 4$

5. $(2x^3 - 3x^2 + x + 5) - (2x^2 + x - 1)$

$= 2x^3 - 3x^2 + x + 5 - 2x^2 - x + 1$

$= 2x^3 - 5x^2 + 6$

Common Error:
Reversing terms: $(2x^2 + x - 1) - (2x^3 - 3x^2 + x + 5)$
Should write: (first) - (second). In this case, $(a) - (b)$.

Common Error:
$2x^3 - 3x^2 + x + 5 - 2x^2 + x - 1$
Should change the sign of each term in the second parentheses: $-(2x^2 + x - 1) = -2x^2 - x + 1$

7.
$$2x^3 - 3x^2 + x + 5$$
$$3x - 2$$
$$\overline{}$$
$$6x^4 - 9x^3 + 3x^2 + 15x$$
$$- 4x^3 + 6x^2 - 2x - 10$$
$$\overline{}$$
$$6x^4 - 13x^3 + 9x^2 + 13x - 10$$

9. $2(x - 1) + 3(2x - 3) - (4x - 5)$
$$= 2x - 2 + 6x - 9 - 4x + 5$$
$$= 4x - 6$$

11. $2y - 3y[4 - 2(y - 1)]$
$$= 2y - 3y[4 - 2y + 2]$$
$$= 2y - 3y[6 - 2y]$$
$$= 2y - 18y + 6y^2$$
$$= -16y + 6y^2$$
$$= 6y^2 - 16y$$

13. $(m - n)(m + n) = m^2 - n^2$

15.

	First Product	Outer Product	Inner Product	Last Product
$(4t - 3)(t - 2) = 4t^2$		$-8t$	$-3t$	$+6$
$= 4t^2 - 11t + 6$				

17.

	First Product	Outer Product	Inner Product	Last Product
$(3x + 2y)(x - 3y) = 3x^2$		$-9xy$	$+2xy$	$-6y^2$
$= 3x^2 - 7xy - 6y^2$				

19. $(2m - 7)(2m + 7) = (2m)^2 - (7)^2 = 4m^2 - 49.$

21.

	First Product	Outer Product	Inner Product	Last Product
$(6x - 4y)(5x + 3y) = 30x^2$		$+18xy$	$-20xy$	$-12y^2$
$= 30x^2 - 2xy - 12y^2$				

23. $(3x - 2y)(3x + 2y) = (3x)^2 - (2y)^2$
$$= 9x^2 - 4y^2$$

25. $(4x - y)^2 = (4x - y)(4x - y)$
$$= (4x)^2 - 2(4x)(y) + (y)^2$$
$$= 16x^2 - 8xy + y^2$$

27.
$$a^2 - ab + b^2$$
$$a + b$$
$$\overline{}$$
$$a^3 - a^2b + ab^2$$
$$a^2b - ab^2 + b^3$$
$$\overline{}$$
$$a^3 + b^3$$

29. $2x - 3\{x + 2[x - (x + 5)] + 1\}$
$$= 2x - 3\{x + 2[x - x - 5] + 1\}$$
$$= 2x - 3\{x + 2[-5] + 1\}$$
$$= 2x - 3\{x - 10 + 1\}$$
$$= 2x - 3\{x - 9\}$$
$$= 2x - 3x + 27$$
$$= -x + 27$$

31. $2\{3[a - 4(1 - a)] - (5 - a)\}$
$$= 2\{3[a - 4 + 4a] - 5 + a\}$$
$$= 2\{3[5a - 4] - 5 + a\}$$
$$= 2\{15a - 12 - 5 + a\}$$
$$= 2\{16a - 17\}$$
$$= 32a - 34$$

33.
$$2x^2 + x - 2$$
$$x^2 - 3x + 5$$
$$\overline{}$$
$$2x^4 + x^3 - 2x^2$$
$$- 6x^3 - 3x^2 + 6x$$
$$10x^2 + 5x - 10$$
$$\overline{}$$
$$2x^4 - 5x^3 + 5x^2 + 11x - 10$$

35.
$$h^2 + hk + k^2$$
$$h^2 - hk + k^2$$
$$\overline{}$$
$$h^4 + h^3k + h^2k^2$$
$$- h^3k - h^2k^2 - hk^3$$
$$h^2k^2 + hk^3 + k^4$$
$$\overline{}$$
$$h^4 + h^2k^2 + k^4$$

37. $(2x - 1)^2 - (3x + 2)(3x - 2) = (2x - 1)^2 - [(3x + 2)(3x - 2)]$
$$= (2x)^2 - 2(2x)(1) + 1^2 - [(3x)^2 - (2)^2]$$
$$= 4x^2 - 4x + 1 - [9x^2 - 4]$$
$$= 4x^2 - 4x + 1 - 9x^2 + 4$$
$$= -5x^2 - 4x + 5$$

39. $(m - 3n)(m + 8n) + (m + 6n)(m + 4n) = m^2 + 8mn - 3mn - 24n^2 + m^2 + 4mn + 6mn + 24n^2$
$$= 2m^2 + 15mn$$

41. $(2m - n)^3 = (2m - n)(2m - n)(2m - n)$
$$= [(2m - n)(2m - n)](2m - n)$$
$$= [(2m)^2 - 2(2m)(n) + n^2](2m - n)$$
$$= [4m^2 - 4mn + n^2](2m - n)$$

The last multiplication is best performed vertically, so we write:

$$4m^2 - 4mn \quad + n^2$$
$$\underline{2m - \quad n}$$
$$8m^3 - \quad 8m^2n + 2mn^2$$
$$\underline{- \quad 4m^2n + 4mn^2 - n^3}$$
$(2m - n)^3 = \ 8m^3 - 12m^2n + 6mn^2 - n^3$

43. $5(x + h) - 4 - (5x - 4) = 5x + 5h - 4 - 5x + 4 = 5h$

45. $3(x + h)^2 + 2(x + h) - (3x^2 + 2x)$
$$= 3(x^2 + 2xh + h^2) + 2x + 2h - 3x^2 - 2x$$
$$= 3x^2 + 6xh + 3h^2 + 2x + 2h - 3x^2 - 2x$$
$$= 6xh + 3h^2 + 2h$$

> **Common Error:**
> $3(x + h)^2 \neq (3x + 3h)^2$
> Exponentiation must be performed before multiplication.

47. $-2(x + h)^2 - 3(x + h) + 7 - (-2x^2 - 3x + 7)$
$$= -2(x^2 + 2xh + h^2) - 3x - 3h + 7 + 2x^2 + 3x - 7$$
$$= -2x^2 - 4xh - 2h^2 - 3x - 3h + 7 + 2x^2 + 3x - 7$$
$$= -4xh - 2h^2 - 3h$$

49. $(x + h)^3 = (x + h)(x + h)(x + h)$
$$= [(x + h)(x + h)](x + h)$$
$$= [x^2 + 2xh + h^2](x + h)$$

The last multiplication is best performed vertically, so we write:
$$x^2 + 2xh + h^2$$
$$\underline{x + \quad h}$$
$$x^3 + 2x^2h + \quad xh^2$$
$$\underline{\quad\quad x^2h + 2xh^2 + h^3}$$
$$x^3 + 3x^2h + 3xh^2 + h^3$$

Therefore $(x + h)^3 - x^3 = x^3 + 3x^2h + 3xh^2 + h^3 - x^3 = 3x^2h + 3xh^2 + h^3$

51. The sum of the first two polynomials:
$(3m^2 - 2m + 5) + (4m^2 - m) = 3m^2 - 2m + 5 + 4m^2 - m = 7m^2 - 3m + 5$

The sum of the last two polynomials:
$(3m^2 - 3m - 2) + (m^3 + m^2 + 2) = 3m^2 - 3m - 2 + m^3 + m^2 + 2 = m^3 + 4m^2 - 3m$

Subtract from the sum of the last two polynomials the sum of the first two.
$$(m^3 + 4m^2 - 3m) - (7m^2 - 3m + 5) = m^3 + 4m^2 - 3m - 7m^2 + 3m - 5$$
$$= m^3 - 3m^2 - 5$$

53. Since $(x - 2)^2 = (x)^2 - 2(x)(2) + (2)^2 = x^2 - 4x + 4$,
$(x - 2)^3 = (x - 2)^2(x - 2) = (x^2 - 4x + 4)(x - 2)$

$$
\begin{array}{r}
x^2 - 4x + 4 \\
x - 2 \\
\hline
x^3 - 4x^2 + 4x \\
-2x^2 + 8x - 8 \\
\hline
x^3 - 6x^2 + 12x - 8
\end{array}
$$

Therefore $2(x - 2)^3 - (x - 2)^2 - 3(x - 2) - 4$
$$= 2(x^3 - 6x^2 + 12x - 8) - (x^2 - 4x + 4) - 3(x - 2) - 4$$
$$= 2x^3 - 12x^2 + 24x - 16 - x^2 + 4x - 4 - 3x + 6 - 4$$
$$= 2x^3 - 13x^2 + 25x - 18$$

55. We will multiply $(x + 2)(x^2 - 3)$ using the FOIL method.

	First Product	Outer Product	Inner Product	Last Product
$(x + 2)(x^2 - 3) = x^3$		$-3x$	$+2x^2$	-6
or $x^3 + 2x^2 - 3x - 6$				

Then $-3x\{x[x - x(2 - x)] - (x + 2)(x^2 - 3)\}$
$$= -3x\{x[x - 2x + x^2] - [x^3 + 2x^2 - 3x - 6]\}$$
$$= -3x\{x[-x + x^2] - [x^3 + 2x^2 - 3x - 6]\}$$
$$= -3x\{-x^2 + x^3 - x^3 - 2x^2 + 3x + 6\}$$
$$= -3x\{-3x^2 + 3x + 6\}$$
$$= 9x^3 - 9x^2 - 18x$$

57. One example is given by choosing $a = 1$ and $b = 1$. Then $(a + b)^2 = (1 + 1)^2 = 2^2 = 4$, but $a^2 + b^2 = 1^2 + 1^2 = 1 + 1 = 2$. Thus, in general, $(a + b)^2 \neq a^2 + b^2$. In fact, since $(a + b)^2 = a^2 + 2ab + b^2$, this quantity can only equal $a^2 + b^2$ if $2ab = 0$. By the properties of 0, either $a = 0$ or $b = 0$. In these cases $(a + b)^2$ would equal $a^2 + b^2$, but only in these.

59. The non-zero term with the highest degree in the polynomial of degree m has degree m. This term is not changed during the addition, since there is no term in the polynomial of degree n with the degree $m (m > n)$. It will still be the highest degree term in the sum, so the sum will be a polynomial of degree m.

61. Now the non-zero term with the highest degree in each polynomial has degree m. It is possible for the coefficients of these terms to be equal in absolute value but opposite in sign, in which case they would add to 0, leaving a term of degree less than m as the term with the highest degree in the sum. Otherwise the terms of degree m will combine to give another term of degree m, which will be the highest degree term in the sum. Thus, $(2x^4 + x^2 + 1) + (-2x^4 + x^2 + 1) = 2x^2 + 2$, but $(2x^4 + x^2 + 1) + (2x^4 + 2x^2 + 1) = 4x^4 + 3x^2 + 2$.
Summarizing, the degree of the sum may be less than or equal to m.

63. There are three quantities in this problem, perimeter, length, and width. They are related by the perimeter formula $P = 2\ell + 2w$. Since x = length of the rectangle, and the width is 5 meters less than the length, $x - 5$ = width of the rectangle. So $P = 2x + 2(x - 5)$ represents the perimeter of the rectangle. Simplifying: $P = 2x + 2x - 10 = 4x - 10$ (meters)

65. There are several quantities involved in this problem. It is important to keep them distinct by using enough words. We write:

$$x = \text{number of nickels}$$
$$x - 5 = \text{number of dimes}$$
$$(x - 5) + 2 = \text{number of quarters}$$

This follows because there are five fewer dimes than nickels $(x - 5)$ and 2 more quarters than dimes (2 more than $x - 5$). Each nickel is worth 5 cents, each dime worth 10 cents, and each quarter worth 25 cents. Hence the value of the nickels is 5 times the number of nickels, the value of the dimes is 10 times the number of dimes, and the value of the quarters is 25 times the number of quarters.

$$\text{value of nickels} = 5x$$
$$\text{value of dimes} = 10(x - 5)$$
$$\text{value of quarters} = 25[(x - 5) + 2]$$

The value of the pile = (value of nickels) + (value of dimes)
$$+ \text{(value of quarters)}$$
$$= 5x + 10(x - 5) + 25[(x - 5) + 2]$$

Simplifying this expression, we get:

$$\begin{aligned}
\text{The value of the pile} &= 5x + 10x - 50 + 25[x - 5 + 2] \\
&= 5x + 10x - 50 + 25[x - 3] \\
&= 5x + 10x - 50 + 25x - 75 \\
&= 40x - 125 \text{ (cents)}
\end{aligned}$$

67. The volume of the plastic shell is equal to the volume of the larger sphere ($V = \frac{4}{3}\pi r^3$) minus the volume of the hole. Since the radius of the hole is x cm and the plastic is 0.3 cm thick, the radius of the larger sphere is $x + 0.3$ cm. Thus, we have

$$\begin{pmatrix} \text{Volume of} \\ \text{shell} \end{pmatrix} = \begin{pmatrix} \text{Volume of} \\ \text{larger sphere} \end{pmatrix} - \begin{pmatrix} \text{Volume of} \\ \text{hole} \end{pmatrix}$$

$$\begin{aligned}
\text{Volume} &= \tfrac{4}{3}\pi(x + 0.3)^3 - \tfrac{4}{3}\pi x^3 \\
&= \tfrac{4}{3}\pi(x + 0.3)(x + 0.3)^2 - \tfrac{4}{3}\pi x^3 \\
&= \tfrac{4}{3}\pi(x + 0.3)(x^2 + 0.6x + 0.09) - \tfrac{4}{3}\pi x^3 \\
&= \tfrac{4}{3}\pi(x^3 + 0.6x^2 + 0.09x + 0.3x^2 + 0.18x + 0.027) - \tfrac{4}{3}\pi x^3 \\
&= \tfrac{4}{3}\pi(x^3 + 0.9x^2 + 0.27x + 0.027) - \tfrac{4}{3}\pi x^3 \\
&= \tfrac{4}{3}\pi x^3 + 1.2\pi x^2 + 0.36\pi x + 0.036\pi - \tfrac{4}{3}\pi x^3 \\
&= 1.2\pi x^2 + 0.36\pi x + 0.036\pi \text{ (cm}^3)
\end{aligned}$$

Exercise 1-3

Key Ideas and Formulas

$$ab + ac = a(b + c)$$
$$u^2 + 2uv + v^2 = (u + v)^2 \qquad \text{Perfect square}$$
$$u^2 - 2uv + v^2 = (u - v)^2 \qquad \text{Perfect square}$$
$$u^2 - v^2 = (u - v)(u + v) \qquad \text{Difference of squares}$$
$$u^3 - v^3 = (u - v)(u^2 + uv + v^2) \qquad \text{Difference of cubes}$$
$$u^3 + v^3 = (u + v)(u^2 - uv + v^2) \qquad \text{Sum of cubes}$$

Common Errors:

Confusing $u^2 + v^2$ (a prime polynomial) with
$$(u + v)^2 = (u + v)(u + v) = u^2 + 2uv + v^2 \text{ (a perfect square).}$$
Confusing $u^2 - v^2 = (u - v)(u + v)$ (difference of squares) with
$$(u - v)^2 = (u - v)(u - v) = u^2 - 2uv + v^2 \text{ (a perfect square).}$$

---

1. $2x^2(3x^2 - 4x - 1)$ **3.** $5xy(2x^2 + 4xy - 3y^2)$ **5.** $(x + 1)(5x - 3)$ **7.** $(y - 2z)(2w - x)$

9. $\begin{aligned}[t] x^2 - 2x + 3x - 6 &= (x^2 - 2x) + (3x - 6) \\ &= x(x - 2) + 3(x - 2) \\ &= (x - 2)(x + 3) \end{aligned}$

11. $\begin{aligned}[t] 6m^2 + 10m - 3m - 5 &= (6m^2 + 10m) - (3m + 5) = 2m(3m + 5) - 1(3m + 5) \\ &= (3m + 5)(2m - 1) \end{aligned}$

13. $\begin{aligned}[t] 2x^2 - 4xy - 3xy + 6y^2 &= (2x^2 - 4xy) - (3xy - 6y^2) = 2x(x - 2y) - 3y(x - 2y) \\ &= (x - 2y)(2x - 3y) \end{aligned}$

15. $8ac + 3bd - 6bc - 4ad = 8ac - 4ad - 6bc + 3bd$
$$= (8ac - 4ad) - (6bc - 3bd)$$
$$= 4a(2c - d) - 3b(2c - d)$$
$$= (2c - d)(4a - 3b)$$

17. $(2x - 1)(x + 3)$ **19.** $(x - 6y)(x + 2y)$ **21.** Prime   **23.** $(5m + 4n)(5m - 4n)$

25. $x^2 + 10xy + 25y^2 = (x + 5y)(x + 5y) = (x + 5y)^2$ **27.** Prime

29. $6x^2 + 48x + 72 = 6(x^2 + 8x + 12) = 6(x + 2)(x + 6)$

31. $2y^3 - 22y^2 + 48y = 2y(y^2 - 11y + 24) = 2y(y - 3)(y - 8)$

33. $16x^2y - 8xy + y = y(16x^2 - 8x + 1) = y(4x - 1)^2$ **35.** $(3s - t)(2s + 3t)$

37. $x^3y - 9xy^3 = xy(x^2 - 9y^2) = xy(x - 3y)(x + 3y)$ **39.** $3m(m^2 - 2m + 5)$

41. $(m + n)(m^2 - mn + n^2)$ **43.** $(c - 1)(c^2 + c + 1)$

45. $6(3x - 5)(2x - 3)^2 + 4(3x - 5)^2(2x - 3) = 2(3x - 5)(2x - 3)[3(2x - 3) + 2(3x - 5)]$
$$= 2(3x - 5)(2x - 3)[6x - 9 + 6x - 10]$$
$$= 2(3x - 5)(2x - 3)(12x - 19)$$

47. $5x^4(9 - x)^4 - 4x^5(9 - x)^3 = x^4(9 - x)^3[5(9 - x) - 4x]$
$$= x^4(9 - x)^3[45 - 5x - 4x]$$
$$= x^4(9 - x)^3(45 - 9x) \quad \text{or} \quad 9x^4(9 - x)^3(5 - x)$$

49. $2(x + 1)(x^2 - 5)^2 + 4x(x + 1)^2(x^2 - 5) = 2(x + 1)(x^2 - 5)[x^2 - 5 + 2x(x + 1)]$
$$= 2(x + 1)(x^2 - 5)[x^2 - 5 + 2x^2 + 2x]$$
$$= 2(x + 1)(x^2 - 5)(3x^2 + 2x - 5)$$
$$= 2(x + 1)(x^2 - 5)(3x + 5)(x - 1)$$

51. $(a - b)^2 - 4(c - d)^2 = (a - b)^2 - [2(c - d)]^2$
$$= [(a - b) - 2(c - d)][(a - b) + 2(c - d)]$$

53. $2am - 3an + 2bm - 3bn = (2am - 3an) + (2bm - 3bn) = a(2m - 3n) + b(2m - 3n)$
$$= (2m - 3n)(a + b)$$

55. Prime

57. $x^3 - 3x^2 - 9x + 27 = (x^3 - 3x^2) - (9x - 27) = x^2(x - 3) - 9(x - 3) = (x - 3)(x^2 - 9)$
$$= (x - 3)(x - 3)(x + 3) = (x - 3)^2(x + 3)$$

59. $a^3 - 2a^2 - a + 2 = (a^3 - 2a^2) - (a - 2) = a^2(a - 2) - 1(a - 2) = (a - 2)(a^2 - 1)$
$$= (a - 2)(a + 1)(a - 1)$$

61. $4(A + B)^2 - 5(A + B) - 6 = [4(A + B) + 3][(A + B) - 2]$

63. $m^4 - n^4 = (m^2)^2 - (n^2)^2 = (m^2 - n^2)(m^2 + n^2) = (m - n)(m + n)(m^2 + n^2)$

65. $s^4t^4 - 8st = st(s^3t^3 - 8) = st[(st)^3 - 2^3] = st(st - 2)[(st)^2 + (st)(2) + (2)^2]$
$$= st(st - 2)[s^2t^2 + 2st + 4]$$

67. $m^2 + 2mn + n^2 - m - n = (m^2 + 2mn + n^2) - (m + n) = (m + n)^2 - 1(m + n)$
$$= (m + n)(m + n - 1)$$

69. $18a^3 - 8a(x^2 + 8x + 16) = 2a[9a^2 - 4(x^2 + 8x + 16)] = 2a\{(3a)^2 - [2(x + 4)]^2\}$
$$= 2a[3a - 2(x + 4)][3a + 2(x + 4)]$$

71. $x^4 + 2x^2 + 1 - x^2 = (x^4 + 2x^2 + 1) - x^2 = (x^2 + 1)^2 - x^2$
$$= (x^2 + 1 - x)(x^2 + 1 + x) = (x^2 - x + 1)(x^2 + x + 1)$$

73. (A) The area of the cardboard can be written as (Original area) - (Removed area),
where the original area = $20^2 = 400$ and the removed area consists of 4 squares
of area x^2 each; thus,
(Original area) - (Removed area) = $400 - 4x^2$ in expanded form.
In factored form, $400 - 4x^2 = 4(100 - x^2) = 4(10 - x)(10 + x)$.

(B) See figure.

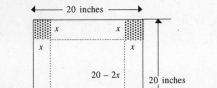

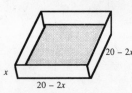

The volume of the box = ℓwh = $x(20 - 2x)(20 - 2x)$
$= 4x(10 - x)(10 - x)$ in factored form

In expanded form, $4x(10 - x)(10 - x) = 4x[10^2 - 2(10)x + x^2]$
$= 4x[100 - 20x + x^2]$
$= 400x - 80x^2 + 4x^3$

Exercise 1-4

Key Ideas and Formulas

Fundamental Property of Fractions

If a, b, and k are real numbers with b, $k \neq 0$, then
$$\frac{ak}{bk} = \frac{a}{b} \qquad \text{Also,} \quad \frac{a}{b} = \frac{ak}{bk}$$

Multiplication and Division

For a, b, c, and d, real numbers,
$$\frac{a}{b} \cdot \frac{c}{d} = \frac{ac}{bd} \qquad\qquad b, \ d \neq 0$$

$$\frac{a}{b} \div \frac{c}{d} = \frac{a}{b} \cdot \frac{d}{c} \qquad\qquad b, \ c, \ d \neq 0$$

Addition and Subtraction

For a, b, and c real numbers
$$\frac{a}{b} + \frac{c}{b} = \frac{a + c}{b} \qquad\qquad b \neq 0$$

$$\frac{a}{b} - \frac{c}{b} = \frac{a - c}{b} \qquad\qquad b \neq 0$$

Common Errors:
There are many possible errors in applying the fundamental property of fractions.
Always be sure that each quantity removed is a factor of the entire numerator and a
factor of the entire denominator. Examples of errors abound, but here are a few
classic types to avoid:

$\left.\begin{array}{l} \dfrac{m + 2}{m - 3} \neq \dfrac{2}{-3} \\[2em] \dfrac{x^2 + 2x - 15}{x^2 + 4x - 5} \neq \dfrac{2x - 15}{4x - 5} \end{array}\right\}$ cannot "cancel" common terms

$\left.\dfrac{x + 7}{ax - bx} \neq \dfrac{7}{a - bx}\right\}$ cannot "cancel" a term against a factor of a term

$\left.\begin{array}{l} \dfrac{x - (x - 4)}{(x - 4)(x + 4)} \neq \dfrac{x}{x + 4} \\[2em] \dfrac{x^2 + 8}{8x^3} \neq \dfrac{x^2}{x^3} \end{array}\right\}$ cannot "cancel" a term against a factor

1. $\left(\dfrac{d^5}{3a} \div \dfrac{d^2}{6a^2}\right) \cdot \dfrac{a}{4d^3} = \left(\dfrac{d^5}{3a} \cdot \dfrac{6a^2}{d^2}\right) \cdot \dfrac{a}{4d^3} = \left(\dfrac{\cancel{d^5}^{d^3}}{\cancel{3a}_1} \cdot \dfrac{\cancel{6a^2}^{2a}}{\cancel{d^2}_1}\right) \cdot \dfrac{a}{4d^3} = \dfrac{2ad^3}{1} \cdot \dfrac{a}{4d^3} = \dfrac{\cancel{2ad^3}^1}{1} \cdot \dfrac{a}{\cancel{4d^3}_2} = \dfrac{a^2}{2}$

3. $\dfrac{2y}{18} - \dfrac{-1}{28} - \dfrac{y}{42} = \dfrac{28y}{252} - \dfrac{-9}{252} - \dfrac{6y}{252} = \dfrac{22y + 9}{252}$ **Note:** $\left.\begin{array}{l} 18 = 2\cdot3\cdot3 \\ 28 = 2\cdot2\cdot7 \\ 42 = 2\cdot3\cdot7 \end{array}\right\}$ so LCD $= 2\cdot2\cdot3\cdot3\cdot7 = 252$

5. $\dfrac{3x + 8}{4x^2} - \dfrac{2x - 1}{x^3} - \dfrac{5}{8x} = \dfrac{2x(3x + 8)}{8x^3} - \dfrac{8(2x - 1)}{8x^3} - \dfrac{x^2(5)}{8x^3}$

$\qquad = \dfrac{6x^2 + 16x - 16x + 8 - 5x^2}{8x^3} = \dfrac{x^2 + 8}{8x^3}$

7. $\dfrac{2x^2 + 7x + 3}{4x^2 - 1} \div (x + 3) = \dfrac{2x^2 + 7x + 3}{4x^2 - 1} \div \dfrac{(x + 3)}{1} = \dfrac{(2x + 1)(x + 3)}{(2x + 1)(2x - 1)} \cdot \dfrac{1}{(x + 3)}$

$\qquad = \dfrac{\cancel{(2x + 1)}^1\,\cancel{(x + 3)}^1}{\cancel{(2x + 1)}_1(2x - 1)} \cdot \dfrac{1}{\cancel{(x + 3)}_1} = \dfrac{1}{2x - 1}$

9. $\dfrac{m + n}{m^2 - n^2} \div \dfrac{m^2 - mn}{m^2 - 2mn + n^2} = \dfrac{m + n}{m^2 - n^2} \cdot \dfrac{m^2 - 2mn + n^2}{m^2 - mn}$

$\qquad = \dfrac{m + n}{(m + n)(m - n)} \cdot \dfrac{(m - n)(m - n)}{m(m - n)}$

$\qquad = \dfrac{\cancel{(m + n)}^1}{\cancel{(m + n)}_1 \cancel{(m - n)}_1} \cdot \dfrac{\cancel{(m - n)}^1 \cancel{(m - n)}^1}{m \cancel{(m - n)}_1} = \dfrac{1}{m}$

11. $\dfrac{1}{a^2 - b^2} + \dfrac{1}{a^2 + 2ab + b^2} = \dfrac{1}{(a + b)(a - b)} + \dfrac{1}{(a + b)(a + b)}$

$\qquad = \dfrac{a + b}{(a + b)(a + b)(a - b)} + \dfrac{a - b}{(a + b)(a + b)(a - b)}$

$\qquad = \dfrac{a + b + a - b}{(a + b)(a + b)(a - b)} = \dfrac{2a}{(a + b)^2(a - b)}$

13. $m - 3 - \dfrac{m - 1}{m - 2} = \dfrac{m - 3}{1} - \dfrac{m - 1}{m - 2} = \dfrac{(m - 3)(m - 2)}{m - 2} - \dfrac{(m - 1)}{m - 2}$

$\qquad = \dfrac{(m^2 - 5m + 6) - (m - 1)}{m - 2} = \dfrac{m^2 - 5m + 6 - m + 1}{m - 2} = \dfrac{m^2 - 6m + 7}{m - 2}$

15. $\dfrac{5}{x - 3} - \dfrac{2}{3 - x} = \dfrac{5}{x - 3} - \dfrac{-2}{x - 3} = \dfrac{5 + 2}{x - 3} = \dfrac{7}{x - 3}$

17. $\dfrac{2}{y + 3} - \dfrac{1}{y - 3} + \dfrac{2y}{y^2 - 9} = \dfrac{2(y - 3)}{(y + 3)(y - 3)} - \dfrac{(y + 3)\cdot 1}{(y + 3)(y - 3)} + \dfrac{2y}{(y + 3)(y - 3)}$

$\qquad = \dfrac{2(y - 3) - (y + 3) + 2y}{(y + 3)(y - 3)} = \dfrac{2y - 6 - y - 3 + 2y}{(y + 3)(y - 3)}$

$\qquad = \dfrac{3y - 9}{(y + 3)(y - 3)} = \dfrac{3\cancel{(y - 3)}^1}{(y + 3)\cancel{(y - 3)}_1} = \dfrac{3}{y + 3}$

19. $\dfrac{1 - \dfrac{y^2}{x^2}}{1 - \dfrac{y}{x}} = \dfrac{x^2\left(1 - \dfrac{y^2}{x^2}\right)}{x^2\left(1 - \dfrac{y}{x}\right)} = \dfrac{x^2 - y^2}{x^2 - xy} = \dfrac{(x + y)\cancel{(x - y)}^{1}}{\underset{1}{x\cancel{(x - y)}}} = \dfrac{x + y}{x}$

21. $\dfrac{6x^3(x^2 + 2)^2 - 2x(x^2 + 2)^3}{x^4} = \dfrac{2x(x^2 + 2)^2[3x^2 - (x^2 + 2)]}{x^4}$

$= \dfrac{2x(x^2 + 2)^2[3x^2 - x^2 - 2]}{x^4}$

$= \dfrac{2\cancel{x}(x^2 + 2)^2(2x^2 - 2)}{\underset{x^3}{\cancel{x^4}}}$

$= \dfrac{2(x^2 + 2)^2(2x^2 - 2)}{x^3}$

$= \dfrac{2(x^2 + 2)^2 \, 2(x^2 - 1)}{x^3}$

$= \dfrac{4(x^2 + 2)^2(x + 1)(x - 1)}{x^3}$

23. $\dfrac{2x(1 - 3x)^3 + 9x^2(1 - 3x)^2}{(1 - 3x)^6} = \dfrac{x(1 - 3x)^2[2(1 - 3x) + 9x]}{(1 - 3x)^6}$

$= \dfrac{x(1 - 3x)^2[2 - 6x + 9x]}{(1 - 3x)^6}$

$= \dfrac{x\cancel{(1 - 3x)^2}^{1}(2 + 3x)}{\underset{(1 - 3x)^4}{\cancel{(1 - 3x)^6}}}$

$= \dfrac{x(2 + 3x)}{(1 - 3x)^4}$

25. $\dfrac{-2x(x + 4)^3 - 3(3 - x^2)(x + 4)^2}{(x + 4)^6} = \dfrac{(x + 4)^2[-2x(x + 4) - 3(3 - x^2)]}{(x + 4)^6}$

$= \dfrac{(x + 4)^2[-2x^2 - 8x - 9 + 3x^2]}{(x + 4)^6}$

$= \dfrac{\cancel{(x + 4)^2}^{1}(x^2 - 8x - 9)}{\underset{(x + 4)^4}{\cancel{(x + 4)^6}}}$

$= \dfrac{x^2 - 8x - 9}{(x + 4)^4}$

$= \dfrac{(x + 1)(x - 9)}{(x + 4)^4}$

27. $\dfrac{y}{y^2 - y - 2} - \dfrac{1}{y^2 + 5y - 14} - \dfrac{2}{y^2 + 8y + 7}$

$= \dfrac{y}{(y - 2)(y + 1)} - \dfrac{1}{(y + 7)(y - 2)} - \dfrac{2}{(y + 1)(y + 7)}$

$= \dfrac{y(y + 7)}{(y + 1)(y - 2)(y + 7)} - \dfrac{1(y + 1)}{(y + 1)(y - 2)(y + 7)} - \dfrac{2(y - 2)}{(y + 1)(y - 2)(y + 7)}$

$= \dfrac{y^2 + 7y - y - 1 - 2y + 4}{(y + 1)(y - 2)(y + 7)} = \dfrac{y^2 + 4y + 3}{(y + 1)(y - 2)(y + 7)} = \dfrac{\cancel{(y + 1)}^{1}(y + 3)}{\cancel{(y + 1)}_{1}(y - 2)(y + 7)}$

$= \dfrac{y + 3}{(y - 2)(y + 7)}$

29. $\dfrac{9 - m^2}{m^2 + 5m + 6} \cdot \dfrac{m + 2}{m - 3} = \dfrac{\cancel{(3 - m)}^{-1}(3 + m)^{1}}{\cancel{(m + 2)}_{1}\cancel{(m + 3)}_{1}} \cdot \dfrac{\cancel{m + 2}^{1}}{\cancel{m - 3}_{1}} = -1$

31. $\dfrac{x + 7}{ax - bx} + \dfrac{y + 9}{by - ay} = \dfrac{x + 7}{x(a - b)} + \dfrac{y + 9}{y(b - a)} = \dfrac{y(x + 7)}{xy(a - b)} + \dfrac{-x(y + 9)}{xy(a - b)}$

$= \dfrac{xy + 7y - xy - 9x}{xy(a - b)} = \dfrac{7y - 9x}{xy(a - b)}$

33. $\dfrac{x^2 - 16}{2x^2 + 10x + 8} \div \dfrac{x^2 - 13x + 36}{x^3 + 1} = \dfrac{x^2 - 16}{2x^2 + 10x + 8} \cdot \dfrac{x^3 + 1}{x^2 - 13x + 36}$

$= \dfrac{\cancel{(x - 4)}^{1}\cancel{(x + 4)}^{1}}{2\cancel{(x + 1)}_{1}\cancel{(x + 4)}_{1}} \cdot \dfrac{\cancel{(x + 1)}^{1}(x^2 - x + 1)}{\cancel{(x - 4)}_{1}(x - 9)} = \dfrac{x^2 - x + 1}{2(x - 9)}$

35. $\dfrac{x^2 - xy}{xy + y^2} + \left(\dfrac{x^2 - y^2}{x^2 + 2xy + y^2} \div \dfrac{x^2 - 2xy + y^2}{x^2y + xy^2} \right) = \dfrac{x^2 - xy}{xy + y^2} \div \left(\dfrac{x^2 - y^2}{x^2 + 2xy + y^2} \cdot \dfrac{x^2y + xy^2}{x^2 - 2xy + y^2} \right)$

$= \dfrac{x^2 - xy}{xy + y^2} \div \left(\dfrac{\cancel{(x - y)}^{1}\cancel{(x + y)}^{1}}{\cancel{(x + y)}_{1}\cancel{(x + y)}_{1}} \cdot \dfrac{xy\cancel{(x + y)}^{1}}{\cancel{(x - y)}_{1}(x - y)} \right)$

$= \dfrac{x^2 - xy}{xy + y^2} \div \dfrac{xy}{x - y} = \dfrac{x^2 - xy}{xy + y^2} \cdot \dfrac{x - y}{xy}$

$= \dfrac{\cancel{x}^{1}(x - y)}{y(x + y)} \cdot \dfrac{x - y}{\cancel{xy}_{1}} = \dfrac{(x - y)^2}{y^2(x + y)}$

37. $\left(\dfrac{x}{x^2 - 16} - \dfrac{1}{x + 4}\right) \div \dfrac{4}{x + 4} = \left(\dfrac{x}{(x - 4)(x + 4)} - \dfrac{1}{x + 4}\right) \div \dfrac{4}{x + 4}$

$= \left(\dfrac{x}{(x - 4)(x + 4)} - \dfrac{(x - 4)}{(x - 4)(x + 4)}\right) \div \dfrac{4}{x + 4}$

$= \dfrac{x - x + 4}{(x - 4)(x + 4)} \div \dfrac{4}{x + 4} = \dfrac{4}{(x - 4)(x + 4)} \div \dfrac{4}{x + 4}$

$= \dfrac{\overset{1}{\cancel{4}}}{(x - 4)\cancel{(x + 4)}} \cdot \dfrac{\cancel{x + 4}^{\,1}}{\underset{1}{\cancel{4}}} = \dfrac{1}{x - 4}$

39. $\dfrac{1 + \frac{2}{x} - \frac{15}{x^2}}{1 + \frac{4}{x} - \frac{5}{x^2}} = \dfrac{x^2\left(1 + \frac{2}{x} - \frac{15}{x^2}\right)}{x^2\left(1 + \frac{4}{x} - \frac{5}{x^2}\right)} = \dfrac{x^2 + 2x - 15}{x^2 + 4x - 5} = \dfrac{\cancel{(x + 5)}^{\,1}(x - 3)}{\underset{1}{\cancel{(x + 5)}}(x - 1)} = \dfrac{x - 3}{x - 1}$

41. $\dfrac{\frac{1}{x + h} - \frac{1}{x}}{h} = \dfrac{\frac{x}{x(x + h)} - \frac{x + h}{x(x + h)}}{h} = \dfrac{\frac{x - x - h}{x(x + h)}}{h} = \dfrac{\frac{-h}{x(x + h)}}{h} = \dfrac{-h}{x(x + h)} \div h$

$= \dfrac{-h}{x(x + h)} \div \dfrac{h}{1} = \dfrac{-\cancel{h}^{\,-1}}{x(x + h)} \cdot \dfrac{1}{\underset{1}{\cancel{h}}} = \dfrac{-1}{x(x + h)}$

43. $\dfrac{\frac{(x + h)^2}{x + h + 2} - \frac{x^2}{x + 2}}{h} = \dfrac{\frac{(x + h)^2(x + 2)}{(x + h + 2)(x + 2)} - \frac{x^2(x + h + 2)}{(x + h + 2)(x + 2)}}{h} = \dfrac{\frac{(x + h)^2(x + 2) - x^2(x + h + 2)}{(x + h + 2)(x + 2)}}{h}$

$= \dfrac{\frac{(x^2 + 2xh + h^2)(x + 2) - x^3 - x^2h - 2x^2}{(x + h + 2)(x + 2)}}{h}$

$= \dfrac{\frac{x^3 + 2x^2h + h^2x + 2x^2 + 4xh + 2h^2 - x^3 - x^2h - 2x^2}{(x + h + 2)(x + 2)}}{h}$

$= \dfrac{\frac{x^2h + h^2x + 4xh + 2h^2}{(x + h + 2)(x + 2)}}{h} = \dfrac{x^2h + h^2x + 4xh + 2h^2}{(x + h + 2)(x + 2)} \div h$

$= \dfrac{x^2h + h^2x + 4xh + 2h^2}{(x + h + 2)(x + 2)} \div \dfrac{h}{1} = \dfrac{\cancel{h}^{\,1}(x^2 + hx + 4x + 2h)}{(x + h + 2)(x + 2)} \cdot \dfrac{1}{\underset{1}{\cancel{h}}}$

$= \dfrac{x^2 + hx + 4x + 2h}{(x + h + 2)(x + 2)}$

45. (A) The solution is incorrect, because in the first step the quantity 4 has been removed from numerator and denominator. But 4 is a term, and only factors can be cancelled. A correct solution would factor numerator and denominator, then cancel common factors, if any.

(B) $\dfrac{x^2 + 5x + 4}{x + 4} = \dfrac{(x + 1)\cancel{(x + 4)}^{\,1}}{\underset{1}{\cancel{x + 4}}} = x + 1$

47. (A) The solution is incorrect, because in the first step the quantity h has been removed from numerator and denominator. Although h is a factor of the denominator it is not a factor of the numerator as shown and only common factors of numerator and denominator can be cancelled. A correct solution would factor numerator and denominator, revealing that h is actually a factor of the numerator in factored form and allowing a correct cancellation of h.

(B) $\dfrac{(x + h)^2 - x^2}{h} = \dfrac{[(x + h) - x][(x + h) + x]}{h}$ using the difference of two squares

$$= \dfrac{[x + h - x][x + h + x]}{h}$$

$$= \dfrac{\overset{1}{\cancel{h}}[2x + h]}{\underset{1}{\cancel{h}}}$$

$$= 2x + h$$

49. (A) The solution is incorrect. In adding a quantity to a fractional expression, the quantity cannot simply be placed in the numerator. A correct solution would rewrite $x - 2$ as a fractional expression, first with denominator 1, and then with denominator equal to the LCD of all denominators in the addition.

(B) $\dfrac{x^2 - 2x}{x^2 - x - 2} + x - 2 = \dfrac{x\overset{1}{\cancel{(x - 2)}}}{(x + 1)\underset{1}{\cancel{(x - 2)}}} + \dfrac{x - 2}{1}$

$$= \dfrac{x}{x + 1} + \dfrac{x - 2}{1}$$

$$= \dfrac{x}{x + 1} + \dfrac{(x - 2)(x + 1)}{x + 1}$$

$$= \dfrac{x + x^2 - x - 2}{x + 1}$$

$$= \dfrac{x^2 - 2}{x - 1}$$

51. (A)(B) The solution is correct.

53. $\dfrac{y - \dfrac{y^2}{y - x}}{1 + \dfrac{x^2}{y^2 - x^2}} = \dfrac{(y^2 - x^2)[y - \dfrac{y^2}{y - x}]}{(y^2 - x^2)[1 + \dfrac{x^2}{y^2 - x^2}]} = \dfrac{(y^2 - x^2)y - \overset{(y + x)}{\cancel{(y^2 - x^2)}}\dfrac{y^2}{\cancel{y - x}}}{y^2 - x^2 + \underset{1}{\cancel{(y^2 - x^2)}}\dfrac{x^2}{\cancel{y^2 - x^2}}}\ 1$

$$= \dfrac{y^3 - x^2y - y^3 - xy^2}{y^2 - x^2 + x^2} = \dfrac{-x^2y - xy^2}{y^2} = \dfrac{-x\overset{1}{\cancel{y}}(x + y)}{\underset{y}{\cancel{y^2}}} = \dfrac{-x(x + y)}{y}$$

55. $2 - \dfrac{1}{1 - \dfrac{2}{a + 2}} = 2 - \dfrac{(a + 2)\cdot 1}{(a + 2)[1 - \dfrac{2}{a + 2}]} = 2 - \dfrac{a + 2}{a + 2 - \underset{1}{\cancel{(a + 2)}}\dfrac{2}{\cancel{a + 2}}} = 2 - \dfrac{a + 2}{a + 2 - 2}$

$$= 2 - \dfrac{a + 2}{a} = \dfrac{2}{1} - \dfrac{a + 2}{a} = \dfrac{2a}{a} - \dfrac{(a + 2)}{a} = \dfrac{2a - a - 2}{a} = \dfrac{a - 2}{a}$$

57. (A) The multiplicative inverse of x is the unique real number $\dfrac{1}{x}$ such that $x(\dfrac{1}{x}) = 1$. Since $\dfrac{c}{d} \cdot \dfrac{d}{c} = \dfrac{cd}{dc} = \dfrac{cd}{cd} = 1$, $\dfrac{d}{c}$ is the multiplicative inverse of $\dfrac{c}{d}$.

(B) By Definition 1 of Section 1-1 of the text $a \div b = a(\dfrac{1}{b}) = a \cdot$ (multiplicative inverse of b). Hence $\dfrac{a}{b} \div \dfrac{c}{d} = \dfrac{a}{b} \cdot$ (multiplicative inverse of $\dfrac{c}{d}$)

$$= \dfrac{a}{b} \cdot \dfrac{d}{c} \text{ by part A.}$$

Exercise 1-5

Key Ideas and Formulas

a^n, n an integer and a real

1. For n a positive integer:

$$a^n = \underbrace{a \cdot a \cdot \cdots \cdot a}_{n \text{ factors of } a}$$

2. For $n = 0$: $a^0 = 1$, $a \neq 0$
0^0 is not defined

3. For n a negative integer:

$$a^n = \frac{1}{a^{-n}} \qquad a \neq 0$$

Properties of Exponents
For m, n and p integers and a
and b real numbers, (division by 0
excluded) then:

1. $a^m a^n = a^{m+n}$

2. $(a^n)^m = a^{mn}$

3. $(ab)^m = a^m b^m$

4. $\left(\dfrac{a}{b}\right)^m = \dfrac{a^m}{b^m} \qquad b \neq 0$

5. $\dfrac{a^m}{a^n} = \begin{cases} a^{m-n} \\ \dfrac{1}{a^{n-m}} \end{cases} \qquad a \neq 0$

6. $a^{-n} = \dfrac{1}{a^n} \qquad a \neq 0$

7. $(a^m b^n)^p = a^{pm} b^{pn}$

8. $\left(\dfrac{a^m}{b^n}\right)^p = \dfrac{a^{pm}}{b^{pn}} \qquad b \neq 0$

9. $\dfrac{a^{-n}}{b^{-m}} = \dfrac{b^m}{a^n} \qquad a, \ b \neq 0$

10. $\left(\dfrac{a}{b}\right)^{-n} = \left(\dfrac{b}{a}\right)^n \qquad a, \ b \neq 0$

---

1. $y^{-5} y^5 = y^{-5+5} = y^0 = 1$

3. $(2x^2)(3x^3)(x^4) = (2 \cdot 3)(x^2 x^3 x^4) = 6x^9$

5. $(3x^3 y^{-2})^2 = 3^2 x^6 y^{-4} = \dfrac{9x^6}{y^4}$

7. $\left(\dfrac{ab^3}{c^2 d}\right)^4 = \dfrac{a^4 (b^3)^4}{(c^2)^4 d^4} = \dfrac{a^4 b^{12}}{c^8 d^4}$

9. $\dfrac{10^{23} \cdot 10^{-11}}{10^{-3} \cdot 10^{-2}} = \dfrac{10^{12}}{10^{-5}} = 10^{12-(-5)} = 10^{17}$

11. $\dfrac{4x^{-2} y^{-3}}{2x^{-3} y^{-1}} = 2x^{-2-(-3)} y^{-3-(-1)} = 2x^1 y^{-2} = \dfrac{2x}{y^2}$ $\boxed{\textbf{Common Error: } \dfrac{x^{-2}}{x^{-3}} \neq x^{-2-3}}$

13. $\left(\dfrac{n^{-3}}{n^{-2}}\right)^{-2} = \dfrac{n^6}{n^4} = n^2$

15. $\dfrac{8 \times 10^3}{2 \times 10^{-5}} = 4 \times 10^{3-(-5)} = 4 \times 10^8$

17. $32,250,000 = 3.2250000. \times 10^7$

7 places left

positive exponent

$= 3.225 \times 10^7$

19. $0.085 = 0.08.5 \times 10^{-2} = 8.5 \times 10^{-2}$

2 places right

negative exponent

21. $0.000\ 000\ 07\ 29 = 7.29 \times 10^{-8}$

8 places right

23. $5 \times 10^{-3} = 0.005. = 0.005$

3 places left

25. $2.69 \times 10^7 = 2.6900\ 000. = 26,900,000$

7 places right

27. $5.9 \times 10^{-10} = 0.000\ 000\ 000\ 5.9 = 0.000\ 000\ 00059$

10 places left

29. $\dfrac{27x^{-5}x^5}{18y^{-6}y^2} = \dfrac{\overset{3}{27}x^{-5+5}}{\underset{2}{18}y^{-6+2}} = \dfrac{3x^0}{2y^{-4}} = \dfrac{3y^4}{2}$ **31.** $\left(\dfrac{x^4y^{-1}}{x^{-2}y^3}\right)^2 = \dfrac{x^8y^{-2}}{x^{-4}y^6} = x^{8-(-4)}y^{-2-6} = x^{12}y^{-8} = \dfrac{x^{12}}{y^8}$

33. $\left(\dfrac{2x^{-3}y^2}{4xy^{-1}}\right)^{-2} = \left(\dfrac{x^{-3}y^2}{2xy^{-1}}\right)^{-2} = \dfrac{x^6y^{-4}}{2^{-2}x^{-2}y^2} = 2^2x^{6-(-2)}y^{-4-2} = 4x^8y^{-6} = \dfrac{4x^8}{y^6}$

35. $\left[\left(\dfrac{u^3v^{-1}w^{-2}}{u^{-2}v^{-2}w}\right)^{-2}\right]^2 = \left(\dfrac{u^3v^{-1}w^{-2}}{u^{-2}v^{-2}w^1}\right)^{-4} = \dfrac{u^{-12}v^4w^8}{u^8v^8w^{-4}} = u^{-20}v^{-4}w^{12} = \dfrac{w^{12}}{u^{20}v^4}$

37. $(x+y)^{-2} = \dfrac{1}{(x+y)^2}$

39. $\dfrac{1+x^{-1}}{1-x^{-2}} = \dfrac{1+\frac{1}{x}}{1-\frac{1}{x^2}} = \dfrac{x^2\left(1+\frac{1}{x}\right)}{x^2\left(1-\frac{1}{x^2}\right)} = \dfrac{x^2+x}{x^2-1} = \dfrac{x\cancel{(x+1)}}{(x-1)\underset{1}{\cancel{(x+1)}}} = \dfrac{x}{x-1}$

41. $\dfrac{x^{-1}-y^{-1}}{x-y} = \dfrac{\frac{1}{x}-\frac{1}{y}}{x-y} = \dfrac{xy\left(\frac{1}{x}-\frac{1}{y}\right)}{xy(x-y)} = \dfrac{\overset{-1}{\cancel{y-x}}}{xy\underset{1}{\cancel{(x-y)}}} = \dfrac{-1}{xy}$

43. $-3(x^3+3)^{-4}(3x^2) = -3\cdot\dfrac{1}{(x^3+3)^4}\cdot 3x^2 = \dfrac{-9x^2}{(x^3+3)^4}$ $\boxed{\text{Common Error: } (x^3+3)^{-4} \neq x^{-12}+3^{-4}}$

45. $2^{3^2} = 64$ on the assumption that 2^3^2 is entered, without parentheses.

47. The identity property for multiplication states that for x in R, $x(1) = x$, and 1 is the only element in R with that property. Therefore, if $a^ma^0 = a^m$, since $a^m(1) = a^m$, a^0 should equal 1.

49. $\dfrac{4x^2-12}{2x} = \dfrac{4x^2}{2x} - \dfrac{12}{2x} = 2x - \dfrac{6}{x} = 2x - 6x^{-1}$

51. $\dfrac{5x^3-2}{3x^2} = \dfrac{5x^3}{3x^2} - \dfrac{2}{3x^2} = \dfrac{5x}{3} - \dfrac{2}{3x^2} = \dfrac{5}{3}x - \dfrac{2}{3}x^{-2}$

53. $\dfrac{2x^3-3x^2+x}{2x^2} = \dfrac{2x^3}{2x^2} - \dfrac{3x^2}{2x^2} + \dfrac{x}{2x^2} = x - \dfrac{3}{2} + \dfrac{1}{2x} = x - \dfrac{3}{2} + \dfrac{1}{2}x^{-1}$

55. $\dfrac{(32.7)(0.000\,000\,008\,42)}{(0.0513)(80,700,000,000)} = \dfrac{(32.7)(8.42\times10^{-9})}{(0.0513)(8.07\times10^{10})} = 6.65\times10^{-17}$

57. $\dfrac{(5,760,000,000)}{(527)(0.000\,007\,09)} = \dfrac{5.76\times10^9}{(527)(7.09\times10^{-6})} = 1.54\times10^{12}$

59. 1.0295×10^{11} **61.** -4.3647×10^{-18}

63. $(9,820,000,000)^3 = (9.82\times10^9)^3 = 9.4697\times10^{29}$

65. $\dfrac{12(a+2b)^{-3}}{6(a+2b)^{-8}} = \dfrac{\overset{2}{\cancel{12}}(a+2b)^{-3-(-8)}}{\underset{1}{\cancel{6}}} = 2(a+2b)^5$

67. $\dfrac{xy^{-2}-yx^{-2}}{y^{-1}-x^{-1}} = \dfrac{\frac{x}{y^2}-\frac{y}{x^2}}{\frac{1}{y}-\frac{1}{x}} = \dfrac{x^2y^2\left(\frac{x}{y^2}-\frac{y}{x^2}\right)}{x^2y^2\left(\frac{1}{y}-\frac{1}{x}\right)} = \dfrac{x^3-y^3}{x^2y-xy^2}$

$= \dfrac{\overset{1}{\cancel{(x-y)}}(x^2+xy+y^2)}{xy\underset{1}{\cancel{(x-y)}}} = \dfrac{x^2+xy+y^2}{xy}$

69. $\left(\dfrac{x^{-1}}{x^{-1}-y^{-1}}\right)^{-1} = \dfrac{x^{-1}-y^{-1}}{x^{-1}} = \dfrac{\frac{1}{x}-\frac{1}{y}}{\frac{1}{x}} = \dfrac{xy\left(\frac{1}{x}-\frac{1}{y}\right)}{xy\left(\frac{1}{x}\right)} = \dfrac{y-x}{y}$

71. mass of earth in pounds
 = mass of earth in grams × number of pounds/gram
 = $6.1 \times 10^{27} \times 2.2 \times 10^{-3}$
 = 13.42×10^{24}
 = 1.342×10^{25}
 = 1.3×10^{25} pounds to two significant digits

73. 1 operation in 10^{-8} seconds means $1 \div 10^{-8}$ operations/second, that is 10^{8} operations in 1 second, that is, 100,000,000 or 100 million. Similarly, 1 operation in 10^{-10} seconds means $1 \div 10^{-10}$, that is 10,000,000,000 or 10 billion operations in 1 second.

Since 1 minute is 60 seconds, multiply each number by 60 to get $60 \times 10^{8} = 6 \times 10^{9}$ or 6,000,000,000 or 6 billion operations in 1 minute and $60 \times 10^{10} = 6 \times 10^{11}$ or 600,000,000,000 or 600 billion operations in 1 minute.

75. average amount of tax paid per person $= \dfrac{\text{amount of tax}}{\text{number of persons}} = \dfrac{349{,}000{,}000{,}000}{242{,}000{,}000}$

$= \dfrac{3.49 \times 10^{11}}{2.42 \times 10^{8}} = 1.44 \times 10^{3}$ dollars per person to three significant digits
$= \$1{,}440$ per person

Exercise 1-6

Key Ideas and Formulas

Note: All properties of exponents treated in Section 1.5 are valid for rational number exponents defined in this section.

For n a natural number and b a real number, $b^{1/n}$ is the principal nth root of b.
1. If n is even and b is positive, then $b^{1/n}$ represents the positive nth root of b.
2. If n is even and b is negative, then $b^{1/n}$ does not represent a real number.
3. If n is odd, then $b^{1/n}$ represents the real nth root of b (there is only one).
4. $0^{1/n} = 0$

For m and n natural numbers and b any real number (except b cannot be negative when n is even):

$b^{m/n} = (b^{1/n})^{m}$ $\qquad$ $b^{m/n} = (b^{m})^{1/n}$

$b^{-m/n} = \dfrac{1}{b^{m/n}}$ $\qquad$ $b \neq 0$

---

1. $16^{1/2} = 4$ $\qquad$ **3.** $16^{3/2} = (16^{1/2})^{3} = 4^{3} = 64$ $\qquad$ **5.** $-36^{1/2} = -(36^{1/2}) = -6$

7. $(-36)^{1/2}$ is not a real number $\qquad$ **9.** $\left(\dfrac{4}{25}\right)^{3/2} = \dfrac{4^{3/2}}{25^{3/2}} = \dfrac{8}{125}$ $\qquad$ **11.** $9^{-3/2} = \dfrac{1}{9^{3/2}} = \dfrac{1}{27}$

13. $y^{1/5}y^{2/5} = y^{1/5+2/5} = y^{3/5}$ $\qquad$ **15.** $d^{2/3}d^{-1/3} = d^{2/3+(-1/3)} = d^{1/3}$

17. $(y^{-8})^{1/16} = y^{(1/16)(-8)} = y^{-8/16} = y^{-1/2} = \dfrac{1}{y^{1/2}}$

19. $(8x^{3}y^{-6})^{1/3} = 8^{1/3}x^{(1/3)(3)}y^{(1/3)(-6)} = 2x^{1}y^{-2} = \dfrac{2x}{y^{2}}$

21. $\left(\dfrac{a^{-3}}{b^{4}}\right)^{1/12} = \dfrac{a^{-3/12}}{b^{4/12}} = \dfrac{a^{-1/4}}{b^{1/3}} = \dfrac{1}{a^{1/4}b^{1/3}}$

23. $\left(\dfrac{4x^{-2}}{y^{4}}\right)^{-1/2} = \dfrac{4^{-1/2}x^{1}}{y^{-2}} = \dfrac{xy^{2}}{4^{1/2}} = \dfrac{xy^{2}}{2}$ $\quad$ | **Common Error:** $\dfrac{4x^{1}}{y^{-2}}$. This is wrong; the exponent $\left(-\frac{1}{2}\right)$ applies to constant as well as variable factors. |

25. $\left(\dfrac{8a^{-4}b^3}{27a^2b^{-3}}\right)^{1/3} = \dfrac{8^{1/3}a^{-4/3}b^1}{27^{1/3}a^{2/3}b^{-1}} = \dfrac{2b^{1-(-1)}}{3a^{2/3-(-4/3)}} = \dfrac{2b^2}{3a^2}$

27. $\dfrac{8x^{-1/3}}{12x^{1/4}} = \dfrac{2}{3x^{1/4-(-1/3)}} = \dfrac{2}{3x^{7/12}}$

29. $\left(\dfrac{a^{2/3}b^{-1/2}}{a^{1/2}b^{1/2}}\right)^2 = \dfrac{a^{4/3}b^{-1}}{a^1b^1} = \dfrac{a^{4/3-1}}{b^{1-(-1)}} = \dfrac{a^{1/3}}{b^2}$

31. $2m^{1/3}(3m^{2/3} - m^6) = 6m^{1/3+2/3} - 2m^{(1/3)+6} = 6m - 2m^{19/3}$

33. $(a^{1/2} + 2b^{1/2})(a^{1/2} - 3b^{1/2}) = a^{1/2}a^{1/2} - 3a^{1/2}b^{1/2} + 2a^{1/2}b^{1/2} - 6b^{1/2}b^{1/2}$
$$= a - a^{1/2}b^{1/2} - 6b$$

35. $(2x^{1/2} - 3y^{1/2})(2x^{1/2} + 3y^{1/2}) = (2x^{1/2})^2 - (3y^{1/2})^2 = 4x - 9y$

37. $(x^{1/2} + 2y^{1/2})^2 = (x^{1/2})^2 + 2(x^{1/2})(2y^{1/2}) + (2y^{1/2})^2 = x + 4x^{1/2}y^{1/2} + 4y$

39. $15^{5/4} = 15^{1.25} = 29.52$

41. $103^{-3/4} = 103^{-0.75} = 0.03093$

43. $2.876^{8/5} = 2.876^{1.6} = 5.421$

45. $(0.000\,000\,077\,35)^{-2/7} = (7.735 \times 10^{-8})^{(-2\div7)} = 107.6$

47. There are many examples; one simple choice is $x = y = 1$. Then $(x + y)^{1/2} = (1 + 1)^{1/2} = 2^{1/2} = \sqrt{2}$, and $x^{1/2} + y^{1/2} = 1^{1/2} + 1^{1/2} = 1 + 1 = 2$; thus, the left side is not equal to the right side.

49. There are many examples; one simple choice is $x = y = 1$. Then $(x + y)^{1/3} = (1 + 1)^{1/3} = 2^{1/3} = \sqrt[3]{2}$ and $\dfrac{1}{(x + y)^3} = \dfrac{1}{(1 + 1)^3} = \dfrac{1}{2^3} = \dfrac{1}{8}$; thus, the left side is not equal to the right side.

51. $\dfrac{12x^{1/2} - 3}{4x^{1/2}} = \dfrac{12x^{1/2}}{4x^{1/2}} - \dfrac{3}{4x^{1/2}} = 3 - \dfrac{3}{4}x^{-1/2}$

53. $\dfrac{3x^{2/3} + x^{1/2}}{5x} = \dfrac{3x^{2/3}}{5x^1} + \dfrac{x^{1/2}}{5x^1} = \dfrac{3}{5}x^{-1/3} + \dfrac{1}{5}x^{-1/2}$

55. $\dfrac{x^2 - 4x^{1/2}}{2x^{1/3}} = \dfrac{x^2}{2x^{1/3}} - \dfrac{4x^{1/2}}{2x^{1/3}} = \dfrac{1}{2}x^{5/3} - 2x^{1/6}$

57. $(a^{3/n}b^{3/m})^{1/3} = a^{(1/3)(3/n)}b^{(1/3)(3/m)} = a^{1/n}b^{1/m}$

59. $(x^{m/4}y^{n/3})^{-12} = x^{-12(m/4)}y^{-12(n/3)} = x^{-3m}y^{-4n} = \dfrac{1}{x^{3m}y^{4n}}$

61. (A) Since $(x^2)^{1/2}$ represents the positive real square root of x^2, it will not equal x if x is negative. So any negative value of x, for example $x = -2$, will make the left side $\neq$ right side. $[(-2)^2]^{1/2} = 4^{1/2} = 2 \neq -2$.

(B) Similarly any positive (or zero) value of x, for example $x = 2$, will make the left side = right side. $[2^2]^{1/2} = 4^{1/2} = 2 = 2$.

(C) Any value of x will make the left side = right side.
$[2^3]^{1/3} = 8^{1/3} = 2$ $[(-2)^3]^{1/3} = [-8]^{1/3} = -2$ $[0^3]^{1/3} = 0^{1/3} = 0$
There is no real value of x such that $(x^3)^{1/3} \neq x$.

63. No. Any negative value of b, for example -4, will make $(b^m)^{1/n}$ not a real number if n is even and m is odd. Thus, $[(-4)^3]^{1/2} = [-64]^{1/2}$ is not a real number.

65.
$$\frac{(2x-1)^{1/2} - (x+2)(\frac{1}{2})(2x-1)^{-1/2}(2)}{(2x-1)}$$

> **Common Error:** Do not "cancel" the $(2x-1)^{1/2}$. It is not a common factor of numerator and denominator.

$$= \frac{(2x-1)^{1/2} - (x+2)(2x-1)^{-1/2}}{(2x-1)^1}$$

$$= \frac{(2x-1)^{1/2} - \frac{x+2}{(2x-1)^{1/2}}}{(2x-1)^1}$$

$$= \frac{(2x-1)^{1/2}}{(2x-1)^{1/2}} \cdot \frac{(2x-1)^{1/2} - \frac{x+2}{(2x-1)^{1/2}}}{(2x-1)^1}$$

$$= \frac{(2x-1)^1 - (x+2)}{(2x-1)^{3/2}}$$

$$= \frac{2x-1-x-2}{(2x-1)^{3/2}} = \frac{x-3}{(2x-1)^{3/2}}$$

67.
$$\frac{2(3x-1)^{1/3} - (2x+1)(\frac{1}{3})(3x-1)^{-2/3}(3)}{(3x-1)^{2/3}}$$

$$= \frac{2(3x-1)^{1/3} - (2x+1)(3x-1)^{-2/3}}{(3x-1)^{2/3}}$$

$$= \frac{2(3x-1)^{1/3} - \frac{(2x+1)}{(3x-1)^{2/3}}}{(3x-1)^{2/3}}$$

$$= \frac{(3x-1)^{2/3}}{(3x-1)^{2/3}} \cdot \frac{2(3x-1)^{1/3} - \frac{(2x+1)}{(3x-1)^{2/3}}}{(3x-1)^{2/3}}$$

$$= \frac{2(3x-1)^1 - (2x+1)}{(3x-1)^{4/3}}$$

$$= \frac{6x-2-2x-1}{(3x-1)^{4/3}}$$

$$= \frac{4x-3}{(3x-1)^{4/3}}$$

69. We are to calculate $N = 10x^{3/4}y^{1/4}$ given $x = 256$ units of labor and $y = 81$ units of capital.

$N = 10(256)^{3/4}(81)^{1/4}$
$\quad = 10(256^{1/4})^3(81)^{1/4}$
$\quad = 10(4^3)(3)$
$\quad = 1,920$ units of finished product.

71. We are to calculate $d = 0.0212v^{7/3}$ given $v = 70$ miles/hour.

$d = 0.0212(70)^{7/3}$
$\quad = 0.0212(70)^{(7 \div 3)}$
$\quad = 428$ feet (to the nearest foot)

Exercise 1-7

Key Ideas and Formulas

$\sqrt[n]{b}$ = the principal nth root of b
$\quad = b^{1/n}$ (b is called the radicand, n the index, in $\sqrt[n]{b}$)

$$b^{m/n} = \begin{cases} \sqrt[n]{b^m} \\ (\sqrt[n]{b})^m \end{cases}$$

For n a natural number > 1, and x and y positive real numbers,

$\sqrt[n]{x^n} = x$

$\sqrt[n]{xy} = \sqrt[n]{x}\,\sqrt[n]{y}$

$\sqrt[n]{\dfrac{x}{y}} = \dfrac{\sqrt[n]{x}}{\sqrt[n]{y}}$

Simplified (Radical) form conditions:

1. No radicand (the expression within the radical sign) contains a factor to a power greater than or equal to the index of the radical.

2. No power of the radicand and the index of the radical have a common factor other than 1.

3. No radical appears in a denominator.

4. No fraction appears within a radical.

1. $m^{2/3} = \sqrt[3]{m^2}$ or $(\sqrt[3]{m})^2$. The first is usually preferred, and we will use this form.

3. $6x^{3/5} = 6(x)^{3/5} = 6\sqrt[5]{x^3}$ **5.** $(4xy^3)^{2/5} = \sqrt[5]{(4xy^3)^2}$ **7.** $(x + y)^{1/2} = \sqrt{x + y}$

9. $\sqrt[5]{b} = b^{1/5}$ **11.** $5\sqrt[4]{x^3} = 5x^{3/4}$ **13.** $\sqrt[5]{(2x^2y)^3} = (2x^2y)^{3/5}$

15. $\sqrt[3]{x} + \sqrt[3]{y} = x^{1/3} + y^{1/3}$ $\boxed{\textbf{Common Error:} \text{ not } (x + y)^{1/3}}$ **17.** $\sqrt[3]{-8} = -2$

19. $\sqrt{9x^8y^4} = \sqrt{9}\sqrt{x^8}\sqrt{y^4} = 3x^4y^2$ **21.** $\sqrt[4]{16m^4n^8} = \sqrt[4]{16}\sqrt[4]{m^4}\sqrt[4]{n^8} = 2mn^2$

23. $\sqrt{8a^3b^5} = \sqrt{4a^2b^4 \cdot 2ab} = \sqrt{4a^2b^4}\sqrt{2ab} = 2ab^2\sqrt{2ab}$

25. $\sqrt[3]{2^4x^4y^7} = \sqrt[3]{2^3x^3y^6 \cdot 2xy} = \sqrt[3]{2^3x^3y^6}\sqrt[3]{2xy} = 2xy^2\sqrt[3]{2xy}$

27. $\sqrt[4]{m^2} = \sqrt[2 \cdot 2]{m^{2 \cdot 1}} = \sqrt[2]{m^1} = \sqrt{m}$ **29.** $\sqrt[5]{\sqrt[3]{xy}} = \sqrt[15]{xy}$

31. $\sqrt[3]{9x^2}\sqrt[3]{9x} = \sqrt[3]{(9x^2)(9x)} = \sqrt[3]{81x^3} = \sqrt[3]{27x^3 \cdot 3} = \sqrt[3]{27x^3}\sqrt[3]{3} = 3x\sqrt[3]{3}$

33. $\dfrac{1}{\sqrt{5}} = \dfrac{1}{\sqrt{5}}\dfrac{\sqrt{5}}{\sqrt{5}} = \dfrac{\sqrt{5}}{5}$ **35.** $\dfrac{6x}{\sqrt{3x}} = \dfrac{6x}{\sqrt{3x}}\dfrac{\sqrt{3x}}{\sqrt{3x}} = \dfrac{6x\sqrt{3x}}{3x} = 2\sqrt{3x}$

37. $\dfrac{2}{\sqrt{2} - 1} = \dfrac{2}{(\sqrt{2} - 1)}\dfrac{(\sqrt{2} + 1)}{(\sqrt{2} + 1)} = \dfrac{2(\sqrt{2} + 1)}{2 - 1} = \dfrac{2(\sqrt{2} + 1)}{1} = 2(\sqrt{2} + 1) = 2\sqrt{2} + 2$

39. $\dfrac{\sqrt{2}}{\sqrt{6} + 2} = \dfrac{\sqrt{2}}{(\sqrt{6} + 2)}\dfrac{(\sqrt{6} - 2)}{(\sqrt{6} - 2)} = \dfrac{\sqrt{2}(\sqrt{6} - 2)}{6 - 4} = \dfrac{\sqrt{2}\sqrt{6} - 2\sqrt{2}}{2} = \dfrac{\sqrt{12} - 2\sqrt{2}}{2} = \dfrac{2\sqrt{3} - 2\sqrt{2}}{2}$

$\qquad = \dfrac{2(\sqrt{3} - \sqrt{2})}{2} = \sqrt{3} - \sqrt{2}$

41. $x\sqrt[5]{3^6x^7y^{11}} = x\sqrt[5]{(3^5x^5y^{10})(3x^2y)} = x\sqrt[5]{3^5x^5y^{10}}\sqrt[5]{3x^2y} = x(3xy^2)\sqrt[5]{3x^2y} = 3x^2y^2\sqrt[5]{3x^2y}$

43. $\dfrac{\sqrt[4]{32m^7n^9}}{2mn} = \dfrac{\sqrt[4]{16m^4n^8 \cdot 2m^3n}}{2mn} = \dfrac{2mn^2\sqrt[4]{2m^3n}}{2mn} = n\sqrt[4]{2m^3n}$

45. $\sqrt[6]{a^4(b - a)^2} = \sqrt[2 \cdot 3]{a^{2 \cdot 2}(b - a)^{2 \cdot 1}} = \sqrt[3]{a^2(b - a)}$ **47.** $\sqrt[3]{\sqrt[4]{a^9b^3}} = \sqrt[3 \cdot 4]{a^{3 \cdot 3}b^{3 \cdot 1}} = \sqrt[4]{a^3b}$

49. $\sqrt[3]{2x^2y^4}\sqrt[3]{3x^5y} = \sqrt[3]{(2x^2y^4)(3x^5y)} = \sqrt[3]{6x^7y^5} = \sqrt[3]{(x^6y^3)(6xy^2)} = \sqrt[3]{x^6y^3}\sqrt[3]{6xy^2} = x^2y\sqrt[3]{6xy^2}$

51. $\sqrt[3]{a^3 + b^3}$ is in simplified form. There is no correct way to simplify this expression further.

53. $\dfrac{\sqrt{2m}\sqrt{5}}{\sqrt{20m}} = \dfrac{\sqrt{10m}}{\sqrt{20m}} = \sqrt{\dfrac{10m}{20m}} = \sqrt{\dfrac{1}{2}} = \sqrt{\dfrac{1}{2} \cdot \dfrac{2}{2}} = \sqrt{\dfrac{2}{4}} = \dfrac{\sqrt{2}}{2}$ or $\dfrac{1}{2}\sqrt{2}$

55. $\dfrac{4a^3b^2}{\sqrt[3]{2ab^2}} = \dfrac{4a^3b^2}{\sqrt[3]{2ab^2}} \cdot \dfrac{\sqrt[3]{2^2a^2b}}{\sqrt[3]{2^2a^2b}} = \dfrac{4a^3b^2\sqrt[3]{4a^2b}}{\sqrt[3]{2^3a^3b^3}} = \dfrac{4a^3b^2\sqrt[3]{4a^2b}}{2ab} = 2a^2b\sqrt[3]{4a^2b}$

Notice that we multiplied numerator and denominator by $\sqrt[3]{2^2a^2b}$ instead of $\sqrt[3]{(2ab^2)^2}$, which would also have rationalized the denominator. We choose the *smallest* exponents that will make the radicand a perfect cube.

57. $\sqrt[4]{\dfrac{3y^3}{4x^3}} = \sqrt[4]{\dfrac{3y^3}{4x^3}\dfrac{4x}{4x}} = \sqrt[4]{\dfrac{12xy^3}{16x^4}} = \dfrac{\sqrt[4]{12xy^3}}{\sqrt[4]{16x^4}} = \dfrac{\sqrt[4]{12xy^3}}{2x}$ or $\dfrac{1}{2x}\sqrt[4]{12xy^3}$

Notice again that we could have multiplied numerator and denominator by $(4x^3)^3$, but the simpler quantity $4x$ is preferred.

59. $\dfrac{3\sqrt{y}}{2\sqrt{y} - 3} = \dfrac{3\sqrt{y}}{(2\sqrt{y} - 3)}\dfrac{(2\sqrt{y} + 3)}{(2\sqrt{y} + 3)} = \dfrac{3\sqrt{y}(2\sqrt{y} + 3)}{(2\sqrt{y})^2 - (3)^2} = \dfrac{6y + 9\sqrt{y}}{4y - 9}$

61. $\dfrac{2\sqrt{5} + 3\sqrt{2}}{5\sqrt{5} + 2\sqrt{2}} = \dfrac{(2\sqrt{5} + 3\sqrt{2})}{(5\sqrt{5} + 2\sqrt{2})}\dfrac{(5\sqrt{5} - 2\sqrt{2})}{(5\sqrt{5} - 2\sqrt{2})}$

$= \dfrac{2\sqrt{5}\cdot 5\sqrt{5} - 2\sqrt{5}\cdot 2\sqrt{2} + 3\sqrt{2}\cdot 5\sqrt{5} - 3\sqrt{2}\cdot 2\sqrt{2}}{(5\sqrt{5})^2 - (2\sqrt{2})^2}$

$= \dfrac{(10)(5) - 4\sqrt{10} + 15\sqrt{10} - (6)(2)}{(25)(5) - (4)(2)} = \dfrac{50 + 11\sqrt{10} - 12}{125 - 8} = \dfrac{38 + 11\sqrt{10}}{117}$

63. $\dfrac{x^2}{\sqrt{x^2 + 9} - 3} = \dfrac{x^2}{(\sqrt{x^2 + 9} - 3)} \cdot \dfrac{(\sqrt{x^2 + 9} + 3)}{(\sqrt{x^2 + 9} + 3)}$

$= \dfrac{x^2(\sqrt{x^2 + 9} + 3)}{(\sqrt{x^2 + 9})^2 - (3)^2} = \dfrac{x^2(\sqrt{x^2 + 9} + 3)}{x^2 + 9 - 9}$

$= \dfrac{\overset{1}{\cancel{x^2}}(\sqrt{x^2 + 9} + 3)}{\underset{1}{\cancel{x^2}}} = \sqrt{x^2 + 9} + 3$

65. Here we want to remove the radical terms from the *numerator*.

$\dfrac{\sqrt{t} - \sqrt{x}}{t - x} = \dfrac{(\sqrt{t} - \sqrt{x})}{t - x} \cdot \dfrac{(\sqrt{t} + \sqrt{x})}{\sqrt{t} + \sqrt{x}} = \dfrac{(\sqrt{t})^2 - (\sqrt{x})^2}{(t - x)(\sqrt{t} + \sqrt{x})}$

$= \dfrac{\overset{1}{\cancel{t - x}}}{\underset{1}{\cancel{(t - x)}}(\sqrt{t} + \sqrt{x})} = \dfrac{1}{\sqrt{t} + \sqrt{x}}$

67. Here we want to remove the radical terms from the *numerator*.

$\dfrac{\sqrt{x + h} - \sqrt{x}}{h} = \dfrac{(\sqrt{x + h} - \sqrt{x})}{h} \cdot \dfrac{(\sqrt{x + h} + \sqrt{x})}{(\sqrt{x + h} + \sqrt{x})} = \dfrac{(\sqrt{x + h})^2 - (\sqrt{x})^2}{h(\sqrt{x + h} + \sqrt{x})} = \dfrac{x + h - x}{h(\sqrt{x + h} + \sqrt{x})}$

$= \dfrac{\overset{1}{\cancel{h}}}{\underset{1}{\cancel{h}}(\sqrt{x + h} + \sqrt{x})} = \dfrac{1}{\sqrt{x + h} + \sqrt{x}}$

69. $\sqrt{0.049\ 375} = 0.2222$

71. $\sqrt[5]{27.0635} = (27.0635)^{1/5} = (27.0635)^{0.2} = 1.934$

73. $\sqrt[7]{0.000\ 000\ 008\ 066} = (8.066 \times 10^{-9})^{1/7}$
$= (8.066 \times 10^{-9})^{(1\div 7)}$
$= 6.979 \times 10^{-2} = 0.06979$

75. $\sqrt[3]{7 + \sqrt[3]{7}} = (7 + 7^{(1\div 3)})^{(1\div 3)} = 2.073$

77. $\sqrt[3]{\sqrt[4]{2}} = (2^{1/4})^{1/3} = (2^{0.25})^{(1\div 3)} = 1.059$

$\sqrt[12]{2} = 2^{1/12} = 2^{(1\div 12)} = 1.059$

79. $\dfrac{1}{\sqrt[3]{4}} = 4^{-1/3} = 4^{(-1\div 3)} = 0.6300$

$\dfrac{\sqrt[3]{2}}{2} = \dfrac{2^{1/3}}{2} = 2^{(1\div 3)} \div 2 = 0.6300$

81. $\sqrt{x^2}$ is the principal square root of x^2, that is, the positive (or zero) square root of x^2. The two square roots of x^2 are x and $-x$. $\sqrt{x^2} = -x$ if and only if $-x$ is positive or zero. Hence $\sqrt{x^2} = -x$ if and only if x is negative or zero. $x \leq 0$.

83. $\sqrt[3]{x^3}$ is the real third root of x^3. This is always equal to x if x is real. All real numbers.

85. (A) $\sqrt{3} + \sqrt{5} \approx 3.968\ 118\ 785$ (B) $\sqrt{2 + \sqrt{3}} + \sqrt{2 - \sqrt{3}} \approx 2.449\ 489\ 743$

(C) $1 + \sqrt{3} \approx 2.732\ 050\ 808$ (D) $\sqrt[3]{10 + 6\sqrt{3}} \approx 2.732\ 050\ 808$

(E) $\sqrt{8 + \sqrt{60}} \approx 3.968\ 118\ 785$ (F) $\sqrt{6} \approx 2.449\ 489\ 743$

Thus, (A) and (E), (B) and (F), and (C) and (D) have the same value to 9 decimal places. To show that they are actually equal:

(A) and (E): $(\sqrt{3} + \sqrt{5})^2 = (\sqrt{3})^2 + 2\sqrt{3}\sqrt{5} + (\sqrt{5})^2 = 3 + \sqrt{4}\sqrt{3}\sqrt{5} + 5 = 8 + \sqrt{60}$

Therefore, since $\sqrt{3} + \sqrt{5}$ is positive, $\sqrt{3} + \sqrt{5}$ is the positive square root of $8 + \sqrt{60}$. Thus, $\sqrt{8 + \sqrt{60}} = \sqrt{3} + \sqrt{5}$.

(B) and (F):

$$\left(\sqrt{2 + \sqrt{3}} + \sqrt{2 - \sqrt{3}}\right)^2 = \left(\sqrt{2 + \sqrt{3}}\right)^2 + 2\sqrt{2 + \sqrt{3}}\sqrt{2 - \sqrt{3}} + \left(\sqrt{2 - \sqrt{3}}\right)^2$$

$$= 2 + \sqrt{3} + 2\sqrt{(2 + \sqrt{3})(2 - \sqrt{3})} + 2 - \sqrt{3}$$

$$= 4 + 2\sqrt{2^2 - (\sqrt{3})^2}$$

$$= 4 + 2\sqrt{4 - 3}$$

$$= 4 + 2 \cdot 1$$

$$= 6$$

Therefore, since $\sqrt{2 + \sqrt{3}} + \sqrt{2 - \sqrt{3}}$ is positive, $\sqrt{2 + \sqrt{3}} + \sqrt{2 - \sqrt{3}}$ is the positive square root of 6. Thus, $\sqrt{2 + \sqrt{3}} + \sqrt{2 - \sqrt{3}} = \sqrt{6}$.

(C) and (D): $(1 + \sqrt{3})^3 = (1 + \sqrt{3})(1 + \sqrt{3})(1 + \sqrt{3})$

$$= (1^2 + 2\sqrt{3} + \sqrt{3}^2)(1 + \sqrt{3})$$

$$= (4 + 2\sqrt{3})(1 + \sqrt{3})$$

$$= 4 + 4\sqrt{3} + 2\sqrt{3} + 2\sqrt{3}\sqrt{3}$$

$$= 10 + 6\sqrt{3}$$

Therefore, $1 + \sqrt{3}$ is the real cube root of $10 + 6\sqrt{3}$. Thus,

$1 + \sqrt{3} = \sqrt[3]{10 + 6\sqrt{3}}$.

87. $\dfrac{1}{\sqrt[3]{a} - \sqrt[3]{b}} = \dfrac{1}{(\sqrt[3]{a} - \sqrt[3]{b})} \dfrac{[(\sqrt[3]{a})^2 + \sqrt[3]{a}\sqrt[3]{b} + (\sqrt[3]{b})^2]}{[(\sqrt[3]{a})^2 + \sqrt[3]{a}\sqrt[3]{b} + (\sqrt[3]{b})^2]}$

$\qquad$ using $(x - y)(x^2 + xy + y^2) = x^3 - y^3$

$\qquad = \dfrac{1[\sqrt[3]{a^2} + \sqrt[3]{ab} + \sqrt[3]{b^2}]}{(\sqrt[3]{a})^3 - (\sqrt[3]{b})^3} = \dfrac{\sqrt[3]{a^2} + \sqrt[3]{ab} + \sqrt[3]{b^2}}{a - b}$

89.
$$\frac{1}{\sqrt{x} - \sqrt{y} + \sqrt{z}} = \frac{1}{[(\sqrt{x} - \sqrt{y}) + \sqrt{z}]} \frac{[(\sqrt{x} - \sqrt{y}) - \sqrt{z}]}{[(\sqrt{x} - \sqrt{y}) - \sqrt{z}]} = \frac{\sqrt{x} - \sqrt{y} - \sqrt{z}}{(\sqrt{x} - \sqrt{y})^2 - (\sqrt{z})^2}$$

$$= \frac{\sqrt{x} - \sqrt{y} - \sqrt{z}}{(\sqrt{x})^2 - 2\sqrt{x}\sqrt{y} + (\sqrt{y})^2 - (\sqrt{z})^2} = \frac{\sqrt{x} - \sqrt{y} - \sqrt{z}}{x - 2\sqrt{xy} + y - z}$$

$$= \frac{\sqrt{x} - \sqrt{y} - \sqrt{z}}{x - 2\sqrt{xy} + y - z} = \frac{\sqrt{x} - \sqrt{y} - \sqrt{z}}{x + y - z - 2\sqrt{xy}}$$

$$= \frac{(\sqrt{x} - \sqrt{y} - \sqrt{z})}{[(x + y - z) - 2\sqrt{xy}]} \frac{[(x + y - z) + 2\sqrt{xy}]}{[(x + y - z) + 2\sqrt{xy}]}$$

$$= \frac{(\sqrt{x} - \sqrt{y} - \sqrt{z})[(x + y - z) + 2\sqrt{xy}]}{(x + y - z)^2 - (2\sqrt{xy})^2}$$

$$= \frac{(\sqrt{x} - \sqrt{y} - \sqrt{z})[(x + y - z) + 2\sqrt{xy}]}{(x + y - z)^2 - 4xy}$$

91.
$$\frac{\sqrt[3]{x + h} - \sqrt[3]{x}}{h} = \frac{(\sqrt[3]{x + h} - \sqrt[3]{x})}{h} \frac{[(\sqrt[3]{x + h})^2 + \sqrt[3]{x + h}\sqrt[3]{x} + (\sqrt[3]{x})^2]}{[(\sqrt[3]{x + h})^2 + \sqrt[3]{x + h}\sqrt[3]{x} + (\sqrt[3]{x})^2]}$$

$$= \frac{(\sqrt[3]{x + h})^3 - (\sqrt[3]{x})^3}{h[\sqrt[3]{(x + h)^2} + \sqrt[3]{x(x + h)} + \sqrt[3]{x^2}]}$$

Using $(a - b)(a^2 + ab + b^2) = a^3 - b^3$

$$= \frac{x + h - x}{h[\sqrt[3]{(x + h)^2} + \sqrt[3]{x(x + h)} + \sqrt[3]{x^2}]}$$

$$= \frac{h}{h[\sqrt[3]{(x + h)^2} + \sqrt[3]{x(x + h)} + \sqrt[3]{x^2}]}$$

$$= \frac{1}{\sqrt[3]{(x + h)^2} + \sqrt[3]{x(x + h)} + \sqrt[3]{x^2}}$$

93. $\sqrt[kn]{x^{km}} = (x^{km})^{1/kn} = x^{km/kn} = x^{m/n} = \sqrt[n]{x^m}$

95.
$$\frac{M_0}{\sqrt{1 - \dfrac{v^2}{c^2}}} = M_0 \div \sqrt{1 - \frac{v^2}{c^2}}$$

$$= M_0 \div \sqrt{\frac{c^2 - v^2}{c^2}} = M_0 \div \frac{\sqrt{c^2 - v^2}}{c}$$

$$= M_0 \cdot \frac{c}{\sqrt{c^2 - v^2}}$$

Now rationalize the denominator

$$= M_0 \cdot \frac{c}{\sqrt{c^2 - v^2}} \frac{\sqrt{c^2 - v^2}}{\sqrt{c^2 - v^2}}$$

$$= \frac{M_0 c\sqrt{c^2 - v^2}}{c^2 - v^2}$$

CHAPTER 1 REVIEW

1. (A) Since 3 is an element of $\{1,2,3,4,5\}$, the statement is true. *T.*

(B) Since 5 is not an element of $\{4,1,2\}$, the statement is true. *T.*

(C) Since B is not an element of $\{1,2,3,4,5\}$, the statement is false. *F.*

(D) Since each element of the set $\{1,2,4\}$ is also an element of the set $\{1,2,3,4,5\}$, the statement is true. *T.*

(E) Since sets B and C have exactly the same elements, the sets are equal. The statement is false. *F.*

(F) The statement is false, since not every element of the set $\{1,2,3,4,5\}$ is an element of the set $\{1,2,4\}$. For example, 3 is an element of the first, but not the second. *F.*   *(1-1)*

2. (A) The commutative property $(\cdot)$ states that, in general, $xy = yx$.
Comparing this with
$x(y + z) = ?$
we see that $x(y + z) = (y + z)x$
$(y + z)x$

(B) The associative property $(+)$ states that, in general,
$(x + y) + z = x + (y + z)$
Comparing this with
$? = 2 + (x + y)$
we see that
$(2 + x) + y = 2 + (x + y)$
$(2 + x) + y$

(C) The distributive property states that, in general,
$(x + y)z = xz + yz$
Comparing this with
$(2 + 3)x = ?$
we see that
$(2 + 3)x = 2x + 3x$
$2x + 3x$ *(1-1)*

3. $(3x - 4) + (x + 2) + (3x^2 + x - 8) + (x^3 + 8)$
$\quad = 3x - 4 + x + 2 + 3x^2 + x - 8 + x^3 + 8$
$\quad = x^3 + 3x^2 + 5x - 2$ *(1-2)*

4. sum of (a) and (c) $= (3x - 4) + (3x^2 + x - 8) = 3x^2 + 4x - 12$
sum of (b) and (d) $= (x + 2) + (x^3 + 8) = x^3 + x + 10$
subtract the sum of (a) and (c) from the sum of (b) and (d)
$= (x^3 + x + 10) - (3x^2 + 4x - 12) = x^3 + x + 10 - 3x^2 - 4x + 12$
$= x^3 - 3x^2 - 3x + 22$ *(1-2)*

5. $3x^2 + x - 8$

$\underline{x^3 + 8}$

$3x^5 + x^4 - 8x^3$

$\underline{\qquad\qquad 24x^2 + 8x - 64}$

$3x^5 + x^4 - 8x^3 + 24x^2 + 8x - 64$ *(1-2)*

6. 3   *(1-2)*

7. 1   *(1-2)*

8. $5x^2 - 3x[4 - 3(x - 2)] = 5x^2 - 3x[4 - 3x + 6]$
$\qquad\qquad\qquad\qquad = 5x^2 - 3x[10 - 3x]$
$\qquad\qquad\qquad\qquad = 5x^2 - 30x + 9x^2$
$\qquad\qquad\qquad\qquad = 14x^2 - 30x$

> **Common Errors:**
> $-3(x - 2) \neq -3x - 6$
> $4 - 3(x - 2) \neq 1(x - 2)$
> The first is incorrect use of the distributive property, the second is incorrect order of operations.

(1-2)

9. $(3m - 5n)(3m + 5n) = (3m)^2 - (5n)^2 = 9m^2 - 25n^2$ *(1-2)*

10.

	First Product	Outer Product	Inner Product	Last Product
$(2x + y)(3x - 4y) = 6x^2$		$-8xy$	$+3xy$	$-4y^2$
$= 6x^2 - 5xy - 4y^2$				

$(1-2)$

11. $(2a - 3b)^2 = (2a)^2 - 2(2a)(3b) + (3b)^2 = 4a^2 - 12ab + 9b^2$ $(1-2)$

12. $(3x - 2)^2$ $(1-3)$

13. Prime $(1-3)$

14. $6n^3 - 9n^2 - 15n = 3n(2n^2 - 3n - 5) = 3n(2n - 5)(n + 1)$ $(1-3)$

15. $\dfrac{2}{5b} - \dfrac{4}{3a^3} - \dfrac{1}{6a^2b^2} = \dfrac{12a^3b}{30a^3b^2} - \dfrac{40b^2}{30a^3b^2} - \dfrac{5a}{30a^3b^2} = \dfrac{12a^3b - 40b^2 - 5a}{30a^3b^2}$ $(1-4)$

16. $\dfrac{3x}{3x^2 - 12x} + \dfrac{1}{6x} = \dfrac{3x}{3x(x - 4)} + \dfrac{1}{6x} = \dfrac{2(3x)}{6x(x - 4)} + \dfrac{1(x - 4)}{6x(x - 4)} = \dfrac{6x + x - 4}{6x(x - 4)} = \dfrac{7x - 4}{6x(x - 4)}$ $(1-4)$

17. $\dfrac{y - 2}{y^2 - 4y + 4} \div \dfrac{y^2 + 2y}{y^2 + 4y + 4} = \dfrac{y - 2}{y^2 - 4y + 4} \cdot \dfrac{y^2 + 4y + 4}{y^2 + 2y}$

$$= \dfrac{\cancel{(y - 2)}^{\,1}}{\cancel{(y - 2)}_{\,1}(y - 2)} \cdot \dfrac{\cancel{(y + 2)}^{\,1}(y + 2)}{y\cancel{(y + 2)}_{\,1}}$$

$$= \dfrac{y + 2}{y(y - 2)}$$

$(1-4)$

18. $\dfrac{u - \dfrac{1}{u}}{1 - \dfrac{1}{u^2}} = \dfrac{u^2\left(u - \dfrac{1}{u}\right)}{u^2\left(1 - \dfrac{1}{u^2}\right)} = \dfrac{u^3 - u}{u^2 - 1} = \dfrac{u\cancel{(u^2 - 1)}^{\,1}}{\cancel{u^2 - 1}_{\,1}} = u$ $(1-4)$

19. $6(xy^3)^5 = 6x^5(y^3)^5 = 6x^5y^{15}$ $(1-5)$

20. $\dfrac{9u^8v^6}{3u^4v^8} = \dfrac{3u^4}{v^2}$ $(1-5)$

21. $(2 \times 10^5)(3 \times 10^{-3}) = (2 \cdot 3) \times 10^{5+(-3)} = 6 \times 10^2$ $(1-5)$

22. $(x^{-3}y^2)^{-2} = (x^{-3})^{-2}(y^2)^{-2} = x^6y^{-4} = \dfrac{x^6}{y^4}$ $(1-6)$

23. $u^{5/3}u^{2/3} = u^{5/3+2/3} = u^{7/3}$ $(1-6)$

24. $(9a^4b^{-2})^{1/2} = 9^{1/2}(a^4)^{1/2}(b^{-2})^{1/2} = 3a^2b^{-1} = \dfrac{3a^2}{b}$ $(1-7)$

25. $3\sqrt[5]{x^2}$ $(1-7)$

26. $-3(xy)^{2/3}$ $(1-7)$

27. $3x\sqrt[3]{x^5y^4} = 3x\sqrt[3]{x^3y^3 \cdot x^2y} = 3x\sqrt[3]{x^3y^3}\sqrt[3]{x^2y} = 3x(xy)\sqrt[3]{x^2y} = 3x^2y\sqrt[3]{x^2y}$ $(1-7)$

28. $\sqrt{2x^2y^5}\sqrt{18x^3y^2} = \sqrt{(2x^2y^5)(18x^3y^2)} = \sqrt{36x^5y^7} = \sqrt{36x^4y^6 \cdot xy} = \sqrt{36x^4y^6}\sqrt{xy}$

$\qquad = 6x^2y^3\sqrt{xy}$ $(1-7)$

29. $\dfrac{6ab}{\sqrt{3a}} = \dfrac{6ab}{\sqrt{3a}}\dfrac{\sqrt{3a}}{\sqrt{3a}} = \dfrac{\cancel{6ab}^{\,2}\sqrt{3a}}{\cancel{3a}} = 2b\sqrt{3a}$ $(1-7)$

1

30. $\dfrac{\sqrt{5}}{3 - \sqrt{5}} = \dfrac{\sqrt{5}}{(3 - \sqrt{5})} \dfrac{(3 + \sqrt{5})}{(3 + \sqrt{5})} = \dfrac{\sqrt{5}(3 + \sqrt{5})}{3^2 - (\sqrt{5})^2} = \dfrac{\sqrt{5}(3 + \sqrt{5})}{9 - 5} = \dfrac{\sqrt{5}(3 + \sqrt{5})}{4}$

$= \dfrac{3\sqrt{5} + 5}{4}$ *(1-7)*

31. $\sqrt[8]{y^6} = \sqrt[2 \cdot 4]{y^{2 \cdot 3}} = \sqrt[4]{y^3}$ *(1-7)*

32. The odd integers between -4 and 2 are -3, -1, and 1. Hence the set is written {-3, -1, 1} *(1-1)*

33. Subtraction.
$(-3) - (-2) = (-3) + [-(-2)]$ is a special case of $a - b = a + (-b)$. *(1-1)*

34. Commutative (+)
$3y + (2x + 5) = (2x + 5) + 3y$ is a special case of $x + y = y + x$. *(1-1)*

35. Distributive
$(2x + 3)(3x + 5) = (2x + 3)3x + (2x + 3)5$ is a special case of
$x(y + z) = xy + xz$. *(1-1)*

36. Associative (·)
$3 \cdot (5x) = (3 \cdot 5)x$ is a special case of $x(yz) = (xy)z$. *(1-1)*

37. Negatives
$\dfrac{a}{-(b - c)} = -\dfrac{a}{b - c}$ is a special case of $\dfrac{a}{-b} = -\dfrac{a}{b}$. (Theorem 1, Part 5) *(1-1)*

38. Identity (+)
$3xy + 0 = 3xy$ is a special case of $x + 0 = x$. *(1-1)*

39. (A) T   (B) F *(1-1)*

40. 0 and -3 are two examples of infinitely many. *(1-1)*

41. (A) a and d   (B) None *(1-2)*

42. $(2x - y)(2x + y) - (2x - y)^2 = (2x)^2 - y^2 - [(2x)^2 - 2(2x)y + y^2]$
$= 4x^2 - y^2 - [4x^2 - 4xy + y^2]$
$= 4x^2 - y^2 - 4x^2 + 4xy - y^2$
$= 4xy - 2y^2$

> **Common Errors:**
> $(2x - y)^2 \neq 4x^2 - y^2$ (exponents do not distribute over subtraction)
> $-(2x - y)^2 \neq (-2x + y)^2$ (square first, then subtract)
> $-(2x - y)^2 \neq -4x^2 - 4xy + y^2$ (must change all signs)

(1-2)

43. A straightforward way to perform this multiplication is vertically:
$m^2 + 2mn - n^2$
$\underline{m^2 - 2mn - n^2}$
$m^4 + 2m^3n - m^2n^2$
$ - 2m^3n - 4m^2n^2 + 2mn^3$
$\underline{ - m^2n^2 - 2mn^3 + n^4}$
$m^4 - 6m^2n^2 + n^4$

Alternately, we can notice
$(m^2 + 2mn - n^2)(m^2 - 2mn - n^2) = [(m^2 - n^2) + 2mn][(m^2 - n^2) - 2mn]$
$= (m^2 - n^2)^2 - (2mn)^2$ (difference of squares)
$= (m^2)^2 - 2m^2n^2 + (n^2)^2 - 4m^2n^2$
$= m^4 - 2m^2n^2 + n^4 - 4m^2n^2$
$= m^4 - 6m^2n^2 + n^4$ *(1-2)*

44. $5(x + h)^2 - 7(x + h) - (5x^2 - 7x) = 5(x^2 + 2xh + h^2) - 7(x + h) - (5x^2 - 7x)$
$$= 5x^2 + 10xh + 5h^2 - 7x - 7h - 5x^2 + 7x$$
$$= 10xh + 5h^2 - 7h \qquad (1\text{-}2)$$

45. $-2x\{(x^2 + 2)(x - 3) - x[x - x(3 - x)]\} = -2x\{(x^2 + 2)(x - 3) - x[x - 3x + x^2]\}$
$$= -2x\{(x^2 + 2)(x - 3) - x[-2x + x^2]\}$$
$$= -2x\{x^3 - 3x^2 + 2x - 6 + 2x^2 - x^3\}$$
$$= -2x\{-x^2 + 2x - 6\}$$
$$= 2x^3 - 4x^2 + 12x \qquad (1\text{-}2)$$

46. $(x - 2y)^3 = (x - 2y)(x - 2y)(x - 2y)$
$$= [(x - 2y)(x - 2y)](x - 2y)$$
$$= [x^2 - 2(x)(2y) + (2y)^2](x - 2y)$$
$$= [x^2 - 4xy + 4y^2](x - 2y)$$

The last multiplication is best performed vertically, so we write

$$
\begin{array}{r}
x^2 - 4xy + 4y^2 \\
x - 2y \\
\hline
x^3 - 4x^2y + 4xy^2 \\
- 2x^2y + 8xy^2 - 8y^3 \\
\hline
x^3 - 6x^2y + 12xy^2 - 8y^3 \qquad (1\text{-}2)
\end{array}
$$

47. $(4x - y)^2 - 9x^2 = (4x - y)^2 - (3x)^2$
$$= [(4x - y) - 3x][(4x - y) + 3x]$$
$$= (x - y)(7x - y) \qquad (1\text{-}3)$$

48. Prime $\qquad (1\text{-}3)$ **49.** $3xy(2x^2 + 4xy - 5y^2) \qquad (1\text{-}3)$

50. $(y - b)^2 - y + b = (y - b)(y - b) - 1(y - b) = (y - b)(y - b - 1) \qquad (1\text{-}3)$

51. $3x^3 + 24y^3 = 3(x^3 + 8y^3) = 3[x^3 + (2y)^3] = 3(x + 2y)[x^2 - x(2y) + (2y)^2]$
$$= 3(x + 2y)(x^2 - 2xy + 4y^2) \qquad (1\text{-}3)$$

52. $y^3 + 2y^2 - 4y - 8 = y^2(y + 2) - 4(y + 2)$
$$= (y + 2)(y^2 - 4) = (y + 2)(y + 2)(y - 2)$$
$$= (y - 2)(y + 2)^2 \qquad (1\text{-}3)$$

53. $2x(x - 4)^3 + 3x^2(x - 4)^2 = x(x - 4)^2[2(x - 4) + 3x]$
$$= x(x - 4)^2[2x - 8 + 3x]$$
$$= x(x - 4)^2(5x - 8) \qquad (1\text{-}3)$$

54. $\dfrac{3x^2(x + 2)^2 - 2x(x + 2)^3}{x^4} = \dfrac{x(x + 2)^2[3x - 2(x + 2)]}{x^4}$
$$= \dfrac{\overset{1}{\cancel{x}}(x + 2)^2[3x - 2x - 4]}{\underset{x^3}{\cancel{x^4}}}$$
$$= \dfrac{(x + 2)^2(x - 4)}{x^3} \qquad (1\text{-}4)$$

55. $\dfrac{m - 1}{m^2 - 4m + 4} + \dfrac{m + 3}{m^2 - 4} + \dfrac{2}{2 - m} = \dfrac{(m - 1)}{(m - 2)(m - 2)} + \dfrac{m + 3}{(m - 2)(m + 2)} + \dfrac{-2}{m - 2}$

$$= \dfrac{(m - 1)(m + 2)}{(m - 2)(m - 2)(m + 2)} + \dfrac{(m + 3)(m - 2)}{(m - 2)(m - 2)(m + 2)}$$
$$+ \dfrac{-2(m - 2)(m + 2)}{(m - 2)(m - 2)(m + 2)}$$

$$= \dfrac{(m - 1)(m + 2) + (m + 3)(m - 2) - 2(m - 2)(m + 2)}{(m - 2)^2(m + 2)}$$

$$= \dfrac{m^2 + m - 2 + m^2 + m - 6 - 2(m^2 - 4)}{(m - 2)^2(m + 2)}$$

$$= \dfrac{2m^2 + 2m - 8 - 2m^2 + 8}{(m - 2)^2(m + 2)} = \dfrac{2m}{(m - 2)^2(m + 2)} \qquad (1\text{-}4)$$

56. $\dfrac{y}{x^2} \div \left(\dfrac{x^2 + 3x}{2x^2 + 5x - 3} \div \dfrac{x^3 y - x^2 y}{2x^2 - 3x + 1} \right) = \dfrac{y}{x^2} \div \left(\dfrac{x^2 + 3x}{2x^2 + 5x - 3} \cdot \dfrac{2x^2 - 3x + 1}{x^3 y - x^2 y} \right)$

$$= \dfrac{y}{x^2} \div \left(\dfrac{x(x + 3)}{(2x - 1)(x + 3)} \cdot \dfrac{(2x - 1)(x - 1)}{x^2 y (x - 1)} \right)$$

$$= \dfrac{y}{x^2} \div \dfrac{1}{xy}$$

$$= \dfrac{y}{x^2} \cdot \dfrac{xy}{1}$$

$$= \dfrac{y^2}{x} \qquad (1\text{-}4)$$

57. $\dfrac{1 - \dfrac{1}{1 + \dfrac{x}{y}}}{1 - \dfrac{1}{1 - \dfrac{x}{y}}} = \dfrac{1 - \dfrac{y(1)}{y\left(1 + \dfrac{x}{y}\right)}}{1 - \dfrac{y(1)}{y\left(1 - \dfrac{x}{y}\right)}} = \dfrac{1 - \dfrac{y}{y + x}}{1 - \dfrac{y}{y - x}} = \dfrac{\dfrac{y + x}{y + x} - \dfrac{y}{y + x}}{\dfrac{y - x}{y - x} - \dfrac{y}{y - x}}$

$$= \dfrac{\dfrac{x}{y + x}}{\dfrac{-x}{y - x}} = \dfrac{x}{y + x} \div \dfrac{-x}{y - x} = \dfrac{x}{y + x} \cdot \dfrac{y - x}{-x} = \dfrac{-1(y - x)}{y + x} = \dfrac{x - y}{x + y} \qquad (1\text{-}4)$$

58. $\dfrac{a^{-1} - b^{-1}}{ab^{-2} - ba^{-2}} = \dfrac{\dfrac{1}{a} - \dfrac{1}{b}}{\dfrac{a}{b^2} - \dfrac{b}{a^2}} = \dfrac{a^2 b^2 \left(\dfrac{1}{a} - \dfrac{1}{b}\right)}{a^2 b^2 \left(\dfrac{a}{b^2} - \dfrac{b}{a^2}\right)}$

$$= \dfrac{ab^2 - a^2 b}{a^3 - b^3} = \dfrac{ab(b - a)}{(a - b)(a^2 + ab + b^2)} = \dfrac{-ab}{a^2 + ab + b^2} \qquad (1\text{-}4,\ 1\text{-}5)$$

59. The solution is incorrect. In adding a quantity to a fractional expression, the quantity cannot simply be placed in the numerator. A correct solution would rewrite $x + 2$ as a fractional expression, first with denominator 1, and then with denominator equal to the LCD of all denominators in the expression. Thus,

$$\dfrac{x^2 + 2x}{x^2 + x - 2} + x + 2 = \dfrac{x(x + 2)}{(x - 1)(x + 2)} + \dfrac{x + 2}{1}$$

$$= \dfrac{x}{x - 1} + \dfrac{x + 2}{1}$$

$$= \dfrac{x}{x - 1} + \dfrac{(x + 2)(x - 1)}{x - 1}$$

$$= \dfrac{x + x^2 + x - 2}{x - 1}$$

$$= \dfrac{x^2 + 2x - 2}{x - 1} \qquad (1\text{-}4)$$

60. $\left(\dfrac{8u^{-1}}{2^2 u^2 v^0}\right)^{-2} \left(\dfrac{u^{-5}}{u^{-3}}\right)^3 = \left(\dfrac{2^3 u^{-1}}{2^2 u^2}\right)^{-2} \left(\dfrac{u^{-5}}{u^{-3}}\right)^3 = (2u^{-3})^{-2} \left(\dfrac{u^{-5}}{u^{-3}}\right)^3 = 2^{-2} u^6 \dfrac{u^{-15}}{u^{-9}}$

$$= \dfrac{1}{2^2} u^6 u^{-15 - (-9)} = \dfrac{1}{2^2} u^6 u^{-6} = \dfrac{1}{4} \qquad (1\text{-}5)$$

61. $\dfrac{5^0}{3^2} + \dfrac{3^{-2}}{2^{-2}} = \dfrac{1}{3^2} + \dfrac{2^2}{3^2} = \dfrac{1}{9} + \dfrac{4}{9} = \dfrac{5}{9} \qquad (1\text{-}5)$

62. $\left(\dfrac{27 x^2 y^{-3}}{8 x^{-4} y^3}\right)^{1/3} = \left(\dfrac{3^3 x^{2 - (-4)} y^{-3 - 3}}{2^3}\right)^{1/3} = \left(\dfrac{3^3 x^6 y^{-6}}{2^3}\right)^{1/3} = \dfrac{3 x^2 y^{-2}}{2} = \dfrac{3x^2}{2y^2} \qquad (1\text{-}6)$

63. $(a^{-1/3}b^{1/4})(9a^{1/3}b^{-1/2})^{3/2} = a^{-1/3}b^{1/4}9^{3/2}a^{1/2}b^{-3/4}$

$$= a^{-1/3+1/2}b^{1/4-3/4}9^{3/2} = 27a^{1/6}b^{-1/2}$$

$$= \frac{27a^{1/6}}{b^{1/2}} \qquad (1\text{-}6)$$

64. $(x^{1/2} + y^{1/2})^2 = (x^{1/2})^2 + 2(x^{1/2})(y^{1/2}) + (y^{1/2})^2$

$$= x + 2x^{1/2}y^{1/2} + y$$

> **Common Error:**
> $(x^{1/2} + y^{1/2})^2 \neq x + y.$
> (Exponents do not distribute over addition.)

$(1\text{-}6)$

65. $(3x^{1/2} - y^{1/2})(2x^{1/2} + 3y^{1/2}) = (3x^{1/2})(2x^{1/2}) + (3x^{1/2})(3y^{1/2})$

$$- (2x^{1/2})(y^{1/2}) - (y^{1/2})(3y^{1/2})$$

$$= 6x + 9x^{1/2}y^{1/2} - 2x^{1/2}y^{1/2} - 3y$$

$$= 6x + 7x^{1/2}y^{1/2} - 3y \qquad (1\text{-}6)$$

66. $\dfrac{0.000\ 000\ 000\ 52}{(1,300)(0.000\ 002)} = \dfrac{5.2 \times 10^{-10}}{(1.3 \times 10^3)(2 \times 10^{-6})}$

$$= \frac{\overset{2}{\cancel{5.2}} \times 10^{-10}}{\underset{1}{\cancel{2.6}} \times 10^{-3}} = 2 \times 10^{-10-(-3)}$$

$$= 2 \times 10^{-7} \qquad (1\text{-}5)$$

67. $\dfrac{(20,410)(0.000\ 003\ 477)}{0.000\ 000\ 022\ 09} = \dfrac{(2.041 \times 10^4)(3.477 \times 10^{-6})}{(2.209 \times 10^{-8})} = 3.213 \times 10^6$

Note that the input is given to 4 significant digits, hence we interpret 4 significant digits of the calculator answer. $(1\text{-}5)$

68. 4.434×10^{-5} $\qquad (1\text{-}5)$

69. -4.541×10^{-6} $\qquad (1\text{-}5)$

70. $82.45^{8/3} = (82.45)^{(8\div3)} = 128,800$ $\qquad (1\text{-}6)$

71. $(0.000\ 000\ 419\ 9)^{2/7} = (4.199 \times 10^{-7})^{(2\div7)} = 0.01507$ $\qquad (1\text{-}6)$

72. $\sqrt[5]{0.006\ 604} = (6.604 \times 10^{-3})^{1/5} = (6.604 \times 10^{-3})^{0.2} = 0.3664$ $\qquad (1\text{-}7)$

73. $\sqrt[3]{3 + \sqrt{2}} = (3 + \sqrt{2})^{(1\div3)} = 1.640$ $\qquad (1\text{-}7)$

74. $\dfrac{2^{-1/2} - 3^{-1/2}}{2^{-1/3} + 3^{-1/3}} = [2^{(-0.5)} - 3^{(-0.5)}] \div [2^{(-1\div3)} + 3^{(-1\div3)}] = 0.08726$ $\qquad (1\text{-}6)$

75. $-2x\sqrt[5]{3^6x^7y^{11}} = -2x\sqrt[5]{3^5x^5y^{10} \cdot 3x^2y} = -2x\sqrt[5]{3^5x^5y^{10}}\sqrt[5]{3x^2y}$

$$= -2x \cdot 3xy^2\sqrt[5]{3x^2y} = -6x^2y^2\sqrt[5]{3x^2y} \qquad (1\text{-}7)$$

76. $\dfrac{2x^2}{\sqrt[3]{4x}} = \dfrac{2x^2}{\sqrt[3]{4x}}\dfrac{\sqrt[3]{2x^2}}{\sqrt[3]{2x^2}} = \dfrac{2x^2\sqrt[3]{2x^2}}{\sqrt[3]{8x^3}} = \dfrac{\overset{x}{\cancel{2x^2}}\sqrt[3]{2x^2}}{\underset{1}{\cancel{2x}}} = x\sqrt[3]{2x^2}$ $\qquad (1\text{-}7)$

77. $\sqrt[5]{\dfrac{3y^2}{8x^2}} = \sqrt[5]{\dfrac{3y^2}{8x^2}\dfrac{4x^3}{4x^3}} = \sqrt[5]{\dfrac{12x^3y^2}{32x^5}} = \dfrac{\sqrt[5]{12x^3y^2}}{\sqrt[5]{32x^5}} = \dfrac{\sqrt[5]{12x^3y^2}}{2x}$ $\qquad (1\text{-}7)$

78. $\sqrt[9]{8x^6y^{12}} = \sqrt[9]{2^3x^6y^{12}} = \sqrt[3 \cdot 3]{2^{3 \cdot 1}x^{3 \cdot 2}y^{3 \cdot 4}} = \sqrt[3]{2x^2y^4} = \sqrt[3]{y^3 \cdot 2x^2y} = \sqrt[3]{y^3}\sqrt[3]{2x^2y} = y\sqrt[3]{2x^2y}$ $(1\text{-}7)$

79. $\sqrt{\sqrt[3]{4x^4}} = \sqrt[2 \cdot 3]{2^{2 \cdot 1}x^{2 \cdot 2}} = \sqrt[3]{2x^2}$ $\qquad (1\text{-}7)$

80. $(2\sqrt{x} - 5\sqrt{y})(\sqrt{x} + \sqrt{y}) = 2\sqrt{x}\sqrt{x} + 2\sqrt{x}\sqrt{y} - 5\sqrt{x}\sqrt{y} - 5\sqrt{y}\sqrt{y}$

$$= 2x - 3\sqrt{x}\sqrt{y} - 5y = 2x - 3\sqrt{xy} - 5y \qquad (1\text{-}7)$$

81. $\dfrac{3\sqrt{x}}{2\sqrt{x} - \sqrt{y}} = \dfrac{3\sqrt{x}}{(2\sqrt{x} - \sqrt{y})} \dfrac{(2\sqrt{x} + \sqrt{y})}{(2\sqrt{x} + \sqrt{y})} = \dfrac{3\sqrt{x}(2\sqrt{x} + \sqrt{y})}{(2\sqrt{x})^2 - (\sqrt{y})^2}$

$\qquad = \dfrac{3\sqrt{x}(2\sqrt{x} + \sqrt{y})}{4x - y} = \dfrac{6x + 3\sqrt{xy}}{4x - y}$ $\hfill (1\text{-}7)$

82. $\dfrac{2\sqrt{u} - 3\sqrt{v}}{2\sqrt{u} + 3\sqrt{v}} = \dfrac{(2\sqrt{u} - 3\sqrt{v})}{(2\sqrt{u} + 3\sqrt{v})} \dfrac{(2\sqrt{u} - 3\sqrt{v})}{(2\sqrt{u} - 3\sqrt{v})} = \dfrac{(2\sqrt{u})^2 - 2(2\sqrt{u})(3\sqrt{v}) + (3\sqrt{v})^2}{(2\sqrt{u})^2 - (3\sqrt{v})^2}$

$\qquad = \dfrac{4u - 12\sqrt{uv} + 9v}{4u - 9v}$ $\hfill (1\text{-}7)$

83. $\dfrac{y^2}{\sqrt{y^2 + 4} - 2} = \dfrac{y^2}{(\sqrt{y^2 + 4} - 2)} \dfrac{(\sqrt{y^2 + 4} + 2)}{(\sqrt{y^2 + 4} + 2)} = \dfrac{y^2(\sqrt{y^2 + 4} + 2)}{(\sqrt{y^2 + 4})^2 - (2)^2} = \dfrac{y^2(\sqrt{y^2 + 4} + 2)}{y^2 + 4 - 4}$

$\qquad = \dfrac{y^2(\sqrt{y^2 + 4} + 2)}{y^2} = \sqrt{y^2 + 4} + 2$ $\hfill (1\text{-}7)$

84. $\dfrac{\sqrt{t} - \sqrt{5}}{t - 5} = \dfrac{(\sqrt{t} - \sqrt{5})}{(t - 5)} \dfrac{(\sqrt{t} + \sqrt{5})}{\sqrt{t} + \sqrt{5}} = \dfrac{(\sqrt{t})^2 - (\sqrt{5})^2}{(t - 5)(\sqrt{t} + \sqrt{5})}$

$\qquad = \dfrac{\overset{1}{\cancel{t - 5}}}{\underset{1}{\cancel{(t - 5)}}(\sqrt{t} + \sqrt{5})} = \dfrac{1}{\sqrt{t} + \sqrt{5}}$ $\hfill (1\text{-}7)$

85. $\dfrac{4\sqrt{x} - 3}{2\sqrt{x}} = \dfrac{4\sqrt{x}}{2\sqrt{x}} - \dfrac{3}{2\sqrt{x}} = 2 - \dfrac{3}{2x^{1/2}} = 2 - \dfrac{3}{2}x^{-1/2} = 2x^0 - \dfrac{3}{2}x^{-1/2}$ $\hfill (1\text{-}7)$

86. Let $\qquad c = 0.54545454\ldots.$ Then

$\qquad 100c = 54.545454\ldots$

So $(100c - c) = (54.545454\ldots) - (0.54545454\ldots)$

$\qquad\qquad = 54$

$\qquad 99c = 54$

$\qquad c = \dfrac{54}{99} = \dfrac{6}{11}$

The number can be written as the quotient of two integers, so it is rational. $\quad (1\text{-}1)$

87. (A) List each element of M. Follow these with each element of N that is not yet listed. $\{-4, -3, 2, 0\}$, or, in increasing order, $\{-4, -3, 0, 2\}$.

(B) List each element of M that is also an element of N. $\{-3, 2\}$. $\hfill (1\text{-}1)$

88. $x^2 - 4x + 1 = (2 - \sqrt{3})^2 - 4(2 - \sqrt{3}) + 1 = (2)^2 - 2(2)\sqrt{3} + (\sqrt{3})^2 - 8 + 4\sqrt{3} + 1$

$\qquad = 4 - 4\sqrt{3} + 3 - 8 + 4\sqrt{3} + 1 = 0$ $\hfill (1\text{-}7)$

89. $x(2x - 1)(x + 3) = x(2x^2 + 6x - x - 3)$

$\qquad\qquad\qquad\quad = x(2x^2 + 5x - 3)$

$\qquad\qquad\qquad\quad = 2x^3 + 5x^2 - 3x$

$\qquad (x - 1)^3 = (x - 1)(x - 1)(x - 1)$

$\qquad\qquad\quad = (x^2 - 2x + 1)(x - 1)$

This multiplication we perform vertically:

$x^2 - 2x + 1$

$\underline{x - 1}$

$x^3 - 2x^2 + x$

$\underline{\quad - x^2 + 2x - 1}$

$x^3 - 3x^2 + 3x - 1$

Hence $x(2x - 1)(x + 3) - (x - 1)^3 = 2x^3 + 5x^2 - 3x - (x^3 - 3x^2 + 3x - 1)$
$$= 2x^3 + 5x^2 - 3x - x^3 + 3x^2 - 3x + 1$$
$$= x^3 + 8x^2 - 6x + 1 \qquad (1\text{-}2)$$

90. $4x(a^2 - 4a + 4) - 9x^3 = x[4(a^2 - 4a + 4) - 9x^2]$
$$= x[2^2(a - 2)^2 - (3x)^2]$$
$$= x[2(a - 2) + 3x][2(a - 2) - 3x]$$
$$= x(2a - 4 + 3x)(2a - 4 - 3x)$$
$$= x(2a + 3x - 4)(2a - 3x - 4) \qquad (1\text{-}3)$$

91. (A) $\sqrt{3 + \sqrt{5}} + \sqrt{3 - \sqrt{5}} \approx 3.162\ 277\ 660$

(B) $\sqrt{4 + \sqrt{15}} + \sqrt{4 - \sqrt{15}} \approx 3.162\ 277\ 660$

(C) $\sqrt{10} \approx 3.162\ 277\ 660$

Thus, (A), (B), and (C) all have the same value to 9 decimal places. To show that they are actually equal:

$$\left(\sqrt{3 + \sqrt{5}} + \sqrt{3 - \sqrt{5}}\right)^2 = \left(\sqrt{3 + \sqrt{5}}\right)^2 + 2\sqrt{3 + \sqrt{5}}\sqrt{3 - \sqrt{5}} + \left(\sqrt{3 - \sqrt{5}}\right)^2$$
$$= 3 + \sqrt{5} + 2\sqrt{(3 + \sqrt{5})(3 - \sqrt{5})} + 3 - \sqrt{5}$$
$$= 6 + 2\sqrt{3^2 - (\sqrt{5})^2}$$
$$= 6 + 2\sqrt{9 - 5}$$
$$= 6 + 4$$
$$= 10$$

Similarly,

$$\left(\sqrt{4 + \sqrt{15}} + \sqrt{4 - \sqrt{15}}\right)^2 = \left(\sqrt{4 + \sqrt{15}}\right)^2 + 2\sqrt{4 + \sqrt{15}}\sqrt{4 - \sqrt{15}} + \left(\sqrt{4 - \sqrt{15}}\right)^2$$
$$= 4 + \sqrt{15} + 2\sqrt{(4 + \sqrt{15})(4 - \sqrt{15})} + 4 - \sqrt{15}$$
$$= 8 + 2\sqrt{4^2 - (\sqrt{15})^2}$$
$$= 8 + 2\sqrt{16 - 15}$$
$$= 8 + 2$$
$$= 10$$

Therefore, since both $\sqrt{3 + \sqrt{5}} + \sqrt{3 - \sqrt{5}}$ and $\sqrt{4 + \sqrt{15}} + \sqrt{4 - \sqrt{15}}$ are positive, they are both equal to the positive square root of 10. Thus,

$$\sqrt{3 + \sqrt{5}} + \sqrt{3 - \sqrt{5}} = \sqrt{4 + \sqrt{15}} + \sqrt{4 - \sqrt{15}} = \sqrt{10} \qquad (1\text{-}7)$$

92. $\dfrac{8(x - 2)^{-3}(x + 3)^2}{12(x - 2)^{-4}(x + 3)^{-2}} = \dfrac{2(x - 2)^{-3-(-4)}(x + 3)^{2-(-2)}}{3} = \dfrac{2(x - 2)^1(x + 3)^4}{3}$
$$= \dfrac{2(x - 2)(x + 3)^4}{3} = \dfrac{2}{3}(x - 2)(x + 3)^4 \qquad (1\text{-}5)$$

93. $\left(\dfrac{a^{-2}}{b^{-1}} + \dfrac{b^{-2}}{a^{-1}}\right)^{-1} = \left(\dfrac{b}{a^2} + \dfrac{a}{b^2}\right)^{-1} = \left(\dfrac{b^3}{a^2 b^2} + \dfrac{a^3}{a^2 b^2}\right)^{-1} = \left(\dfrac{b^3 + a^3}{a^2 b^2}\right)^{-1} = \dfrac{a^2 b^2}{b^3 + a^3}$ or $\dfrac{a^2 b^2}{a^3 + b^3}$ $(1\text{-}5)$

94. $(x^{1/3} - y^{1/3})(x^{2/3} + x^{1/3}y^{1/3} + y^{2/3}) = (x^{1/3})^3 - (y^{1/3})^3 = x - y \qquad (1\text{-}6)$

95. $\left(\dfrac{x^{m^2}}{x^{2m-1}}\right)^{1/(m-1)} = (x^{m^2-(2m-1)})^{1/(m-1)} = (x^{m^2-2m+1})^{1/(m-1)} = x^{(m^2-2m+1)/(m-1)} = x^{(m-1)^2/(m-1)}$
$$= x^{\overset{m-1}{\cancel{(m-1)^2}}/\underset{1}{\cancel{m-1}}} = x^{m-1} \qquad (1\text{-}6)$$

96.
$$\frac{1}{1 - \sqrt[3]{x}} = \frac{1}{(1 - \sqrt[3]{x})} \frac{[1 + \sqrt[3]{x} + (\sqrt[3]{x})^2]}{[1 + \sqrt[3]{x} + (\sqrt[3]{x})^2]} = \frac{1 + \sqrt[3]{x} + \sqrt[3]{x^2}}{(1)^3 - (\sqrt[3]{x})^3}$$

Common Error:
$(1 - \sqrt[3]{x})(1 + \sqrt[3]{x}) \neq 1 - x$
This would not be a correct application of
$(a - b)(a + b) = a^2 - b^2$

$$= \frac{1 + \sqrt[3]{x} + \sqrt[3]{x^2}}{1 - x}$$

(1-7)

97.
$$\frac{\sqrt[3]{t} - \sqrt[3]{5}}{t - 5} = \frac{(\sqrt[3]{t} - \sqrt[3]{5})}{(t - 5)} \frac{[(\sqrt[3]{t})^2 + \sqrt[3]{t}\sqrt[3]{5} + (\sqrt[3]{5})^2]}{[(\sqrt[3]{t})^2 + \sqrt[3]{t}\sqrt[3]{5} + (\sqrt[3]{5})^2]} = \frac{(\sqrt[3]{t})^3 - (\sqrt[3]{5})^3}{(t - 5)[\sqrt[3]{t^2} + \sqrt[3]{5t} + \sqrt[3]{5^2}]}$$

$$= \frac{\overset{1}{\cancel{t - 5}}}{\underset{1}{\cancel{(t - 5)}}[\sqrt[3]{t^2} + \sqrt[3]{5t} + \sqrt[3]{25}]} = \frac{1}{\sqrt[3]{t^2} + \sqrt[3]{5t} + \sqrt[3]{25}}$$

(1-7)

98.
$$\sqrt[(n+1)]{x^{n^2}x^{2n+1}} = \sqrt[n+1]{x^{n^2+2n+1}}$$

$$= (x^{n^2+2n+1})^{1/(n+1)} = x^{(n^2+2n+1)/(n+1)} = x^{(n+1)^2/(n+1)} = x^{\overset{n+1}{\cancel{(n+1)}^2/\cancel{n+1}}} = x^{n+1}$$

(1-7)

99. The volume of the concrete wall is equal to the volume of the outer cylinder ($V = \pi r^2 h$) minus the volume of the basin. Since the radius of the basin is x ft and the concrete is 2 ft thick, the radius of the outer cylinder is $x + 2$ ft. Thus, we have

$$\begin{pmatrix} \text{Volume of} \\ \text{concrete} \\ \text{wall} \end{pmatrix} = \begin{pmatrix} \text{Volume of} \\ \text{outer} \\ \text{cylinder} \end{pmatrix} - \begin{pmatrix} \text{Volume} \\ \text{of} \\ \text{basin} \end{pmatrix}$$

$$\begin{aligned} \text{Volume} &= \pi(x + 2)^2 3 - \pi x^2 3 \\ &= \pi(x^2 + 4x + 4)3 - \pi x^2 3 \\ &= 3\pi x^2 + 12\pi x + 12\pi - 3\pi x^2 \\ &= 12\pi x + 12\pi \ (\text{ft}^3) \end{aligned}$$

(1-2)

100. average energy consumption per person $= \dfrac{\text{total energy consumption}}{\text{number of persons}}$

$$= \frac{2,257,000,000,000}{235,000,000}$$

$$= \frac{2.257 \times 10^{12}}{2.35 \times 10^8}$$

$$= 9.60 \times 10^3 \text{ kilograms per person}$$

$$= 9,600 \text{ kg per person}$$

(1-5)

101. (A) We are to estimate $N = 20x^{1/2}y^{1/2}$ given $x = 1,600$ units of capital and $y = 900$ units of labor.
$N = 20(1,600)^{1/2}(900)^{1/2} = 20(40)(30) = 24,000$ units produced.

(B) Given $x = 3,200$ units of capital and $y = 1,800$ units of labor, then
$$\begin{aligned} N = 20(3,200)^{1/2}(1,800)^{1/2} &= 20(2 \cdot 1,600)^{1/2}(2 \cdot 900)^{1/2} \\ &= 20 \cdot 2^{1/2}(1,600)^{1/2} \, 2^{1/2}(900)^{1/2} \\ &= 20 \cdot 2^{1/2}(40) \cdot 2^{1/2}(30) \\ &= 20(40)(30) \cdot 2^{1/2} \cdot 2^{1/2} \\ &= 24,000 \cdot 2^1 \\ &= 48,000 \text{ units produced} \end{aligned}$$

(C) The effect of raising x to $2x$ and y to $2y$ is to replace N by
$$\begin{aligned} 20(2x)^{1/2}(2y)^{1/2} &= 20 \cdot 2^{1/2}x^{1/2} 2^{1/2}y^{1/2} \\ &= 20 \cdot 2^{1/2} \cdot 2^{1/2}x^{1/2}y^{1/2} \\ &= 2^1 \cdot 20x^{1/2}y^{1/2} \\ &= 2N \end{aligned}$$

Thus, the production is doubled at any production level.

(1-6)

102. $\dfrac{1}{\dfrac{1}{R_1} + \dfrac{1}{R_2} + \dfrac{1}{R_3}} = \dfrac{R_1 R_2 R_3 \cdot 1}{R_1 R_2 R_3 \left(\dfrac{1}{R_1} + \dfrac{1}{R_2} + \dfrac{1}{R_3} \right)} = \dfrac{R_1 R_2 R_3}{R_2 R_3 + R_1 R_3 + R_1 R_2}$　　*(1-4)*

103. (A) The area of the cardboard can be written as (Original area) - (Removed area), where the original area = 16 × 30 = 480 and the removed area consists of 6 squares of area x^2 each; thus

　　　(Original area) - (Removed area) = $480 - 6x^2$ in expanded form.

In factored form, $480 - 6x^2 = 6(80 - x^2)$

(B) See figure.

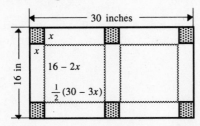 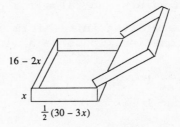

The volume of the box = $\ell wh = x(16 - 2x)\dfrac{1}{2}(30 - 3x)$

　　　　　　　　　　　$= x(16 - 2x)(15 - 1.5x)$
　　　　　　　　　　　　or
　　　　　　　　　$3x(8 - x)(10 - x)$ in factored form.

In expanded form, $3x(8 - x)(10 - x) = 3x(80 - 18x + x^2)$
　　　　　　　　　　　　　　　　　$= 240x - 54x^2 + 3x^3$　　*(1-3)*

CHAPTER 2

Exercise 2-1

Key Ideas and Formulas

The **solution set** for an equation is defined to be the set of elements in the domain of the variable that makes the equation true. Each element of the solution set is called a **solution**, or **root**, of the equation. To **solve an equation** is to find the solution set for the equation.

An equation is called an **identity** if the equation is true for all elements from the domain of the variable. An equation is called a **conditional equation** if it is true for certain domain values and false for others.

Any equation that can be written in the form $ax + b = 0$ $a \neq 0$ is called a linear, or first degree, equation. To solve a linear equation, use properties of equality:

If $a = b$, then $a + c = b + c$

If $a = b$, then $a - c = b - c$

If $a = b$, then $ca = cb$ $c \neq 0$

If $a = b$, then $\dfrac{a}{c} = \dfrac{b}{c}$ $c \neq 0$

If $a = b$, then $b = a$

If $a = b$, then either may replace the other in any statement without changing the truth or falsity of the statement.

To solve equations involving variables in the denominator, we must exclude any value of the variable that will make a denominator 0. With these values excluded, we may multiply through by the LCD.

---

1. $3(x + 2) = 5(x - 6)$
$3x + 6 = 5x - 30$
$-2x = -36$
$x = 18$
Solution: 18

3. $5 + 4(t - 2) = 2(t + 7) + 1$
$5 + 4t - 8 = 2t + 14 + 1$
$4t - 3 = 2t + 15$
$2t = 18$
$t = 9$
Solution: 9

5. $3 - \dfrac{2x - 3}{3} = \dfrac{5 - x}{2}$ LCD = 6

$6 \cdot 3 - 6\dfrac{(2x - 3)}{3} = 6\dfrac{(5 - x)}{2}$

$18 - 2(2x - 3) = 3(5 - x)$

$18 - 4x + 6 = 15 - 3x$

$-4x + 24 = 15 - 3x$

$-x = -9$

$x = 9$

Solution: 9

> **Common Error:**
> After line 2, students often write
> $18 - \overset{2}{\cancel{6}}\dfrac{2x - 3}{\cancel{3}} = \ldots$
> $18 - 4x - 3 = \ldots$
> forgetting to distribute the -2. Put compound numerators in parentheses to avoid this.

7.

$$5 - \frac{2x - 1}{4} = \frac{x + 2}{3} \quad \text{LCD} = 12$$

$$12 \cdot 5 - 3(2x - 1) = 4(x + 2)$$
$$60 - 6x + 3 = 4x + 8$$
$$-6x + 63 = 4x + 8$$
$$-10x = -55$$
$$x = \frac{-55}{-10}$$

Solution: $\frac{11}{2}$ or 5.5

9.

$$0.1(x - 7) + 0.05x = 0.8$$
$$0.1x - 0.7 + 0.05x = 0.8$$
$$0.15x - 0.7 = 0.8$$
$$0.15x = 1.5$$
$$x = \frac{1.5}{0.15}$$
$$x = 10$$

Solution: 10

Problems of this type can also be solved by elimination of decimals. Thus we could multiply every term on both sides by 100 to get $10(x - 7) + 5x = 80$ and proceed from here.

11.

$$0.3x - 0.04(x + 1) = 2.04$$
$$0.3x - 0.04x - 0.04 = 2.04$$
$$0.26x - 0.04 = 2.04$$
$$0.26x = 2.08$$
$$x = \frac{2.08}{0.26}$$
$$x = 8$$

Solution: 8

13.

$$\frac{1}{m} - \frac{1}{9} = \frac{4}{9} - \frac{2}{3m}$$

Excluded value: $m \neq 0$ LCD = $9m$

$$9m \cdot \frac{1}{m} - 9m \cdot \frac{1}{9} = 9m \cdot \frac{4}{9} - 9m \cdot \frac{2}{3m}$$
$$9 - m = 4m - 6$$
$$-5m = -15$$
$$m = 3$$

Solution: 3

15.

$$\frac{5x}{x + 5} = 2 - \frac{25}{x + 5}$$

Excluded value: $x \neq -5$ LCD = $x + 5$

$$(x + 5)\frac{5x}{x + 5} = (x + 5)2 - (x + 5)\frac{25}{x + 5}$$
$$5x = 2x + 10 - 25$$
$$5x = 2x - 15$$
$$3x = -15$$
$$x = -5$$

No Solution: -5 is excluded

> **Common Error:** Do not accept the proposed solution (-5) without checking the list of excluded values.

17.

$$\frac{2x}{10} - \frac{3 - x}{14} = \frac{2 + x}{5} - \frac{1}{2} \quad \text{LCD} = 70$$

$$70 \cdot \frac{2x}{10} - 70 \cdot \frac{(3 - x)}{14} = 70 \cdot \frac{(2 + x)}{5} - 70 \cdot \frac{1}{2}$$
$$14x - 5(3 - x) = 14(2 + x) - 35$$
$$14x - 15 + 5x = 28 + 14x - 35$$
$$19x - 15 = 14x - 7$$
$$5x = 8$$
$$x = \frac{8}{5}$$

Solution: $\frac{8}{5}$

19.

$$\frac{1}{3} - \frac{s - 2}{2s + 4} = \frac{s + 2}{3s + 6} \quad \text{Excluded value: } s \neq -2$$

$$\frac{1}{3} - \frac{s - 2}{2(s + 2)} = \frac{s + 2}{3(s + 2)} \quad \text{LCD} = 6(s + 2)$$

$$6(s + 2) \cdot \frac{1}{3} - 6(s + 2)\frac{(s - 2)}{2(s + 2)} = 6(s + 2)\frac{(s + 2)}{3(s + 2)}$$
$$2(s + 2) - 3(s - 2) = 2(s + 2)$$
$$2s + 4 - 3s + 6 = 2s + 4$$
$$-s + 10 = 2s + 4$$
$$-3s = -6$$
$$s = 2$$

Solution: 2

21.
$$\frac{3x}{2 - x} + \frac{6}{x - 2} = 3 \quad \text{Excluded value: } x \neq 2 \quad \text{LCD} = x - 2$$

$$(x - 2)\frac{3x}{2 - x} + (x - 2)\frac{6}{x - 2} = (x - 2)3$$

$$-3x + 6 = (x - 2)3$$
$$-3x + 6 = 3x - 6$$
$$-6x = -12$$
$$x = 2$$

No solution: 2 is excluded

23.
$$\frac{5t - 22}{t^2 - 6t + 9} - \frac{11}{t^2 - 3t} - \frac{5}{t} = 0$$

$$\frac{5t - 22}{(t - 3)^2} - \frac{11}{t(t - 3)} - \frac{5}{t} = 0 \quad \begin{array}{l}\text{Excl. values: } t \neq 0, 3 \\ \text{LCD} = t(t - 3)^2\end{array}$$

$$t(t - 3)^2\frac{(5t - 22)}{(t - 3)^2} - t(t - 3)^2\frac{11}{t(t - 3)} - t(t - 3)^2\frac{5}{t} = 0$$

$$t(5t - 22) - (t - 3)11 - (t - 3)^2 5 = 0$$
$$5t^2 - 22t - 11t + 33 - 5(t^2 - 6t + 9) = 0$$
$$5t^2 - 33t + 33 - 5t^2 + 30t - 45 = 0$$
$$-3t - 12 = 0$$
$$-3t = 12$$
$$t = -4$$

Solution: -4

25.
$$3.142x - 0.4835(x - 4) = 6.795$$
$$3.142x - 0.4835x + 1.934 = 6.795$$
$$2.6585x + 1.934 = 6.795$$
$$2.6585x = 4.861$$
$$x = \frac{4.861}{2.6585}$$
$$x = 1.83 \text{ to 3 significant digits}$$

Solution: 1.83

27.
$$\frac{2.32x}{x - 2} - \frac{3.76}{x} = 2.32$$

Excluded values : $x \neq 0, 2$ \quad LCD $= x(x - 2)$

$$x(x - 2)\frac{2.32x}{x - 2} - x(x - 2)\frac{3.76}{x} = 2.32x(x - 2)$$
$$2.32x^2 - 3.76(x - 2) = 2.32x(x - 2)$$
$$2.32x^2 - 3.76x + 7.52 = 2.32x^2 - 4.64x$$
$$-3.76x + 7.52 = -4.64x$$
$$7.52 = -0.88x$$
$$x = -8.55$$

Solution: -8.55

29.
$$a_n = a_1 + (n - 1)d$$
$$a_1 + (n - 1)d = a_n$$
$$(n - 1)d = a_n - a_1$$
$$d = \frac{a_n - a_1}{n - 1}$$

31.
$$\frac{1}{f} = \frac{1}{d_1} + \frac{1}{d_2} \quad \text{LCD} = d_1 d_2 f$$

$$d_1 d_2 f\frac{1}{f} = d_1 d_2 f\frac{1}{d_1} + d_1 d_2 f\frac{1}{d_2}$$
$$d_1 d_2 = d_2 f + d_1 f$$
$$d_2 f + d_1 f = d_1 d_2$$
$$(d_2 + d_1)f = d_1 d_2$$
$$f = \frac{d_1 d_2}{d_2 + d_1}$$

33.
$$A = 2ab + 2ac + 2bc$$
$$2ab + 2ac + 2bc = A$$
$$2ab + 2ac = A - 2bc$$
$$a(2b + 2c) = A - 2bc$$
$$a = \frac{A - 2bc}{2b + 2c}$$

35.
$$y = \frac{2x - 3}{3x + 5}$$
$$(3x + 5)y = 2x - 3$$
$$3xy + 5y = 2x - 3$$
$$5y + 3 = 2x - 3xy$$
$$5y + 3 = x(2 - 3y)$$
$$\frac{5y + 3}{2 - 3y} = x$$
$$x = \frac{5y + 3}{2 - 3y}$$

37. The "solution" is incorrect. Although 3 is a solution of the two last equations, they are not equivalent to the first equation because both sides have been multiplied by $x - 3$, which is zero when $x = 3$. It is not permitted to multiply both sides of an equation by zero. When $x = 3$, the first equation involves division by zero. Since 3, the only possible solution, is not a solution, the given (first) equation has no solution.

39.
$$\frac{x - \frac{1}{x}}{1 + \frac{1}{x}} = 3 \quad \text{Excl. val.: } x \neq 0$$

$$\frac{x(x - \frac{1}{x})}{x(1 + \frac{1}{x})} = 3$$

$$\frac{x^2 - 1}{x + 1} = 3 \quad \text{Excl. val.: } x \neq -1$$

$$\frac{(x - 1)\overset{1}{\cancel{(x + 1)}}}{\underset{1}{\cancel{x + 1}}} = 3$$

$$x - 1 = 3$$
$$x = 4$$

Solution: 4

41.
$$\frac{x + 1 - \frac{2}{x}}{1 - \frac{1}{x}} = x + 2 \quad \text{Excl. val.: } x \neq 0$$

$$\frac{x(x + 1 - \frac{2}{x})}{x(1 - \frac{1}{x})} = x + 2$$

$$\frac{x^2 + x - 2}{x - 1} = x + 2 \quad \text{Excl. val.: } x \neq 1$$

$$\frac{\overset{1}{\cancel{(x - 1)}}(x + 2)}{\underset{1}{\cancel{x - 1}}} = x + 2$$

$$x + 2 = x + 2$$

Solution: All real numbers except the excluded numbers 0 and 1.

43.
$$y = \frac{a}{1 + \frac{b}{x + c}}$$

$$y = \frac{a(x + c)}{(x + c)(1 + \frac{b}{x + c})}$$

$$y = \frac{a(x + c)}{x + c + b}$$

$$y = \frac{ax + ac}{x + c + b}$$

$$y(x + c + b) = ax + ac$$
$$xy + cy + by = ax + ac$$
$$cy + by - ac = ax - xy$$
$$cy + by - ac = x(a - y)$$
$$\frac{cy + by - ac}{a - y} = x$$

$$x = \frac{cy + by - ac}{a - y}$$

45. Let x = the number,
Then 10 less than two thirds the number is one fourth the number
$$\frac{2}{3}x - 10 \quad = \quad \frac{1}{4}x$$

$$\frac{2}{3}x - 10 = \frac{1}{4}x$$

$$12\left(\frac{2}{3}x\right) - 12(10) = 12\left(\frac{1}{4}x\right)$$

$$8x - 120 = 3x$$
$$-120 = -5x$$
$$x = 24$$

The number is 24.

47.
Let x = first of the consecutive even numbers
$x + 2$ = second of the numbers
$x + 4$ = third of the numbers
$x + 6$ = fourth of the numbers
first + second + third = 2 more than twice fourth
$$x + x + 2 + x + 4 = 2 + 2(x + 6)$$
$$3x + 6 = 2 + 2x + 12$$
$$3x + 6 = 2x + 14$$
$$x = 8$$
The four consecutive numbers are 8, 10, 12, 14.

49. Let w = width of rectangle
$2w - 3$ = length of rectangle
We use the perimeter formula
$$P = 2a + 2b.$$
$54 = 2w + 2(2w - 3)$
$54 = 2w + 4w - 6$ $10 = w$
$54 = 6w - 6$ $17 = 2w - 3$
$60 = 6w$ dimensions:
                           17 meters × 10 meters

51. Let P = perimeter of triangle
16 = length of one side
$\dfrac{2}{7}P$ = length of second side
$\dfrac{1}{3}P$ = length of third side
We use the perimeter formula
$$P = a + b + c$$
$$P = 16 + \dfrac{2}{7}P + \dfrac{1}{3}P$$
$$21P = 21(16) + 21\left(\dfrac{2}{7}P\right) + 21\left(\dfrac{1}{3}P\right)$$
$21P = 336 + 6P + 7P$
$21P = 336 + 13P$
$8P = 336$
$P = 42$ feet

53.
Let P = price before discount
$0.20P$ = 20 percent discount on P
Then price before discount - discount = price after discount
$$P - 0.20P = 72$$
$$0.8P = 72$$
$$P = \dfrac{72}{0.8}$$
$$P = \$90$$

55. Let x = sales of employee
Then $x - 7,000$ = sales on which 8% commission is paid
$0.08(x - 7,000)$ = (rate of commission) × (sales) = (amount of commission)
$2,150 + 0.08(x - 7,000)$ = (base salary) + (amount of commission) = earnings
                Earnings = 3,170
$2,150 + 0.08(x - 7,000) = 3,170$
$2,150 + 0.08x - 560 = 3,170$
$0.08x + 1,590 = 3,170$
$0.08x = 1,580$
$$x = \dfrac{1,580}{0.08}$$
$x = \$19,750$

57. (A) We note: The temperature increased 2.5°C for each additional 100 meters of depth. Hence, the temperature increased 25 degrees for each additional kilometer of depth.

Let x = the depth (in kilometers), then $x - 3$ = the depth beyond 3 kilometers.
$25(x - 3)$ = the temperature increase for $x - 3$ kilometers of depth.
T = temperature at 3 kilometers + temperature increase.
$T = 30 + 25(x - 3)$

(B) We are to find T when $x = 15$. We use the above relationship as a formula.
$T = 30 + 25(15 - 3)$
$= 330°C$

(C) We are to find x when $T = 280$. We use the above relationship as an equation.
$280 = 30 + 25(x - 3)$
$280 = 30 + 25x - 75$
$280 = -45 + 25x$
$325 = 25x$
$x = 13$ kilometers

59. Let D = distance from earthquake to station

Then $\dfrac{D}{5}$ = time of primary wave.

$\dfrac{D}{3}$ = time of secondary wave.

Time difference = time of *slower* secondary wave - time of *faster* primary wave

$12 = \dfrac{D}{3} - \dfrac{D}{5}$

$15(12) = 15\left(\dfrac{D}{3}\right) - 15\left(\dfrac{D}{5}\right)$

$\qquad 180 = 5D - 3D$

$\qquad 180 = 2D$

$\qquad\quad D = 90$ miles

> **Common Error:**
> Time difference is *not* time of fast wave (short) - time of slow wave (long)

61. We set up the proportion

$$\dfrac{\text{marked trout in second sample}}{\text{total number in second sample}} = \dfrac{\text{marked trout in first sample}}{\text{total trout population}}$$

Let x = total population

$\dfrac{8}{200} = \dfrac{200}{x}$

$200x\left(\dfrac{8}{200}\right) = 200x\left(\dfrac{200}{x}\right)$

$\qquad 8x = 40{,}000$

$\qquad\ x = 5{,}000$ trout

63.
$\qquad\qquad$ Let x = amount of distilled water

$\qquad\qquad\qquad 50$ = amount of 30% solution

$\qquad\qquad$ Then $50 + x$ = amount of 25% solution

acid in 30% solution + acid in distilled water = acid in 25% solution

$\qquad\quad 0.3(50) + 0 = 0.25(50 + x)$

$\qquad\qquad 0.3(50) = 0.25(50 + x)$

$\qquad\qquad\qquad 15 = 12.5 + 0.25x$

$\qquad\qquad\quad 2.5 = 0.25x$

$\qquad\qquad\quad\ x = 10$ gallons

65.
$\qquad\qquad$ Let x = amount of 50% solution

$\qquad\qquad\qquad 5$ = amount of distilled water

$\qquad\qquad$ Then $x - 5$ = amount of 90% solution

acid in 90% solution + acid in distilled water = acid in 50% solution

$\quad 0.9(x - 5) + 0 = 0.5x$

$\qquad 0.9x - 4.5 = 0.5x$

$\qquad\qquad -4.5 = -0.4x$

$\qquad\qquad\quad\ x = 11.25$ liters

67.
$\qquad$ Let t = time for both computers to finish the job

Then $t + 1$ = time worked by old computer

$\qquad\quad t$ = time worked by new computer

Since the old computer can do 1 job in 5 hours, it works at a rate

(1 job) ÷ (5 hours) = $\dfrac{1}{5}$ job per hour

Similarly the new computer works at a rate of $\dfrac{1}{3}$ job per hour.

Part of job completed by old computer in $t + 1$ hours	+	Part of job completed by new computer in t hours	= 1 whole job.
(Rate of old)(time of old)	+	(Rate of new)(Time of new)	= 1
$\frac{1}{5}(t + 1)$	+	$\frac{1}{3}(t)$	= 1

$$15\left(\frac{1}{5}\right)(t + 1) + 15\left(\frac{1}{3}t\right) = 15$$
$$3(t + 1) + 5t = 15$$
$$3t + 3 + 5t = 15$$
$$8t + 3 = 15$$
$$8t = 12$$
$$t = 1.5 \text{ hours}$$

69. Let d = distance flown north

(A) Using $t = \frac{d}{r}$, we note:

$$\text{rate flying north} = 150 - 30 = 120 \text{ miles per hour}$$
$$\text{rate flying south} = 150 + 30 = 180 \text{ miles per hour}$$

time flying north + time flying south = 3 hours

$$\frac{d}{120} + \frac{d}{180} = 3$$
$$360\frac{d}{120} + 360\frac{d}{180} = 3(360)$$
$$3d + 2d = 1080$$
$$5d = 1080$$
$$d = 216 \text{ miles}$$

(B) We still use the above ideas, except that rate flying north = rate flying south = 150 miles per hour.

$$\frac{d}{150} + \frac{d}{150} = 3$$
$$\frac{2d}{150} = 3$$
$$\frac{d}{75} = 3$$
$$d = 225 \text{ miles}$$

71. Let x = frequency of second note
y = frequency of third note

$$\frac{264}{4} = \frac{x}{5} = \frac{y}{6}$$
$$\frac{264}{4} = \frac{x}{5} \qquad \frac{264}{4} = \frac{y}{6}$$
$$66 = \frac{x}{5} \qquad 66 = \frac{y}{6}$$
$$x = 330 \text{ hertz} \quad y = 396 \text{ hertz}$$

73. We are to find d when $p = 40$.

$$40 = -\frac{1}{5}d + 70$$
$$5(40) = 5\left(-\frac{1}{5}d\right) + 5(70)$$
$$200 = -d + 350$$
$$-150 = -d$$
$$d = 150 \text{ centimeters}$$

75. total height = height in sand + height in water + height in air

Let h = total height

$$h = \frac{1}{5}h + 20 + \frac{2}{3}h$$
$$15h = 3h + 300 + 10h$$
$$15h = 13h + 300$$
$$2h = 300$$
$$h = 150 \text{ feet}$$

77. The hands will meet when the minute hand has made exactly 1 more revolution than the hour hand.

$$\text{Let } t = \text{time after 12 o'clock in hours}$$
$$\text{rate of hour hand} = \tfrac{1}{12} \text{ revolution per hour}$$
$$\text{rate of minute hand} = 1 \text{ revolution per hour}$$
$$\text{distance of minute hand} = 1 + \text{distance of hour hand}$$
$$1(t) = 1 + \tfrac{1}{12}(t)$$
$$t = 1 + \tfrac{1}{12}t$$
$$12t = 12 + t$$
$$11t = 12$$
$$t = \tfrac{12}{11} \text{ hours}$$

That clock will read $12 + \tfrac{12}{11}$ o'clock, that is $1\tfrac{1}{11}$ hours after 12, or 1 hour $\tfrac{60}{11}$ minutes after 12. This will be $\tfrac{60}{11}$ minutes, or $5\tfrac{5}{11}$ minutes after 1 PM.

Exercise 2-2

Key Ideas and Formulas

A **system** of two linear equations in two variables can be written in the form:

$$ax + by = h$$
$$cx + dy = k$$

where x and y are variables and a, b, c, d, h, and k are real constants. A pair of numbers $x = x_0$ and $y = y_0$ is a **solution** of this system if each equation is satisfied by the pair. The set of all such pairs of numbers is called the **solution set** for the system. To **solve** a system is to find its solution set.

Such systems can be solved by the method of substitution (see text).

---

1. $y = 2x + 3$
$y = 3x - 5$

Substitute y from the first equation into the second equation to eliminate y.

$2x + 3 = 3x - 5$
$-x + 3 = -5$
$\quad -x = -8$
$\quad\ x = 8$

Now replace x with 8 in the first equation to find y.

$y = 2 \cdot 8 + 3$
$y = 19$

Solution: $x = 8$, $y = 19$

3. $x - y = 4$
$x + 3y = 12$

Solve the first equation for x in terms of y.
$x = 4 + y$

Substitute into the second equation to eliminate x.

$(4 + y) + 3y = 12$
$\qquad\quad 4y = 8$
$\qquad\quad\ y = 2$

Now replace y with 2 in the first equation to find x.

$x - 2 = 4$
$\qquad x = 6$
Solution: $x = 6$, $y = 2$

5. $3x - y = 7$
$2x + 3y = 1$
Solve the first equation for y in terms of x.
$-y = 7 - 3x$
$y = -7 + 3x$
Substitute into the second equation to eliminate y.
$2x + 3(-7 + 3x) = 1$
$2x - 21 + 9x = 1$
$11x = 22$
$x = 2$
Now replace x with 2 in the first equation to find y.
$3 \cdot 2 - y = 7$
$6 - y = 7$
$y = -1$
Solution: $x = 2$, $y = -1$

7. $4x + 3y = 26$
$3x - 11y = -7$
Solve the second equation for x in terms of y.
$3x = 11y - 7$
$x = \dfrac{11y - 7}{3}$
Substitute into the first equation to eliminate x.
$4\left(\dfrac{11y - 7}{3}\right) + 3y = 26$
$\dfrac{44y - 28}{3} + 3y = 26$
$44y - 28 + 9y = 78$
$53y = 106$
$y = 2$
$x = \dfrac{11 \cdot 2 - 7}{3}$
$x = 5$
Solution: $x = 5$, $y = 2$

9. $7m + 12n = -1$
$5m - 3n = 7$
Solve the first equation for n in terms of m.
$12n = -1 - 7m$
$n = \dfrac{-1 - 7m}{12}$
Substitute into the second equation to eliminate n.
$5m - 3\left(\dfrac{-1 - 7m}{12}\right) = 7$
$5m - \dfrac{-1 - 7m}{4} = 7$
$20m + 1 + 7m = 28$
$27m = 27$
$m = 1$
$n = \dfrac{-1 - 7(1)}{12}$
$n = -\dfrac{2}{3}$
Solution: $m = 1$, $n = -\dfrac{2}{3}$

11. $y = 0.08x$
$y = 100 + 0.04x$
Substitute y from the first equation into the second equation to eliminate y.
$0.08x = 100 + 0.04x$
$0.04x = 100$
$x = 2,500$
$y = 0.08(2,500)$
$y = 200$
Solution: $x = 2,500$, $y = 200$

13. $0.2u - 0.5v = 0.07$
$0.8u - 0.3v = 0.79$

For convenience, eliminate decimals by multiplying both sides of each equation by 100.

$20u - 50v = 7$
$80u - 30v = 79$

| **Common Error**: $2u - 5v \neq 7$ |
| $8u - 3v \neq 79$ |

Solve the first equation for u in terms of v and substitute into the second equation to eliminate u.

$$20u = 50v + 7$$

$$u = \frac{50v + 7}{20}$$

$$80\left(\frac{50v + 7}{20}\right) - 30v = 79$$

$$4(50v + 7) - 30v = 79$$

$$200v + 28 - 30v = 79$$

$$170v = 51$$

$$v = 0.3$$

$$u = \frac{50(0.3) + 7}{20}$$

$$u = 1.1$$

Solution: $u = 1.1$, $v = 0.3$

15. $\dfrac{2}{5}x + \dfrac{3}{2}y = 2$

$\dfrac{7}{3}x - \dfrac{5}{4}y = -5$

Eliminate fractions by multiplying both sides of the first equation by 10 and both sides of the second equation by 12.

$$10\left(\frac{2}{5}x + \frac{3}{2}y\right) = 20$$

$$4x + 15y = 20$$

$$12\left(\frac{7}{3}x - \frac{5}{4}y\right) = -60$$

$$28x - 15y = -60$$

Solve the first equation for y in terms of x and substitute into the second equation to eliminate y.

$$15y = 20 - 4x$$

$$y = \frac{20 - 4x}{15}$$

$$28x - 15\left(\frac{20 - 4x}{15}\right) = -60$$

$$28x - (20 - 4x) = -60$$

$$28x - 20 + 4x = -60$$

$$32x = -40$$

$$x = -\frac{5}{4}$$

$$y = \frac{20 - 4\left(-\frac{5}{4}\right)}{15}$$

$$y = \frac{20 + 5}{15}$$

$$y = \frac{5}{3}$$

Solution: $x = -\dfrac{5}{4}$, $y = \dfrac{5}{3}$

17. If a contradiction such as $0 = 1$ is encountered, this indicates that the system has no solutions. In the example, if we solve the first equation for x in terms of y we obtain $x = 2y - 3$. Substituting this into the second equation yields

$$-2(2y - 3) + 4y = 7$$

$$-4y + 6 + 4y = 7$$

$$6 = 7$$

Thus, this system has no solutions.

19. $x = 2 + p - 2q$

$y = 3 - p + 3q$

Solve the first equation for p in terms of q, x, and y and substitute into the second equation to eliminate p, then solve for q in terms of x and y.

$$p = x - 2 + 2q$$
$$y = 3 - (x - 2 + 2q) + 3q$$
$$y = 3 - x + 2 - 2q + 3q$$
$$y = 5 - x + q$$
$$q = x + y - 5$$

Now substitute this expression for q into $p = x - 2 + 2q$ to find p in terms of x and y.

$$p = x - 2 + 2(x + y - 5)$$
$$p = x - 2 + 2x + 2y - 10$$
$$p = 3x + 2y - 12$$

Solution: $p = 3x + 2y - 12$, $q = x + y - 5$

To check this solution substitute into the original equations to see if true statements result:

$$x = 2 + p - 2q \qquad\qquad y = 3 - p + 3q$$
$$x \overset{?}{=} 2 + (3x + 2y - 12) - 2(x + y - 5) \quad y \overset{?}{=} 3 - (3x + 2y - 12) + 3(x + y - 5)$$
$$x \overset{?}{=} 2 + 3x + 2y - 12 - 2x - 2y + 10 \quad y \overset{?}{=} 3 - 3x - 2y + 12 + 3x + 3y - 15$$
$$x \overset{\surd}{=} x \qquad\qquad\qquad\qquad y \overset{\surd}{=} y$$

21. $ax + by = h$
$cx + dy = k$

Solve the first equation for x in terms of y and the constants.

$$ax = h - by$$
$$x = \frac{h - by}{a} \qquad (a \neq 0)$$

Substitute this expression into the second equation to eliminate x.

$$c\left(\frac{h - by}{a}\right) + dy = k$$

$$ac\left(\frac{h - by}{a}\right) + ady = ak$$
$$c(h - by) + ady = ak$$
$$ch - bcy + ady = ak$$
$$(ad - bc)y = ak - ch$$
$$y = \frac{ak - ch}{ad - bc} \qquad ad - bc \neq 0$$

Similarly, solve the first equation for y in terms of x and the constants.

$$by = h - ax$$
$$y = \frac{h - ax}{b} \qquad (b \neq 0)$$

Substitute this expression into the second equation to eliminate y.

$$cx + d\left(\frac{h - ax}{b}\right) = k$$

$$bcx + bd\left(\frac{h - ax}{b}\right) = bk$$
$$bcx + d(h - ax) = bk$$
$$bcx + dh - adx = bk$$
$$(bc - ad)x = bk - dh$$
$$x = \frac{bk - dh}{bc - ad} \qquad bc - ad \neq 0$$

or, for consistency with the expression for y,

$$x = \frac{dh - bk}{ad - bc}$$

Solution: $x = \dfrac{dh - bk}{ad - bc}$, $y = \dfrac{ak - ch}{ad - bc} \qquad ad - bc \neq 0$

23. Let x = airspeed of the plane
     y = rate at which wind is blowing
Then
     $x - y$ = ground speed flying from Atlanta to Los Angeles (head wind)
     $x + y$ = ground speed flying from Los Angeles to Atlanta (tail wind)
Then, applying Distance = Rate × Time, we have
     $2,100 = 8.75(x - y)$
     $2,100 = 5(x + y)$

After simplification, we have
     $x - y = 240$
     $x + y = 420$
Solve the first equation for x in terms of y and substitute into the second equation.
$$x = 240 + y$$
$$240 + y + y = 420$$
$$2y = 180$$
$$y = 90 \text{ mph} = \text{wind rate}$$
$$x = 240 + y$$
$$x = 240 + 90$$
$$x = 330 \text{ mph} = \text{airspeed}$$

25. Let x = time rowed upstream
     y = time rowed downstream
Then $x + y = \dfrac{1}{4}$ (15 min = $\dfrac{1}{4}$ hr.)
Since   rate upstream = 20 - 2 = 18 mph and
       rate downstream = 20 + 2 = 22 mph,
applying Distance = Rate × Time to the equal distances upstream and downstream, we have
     $18x = 22y$
Solve the first equation for y in terms of x and substitute into the second equation.
$$y = \frac{1}{4} - x$$
$$18x = 22\left(\frac{1}{4} - x\right)$$
$$18x = 5.5 - 22x$$
$$40x = 5.5$$
$$x = 0.1375 \text{ hr.}$$
Then the distance rowed upstream = $18x = 18(0.1375) = 2.475$ km.

27. Let x = amount of 50% solution
     y = amount of 80% solution
100 milliliters are required, hence
     $x + y = 100$
68% of the 100 milliliters must be acid, hence
     $0.50x + 0.80y = 0.68(100)$

Solve the first equation for y in terms of x and substitute into the second equation.
$$y = 100 - x$$
$$0.50x + 0.80(100 - x) = 0.68(100)$$
$$0.5x + 80 - 0.8x = 68$$
$$-0.3x = -12$$
$$x = 40 \text{ milliliters of 50\% solution}$$
$$y = 100 - x$$
$$y = 100 - 40 = 60 \text{ milliliters of 80\% solution}$$

29. "Break even" means Cost = Revenue.

Let y = Cost = Revenue.

Let x = number of records sold

$\quad y$ = Revenue = number of records sold $\times$ price per record

$\quad y = x(8.00)$

$\quad y$ = Cost = Fixed Cost + Variable Cost

$\qquad\qquad$ = 17,680 + number of records $\times$ cost per record

$\qquad\quad y = 17,680 + x(4.60)$

Substitute y from the first equation into the second equation to eliminate y.

$\quad 8.00x = 17,680 + 4.60x$

$\quad 3.40x = 17,680$

$\qquad x = \dfrac{17,680}{3.40}$

$\qquad x = 5,200$ records

31. Let x = amount invested at 10% $\quad 0.1x$ = yield on amount invested at 10%

$\quad\ y$ = amount invested at 15% $\quad 0.15y$ = yield on amount invested at 15%

Then

$\qquad\qquad x + y = 12,000 \quad$ (amount invested)

$\quad 0.1x + 0.15y = 0.12(12,000) \quad$ (total yield)

Solve the first equation for y in terms of x and substitute into the second equation.

$\quad 0.1x + 0.15(12,000 - x) \qquad\quad = 0.12(12,000)$

$\quad 0.1x + 0.15(12,000) - 0.15x = 0.12(12,000)$

$\quad 0.1x + 1,800 - 0.15x \qquad\quad = 1,440$

$\qquad\qquad\quad -0.05x + 1,800 = 1,440$

$\qquad\qquad\quad -0.05x \qquad\qquad = -360$

$\qquad\qquad\qquad\qquad\quad x = \$7,200$ invested at 10%

$\qquad\qquad y = 12,000 - x = \$4,800$ invested at 15%

33. Let x = number of hours Mexico plant is operated

$\quad\ y$ = number of hours Taiwan plant is operated

Then (Production at Mexico plant) + (Production at Taiwan plant) = (Total Production)

$\qquad\quad 40x \qquad\qquad + \qquad\quad 20y \qquad\qquad = 4000$ (keyboards)

$\qquad\quad 32x \qquad\qquad + \qquad\quad 32y \qquad\qquad = 4000$ (screens)

Solve the first equation for y in terms of x and substitute into the second equation.

$\qquad\quad 20y = 4,000 - 40x$

$\qquad\qquad y = 200 - 2x$

$\qquad 32x + 32(200 - 2x) = 4,000$

$\qquad 32x + 6,400 - 64x = 4,000$

$\qquad\qquad\qquad -32x = -2,400$

$\qquad\qquad\qquad\quad x = 75$ hours Mexico plant

$\qquad\qquad\quad y = 200 - 2x = 200 - 2(75)$

$\qquad\qquad\qquad\qquad = 50$ hours Taiwan plant

35. Let x = number of grams of Mix A

$\quad\ y$ = number of grams of Mix B

Then (Nutrition from Mix A) + (Nutrition for Mix B) = (Total Nutrition)

$\qquad\qquad 0.10x \qquad + \qquad 0.20y \qquad = \qquad 20$ (Total protein)

$\qquad\qquad 0.06x \qquad + \qquad 0.02y \qquad = \qquad 6$ (Total fat)

For convenience, eliminate decimals by multiplying both sides of the first equation by 10 and the second equation by 100.

$\qquad\qquad x + 2y = 200$

$\qquad\quad 6x + 2y = 600$

Solve the first equation for x in terms of y and substitute into the second equation.

$$x = 200 - 2y$$
$$6(200 - 2y) + 2y = 600$$
$$1200 - 12y + 2y = 600$$
$$-10y = -600$$
$$y = 60 \text{ grams Mix } A$$
$$x = 200 - 2y = 200 - 2(60)$$
$$= 80 \text{ grams Mix } B$$

37. (A) Following the hint, write $p = aq + b$.
Since $p = 0.60$ corresponds to supply $q = 450$,
$$0.60 = 450a + b$$
Since $p = 0.90$ corresponds to supply $q = 750$,
$$0.90 = 750a + b$$

Solve the first equation for b in terms of a and substitute into the second equation.
$$b = 0.60 - 450a$$
$$0.90 = 750a + 0.60 - 450a$$
$$0.30 = 300a$$
$$a = 0.001$$
$$b = 0.60 - 450a = 0.60 - 450(0.001)$$
$$= 0.15$$
Thus, the supply equation is $p = 0.001q + 0.15$.

(B) Use the hint for part (A) to write $p = cq + d$.
Since $p = 0.60$ corresponds to demand $q = 645$,
$$0.60 = 645c + d$$
Since $p = 0.90$ corresponds to demand $q = 495$,
$$0.90 = 495c + d$$
Solve the first equation for d in terms of c and substitute into the second equation.
$$d = 0.60 - 645c$$
$$0.90 = 495c + 0.60 - 645c$$
$$0.30 = -150c$$
$$c = -0.002$$
$$d = 0.60 - 645c = 0.60 - 645(-0.002)$$
$$= 1.89$$
Thus, the demand equation is $p = -0.002q + 1.89$

(C) Solve the system of equations
$$p = 0.001q + 0.15$$
$$p = -0.002q + 1.89$$
Substitute p from the first equation into the second equation to eliminate p.
$$0.001q + 0.15 = -0.002q + 1.89$$
$$0.003q = 1.74$$
$$q = 580 \text{ bushels} = \text{equilibrium quantity}$$
$$p = 0.001q + 0.15 = 0.001(580) + 0.15$$
$$= \$0.73 \text{ equilibrium price}$$

63. $s = a + bt^2$

(A) We are given: When $t = 1$, $s = 180$
$\qquad\qquad\qquad\quad$ When $t = 2$, $s = 132$
Substituting these values in the given equation, we have
$$180 = a + b(1)^2$$
$$132 = a + b(2)^2 \quad \text{or}$$
$$180 = a + b$$
$$132 = a + 4b$$

Solve the first equation for a in terms of b and substitute into the second equation to eliminate a.

$$a = 180 - b$$
$$132 = 180 - b + 4b$$
$$132 = 180 + 3b$$
$$-48 = 3b$$
$$b = -16$$
$$180 = a - 16$$
$$a = 196$$

(B) The height of the building is represented by s, the distance of the object above the ground, when $t = 0$. Since we now know
$$s = 196 - 16t^2$$
from part (A), when $t = 0$, $s = 196$ feet is the height of the building.

(C) The object falls until s, its distance above the ground, is zero. Since
$$s = 196 - 16t^2$$
we substitute $s = 0$ and solve for t.
$$0 = 196 - 16t^2$$
$$16t^2 = 196$$
$$t^2 = \frac{196}{16}$$
$$t = \frac{14}{4} \quad \text{(discarding the negative solution)}$$
$$t = 3.5 \text{ seconds}$$

41. Let p = time of primary wave
$\quad\quad s$ = time for secondary wave
We know
$s - p = 16$ (time difference)
To find a second equation, we have to use Distance = Rate × Time
$\quad 5p$ = distance for primary wave
$\quad 3s$ = distance for secondary wave
These distances are equal, hence
$\quad 5p = 3s$

Solve the first equation for s in terms of p and substitute into the second equation to eliminate s.

$$s = p + 16$$
$$5p = 3(p + 16)$$
$$5p = 3p + 48$$
$$2p = 48$$
$$p = 24 \text{ seconds}$$
$$s = 24 + 16$$
$$s = 40 \text{ seconds}$$

The distance traveled = $5p = 3s = 120$ miles

Exercise 2-3

Key Ideas and Formulas

a is less than b ($a < b$) or b is greater than a ($b > a$) if there exists a positive real number p such that $a + p = b$.

For any two real numbers a and b, $a < b$, $a > b$, or $a = b$. (Trichotomy principle)

For any real numbers a, b, and c with $a < b$, then:
1. $a + c < b + c$
2. $a - c < b - c$

3. If c is positive, then $ac < bc$ and $\dfrac{a}{c} < \dfrac{b}{c}$

4. If c is negative, then $ac > bc$ and $\dfrac{a}{c} > \dfrac{b}{c}$

Similar properties hold if $a > b$, then $a + c > b + c$ and so on. The order of an inequality reverses if we multiply or divide both sides of an inequality statement by a negative number.

1. $-8 \leq x \leq 7$

3. $-6 \leq x < 6$

5. $x \geq -6$

7. $(-2, 6]$

9. $(-7, 8)$

11. $(-\infty, -2]$

13. $[-7, 2); -7 \leq x < 2$

15. $(-\infty, 0]; x \leq 0$

17. $7x - 8 < 4x + 7$
$\qquad 3x < 15$
$\qquad x < 5 \text{ or } (-\infty, 5)$

19. $3 - x \geq 5(3 - x)$
$\quad 3 - x \geq 15 - 5x$
$\qquad 4x \geq 12$
$\qquad x \geq 3 \text{ or } [3, \infty)$

21. $\dfrac{N}{-2} > 4$
$\qquad N < -8 \text{ or } (-\infty, -8)$

23. $-5t < -10$
$\qquad t > 2 \text{ or } (2, \infty)$

25. $3 - m < 4(m - 3)$
$\quad 3 - m < 4m - 12$
$\quad -5m < -15$
$\qquad m > 3 \text{ or } (3, \infty)$

Common Error:
Neglecting to reverse the order after division by -5.

27. $-2 - \dfrac{B}{4} \leq \dfrac{1 + B}{3}$
$$12\left(-2 - \frac{B}{4}\right) \leq 12\frac{(1 + B)}{3}$$
$\quad -24 - 3B \leq 4(1 + B)$
$\quad -24 - 3B \leq 4 + 4B$
$\qquad -7B \leq 28$
$\qquad B \geq -4 \text{ or } [-4, \infty)$

29. $-4 < 5t + 6 \leq 21$
$\quad -10 < 5t \leq 15$
$\quad\; -2 < t \leq 3 \text{ or } (-2, 3]$

31.

$[4, 7]$

$(-5, 5)$

$(-5, 5) \cup [4, 7] = (-5, 7]$ **$-5 < x \leq 7$**

33.

$[-1, 4)$

$(2, 6]$

$[-1, 4) \cap (2, 6] = (2, 4)$ **$2 < x < 4$**

35.

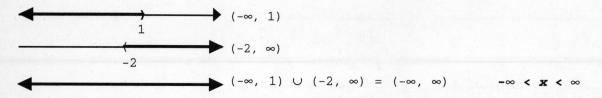

$(-\infty, 1)$

$(-2, \infty)$

$(-\infty, 1) \cup (-2, \infty) = (-\infty, \infty)$ $-\infty < x < \infty$

37.

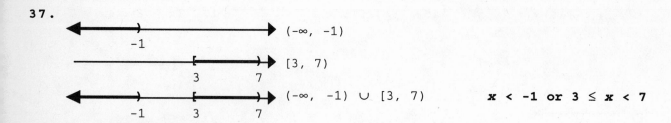

$(-\infty, -1)$

$[3, 7)$

$(-\infty, -1) \cup [3, 7)$ $x < -1$ or $3 \leq x < 7$

39.

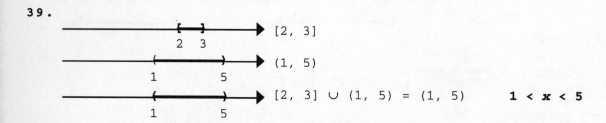

$[2, 3]$

$(1, 5)$

$[2, 3] \cup (1, 5) = (1, 5)$ $1 < x < 5$

41.

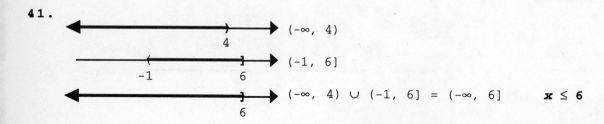

$(-\infty, 4)$

$(-1, 6]$

$(-\infty, 4) \cup (-1, 6] = (-\infty, 6]$ $x \leq 6$

43. $\dfrac{q}{7} - 3 > \dfrac{q - 4}{3} + 1$

$21\left(\dfrac{q}{7} - 3\right) > 21\left(\dfrac{(q - 4)}{3} + 1\right)$

$3q - 63 > 7(q - 4) + 21$
$3q - 63 > 7q - 28 + 21$
$3q - 63 > 7q - 7$
$\quad\;\; -4q > 56$
$\qquad q < -14$ or $(-\infty, -14)$

45. $\dfrac{2x}{5} - \dfrac{1}{2}(x - 3) \leq \dfrac{2x}{3} - \dfrac{3}{10}(x + 2)$ LCD = 30

$12x - 15(x - 3) \leq 20x - 9(x + 2)$
$12x - 15x + 45 \leq 20x - 9x - 18$
$-3x + 45 \leq 11x - 18$
$-14x \leq -63$
$x \geq 4.5$ or $[4.5, \infty)$

47.
$$-4 \le \frac{9}{5}x + 32 \le 68$$
$$-36 \le \frac{9}{5}x \le 36$$
$$\frac{5}{9}(-36) \le x \le \frac{5}{9}(36)$$
$$-20 \le x \le 20 \text{ or } [-20, 20]$$

49.
$$-12 < \frac{3}{4}(2 - x) \le 24$$
$$\frac{4}{3}(-12) < 2 - x \le \frac{4}{3}(24)$$
$$-16 < 2 - x \le 32$$
$$-18 < -x \le 30$$
$$18 > x \ge -30$$
$$-30 \le x < 18 \text{ or } [-30, 18)$$

51.
$$16 < 7 - 3x \le 31$$
$$9 < -3x \le 24$$
$$-3 > x \ge -8$$
$$-8 \le x < -3 \text{ or } [-8, -3)$$

53.
$$-6 < -\frac{2}{5}(1 - x) \le 4$$
$$-30 < -2(1 - x) \le 20$$
$$-30 < -2 + 2x \le 20$$
$$-28 < 2x \le 22$$
$$-14 < x \le 11 \text{ or } (-14, 11]$$

55.
$$5.23(x - 0.172) \le 6.02x - 0.427$$
$$5.23x - 0.89956 \le 6.02x - 0.427$$
$$5.23x - 6.02x - 0.89956 \le -0.427$$
$$-0.79x - 0.89956 \le -0.427$$
$$-0.79x \le -0.427 + 0.89956$$
$$-0.79x \le 0.47256$$
$$x \ge -0.60$$

57.
$$-0.703 < 0.112 - 2.28x < 0.703$$
$$-0.703 - 0.112 < -2.28x < 0.703 - 0.112$$
$$-0.815 < -2.28x < 0.591$$
$$0.357 > x > -0.259$$
$$-0.259 < x < 0.357$$

59. $\sqrt{1 - x}$ represents a real number exactly when $1 - x$ is positive or zero. We can write this as an inequality statement and solve for x.

$$1 - x \ge 0$$
$$-x \ge -1$$
$$x \le 1$$

61. $\sqrt{3x + 5}$ represents a real number exactly when $3x + 5$ is positive or zero. We can write this as an inequality statement and solve for x.

$$3x + 5 \ge 0$$
$$3x \ge -5$$
$$x \ge -\frac{5}{3}$$

63. $\dfrac{1}{\sqrt[4]{2x + 3}}$ represents a real number exactly when $2x + 3$ is positive. (*not* zero).
We can write this as an inequality statement and solve for x.

$$2x + 3 > 0$$
$$2x > -3$$
$$x > -\frac{3}{2}$$

65. (A) For $ab > 0$, ab must be positive, hence a and b must have the same sign.
Either
1. $a > 0$ and $b > 0$ or
2. $a < 0$ and $b < 0$

(B) For $ab < 0$, ab must be negative, hence a and b must have opposite signs.
Either
1. $a > 0$ and $b < 0$ or
2. $a < 0$ and $b > 0$

(C) For $\dfrac{a}{b} > 0$, $\dfrac{a}{b}$ must be positive, hence a and b must have the same sign.
Answer as in (A).

(D) For $\frac{a}{b} < 0$, $\frac{a}{b}$ must be negative, hence a and b must have opposite signs. Answer as in (B).

67. (A) If $a - b = 1$, then $a = b + 1$. Therefore, a is greater than b. >

(B) If $u - v = -2$, then $v = u + 2$. Therefore, u is less than v. <

69. If $\frac{b}{a}$ is greater than 1

$$\frac{b}{a} > 1$$

$a \cdot \frac{b}{a} < a \cdot 1$. (since a is negative.)

$b < a$

$0 < a - b$

$a - b$ is positive

71. (A) F   (B) T   (C) T

73. If $a < b$, then by definition of <, there exists a positive number p such that $a + p = b$. Then, adding c to both sides, we obtain $(a + c) + p = b + c$, where p is positive. Hence, by definition of <, we have
$a + c < b + c$

75. (A) If $a < b$, then by definition of <, there exists a positive number p such that $a + p = b$. If we multiply both sides of this by the positive number c, we obtain $(a + p)c = bc$, or $ac + pc = bc$, where pc is positive. Hence, by definition of <, we have $ac < bc$.

(B) If $a < b$, then by definition of <, there exists a positive number p such that $a + p = b$. If we multiply both sides of this by the negative number c, we obtain $(a + p)c = bc$, or $ac + pc = bc$, where pc is negative. Hence, by definition of <, we have $ac > bc$.

77. We want $200 \leq T \leq 300$. We are given $T = 30 + 25(x - 3)$. Substituting, we must solve
$200 \leq 30 + 25(x - 3) \leq 300$
$200 \leq 30 + 25x - 75 \leq 300$
$200 \leq -45 + 25x \leq 300$
$245 \leq 25x \leq 345$
$9.8 \leq x \leq 13.8$
(from 9.8 km to 13.8 km)

79. Let x = number of calculators sold
Then Revenue = (price per calculator) × (number of calculators sold) = $63x$
Cost = fixed cost + variable cost
= 650,000 + (Cost per calculator) × (number sold)
= 650,000 + 47x

(A) We want Revenue > Cost
$63x > 650,000 + 47x$
$16x > 650,000$
$x > 40,625$
More than 40,625 calculators must be sold for the company to make a profit.

(B) We want Revenue = Cost
$63x = 650,000 + 47x$
$16x = 650,000$
$x = 40,625$

(C) 40,625 calculators sold represents the break-even point, the boundary between profit and loss.

81. (A) The company might try to increase sales and keep the price the same (see part B). It might try to increase the price and keep the sales the same (see part C). Either of these strategies would need further analysis and implementation that are out of place in a discussion here.

(B) Here the cost has been changed to $650,000 + 50.5x$, but the revenue is still $63x$.

   Revenue > Cost

$$63x > 650,000 + 50.5x$$
$$12.5x > 650,000$$
$$x > 52,000 \text{ calculators}$$

(C) Let p = the new price. Here the cost is still $650,000 + 50.5x$ as in part (B) where x is now known to be $40,625$. Thus, cost = $650,000 + 50.5(40,625)$. The revenue is (price per calculator) × (number of calculators) = $p(40,625)$.

   Revenue > Cost

$$p(40,625) > 650,000 + 50.5(40,625)$$
$$p > \frac{650,000 + 50.5(40,625)}{40,625}$$
$$p > 66.50$$

The price could be raised by $3.50 to $66.50.

83. We want $220 \leq W \leq 2,750$. We are given $W = 110I$. Substituting, we must solve
$$220 \leq 110I \leq 2,750$$
$$2 \leq I \leq 25 \text{ or } [2, 25].$$

85. We are given that the benefit reduction (B) is $1 for each $2 that earnings (E) exceed $8,880. Thus, $B = \frac{1}{2}(E - 8,880)$.

In this situation, we are dealing with the range of values for E of between $13,000 and $16,000, that is:
$$13,000 \leq E \leq 16,000$$
Then
$$13,000 - 8,880 \leq E - 8,880 \leq 16,000 - 8,880$$
$$4,120 \leq E - 8,880 \leq 7,120$$
$$\frac{1}{2}(4,120) \leq \frac{1}{2}(E - 8,880) \leq \frac{1}{2}(7,120)$$
$$2,060 \leq B \leq 3,560$$
Thus, the benefit reduction is in the range between $2,060 and $3,560.

Exercise 2-4
Key Ideas and Formulas

$$|x| = \begin{cases} x \text{ if } x \geq 0 \\ -x \text{ if } x < 0 \end{cases}$$ [Note: $-x$ is positive if x is negative]

The distance between two points A and B on a number line with coordinates a and b respectively is given by
$$d(A, B) = |b - a| = |a - b|$$

Geometric Interpretation of Absolute Value Equations and Inequalities

Form ($d > 0$)	Geometric Interpretation	Solution	Graph
$\lvert x - c \rvert = d$	Distance between x and c is equal to d.	$\{c - d, c + d\}$	
$\lvert x - c \rvert < d$	Distance between x and c is less than d.	$(c - d, c + d)$	
$0 < \lvert x - c \rvert < d$	Distance between x and c is less than d but $x \neq c$.	$(c - d, c) \cup (c, c + d)$	
$\lvert x - c \rvert > d$	Distance between x and c is greater than d.	$(-\infty, c - d) \cup (c + d, \infty)$	

For $p > 0$

$\quad |x| = p$ is equivalent to $x = p$ or $x = -p$

$\quad |x| < p$ is equivalent to $-p < x < p$

$\quad |x| > p$ is equivalent to $x < -p$ or $x > p$

Common Error:
Writing: $p < x < -p$
This is true for no real value of x,
since $p < -p$ is false.

To solve: (p assumed positive)

$\quad |ax + b| = p$,   solve $ax + b = p$ or $ax + b = -p$

$\quad |ax + b| < p$,   solve $-p < ax + b < p$

$\quad |ax + b| > p$,   solve $ax + b < -p$ or $ax + b > p$

---

1. $\sqrt{5}$

3. $|(-6) - (-2)| = |-4| = 4$

5. $|5 - \sqrt{5}| = 5 - \sqrt{5}$ since $5 - \sqrt{5}$ is positive.

7. $|\sqrt{5} - 5| = -(\sqrt{5} - 5) = 5 - \sqrt{5}$ since $\sqrt{5} - 5$ is negative.

9. $d(A,B) = |b - a|$
$\quad\quad\quad\quad = |5 - (-7)|$
$\quad\quad\quad\quad = |12|$
$\quad\quad\quad\quad = 12$

11. $d(A,B) = |b - a|$
$\quad\quad\quad\quad = |-7 - 5|$
$\quad\quad\quad\quad = |-12|$
$\quad\quad\quad\quad = 12$

13. $d(B,0) = |0 - (-4)|$
$\quad\quad\quad\quad = |4|$
$\quad\quad\quad\quad = 4$

15. $d(0,B) = |-4 - 0|$
$\quad\quad\quad\quad = |-4|$
$\quad\quad\quad\quad = 4$

17. $d(B,C) = |5 - (-4)|$
$\quad\quad\quad\quad = |9|$
$\quad\quad\quad\quad = 9$

19. The distance between x and 3 is equal to 4.
$\quad |x - 3| = 4$

21. The distance between m and -2 is equal to 5.
$\quad |m - (-2)| = 5$
$\quad |m + 2| = 5$

23. The distance between x and 3 is less than 5.
$\quad |x - 3| < 5$

25. The distance between p and -2 is more than 6.
$\quad |p - (-2)| > 6$
$\quad |p + 2| > 6$

27. The distance between q and 1 is not less than 2.
$\quad |q - 1| \geq 2$

29. x is no more than 7 units from the origin.
$-7 \leq x \leq 7$ $[-7, 7]$

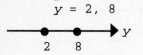

31. x is at least 7 units from the origin.
$x \leq -7$ or $x \geq 7$
$(-\infty, -7] \cup [7, \infty)$

Common Error: $7 \leq x \leq -7$
This is meaningless

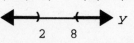

33. y is 3 units from 5.
$\quad |y - 5| = 3$
$\quad\quad y - 5 = \pm 3$
$\quad\quad\quad y = 5 \pm 3$
$\quad\quad\quad y = 2, 8$

35. y is less than 3 units from 5.
$\quad |y - 5| < 3$
$-3 < y - 5 < 3$
$\quad 2 < y < 8$
$(2, 8)$

37. y is more than 3 units from 5.
$\quad |y - 5| > 3$
$\quad y - 5 < -3$ or $y - 5 > 3$
$\quad\quad y < 2$ or $y > 8$
$(-\infty, 2) \cup (8, \infty)$

39. $|u - (-8)| = 3$
u is 3 units from -8.
$|u + 8| = 3$
$u + 8 = \pm 3$
$u = -8 \pm 3$
$u = -11$ or -5

$-11 \quad -5$

41. $|u - (-8)| \leq 3$
u is no more than 3
units from -8.
$|u + 8| \leq 3$
$-3 \leq u + 8 \leq 3$
$-11 \leq u \leq -5$
$[-11, -5]$

$-11 \quad -5$

43. $|u - (-8)| \geq 3$
u is at least 3 units
from -8.
$|u + 8| \geq 3$
$u + 8 \leq -3$ or $u + 8 \geq 3$
$u \leq -11$ or $u \geq -5$
$(-\infty, -11] \cup [-5, \infty)$

$-11 \quad -5$

45. $|5x - 3| \leq 12$
$-12 \leq 5x - 3 \leq 12$
$-9 \leq 5x \leq 15$
$-\dfrac{9}{5} \leq x \leq 3$

$\left[-\dfrac{9}{5},\ 3\right]$

47. $|2y - 8| > 2$
$2y - 8 < -2$ or $2y - 8 > 2$
$2y < 6$ or $2y > 10$
$y < 3$ or $y > 5$
$(-\infty,\ 3) \cup (5,\ \infty)$

49. $|5t - 7| = 11$
$5t - 7 = \pm 11$
$5t = 7 \pm 11$
$t = \dfrac{7 \pm 11}{5}$
$t = -\dfrac{4}{5},\ \dfrac{18}{5}$

51. $|9 - 7u| < 14$
$-14 < 9 - 7u < 14$
$-23 < -7u < 5$
$\dfrac{23}{7} > u > -\dfrac{5}{7}$
$-\dfrac{5}{7} < u < \dfrac{23}{7}$

$\left(-\dfrac{5}{7},\ \dfrac{23}{7}\right)$

53. $\left|1 - \dfrac{2}{3}x\right| \geq 5$

$1 - \dfrac{2}{3}x \leq -5$ or $1 - \dfrac{2}{3}x \geq 5$

$-\dfrac{2}{3}x \leq -6$ or $-\dfrac{2}{3}x \geq 4$

$x \geq 9$ or $x \leq -6$

$(-\infty,\ -6] \cup [9,\ \infty)$

55. $\left|\dfrac{9}{5}C + 32\right| < 31$

$-31 < \dfrac{9}{5}C + 32 < 31$

$-63 < \dfrac{9}{5}C < -1$

$-35 < C < -\dfrac{5}{9}$

$\left(-35,\ -\dfrac{5}{9}\right)$

57. $\sqrt{x^2} < 2$
$|x| < 2$
$-2 < x < 2$
$(-2,\ 2)$

59. $\sqrt{(1 - 3t)^2} \leq 2$
$|1 - 3t| \leq 2$
$-2 \leq 1 - 3t \leq 2$
$-3 \leq -3t \leq 1$
$1 \geq t \geq -\dfrac{1}{3}$
$-\dfrac{1}{3} \leq t \leq 1$

$\left[-\dfrac{1}{3},\ 1\right]$

61. $\sqrt{(2t - 3)^2} > 3$
$|2t - 3| > 3$
$2t - 3 < -3$ or $2t - 3 > 3$
$2t < 0$ $2t > 6$
$t < 0$ $t > 3$
$(-\infty,\ 0) \cup (3,\ \infty)$

63. $0 < |x - 3| < 0.1$
The distance between x and 3 is
less than 0.1 but $x \neq 3$.
$-0.1 < x - 3 < 0.1$ except $x \neq 3$
$2.9 < x < 3.1$ but $x \neq 3$
$2.9 < x < 3$ or $3 < x < 3.1$
$(2.9,\ 3) \cup (3,\ 3.1)$

$2.9 \quad 3 \quad 3.1$

65. $0 < |x - c| < d$
The distance between x and c is less
than d but $x \neq c$.
$-d < x - c < d$ except $x \neq c$
$c - d < x < c + d$ but $x \neq c$
$c - d < x < c$ or $c < x < c + d$
$(c - d,\ c) \cup (c,\ c + d)$

$c - d \quad c \quad c + d$

67. $|x| = x$ if x is positive or zero. So $|x - 5| = x - 5$ if $x - 5$ is positive or
zero. We must solve $x - 5 \geq 0$. The solution is $x \geq 5$.

69. $|x| = -x$ if x is negative or zero. So $|x + 8| = -(x + 8)$ if $x + 8$ is negative or zero. We must solve $x + 8 \leq 0$. The solution is $x \leq -8$.

71. As in 67, we must solve $4x + 3 \geq 0$.
$$4x + 3 \geq 0$$
$$4x \geq -3$$
$$x \geq -\frac{3}{4}$$

73. As in 69, we must solve $5x - 2 \leq 0$.
$$5x - 2 \leq 0$$
$$5x \leq 2$$
$$x \leq \frac{2}{5}$$

75. We consider four cases for $|3 - x| + 3 = |2 - 3x|$

Case 1: $3 - x > 0$ and $2 - 3x > 0$, that is $x < 3$ and $x < \frac{2}{3}$ (or simply $x < \frac{2}{3}$)

$|3 - x| = 3 - x$ and $|2 - 3x| = 2 - 3x$
Hence: $3 - x + 3 = 2 - 3x$
$$-x + 6 = 2 - 3x$$
$$2x = -4$$
$$x = -2$$
which is a possible value for x in this case.

Case 2: $3 - x > 0$ and $2 - 3x < 0$, that is $x < 3$ and $x > \frac{2}{3}$ (or simply $\frac{2}{3} < x < 3$)

$|3 - x| = 3 - x$ and $|2 - 3x| = -(2 - 3x) = 3x - 2$
Hence: $3 - x + 3 = 3x - 2$
$$-x + 6 = 3x - 2$$
$$-4x = -8$$
$$x = 2$$
which is a possible value for x in this case.

Case 3: $3 - x < 0$ and $2 - 3x < 0$, that is $x > 3$ and $x > \frac{2}{3}$ (or simply $x > 3$)

$|3 - x| = -(3 - x) = x - 3$ and $|2 - 3x| = -(2 - 3x) = 3x - 2$
Hence: $x - 3 + 3 = 3x - 2$
$$x = 3x - 2$$
$$-2x = -2$$
$$x = 1$$
which is not a possible value for x in this case. ($x > 3$).

Case 4: $3 - x < 0$ and $2 - 3x > 0$, that is $x > 3$ and $x < \frac{2}{3}$. These are mutually contradictory, so no solution is possible in this case.
Solution: $-2, 2$

77. *Case 1:* $x > 0$. Then $|x| = x$. Hence $\dfrac{x}{|x|} = \dfrac{x}{x} = 1$.

Case 2: $x = 0$. Then $|x| = 0$. Hence $\dfrac{x}{|x|}$ is not defined.

Case 3: $x < 0$. Then $|x| = -x$. Hence $\dfrac{x}{|x|} = \dfrac{x}{-x} = -1$.

Thus, the possible values of $\dfrac{x}{|x|}$ are 1 and -1.

79. There are three possible relations between real numbers a and b; either $a = b$, $a > b$, or $a < b$. We examine each case separately.
Case 1: $a = b$
$|b - a| = |0| = 0$; $|a - b| = |0| = 0$
Case 2: $a > b$
$|b - a| = -(b - a) = a - b$
$|a - b| = a - b$
Case 3: $b > a$
$|b - a| = b - a$
$|a - b| = -(a - b) = b - a$
Thus in all three cases $|b - a| = |a - b|$.

81. If $m < n$, then $m + m < m + n$ (adding m to both sides)
Also, $m + n < n + n$ (adding n to both sides). Hence,
$m + m < m + n < n + n$
$\quad 2m < m + n < 2n$
$\quad\quad m < \dfrac{m + n}{2} < n$

83. *Case 1.* $m > 0$. Then $|m| = m$ $|-m| = -(-m) = m$. Hence $|m| = |-m|$
Case 2. $m < 0$. Then $|m| = -m$ $|-m| = -m$. Hence $|m| = |-m|$
Case 3. $m = 0$. Then $0 = m = -m$, hence $|m| = |-m| = 0$.

85. If $n \neq 0$, $n > 0$ or $n < 0$.
Case 1. $n > 0$. If $m \geq 0$ $|m| = m$, $\dfrac{m}{n} \geq 0$ $\left|\dfrac{m}{n}\right| = \dfrac{m}{n}$ $|n| = n$.
Hence: $\left|\dfrac{m}{n}\right| = \dfrac{m}{n} = \dfrac{|m|}{|n|}$

If $m < 0$ $|m| = -m$, $\dfrac{m}{n} < 0$ $\left|\dfrac{m}{n}\right| = -\dfrac{m}{n}$ $|n| = n$.
Hence: $\left|\dfrac{m}{n}\right| = -\dfrac{m}{n} = \dfrac{-m}{n} = \dfrac{|m|}{|n|}$

Case 2. $n < 0$. If $m > 0$ $|m| = m$, $\dfrac{m}{n} < 0$ $\left|\dfrac{m}{n}\right| = -\dfrac{m}{n}$ $|n| = -n$
Hence: $\left|\dfrac{m}{n}\right| = -\dfrac{m}{n} = \dfrac{m}{-n} = \dfrac{|m|}{|n|}$

If $m \leq 0$ $|m| = -m$ $\dfrac{m}{n} > 0$ $\left|\dfrac{m}{n}\right| = \dfrac{m}{n}$ $|n| = -n$
Hence: $\left|\dfrac{m}{n}\right| = \dfrac{m}{n} = \dfrac{-m}{-n} = \dfrac{|m|}{|n|}$

87. First note that $a \leq b$ is true if $a < b$ or if $a = b$. Hence $a < b$ implies $a \leq b$.
Also $a = b$ implies $a \leq b$. Now consider three cases ($m > 0$, $m = 0$, $m < 0$).
Case 1. $m > 0$. Then $|m| = m$. Also $-|m| < 0$
Hence $-|m| < 0 < m = |m|$
$\quad\quad -|m| < m = |m|$
$\quad\quad -|m| \leq m \leq |m|$
Case 2. $m = 0$. Then $-|m| = m = |m| = 0$.
Hence $-|m| \leq m \leq |m|$
Case 3. $m < 0$. Then $|m| = -m$, hence $-|m| = m$. Also $|m| > 0$.
Hence $-|m| = m < 0 < |m|$
$\quad\quad -|m| = m < |m|$
$\quad\quad -|m| \leq m \leq |m|$

89. Consider three cases ($a < b$, $a = b$, $a > b$)
Case 1. $a < b$. Then $\max(a,b) = b$. Also $a - b < 0$, hence $|a - b| = b - a$.
Hence $\dfrac{1}{2}[a + b + |a - b|] = \dfrac{1}{2}[a + b + b - a] = \dfrac{1}{2}(2b) = b$.

Thus $\max(a,b) = b = \dfrac{1}{2}[a + b + |a - b|]$

Case 2. $a = b$. Then $\max(a,b) = a = b$. Also $a - b = 0$, hence $|a - b| = 0$.
Hence $\dfrac{1}{2}[a + b + |a - b|] = \dfrac{1}{2}[a + b] = \dfrac{1}{2}[2a] = a$ or $\dfrac{1}{2}[2b] = b$.

Thus $\max(a,b) = a = b = \dfrac{1}{2}[a + b + |a - b|]$

Case 3. $a > b$. Then $\max(a,b) = a$. Also $a - b > 0$, hence $|a - b| = a - b$
Hence $\dfrac{1}{2}[a + b + |a - b|] = \dfrac{1}{2}[a + b + a - b] = \dfrac{1}{2}(2a) = a$.

Thus $\max(a,b) = a = \dfrac{1}{2}[a + b + |a - b|]$

91. $\left| \dfrac{x - 45.4}{3.2} \right| < 1$

$$-1 < \dfrac{x - 45.4}{3.2} < 1$$
$$-3.2 < x - 45.4 < 3.2$$
$$42.2 < x < 48.6$$

93. The difference between P and 500 has an absolute value of no more than 20.
$|P - 500| \le 20$

95. The difference between A and 12.436 has an absolute value of less than the error of 0.001.
$|A - 12.436| < 0.001$
$-0.001 < A - 12.436 < 0.001$
$12.435 < A < 12.437$ or, in interval notation, (12.435, 12.437)

97. The difference between N and 2.37 has an absolute value of no more than 0.005.
$|N - 2.37| \le 0.005$.

Exercise 2-5

Key Ideas and Formulas

For a, b real numbers.

$$i = \text{the imaginary unit} = \sqrt{-1}$$
$$a + bi = \text{complex number (or imaginary number) in standard form}$$
$$0 + bi = \text{pure imaginary number}$$
$$a = a + 0i = \text{real number}$$
$$0 + 0i = \text{zero}$$
$$a - bi = \text{conjugate of } a + bi$$
$$a + bi = c + di \text{ if and only if } a = c \text{ and } b = d$$
$$(a + bi) + (c + di) = (a + c) + (b + d)i$$
$$(a + bi)(c + di) = (ac - bd) + (ad + bc)i$$
$$(a + bi)(a - bi) = a^2 + b^2$$
$$\dfrac{a + bi}{c} = \dfrac{a}{c} + \dfrac{b}{c}i$$
$$\dfrac{a + bi}{c + di} = \dfrac{(a + bi)}{(c + di)}\dfrac{(c - di)}{(c - di)} \text{ which is simplified to}$$
$$\dfrac{ac + bd}{c^2 + d^2} + \dfrac{bc - ad}{c^2 + d^2}i$$

The principal square root of a negative real number, denoted by $\sqrt{-a}$ where a is positive, is defined by $\sqrt{-a} = i\sqrt{a}$

$a + bi$, where a and b are real numbers and i is the imaginary unit, is called a complex number in standard form.

---

1. $(2 + 4i) + (5 + i) = 2 + 4i + 5 + i = 7 + 5i$

3. $(-2 + 6i) + (7 - 3i) = -2 + 6i + 7 - 3i = 5 + 3i$

5. $(6 + 7i) - (4 + 3i) = 6 + 7i - 4 - 3i = 2 + 4i$

7. $(3 + 5i) - (-2 - 4i) = 3 + 5i + 2 + 4i = 5 + 9i$

9. $(4 - 5i) + 2i = 4 - 5i + 2i = 4 - 3i$

11. $(4i)(6i) = 24i^2 = 24(-1) = -24$ or $-24 + 0i$

13. $-3i(2 - 4i) = -6i + 12i^2 = -6i + 12(-1) = -12 - 6i$

15. $(3 + 3i)(2 - 3i) = 6 - 9i + 6i - 9i^2 = 6 - 3i - 9(-1) = 6 - 3i + 9 = 15 - 3i$

17. $(2 - 3i)(7 - 6i) = 14 - 12i - 21i + 18i^2 = 14 - 33i - 18 = -4 - 33i$

19. $(7 + 4i)(7 - 4i) = 49 - 16i^2 = 49 + 16 = 65$ or $65 + 0i$

21. $\dfrac{1}{2 + i} = \dfrac{1}{(2 + i)} \dfrac{(2 - i)}{(2 - i)} = \dfrac{2 - i}{4 - i^2} = \dfrac{2 - i}{4 + 1} = \dfrac{2 - i}{5} = \dfrac{2}{5} - \dfrac{1}{5}i$

23. $\dfrac{3 + i}{2 - 3i} = \dfrac{(3 + i)}{(2 - 3i)} \dfrac{(2 + 3i)}{(2 + 3i)} = \dfrac{6 + 11i + 3i^2}{4 - 9i^2} = \dfrac{6 + 11i - 3}{4 + 9} = \dfrac{3 + 11i}{13} = \dfrac{3}{13} + \dfrac{11}{13}i$

25. $\dfrac{13 + i}{2 - i} = \dfrac{(13 + i)}{(2 - i)} \dfrac{(2 + i)}{(2 + i)} = \dfrac{26 + 15i + i^2}{4 - i^2} = \dfrac{26 + 15i - 1}{4 + 1} = \dfrac{25 + 15i}{5} = 5 + 3i$

27. $(2 - \sqrt{-4}) + (5 - \sqrt{-9}) = (2 - i\sqrt{4}) + (5 - i\sqrt{9}) = 2 - 2i + 5 - 3i = 7 - 5i$

29. $(9 - \sqrt{-9}) - (12 - \sqrt{-25}) = (9 - i\sqrt{9}) - (12 - i\sqrt{25})$
$$= (9 - 3i) - (12 - 5i) = 9 - 3i - 12 + 5i = -3 + 2i$$

31. $(3 - \sqrt{-4})(-2 + \sqrt{-49}) = (3 - i\sqrt{4})(-2 + i\sqrt{49})$
$$= (3 - 2i)(-2 + 7i) = -6 + 25i - 14i^2 = -6 + 25i + 14 = 8 + 25i$$

33. $\dfrac{5 - \sqrt{-4}}{7} = \dfrac{5 - i\sqrt{4}}{7} = \dfrac{5 - 2i}{7} = \dfrac{5}{7} - \dfrac{2}{7}i$

35. $\dfrac{1}{2 - \sqrt{-9}} = \dfrac{1}{2 - i\sqrt{9}} = \dfrac{1}{2 - 3i} = \dfrac{1}{(2 - 3i)} \dfrac{(2 + 3i)}{(2 + 3i)} = \dfrac{2 + 3i}{4 - 9i^2}$
$$= \dfrac{2 + 3i}{4 + 9} = \dfrac{2 + 3i}{13} = \dfrac{2}{13} + \dfrac{3}{13}i$$

37. $\dfrac{2}{5i} = \dfrac{2}{5i} \cdot \dfrac{i}{i} = \dfrac{2i}{5i^2} = \dfrac{2i}{-5} = -\dfrac{2}{5}i$ or $0 - \dfrac{2}{5}i$

39. $\dfrac{1 + 3i}{2i} = \dfrac{(1 + 3i)}{(2i)} \cdot \dfrac{i}{i} = \dfrac{i + 3i^2}{2i^2} = \dfrac{i - 3}{-2} = \dfrac{3}{2} - \dfrac{1}{2}i$

41. $(2 - 3i)^2 - 2(2 - 3i) + 9 = 4 - 12i + 9i^2 - 4 + 6i + 9$
$$= 4 - 12i - 9 - 4 + 6i + 9 = -6i \text{ or } 0 - 6i$$

43. $x^2 - 2x + 2 = (1 - i)^2 - 2(1 - i) + 2 = 1 - 2i + i^2 - 2 + 2i + 2$
$$= 1 - 2i - 1 - 2 + 2i + 2 = 0 \text{ or } 0 + 0i$$

45. $i^{18} = i^{16} \cdot i^2 = (i^4)^4 \cdot i^2 = 1^4(-1) = -1$
   largest integer in 18 exactly divisible by 4
$i^{32} = (i^4)^8 = 1^8 = 1$
$i^{67} = i^{64} \cdot i^3 = (i^4)^{16} \cdot i^2 \cdot i = 1^{16}(-1)i = -i$
   largest integer in 67 exactly divisible by 4

47. According to the definition of equality for complex numbers
   $(2x - 1) + (3y + 2)i = 5 - 4i = 5 + (-4)i$
   if and only if
   $2x - 1 = 5$ and $3y + 2 = -4$
   So $2x = 6$ and $3y = -6$
   $x = 3$ $y = -2$

49. $\sqrt{3 - x}$ represents an imaginary number when $3 - x$ is negative. We solve:
   $3 - x < 0$
   $3 < x$ or $x > 3$

51. $\sqrt{2 - 3x}$ represents an imaginary number when $2 - 3x$ is negative. We solve:
   $2 - 3x < 0$
   $-3x < -2$
   $x > \dfrac{2}{3}$

53. $(3.17 - 4.08i)(7.14 + 2.76i)$
$$= (3.17)(7.14) + (3.17)(2.76i) - (4.08i)(7.14) - (4.08i)(2.76i)$$
$$= 22.6338 + 8.7492i - 29.1312i - 11.2608i^2$$
$$= 22.6338 - 20.382i + 11.2608$$
$$= 33.89 - 20.38i$$

55. $\dfrac{8.14 + 2.63i}{3.04 + 6.27i} = \dfrac{(8.14 + 2.63i)}{(3.04 + 6.27i)} \cdot \dfrac{(3.04 - 6.27i)}{(3.04 - 6.27i)}$

$$= \dfrac{(8.14)(3.04) - (8.14)(6.27i) + (2.63i)(3.04) - (2.63i)(6.27i)}{(3.04)^2 + (6.27)^2}$$

$$= \dfrac{24.7456 - 51.0378i + 7.9952i - 16.4901i^2}{9.2416 + 39.3129}$$

$$= \dfrac{24.7456 - 43.0426i + 16.4901}{48.55455} = \dfrac{41.2357 - 43.0426i}{48.5545} = 0.85 - 0.89i$$

57. $(a + bi) + (c + di) = a + bi + c + di = a + c + bi + di = (a + c) + (b + d)i$

59. $(a + bi)(a - bi) = a^2 - b^2i^2 = a^2 + b^2$ or $(a^2 + b^2) + 0i$

61. $(a + bi)(c + di) = ac + adi + bci + bdi^2 = ac + (ad + bc)i - bd$
$$= (ac - bd) + (ad + bc)i$$

63. $i^{4k} = (i^4)^k = (i^2 \cdot i^2)^k = [(-1)(-1)]^k = 1^k = 1$

65. 1. Definition of addition
2. Commutative property for addition of real numbers.
3. Definition of addition (read from right to left).

67. The product of a complex number and its conjugate is a real number.
$$z\bar{z} = (x + yi)(x - yi)$$
$$= x^2 - (yi)^2$$
$$= x^2 - y^2i^2$$
$$= x^2 + y^2 \text{ or}$$
$$(x^2 + y^2) + 0i.$$
This is a real number.

69. The conjugate of a complex number is equal to the complex number if and only if the number is real. To prove a theorem containing the phrase "if and only if", it is often helpful to prove two parts separately. Thus: $\bar{z} = z$ if z is real; $\bar{z} = z$ only if z is real

Hypothesis: z is real
Conclusion: $\bar{z} = z$
Proof: Assume z is real, then
               $z = x + 0i = x$
               $\bar{z} = x - 0i = x$

               Hence $z = \bar{z}$.

Hypothesis: $\bar{z} = z$
Conclusion: z is real
Proof: Assume $\bar{z} = z$,
               that is, $x - yi = x + yi$
               Then by the definition of equality
               $x = x$ $-y = y$
                          $-2y = 0$
                           $y = 0$
               Hence $z = x + 0i$, that is, z is real.

71. The conjugate of the sum of two complex numbers is equal to the sum of their conjugates.

$$\overline{z + w} = \overline{(x + yi) + (u + vi)}$$
$$= \overline{x + yi + u + vi}$$
$$= \overline{x + u + (y + v)i}$$
$$= (x + u) - (y + v)i$$
$$= x + u - yi - vi$$
$$= (x - yi) + (u - vi)$$
$$= \overline{z} + \overline{w}$$

73. The conjugate of the product of two complex numbers is equal to the product of their conjugates.

$$\overline{zw} = \overline{(x + yi)(u + vi)}$$
$$= \overline{xu + xvi + yui + yvi^2}$$
$$= \overline{xu + (xv + yu)i - yv}$$
$$= xu - yv - (xv + yu)i$$
$$= xu - xvi - yv - yui$$
$$= x(u - vi) - yui + yv(-1)$$
$$= x(u - vi) - yui + yvi^2$$
$$= x(u - vi) - yi(u - vi)$$
$$= (x - yi)(u - vi)$$
$$= \overline{z}\ \overline{w}$$

Exercise 2-6

Key Ideas and Formulas

A quadratic equation in one variable is any equation that can be written in the form
$$ax^2 + bx + c = 0 \qquad a \neq 0 \qquad \text{(standard form)}$$
where x is a variable and a, b, c are constants.

To solve a quadratic equation in standard form, check if it can be solved using factoring and the zero property:
$$m \cdot n = 0 \text{ if and only if } m = 0 \text{ or } n = 0 \text{ or both}$$
If not, generally the quadratic formula is applied:
$$x = \frac{-b \pm \sqrt{b^2 - 4ac}}{2a} \qquad (a \neq 0)$$

Occasionally the square root property
$$\text{if } A^2 = C, \text{ then } A = \pm\sqrt{C}$$
can be useful, especially if the equation is in the form
$$ax^2 = b \text{ or } (ax + b)^2 = d$$

To complete the square of a quadratic of the form $x^2 + bx$, add the square of one-half the coefficient of x; that is, add $(b/2)^2$. Thus,
$$x^2 + bx$$
$$x^2 + bx + \left(\frac{b}{2}\right)^2 = \left(x + \frac{b}{2}\right)^2$$

To solve a quadratic equation by completing the square, then, put the equation in the form $x^2 + bx = m$, add $(b/2)^2$ to both sides and use the square root property to complete the solution.

The quantity $b^2 - 4ac$, appearing in the quadratic formula, is called the discriminant. If the discriminant is positive, the equation has two distinct real roots. If the discriminant is 0, the equation has one real root (a double root). If the discriminant is negative, the equation has two imaginary roots, one the conjugate of the other.

1.
$$4u^2 = 8u$$
$$4u^2 - 8u = 0$$
$$4u(u - 2) = 0$$
$$4u = 0 \text{ or } u - 2 = 0$$
$$u = 0 \qquad u = 2$$

3.
$$9y^2 = 12y - 4$$
$$9y^2 - 12y + 4 = 0$$
$$(3y - 2)^2 = 0$$
$$3y - 2 = 0$$
$$3y = 2$$
$$y = \frac{2}{3} \text{ (double root)}$$

5.
$$11x = 2x^2 + 12$$
$$0 = 2x^2 - 11x + 12$$
$$2x^2 - 11x + 12 = 0$$
$$(2x - 3)(x - 4) = 0$$
$$2x - 3 = 0 \text{ or } x - 4 = 0$$
$$2x = 3 \text{ or } x = 4$$
$$x = \frac{3}{2}$$

7.
$$m^2 - 12 = 0$$
$$m^2 = 12$$
$$m = \pm\sqrt{12}$$
$$m = \pm 2\sqrt{3}$$

9.
$$x^2 + 25 = 0$$
$$x^2 = -25$$
$$x = \pm 5i$$

11.
$$9y^2 - 16 = 0$$
$$9y^2 = 16$$
$$y^2 = \frac{16}{9}$$
$$y = \pm\sqrt{\frac{16}{9}}$$
$$y = \pm\frac{4}{3}$$

13.
$$4x^2 + 25 = 0$$
$$4x^2 = -25$$
$$x^2 = -\frac{25}{4}$$
$$x = \pm\sqrt{-\frac{25}{4}}$$
$$x = \pm\frac{\sqrt{-25}}{\sqrt{4}}$$
$$x = \pm\frac{5i}{2} \text{ or } \pm\frac{5}{2}i$$

15.
$$(n + 5)^2 = 9$$
$$n + 5 = \pm\sqrt{9}$$
$$n + 5 = \pm 3$$
$$n = -5 \pm 3$$
$$n = -5 + 3 \text{ or } -5 - 3$$
$$n = -2, -8$$

17.
$$(d - 3)^2 = -4$$
$$d - 3 = \pm\sqrt{-4}$$
$$d - 3 = \pm 2i$$
$$d = 3 \pm 2i$$

19. $x^2 - 10x - 3 = 0$
$$x = \frac{-b \pm \sqrt{b^2 - 4ac}}{2a} \qquad a = 1, \ b = -10, \ c = -3$$
$$x = \frac{-(-10) \pm \sqrt{(-10)^2 - 4(1)(-3)}}{2(1)}$$
$$x = \frac{10 \pm \sqrt{112}}{2} = \frac{10 \pm 4\sqrt{7}}{2}$$
$$= 5 \pm 2\sqrt{7}$$

> **Common Error:**
> It is incorrect to "cancel" this way:
> $$\frac{\cancel{10} \pm \sqrt{112}}{2} \neq 5 \pm \sqrt{112}$$

21.
$$x^2 + 8 = 4x$$
$$x^2 - 4x + 8 = 0$$
$$x = \frac{-b \pm \sqrt{b^2 - 4ac}}{2a} \qquad a = 1, \ b = -4, \ c = 8$$
$$x = \frac{-(-4) \pm \sqrt{(-4)^2 - 4(1)(8)}}{2(1)}$$
$$x = \frac{4 \pm \sqrt{-16}}{2}$$
$$x = \frac{4 \pm i\sqrt{16}}{2}$$
$$x = \frac{4 \pm 4i}{2}$$
$$x = 2 \pm 2i$$

23.
$$2x^2 + 1 = 4x$$
$$2x^2 - 4x + 1 = 0$$

$$x = \frac{-b \pm \sqrt{b^2 - 4ac}}{2a} \qquad a = 2, \; b = -4, \; c = 1$$

$$= \frac{-(-4) \pm \sqrt{(-4)^2 - 4(2)(1)}}{2(2)}$$

$$x = \frac{4 \pm \sqrt{8}}{4}$$

$$x = \frac{4 \pm 2\sqrt{2}}{4}$$

$$x = \frac{2 \pm \sqrt{2}}{2}$$

> **Common Errors:**
> $$\frac{2 \pm \sqrt{2}}{2} \neq \pm\sqrt{2}$$
> $$\neq 1 \pm \sqrt{2}$$
> These involve incorrect "cancelling".

25.
$$5x^2 + 2 = 2x$$
$$5x^2 - 2x + 2 = 0$$

$$x = \frac{-b \pm \sqrt{b^2 - 4ac}}{2a} \qquad a = 5, \; b = -2, \; c = 2$$

$$x = \frac{-(-2) \pm \sqrt{(-2)^2 - 4(5)(2)}}{2(5)}$$

$$x = \frac{2 \pm \sqrt{-36}}{10}$$

$$x = \frac{2 \pm 6i}{10}$$

$$x = \frac{1}{5} \pm \frac{3}{5}i$$

27.
$$x^2 - 6x - 3 = 0$$
$$x^2 - 6x = 3$$
$$x^2 - 6x + 9 = 12$$
$$(x - 3)^2 = 12$$
$$x - 3 = \pm\sqrt{12}$$
$$x = 3 \pm \sqrt{12}$$
$$x = 3 \pm 2\sqrt{3}$$

29.
$$2y^2 - 6y + 3 = 0$$
$$y^2 - 3y + \frac{3}{2} = 0$$
$$y^2 - 3y = -\frac{3}{2}$$
$$y^2 - 3y + \frac{9}{4} = -\frac{3}{2} + \frac{9}{4}$$
$$\left(y - \frac{3}{2}\right)^2 = \frac{3}{4}$$
$$y - \frac{3}{2} = \pm\sqrt{\frac{3}{4}}$$
$$y = \frac{3}{2} \pm \frac{\sqrt{3}}{\sqrt{4}}$$
$$y = \frac{3}{2} \pm \frac{\sqrt{3}}{2}$$
$$y = \frac{3 \pm \sqrt{3}}{2}$$

31.
$$3x^2 - 2x - 2 = 0$$
$$x^2 - \frac{2}{3}x - \frac{2}{3} = 0$$
$$x^2 - \frac{2}{3}x = \frac{2}{3}$$
$$x^2 - \frac{2}{3}x + \frac{1}{9} = \frac{2}{3} + \frac{1}{9}$$
$$\left(x - \frac{1}{3}\right)^2 = \frac{7}{9}$$
$$x - \frac{1}{3} = \pm\sqrt{\frac{7}{9}}$$
$$x - \frac{1}{3} = \pm\frac{\sqrt{7}}{3}$$
$$x = \frac{1}{3} \pm \frac{\sqrt{7}}{3}$$
$$x = \frac{1 \pm \sqrt{7}}{3}$$

33.
$$x^2 + mx + n = 0$$
$$x^2 + mx = -n$$
$$x^2 + mx + \frac{m^2}{4} = \frac{m^2}{4} - n$$
$$\left(x + \frac{m}{2}\right)^2 = \frac{m^2 - 4n}{4}$$
$$x + \frac{m}{2} = \pm\sqrt{\frac{m^2 - 4n}{4}}$$
$$x = -\frac{m}{2} \pm \frac{\sqrt{m^2 - 4n}}{2}$$
$$x = \frac{-m \pm \sqrt{m^2 - 4n}}{2}$$

35.
$$12x^2 + 7x = 10$$
$$12x^2 + 7x - 10 = 0$$
$(4x + 5)(3x - 2) = 0$ Polynomial is factorable.
$$4x + 5 = 0 \text{ or } 3x - 2 = 0$$
$$4x = -5 \qquad 3x = 2$$
$$x = -\frac{5}{4} \qquad x = \frac{2}{3}$$

37. $(2y - 3)^2 = 5$ Format for the square root method.
$$2y - 3 = \pm\sqrt{5}$$
$$2y = 3 \pm \sqrt{5}$$
$$y = \frac{3 \pm \sqrt{5}}{2}$$

39.
$$x^2 = 3x + 1$$
$x^2 - 3x - 1 = 0$ Polynomial is not factorable, use quadratic formula.
$$x = \frac{-b \pm \sqrt{b^2 - 4ac}}{2a} \quad a = 1, b = -3, c = -1$$
$$x = \frac{-(-3) \pm \sqrt{(-3)^2 - 4(1)(-1)}}{2(1)}$$
$$x = \frac{3 \pm \sqrt{13}}{2}$$

41.
$$7n^2 = -4n$$
$$7n^2 + 4n = 0$$
$n(7n + 4) = 0$ Polynomial is factorable.
$$n = 0 \text{ or } 7n + 4 = 0$$
$$7n = -4$$
$$n = -\frac{4}{7}$$

43.
$1 + \dfrac{8}{x^2} = \dfrac{4}{x}$ Excluded value: $x \neq 0$
$$x^2 + 8 = 4x$$
$x^2 - 4x + 8 = 0$ Polynomial is not factorable, use quadratic formula, or complete the square.
$$x^2 - 4x = -8$$
$$x^2 - 4x + 4 = -4$$
$$(x - 2)^2 = -4$$
$$x - 2 = \pm\sqrt{-4}$$
$$x - 2 = \pm i\sqrt{4}$$
$$x - 2 = \pm 2i$$
$$x = 2 \pm 2i$$

45. $\dfrac{24}{10 + m} + 1 = \dfrac{24}{10 - m}$ Excluded value: $m \neq -10, 10$: LCD is $(10 + m)(10 - m)$
$$(10 + m)(10 - m)\frac{24}{10 + m} + (10 + m)(10 - m) = (10 + m)(10 - m)\frac{24}{10 - m}$$
$$24(10 - m) + 100 - m^2 = 24(10 + m)$$
$$240 - 24m + 100 - m^2 = 240 + 24m$$
$$340 - 24m - m^2 = 240 + 24m$$
$$0 = m^2 + 48m - 100$$
$m^2 + 48m - 100 = 0$ Polynomial is factorable.
$$(m + 50)(m - 2) = 0$$
$$m + 50 = 0 \text{ or } m - 2 = 0$$
$$m = -50 \qquad m = 2$$

47. $\dfrac{2}{x - 2} = \dfrac{4}{x - 3} - \dfrac{1}{x + 1}$ Excluded values: $x \neq 2,\ 3,\ -1$

$$(x - 2)(x - 3)(x + 1)\dfrac{2}{x - 2} = (x - 2)(x - 3)(x + 1)\dfrac{4}{x - 3} - (x - 2)(x - 3)(x + 1)\dfrac{1}{x + 1}$$

$$2(x - 3)(x + 1) = 4(x - 2)(x + 1) - (x - 2)(x - 3)$$
$$2(x^2 - 2x - 3) = 4(x^2 - x - 2) - (x^2 - 5x + 6)$$
$$2x^2 - 4x - 6 = 4x^2 - 4x - 8 - x^2 + 5x - 6$$
$$2x^2 - 4x - 6 = 3x^2 + x - 14$$
$$0 = x^2 + 5x - 8$$

$x^2 + 5x - 8 = 0$ Polynomial is not factorable, use quadratic formula.

$$x = \dfrac{-b \pm \sqrt{b^2 - 4ac}}{2a} \quad a = 1,\ b = 5,\ c = -8$$

$$x = \dfrac{-5 \pm \sqrt{(5)^2 - 4(1)(-8)}}{2(1)} = \dfrac{-5 \pm \sqrt{57}}{2}$$

49. $\dfrac{x + 2}{x + 3} - \dfrac{x^2}{x^2 - 9} = 1 - \dfrac{x - 1}{3 - x}$ Excluded values: $x \neq 3,\ -3$

$$(x - 3)(x + 3)\dfrac{(x + 2)}{x + 3} - (x - 3)(x + 3)\dfrac{x^2}{x^2 - 9} = (x - 3)(x + 3) - (x - 3)(x + 3)\dfrac{x - 1}{3 - x}$$

$$(x - 3)(x + 2) - x^2 = x^2 - 9 + (x - 1)(x + 3)$$
$$x^2 - x - 6 - x^2 = x^2 - 9 + x^2 + 2x - 3$$
$$-x - 6 = 2x^2 + 2x - 12$$
$$0 = 2x^2 + 3x - 6$$

$2x^2 + 3x - 6 = 0$ Polynomial is not factorable, use quadratic formula.

$$x = \dfrac{-b \pm \sqrt{b^2 - 4ac}}{2a} \quad a = 2,\ b = 3,\ c = -6$$

$$x = \dfrac{-3 \pm \sqrt{(3)^2 - 4(2)(-6)}}{2(2)}$$

$$x = \dfrac{-3 \pm \sqrt{57}}{4}$$

51. According to Theorem 3, Section 2-4, $|3u - 2| = u^2$ is equivalent to $3u - 2 = u^2$ or $3u - 2 = -u^2$. Hence we must solve:

$3u - 2 = u^2$ or $3u - 2 = -u^2$

$0 = u^2 - 3u + 2$ $u^2 + 3u - 2 = 0$

$0 = (u - 1)(u - 2)$ $u = \dfrac{-b \pm \sqrt{b^2 - 4ac}}{2a} \quad a = 1,\ b = 3,\ c = -2$

$u - 1 = 0$ or $u - 2 = 0$

$u = 1$ $u = 2$

$$u = \dfrac{-3 \pm \sqrt{(3)^2 - 4(1)(-2)}}{2(1)}$$

$$u = \dfrac{-3 \pm \sqrt{17}}{2}$$

53. $s = \dfrac{1}{2}gt^2$

$\dfrac{1}{2}gt^2 = s$

$gt^2 = 2s$

$t^2 = \dfrac{2s}{g}$

$t = \sqrt{\dfrac{2s}{g}}$

55. $P = EI - RI^2$

$RI^2 - EI + P = 0$

$$I = \dfrac{-b \pm \sqrt{b^2 - 4ac}}{2a} \quad a = R,\ b = -E,\ c = P$$

$$I = \dfrac{-(-E) \pm \sqrt{(-E)^2 - 4(R)(P)}}{2(R)}$$

$$I = \dfrac{E + \sqrt{E^2 - 4RP}}{2R} \text{ (positive square root)}$$

57. $2.07x^2 - 3.79x + 1.34 = 0$

$$x = \frac{-b \pm \sqrt{b^2 - 4ac}}{2a} \quad a = 2.07, \ b = -3.79, \ c = 1.34$$

$$x = \frac{-(-3.79) \pm \sqrt{(-3.79)^2 - 4(2.07)(1.34)}}{2(2.07)} = \frac{3.79 \pm 1.81}{4.14}$$

$$x = 1.35, \ 0.48$$

59. $4.83x^2 + 2.04x - 3.18 = 0$

$$x = \frac{-b \pm \sqrt{b^2 - 4ac}}{2a} \quad a = 4.83, \ b = 2.04, \ c = -3.18$$

$$x = \frac{-2.04 \pm \sqrt{(2.04)^2 - 4(4.83)(-3.18)}}{2(4.83)} = \frac{-2.04 \pm 8.10}{9.66}$$

$$x = -1.05, \ 0.63$$

61. In this problem, $a = 1$, $b = 4$, $c = c$. Thus, the discriminant $b^2 - 4ac = (4)^2 - 4(1)(c) = 16 - 4c$. Hence,
if $16 - 4c > 0$, thus $16 > 4c$ or $c < 4$, there are two distinct real roots.
if $16 - 4c = 0$, thus $c = 4$, there is one real double root,
and if $16 - 4c < 0$, thus $16 < 4c$ or $c > 4$, there are two distinct imaginary roots.

63. $0.0134x^2 + 0.0414x + 0.0304 = 0$
The discriminant is $b^2 - 4ac$, where $a = 0.0134$, $b = 0.0414$, $c = 0.0304$.
$b^2 - 4ac = (0.0414)^2 - 4(0.0134)(0.0304) = 8.45 \times 10^{-5} > 0$
Since the discriminant is positive, the equation has real solutions.

65. $0.0134x^2 + 0.0214x + 0.0304 = 0$
Here $a = 0.0134$, $b = 0.0214$, $c = 0.0304$.
$b^2 - 4ac = (0.0214)^2 - 4(0.0134)(0.0304) = -1.17 \times 10^{-3} < 0$
Since $b^2 - 4ac$, the discriminant, is negative, the equation has no real solutions.

67. $\sqrt{3}x^2 = 8\sqrt{2}x - 4\sqrt{3}$

$\sqrt{3}x^2 - 8\sqrt{2}x + 4\sqrt{3} = 0$

$$x = \frac{-b \pm \sqrt{b^2 - 4ac}}{2a} \quad a = \sqrt{3},$$
$$b = -8\sqrt{2}$$
$$c = 4\sqrt{3}$$

$$x = \frac{-(-8\sqrt{2}) \pm \sqrt{(-8\sqrt{2})^2 - 4(\sqrt{3})(4\sqrt{3})}}{2(\sqrt{3})}$$

$$x = \frac{8\sqrt{2} \pm \sqrt{(64)(2) - (16)(3)}}{2\sqrt{3}}$$

$$x = \frac{8\sqrt{2} \pm \sqrt{80}}{2\sqrt{3}}$$

$$x = \frac{8\sqrt{2} \pm 4\sqrt{5}}{2\sqrt{3}}$$

$$x = \frac{2(4\sqrt{2} \pm 2\sqrt{5})}{2\sqrt{3}} \cdot \frac{\sqrt{3}}{\sqrt{3}}$$

$$x = \frac{4\sqrt{6} \pm 2\sqrt{15}}{3} \text{ or } \frac{4}{3}\sqrt{6} \pm \frac{2}{3}\sqrt{15}$$

69. $x^2 + 2ix = 3$

$x^2 + 2ix - 3 = 0$

$$x = \frac{-b \pm \sqrt{b^2 - 4ac}}{2a} \quad a = 1$$
$$b = 2i$$
$$c = -3$$

$$x = \frac{-2i \pm \sqrt{(2i)^2 - 4(1)(-3)}}{2(1)}$$

$$x = \frac{-2i \pm \sqrt{-4 + 12}}{2}$$

$$x = \frac{-2i \pm \sqrt{8}}{2}$$

$$x = \frac{-2i \pm 2\sqrt{2}}{2}$$

$$x = \frac{2(-i \pm \sqrt{2})}{2}$$

$$x = -i \pm \sqrt{2}$$

$$x = \sqrt{2} - i, \ -\sqrt{2} - i$$

71. $x^3 - 1 = 0$

$(x - 1)(x^2 + x + 1) = 0$

$x - 1 = 0$ or $x^2 + x + 1 = 0$

$x = 1$ $x = \dfrac{-b \pm \sqrt{b^2 - 4ac}}{2a}$ $a = 1,\ b = 1,\ c = 1$

$x = \dfrac{-1 \pm \sqrt{(1)^2 - 4(1)(1)}}{2(1)}$

$x = \dfrac{-1 \pm \sqrt{1 - 4}}{2}$

$x = \dfrac{-1 \pm \sqrt{-3}}{2}$

$x = \dfrac{-1 \pm i\sqrt{3}}{2}$ or $-\dfrac{1}{2} \pm \dfrac{1}{2} i\sqrt{3}$

73. If a quadratic equation has two roots, they are $\dfrac{-b + \sqrt{b^2 - 4ac}}{2a}$ and

$\dfrac{-b - \sqrt{b^2 - 4ac}}{2a}$. If a, b, c are rational, then so are $-b$, $2a$, and $b^2 - 4ac$.

Then, *either* $\sqrt{b^2 - 4ac}$ is rational, hence $\dfrac{-b + \sqrt{b^2 - 4ac}}{2a}$ and $\dfrac{-b - \sqrt{b^2 - 4ac}}{2a}$ are

both rational, *or*, $\sqrt{b^2 - 4ac}$ is irrational, hence $\dfrac{-b + \sqrt{b^2 - 4ac}}{2a}$ and

$\dfrac{-b - \sqrt{b^2 - 4ac}}{2a}$ are both irrational , *or*, $\sqrt{b^2 - 4ac}$ is imaginary, hence

$\dfrac{-b + \sqrt{b^2 - 4ac}}{2a}$ and $\dfrac{-b + \sqrt{b^2 - 4ac}}{2a}$ are both imaginary. There is no other

possibility; hence, one root cannot be rational while the other is irrational.

75. $r_1 = \dfrac{-b + \sqrt{b^2 - 4ac}}{2a}$ $r_2 = \dfrac{-b - \sqrt{b^2 - 4ac}}{2a}$

$r_1 r_2 = \dfrac{(-b + \sqrt{b^2 - 4ac})}{2a} \dfrac{(-b - \sqrt{b^2 - 4ac})}{2a}$

$= \dfrac{(-b)^2 - (\sqrt{b^2 - 4ac})^2}{4a^2} = \dfrac{b^2 - (b^2 - 4ac)}{4a^2} = \dfrac{b^2 - b^2 + 4ac}{4a^2} = \dfrac{4ac}{4a^2} = \dfrac{c}{a}$

77. The $\pm$ in front still yields the same two numbers even if a is negative.

79. Let x = one number.
Since their sum is 21,
$21 - x$ = other number
Then, since their product is 104,
$$x(21 - x) = 104$$
$$21x - x^2 = 104$$
$$0 = x^2 - 21x + 104$$
$$x^2 - 21x + 104 = 0$$
$$(x - 13)(x - 8) = 0$$
$$x - 13 = 0 \quad \text{or} \quad x - 8 = 0$$
$$x = 13 \qquad\qquad x = 8$$
The numbers are 8 and 13.

81. Let x = first of the two
consecutive even integers.
Then $x + 2$ = second of these integers
Since their product is 168,
$$x(x + 2) = 168$$
$$x^2 + 2x = 168$$
$$x^2 + 2x - 168 = 0$$
$$(x - 12)(x + 14) = 0$$
$$x - 12 = 0 \quad \text{or} \quad x + 14 = 0$$
$$x = 12 \qquad\qquad x = -14$$
If $x = 12$, the two consecutive positive
even integers must be 12 and 14. We
discard the other solution, since the
numbers must be positive.

83. Let x = amount of increase

Then $x + 4$ = length of new rectangle

$x + 2$ = width of new rectangle

Using $A = ab$, we have

Area of new rectangle = 2 × Area of old rectangle.

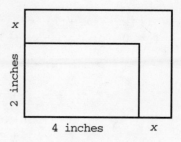

$$(x + 4)(x + 2) = 2(4)(2)$$
$$x^2 + 6x + 8 = 16$$
$$x^2 + 6x - 8 = 0$$
$$x^2 + 6x = 8$$
$$x^2 + 6x + 9 = 17$$
$$(x + 3)^2 = 17$$
$$x + 3 = \pm\sqrt{17}$$
$$x = -3 \pm \sqrt{17}$$

We discard the negative solution and take $x = -3 + \sqrt{17} = 1.12$. Then dimensions of new rectangle are $x + 4$ by $x + 2$ or 5.12 by 3.12 inches.

85. Let r = interest rate. Applying the given formula, we have

$$1440 = 1000(1 + r)^2$$
$$1.44 = (1 + r)^2$$
$$(1 + r)^2 = 1.44$$
$$1 + r = \pm\sqrt{1.44}$$
$$1 + r = \pm1.2$$
$$r = -1 \pm 1.2$$

Discarding the negative root, we have
$r = .2$ or 20%.

87. Let r = rate of slow plane.

Then $r + 140$ = rate of fast plane.

After 1 hour $r(1) = r$ = distance traveled by slow plane.

$(r + 140)(1) = r + 140$ = distance traveled by fast plane.

Applying the Pythagorean theorem, we have
$$r^2 + (r + 140)^2 = 260^2$$
$$r^2 + r^2 + 280r + 19,600 = 67,600$$
$$2r^2 + 280r - 48,000 = 0$$
$$r^2 + 140r - 24,000 = 0$$
$$(r + 240)(r - 100) = 0$$
$$r + 240 = 0 \quad \text{or} \quad r - 100 = 0$$
$$r = -240 \qquad\qquad r = 100$$

Discarding the negative solution, we have

$r = 100$ miles per hour = rate of slow plane.

$r + 140 = 240$ miles per hour = rate of fast plane.

89. Let t = time for smaller pipe to fill tank alone

$t - 5$ = time for larger pipe to fill tank alone

5 = time for both pipes to fill tank together

Then $\dfrac{1}{t}$ = rate for smaller pipe

$\dfrac{1}{t - 5}$ = rate for larger pipe

$$\begin{pmatrix} \text{Part of job} \\ \text{completed by} \\ \text{smaller pipe} \end{pmatrix} + \begin{pmatrix} \text{Part of job} \\ \text{completed by} \\ \text{larger pipe} \end{pmatrix} = 1 \text{ whole job}$$

$$\frac{1}{t}(5) \quad + \quad \frac{1}{t-5}(5) \quad = 1$$

$$\frac{5}{t} \quad + \quad \frac{5}{t-5} \quad = 1 \quad \text{Excluded values: } t \neq 0, 5$$

$$t(t-5)\frac{5}{t} \quad + \quad t(t-5)\frac{5}{t-5} = t(t-5)$$

$$5(t-5) + 5t = t(t-5)$$
$$5t - 25 + 5t = t^2 - 5t$$
$$10t - 25 = t^2 - 5t$$
$$0 = t^2 - 15t + 25$$
$$t^2 - 15t + 25 \qquad = 0$$

$$t = \frac{-b \pm \sqrt{b^2 - 4ac}}{2a} \quad a = 1, \ b = -15, \ c = 25$$

$$t = \frac{-(-15) \pm \sqrt{(-15)^2 - 4(1)(25)}}{2(1)} = \frac{15 \pm \sqrt{125}}{2}$$

$$t = 13.09, \ 1.91$$
$$t - 5 = 8.09, \ -3.09$$

Discarding the answer for t which results in a negative answer for $t - 5$, we have 13.09 hours for smaller pipe alone, 8.09 hours for larger pipe alone.

91. Let v = speed of car. Applying the given formula, we have

$$165 = 0.044v^2 + 1.1v$$
$$0 = 0.044v^2 + 1.1v - 165$$
$$0.044v^2 + 1.1v - 165 = 0$$

$$v = \frac{-b \pm \sqrt{b^2 - 4ac}}{2a} \quad a = 0.044, \ b = 1.1, \ c = -165$$

$$v = \frac{-1.1 \pm \sqrt{(1.1)^2 - 4(0.044)(-165)}}{2(0.044)} = \frac{-1.1 \pm 5.5}{0.088}$$

$$v = -75 \text{ or } 50$$

Discarding the negative answer, we have $v = 50$ miles per hour.

93. Let ℓ = length of building
w = width of building.
Then, using the hint, in the similar triangles
ABC and CDE, we have

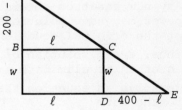

$$\frac{200 - w}{w} = \frac{\ell}{400 - \ell}$$

Therefore $w(400 - \ell)\dfrac{200 - w}{w} = w(400 - \ell)\dfrac{\ell}{400 - \ell}$

$$(400 - \ell)(200 - w) = w\ell$$
$$80,000 - 400w - 200\ell + w\ell = w\ell$$
$$80,000 - 400w = 200\ell$$
$$400 - 2w = \ell$$

Since the cross-sectional area of the building is given as 15,000 ft^2, we have

$$\ell w = 15,000$$

Substituting, we get,
$$(400 - 2w)w = 15,000$$
$$400w - 2w^2 = 15,000$$
$$0 = 2w^2 - 400w + 15,000$$
$$0 = w^2 - 200w + 7,500$$
$$0 = (w - 50)(w - 150)$$
$$w - 50 = 0 \quad \text{or} \quad w - 150 = 0$$
$$w = 50 \qquad\qquad w = 150$$
$$\ell = 400 - 2w \qquad \ell = 400 - 2w$$
$$\quad = 400 - 2(50) \qquad = 400 - 2(150)$$
$$\quad = 300 \qquad\qquad = 100$$

Thus, there are two solutions: the building is 50 ft wide and 300 ft long or 150 ft wide and 100 ft long.

95. Let x = distance from the warehouse to Factory A. Since the distance from the warehouse to Factory B via Factory A is known (it is the difference in odometer readings: 52937–52846) to be 91 miles, then

$91 - x$ = distance from Factory A to Factory B.

The distance from Factory B to the warehouse is known (it is the difference in odometer readings: 53002–52937) to be 65 miles. Applying the Pythagorean theorem, we have

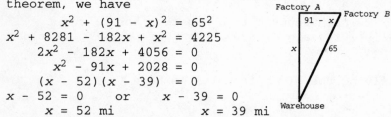

$$x^2 + (91 - x)^2 = 65^2$$
$$x^2 + 8281 - 182x + x^2 = 4225$$
$$2x^2 - 182x + 4056 = 0$$
$$x^2 - 91x + 2028 = 0$$
$$(x - 52)(x - 39) = 0$$
$$x - 52 = 0 \quad \text{or} \quad x - 39 = 0$$
$$x = 52 \text{ mi} \qquad\quad x = 39 \text{ mi}$$

Since we are told that the distance from the warehouse to Factory A was greater than the distance from Factory A to Factory B, we discard the solution $x = 39$, which would lead to $91 - x = 52$ miles, a contradiction.

52 miles.

Exercise 2-7

Key Ideas and Formulas

1. If both sides of an equation are squared, then the solution set of the original equation is a subset of the solution set of the new equation.

   Any new equation obtained by raising both members of an equation to the same power may have solutions (called *extraneous solutions*) that are not solutions of the original equation.

   Thus, every solution of the new equation must be checked in the original equation to eliminate extraneous solutions.

2. To solve an equation that is not quadratic, it may be possible to transform it to the form

$$au^2 + bu + c = 0$$

where u is an expression in some other variable.

1. $\sqrt[3]{x + 5} = 3$

$x + 5 = 27$

$x = 22$

Check:

$\sqrt[3]{22 + 5} \overset{?}{=} 3$

$\sqrt[3]{27} \overset{?}{=} 3$

$3 \overset{\checkmark}{=} 3$

Solution: 22

3. $\sqrt{5n + 9} = n - 1$

$5n + 9 = n^2 - 2n + 1$

$0 = n^2 - 7n - 8$

$n^2 - 7n - 8 = 0$

$(n - 8)(n + 1) = 0$

$n = 8, -1$

Check: $\sqrt{5(8) + 9} \overset{?}{=} 8 - 1$

$7 \overset{\checkmark}{=} 7$

$\sqrt{5(-1) + 9} \overset{?}{=} -1 - 1$

$2 \neq -2$

Solution: 8

5. $\sqrt{x + 5} + 7 = 0$

$\sqrt{x + 5} = -7$

Since the left side is non-negative for real x, there is no solution. However, if we don't notice this, we proceed as follows:

$x + 5 = 49$

$x = 44$

But this is an extraneous solution, since checking gives

$\sqrt{44 + 5} + 7 \overset{?}{=} 0$

$14 \neq 0$

No solution.

7. $\sqrt{3x + 4} = 2 + \sqrt{x}$

$3x + 4 = 4 + 4\sqrt{x} + x$

$2x = 4\sqrt{x}$

$x = 2\sqrt{x}$

$x^2 = 4x$

$x^2 - 4x = 0$

$x(x - 4) = 0$

$x = 0, 4$

Check: $\sqrt{3(0) + 4} \overset{?}{=} 2 + \sqrt{0}$

$2 \overset{\checkmark}{=} 2$

$\sqrt{3(4) + 4} \overset{?}{=} 2 + \sqrt{4}$

$4 \overset{\checkmark}{=} 4$

Solution: 0, 4

9. $y^4 - 2y^2 - 8 = 0$

Let $u = y^2$, then

$u^2 - 2u - 8 = 0$

$(u - 4)(u + 2) = 0$

$u = 4, -2$

$y^2 = 4 \qquad y^2 = -2$

$y = \pm 2 \qquad y = \pm i\sqrt{2}$

11. $x^{10} + 3x^5 - 10 = 0$

Let $u = x^5$, then

$u^2 + 3u - 10 = 0$

$(u + 5)(u - 2) = 0$

$u = -5, 2$

$x^5 = -5 \qquad x^5 = 2$

$x = \sqrt[5]{-5} \text{ or } -\sqrt[5]{5}, \ x = \sqrt[5]{2}$

13. $2x^{2/3} + 3x^{1/3} - 2 = 0$

Let $u = x^{1/3}$, then

$2u^2 + 3u - 2 = 0$

$(2u - 1)(u + 2) = 0$

$u = \dfrac{1}{2}, -2$

$x^{1/3} = \dfrac{1}{2} \qquad x^{1/3} = -2$

$x = \dfrac{1}{8} \qquad x = -8$

15. $(m^2 - m)^2 - 4(m^2 - m) = 12$

Let $u = m^2 - m$, then

$u^2 - 4u = 12$

$u^2 - 4u - 12 = 0$

$(u - 6)(u + 2) = 0$

$u = 6, -2$

$m^2 - m = 6 \qquad m^2 - m = -2$

$m^2 - m - 6 = 0 \qquad m^2 - m + 2 = 0$

$(m - 3)(m + 2) = 0 \qquad m = \dfrac{-(-1) \pm \sqrt{(-1)^2 - 4(1)(2)}}{2(1)}$

$m = 3, -2 \qquad m = \dfrac{1 \pm \sqrt{-7}}{2} \text{ or } \dfrac{1 \pm i\sqrt{7}}{2}$

$m = \dfrac{1}{2} \pm \dfrac{\sqrt{7}}{2} i$

17.

$$\sqrt{u - 2} = 2 + \sqrt{2u + 3}$$
$$u - 2 = 4 + 4\sqrt{2u + 3} + 2u + 3$$
$$u - 2 = 2u + 7 + 4\sqrt{2u + 3}$$
$$-u - 9 = 4\sqrt{2u + 3}$$
$$u^2 + 18u + 81 = 16(2u + 3)$$
$$u^2 + 18u + 81 = 32u + 48$$
$$u^2 - 14u + 33 = 0$$
$$(u - 3)(u - 11) = 0$$
$$u = 3, 11$$

Check: $\sqrt{3 - 2} \overset{?}{=} 2 + \sqrt{2(3) + 3}$
$$1 \neq 5$$
$$\sqrt{11 - 2} \overset{?}{=} 2 + \sqrt{2(11) + 3}$$
$$3 \neq 7$$

No solution.

> **Common Error:**
> $u - 2 = 4 + 2u + 3$ is not an equivalent equation to the given equation.
> $(2 + \sqrt{2u + 3})^2 \neq 4 + 2u + 3$

19.

$$\sqrt{3y - 2} = 3 - \sqrt{3y + 1}$$
$$3y - 2 = 9 - 6\sqrt{3y + 1} + 3y + 1$$
$$3y - 2 = 3y + 10 - 6\sqrt{3y + 1}$$
$$-12 = -6\sqrt{3y + 1}$$
$$2 = \sqrt{3y + 1}$$
$$4 = 3y + 1$$
$$3 = 3y$$
$$y = 1$$

Check: $\sqrt{3(1) - 2} \overset{?}{=} 3 - \sqrt{3(1) + 1}$
$$1 \overset{\checkmark}{=} 1$$

Solution: 1

21.

$$\sqrt{7x - 2} - \sqrt{x + 1} = \sqrt{3}$$
$$\sqrt{7x - 2} = \sqrt{x + 1} + \sqrt{3}$$
$$7x - 2 = x + 1 + 2\sqrt{3}\sqrt{x + 1} + 3$$
$$7x - 2 = x + 4 + 2\sqrt{3(x + 1)}$$
$$6x - 6 = 2\sqrt{3(x + 1)}$$
$$3x - 3 = \sqrt{3x + 3}$$
$$9x^2 - 18x + 9 = 3x + 3$$
$$9x^2 - 21x + 6 = 0$$
$$3x^2 - 7x + 2 = 0$$
$$(3x - 1)(x - 2) = 0$$
$$x = \tfrac{1}{3}, 2$$

Check: $\sqrt{7(\tfrac{1}{3}) - 2} - \sqrt{\tfrac{1}{3} + 1} \overset{?}{=} \sqrt{3}$
$$\sqrt{\tfrac{1}{3}} - \sqrt{\tfrac{4}{3}} \overset{?}{=} \sqrt{3}$$
$$-\tfrac{1}{3}\sqrt{3} \neq \sqrt{3}$$
$$\sqrt{7(2) - 2} - \sqrt{2 + 1} \overset{?}{=} \sqrt{3}$$
$$\sqrt{12} - \sqrt{3} \overset{?}{=} \sqrt{3}$$
$$\sqrt{3} \overset{\checkmark}{=} \sqrt{3}$$

Solution: 2

23. $3n^{-2} - 11n^{-1} - 20 = 0$
Let $u = n^{-1}$, then
$$3u^2 - 11u - 20 = 0$$
$$(3u + 4)(u - 5) = 0$$
$$u = -\tfrac{4}{3}, 5$$
$$n^{-1} = -\tfrac{4}{3} \qquad n^{-1} = 5$$
$$n = -\tfrac{3}{4} \qquad n = \tfrac{1}{5}$$

25. $9y^{-4} - 10y^{-2} + 1 = 0$
Let $u = y^{-2}$, then
$$9u^2 - 10u + 1 = 0$$
$$(9u - 1)(u - 1) = 0$$
$$u = \tfrac{1}{9}, 1$$
$$y^{-2} = \tfrac{1}{9} \qquad y^{-2} = 1$$
$$y^2 = 9 \qquad y^2 = 1$$
$$y = \pm 3 \qquad y = \pm 1$$

27. $y^{1/2} - 3y^{1/4} + 2 = 0$
Let $u = y^{1/4}$, then
$$u^2 - 3u + 2 = 0$$
$$(u - 1)(u - 2) = 0$$
$$u = 1, 2$$
$$y^{1/4} = 1 \qquad y^{1/4} = 2$$
$$y = 1 \qquad y = 16$$

> **Common Error:**
> $y \neq 2^{1/4}$
> Both sides must be raised to the fourth power to eliminate the $y^{1/4}$.

29. $(m - 5)^4 + 36 = 13(m - 5)^2$

Let $u = (m - 5)^2$, then

$$u^2 + 36 = 13u$$

$$u^2 - 13u + 36 = 0$$

$$(u - 4)(u - 9) = 0$$

$$u = 4, 9$$

$(m - 5)^2 = 4 \qquad (m - 5)^2 = 9$

$m - 5 = \pm 2 \qquad m - 5 = \pm 3$

$m = 5 \pm 2 \qquad m = 5 \pm 3$

$m = 3, 7 \qquad m = 2, 8$

31. $\sqrt{5 - 2x} - \sqrt{x + 6} = \sqrt{x + 3}$

$$\sqrt{5 - 2x} = \sqrt{x + 6} + \sqrt{x + 3}$$

$$5 - 2x = x + 6 + 2\sqrt{x + 6}\sqrt{x + 3} + x + 3$$

$$5 - 2x = 2x + 9 + 2\sqrt{x + 6}\sqrt{x + 3}$$

$$-4x - 4 = 2\sqrt{x + 6}\sqrt{x + 3}$$

$$-2x - 2 = \sqrt{(x + 6)(x + 3)}$$

$$4x^2 + 8x + 4 = (x + 6)(x + 3)$$

$$4x^2 + 8x + 4 = x^2 + 9x + 18$$

$$3x^2 - x - 14 = 0$$

$$(3x - 7)(x + 2) = 0$$

$$x = \tfrac{7}{3}, -2$$

Check:

$$\sqrt{5 - 2(\tfrac{7}{3})} - \sqrt{\tfrac{7}{3} + 6} \stackrel{?}{=} \sqrt{\tfrac{7}{3} + 3}$$

$$\sqrt{\tfrac{1}{3}} - \sqrt{\tfrac{25}{3}} \stackrel{?}{=} \sqrt{\tfrac{16}{3}}$$

$$-\frac{4\sqrt{3}}{3} \neq \frac{4\sqrt{3}}{3}$$

$$\sqrt{5 - 2(-2)} - \sqrt{-2 + 6} \stackrel{?}{=} \sqrt{-2 + 3}$$

$$1 = 1$$

Solution: -2

33. $2 + 3y^{-4} = 6y^{-2}$

$2y^4 + 3 = 6y^2$ Multiply both members by y^4 $\quad y \neq 0$

$2y^4 - 6y^2 + 3 = 0$

Let $u = y^2$, then

$2u^2 - 6u + 3 = 0$

$$u = \frac{-b \pm \sqrt{b^2 - 4ac}}{2a} \quad a = 2, \ b = -6, \ c = 3$$

$$u = \frac{-(-6) \pm \sqrt{(-6)^2 - 4(2)(3)}}{2(2)}$$

$$u = \frac{6 \pm \sqrt{36 - 24}}{4}$$

$$u = \frac{6 \pm \sqrt{12}}{4}$$

$$u = \frac{2(3 \pm \sqrt{3})}{4}$$

$$u = \frac{3 \pm \sqrt{3}}{2}$$

$$y^2 = \frac{3 \pm \sqrt{3}}{2}$$

$$y = \pm\sqrt{\frac{3 \pm \sqrt{3}}{2}} \text{ (four roots)}$$

35. By squaring: $m - 7\sqrt{m} + 12 = 0$

$$m + 12 = 7\sqrt{m}$$
$$m^2 + 24m + 144 = 49m$$
$$m^2 - 25m + 144 = 0$$
$$(m - 9)(m - 16) = 0$$
$$m = 9, \; 16$$

Check: $9 - 7\sqrt{9} + 12 \overset{?}{=} 0$

$$0 \overset{\sqrt{}}{=} 0$$

$$16 - 7\sqrt{16} + 12 \overset{?}{=} 0$$

$$0 \overset{\sqrt{}}{=} 0$$

Solution: 9, 16

By substitution: $m - 7\sqrt{m} + 12 = 0$

Let $u = \sqrt{m}$, then

$$u^2 - 7u + 12 = 0$$
$$(u - 4)(u - 3) = 0$$
$$u = 3, \; 4$$

$$\sqrt{m} = 3 \qquad\qquad \sqrt{m} = 4$$
$$m = 9 \qquad\qquad m = 16$$

These answers have already been checked.

37. By squaring: $t - 11\sqrt{t} + 18 = 0$

$$t + 18 = 11\sqrt{t}$$
$$t^2 + 36t + 324 = 121t$$
$$t^2 - 85t + 324 = 0$$
$$(t - 4)(t - 81) = 0$$
$$t = 4, \; 81$$

Check: $4 - 11\sqrt{4} + 18 \overset{?}{=} 0$

$$0 \overset{\sqrt{}}{=} 0$$

$$81 - 11\sqrt{81} + 18 \overset{?}{=} 0$$

$$0 \overset{\sqrt{}}{=} 0$$

Solution 4, 81

By substitution: $t = 11\sqrt{t} + 18 = 0$

Let $u = \sqrt{t}$, then

$$u^2 - 11u + 18 = 0$$
$$(u - 9)(u - 2) = 0$$
$$u = 2, \; 9$$

$$u = 2 \qquad u = 9$$
$$\sqrt{t} = 2 \qquad \sqrt{t} = 9$$
$$t = 4 \qquad t = 81$$

These answers have already been checked.

39. Let x = width of cross-section of the beam

y = depth of cross-section of the beam

From the Pythagorean theorem

$$x^2 + y^2 = 16^2$$

Thus,

$$y = \sqrt{256 - x^2}$$

Since the area of the rectangle is given by xy, we have

$$xy = 120$$
$$x\sqrt{256 - x^2} = 120$$
$$x^2(256 - x^2) = 14{,}400$$
$$256x^2 - x^4 = 14{,}400$$
$$-x^4 + 256x^2 - 14{,}400 = 0$$
$$(x^2)^2 - 256x^2 + 14{,}400 = 0$$

$$x^2 = \frac{-b \pm \sqrt{b^2 - 4ac}}{2a} \quad a = 1, \; b = -256, \; c = 14{,}400$$

$$x^2 = \frac{-(-256) \pm \sqrt{(-256)^2 - 4(1)(14{,}400)}}{2(1)}$$

$$x^2 = \frac{256 \pm \sqrt{65{,}536 - 57{,}600}}{2}$$

$$x^2 = \frac{256 \pm \sqrt{7{,}936}}{2}$$

$$x^2 = 128 \pm \sqrt{1{,}984}$$

$$x = \sqrt{128 \pm \sqrt{1{,}984}}$$

If $x = \sqrt{128 + \sqrt{1{,}984}} \approx 13.1$ then

$$y = \sqrt{256 - x^2}$$
$$= \sqrt{256 - (128 + \sqrt{1{,}984})}$$
$$= \sqrt{128 - \sqrt{1{,}984}} \approx 9.1$$

Thus the dimensions of the rectangle are 13.1 inches by 9.1 inches. Notice that if $x = \sqrt{128 - \sqrt{1,984}}$, then $y = \sqrt{128 + \sqrt{1,984}}$ and the dimensions are still 13.1 inches by 9.1 inches.

41.

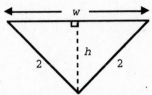

Let w = width of trough
h = altitude of triangular end

Examining the triangular end of the trough sketched above, we see that $h^2 + \left(\frac{1}{2}w\right)^2 = 2^2$. The area of this end, $A = \frac{1}{2}wh$. Since the volume of the trough V is given by $V = A \cdot 6$, we have

$$9 = 6A$$

$$9 = 6\left(\frac{1}{2}wh\right)$$

$$9 = 3wh$$

$$3 = wh$$

Since $h^2 = 2^2 - \left(\frac{1}{2}w\right)^2$

$$h^2 = 2^2 - \frac{1}{4}w^2$$

$$h = \sqrt{4 - \frac{1}{4}w^2}$$

Hence we solve

$$3 = w\sqrt{4 - \frac{1}{4}w^2}$$

$$9 = w^2\left(4 - \frac{1}{4}w^2\right)$$

$$9 = 4w^2 - \frac{1}{4}(w^2)^2$$

$$36 = 16w^2 - (w^2)^2$$

$$(w^2)^2 - 16w^2 + 36 = 0$$

$$w^2 = \frac{-b \pm \sqrt{b^2 - 4ac}}{2a} \quad a = 1, \ b = -16, \ c = 36$$

$$w^2 = \frac{-(-16) \pm \sqrt{(-16)^2 - 4(1)(36)}}{2(1)}$$

$$w^2 = \frac{16 \pm \sqrt{256 - 144}}{2}$$

$$w^2 = \frac{16 \pm \sqrt{112}}{2}$$

$$w^2 = 8 \pm 2\sqrt{7}$$

$$w = \sqrt{8 \pm 2\sqrt{7}}$$

$$w = 1.65 \text{ ft or } 3.65 \text{ ft}$$

43. $p = 14 + 0.01q$

$p = 50 - 0.5\sqrt{q}$

Substitute p from the first equation into the second to eliminate p.

$$14 + 0.01q = 50 - 0.5\sqrt{q}$$

$$-36 + 0.01q = -0.5\sqrt{q}$$

$$(-36 + 0.01q)^2 = (-0.5\sqrt{q})^2$$

$$1296 - 0.72q + 0.0001q^2 = 0.25q$$

$$0.0001q^2 - 0.97q + 1296 = 0$$

$$q = \frac{-b \pm \sqrt{b^2 - 4ac}}{2a} \qquad a = 0.0001, \; b = -0.97, \; c = 1296$$

$$q = \frac{-(-0.97) \pm \sqrt{(-0.97)^2 - 4(0.0001)(1296)}}{2(0.0001)}$$

$$q = \frac{0.97 \pm \sqrt{0.9409 - 0.5184}}{0.0002}$$

$$q = \frac{0.97 \pm \sqrt{0.4225}}{0.0002}$$

$$q = \frac{0.97 \pm 0.65}{0.0002}$$

$$q = 1600 \text{ or } 8100$$

Check: $14 + 0.01(1600) \overset{?}{=} 50 - 0.5\sqrt{1600}$ $\qquad$ $14 + 0.01(8100) \overset{?}{=} 50 - 0.5\sqrt{8100}$

$\qquad\qquad\qquad\quad 30 \overset{\surd}{=} 30$ $\qquad\qquad\qquad\qquad\qquad\qquad 95 \neq 5$

$\qquad q = 1600$ telephones

$\qquad p = 14 + 0.01q = 14 + 0.01(1600)$

$\qquad\qquad = \$30$

Exercise 2-8
Key Ideas and Formulas

A non-zero polynomial will have a constant sign (either always positive or always negative) within each interval determined by its real zeros plotted on a number line. If a polynomial has no real zeros, then the polynomial is either positive over the whole line or negative over the whole line.

The rational expression P/Q, where P and Q are non-zero polynomials, will have a constant sign (either always positive or always negative) within each interval determined by the real zeros of P and Q plotted on a number line. If neither P nor Q have real zeros, then the rational expression P/Q is either positive over the whole line or negative over the whole line.

To solve a polynomial inequality:

Step 1: Write the polynomial inequality in standard form.

Step 2: Find all real zeros of the polynomial.

Step 3: Plot the real zeros on a number line, dividing the number line into intervals.

Step 4: Choose a test number (that is easy to compute with) in each interval and evaluate the polynomial for each number.

Step 5: Using the results of Step 4 construct a sign chart, showing the sign of the polynomial in each interval.

Step 6: From the sign chart, write down the solution (and draw the graph, if required) of the original polynomial inequality.

A similar sequence of steps is required to solve a rational inequality.

1.
$$x^2 < 10 - 3x$$
$$x^2 + 3x - 10 < 0$$
$$(x + 5)(x - 2) < 0$$
Zeros: -5, 2

$(-\infty, -5)$ $(-5, 2)$ $(2, \infty)$

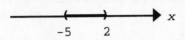

-5   2

$x^2 + 3x - 10 = (x + 5)(x - 2)$			
Test Number	-6	0	3
Value of Polynomial for Test Number	8	-10	8
Sign of Polynomial in Interval	+	-	+
Interval	$(-\infty, -5)$	$(-5, 2)$	$(2, \infty)$

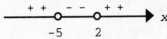

-5   2

$x^2 + 3x - 10$ is negative within the interval $(-5, 2)$. $-5 < x < 2$.

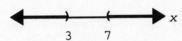

-5   2

3.
$$x^2 + 21 > 10x$$
$$x^2 - 10x + 21 > 0$$
$$(x - 3)(x - 7) > 0$$
Zeros: 3, 7

$(-\infty, 3)$ $(3, 7)$ $(7, \infty)$

3   7

$x^2 - 10x + 21 = (x - 3)(x - 7)$			
Test Number	2	5	8
Value of Polynomial for Test Number	5	-4	5
Sign of Polynomial in Interval	+	-	+
Interval	$(-\infty, 3)$	$(3, 7)$	$(7, \infty)$

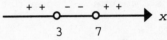

3   7

$x^2 - 10x + 21$ is positive within the intervals $(-\infty, 3)$ and $(7, \infty)$.
$x < 3$ or $x > 7$

3   7

5.
$$x^2 \leq 8x$$
$$x^2 - 8x \leq 0$$
$$x(x - 8) \leq 0$$
Zeros: 0,8

0,8 are part of the solution set, so we use solid dots.

$(-\infty, 0)$ $(0, 8)$ $(8, \infty)$

0   8

$x^2 - 8x = x(x - 8)$			
Test Number	-1	4	9
Value of Polynomial for Test Number	9	-16	9
Sign of Polynomial in Interval	+	-	+
Interval	$(-\infty, 0)$	$(0, 8)$	$(8, \infty)$

+ +   - -   + +

0   8

$x^2 - 8x$ is non-positive within the interval $[0,8]$. $0 \leq x \leq 8$

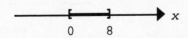

0   8

7.
$$x^2 + 5x \leq 0$$
$$x(x + 5) \leq 0$$
Zeros: -5, 0

-5, 0 are part of the solution set, so we use solid dots.

$(-\infty, -5)$ $(-5, 0)$ $(0, \infty)$

-5   0

$x^2 + 5x = x(x + 5)$			
Test Number	-6	-1	1
Value of Polynomial for Test Number	6	-4	6
Sign of Polynomial in Interval	+	-	+
Interval	$(-\infty, -5)$	$(-5, 0)$	$(0, \infty)$

+ +   - -   + +

-5   0

$x^2 + 5x$ is non-positive within the interval $[-5,0]$. $-5 \leq x \leq 0$

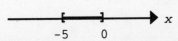

-5   0

9.

$$x^2 > 4$$
$$x^2 - 4 > 0$$
$$(x + 2)(x - 2) > 0$$

Zeros: -2, 2

$(-\infty, -2)$ $(-2, 2)$ $(2, \infty)$

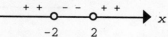

$x^2 - 4 = (x + 2)(x - 2)$			
Test Number	-3	0	3
Value of Polynomial for Test Number	5	-4	5
Sign of Polynomial in Interval	$+$	$-$	$+$
Interval	$(-\infty, -2)$	$(-2, 2)$	$(2, \infty)$

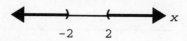

$x^2 - 4$ is positive within the intervals $(-\infty, -2)$ and $(2, \infty)$. $x < -2$ or $x > 2$

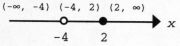

11. $\dfrac{x - 2}{x + 4} \le 0$

Zeros of P, Q: -4, 2

2 is a part of the solution set, so we use a solid dot there. -4 is not part of the solution set, ($\frac{P}{Q}$ is not defined there) so we use an open dot there.

$(-\infty, -4)$ $(-4, 2)$ $(2, \infty)$

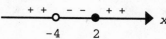

$\dfrac{P}{Q} = \dfrac{x - 2}{x + 4}$			
Test Number	-5	0	3
Value of $\frac{P}{Q}$	7	$-\frac{1}{2}$	$\frac{1}{7}$
Sign of $\frac{P}{Q}$	$+$	$-$	$+$
Interval	$(-\infty, -4)$	$(-4, 2)$	$(2, \infty)$

$\dfrac{x - 2}{x + 4}$ is non-positive within the interval $(-4, 2]$. $-4 < x \le 2$

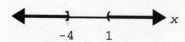

13. $\dfrac{x + 4}{1 - x} \le 0$

Zeros of P, Q: -4, 1

-4 is a part of the solution set, so we use a solid dot there. 1 is not part of the solution set, ($\frac{P}{Q}$ is not defined there) so we use an open dot there.

$(-\infty, -4)$ $(-4, 1)$ $(1, \infty)$

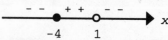

$\dfrac{P}{Q} = \dfrac{x + 4}{1 - x}$			
Test Number	-5	0	2
Value of $\frac{P}{Q}$	$-\frac{1}{6}$	4	-6
Sign of $\frac{P}{Q}$	$-$	$+$	$-$
Interval	$(-\infty, -4)$	$(-4, 1)$	$(1, \infty)$

$\dfrac{x + 4}{1 - x}$ is non-positive within the intervals $(-\infty, -4]$ and $(1, \infty)$. $x \le -4$ or $x > 1$

15. $\dfrac{x^2 + 5x}{x - 3} \geq 0$

$\dfrac{x(x + 5)}{x - 3} \geq 0$

Zeros of P, Q: -5, 0, 3
-5 and 0 are part of the
solution set, so we use solid
dots there. 3 is not part of
the solution set, ($\frac{P}{Q}$ is not
defined there) so we use an
open dot there.

(-∞, -5) (-5, 0) (0, 3) (3, ∞)

	$\frac{P}{Q} = \frac{x(x + 5)}{x - 3}$			
Test Number	-6	-1	1	4
Value of $\frac{P}{Q}$	$-\frac{1}{9}$	1	-3	36
Sign of $\frac{P}{Q}$	-	+	-	+
Interval	(-∞, -5)	(-5, 0)	(0, 3)	(3, ∞)

$\dfrac{x^2 + 5x}{x - 3}$ is non-negative within the intervals [-5, 0] and (3, ∞).

$-5 \leq x \leq 0$ or $3 < x$.

17. $\dfrac{(x + 1)^2}{x^2 + 2x - 3} \leq 0$

$\dfrac{(x + 1)(x + 1)}{(x - 1)(x + 3)} \geq 0$

Zeros of P, Q: -1, 1, -3
-1 is part of the solution set,
so we use a solid dot there.
-3 and 1 are not part of the
solution set, ($\frac{P}{Q}$ is not defined
there) so we use open dots
there.

(-∞, -3) (-3, -1) (-1, 1) (1, ∞)

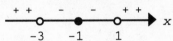

	$\frac{P}{Q} = \frac{(x + 1)(x + 1)}{(x - 1)(x + 3)}$			
Test Number	-4	-2	0	2
Value of $\frac{P}{Q}$	$\frac{9}{5}$	$-\frac{1}{3}$	$-\frac{1}{3}$	$\frac{9}{5}$
Sign of $\frac{P}{Q}$	+	-	-	+
Interval	(-∞, -3)	(-3, -1)	(-1, 1)	(1, ∞)

$\dfrac{(x + 1)^2}{x^2 + 2x - 3}$ is non-positive within the intervals (-3, -1) and (-1, 1) as well
as at $x = -1$. More simply, we can write $-3 < x < 1$ or $(-3, 1)$.

19. $\dfrac{1}{x} < 4$

$\dfrac{1}{x} - 4 < 0$

$\dfrac{1 - 4x}{x} < 0$

Zeros of P, Q: 0, $\dfrac{1}{4}$

(-∞, 0) (0, 1/4) (1/4, ∞)

	$\frac{P}{Q} = \frac{1 - 4x}{x}$		
Test Number	-1	0.1	1
Value of $\frac{P}{Q}$	-5	6	-3
Sign of $\frac{P}{Q}$	-	+	-
Interval	(-∞, 0)	(0, $\frac{1}{4}$)	($\frac{1}{4}$, ∞)

$\dfrac{1 - 4x}{x} < 0$ and $\dfrac{1}{x} < 4$ within the intervals $(-∞, 0)$ and $\left(\dfrac{1}{4}, ∞\right)$. $x < 0$ or $\dfrac{1}{4} < x$

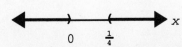

21.

$$\frac{3x + 1}{x + 4} \leq 1$$

$$\frac{3x + 1}{x + 4} - 1 \leq 0$$

$$\frac{3x + 1 - (x + 4)}{x + 4} \leq 0$$

$$\frac{2x - 3}{x + 4} \leq 0$$

> **Common Error:**
> $3x + 1 \leq x + 4$
> This is not equivalent
> to the given inequality.

> **Common Error:**
> $3x + 1 - x + 4$ is an
> incorrect numerator.

$(-\infty, -4)$ $(-4, 3/2)$ $(3/2, \infty)$

-4 $\quad \frac{3}{2}$

Zeros of P, Q: -4, $\frac{3}{2}$

$\frac{3}{2}$ is part of the solution set so we use a solid dot there. -4 is not part of the solution set ($\frac{P}{Q}$ is not defined there) so we use an open dot there.

	$\frac{P}{Q} = \frac{2x - 3}{x + 4}$		
Test Number	-5	0	2
Value of $\frac{P}{Q}$	13	$-\frac{3}{4}$	$\frac{1}{6}$
Sign of $\frac{P}{Q}$	$+$	$-$	$+$
Interval	$(-\infty, -4)$	$(-4, \frac{3}{2})$	$(\frac{3}{2}, \infty)$

$+$ $+$ $\quad$ $-$ $-$ $\quad$ $+$ $+$

-4 $\quad \frac{3}{2}$

$\frac{2x - 3}{x + 4} \leq 0$ and $\frac{3x + 1}{x + 4} \leq 1$ within the interval $\left(-4, \frac{3}{2}\right]$. $-4 < x \leq \frac{3}{2}$

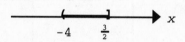

-4 $\quad \frac{3}{2}$

23.

$$\frac{2}{x + 1} \geq \frac{1}{x - 2}$$

$$\frac{2}{x + 1} - \frac{1}{x - 2} \geq 0$$

$$\frac{2(x - 2) - (x + 1)}{(x + 1)(x - 2)} \geq 0$$

$$\frac{2x - 4 - x - 1}{(x + 1)(x - 2)} \geq 0$$

$$\frac{x - 5}{(x + 1)(x - 2)} \geq 0$$

Zeros of P, Q: 5, -1, 2

5 is a part of the solution set, so we use a solid dot there. -1 and 2 are not part of the solution set, ($\frac{P}{Q}$ is not defined there) so we use an open dot there.

$(-\infty, -1)$ $(-1, 2)$ $(2, 5)$ $(5, \infty)$

-1 $\quad$ 2 $\quad$ 5

$\frac{x - 5}{(x - 1)(x - 2)} \geq 0$ and

$\frac{2}{x + 1} \geq \frac{1}{x - 2}$ within the intervals $(-1, 2)$ and $[5, \infty)$.

$-1 < x < 2$ or $5 \leq x$

-1 $\quad$ 2 $\quad$ 5

	$\frac{P}{Q} = \frac{x - 5}{(x + 1)(x - 2)}$			
Test Number	-2	0	3	6
Value of $\frac{P}{Q}$	$-\frac{7}{4}$	$\frac{5}{2}$	$-\frac{1}{2}$	$\frac{1}{28}$
Sign of $\frac{P}{Q}$	$-$	$+$	$-$	$+$
Interval	$(-\infty, -1)$	$(-1, 2)$	$(2, 5)$	$(5, \infty)$

$-$ $-$ $\quad$ $+$ $+$ $\quad$ $-$ $-$ $\quad$ $+$ $+$

-1 $\quad$ 2 $\quad$ 5

25.
$$x^3 + 2x^2 \leq 8x$$
$$x^3 + 2x^2 - 8x \leq 0$$
$$x(x^2 + 2x - 8) \leq 0$$
$$x(x + 4)(x - 2) \leq 0$$
Zeros: 0, -4, 2
0, -4, 2 are part of the solution set, so we use solid dots there.

$x^3 + 2x^2 - 8x = x(x + 4)(x - 2)$				
Test Number	-5	-1	1	3
Value of Polynomial for Test Number	-35	9	-5	21
Sign of Polynomial in Interval	-	+	-	+
Interval	$(-\infty, -4)$	$(-4, 0)$	$(0, 2)$	$(2, \infty)$

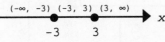

$x(x + 4)(x - 2) \leq 0$ and $x^3 + 2x^2 \leq 8x$ within the intervals $(-\infty, -4]$ and $[0, 2]$.
$x \leq -4$ or $0 \leq x \leq 2$

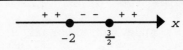

27. $\sqrt{x^2 - 9}$ will represent a real number when $x^2 - 9 \geq 0$.
$$x^2 - 9 \geq 0$$
$$(x + 3)(x - 3) \geq 0$$
Zeros of P, Q: -3, 3
-3 and 3 are part of the solution set, so we use solid dots.

$x^2 - 9 = (x + 3)(x - 3)$			
Test Number	-4	0	4
Value of Polynomial for Test Number	7	-9	7
Sign of Polynomial in Interval	+	-	+
Interval	$(-\infty, -3)$	$(-3, 3)$	$(3, \infty)$

$x^2 - 9 \geq 0$ and $\sqrt{x^2 - 9}$ will represent a real number within the intervals $(-\infty, -3]$ and $[3, \infty)$. $x \leq -3$ or $x \geq 3$

29. $\sqrt{2x^2 + x - 6}$ will represent a real number when $2x^2 + x - 6 \geq 0$
$$2x^2 + x - 6 \geq 0$$
$$(2x - 3)(x + 2) \geq 0$$
Zeros: $-2, \dfrac{3}{2}$
-2 and $\dfrac{3}{2}$ are part of the solution set, so we use solid dots.

$2x^2 + x - 6 = (2x - 3)(x + 2)$			
Test Number	-3	0	2
Value of Polynomial for Test Number	9	-6	4
Sign of Polynomial in Interval	+	-	+
Interval	$(-\infty, -2)$	$(-2, \frac{3}{2})$	$(\frac{3}{2}, \infty)$

$2x^2 + x - 6 \geq 0$ and $\sqrt{2x^2 + x - 6}$ will represent a real number within the intervals $(-\infty, -2]$ and $\left[\dfrac{3}{2}, \infty\right)$. $x \leq -2$ or $x \geq \dfrac{3}{2}$

31. $\sqrt{\dfrac{x + 7}{3 - x}}$ will represent a real number when $\dfrac{x + 7}{3 - x} \geq 0$.

$\dfrac{x + 7}{3 - x} \geq 0$

Zeros of $\dfrac{P}{Q}$: -7, 3

-7 is a part of the solution set, so we use a solid dot there. 3 is not part of the solution set, ($\frac{P}{Q}$ is not defined there) so we use an open dot there.

	$\dfrac{P}{Q} = \dfrac{x + 7}{3 - x}$		
Test Number	-8	0	4
Value of $\dfrac{P}{Q}$	$-\dfrac{1}{11}$	$\dfrac{7}{3}$	-11
Sign of $\dfrac{P}{Q}$	$-$	$+$	$-$
Interval	$(-\infty,\ -7)$	$(-7,\ 3)$	$(3,\ \infty)$

$\dfrac{x + 7}{3 - x} \geq 0$ and $\sqrt{\dfrac{x + 7}{3 - x}}$ will represent a real number within the interval $[-7,\ 3)$.

$-7 \leq x < 3$

33. Since r_1 and r_2 are distinct real roots of $ax^2 + bx + c = 0$, the inequality can be written:

$a(x - r_1)(x - r_2) > 0$

If $a > 0$, this is equivalent to $(x - r_1)(x - r_2) > 0$

If $a < 0$, this is equivalent to $(x - r_1)(x - r_2) < 0$

Zeros: r_1, r_2

	$(x - r_1)(x - r_2)$		
Test Number	$t_1 < r_1$	$r_1 < t_2 < r_2$	$r_2 < t_3$
Sign of Polynomial in Interval	$+$	$-$	$+$
Interval	$(-\infty,\ r_1)$	$(r_1,\ r_2)$	$(r_2,\ \infty)$

Thus if $a > 0$, $(x - r_1)(x - r_2) > 0$ and $ax^2 + bx + c > 0$ within the intervals $(-\infty,\ r_1)$ and $(r_2,\ \infty)$.

Example: $x^2 - 5x - 6 > 0$ or $(x + 1)(x - 6) > 0$. Solution: $(-\infty,\ -1) \cup (6,\ \infty)$

If $a < 0$, $(x - r_1)(x - r_2) < 0$ and $ax^2 + bx + c > 0$ within the interval $(r_1,\ r_2)$

Example: $6 - x - x^2 > 0$ or $(x + 3)(x - 2) < 0$. Solution: $(-3,\ 2)$

35. Since r is a double root of $ax^2 + bx + c = 0$, the inequality can be written

$a(x - r)^2 \geq 0$

If $a > 0$, this is true for all real x, and the solution set is R, the set of all real numbers.

If $a < 0$, this is equivalent to $(x - r)^2 \leq 0$, which is true only if $x = r$. The solution set is $\{r\}$.

37. One example is $x^2 \geq 0$.

39.
$$x^2 + 1 < 2x$$
$$x^2 - 2x + 1 < 0$$
$$(x - 1)^2 < 0$$

Since the square of no real number is negative, these statements are never true for any real number x. No solution (and no graph). $\varnothing$ is the solution set.

41. $x^2 < 3x - 3$
$x^2 - 3x + 3 < 0$
We attempt to find all real zeros of the polynomial.
$x^2 - 3x + 3 = 0$

$$x = \frac{-b \pm \sqrt{b^2 - 4ac}}{2a} \quad a = 1, \; b = -3, \; c = 3$$

$$x = \frac{-(-3) \pm \sqrt{(-3)^2 - 4(1)(3)}}{2(1)}$$

$$x = \frac{3 \pm \sqrt{-3}}{2}$$

The polynomial has no real zeros. Hence the statement is either true for all real x or for no real x. To determine which, we choose a test number, say 0.
$x^2 < 3x - 3$
$0^2 \overset{?}{<} 3(0) - 3$
$0 \overset{?}{<} -3$ False.
The statement is never true for any real number x. No solution (and no graph). $\varnothing$ is the solution set.

43. $x^2 - 1 \geq 4x$
$x^2 - 4x - 1 \geq 0$
Find all real zeros of the polynomial.
$x^2 - 4x - 1 = 0$

$$x = \frac{-b \pm \sqrt{b^2 - 4ac}}{2a} \quad a = 1, \; b = -4, \; c = -1$$

$$x = \frac{-(-4) \pm \sqrt{(-4)^2 - 4(1)(-1)}}{2(1)}$$

$$x = \frac{4 \pm \sqrt{16 + 4}}{2}$$

$$x = \frac{4 \pm \sqrt{20}}{2}$$

$$x = 2 \pm \sqrt{5}$$
$$\approx -0.236, \; 4.236$$

> **Common Error:**
> $x \neq 2 \pm \sqrt{20}$

Plot the real zeros on a number line.

Polynomial $x^2 - 4x - 1$			
Test Number	-1	0	5
Value of Polynomial for Test Number	4	-1	4
Sign of Polynomial in Interval	+	-	+
Interval	$(-\infty, 2 - \sqrt{5})$	$(2 - \sqrt{5}, 2 + \sqrt{5})$	$(2 + \sqrt{5}, \infty)$

$x^2 - 4x - 1 \geq 0$ and $x^2 - 1 \geq 4x$ within the intervals $(-\infty, 2 - \sqrt{5}]$ and $[2 + \sqrt{5}, \infty)$.
$x \leq 2 - \sqrt{5}$ or $x \geq 2 + \sqrt{5}$

45.
$$x^3 > 2x^2 + x$$
$$x^3 - 2x^2 - x > 0$$

Find all real zeros of the polynomial.
$$x^3 - 2x^2 - x = 0$$
$$x(x^2 - 2x - 1) = 0$$

$$x = 0 \text{ or } x^2 - 2x - 1 = 0$$

$$x = \frac{-b \pm \sqrt{b^2 - 4ac}}{2a} \quad a = 1, \ b = -2, \ c = -1$$

$$x = \frac{-(-2) \pm \sqrt{(-2)^2 - 4(1)(-1)}}{2(1)}$$

$$x = \frac{2 \pm \sqrt{8}}{2}$$

$$x = 1 \pm \sqrt{2}$$

$$\approx -0.414, \ 2.414$$

Plot the real zeros on a number line.

Polynomial $x^3 - 2x^2 - x$				
Test Number	-1	-0.1	1	3
Value of Polynomial for Test Number	-2	0.079	-2	6
Sign of Polynomial in Interval	$-$	$+$	$-$	$+$
Interval	$(-\infty, 1 - \sqrt{2})$	$(1 - \sqrt{2}, 0)$	$(0, 1 + \sqrt{2})$	$(1 + \sqrt{2}, \infty)$

$x^3 - 2x^2 - x > 0$ and $x^3 > 2x^2 + x$ within the intervals $(1 - \sqrt{2}, 0)$, and

$(1 + \sqrt{2}, \infty)$. $1 - \sqrt{2} < x < 0$ or $x > 1 + \sqrt{2}$

47.
$$4x^4 + 4 \le 17x^2$$
$$4x^4 - 17x^2 + 4 \le 0$$
$$(4x^2 - 1)(x^2 - 4) \le 0$$
$$(2x - 1)(2x + 1)(x - 2)(x + 2) \le 0$$

Zeros: $-2, \ -\dfrac{1}{2}, \ \dfrac{1}{2}, \ 2$

$-2, \ -\dfrac{1}{2}, \ \dfrac{1}{2},$ and 2 are part of the solution set, so we use solid dots.

$4x^4 - 17x^2 + 4 = (2x - 1)(2x + 1)(x - 2)(x + 2)$					
Test Number	-3	-1	0	1	3
Value of Polynomial for Test Number	175	-9	4	-9	175
Sign of Polynomial in Interval	$+$	$-$	$+$	$-$	$+$
Interval	$(-\infty, -2)$	$(-2, -\frac{1}{2})$	$(-\frac{1}{2}, \frac{1}{2})$	$(\frac{1}{2}, 2)$	$(2, \infty)$

$4x^4 - 17x^2 + 4 \le 0$ and $4x^4 + 4 \le 17x^2$ within the intervals $\left[-2, \ -\dfrac{1}{2}\right]$ and $\left[\dfrac{1}{2}, \ 2\right]$.

$-2 \le x \le -\dfrac{1}{2}$ or $\dfrac{1}{2} \le x \le 2$.

49. $|x^2 - 1| \leq 3$ According to Theorem 2, Section 2-4, we must consider the equations $x^2 - 1 = 3$ and $x^2 - 1 = -3$ to solve the equality part of the inequality statement, and the double inequality $-3 < x^2 - 1 < 3$ to solve the inequality part.

Case 1:

$x^2 - 1 = 3$

$\quad x^2 = 4$

There are two solutions: -2 and 2.

Case 2:

$x^2 - 1 = -3$

$\quad x^2 = -2$

There are no real solutions.

Case 3:

$-3 < x^2 - 1 < 3$

To satisfy these relations, both statements $-3 < x^2 - 1$ and $x^2 - 1 < 3$ must hold. But $-3 < x^2 - 1$ or $-4 < x^2$ holds for all real x, so we need only examine $x^2 - 1 < 3$ or $x^2 - 4 < 0$. $(x - 2)(x + 2) < 0$

Zeros: -2, 2

	$x^2 - 4 = (x - 2)(x + 2)$		
Test Number	-3	0	3
Value of Polynomial for Test Number	5	-4	5
Sign of Polynomial in Interval	+	–	+
Interval	$(-\infty, -2)$	$(-2, 2)$	$(2, \infty)$

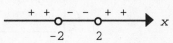

$(x - 2)(x + 2)$ is negative within the interval $(-2, 2)$. Combining the solutions from the separate cases, $|x^2 - 1| \leq 3$ when $-2 \leq x \leq 2$. $[-2, 2]$

51. (A) A profit will result if cost is less than revenue; that is, if

$\qquad C < R$

$\qquad 28 - 2p < 9p - p^2$

$p^2 - 11p + 28 < 0$

$(p - 7)(p - 4) < 0$

Zeros: 4, 7

	$p^2 - 11p + 28 = (p - 7)(p - 4)$		
Test Number	3	5	8
Value of Polynomial for Test Number	4	-2	4
Sign of Polynomial in Interval	+	–	+
Interval	$(-\infty, 4)$	$(4, 7)$	$(7, \infty)$

$p^2 - 11p + 28 < 0$ and a profit will occur $(C < R)$, for $\$4 < p < \$7$ or $(\$4, \$7)$

(B) A loss will result if cost is greater than revenue; that is, if

$\qquad C > R$

$\qquad 28 - 2p > 9p - p^2$

$p^2 - 11p + 28 \qquad > 0$

Referring to the sign chart in part (A), we see that $p^2 - 11p + 28 > 0$, and a loss will occur $(C > R)$, for $p < \$4$ or $p > \$7$.

Since a negative price doesn't make sense we delete any number to the left of 0. Thus, a loss will occur for $\$0 \leq p < \$4$ or $p > \$7$. $[\$0, \$4) \cup (\$7, \infty)$.

53. The object will be 160 feet or higher while $160 \leq d$.

$$160 \leq 112t - 16t^2$$
$$16t^2 - 112t + 160 \leq 0$$
$$t^2 - 7t + 10 \leq 0 \quad \text{(dividing both sides by 16)}$$
$$(t - 2)(t - 5) \leq 0$$

Zeros: 2, 5

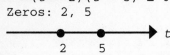

$t^2 - 7t + 10 = (t - 2)(t - 5)$			
Test Number	0	3	6
Value of Polynomial for Test Number	10	-2	4
Sign of Polynomial in Interval	+	-	+
Interval	$(-\infty, 2)$	$(2, 5)$	$(5, \infty)$

$$\begin{array}{ccc} + + & - - & + + \\ \end{array}$$
2   5

$t^2 - 7t + 10 \leq 0$, and the object will be 160 feet or higher, for $2 \leq t \leq 5$. [2, 5]

55. It will take the car more than 330 feet to stop when $d > 330$.

$$0.044v^2 + 1.1v > 330$$
$$0.044v^2 + 1.1v - 330 > 0$$
$$v^2 + 25v - 7{,}500 > 0 \quad \text{(dividing both sides by 0.044)}$$
$$(v - 75)(v + 100) > 0$$

Zeros: -100, 75

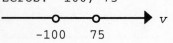

$v^2 + 25v - 7{,}500 = (v - 75)(v + 100)$			
Test Number	-200	0	100
Value of Polynomial for Test Number	27,500	-7,500	5,000
Sign of Polynomial in Interval	+	-	+
Interval	$(-\infty, -100)$	$(-100, 75)$	$(75, \infty)$

$$\begin{array}{ccc} + + & - - & + + \\ \end{array} \quad v$$
-100   75

$v^2 + 25v - 7{,}500 > 0$, and it will take the car more than 330 feet to stop, for $v < -100$ or $v > 75$.

Since a negative speed doesn't make sense we delete any number to the left of 0. Thus, $v > 75$ miles per hour.

57. Sales will be 8,000 units or more when $8 \leq S$.

$$8 \leq \frac{200t}{t^2 + 100}$$
$$8 - \frac{200t}{t^2 + 100} \leq 0$$
$$\frac{8(t^2 + 100) - 200t}{t^2 + 100} \leq 0$$
$$\frac{8t^2 - 200t + 800}{t^2 + 100} \leq 0$$

Find all real zeros of P and Q,

$$\frac{P}{Q} = \frac{8t^2 - 200t + 800}{t^2 + 100}.$$

$$8t^2 - 200t + 800 = 0$$
$$t^2 - 25t + 100 = 0$$
$$(t - 5)(t - 20) = 0$$
$$t = 5, 20 \quad \text{(Zeros of } P\text{)}$$

$t^2 + 100$ has no real zero.

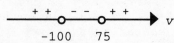

$\frac{P}{Q} = \frac{8t^2 - 200t + 800}{t^2 + 100}$			
Test Number	0	10	30
Value of $\frac{P}{Q}$	8	-2	2
Sign of $\frac{P}{Q}$	+	-	+
Interval	$(-\infty, 5)$	$(5, 20)$	$(20, \infty)$

$$\begin{array}{ccc} + + & - - & + \\ \end{array} \quad t$$
5   20

$\dfrac{8t^2 - 200t + 800}{t^2 + 100} \leq 0$ and sales will be 8 thousand units or more, for $5 \leq t \leq 20$.

CHAPTER 2 REVIEW

1. $0.05x + 0.25(30 - x) = 3.3$
$0.05x + 7.5 - 0.25x = 3.3$
$-0.2x + 7.5 = 3.3$
$-0.2x = -4.2$
$x = \dfrac{-4.2}{-0.2}$
$x = 21$ *(2-1)*

2. $\dfrac{5x}{3} - \dfrac{4 + x}{2} = \dfrac{x - 2}{4} + 1$
$12\dfrac{5x}{3} - 12\dfrac{(4 + x)}{2} = 12\dfrac{(x - 2)}{4} + 12$
$20x - 6(4 + x) = 3(x - 2) + 12$
$20x - 24 - 6x = 3x - 6 + 12$
$14x - 24 = 3x + 6$
$11x = 30$
$x = \dfrac{30}{11}$ *(2-1)*

3. $y = 4x - 9$
$y = -x + 6$
Substitute y from the first equation into the second equation to eliminate y.
$4x - 9 = -x + 6$
$5x - 9 = 6$
$5x = 15$
$x = 3$
$y = -x + 6 = -3 + 6 = 3$
$x = 3, y = 3$ *(2-2)*

4. $3(2 - x) - 2 \leq 2x - 1$
$6 - 3x - 2 \leq 2x - 1$
$-3x + 4 \leq 2x - 1$
$-5x \leq -5$
$x \geq 1$
$[1, \infty)$

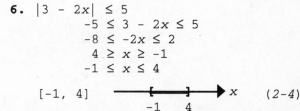

(2-3)

5. $|y + 9| < 5$
$-5 < y + 9 < 5$
$-14 < y < -4$
$(-14, -4)$

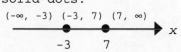

-14   -4   *(2-4)*

6. $|3 - 2x| \leq 5$
$-5 \leq 3 - 2x \leq 5$
$-8 \leq -2x \leq 2$
$4 \geq x \geq -1$
$-1 \leq x \leq 4$
$[-1, 4]$

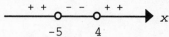

-1   4   *(2-4)*

7. $x^2 + x < 20$
$x^2 + x - 20 < 0$
$(x + 5)(x - 4) < 0$
Zeros: -5, 4
$(-\infty, -5)$ $(-5, 4)$ $(4, \infty)$

	$x^2 + x - 20 = (x + 5)(x - 4)$		
Test Number	-6	0	5
Value of Polynomial for Test Number	10	-20	10
Sign of Polynomial in Interval	$+$	$-$	$+$
Interval	$(-\infty, -5)$	$(-5, 4)$	$(4, \infty)$

$x^2 + x - 20$ is negative within the interval $(-5, 4)$. $-5 < x < 4$.

(2-8)

8. $x^2 \geq 4x + 21$
$x^2 - 4x - 21 \geq 0$
$(x + 3)(x - 7) \geq 0$
Zeros: -3, 7
-3, 7 are part of the solution set, so we use solid dots.
$(-\infty, -3)$ $(-3, 7)$ $(7, \infty)$

	$x^2 - 4x - 21 = (x + 3)(x - 7)$		
Test Number	-4	0	8
Value of Polynomial for Test Number	11	-21	11
Sign of Polynomial in Interval	$+$	$-$	$+$
Interval	$(-\infty, -3)$	$(-3, 7)$	$(7, \infty)$

$x^2 - 4x - 21 \geq 0$ is non-negative within the intervals $(-\infty, -3]$ and $[7, \infty)$.
$x \leq -3$ or $x \geq 7$

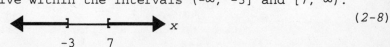

(2-8)

9. (A) $(-3 + 2i) + (6 - 8i) = -3 + 2i + 6 - 8i = 3 - 6i$

(B) $(3 - 3i)(2 + 3i) = 6 + 3i - 9i^2 = 6 + 3i + 9 = 15 + 3i$

(C) $\dfrac{13 - i}{5 - 3i} = \dfrac{(13 - i)}{(5 - 3i)} \dfrac{(5 + 3i)}{(5 + 3i)}$

$= \dfrac{65 + 34i - 3i^2}{25 - 9i^2} = \dfrac{65 + 34i + 3}{25 + 9} = \dfrac{68 + 34i}{34} = 2 + i$ $\qquad (2\text{-}5)$

10. $2x^2 - 7 = 0$

$2x^2 = 7$

$x^2 = \dfrac{7}{2}$

$x = \pm\sqrt{\dfrac{7}{2}}$

$x = \pm\dfrac{\sqrt{14}}{2}$ $\quad (2\text{-}6)$

11. $2x^2 = 4x$

$2x^2 - 4x = 0$

$2x(x - 2) = 0$

$2x = 0 \quad x - 2 = 0$

$x = 0 \qquad x = 2$

$(2\text{-}6)$

12. $2x^2 = 7x - 3$

$2x^2 - 7x + 3 = 0$

$(2x - 1)(x - 3) = 0$

$2x - 1 = 0 \qquad x - 3 = 0$

$x = \dfrac{1}{2} \qquad x = 3$

$(2\text{-}6)$

13. $m^2 + m + 1 = 0$

$m = \dfrac{-b \pm \sqrt{b^2 - 4ac}}{2a}$ $\quad a = 1$

$b = 1$

$c = 1$

$m = \dfrac{-1 \pm \sqrt{(1)^2 - 4(1)(1)}}{2(1)}$

$m = \dfrac{-1 \pm \sqrt{-3}}{2}$

$m = \dfrac{-1 \pm i\sqrt{3}}{2}$

$m = -\dfrac{1}{2} \pm \dfrac{\sqrt{3}}{2}i$ $\qquad (2\text{-}6)$

14. $y^2 = \dfrac{3}{2}(y + 1)$

$2y^2 = 3(y + 1)$

$2y^2 = 3y + 3$

$2y^2 - 3y - 3 = 0$

$y = \dfrac{-b \pm \sqrt{b^2 - 4ac}}{2a}$ $\quad a = 2$

$b = -3$

$c = -3$

$y = \dfrac{-(-3) \pm \sqrt{(-3)^2 - 4(2)(-3)}}{2(2)}$

$y = \dfrac{3 \pm \sqrt{33}}{4}$

$(2\text{-}6)$

15. $\sqrt{5x - 6} - x = 0$

$\sqrt{5x - 6} = x$

$5x - 6 = x^2$

$0 = x^2 - 5x + 6$

$x^2 - 5x + 6 = 0$

$(x - 3)(x - 2) = 0$

$x = 2, 3$

Check: $\sqrt{5(2) - 6} - 2 \overset{?}{=} 0$

$0 \overset{\surd}{=} 0$

$\sqrt{5(3) - 6} - 3 \overset{?}{=} 0$

$0 \overset{\surd}{=} 0$

Solution: 2, 3 $\qquad (2\text{-}7)$

16. $3x + 2y = 5$

$4x - y = 14$

Solve the first equation for y in terms of x and substitute into the second equation

$2y = 5 - 3x$

$y = \dfrac{5 - 3x}{2}$

$4x - \left(\dfrac{5 - 3x}{2}\right) = 14$

$8x - \dfrac{2}{1}\left(\dfrac{5 - 3x}{2}\right) = 28$

$8x - (5 - 3x) = 28$

$8x - 5 + 3x = 28$

$11x = 33$

$x = 3$

$y = \dfrac{5 - 3x}{2} = \dfrac{5 - 3 \cdot 3}{2}$

$y = -2$

$x = 3, \ y = -2$ $\qquad (2\text{-}2)$

17. $\sqrt{3 - 5x}$ represents a real number exactly when $3 - 5x$ is positive or zero. We can write this as an inequality statement and solve for x.

$$3 - 5x \geq 0$$
$$-5x \geq -3$$
$$x \leq \frac{3}{5} \qquad (2\text{-}3)$$

18.
$$\frac{7}{2 - x} = \frac{10 - 4x}{x^2 + 3x - 10}$$

$$\frac{7}{2 - x} = \frac{10 - 4x}{(x - 2)(x + 5)} \qquad \text{Excluded values: } x \neq 2, \ -5$$

$$(\overset{-1}{\cancel{x - 2}})(x + 5) \ \frac{7}{\underset{1}{\cancel{2 - x}}} = (x - 2)(x + 5) \ \frac{10 - 4x}{(x - 2)(x + 5)}$$

$$-7(x + 5) = 10 - 4x$$
$$-7x - 35 = 10 - 4x$$
$$-3x = 45$$
$$x = -15 \qquad (2\text{-}1)$$

19.
$$\frac{u - 3}{2u - 2} = \frac{1}{6} - \frac{1 - u}{3u - 3}$$

$$\frac{u - 3}{2(u - 1)} = \frac{1}{6} - \frac{1 - u}{3(u - 1)} \qquad \text{Excluded value: } u \neq 1$$

$$6(u - 1) \ \frac{(u - 3)}{2(u - 1)} = 6(u - 1) \ \frac{1}{6} - 6(u - 1) \ \frac{(1 - u)}{3(u - 1)}$$

$$3(u - 3) = u - 1 - 2(1 - u)$$
$$3u - 9 = u - 1 - 2 + 2u$$
$$3u - 9 = 3u - 3$$
$$-9 = -3$$

No solution $\qquad (2\text{-}1)$

20. $5m + 6n = 2$
$4m - 9n = 20$
Solve the first equation for m in terms of n and substitute into the second equation.

$$5m = 2 - 6n$$
$$m = \frac{2 - 6n}{5}$$

$$4\left(\frac{2 - 6n}{5}\right) - 9n = 20$$

$$\frac{8 - 24n}{5} - 9n = 20$$

$$\frac{5}{1}\left(\frac{8 - 24n}{5}\right) - 45n = 100$$

$$8 - 24n - 45n = 100$$
$$-69n = 92$$
$$n = -\frac{4}{3}$$

$$m = \frac{2 - 6\left(-\frac{4}{3}\right)}{5} = \frac{2 + 8}{5} = 2$$

$$m = 2, \ n = -\frac{4}{3} \qquad (2\text{-}2)$$

21.
$$\frac{x + 3}{8} \leq 5 - \frac{2 - x}{3}$$

$$24\frac{(x + 3)}{8} \leq 120 - 24\frac{(2 - x)}{3}$$

$$3(x + 3) \leq 120 - 8(2 - x)$$
$$3x + 9 \leq 120 - 16 + 8x$$
$$3x + 9 \leq 8x + 104$$
$$-5x \leq 95$$
$$x \geq -19$$

$[-19, \infty)$

$-19 \qquad (2\text{-}3)$

22. $|3x - 8| > 2$

$$3x - 8 < -2 \quad \text{or} \quad 3x - 8 > 2$$
$$3x < 6 \qquad\qquad\qquad 3x > 10$$
$$x < 2 \quad \text{or} \qquad\quad x > \frac{10}{3}$$

$(-\infty, \ 2) \cup \left(\frac{10}{3}, \ \infty\right)$

$2 \qquad \frac{10}{3}$

$\qquad\qquad\qquad\qquad (2\text{-}4)$

23.
$$\frac{1}{x} < 2$$

$$\frac{1}{x} - 2 < 0$$

$$\frac{1 - 2x}{x} < 0$$

Zeros of P, Q: 0, $\frac{1}{2}$

	$\frac{P}{Q} = \frac{1 - 2x}{x}$		
Test Number Value of $\frac{P}{Q}$	-1 -3	0.1 8	1 -1
Sign of $\frac{P}{Q}$	$-$	$+$	$-$
Interval	$(-\infty, 0)$	$(0, \frac{1}{2})$	$(\frac{1}{2}, \infty)$

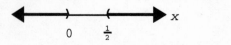

$\frac{1 - 2x}{x} < 0$ and $\frac{1}{x} < 2$ within the intervals $(-\infty, 0)$ and $\left(\frac{1}{2}, \infty\right)$. $x < 0$ or $x > \frac{1}{2}$

(2-8)

24.
$$\frac{3}{x - 4} \leq \frac{2}{x - 3}$$

$$\frac{3}{x - 4} - \frac{2}{x - 3} \leq 0$$

$$\frac{3(x - 3) - 2(x - 4)}{(x - 4)(x - 3)} \leq 0$$

$$\frac{3x - 9 - 2x + 8}{(x - 4)(x - 3)} \leq 0$$

$$\frac{x - 1}{(x - 4)(x - 3)} \leq 0$$

Zeros of P, Q: 1, 3, 4
1 is part of the solution set so we use a solid dot there. 3 and 4 are not part of the solution set ($\frac{P}{Q}$ is not defined there) so we use open dots there.

	$\frac{P}{Q} = \frac{x - 1}{(x - 4)(x - 3)}$			
Test Number Value of $\frac{P}{Q}$	0 $-\frac{1}{12}$	2 $\frac{1}{2}$	3.5 -10	5 2
Sign of $\frac{P}{Q}$	$-$	$+$	$-$	$+$
Interval	$(-\infty, 1)$	$(1, 3)$	$(3, 4)$	$(4, \infty)$

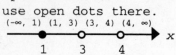

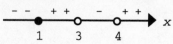

$\frac{x - 1}{(x - 4)(x - 3)} \leq 0$ and $\frac{3}{x - 4} \leq \frac{2}{x - 3}$ within the intervals $(-\infty, 1]$ and $(3, 4)$.
$x \leq 1$ or $3 < x < 4$

(2-8)

25.
$$\sqrt{(1 - 2m)^2} \leq 3$$
$$|1 - 2m| \leq 3$$
$$-3 \leq 1 - 2m \leq 3$$
$$-4 \leq -2m \leq 2$$

$$2 \geq m \geq -1$$
$$-1 \leq m \leq 2$$

$[-1, 2]$

(2-4)

26. $\sqrt{\dfrac{x + 4}{2 - x}}$ will represent a real number when $\dfrac{x + 4}{2 - x} \geq 0$.

$\dfrac{x + 4}{2 - x} \geq 0$

Zeros of P, Q: -4, 2

-4 is part of the solution set so we use a solid dot there. 2 is not part of the solution set ($\frac{P}{Q}$ is not defined there) so we use an open dot there.

	\multicolumn{3}{c}{$\dfrac{P}{Q} = \dfrac{x + 4}{2 - x}$}
Test Number	-5	0	3
Value of $\frac{P}{Q}$	$-\frac{1}{7}$	2	-7
Sign of $\frac{P}{Q}$	$-$	$+$	$-$
Interval	$(-\infty, -4)$	$(-4, 2)$	$(2, \infty)$

$\dfrac{x + 4}{2 - x} \geq 0$ and $\sqrt{\dfrac{x + 4}{2 - x}}$ will represent a real number within the interval $[-4, 2)$.

$-4 \leq x < 2$

(2-8)

27. (A) $d(A,B) = |-2 - (-8)| = |6| = 6$ (B) $d(B,A) = |-8 - (-2)| = |-6| = 6$ *(2-4)*

28. (A) $(3 + i)^2 - 2(3 + i) + 3 = 9 + 6i + i^2 - 6 - 2i + 3$
$$= 9 + 6i - 1 - 6 - 2i + 3$$
$$= 5 + 4i$$

(B) $i^{27} = i^{26}i = (i^2)^{13}i = (-1)^{13}i = (-1)i = -i$ *(2-5)*

29. (A) $(2 - \sqrt{-4}) - (3 - \sqrt{-9}) = (2 - i\sqrt{4}) - (3 - i\sqrt{9}) = (2 - 2i) - (3 - 3i)$
$$= 2 - 2i - 3 + 3i = -1 + i$$

(B) $\dfrac{2 - \sqrt{-1}}{3 + \sqrt{-4}} = \dfrac{2 - i\sqrt{1}}{3 + i\sqrt{4}} = \dfrac{2 - i}{3 + 2i} = \dfrac{(2 - i)}{(3 + 2i)}\dfrac{(3 - 2i)}{(3 - 2i)} = \dfrac{6 - 7i + 2i^2}{9 - 4i^2} = \dfrac{6 - 7i - 2}{9 + 4}$

$\qquad = \dfrac{4 - 7i}{13} = \dfrac{4}{13} - \dfrac{7}{13}i$

(C) $\dfrac{4 + \sqrt{-25}}{\sqrt{-4}} = \dfrac{4 + i\sqrt{25}}{i\sqrt{4}} = \dfrac{4 + 5i}{2i} = \dfrac{4 + 5i}{2i}\dfrac{i}{i} = \dfrac{4i + 5i^2}{2i^2} = \dfrac{4i - 5}{-2} = \dfrac{5}{2} - 2i$ *(2-5)*

30. $\left(u + \dfrac{5}{2}\right)^2 = \dfrac{5}{4}$

$\qquad u + \dfrac{5}{2} = \pm\sqrt{\dfrac{5}{4}}$

$\qquad u + \dfrac{5}{2} = \pm\dfrac{\sqrt{5}}{2}$

$\qquad u = -\dfrac{5}{2} \pm \dfrac{\sqrt{5}}{2}$

$\qquad u = \dfrac{-5 \pm \sqrt{5}}{2}$ *(2-6)*

31. $1 + \dfrac{3}{u^2} = \dfrac{2}{u}$ Excluded value: $u \neq 0$

$\qquad u^2 + 3 = 2u$

$\qquad u^2 - 2u = -3$

$\qquad u^2 - 2u + 1 = -2$

$\qquad (u - 1)^2 = -2$

$\qquad u - 1 = \pm\sqrt{-2}$

$\qquad u = 1 \pm \sqrt{-2}$

$\qquad u = 1 \pm i\sqrt{2}$ *(2-6)*

32.

$$\frac{x}{x^2 - x - 6} - \frac{2}{x - 3} = 3$$

$$\frac{x}{(x - 3)(x + 2)} - \frac{2}{x - 3} = 3 \quad \text{Excluded values: } x \neq 3, -2$$

$$(x - 3)(x + 2)\frac{x}{(x - 3)(x + 2)} - (x - 3)(x + 2)\frac{2}{x - 3} = 3(x - 3)(x + 2)$$

$$x - 2(x + 2) = 3(x - 3)(x + 2)$$

$$x - 2x - 4 = 3(x^2 - x - 6)$$

$$-x - 4 = 3x^2 - 3x - 18$$

$$0 = 3x^2 - 2x - 14$$

$$3x^2 - 2x - 14 = 0$$

$$x = \frac{-b \pm \sqrt{b^2 - 4ac}}{2a} \qquad a = 3$$

$$b = -2$$
$$c = -14$$

$$x = \frac{-(-2) \pm \sqrt{(-2)^2 - 4(3)(-14)}}{2(3)}$$

$$x = \frac{2 \pm \sqrt{172}}{6}$$

$$x = \frac{2 \pm 2\sqrt{43}}{6}$$

$$x = \frac{1 \pm \sqrt{43}}{3} \qquad (2\text{-}6)$$

33. $2x^{2/3} - 5x^{1/3} - 12 = 0$

Let $u = x^{1/3}$, then

$$2u^2 - 5u - 12 = 0$$

$$(2u + 3)(u - 4) = 0$$

$$u = -\frac{3}{2},\ 4$$

$$x^{1/3} = -\frac{3}{2} \qquad x^{1/3} = 4$$

$$x = -\frac{27}{8} \qquad x = 64 \qquad (2\text{-}7)$$

34. $m^4 + 5m^2 - 36 = 0$

Let $u = m^2$, then

$$u^2 + 5u - 36 = 0$$

$$(u + 9)(u - 4) = 0$$

$$u = -9,\ 4$$

$$m^2 = -9 \qquad m^2 = 4$$

$$m = \pm 3i \qquad m = \pm 2 \qquad (2\text{-}7)$$

35. $\sqrt{y - 2} - \sqrt{5y + 1} = -3$

$$-\sqrt{5y + 1} = -3 - \sqrt{y - 2}$$

$$5y + 1 = 9 + 6\sqrt{y - 2} + y - 2$$

$$5y + 1 = y + 7 + 6\sqrt{y - 2}$$

$$4y - 6 = 6\sqrt{y - 2}$$

$$2y - 3 = 3\sqrt{y - 2}$$

$$4y^2 - 12y + 9 = 9(y - 2)$$

$$4y^2 - 12y + 9 = 9y - 18$$

$$4y^2 - 21y + 27 = 0$$

$$(4y - 9)(y - 3) = 0$$

$$y = \frac{9}{4},\ 3$$

Common Error:
$y - 2 - 5y + 1 = 9$
is not equivalent to the
equation formed by
squaring both members
of the given equation.

Check: $\sqrt{\frac{9}{4} - 2} - \sqrt{5\left(\frac{9}{4}\right) + 1} \overset{?}{=} -3$

$$\sqrt{\frac{1}{4}} - \sqrt{\frac{49}{4}} \overset{?}{=} -3$$

$$-3 \overset{\checkmark}{=} -3$$

$$\sqrt{3 - 2} - \sqrt{5(3) + 1} \overset{?}{=} -3$$

$$-3 \overset{\checkmark}{=} -3$$

Solution: $\frac{9}{4},\ 3$ $\qquad (2\text{-}7)$

36. $2.15x - 3.73(x - 0.93) = 6.11x$

$$2.15x - 3.73x + 3.4689 = 6.11x$$

$$-1.58x + 3.4689 = 6.11x$$

$$3.4689 = 7.69x$$

$$x = 0.45 \quad (2\text{-}1)$$

37. $-1.52 \leq 0.77 - 2.04x \leq 5.33$

$$-2.29 \leq -2.04x \leq 4.56$$

$$1.12 \geq x \geq -2.24$$

$$-2.24 \leq x \leq 1.12 \text{ or } [-2.24, 1.12]$$

$$(2\text{-}3)$$

38. $\dfrac{3.77 - 8.47i}{6.82 - 7.06i} = \dfrac{(3.77 - 8.47i)}{(6.82 - 7.06i)}\dfrac{(6.82 + 7.06i)}{(6.82 + 7.06i)}$

$$= \dfrac{(3.77)(6.82) + (3.77)(7.06i) - (6.82)(8.47i) - (8.47)(7.06)i^2}{(6.82)^2 + (7.06)^2}$$

$$= \dfrac{25.7114 + 26.6162i - 57.7654i + 59.7982}{46.5124 + 49.8436}$$

$$= \dfrac{85.5096 - 31.1492i}{96.356}$$

$$= 0.89 - 0.32i \qquad\qquad (2\text{-}5)$$

39. $6.09x^2 + 4.57x - 8.86 = 0$

$$x = \dfrac{-b \pm \sqrt{b^2 - 4ac}}{2a} \quad a = 6.09,\ b = 4.57,\ c = -8.86$$

$$x = \dfrac{-4.57 \pm \sqrt{(4.57)^2 - 4(6.09)(-8.86)}}{2(6.09)}$$

$$x = \dfrac{-4.57 \pm \sqrt{236.7145}}{12.18}$$

$$x = \dfrac{-4.57 \pm 15.3855}{12.18}$$

$$x = -1.64,\ 0.89 \qquad\qquad (2\text{-}6)$$

40. $15.2x + 5.6y = 20$
$2.5x + 7.5y = 10$
Solve the first equation for y in terms of x and substitute into the second equation.

$$5.6y = 20 - 15.2x$$

$$y = \dfrac{20 - 15.2x}{5.6}$$

$$2.5x + 7.5\left(\dfrac{20 - 15.2x}{5.6}\right) = 10$$

$$2.5x + \dfrac{150 - 114x}{5.6} = 10$$

$$5.6(2.5x) + \dfrac{5.6}{1}\left(\dfrac{150 - 114x}{5.6}\right) = 56$$

$$14x + 150 - 114x = 56$$

$$-100x = -94$$

$$x = 0.94$$

$$y = \dfrac{20 - 15.2(0.94)}{5.6}$$

$$y = 1.02$$

$$x = 0.94,\ y = 1.02 \qquad\qquad (2\text{-}2)$$

41.
$$P = M - Mdt$$
$$M - Mdt = P$$
$$M(1 - dt) = P$$
$$M = \dfrac{P}{1 - dt} \qquad (2\text{-}1)$$

42.
$$P = EI - RI^2$$
$$RI^2 - EI + P = 0$$
$$I = \dfrac{-b \pm \sqrt{b^2 - 4ac}}{2a} \quad \begin{array}{l} a = R \\ b = -E \\ c = P \end{array}$$

$$I = \dfrac{-(-E) \pm \sqrt{(-E)^2 - 4(R)(P)}}{2(R)}$$

$$I = \dfrac{E \pm \sqrt{E^2 - 4PR}}{2R} \qquad\qquad (2\text{-}6)$$

43.

$$x = \frac{4y + 5}{2y + 1}$$

$$x(2y + 1) = (2y + 1)\frac{4y + 5}{2y + 1}$$

$$2xy + x = 4y + 5$$

$$2xy + x - 4y = 5$$

$$2xy - 4y = 5 - x$$

$$y(2x - 4) = 5 - x$$

$$y = \frac{5 - x}{2x - 4} \qquad (2\text{-}1)$$

44. The original equation can be rewritten as

$$\frac{4}{(x - 1)(x - 3)} = \frac{3}{(x - 1)(x - 2)}$$

Thus, $x = 1$ cannot be a solution of this equation. This extraneous solution was introduced when both sides were multiplied by $x - 1$ in the second line. $x = 1$ must be discarded and the only correct solution is $x = -1$. $\qquad (2\text{-}1)$

45. In this problem, $a = 1$, $b = -6$, $c = c$. Thus, the discriminant $b^2 - 4ac = (-6)^2 - 4(1)(c) = 36 - 4c$. Hence,

if $36 - 4c > 0$, thus $36 > 4c$ or $c < 9$, there are two distinct real roots.

if $36 - 4c = 0$, thus $c = 9$, there is one real double root.

if $36 - 4c < 0$, thus $36 < 4c$ or $c > 9$, there are two distinct imaginary roots. $(2\text{-}6)$

46. The given inequality $a + b < b - a$ is equivalent to, successively,

$a < -a$

$2a < 0$

$a < 0$

Thus its truth is independent of the value of b, and dependent on a being negative. True for all real b and all negative a. $\qquad (2\text{-}3)$

47. If $a > b$ and b is negative, then $\frac{a}{b} < \frac{b}{b}$, that is, $\frac{a}{b} < 1$, since dividing both

sides by b reverses the order of the inequality. $\frac{a}{b}$ is less than 1. $\qquad (2\text{-}3)$

48.

$$y = \frac{1}{1 - \frac{1}{1 - x}}$$

$$y = \frac{1(1 - x)}{(1 - x)1 - (1 - x)\frac{1}{1 - x}}$$

$$y = \frac{1 - x}{1 - x - 1}$$

$$y = \frac{1 - x}{-x}$$

$$-xy = 1 - x$$

$$x - xy = 1$$

$$x(1 - y) = 1$$

$$x = \frac{1}{1 - y} \qquad (2\text{-}1)$$

49. $0 < |x - 6| < d$ means: the distance between x and 6 is less than d but $x \neq 6$.

$-d < x - 6 < d$ except $x \neq 6$

$6 - d < x < 6 + d$ but $x \neq 6$

$6 - d < x < 6$ or $6 < x < 6 + d$

$(6 - d, 6) \cup (6, 6 + d)$

$(2\text{-}4)$

50.

$$2x^2 = \sqrt{3}x - \frac{1}{2}$$

$$4x^2 = 2\sqrt{3}x - 1$$

$$4x^2 - 2\sqrt{3}x + 1 = 0$$

$$x = \frac{-b \pm \sqrt{b^2 - 4ac}}{2a} \quad a = 4, \ b = -2\sqrt{3}, \ c = 1$$

$$x = \frac{-(-2\sqrt{3}) \pm \sqrt{(-2\sqrt{3})^2 - 4(4)(1)}}{2(4)}$$

$$x = \frac{2\sqrt{3} \pm \sqrt{-4}}{8}$$

$$x = \frac{2\sqrt{3} \pm 2i}{8}$$

$$x = \frac{\sqrt{3} \pm i}{4} \quad \text{or} \quad \frac{\sqrt{3}}{4} \pm \frac{1}{4}i \qquad (2\text{-}6)$$

51.

$$4 = 8x^{-2} + x^{-4}$$

$$4x^4 = 8x^2 + 1 \quad \text{Multiply both members by } x^4 \quad x \neq 0$$

$$4x^4 - 8x^2 - 1 = 0$$

Let $u = x^2$, then

$$4u^2 - 8u - 1 = 0$$

$$u = \frac{-b \pm \sqrt{b^2 - 4ac}}{2a} \quad a = 4, \ b = -8, \ c = -1$$

$$u = \frac{-(-8) \pm \sqrt{(-8)^2 - 4(4)(-1)}}{2(4)}$$

$$u = \frac{8 \pm \sqrt{64 + 16}}{8}$$

$$u = \frac{8 \pm \sqrt{80}}{8}$$

$$u = \frac{8 \pm 4\sqrt{5}}{8}$$

> **Common Error:**
> It is incorrect to "cancel" the 8's at this point.

$$u = \frac{4(2 \pm \sqrt{5})}{8}$$

$$u = \frac{2 \pm \sqrt{5}}{2}$$

$$x^2 = \frac{2 \pm \sqrt{5}}{2}$$

$$x = \pm\sqrt{\frac{2 + \sqrt{5}}{2}} \quad \text{(two real roots)} \qquad (2\text{-}7)$$

52.

$$(a + bi)\left(\frac{a}{a^2 + b^2} - \frac{b}{a^2 + b^2}i\right)$$

$$= \frac{(a + bi)}{1}\left(\frac{a}{a^2 + b^2} - \frac{bi}{a^2 + b^2}\right)$$

$$= \frac{a(a + bi)}{a^2 + b^2} - \frac{bi(a + bi)}{a^2 + b^2}$$

$$= \frac{a^2 + abi - abi - b^2 i^2}{a^2 + b^2}$$

$$= \frac{a^2 + b^2}{a^2 + b^2} = 1 \qquad (2\text{-}5)$$

53.

$$2x > \frac{x^2}{5} + 5$$

$$10x > x^2 + 25$$

$$0 > x^2 - 10x + 25$$

$$0 > (x - 5)^2$$

Since the right side is never negative for x a real number, this statement is satisfied by no real number. $\qquad (2\text{-}8)$

54.
$$\frac{x^2}{4} + 4 \geq 2x$$
$$x^2 + 16 \geq 8x$$
$$x^2 - 8x + 16 \geq 0$$
$$(x - 4)^2 \geq 0$$

Since the left side is always positive or zero for x a real number, this statement is satisfied by all real numbers x. *(2-8)*

55. $\left| x - \dfrac{8}{x} \right| \geq 2$

There are two ways for this statement to hold. The solution set is the union of the solution sets for:

$x - \dfrac{8}{x} \leq -2$ and $x - \dfrac{8}{x} \geq 2$

Case 1: $x - \dfrac{8}{x} \leq -2$

$$x - \frac{8}{x} + 2 \leq 0$$
$$\frac{x^2 - 8 + 2x}{x} \leq 0$$
$$\frac{x^2 + 2x - 8}{x} \leq 0$$
$$\frac{(x - 2)(x + 4)}{x} \leq 0$$

Zeros of P, Q: 2, -4, 0

2 and -4 are part of the solution set, so we use solid dots there. 0 is not part of the solution set, so we use an open dot there.

$\dfrac{P}{Q} = \dfrac{(x - 2)(x + 4)}{x}$				
Test Number	-5	-1	1	3
Value of $\dfrac{P}{Q}$	$-\dfrac{7}{5}$	9	-5	$\dfrac{7}{3}$
Sign of $\dfrac{P}{Q}$	–	+	–	+
Interval	$(-\infty, -4)$	$(-4, 0)$	$(0, 2)$	$(2, \infty)$

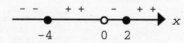

$\dfrac{(x - 2)(x + 4)}{x} \leq 0$ and $x - \dfrac{8}{x} \leq -2$ within the intervals $(-\infty, -4]$ and $(0, 2]$.

Case 2: $x - \dfrac{8}{x} \geq 2$

$$x - \frac{8}{x} - 2 \geq 0$$
$$\frac{x^2 - 8 - 2x}{x} \geq 0$$
$$\frac{x^2 - 2x - 8}{x} \geq 0$$
$$\frac{(x - 4)(x + 2)}{x} \geq 0$$

Zeros of P, Q: -2, 4, 0. -2 and 4 are part of the solution set, so we use solid dots there. 0 is not part of the solution set, so we use an open dot there.

$\dfrac{P}{Q} = \dfrac{(x - 4)(x + 2)}{x}$				
Test Number	-3	-1	1	5
Value of $\dfrac{P}{Q}$	$-\dfrac{7}{3}$	5	-9	$\dfrac{7}{5}$
Sign of $\dfrac{P}{Q}$	–	+	–	+
Interval	$(-\infty, -2)$	$(-2, 0)$	$(0, 4)$	$(4, \infty)$

$\dfrac{(x - 4)(x + 2)}{x} \geq 0$ and $x - \dfrac{8}{x} \geq 2$ within the intervals $[-2, 0)$ and $[4, \infty)$.

Combining Case 1 and Case 2, we have: $x \leq -4$ or $-2 \leq x < 0$ or $0 < x \leq 2$ or $x \geq 4$
$(-\infty, -4] \cup [-2, 0) \cup (0, 2] \cup [4, \infty)$ *(2-8)*

56. $x = 2 + 3u + 7v$
$y = -3 + 2u + 5v$

Solve the first equation for u in terms of the other variables and substitute into the second equation.

$x - 2 - 7v = 3u$

$$u = \frac{x - 2 - 7v}{3}$$

$$y = -3 + 2\left(\frac{x - 2 - 7v}{3}\right) + 5v$$

$$y = -3 + \frac{2x - 4 - 14v}{3} + 5v$$

$$3y = -9 + \frac{3}{1}\left(\frac{2x - 4 - 14v}{3}\right) + 15v$$

$$3y = -9 + 2x - 4 - 14v + 15v$$

$13 - 2x + 3y = v$

$$u = \frac{x - 2 - 7v}{3} = \frac{x - 2 - 7(13 - 2x + 3y)}{3} = \frac{x - 2 - 91 + 14x - 21y}{3}$$

$$= \frac{15x - 21y - 93}{3} = 5x - 7y - 31$$

The checking steps are omitted for lack of space.　　　　　　　　　　　　　　　　*(2-2)*

57. (A) $2x - y = -5$
$-6x + 3y = 15$

Solve the first equation for y in terms of x and substitute into the second equation.

$-y = -2x - 5$
$y = 2x + 5$
$-6x + 3(2x + 5) = 15$
　　　　　　$15 = 15$

Thus, the system has an infinite number of solutions.

(B) $2x - y = -5$
$-6x + 3y = 10$

Solve the first equation for y in terms of x and substitute into the second equation.

$-y = -2x - 5$
$y = 2x + 5$
$-6x + 3(2x + 5) = 10$
　　　　　　$15 = 10$

Thus, the system has no solution.　　　　*(2-2)*

58. Let x = the number
$\frac{1}{x}$ = its reciprocal

Then　　　$x - \frac{1}{x} = \frac{16}{15}$　Excluded value: $x \neq 0$

$15x^2 - 15 = 16x$
$15x^2 - 16x - 15 = 0$
$(5x + 3)(3x - 5) = 0$
$5x + 3 = 0$ or $3x - 5 = 0$
　　　$x = -\frac{3}{5}$　　　　　$x = \frac{5}{3}$　　*(2-6)*

59. (A) $H = 0.7(220 - A)$

(B) We are to find H when $A = 20$.
$H = 0.7(220 - 20)$
$H = 140$ beats per minute.
(C) We are to find A when $H = 126$.
$126 = 0.7(220 - A)$
$126 = 154 - 0.7A$
$-28 = -0.7A$
$A = 40$ years old.　　　　　　*(2-2)*

60. Let　　　x = amount of 80% solution
Then $50 - x$ = amount of 30% solution
since　　50 = amount of 60% solution

acid in		acid in		acid in
80% solution	+	30% solution	=	60% solution

$0.8(x) + 0.3(50 - x) = 0.6(50)$
$0.8x + 15 - 0.3x = 30$
$15 + 0.5x = 30$
$0.5x = 15$
$x = \dfrac{15}{0.5}$
$x = 30$ milliliters of 80% solution
$50 - x = 20$ milliliters of 30% solution　　*(2-2)*

61. Let x = the rate of the current
Then $12 - x$ = the rate of the boat upstream
$12 + x$ = the rate of the boat downstream

Solving $d = rt$ for t, we have $t = \dfrac{d}{r}$. We use this formula, together with

time upstream = 2 + time downstream

$$\text{time upstream} = \frac{\text{distance upstream}}{\text{rate upstream}} = \frac{45}{12 - x}$$

$$\text{time downstream} = \frac{\text{distance downstream}}{\text{rate downstream}} = \frac{45}{12 + x}$$

So, $\quad \dfrac{45}{12 - x} = 2 + \dfrac{45}{12 + x}$ Excluded values: $x \neq -12,\ 12$

$$45(12 + x) = (12 + x)(12 - x)2 + 45(12 - x)$$
$$540 + 45x = 288 - 2x^2 + 540 - 45x$$
$$2x^2 + 90x - 288 = 0$$
$$2(x + 48)(x - 3) = 0$$
$$x = -48,\ 3$$

Discarding the negative answer, we have rate = 3 miles per hour. $\qquad$ $(2\text{-}6)$

62. (A) Let x = distance rowed
then $15 - 3 = 12$ km/hr = the rate rowed upstream
$15 + 3 = 18$ km/hr = the rate rowed downstream

Using $t = \dfrac{d}{r}$ as in the previous problem yields

$$\text{time upstream} = \frac{x}{12}$$

$$\text{time downstream} = \frac{x}{18}$$

So $\quad \dfrac{x}{12} + \dfrac{x}{18} = \dfrac{25}{60}$ $\quad$ LCD = 180

$$\frac{180}{1} \cdot \frac{x}{12} + \frac{180}{1} \cdot \frac{x}{18} = \frac{180}{1} \cdot \frac{25}{60}$$
$$15x + 10x = 75$$
$$25x = 75$$
$$x = 3 \text{ km}$$

(B) Now let x = still-water speed
$x - 3$ = the rate rowed upstream
$x + 3$ = the rate rowed downstream

$$\text{time upstream} = \frac{3}{x - 3}$$

$$\text{time downstream} = \frac{3}{x + 3}$$

So $\quad \dfrac{3}{x - 3} + \dfrac{3}{x + 3} = \dfrac{23}{60}$ $\quad$ Excluded values: $x = 3,\ -3$

$$60(x + 3)3 + 60(x - 3)3 = 23(x + 3)(x - 3)$$
$$180x + 540 + 180x - 540 = 23x^2 - 207$$
$$0 = 23x^2 - 360x - 207$$
$$x = \frac{-b \pm \sqrt{b^2 - 4ac}}{2a} \quad a = 23,\ b = -360,\ c = -207$$
$$x = \frac{-(-360) \pm \sqrt{(-360)^2 - 4(23)(-207)}}{2(23)}$$
$$x = 16.2 \text{ or } -0.6$$

Discarding the negative answer, we have $x = 16.2$ km/hr.

(C) Now 18 - 3 = 15 km/hr = the rate rowed upstream
           18 + 3 = 21 km/hr = the rate rowed downstream

So $\dfrac{3}{15} + \dfrac{3}{21}$ = round trip time

               = 0.343 hr
               = 0.343 × 60 min
               = 20.6 min *(2-1, 2-6)*

63. Let x = number of bags of brand A
     y = number of bags of brand B
Then (Nutrition from brand A) + (Nutrition from brand B) = (Total nutrition)
               $8x$ + $4y$ =   860 (Total nitrogen)
               $9x$ + $7y$ = 1,080 (Total phosphoric
                                                                 acid)

Solve the first equation for y in terms of x and substitute into the second equation.
           $4y = 860 - 8x$
             $y = 215 - 2x$
     $9x + 7(215 - 2x) = 1,080$
     $9x + 1,505 - 14x = 1,080$
               $-5x = -425$
                 $x = 85$ bags of brand A
                 $y = 215 - 2(85)$
                 $y = 45$ bags of brand B *(2-2)*

64. (A) Apply the given formula with $C = 15$.
           $15 = x^2 - 10x + 31$
            $0 = x^2 - 10x + 16$
    $x^2 - 10x + 16 = 0$
    $(x - 8)(x - 2) = 0$
           $x = 2$ or 8
Thus the output could be either 2000 or 8000 units.

(B) Apply the given formula with $C = 6$.
          $6 = x^2 - 10x + 31$
          $0 = x^2 - 10x + 25$
    $x^2 - 10x + 25 = 0$
       $(x - 5)^2 = 0$
           $x = 5$
Thus the output must be 5000 units. *(2-6)*

65. The break-even points are defined by $C = R$ (cost = revenue). Applying the formulas in this problem and the previous one, we have
$x^2 - 10x + 31 = 3x$
$x^2 - 13x + 31 = 0$

$$x = \frac{-b \pm \sqrt{b^2 - 4ac}}{2a} \quad a = 1,\ b = -13,\ c = 31$$

$$x = \frac{-(-13) \pm \sqrt{(-13)^2 - 4(1)(31)}}{2(1)}$$

$$x = \frac{13 \pm \sqrt{45}}{2} \text{ thousand or approximately 3,146 and 9,854 units}\quad (2\text{-}6)$$

66. A profit will result if $R > C$, that is, if
        $3x > x^2 - 10x + 31$
$x^2 - 10x + 31 < 3x$
$x^2 - 13x + 31 < 0$
The zeros of this polynomial were determined in problem 65 to be $\dfrac{13 \pm \sqrt{45}}{2}$ (in thousands)

$x^2 - 13x + 31$			
Test Number	0	5	10
Value of Polynomial for Test Number	31	-9	1
Sign of Polynomial in Interval	+	−	+
Interval	$(-\infty, \frac{13 - \sqrt{45}}{2})$	$(\frac{13 - \sqrt{45}}{2}, \frac{13 + \sqrt{45}}{2})$	$(\frac{13 + \sqrt{45}}{2}, \infty)$
approximately	$(-\infty, 3.146)$	$(3.146, 9.854)$	$(9.854, \infty)$

$x^2 - 13x + 31 < 0$, and a profit will result if $\dfrac{13 - \sqrt{45}}{2} < x < \dfrac{13 + \sqrt{45}}{2}$, or, approximately, $3.146 < x < 9.854$, in thousands. (2-8)

67. The distance of T from 110 must be no greater than 5. $|T - 110| \leq 5$ (2-4)

68. Let x = width of page
y = height of page

Then $xy = 480$, thus $y = \dfrac{480}{x}$.

Since the printed portion is surrounded by margins of 2 cm on each side, we have
$x - 4$ = width of printed portion
$y - 4$ = height of printed portion
Hence
$(x - 4)(y - 4) = 320$, that is

$$(x - 4)\left(\dfrac{480}{x} - 4\right) = 320$$

Solving this, we obtain:

$$x\left(\dfrac{480}{x}\right) - 4x - 4\left(\dfrac{480}{x}\right) + 16 = 320$$

$$480 - 4x - \dfrac{1{,}920}{x} + 16 = 320$$

$$-4x - \dfrac{1{,}920}{x} = -176 \quad \text{LCD: } x \quad x \neq 0$$

$$-4x^2 - 1{,}920 = -176x$$

$$0 = 4x^2 - 176x + 1{,}920$$

$$0 = x^2 - 44x + 480$$

$$0 = (x - 20)(x - 24)$$

$$x - 20 = 0 \quad \text{or} \quad x - 24 = 0$$

$$x = 20 \qquad\qquad x = 24$$

$$\dfrac{480}{x} = 24 \qquad\qquad \dfrac{480}{x} = 20$$

Thus, the dimensions of the page are 20 cm by 24 cm. (2-6)

69.

In the isosceles triangle we note:

$$\frac{1}{2}Bh = A = 24$$

Hence

$$h = \frac{48}{B}$$

Applying the Pythagorean theorem, we have

$$h^2 + \left(\frac{B}{2}\right)^2 = 8^2$$

$$\left(\frac{48}{B}\right)^2 + \left(\frac{B}{2}\right)^2 = 8^2$$

$$\frac{2,304}{B^2} + \frac{B^2}{4} = 64$$

$$4B^2\left(\frac{2,304}{B^2}\right) + 4B^2\left(\frac{B^2}{4}\right) = 4B^2(64)$$

$$9,216 + B^4 = 256B^2$$

$$(B^2)^2 - 256B^2 + 9,216 = 0$$

$$B^2 = \frac{-b \pm \sqrt{b^2 - 4ac}}{2a} \quad a = 1, \ b = -256, \ c = 9,216$$

$$B^2 = \frac{-(-256) \pm \sqrt{(-256)^2 - 4(1)(9,216)}}{2(1)}$$

$$B^2 = \frac{256 \pm \sqrt{65,536 - 36,864}}{2}$$

$$B^2 = \frac{256 \pm \sqrt{28,672}}{2}$$

$$B^2 = 128 \pm 32\sqrt{7}$$

$$B = \sqrt{128 \pm 32\sqrt{7}}$$

$$B = 14.58 \text{ ft or } 6.58 \text{ ft}$$

(2-7)

CUMULATIVE REVIEW EXERCISE (Chapters 1 and 2)

1. $\dfrac{7x}{5} - \dfrac{3 + 2x}{2} = \dfrac{x - 10}{3} + 2$

$30\dfrac{7x}{5} - 30\dfrac{(3 + 2x)}{2} = 30\dfrac{(x - 10)}{3} + 2(30)$

$42x - 15(3 + 2x) = 10(x - 10) + 60$

$42x - 45 - 30x = 10x - 100 + 60$

$12x - 45 = 10x - 40$

$2x = 5$

$x = \dfrac{5}{2}$ $(2\text{-}1)$

2. (A) The distributive property states that, in general,

$x(y + z) = xy + xz$

Comparing this with

$c(a + b) = ?$

we see that

$c(a + b) = ca + cb$

$ca + cb$

(B) The associative property (+) states that, in general,

$(x + y) + z = x + (y + z)$

Comparing this with

$(a + b) + c = ?$

we see that

$(a + b) + c = a + (b + c)$

$a + (b + c)$

(C) The commutative property ($\cdot$) states that, in general,

$xy = yx$

Comparing this with

$(a + b)c = ?$

we see that

$(a + b)c = c(a + b)$

$c(a + b)$ $(1\text{-}1)$

3. $2x - 3y = 8$

$4x + y = 2$

Solve the second equation for y in terms of x and substitute into the first equation.

$y = 2 - 4x$

$2x - 3(2 - 4x) = 8$

$2x - 6 + 12x = 8$

$14x = 14$

$x = 1$

$y = 2 - 4x = 2 - 4(1) = -2$

$x = 1, \ y = -2$ $(2\text{-}2)$

4. $4x(x - 3) + 2(3x + 5)$

$= 4x^2 - 12x + 6x + 10$

$= 4x^2 - 6x + 10$ $(1\text{-}2)$

5. $3x(x - 7) - (5x - 9)$

$= 3x^2 - 21x - 5x + 9$

$= 3x^2 - 26x + 9$ $(1\text{-}2)$

6.

	First Product	Outer Product	Inner Product	Last Product
$(2x - 5y)(3x + 4y)$	$= 6x^2$	$+8xy$	$-15xy$	$-20y^2$
	$= 6x^2 - 7xy - 20y^2$			

$(1\text{-}2)$

7. $(2a - 3b)(2a + 3b) = (2a)^2 - (3b)^2 = 4a^2 - 9b^2$ $(1\text{-}2)$

8. $(5m + 2n)^2 = (5m)^2 + 2(5m)(2n) + (2n)^2$

$= 25m^2 + 20mn + 4n^2$ $(1\text{-}2)$

9. $2(3 - y) + 4 \leq 5 - y$

$6 - 2y + 4 \leq 5 - y$

$-2y + 10 \leq 5 - y$

$-y \leq -5$

$y \geq 5$

$[5, \infty)$

$(2\text{-}3)$

10. $|x - 2| < 7$

$-7 < x - 2 < 7$

$-5 < x < 9$

$(-5, 9)$

$(2\text{-}4)$

11.
$$x^2 + 3x \geq 10$$
$$x^2 + 3x - 10 \geq 0$$
$$(x + 5)(x - 2) \geq 0$$
Zeros: -5, 2
-5, 2 are part of the solution set, so we use solid dots.

$x^2 + 3x - 10 = (x + 5)(x - 2)$			
Test Number	-6	0	3
Value of Polynomial for Test Number	8	-10	8
Sign of Polynomial in Interval	+	-	+
Interval	$(-\infty, -5)$	$(-5, 2)$	$(2, \infty)$

$x^2 + 3x - 10$ is non-negative within the intervals $(-\infty, -5]$ and $[2, \infty)$.
$(-\infty, -5] \cup [2, \infty)$

$(2-8)$

12. Prime $(1-3)$ **13.** $(3t + 5)(2t - 1)$ $(1-3)$

14.
$$\frac{6}{x^2 - 3x} - \frac{4}{x^2 - 2x} = \frac{6}{x(x - 3)} - \frac{4}{x(x - 2)}$$
$$= \frac{6(x - 2)}{x(x - 2)(x - 3)} - \frac{4(x - 3)}{x(x - 2)(x - 3)}$$
$$= \frac{6x - 12 - 4x + 12}{x(x - 2)(x - 3)}$$
$$= \frac{2\overset{1}{\cancel{x}}}{\underset{1}{\cancel{x}}(x - 2)(x - 3)}$$
$$= \frac{2}{(x - 2)(x - 3)}$$

Common Error:
$6(x^2 - 2x) - 4(x^2 - 3x)$.
Multiplying the given expression
is incorrect. The LCD should be
found, then used to build each
fractional expression to higher terms.

$(1-4)$

15. $\dfrac{\frac{4}{y} - y}{\frac{2}{y^2} - \frac{1}{2}} = \dfrac{2y^2\left(\frac{4}{y} - y\right)}{2y^2\left(\frac{2}{y^2} - \frac{1}{2}\right)} = \dfrac{8y - 2y^3}{4 - y^2} = \dfrac{2y\overset{1}{(\cancel{4 - y^2})}}{\underset{1}{\cancel{4 - y^2}}} = 2y$ $(1-4)$

16. (A) $(2 - 3i) - (-5 + 7i) = 2 - 3i + 5 - 7i = 7 - 10i$

(B) $(1 + 4i)(3 - 5i) = 3 + 7i - 20i^2 = 3 + 7i + 20 = 23 + 7i$

(C) $\dfrac{5 + i}{2 + 3i} = \dfrac{(5 + i)}{(2 + 3i)}\dfrac{(2 - 3i)}{(2 - 3i)} = \dfrac{10 - 13i - 3i^2}{4 - 9i^2}$

$\qquad = \dfrac{10 - 13i + 3}{4 + 9} = \dfrac{13 - 13i}{13} = 1 - i$ $(2-5)$

17. $3x^2(xy^2)^5 = 3x^2x^5(y^2)^5 = 3x^7y^{10}$ $(1-5)$ **18.** $\dfrac{(2xy^2)^3}{4x^4y^4} = \dfrac{8x^3y^6}{4x^4y^4} = \dfrac{2y^2}{x}$ $(1-5)$

19. $(a^4b^{-6})^{3/2} = (a^4)^{3/2}(b^{-6})^{3/2}$
$\qquad = a^6b^{-9}$
$\qquad = \dfrac{a^6}{b^9}$ $(1-6)$

20.
$$3x^2 = -12x$$
$$3x^2 + 12x = 0$$
$$3x(x + 4) = 0$$
$$3x = 0 \quad \text{or} \quad x + 4 = 0$$
$$x = 0 \qquad\qquad x = -4$$
$(2-6)$

21. $4x^2 - 20 = 0$
$$4x^2 = 20$$
$$x^2 = 5$$
$$x = \pm\sqrt{5} \qquad (2\text{-}6)$$

22. $x^2 - 6x + 2 = 0$
$$x^2 - 6x = -2$$
$$x^2 - 6x + 9 = 7$$
$$(x - 3)^2 = 7$$
$$x - 3 = \pm\sqrt{7}$$
$$x = 3 \pm \sqrt{7} \qquad (2\text{-}6)$$

23. $x - \sqrt{12 - x} = 0$
$$x = \sqrt{12 - x}$$
$$x^2 = 12 - x$$
$$x^2 + x - 12 = 0$$
$$(x + 4)(x - 3) = 0$$
$$x = -4, \ 3$$

Check: $-4 - \sqrt{12 - (-4)} \overset{?}{=} 0$
$$-4 - 4 \overset{?}{=} 0$$
$$-8 \neq 0$$
$$3 - \sqrt{12 - 3} \overset{?}{=} 0$$
$$3 - 3 \overset{?}{=} 0$$
$$0 \overset{\sqrt{}}{=} 0$$

Solution: 3 $\qquad (2\text{-}7)$

24. $5\sqrt[4]{a^3}$ $\qquad (1\text{-}7)$

25. $2x^{2/5}y^{3/5}$ $\qquad (1\text{-}7)$

26. $xy^2\sqrt[3]{x^4y^8} = xy^2\sqrt[3]{x^3y^6 \cdot xy^2} = xy^2\sqrt[3]{x^3y^6}\sqrt[3]{xy^2}$
$$= xy^2xy^2\sqrt[3]{xy^2} = x^2y^4\sqrt[3]{xy^2} \qquad (1\text{-}7)$$

27. $\sqrt{2xy^3}\sqrt{6x^2y} = \sqrt{(2xy^3)(6x^2y)} = \sqrt{12x^3y^4} = \sqrt{4x^2y^4 \cdot 3x} = \sqrt{4x^2y^4}\sqrt{3x} = 2xy^2\sqrt{3x}$ $\quad (1\text{-}7)$

28. $\dfrac{3 + \sqrt{2}}{2 + 3\sqrt{2}} = \dfrac{(3 + \sqrt{2})}{(2 + 3\sqrt{2})}\dfrac{(2 - 3\sqrt{2})}{(2 - 3\sqrt{2})} = \dfrac{6 - 9\sqrt{2} + 2\sqrt{2} - 3\sqrt{2}\sqrt{2}}{2^2 - (3\sqrt{2})^2}$

$$= \dfrac{6 - 7\sqrt{2} - 6}{4 - 18}$$
$$= \dfrac{-7\sqrt{2}}{-14}$$
$$= \dfrac{\sqrt{2}}{2} \text{ or } \dfrac{1}{2}\sqrt{2} \qquad (1\text{-}7)$$

29. $\sqrt{2 + 3x}$ represents a real number exactly when $2 + 3x$ is positive or zero. We can write this as an inequality statement and solve for x.
$$2 + 3x \geq 0$$
$$3x \geq -2$$
$$x \geq -\dfrac{2}{3} \text{ or } \left[-\dfrac{2}{3}, \infty\right) \qquad (2\text{-}3)$$

30. The prime factors of 60 are 2, 3, and 5. Hence the set is written {2, 3, 5}.
$\qquad (1\text{-}1, \ 1\text{-}2)$

31. (A) This is false. For example, if $a = 2$ and $b = \dfrac{1}{2}$, then $ab = 1$. *F*

(B) This is true. This restates the zero property of real numbers. *T*

(C) This is false. In fact, a real number with a repeating decimal expansion is a *rational* number. *F* $\qquad (1\text{-}1)$

32.
$$\frac{x+3}{2x+2} + \frac{5x+2}{3x+3} = \frac{5}{6}$$

$$\frac{(x+3)}{2(x+1)} + \frac{(5x+2)}{3(x+1)} = \frac{5}{6} \quad \text{LCD: } 6(x+1) \quad \text{Excluded value: } x \neq -1$$

$$6(x+1)\frac{(x+3)}{2(x+1)} + 6(x+1)\frac{(5x+2)}{3(x+1)} = 6(x+1)\frac{5}{6}$$

$$3(x+3) + 2(5x+2) = 5(x+1)$$

$$3x+9 + 10x + 4 = 5x+5$$

$$13x + 13 = 5x + 5$$

$$8x = -8$$

$$x = -1$$

No solution: -1 is excluded   *(2-1)*

33.
$$\frac{3}{x} = \frac{6}{x+1} - \frac{1}{x-1} \quad \text{Excluded values: } x \neq 0, 1, -1$$

$$x(x+1)(x-1)\frac{3}{x} = x(x+1)(x-1)\frac{6}{x+1} - x(x+1)(x-1)\frac{1}{x-1}$$

$$3(x+1)(x-1) = 6x(x-1) - x(x+1)$$

$$3(x^2-1) = 6x^2 - 6x - x^2 - x$$

$$3x^2 - 3 = 5x^2 - 7x$$

$$0 = 2x^2 - 7x + 3$$

$$0 = (2x-1)(x-3)$$

$$2x - 1 = 0 \quad \text{or} \quad x - 3 = 0$$

$$2x = 1 \qquad\qquad x = 3$$

$$x = \frac{1}{2}$$

(2-6)

34.
$$2x + 1 = 3\sqrt{2x-1}$$

$$(2x+1)^2 = 9(2x-1)$$

$$4x^2 + 4x + 1 = 18x - 9$$

$$4x^2 - 14x + 10 = 0$$

$$2x^2 - 7x + 5 = 0$$

$$(2x-5)(x-1) = 0$$

$$2x - 5 = 0 \quad \text{or} \quad x - 1 = 0$$

$$x = \frac{5}{2} \qquad\qquad x = 1$$

Check: $2\left(\frac{5}{2}\right) + 1 \overset{?}{=} 3\sqrt{2\left(\frac{5}{2}\right) - 1}$

$$5 + 1 \overset{?}{=} 3\sqrt{5 - 1}$$

$$6 \overset{\checkmark}{=} 6$$

$$2 \cdot 1 + 1 \overset{?}{=} 3\sqrt{2 \cdot 1 - 1}$$

$$3 \overset{\checkmark}{=} 3$$

Solution: $1, \frac{5}{2}$ *(2-7)*

35. $2x - 3y = 9$
$4x + 2y = 23$
Solve the first equation for x in terms of y and substitute into the second equation.

$$2x = 3y + 9$$

$$x = \frac{3y+9}{2}$$

$$\frac{4}{1}\left(\frac{3y+9}{2}\right) + 2y = 23$$

$$2(3y+9) + 2y = 23$$

$$6y + 18 + 2y = 23$$

$$8y = 5$$

$$y = \frac{5}{8}$$

$$x = \frac{3y+9}{2} = \frac{3\left(\frac{5}{8}\right)+9}{2} = \frac{87}{16}$$

$$x = \frac{87}{16}, \ y = \frac{5}{8}$$

(2-2)

36. (a) is a polynomial with degree 2

(b) is not a polynomial

(c) is a polynomial with degree 4

(d) is not a polynomial   *(1-2)*

37. $(a + 2b)(2a + b) - (a - 2b)(2a - b) = 2a^2 + ab + 4ab + 2b^2 - (2a^2 - ab - 4ab + 2b^2)$

$$= 2a^2 + 5ab + 2b^2 - (2a^2 - 5ab + 2b^2)$$

$$= 2a^2 + 5ab + 2b^2 - 2a^2 + 5ab - 2b^2$$

$$= 10ab \qquad (1-2)$$

38. $3(x + h)^2 - 4(x + h) - (3x^2 - 4x) = 3(x^2 + 2xh + h^2) - 4x - 4h - 3x^2 + 4x$

$$= 3x^2 + 6xh + 3h^2 - 4x - 4h - 3x^2 + 4x$$

$$= 6xh + 3h^2 - 4h \qquad (1-2)$$

39. $(4m + 3n)^3 = (4m + 3n)(4m + 3n)(4m + 3n)$

$$= [(4m + 3n)(4m + 3n)](4m + 3n)$$

$$= [(4m)^2 + 2(4m)(3n) + (3n)^2](4m + 3n)$$

$$= [16m^2 + 24mn + 9n^2](4m + 3n)$$

The last multiplication is best performed vertically, so we write

$$\begin{array}{r} 16m^2 + 24mn + 9n^2 \\ \underline{4m + 3n} \\ 64m^3 + 96m^2n + 36mn^2 \\ \underline{48m^2n + 72mn^2 + 27n^3} \\ 64m^3 + 144m^2n + 108mn^2 + 27n^3 \end{array} \qquad (1-2)$$

40. $(2y + 4)^2 - y^2 = [(2y + 4) - y][(2y + 4) + y]$

$$= (2y + 4 - y)(2y + 4 + y)$$

$$= (y + 4)(3y + 4) \qquad (1-3)$$

41. $a^3 + 3a^2 - 4a - 12 = (a^3 + 3a^2) - (4a + 12)$

$$= a^2(a + 3) - 4(a + 3)$$

$$= (a + 3)(a^2 - 4)$$

$$= (a + 3)(a + 2)(a - 2) \qquad (1-3)$$

42. $3x^4(x + 1)^2 + 4x^3(x + 1)^3 = x^3(x + 1)^2[3x + 4(x + 1)]$

$$= x^3(x + 1)^2[3x + 4x + 4]$$

$$= x^3(x + 1)^2(7x + 4) \qquad (1-3)$$

43. $\dfrac{-3x^4(x + 1)^2 + 4x^3(x + 1)^3}{(x + 1)^6} = \dfrac{x^3(x + 1)^2[-3x + 4(x + 1)]}{(x + 1)^6}$

$$= \frac{x^3\cancel{(x + 1)^2}^{\,1}[-3x + 4x + 4]}{\cancel{(x + 1)^6}_{(x + 1)^4}}$$

$$= \frac{x^3(x + 4)}{(x + 1)^4} \qquad (1-4)$$

44. $\dfrac{2a + 4b}{a^2 - b^2} + \dfrac{3a}{a^2 - 3ab + 2b^2} - \dfrac{3b}{a^2 - ab - 2b^2}$

$$= \dfrac{2a + 4b}{(a + b)(a - b)} + \dfrac{3a}{(a - b)(a - 2b)} - \dfrac{3b}{(a - 2b)(a + b)}$$

$$= \dfrac{(2a + 4b)(a - 2b)}{(a + b)(a - b)(a - 2b)} + \dfrac{3a(a + b)}{(a + b)(a - b)(a - 2b)} - \dfrac{3b(a - b)}{(a + b)(a - b)(a - 2b)}$$

$$= \dfrac{(2a + 4b)(a - 2b) + 3a(a + b) - 3b(a - b)}{(a + b)(a - b)(a - 2b)}$$

$$= \dfrac{2a^2 - 4ab + 4ab - 8b^2 + 3a^2 + 3ab - 3ab + 3b^2}{(a + b)(a - b)(a - 2b)}$$

$$= \dfrac{5a^2 - 5b^2}{(a + b)(a - b)(a - 2b)}$$

$$= \dfrac{5 \overset{1}{\cancel{(a + b)}}\, \overset{1}{\cancel{(a - b)}}}{\underset{1}{\cancel{(a + b)}}\, \underset{1}{\cancel{(a - b)}}\,(a - 2b)}$$

$$= \dfrac{5}{a - 2b} \qquad\qquad (1\text{-}4)$$

45. $\dfrac{2 - \dfrac{4}{2 - \dfrac{x}{y}}}{2 - \dfrac{4}{2 + \dfrac{x}{y}}} = \dfrac{2 - \dfrac{y(4)}{y\left(2 - \dfrac{x}{y}\right)}}{2 - \dfrac{y(4)}{y\left(2 + \dfrac{x}{y}\right)}}$

$$= \dfrac{2 - \dfrac{4y}{2y - x}}{2 - \dfrac{4y}{2y + x}}$$

$$= \dfrac{\dfrac{2(2y - x)}{2y - x} - \dfrac{4y}{2y - x}}{\dfrac{2(2y + x)}{2y + x} - \dfrac{4y}{2y + x}}$$

$$= \dfrac{\dfrac{4y - 2x - 4y}{2y - x}}{\dfrac{4y + 2x - 4y}{2y + x}}$$

$$= \dfrac{\dfrac{-2x}{2y - x}}{\dfrac{2x}{2y + x}}$$

$$= -\dfrac{2x}{2y - x} + \dfrac{2x}{2y + x}$$

$$= \dfrac{\overset{1}{\cancel{2x}}}{x - 2y} \cdot \dfrac{x + 2y}{\underset{1}{\cancel{2x}}}$$

$$= \dfrac{x + 2y}{x - 2y} \qquad\qquad (1\text{-}4)$$

46. The solution is incorrect, because in the first step the quantity -2 has been removed from numerator and denominator. But -2 is a term, and only factors can be cancelled. A correct solution would factor numerator and denominator, then cancel common factors, as follows:

$$\dfrac{x^2 - x - 2}{x - 2} = \dfrac{\overset{1}{\cancel{(x - 2)}}(x + 1)}{\underset{1}{\cancel{x - 2}}}$$

$$= x + 1 \qquad (1\text{-}4)$$

47. The original equation can be rewritten as

$$\dfrac{5}{(x + 2)(x + 3)} = \dfrac{3}{(x + 1)(x + 2)}$$

Thus, $x = -2$ cannot be a solution of this equation. This extraneous solution was introduced when both sides were multiplied by $x + 2$ in the second line. $x = -2$ must be discarded and the only correct solution is $x = 2$. $\qquad (2\text{-}6)$

48. $|4x - 9| > 3$

$\quad 4x - 9 < -3 \quad$ or $\quad 4x - 9 > 3$

$\qquad 4x < 6 \qquad\qquad\qquad 4x > 12$

$\qquad x < \dfrac{3}{2} \quad$ or $\qquad\quad x > 3$

$\left(-\infty, \dfrac{3}{2}\right) \cup (3, \infty)$

$\qquad\qquad\qquad\qquad\qquad\qquad (2\text{-}4)$

49. $\sqrt{(3m - 4)^2} \le 2$

$\qquad |3m - 4| \le 2$

$\quad -2 \le 3m - 4 \le 2$

$\qquad\quad 2 \le 3m \le 6$

$\qquad\quad \dfrac{2}{3} \le m \le 2$

$\left[\dfrac{2}{3}, \; 2\right]$

$\qquad\qquad\qquad\qquad\qquad (2\text{-}4)$

50. $\qquad\qquad \dfrac{2}{x + 1} \ge \dfrac{1}{x - 2}$

Common Error:
$2(x - 2) \ge x + 1$ is not equivalent to the given inequality.

$\qquad \dfrac{2}{x + 1} - \dfrac{1}{x - 2} \ge 0$

$\dfrac{2(x - 2) - (x + 1)}{(x + 1)(x - 2)} \ge 0$

$\dfrac{2x - 4 - x - 1}{(x + 1)(x - 2)} \ge 0$

$\qquad \dfrac{x - 5}{(x + 1)(x - 2)} \ge 0$

Zeros of P, Q: $-1, 2, 5$
5 is part of the solution set
so we use a solid dot there.
-1 and 2 are not part of the
solution set ($\frac{P}{Q}$ is not defined
there) so we use open dots
there.

$(-\infty, -1) \; (-1, 2) \; (2, 5) \; (5, \infty)$

	$\dfrac{P}{Q} = \dfrac{x - 5}{(x + 1)(x - 2)}$			
Test Number	-2	0	3	6
Value of $\frac{P}{Q}$	$-\frac{7}{4}$	$\frac{5}{2}$	$-\frac{1}{2}$	$\frac{1}{28}$
Sign of $\frac{P}{Q}$	$-$	$+$	$-$	$+$
Interval	$(-\infty, -1)$	$(-1, 2)$	$(2, 5)$	$(5, \infty)$

$\dfrac{x - 5}{(x + 1)(x - 2)} \ge 0$ and $\dfrac{2}{x + 1} \ge \dfrac{1}{x - 2}$ within the intervals $(-1, 2)$ and $[5, \infty)$.

$-1 < x < 2$ or $5 \le x$

$\qquad\qquad\qquad\qquad\qquad\qquad\qquad (2\text{-}8)$

51. $\dfrac{\sqrt{x - 2}}{x - 4}$ represents a real number if $x - 2$ is positive or zero, except if $x = 4$.

Thus, $x \ge 2$, $x \ne 4$, or $[2, 4) \cup (4, \infty)$.

$\qquad\qquad\qquad\qquad\qquad\qquad\qquad (2\text{-}3)$

52. (A) $(2 - 3i)^2 - (4 - 5i)(2 - 3i) - (2 + 10i)$

$\quad = (2)^2 - 2(2)(3i) + (3i)^2 - (8 - 12i - 10i + 15i^2) - 2 - 10i$

$\quad = 4 - 12i + 9i^2 - (8 - 22i + 15i^2) - 2 - 10i$

$\quad = 4 - 12i + 9i^2 - 8 + 22i - 15i^2 - 2 - 10i$

$\quad = 4 - 12i - 9 - 8 + 22i + 15 - 2 - 10i$

$\quad = 0 + 0i$ or 0

(B) $\dfrac{3}{5} + \dfrac{4}{5}i + \dfrac{1}{\frac{3}{5} + \frac{4}{5}i} = \dfrac{3}{5} + \dfrac{4}{5}i + \dfrac{1}{\left(\frac{3}{5} + \frac{4}{5}i\right)} \dfrac{\left(\frac{3}{5} - \frac{4}{5}i\right)}{\left(\frac{3}{5} - \frac{4}{5}i\right)}$

$\qquad\qquad = \dfrac{3}{5} + \dfrac{4}{5}i + \dfrac{\frac{3}{5} - \frac{4}{5}i}{\frac{9}{25} - \frac{16}{25}i^2}$

$\qquad\qquad = \dfrac{3}{5} + \dfrac{4}{5}i + \dfrac{\frac{3}{5} - \frac{4}{5}i}{\frac{9}{25} + \frac{16}{25}}$

$\qquad\qquad = \dfrac{3}{5} + \dfrac{4}{5}i + \dfrac{3}{5} - \dfrac{4}{5}i$

$\qquad\qquad = \dfrac{6}{5}$

(C) $i^{35} = i^{32}i^3 = (i^4)^8(-i) = 1^8(-i) = -i$ $\qquad\qquad$ (2-5)

53. (A) $(5 + 2\sqrt{-9}) - (2 - 3\sqrt{-16}) = (5 + 2i\sqrt{9}) - (2 - 3i\sqrt{16})$
$\qquad\qquad\qquad\qquad\qquad\qquad = (5 + 6i) - (2 - 12i)$
$\qquad\qquad\qquad\qquad\qquad\qquad = 5 + 6i - 2 + 12i$
$\qquad\qquad\qquad\qquad\qquad\qquad = 3 + 18i$

(B) $\dfrac{2 + 7\sqrt{-25}}{3 - \sqrt{-1}} = \dfrac{2 + 7i\sqrt{25}}{3 - i}$

$\qquad\qquad\quad = \dfrac{2 + 35i}{3 - i}$

$\qquad\qquad\quad = \dfrac{(2 + 35i)}{(3 - i)} \dfrac{(3 + i)}{(3 + i)}$

$\qquad\qquad\quad = \dfrac{6 + 107i + 35i^2}{9 - i^2}$

$\qquad\qquad\quad = \dfrac{6 + 107i - 35}{9 + 1}$

$\qquad\qquad\quad = \dfrac{-29 + 107i}{10}$

$\qquad\qquad\quad = -2.9 + 10.7i$

(C) $\dfrac{12 - \sqrt{-64}}{\sqrt{-4}} = \dfrac{12 - i\sqrt{64}}{i\sqrt{4}} = \dfrac{12 - 8i}{2i} = \dfrac{12 - 8i}{2i} \dfrac{i}{i}$

$\qquad\quad = \dfrac{12i - 8i^2}{2i^2} = \dfrac{12i + 8}{-2} = -4 - 6i$ $\qquad\qquad$ (2-5)

54. (A) 0.1767

(B) $\sqrt[7]{12.47} = (12.47)^{1/7} = (12.47)^{(1\div 7)} = 1.434$

(C) $\sqrt[5]{\sqrt[3]{9} + 4} = (9^{1/3} + 4)^{1/5} = (9^{(1\div 3)} + 4)^{(1\div 5)} = 1.435$

(D) $\dfrac{(52{,}180{,}000{,}000)(0.000\,000\,002\,973)}{0.000\,000\,000\,000\,271} = \dfrac{(5.218 \times 10^{10})(2.973 \times 10^{-9})}{2.71 \times 10^{-13}}$
$\qquad\qquad\qquad\qquad\qquad\qquad\qquad\qquad\qquad = 5.724 \times 10^{14}$ $\qquad\qquad$ (1-5, 1-7)

55. $3xy^2\sqrt[3]{16x^5y^7} = 3xy^2\sqrt[3]{2^3x^3y^6 \cdot 2x^2y}$

$\qquad\qquad\qquad\quad = 3xy^2\sqrt[3]{2^3x^3y^6}\sqrt[3]{2x^2y}$

$\qquad\qquad\qquad\quad = 3xy^2 \cdot 2xy^2\sqrt[3]{2x^2y}$

$\qquad\qquad\qquad\quad = 6x^2y^4\sqrt[3]{2x^2y}$ $\qquad\qquad$ (1-7)

56. $\sqrt[5]{\dfrac{7a^3}{16b^4}} = \sqrt[5]{\dfrac{7a^3}{16b^4}\dfrac{2b}{2b}} = \sqrt[5]{\dfrac{14a^3b}{32b^5}} = \dfrac{\sqrt[5]{14a^3b}}{\sqrt[5]{32b^5}} = \dfrac{\sqrt[5]{14a^3b}}{2b}$ $\qquad$ (1-7)

57. $\sqrt[3]{\sqrt{8y^9}} = \sqrt[2\cdot3]{2^{3\cdot1}y^{3\cdot3}} = \sqrt{2y^3} = \sqrt{y^2\cdot2y} = \sqrt{y^2}\sqrt{2y} = y\sqrt{2y}$ $\qquad$ (1-7)

58. $\dfrac{t}{\sqrt{t+9}-3} = \dfrac{t}{(\sqrt{t+9}-3)}\dfrac{(\sqrt{t+9}+3)}{(\sqrt{t+9}+3)}$

$\qquad\qquad = \dfrac{t(\sqrt{t+9}+3)}{(\sqrt{t+9})^2-3^2}$

$\qquad\qquad = \dfrac{t(\sqrt{t+9}+3)}{t+9-9}$

$\qquad\qquad = \dfrac{\overset{1}{\cancel{t}}(\sqrt{t+9}+3)}{\underset{1}{\cancel{t}}}$

$\qquad\qquad = \sqrt{t+9}+3$ $\qquad\qquad\qquad$ (1-7)

59. (A) $\sqrt{7+4\sqrt{3}} \approx 3.732\,050\,808$ $\qquad$ (B) $2+\sqrt{3} \approx 3.732\,050\,808$

(C) $\sqrt[3]{26+15\sqrt{3}} \approx 3.732\,050\,808$

Thus, (A), (B) and (C) all have the same value to 9 decimal places. To show that they are actually equal:

$\qquad (2+\sqrt{3})^2 = 2^2 + 2\cdot2\cdot\sqrt{3} + (\sqrt{3})^2 = 4 + 4\sqrt{3} + 3 = 7 + 4\sqrt{3}$

Therefore, since $2+\sqrt{3}$ is positive, $2+\sqrt{3}$ is the positive square root of $7+4\sqrt{3}$.

Thus, $\sqrt{7+4\sqrt{3}} = 2+\sqrt{3}$

$\qquad (2+\sqrt{3})^3 = (2+\sqrt{3})^2(2+\sqrt{3})$

$\qquad\qquad\qquad = (7+4\sqrt{3})(2+\sqrt{3})$

$\qquad\qquad\qquad = 14 + 7\sqrt{3} + 8\sqrt{3} + 4\sqrt{3}\sqrt{3}$

$\qquad\qquad\qquad = 26 + 15\sqrt{3}$

Therefore, $2+\sqrt{3}$ is the real cube root of $26+15\sqrt{3}$. Thus, $2+\sqrt{3} = \sqrt[3]{26+15\sqrt{3}}$.

$\qquad\qquad\qquad\qquad\qquad\qquad\qquad\qquad\qquad\qquad\qquad$ (1-7)

60. (A) $\dfrac{2x^3+10}{5x^2} = \dfrac{2x^3}{5x^2} + \dfrac{10}{5x^2} = \dfrac{2x}{5} + \dfrac{2}{x^2} = \dfrac{2}{5}x + 2x^{-2}$

(B) $\dfrac{2\sqrt{x}-5}{4\sqrt{x}} = \dfrac{2\sqrt{x}}{4\sqrt{x}} - \dfrac{5}{4\sqrt{x}} = \dfrac{1}{2} - \dfrac{5}{4\sqrt{x}} = \dfrac{1}{2} - \dfrac{5}{4}x^{-1/2}$ or $\dfrac{1}{2}x^0 - \dfrac{5}{4}x^{-1/2}$ $\quad$ (1-5, 1-7)

61. $\qquad 1 + \dfrac{14}{y^2} = \dfrac{6}{y}$ Excluded value: $y \neq 0$

$\qquad y^2(1) + y^2\dfrac{14}{y^2} = y^2\dfrac{6}{y}$

$\qquad\qquad y^2 + 14 = 6y$

$\qquad y^2 - 6y + 14 = 0$

$\qquad\qquad y^2 - 6y = -14$

$\qquad y^2 - 6y + 9 = -5$

$\qquad\qquad (y-3)^2 = -5$

$\qquad\qquad\quad y - 3 = \pm\sqrt{-5}$

$\qquad\qquad\qquad y = 3 \pm \sqrt{-5}$

$\qquad\qquad\qquad y = 3 \pm i\sqrt{5}$ $\qquad\qquad\qquad$ (2-6)

62. $4x^{2/3} - 4x^{1/3} - 3 = 0$
Let $u = x^{1/3}$, then
$$4u^2 - 4u - 3 = 0$$
$$(2u - 3)(2u + 1) = 0$$
$$u = \frac{3}{2}, \; -\frac{1}{2}$$

$x^{1/3} = \frac{3}{2} \qquad x^{1/3} = -\frac{1}{2}$

$x = \frac{27}{8} \qquad x = -\frac{1}{8} \qquad$ (2-7)

63. $u^4 + u^2 - 12 = 0$
Let $w = u^2$, then
$$w^2 + w - 12 = 0$$
$$(w + 4)(w - 3) = 0$$
$$w = -4, \; 3$$
$u^2 = -4 \qquad u^2 = 3$

$u = \pm 2i \qquad u = \pm\sqrt{3} \qquad$ (2-7)

64. $\sqrt{8t - 2} - 2\sqrt{t} = 1$
$$\sqrt{8t - 2} = 2\sqrt{t} + 1$$
$$8t - 2 = 4t + 4\sqrt{t} + 1$$
$$4t - 3 = 4\sqrt{t}$$
$$16t^2 - 24t + 9 = 16t$$
$$16t^2 - 40t + 9 = 0$$
$$(4t - 1)(4t - 9) = 0$$
$$t = \tfrac{1}{4}, \; \tfrac{9}{4}$$

Check: $\sqrt{8(\frac{1}{4}) - 2} - 2\sqrt{\frac{1}{4}} \overset{?}{=} 1$

$0 - 2(\tfrac{1}{2}) \overset{?}{=} 1$

$-1 \neq 1$

$\sqrt{8(\frac{9}{4}) - 2} - 2\sqrt{\frac{9}{4}} \overset{?}{=} 1$

$\sqrt{16} - 2\sqrt{\frac{9}{4}} \overset{?}{=} 1$

$4 - 2(\tfrac{3}{2}) \overset{?}{=} 1$

$1 = 1 \quad \checkmark$

Solution: $\frac{9}{4}$ $\qquad$ (2-7)

65. $-3.45 < 1.86 - 0.33x \leq 7.92$
$-5.31 < -0.33x \leq 6.06$
$16.09 > x \geq -18.36$
$-18.36 \leq x < 16.09$ or $[-18.36, 16.09)$ $\qquad$ (2-3)

66. $2.35x^2 + 10.44x - 16.47 = 0$
$$x = \frac{-b \pm \sqrt{b^2 - 4ac}}{2a} \quad a = 2.35, \; b = 10.44, \; c = -16.47$$
$$x = \frac{-10.44 \pm \sqrt{(10.44)^2 - 4(2.35)(-16.47)}}{2(2.35)}$$
$$= \frac{-10.44 \pm \sqrt{263.8116}}{4.70}$$
$$= \frac{-10.44 \pm 16.2423}{4.70}$$
$$= -5.68, \; 1.23 \qquad (2-6)$$

67. $12.5x + 2.5y = 20$
$3.5x + 8.7y = 10$
Solve the first equation for y in terms of x and substitute into the second equation.
$$2.5y = 20 - 12.5x$$
$$y = \frac{20 - 12.5x}{2.5}$$
$$y = 8 - 5x$$
$$3.5x + 8.7(8 - 5x) = 10$$
$$3.5x + 69.6 - 43.5x = 10$$
$$-40x = -59.6$$
$$x = 1.49$$
$$y = 8 - 5(1.49) = 0.55$$
$x = 1.49, \; y = 0.55 \qquad (2-2)$

68.
$$\frac{x - 2}{x + 1} = \frac{2y + 1}{y - 2}$$

$$(x + 1)(y - 2)\frac{x - 2}{x + 1} = (x + 1)(y - 2)\frac{2y + 1}{y - 2}$$

$$(y - 2)(x - 2) = (x + 1)(2y + 1)$$

$$xy - 2y - 2x + 4 = 2xy + x + 2y + 1$$

$$3 - 3x = xy + 4y$$

$$3 - 3x = y(x + 4)$$

$$y = \frac{3 - 3x}{x + 4} \qquad (2\text{-}1)$$

69. $x = -1 + 5s + 2t$
$y = 2 + 2s + t$
Solve the second equation for t in terms of the other variables and substitute into the first equation.

$t = y - 2 - 2s$

$x = -1 + 5s + 2(y - 2 - 2s)$

$x = -1 + 5s + 2y - 4 - 4s$

$x = -5 + s + 2y$

$s = 5 + x - 2y$

$t = y - 2 - 2s = y - 2 - 2(5 + x - 2y) = y - 2 - 10 - 2x + 4y$

$t = -12 - 2x + 5y$

The checking steps are omitted for lack of space. $\qquad (2\text{-}2)$

70. $x^2 - x - 1 = \left(\frac{1}{2} - \frac{\sqrt{5}}{2}\right)^2 - \left(\frac{1}{2} - \frac{\sqrt{5}}{2}\right) - 1$

$$= \left(\frac{1}{2}\right)^2 - 2\left(\frac{1}{2}\right)\left(\frac{\sqrt{5}}{2}\right) + \left(\frac{\sqrt{5}}{2}\right)^2 - \frac{1}{2} + \frac{\sqrt{5}}{2} - 1$$

$$= \frac{1}{4} - \frac{\sqrt{5}}{2} + \frac{5}{4} - \frac{1}{2} + \frac{\sqrt{5}}{2} - 1$$

$$= \frac{6}{4} - \frac{3}{2}$$

$$= 0 \qquad (1\text{-}7)$$

71. $x^2 - x + 2 = \left(\frac{1}{2} - \frac{i}{2}\sqrt{7}\right)^2 - \left(\frac{1}{2} - \frac{i}{2}\sqrt{7}\right) + 2$

$$= \left(\frac{1}{2}\right)^2 - 2\left(\frac{1}{2}\right)\left(\frac{i}{2}\sqrt{7}\right) + \left(\frac{i}{2}\sqrt{7}\right)^2 - \frac{1}{2} + \frac{i}{2}\sqrt{7} + 2$$

$$= \frac{1}{4} - \frac{i}{2}\sqrt{7} + \frac{7i^2}{4} - \frac{1}{2} + \frac{i}{2}\sqrt{7} + 2$$

$$= \frac{1}{4} - \frac{7}{4} - \frac{1}{2} + 2$$

$$= 0 \qquad (2\text{-}5)$$

72. $(x + 1)^3 - (x - 1)^3 = (x + 1)(x + 1)(x + 1) - (x - 1)(x - 1)(x - 1)$

$$= [(x + 1)(x + 1)](x + 1) - [(x - 1)(x - 1)](x - 1)$$

$$= [x^2 + 2x + 1](x + 1) - [x^2 - 2x + 1](x - 1)$$

$$= x^3 + 2x^2 + x + x^2 + 2x + 1 - (x^3 - 2x^2 + x - x^2 + 2x - 1)$$

$$= x^3 + 3x^2 + 3x + 1 - (x^3 - 3x^2 + 3x - 1)$$

$$= x^3 + 3x^2 + 3x + 1 - x^3 + 3x^2 - 3x + 1$$

$$= 6x^2 + 2$$

> **Common Error:**
>
> $(x + 1)^3 - (x - 1)^3 \neq (x + 1 - (x - 1))^3$
> $\neq (x + 1 - x + 1)^3$
> First cube, then subtract.

$\qquad (1\text{-}2)$

73. The given inequality
$$a - b < b - a$$
is equivalent to, successively,
$$2a - b < b$$
$$2a < 2b$$
$$a < b$$
Thus it is true for all real a and b such that $a < b$.   *(2-3)*

74. $9b^4 - 16b^2(a^2 - 2a + 1)$
$= b^2[9b^2 - 16(a^2 - 2a + 1)]$
$= b^2[9b^2 - 16(a - 1)^2]$
$= b^2\{(3b)^2 - [4(a - 1)]^2\}$
$= b^2[3b - 4(a - 1)][3b + 4(a - 1)]$
$= b^2(3b - 4a + 4)(3b + 4a - 4)$ *(1-3)*

75.
$$\dfrac{1 - \dfrac{1}{1 + \dfrac{1}{1 + m}}}{\dfrac{m}{2 - m} + \dfrac{m}{2 + m}} = \dfrac{1 - \dfrac{1(1 + m)}{1(1 + m) + (1 + m)\dfrac{1}{1 + m}}}{\dfrac{m}{2 - m} + \dfrac{m}{2 + m}}$$

$$= \dfrac{1 - \dfrac{1 + m}{1 + m + 1}}{\dfrac{m}{2 - m} + \dfrac{m}{2 + m}}$$

$$= \dfrac{1 - \dfrac{1 + m}{2 + m}}{\dfrac{m}{2 - m} + \dfrac{m}{2 + m}}$$

$$= \dfrac{(2 - m)(2 + m)}{(2 - m)(2 + m)} \dfrac{1 - \dfrac{1 + m}{2 + m}}{\dfrac{m}{2 - m} + \dfrac{m}{2 + m}}$$

$$= \dfrac{(2 - m)(2 + m) - (2 - m)(1 + m)}{(2 + m)m + (2 - m)m}$$

$$= \dfrac{4 - m^2 - (2 + m - m^2)}{2m + m^2 + 2m - m^2}$$

$$= \dfrac{4 - m^2 - 2 - m + m^2}{4m}$$

$$= \dfrac{2 - m}{4m} \qquad (1-4)$$

76.
$$\dfrac{x + y}{y - \dfrac{x + y}{x - y}} = 1$$

$$\dfrac{(x + y)(x - y)}{y(x - y) - (x + y)} = 1$$

$$\dfrac{x^2 - y^2}{xy - y^2 - x - y} = 1$$

$$x^2 - y^2 = xy - y^2 - x - y$$

$$x^2 + x = xy - y$$

$$x^2 + x = y(x - 1)$$

$$y = \dfrac{x^2 + x}{x - 1} \qquad (2-1)$$

77.
$$3x^2 = 2\sqrt{2}x - 1$$
$$3x^2 - 2\sqrt{2}x + 1 = 0$$
$$x = \dfrac{-b \pm \sqrt{b^2 - 4ac}}{2a} \qquad \begin{array}{l} a = 3 \\ b = -2\sqrt{2} \\ c = 1 \end{array}$$

$$x = \dfrac{-(-2\sqrt{2}) \pm \sqrt{(-2\sqrt{2})^2 - 4(3)(1)}}{2(3)}$$

$$x = \dfrac{2\sqrt{2} \pm \sqrt{8 - 12}}{6}$$

$$x = \dfrac{2\sqrt{2} \pm \sqrt{-4}}{6}$$

$$x = \dfrac{2\sqrt{2} \pm 2i}{6}$$

$$x = \dfrac{2(\sqrt{2} \pm i)}{6}$$

$$x = \dfrac{\sqrt{2} \pm i}{3} \qquad (2-6)$$

78. In this problem, $a = 1$, $b = b$, $c = 1$. Thus, the discriminant $b^2 - 4ac = b^2 - 4 \cdot 1 \cdot 1 = b^2 - 4$. Hence, the number and types of roots depend on the sign of $b^2 - 4 = (b + 2)(b - 2)$. Charting the sign behavior of $b^2 - 4 = (b - 2)(b + 2)$, note:

Zeros: -2, 2

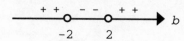

$b^2 - 4 = (b + 2)(b - 2)$			
Test Number	-3	0	3
Value of Polynomial for Test Number	5	-4	5
Sign of Polynomial in Interval	$+$	$-$	$+$
Interval	$(-\infty, -2)$	$(-2, 2)$	$(2, \infty)$

Hence,

if $b^2 - 4 > 0$, there are two distinct real roots. This occurs if $b < -2$ or $b > 2$.
if $b^2 - 4 = 0$, there is one real double root. This occurs if $b = -2$ or $b = 2$.
if $b^2 - 4 < 0$, there are two distinct imaginary roots. This occurs if $-2 < b < 2$.

(2-6, 2-8)

79.
$$1 = 6x^{-2} + 9x^{-4}$$
$$1 = \frac{6}{x^2} + \frac{9}{x^4} \quad \text{Excluded value: } x \neq 0$$
$$x^4 = 6x^2 + 9$$
$$x^4 - 6x^2 - 9 = 0$$
$$x^4 - 6x^2 = 9$$
$$x^4 - 6x^2 + 9 = 18$$
$$(x^2 - 3)^2 = 18$$
$$x^2 - 3 = \pm\sqrt{18}$$
$$x^2 = 3 \pm \sqrt{18}$$
$$x^2 = 3 \pm 3\sqrt{2}$$

$3 - 3\sqrt{2}$ is negative; there can be no real solutions of $x^2 = 3 - 3\sqrt{2}$. Since we are asked to find all real roots of the original equation, we discard $x^2 = 3 - 3\sqrt{2}$.

$$x^2 = 3 + 3\sqrt{2}$$
$$x = \pm\sqrt{3 + 3\sqrt{2}}$$

(2-7)

80. (A) $(a^{1/4} - b^{1/4})(a^{1/4} + b^{1/4})(a^{1/2} + b^{1/2}) = [(a^{1/4})^2 - (b^{1/4})^2](a^{1/2} + b^{1/2})$
$$= (a^{1/2} - b^{1/2})(a^{1/2} + b^{1/2})$$
$$= (a^{1/2})^2 - (b^{1/2})^2$$
$$= a - b$$

Common Error: $(a^{1/4})^2 \neq a^{1/16}$

(B) $\left(\dfrac{x^{-1}y + xy^{-1}}{xy^{-1} - x^{-1}y}\right)^{-1} = \left(\dfrac{\frac{y}{x} + \frac{x}{y}}{\frac{x}{y} - \frac{y}{x}}\right)^{-1}$

> **Common Error:**
> "Bringing in" the exponent -1.
> $$\left(\frac{x^{-1}y + xy^{-1}}{xy^{-1} - x^{-1}y}\right)^{-1} \neq \frac{xy^{-1} + x^{-1}y}{x^{-1}y - xy^{-1}}$$
> Exponentiation does not distribute over $+$ or $-$.

$$= \left(\dfrac{xy\,\frac{y}{x} + \frac{x}{y}}{xy\,\frac{x}{y} - \frac{y}{x}}\right)^{-1}$$

$$= \left(\dfrac{y^2 + x^2}{x^2 - y^2}\right)^{-1}$$

$$= \dfrac{x^2 - y^2}{x^2 + y^2}$$

(1-5, 1-7)

81. $\dfrac{\sqrt[3]{8+h}-2}{h} = \dfrac{\sqrt[3]{8+h}-2}{h} \cdot \dfrac{(\sqrt[3]{8+h})^2 + 2\sqrt[3]{8+h} + 2^2}{(\sqrt[3]{8+h})^2 + 2\sqrt[3]{8+h} + 2^2}$

$\qquad = \dfrac{(\sqrt[3]{8+h})^3 - 2^3}{h[(\sqrt[3]{8+h})^2 + 2\sqrt[3]{8+h} + 4]}$

$\qquad = \dfrac{8+h-8}{h[(8+h)^{2/3} + 2(8+h)^{1/3} + 4]}$

$\qquad = \dfrac{\overset{1}{\cancel{h}}}{\underset{1}{\cancel{h}}[(8+h)^{2/3} + 2(8+h)^{1/3} + 4]}$

$\qquad = \dfrac{1}{(8+h)^{2/3} + 2(8+h)^{1/3} + 4}$ $(1\text{-}7)$

82. $\dfrac{a+bi}{a-bi} = \dfrac{(a+bi)}{(a-bi)} \cdot \dfrac{(a+bi)}{(a+bi)} = \dfrac{(a+bi)^2}{a^2 - (bi)^2} = \dfrac{a^2 + 2abi + b^2 i^2}{a^2 - b^2 i^2}$

$\qquad\qquad = \dfrac{a^2 + 2abi - b^2}{a^2 + b^2}$

$\qquad\qquad = \dfrac{a^2 - b^2 + 2abi}{a^2 + b^2}$

$\qquad\qquad = \dfrac{a^2 - b^2}{a^2 + b^2} + \dfrac{2ab}{a^2 + b^2}\,i$ $(2\text{-}5)$

83. $\left|\dfrac{x+4}{x}\right| < 3$

$-3 < \dfrac{x+4}{x} < 3$

It is difficult (and a rich source of possible errors) to solve this double inequality as a double inequality. We choose to solve the inequalities $-3 < \dfrac{x+4}{x}$ and $\dfrac{x+4}{x} < 3$ separately. Then the solution set of the given inequality is the intersection of the solution sets of the two inequalities.

Part 1: $-3 < \dfrac{x+4}{x}$

$\qquad 0 < \dfrac{x+4}{x} + 3$

$\qquad 0 < \dfrac{x+4+3x}{x}$

$\qquad 0 < \dfrac{4x+4}{x}$

Zeros of $\dfrac{P}{Q}$: -1, 0

(−∞, −1) (−1, 0) (0, ∞)

Part 2: $\dfrac{x+4}{x} < 3$

$\qquad \dfrac{x+4}{x} - 3 < 0$

$\qquad \dfrac{x+4-3x}{x} < 0$

$\qquad \dfrac{4-2x}{x} < 0$

Zeros of $\dfrac{P}{Q}$: 0, 2

(−∞, 0) (0, 2) (2, ∞)

	$\dfrac{P}{Q} = \dfrac{4x+4}{x}$		
Test Number	−2	−$\frac{1}{2}$	1
Value of $\dfrac{P}{Q}$	2	−4	8
Sign of $\dfrac{P}{Q}$	+	−	+
Interval	(−∞, −1)	(−1, 0)	(0, ∞)

	$\dfrac{P}{Q} = \dfrac{4-2x}{x}$		
Test Number	−1	1	3
Value of $\dfrac{P}{Q}$	−6	2	−$\frac{2}{3}$
Sign of $\dfrac{P}{Q}$	−	+	−
Interval	(−∞, 0)	(0, 2)	(2, ∞)

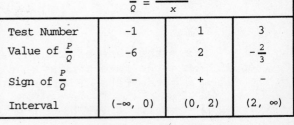

$0 < \dfrac{4x + 4}{4}$ and $-3 < \dfrac{x + 4}{x}$ within the intervals $(-\infty, -1)$ and $(0, \infty)$. $\qquad$ $\dfrac{4 - 2x}{x} < 0$ and $\dfrac{x + 4}{x} < 3$ within the intervals $(-\infty, 0)$ and $(2, \infty)$.

Combining the results of parts 1 and 2, the solution set of $\left| \dfrac{x + 4}{x} \right| < 3$ is the intersection of the two subsets of the real number line, all points which lie in *both* sets.

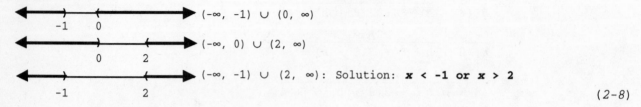

$\qquad\qquad\qquad\qquad\qquad\qquad\qquad\qquad\qquad\qquad\qquad\qquad\qquad\qquad$ *(2-8)*

84. $\sqrt[n]{\left(\dfrac{x^{2n^2}}{x^{4n}} \right)^{1/(n-2)}}$

$= \sqrt[n]{\left(x^{2n^2 - 4n} \right)^{1/(n-2)}}$

$= \sqrt[n]{x^{(2n^2 - 4n)/(n-2)}}$

$= \sqrt[n]{x^{[2n(n-2)]/(n-2)}}$

$= \sqrt[n]{x^{2n}}$

$= x^2 \qquad\qquad$ *(1-7)*

85. Let $x =$ the number

$\dfrac{1}{x} =$ its reciprocal

Then

$\qquad x - \dfrac{1}{x} = \dfrac{3}{2} \quad$ Excl. value: $x \neq 0$

$\qquad\qquad 2x^2 - 2 = 3x$

$\qquad 2x^2 - 3x - 2 = 0$

$\qquad (2x + 1)(x - 2) = 0$

$\qquad 2x + 1 = 0 \quad$ or $\quad x - 2 = 0$

$\qquad\qquad x = -\dfrac{1}{2} \qquad\qquad x = 2 \quad$ *(2-6)*

86. Let $\quad t =$ time for newer belt to fill car alone

$\qquad 14 =$ time for older belt to fill car alone

$\qquad 6 =$ time for both belts to fill car together

Then $\dfrac{1}{t} =$ rate for newer belt

$\dfrac{1}{14} =$ rate for older belt

$\begin{pmatrix} \text{Part of job} \\ \text{completed by} \\ \text{newer belt} \end{pmatrix} + \begin{pmatrix} \text{Part of job} \\ \text{completed by} \\ \text{older belt} \end{pmatrix} = 1$ whole job

$\qquad \dfrac{1}{t}(6) \qquad + \qquad \dfrac{1}{14}(6) \qquad = 1$

$\qquad\qquad\qquad \dfrac{6}{t} + \dfrac{6}{14} = 1 \quad$ Excluded value: $t \neq 0$

$\qquad\qquad 14t\dfrac{6}{t} + 14t\dfrac{6}{14} = 14t$

$\qquad\qquad\qquad 84 + 6t = 14t$

$\qquad\qquad\qquad\qquad 84 = 8t$

$\qquad\qquad\qquad\qquad t = 10.5$ minutes $\qquad$ *(2-2)*

87. Let $\qquad x =$ the rate of the current

Then $15 - x =$ the rate of the boat upstream

$\qquad 15 + x =$ the rate of the boat downstream

Solving $d = rt$ for t, we have $t = \dfrac{d}{r}$. We use this formula, together with

time upstream + time downstream = 4.8.

$$\text{time upstream} = \frac{\text{distance upstream}}{\text{rate upstream}} = \frac{35}{15 - x}$$

$$\text{time downstream} = \frac{\text{distance downstream}}{\text{rate downstream}} = \frac{35}{15 + x}$$

So,

$$\frac{35}{15 - x} + \frac{35}{15 + x} = 4.8 \quad \text{Excluded values: } x \neq 15, -15$$

$$35(15 + x) + 35(15 - x) = 4.8(15 + x)(15 - x)$$

$$525 + 35x + 525 - 35x = 4.8(225 - x^2)$$

$$1050 = 1080 - 4.8x^2$$

$$-30 = -4.8x^2$$

$$6.25 = x^2$$

$$x = 2.5 \text{ miles per hour (discarding the negative answer)}$$

(2-6)

88. Let $\quad x$ = amount of distilled water (0% acid)

$\quad\quad\quad$ 24 = amount of 90% solution

Then $x + 24$ = amount of 60% solution

$$\begin{array}{ccc} \text{acid in} \\ \text{distilled} & + & \text{acid in} \\ \text{water} & & \text{90\% solution} \end{array} = \begin{array}{c} \text{acid in} \\ \text{60\% solution} \end{array}$$

$$0 \quad + \quad 0.9(24) \quad = 0.6(x + 24)$$

$$21.6 = 0.6x + 14.4$$

$$7.2 = 0.6x$$

$$x = 12 \text{ gallons} \quad\quad (2-2)$$

89. "Break even" means Cost = Revenue

Let x = number of books sold

Revenue = number of books sold × price per book

$$= x(9.65)$$

$\quad$ Cost = Fixed Cost + Variable Cost

$$= 41,800 + \text{number of books} \times \text{cost per book}$$

$$= 41,800 + x(4.90)$$

$$9.65x = 41,800 + 4.90x$$

$$4.75x = 41,800$$

$$x = \frac{41,800}{4.75}$$

$$x = 8,800 \text{ books} \quad\quad\quad (2-2)$$

90. The distance of p from 200 must be no greater than 10.

$$|p - 200| \leq 10 \quad\quad\quad (2-4)$$

91. Solve the system of equations

$$p = 5.5 + 0.002q$$

$$p = 22 - 0.001q$$

Substitute p from the first equation into the second equation to eliminate p.

$$5.5 + 0.002q = 22 - 0.001q$$

$$0.003q = 16.5$$

$$q = 5,500 \text{ cheeseheads} = \text{equilibrium quantity}$$

$$p = 5.5 + 0.002q = 5.5 + 0.002(5,500)$$

$$= \$16.50 \text{ equilibrium price} \quad\quad\quad (2-2)$$

92. The volume of the concrete wall is equal to the volume of the outer box ($V = \ell wh$) minus the volume of the inner sandbox space. Since the length and width of the outer box are x ft and the concrete is $\frac{1}{2}$ foot thick, the length and the width of the inner box are $x - 1$ ft. Thus, we have

$$\begin{pmatrix} \text{Volume of} \\ \text{outer box} \end{pmatrix} - \begin{pmatrix} \text{Volume of} \\ \text{inner box} \end{pmatrix} = \begin{pmatrix} \text{Volume of} \\ \text{concrete} \end{pmatrix}$$

$$x \cdot x \cdot 1 - (x - 1)(x - 1)1 = V$$
$$x^2 - (x - 1)^2 = V$$
$$x^2 - (x^2 - 2x + 1) = V$$
$$x^2 - x^2 + 2x - 1 = V$$
$$V = 2x - 1 \text{ cubic feet} \qquad (1\text{-}2)$$

93. (A) A profit will result if revenue is greater than cost; that is, if

$$R > C$$
$$15p - 2p^2 > 88 - 12p$$
$$-88 + 27p - 2p^2 > 0$$
$$2p^2 - 27p + 88 < 0$$
$$(2p - 11)(p - 8) < 0$$

Zeros: 5.5, 8

$(-\infty, 5.5)$ $(5.5, 8)$ $(8, \infty)$

5.5   8 $\to$ p

$2p^2 - 27p + 88 = (2p - 11)(p - 8)$			
Test Number	0	6	9
Value of Polynomial for Test Number	88	-2	7
Sign of Polynomial in Interval	+	−	+
Interval	$(-\infty, 5.5)$	$(5.5, 8)$	$(8, \infty)$

+ +   − −   + + $\to$ p
5.5   8

$2p^2 - 27p + 88 < 0$ and a profit will occur ($R > C$) for $\$5.5 < p < \$8$ or ($\$5.5, \$8$).

(B) A loss will result if cost is greater than revenue, that is if

$$C > R$$
$$88 - 12p > 15p - 2p^2$$
$$2p^2 - 27p + 88 > 0$$

Referring to the sign chart in part (A), we see that $2p^2 - 27p + 88 > 0$, and a loss will occur ($C > R$), for $p < \$5.5$ or $p > \$8$. Since a negative price doesn't make sense, we delete any number to the left of 0. Thus, a loss will occur for $\$0 \le p < \$5.5$ or $p > \$8$. [$\$0, \$5.5) \cup (\$8, \infty)$ $\qquad (2\text{-}8)$

94.   Let x = the distance from port A to port B

Then $115 - x$ = the distance from port B to port C

Applying the Pythagorean theorem, we have

$$x^2 + (115 - x)^2 = 85^2$$
$$x^2 + 13,225 - 230x + x^2 = 7,225$$
$$2x^2 - 230x + 6,000 = 0$$
$$x^2 - 115x + 3,000 = 0$$
$$(x - 40)(x - 75) = 0$$
$$x = 40, 75$$
$$115 - x = 75, 40$$

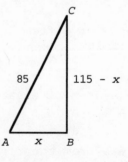

Based on the given information, there are two possible solutions:
40 miles from A to B and 75 miles from B to C *or*
75 miles from A to B and 40 miles from B to C $\qquad (2\text{-}6)$

95. $s = a + bt^2$

(A) We are given: When $t = 5$, $s = 2,100$

When $t = 10$, $s = 900$

Substituting these values into the given equation, we have

$2,100 = a + b(5)^2$

$900 = a + b(10)^2$ or

$2,100 = a + 25b$

$900 = a + 100b$

Solve the first equation for a in terms of b and substitute into the second equation to eliminate a.

$a = 2,100 - 25b$

$900 = 2,100 - 25b + 100b$

$-1,200 = 75b$

$b = -16$

$a = 2,100 - 25b = 2,100 - 25(-16) = 2,500$

(B) The height of the balloon is represented by s, the distance of the object above the ground, when $t = 0$. Since we now know

$s = 2,500 - 16t^2$

from part (A), when $t = 0$, $s = 2,500$ feet is the height of the balloon.

(C) The object falls until s, its distance above the ground, is zero. Since

$s = 2,500 - 16t^2$

we substitute $s = 0$ and solve for t.

$0 = 2,500 - 16t^2$

$16t^2 = 2,500$

$t^2 = \dfrac{2,500}{16}$

$t = \dfrac{50}{4}$ (discarding the negative solution)

$t = 12.5$ seconds

(2-1, 2-2, 2-6)

CHAPTER 3
Exercise 3-1
Key Ideas and Formulas

There is a one-to-one correspondence between the points in a plane and the elements in the set of all ordered pairs of real numbers.

The graph of an equation in two variables is the graph of its solution set, the set of all ordered pairs of real numbers (a, b) that satisfy the equation.

To sketch the graph of an equation, we include enough points from its solution set so that the total graph is apparent. We may use point-to-point plotting and aids to graphing, such as symmetry.

A graph is symmetric with respect to the y axis if $(-a, b)$ is on the graph whenever (a, b) is on the graph.

A graph is symmetric with respect to the x axis if $(a, -b)$ is on the graph whenever (a, b) is on the graph.

A graph is symmetric with respect to the origin if $(-a, -b)$ is on the graph whenever (a, b) is on the graph.

Test for symmetry with respect to the	if equation is equivalent when
y axis	x is replaced with $-x$
x axis	y is replaced with $-y$
origin	x and y are replaced with $-x$ and $-y$.

Distance Formula:

Distance between $P_1(x_1, y_1)$ and $P_2(x_2, y_2)$: $d(P_1, P_2) = \sqrt{(x_2 - x_1)^2 + (y_2 - y_1)^2}$

Standard Equations of a Circle

1. Circle with radius r and center at (h, k): $(x - h)^2 + (y - k)^2 = r^2$ $\quad r > 0$
2. Circle with radius r and center at $(0, 0)$: $x^2 + y^2 = r^2$ $\quad\quad\quad r > 0$

---

1. $y = 2x - 4$ $\quad$ y axis symmetry? $y = 2(-x) - 4$ $\quad$ $y = -2x - 4$ $\quad$ No, not equivalent
$\quad\quad\quad\quad\quad$ x axis symmetry? $-y = 2x - 4$ $\quad\quad$ $y = 4 - 2x$ $\quad\quad$ No, not equivalent
$\quad\quad\quad\quad\quad$ origin symmetry? $-y = 2(-x) - 4$ $\quad$ $-y = -2x - 4$
$\quad\quad\quad\quad\quad\quad\quad\quad\quad\quad\quad\quad\quad\quad\quad\quad\quad\quad\quad\quad$ $y = 2x + 4$ $\quad\quad$ No, not equivalent

x	y
0	-4
2	0
4	4

The graph has none of the three symmetry properties.

3. $y = \frac{1}{2}x$ y axis symmetry? $y = \frac{1}{2}(-x)$ $y = -\frac{1}{2}x$ No, not equivalent

x axis symmetry? $-y = \frac{1}{2}x$ $y = -\frac{1}{2}x$ No, not equivalent

origin symmetry? $-y = \frac{1}{2}(-x)$ $y = \frac{1}{2}x$ Yes, equivalent

x	y
0	0
4	2

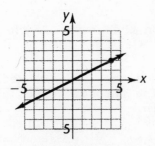

The graph has symmetry with respect to the origin.

We reflect the portion of the graph in quadrant I through the origin, using the origin symmetry.

5. $|y| = x$ y axis symmetry? $|y| = -x$ No, not equivalent

x axis symmetry? $|-y| = x$ $|y| = x$,   Yes, equivalent

origin symmetry? $|-y| = -x$ $|y| = -x$ No, not equivalent

x	y
0	0
4	4

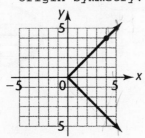

The graph has symmetry with respect to the x axis.

We reflect the portion of the graph in quadrant I through the x axis, using the x axis symmetry.

7. $|x| = |y|$ y axis symmetry? $|-x| = |y|$ $|x| = |y|$ Yes, equivalent

x axis symmetry? $|x| = |-y|$ $|x| = |y|$ Yes, equivalent

origin symmetry follows automatically.

x	y
0	0
4	4

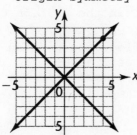

The graph has all three symmetries.

We reflect the portion of the graph in quadrant I through the y axis, and the x axis, and the origin, using all three symmetries.

9. $d = \sqrt{[3 - (-6)]^2 + [4 - (-4)]^2}$

$= \sqrt{(9)^2 + (8)^2} = \sqrt{145}$

Common Error: *not 9 + 8.*

11. $d = \sqrt{[(4) - (6)]^2 + [(-2) - (6)]^2}$

$= \sqrt{(-2)^2 + (-8)^2} = \sqrt{68}$

13. $(x - 0)^2 + (y - 0)^2 = 7^2$
$x^2 + y^2 = 49$

15. $(x - 2)^2 + (y - 3)^2 = 6^2$
$(x - 2)^2 + (y - 3)^2 = 36$

17. $[x - (-4)]^2 + (y - 1)^2 = (\sqrt{7})^2$
$(x + 4)^2 + (y - 1)^2 = 7$

Common Error: *not $(x - 4)^2$*

19. (A) When $x = -3$, the corresponding y value on the graph is 3, to the nearest integer.

(B) When $x = 2$, the corresponding y value on the graph is -2, to the nearest integer.

(C) Three values of x correspond to $y = 3$ on the graph. To the nearest integer they are -3, -1, and 4.

(D) Three values of x correspond to $y = -1$ on the graph. To the nearest integer, they are -4, 1, and 3.

21. (A) Reflect the given graph across the x axis.

(B) Reflect the given graph across the y axis.

(C) Reflect the given graph through the origin.

(D) Reflect the given graph across the y axis, then reflect the resulting curve across the x axis.

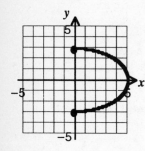
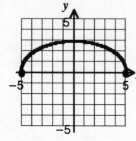
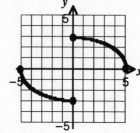
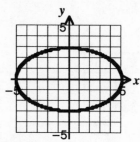

23. $y^2 = x + 2$ y axis symmetry? $y^2 = (-x) + 2$ $y^2 = 2 - x$ No, not equivalent

x axis symmetry? $(-y)^2 = x + 2$ $y^2 = x + 2$ Yes, equivalent

origin symmetry? $(-y)^2 = (-x) + 2$ $y^2 = 2 - x$ No, not equivalent

x	y
-2	0
-1	1
2	2

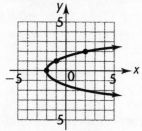

The graph has x axis symmetry.
To obtain the portion of the graph for $y \geq 0$, we sketch $y = \sqrt{x + 2}$, $x \geq -2$. We reflect the portion of the graph in quadrant I across the x axis, using the x axis symmetry.

25. $y = x^2 + 1$ y axis symmetry? $y = (-x)^2 + 1$ $y = x^2 + 1$ Yes, equivalent

x axis symmetry? $-y = x^2 + 1$ $y = -x^2 - 1$ No, not equivalent

origin symmetry? $-y = (-x)^2 + 1$ $y = -x^2 - 1$ No, not equivalent

x	y
0	1
1	2
2	5

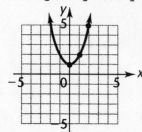

The graph has y axis symmetry.
We reflect the portion of the graph in quadrant I across the y axis, using the y axis symmetry.

27. $x^2 + 4y^2 = 4$ y axis symmetry? $(-x)^2 + 4y^2 = 4$ $x^2 + 4y^2 = 4$ Yes, equivalent

$4y^2 = 4 - x^2$ x axis symmetry? $x^2 + 4(-y)^2 = 4$ $x^2 + 4y^2 = 4$ Yes, equivalent

$y^2 = \dfrac{4 - x^2}{4}$ origin symmetry follows automatically.

$y = \pm\dfrac{1}{2}\sqrt{4 - x^2}$

x	y
0	1
1	$\dfrac{\sqrt{3}}{2} \approx .9$
2	0

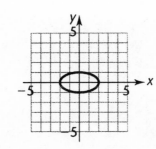

The graph has all three symmetries.
To obtain the quadrant I portion of the graph, we sketch $y = \dfrac{1}{2}\sqrt{4 - x^2}$,
$0 \le x \le 2$. We reflect this graph across the y axis, then reflect everything across the x axis.

29. $4y^2 - x^2 = 1$ y axis symmetry? $4y^2 - (-x)^2 = 1$ $4y^2 - x^2 = 1$ Yes, equivalent

$4y^2 = x^2 + 1$ x axis symmetry? $4(-y)^2 - x^2 = 1$ $4y^2 - x^2 = 1$ Yes, equivalent

$y^2 = \dfrac{x^2 + 1}{4}$ origin symmetry follows automatically.

$y = \pm\dfrac{1}{2}\sqrt{x^2 + 1}$

x	y
0	$\dfrac{1}{2}$
2	$\dfrac{1}{2}\sqrt{5} \approx 1.1$
4	$\dfrac{1}{2}\sqrt{17} \approx 2.0$

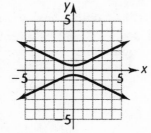

The graph has all three symmetries.
To obtain the quadrant I portion of the graph, we sketch $y = \dfrac{1}{2}\sqrt{x^2 + 1}$, $x \ge 0$.
We reflect this graph across the y axis, then reflect everything across the x axis.

31. $y^3 = x$ y axis symmetry? $y^3 = -x$ No, not equivalent

x axis symmetry? $(-y)^3 = x$ $y^3 = -x$ No, not equivalent

origin symmetry? $(-y)^3 = -x$ $y^3 = x$ Yes, equivalent

x	y
0	0
1	1
8	2

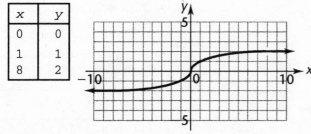

The graph has symmetry with respect to the origin.

We reflect the portion of the graph in quadrant I through the origin, using the origin symmetry.

33. $y = 0.6x^2 - 4.5$ y axis symmetry? $y = 0.6(-x)^2 - 4.5$ $y = 0.6x^2 - 4.5$ Yes, equivalent

x axis symmetry? $-y = 0.6x^2 - 4.5$ $y = -0.6x^2 + 4.5$ No, not equivalent

origin symmetry? $-y = 0.6(-x^2) - 4.5$ $y = -0.6x^2 + 4.5$ No, not equivalent

x	y
0	-4.5
1	-3.9
2	-2.1
3	0.9
4	5.1

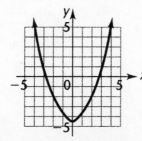

The graph has y axis symmetry.

We reflect the portion of the graph for $x \ge 0$ across the y axis, using the y axis symmetry.

35. $y = \sqrt{17 - x^2}$ y axis symmetry? $y = \sqrt{17 - (-x)^2}$ $y = \sqrt{17 - x^2}$ Yes, equivalent

x axis symmetry? $-y = \sqrt{17 - x^2}$ $y = -\sqrt{17 - x^2}$ No, not equivalent

origin symmetry? $-y = \sqrt{17 - (-x)^2}$ $y = -\sqrt{17 - x^2}$ No, not equivalent

x	y
0	$\sqrt{17} \approx 4.1$
1	$\sqrt{16} = 4$
2	$\sqrt{13} \approx 3.6$
3	$\sqrt{8} \approx 2.8$
4	$\sqrt{1} = 1$
$\sqrt{17}$	0

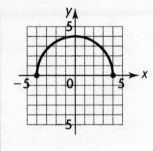

The graph has y axis symmetry.

We reflect the portion of the graph in quadrant I across the y axis, using the y axis symmetry.

37. $y = x^{2/3}$ y axis symmetry? $y = (-x)^{2/3}$ $y = x^{2/3}$ Yes, equivalent

x axis symmetry? $-y = x^{2/3}$ $y = -(x^{2/3})$ No, not equivalent

origin symmetry? $-y = (-x)^{2/3}$ $y = -(x^{2/3})$ No, not equivalent

x	y
0	0
1	1
2	$2^{2/3} \approx 1.6$
3	$3^{2/3} \approx 2.1$

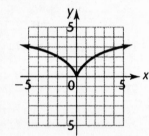

The graph has y axis symmetry.

We reflect the portion of the graph in quadrant I across the y axis, using the y axis symmetry.

39. The lengths of the three sides are found from the distance formula to be:

$$\sqrt{[1 - (-3)]^2 + [(-2) - (2)]^2} = \sqrt{4^2 + (-4)^2} = \sqrt{32}$$

$$\sqrt{[8 - 1]^2 + [5 - (-2)]^2} = \sqrt{7^2 + 7^2} = \sqrt{98}$$

$$\sqrt{[8 - (-3)]^2 + [5 - 2]^2} = \sqrt{11^2 + 3^2} = \sqrt{130}$$

Since $(\sqrt{130})^2 = (\sqrt{32})^2 + (\sqrt{98})^2$, the triangle is a right triangle.

> **Common Error:**
> Confusing these true statements with the *false* statement
> $\sqrt{130} = \sqrt{32} + \sqrt{98}$.

41. Perimeter = sum of lengths of all three sides

$= \sqrt{[1 - (-3)]^2 + [(-2) - 1]^2} + \sqrt{(4 - 1)^2 + [3 - (-2)]^2} + \sqrt{[4 - (-3)]^2 + (3 - 1)^2}$

$= \sqrt{16 + 9} + \sqrt{9 + 25} + \sqrt{49 + 4}$

$= \sqrt{25} + \sqrt{34} + \sqrt{53}$

$= 18.11$

43. The distance formula requires that $\sqrt{(x - 2)^2 + [8 - (-4)]^2} = 13$

Solving, we have

$$\sqrt{(x - 2)^2 + (12)^2} = 13$$
$$(x - 2)^2 + (12)^2 = (13)^2$$
$$(x - 2)^2 + 144 = 169$$
$$(x - 2)^2 = 25$$
$$x - 2 = \pm 5$$
$$x = 2 \pm 5$$
$$x = -3, 7$$

45. $(x + 4)^2 + (y - 2)^2 = 7$
$[x - (-4)]^2 + (y - 2)^2 = (\sqrt{7})^2$
Center $(-4, 2)$; Radius $= \sqrt{7}$

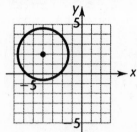

47. $x^2 + y^2 - 6x - 4y = 36$
$x^2 - 6x + y^2 - 4y = 36$
$x^2 - 6x + 9 + y^2 - 4y + 4 = 36 + 9 + 4$
$(x - 3)^2 + (y - 2)^2 = 49$
Center $(3, 2)$; Radius $= 7$

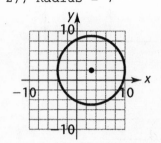

49. $x^2 + y^2 + 8x - 6y + 8 = 0$
$x^2 + 8x + y^2 - 6y = -8$
$x^2 + 8x + 16 + y^2 - 6y + 9 = -8 + 16 + 9$
$(x + 4)^2 + (y - 3)^2 = 17$
Center $(-4, 3)$; Radius $= \sqrt{17}$

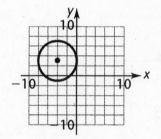

51. (A) & (B)

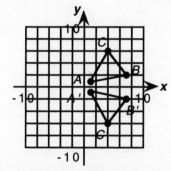

(C) Each point, and therefore the whole graph, is transformed into its reflection across the x axis.

53. (A) & (B)

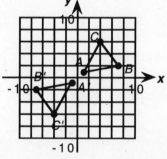

(C) Each point, and therefore the whole graph, is transformed into its reflection across the origin.

55. $x^2 + y^2 = 3$
$y^2 = 3 - x^2$
$y = \pm\sqrt{3 - x^2}$

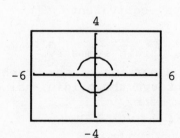

57. $(x + 3)^2 + (y + 1)^2 = 2$
$(y + 1)^2 = 2 - (x + 3)^2$
$y + 1 = \pm\sqrt{2 - (x + 3)^2}$
$y = -1 \pm \sqrt{2 - (x + 3)^2}$

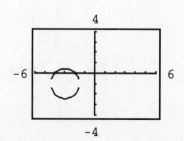

59.

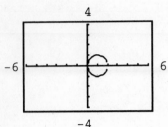

From the graph, the center of the circle is at $(1, 0)$ and its radius is 1. The equation of the circle is therefore $(x - 1)^2 + (y - 0)^2 = 1^2$ or
$$(x - 1)^2 + y^2 = 1$$
To check this observation, solve this equation for y in terms of x and note that this gives the original equations for $0 \le x \le 2$.

Solving for y, we obtain
$$y^2 = 1 - (x - 1)^2$$
$$y^2 = 1 - (x^2 - 2x + 1)$$
$$y^2 = 1 - x^2 + 2x - 1$$
$$y^2 = 2x - x^2$$
$$y = \pm\sqrt{2x - x^2}$$

61.

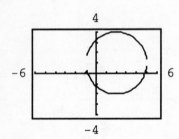

From the graph, the center of the circle is at $(2, 1)$ and its radius is 3. The equation of the circle is therefore $(x - 2)^2 + (y - 1)^2 = 3^2$ or
$$(x - 2)^2 + (y - 1)^2 = 9$$
To check this observation, solve this equation for y in terms of x and note that this gives the original equations for $-1 \le x \le 5$.

Solving for y, we obtain
$$(y - 1)^2 = 9 - (x - 2)^2$$
$$(y - 1)^2 = 9 - (x^2 - 4x + 4)$$
$$(y - 1)^2 = 9 - x^2 + 4x - 4$$
$$(y - 1)^2 = 5 + 4x - x^2$$
$$y - 1 = \pm\sqrt{5 + 4x - x^2}$$
$$y = 1 \pm \sqrt{5 + 4x - x^2}$$

63. $y^3 = |x|$ y axis symmetry? $y^3 = |-x|$ $y^3 = |x|$ Yes, equivalent

   x axis symmetry? $(-y)^3 = |x|$ $-y^3 = |x|$ No, not equivalent

   origin symmetry? $(-y)^3 = |-x|$ $-y^3 = |x|$ No, not equivalent

x	y
0	0
1	1
8	2

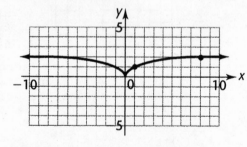

The graph has y axis symmetry.

We reflect the portion of the graph in quadrant I across the y axis, using the y axis symmetry.

65. $xy = 1$ y axis symmetry? $-xy = 1$ No, not equivalent

   x axis symmetry? $x(-y) = 1$ $-xy = 1$ No, not equivalent

   origin symmetry? $(-x)(-y) = 1$ $xy = 1$ Yes, equivalent

x	y
1	1
2	$\frac{1}{2}$
3	$\frac{1}{3}$
$\frac{1}{2}$	2
$\frac{1}{3}$	3

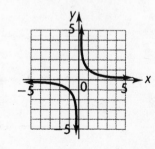

The graph has origin symmetry.

We reflect the portion of the graph in quadrant I through the origin, using the origin symmetry.

67. $y = 6x - x^2$ y axis symmetry? $y = 6(-x) - (-x)^2$ $y = -6x - x^2$ No, not equivalent
x axis symmetry? $-y = 6x - x^2$ $y = x^2 - 6x$ No, not equivalent
origin symmetry? $-y = 6(-x) - (-x)^2$ $y = x^2 + 6x$ No, not equivalent

x	y
-1	-7
0	0
1	5
2	8
3	9
4	8
5	5
6	0
7	-7

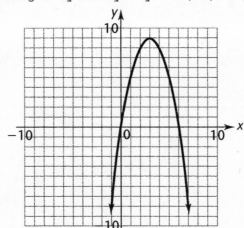

The graph has none of the three symmetries.

A larger table of values is needed since we have no symmetry information.

69. From geometry we know that the perpendicular bisector of a line segment is the set of points whose distances to the end points of the segment are equal. Using the distance formula, let (x, y) be a typical point on the perpendicular bisector. Then the distance from (x, y) to $(-6, -2)$ equals the distance from (x, y) to $(4, 4)$.

$$\sqrt{[x - (-6)]^2 + [y - (-2)]^2} = \sqrt{(x - 4)^2 + (y - 4)^2}$$
$$(x + 6)^2 + (y + 2)^2 = (x - 4)^2 + (y - 4)^2$$
$$x^2 + 12x + 36 + y^2 + 4y + 4 = x^2 - 8x + 16 + y^2 - 8y + 16$$
$$20x + 12y = -8$$
$$5x + 3y = -2$$

71. The center of the circle must be the midpoint of the diameter segment. Using the midpoint formula from problem 70, we have: center(= midpoint) has coordinates $\left(\dfrac{7 + 1}{2}, \dfrac{-3 + 7}{2}\right) = (4, 2)$. The radius of the circle must be the distance from the center to one of the endpoints of the diameter. It does not matter which endpoint we choose. For example,
$r = \sqrt{(7 - 4)^2 + [-3 - 2]^2} = \sqrt{3^2 + (-5)^2} = \sqrt{34}$.
The equation of the circle is
$(x - 4)^2 + (y - 2)^2 = (\sqrt{34})^2$ or $(x - 4)^2 + (y - 2)^2 = 34$.

73. (A) $6.00 on the price scale corresponds to 3,000 cases on the demand scale.

(B) The demand decreases from 3,000 to 2,600 cases, that is, by 400 cases.

(C) The demand increases from 3,000 to 3,600 cases, that is, by 600 cases.

(D) Demand decreases with increasing price and increases with decreasing price. To increase demand from 2,000 to 4,000 cases, a price decrease from $6.90 to $5.60 is necessary.

75. (A) 9:00 is halfway from 6 AM to noon, and corresponds to a temperature of 53°.

(B) The highest temperature occurs halfway from noon to 6 PM, at 3 PM. This temperature is 68°.

(C) This temperature occurs at 1 AM, 7 AM, and 11 PM.

77. (A) There is no obvious symmetry. A table of values yields the following approximate values:

x	0	0.5	1	1.5	2
v	0.7	0.6	0.5	0.35	0

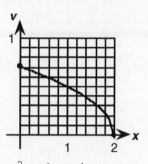

(C) For a displacement of 2 cm, the ball is stationary ($v = 0$). As the vertical displacement approaches 0, the ball gathers speed until $v = 0.5\sqrt{2} \approx 0.7$ m/sec. The velocity carries the ball through the equilibrium position ($x = 0$) to rise again to displacement 2 cm and velocity 0.

79. Using the hint, we note that $(2, r - 1)$ must satisfy $x^2 + y^2 = r^2$, that is

$$2^2 + (r - 1)^2 = r^2$$
$$4 + r^2 - 2r + 1 = r^2$$
$$-2r + 5 = 0$$
$$r = \frac{5}{2} \text{ or } 2.5 \text{ ft.}$$

81. (A) From the drawing, we can write:

$$\left(\begin{array}{c}\text{Distance from tower} \\ \text{to town } B\end{array}\right) = 2 \times \left(\begin{array}{c}\text{Distance from tower} \\ \text{to town } A\end{array}\right)$$

$$\left(\begin{array}{c}\text{Distance from } (x, y) \\ \text{to } (36, 15)\end{array}\right) = 2 \times \left(\begin{array}{c}\text{Distance from } (x, y) \\ \text{to } (0, 0)\end{array}\right)$$

$$\sqrt{(36 - x)^2 + (15 - y)^2} = 2\sqrt{(0 - x)^2 + (0 - y)^2}$$
$$\sqrt{(36 - x)^2 + (15 - y)^2} = 2\sqrt{x^2 + y^2}$$
$$(36 - x)^2 + (15 - y)^2 = 4(x^2 + y^2)$$
$$1,296 - 72x + x^2 + 225 - 30y + y^2 = 4x^2 + 4y^2$$
$$1,521 = 3x^2 + 3y^2 + 72x + 30y$$
$$507 = x^2 + y^2 + 24x + 10y$$
$$144 + 25 + 507 = x^2 + 24x + 144 + y^2 + 10y + 25$$
$$676 = (x + 12)^2 + (y + 5)^2$$

The circle has center $(-12, -5)$ and radius 26.

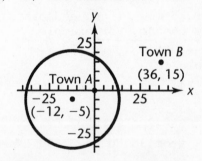

(B) All points due east of Town A have y coordinate 0 in this coordinate system. The points on the circle for which $y = 0$ are found by substituting $y = 0$ into the equation of the circle and solving for x.

$$(x + 12)^2 + (y + 5)^2 = 676$$
$$(x + 12)^2 + 25 = 676$$
$$(x + 12)^2 = 651$$
$$x + 12 = \pm\sqrt{651}$$
$$x = -12 \pm \sqrt{651}$$

Choosing the positive square root so that x is greater than -12 (east rather than west) we have $x = -12 + \sqrt{651} \approx 13.5$ miles.

Exercise 3-2

Key Ideas and Formulas

The graph of the equation

$$Ax + By = C$$

where A, B, and C are constants (A and B not both 0), is a straight line. Any straight line in a rectangular coordinate system has an equation of this form. If a line passes through two distinct points $P_1(x_1, y_1)$ and $P_2(x_2, y_2)$, its slope is given by

$$m = \frac{y_2 - y_1}{x_2 - x_1} \quad x_2 \neq x_1$$

$$= \frac{\text{vertical change}}{\text{horizontal change}} = \frac{\text{rise}}{\text{run}}.$$

For a vertical line, $x_2 = x_1$, hence slope is not defined. Equation: $x = a$. For a horizontal line, $y_2 = y_1$, slope = 0. Equation: $y = b$.

Slope-intercept form of an equation of a non-vertical line:

$$y = mx + b \quad m = \frac{\text{Rise}}{\text{Run}} = \text{Slope} \quad b = y \text{ intercept}$$

Point-slope form of an equation of a non-vertical line through (x_1, y_1) with slope m:
$$y - y_1 = m(x - x_1)$$

Given two nonvertical lines L_1 and L_2 with slopes m_1 and m_2, respectively, then

$$L_1 \text{ is parallel to } L_2 \ (L_1 \parallel L_2) \text{ if and only if } m_1 = m_2$$

$$L_1 \text{ is perpendicular to } L_2 \ (L_1 \perp L_2) \text{ if and only if } m_1 m_2 = -1. \ (m_1 = -\frac{1}{m_2})$$

---

1. The x intercept is -2. The y intercept is 2. From the point (-2, 0) to the point (0, 2), the value of y increases by 2 units as the value of x increases by 2 units. Thus slope = $\frac{\text{change in } y}{\text{change in } x} = \frac{2}{2} = 1$.

3. The x intercept is -2. The y intercept is -4. From the point (-2, 0) to the point (0, -4) the value of y decreases by 4 units as the value of x increases by 2 units. Thus, the slope = $\frac{\text{change in } y}{\text{change in } x} = \frac{-4}{2} = -2$.

5. The x intercept is 3. The y intercept is -1. From the point (0, -1) to the point (3, 0) the value of y increases by 1 unit as the value of x increases by 3 units. Thus, the slope = $\frac{\text{change in } y}{\text{change in } x} = \frac{1}{3}$.

7. $y = -\frac{3}{5}x + 4$ slope $-\frac{3}{5}$

x	y
0	4
5	1
-5	7

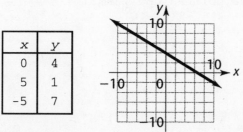

9. $y = -\frac{3}{4}x$ slope $-\frac{3}{4}$

x	y
0	0
4	-3
-4	3

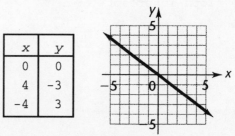

11. $2x - 3y = 15$

$\qquad -3y = -2x + 15$

$\qquad y = \frac{2}{3}x - 5$ slope $\frac{2}{3}$

x	y
0	-5
3	-3
-3	-7

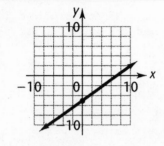

13. $4x - 5y = -24$

$\qquad -5y = -4x - 24$

$\qquad y = \frac{4}{5}x + \frac{24}{5}$ slope $\frac{4}{5}$

x	y
-1	4
4	8
9	12

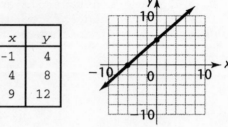

15. $\frac{y}{8} - \frac{x}{4} = 1$

$\qquad \frac{y}{8} = \frac{x}{4} + 1$

$\qquad y = 2x + 8$ slope 2

x	y
0	8
-4	0
-8	-8

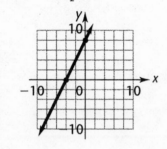

17. $x = -3$ slope not defined
vertical line

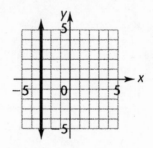

19. $y = 3.5$ slope 0
horizontal line

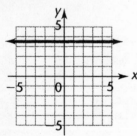

21. Slope and y intercept are given; we
use slope-intercept form.
$y = 1x + 0$
$y = x$

23. Slope and y intercept are given; we
use slope-intercept form.

$y = -\frac{2}{3}x + (-4)$

$y = -\frac{2}{3}x - 4$

25. A point and the slope are given; we
use point-slope form.
$y - 4 = -3(x - 0)$
$y - 4 = -3x$
$\qquad y = -3x + 4$

27. A point and the slope are given; we use point-slope form.

$$y - 4 = -\frac{2}{5}[x - (-5)]$$

$$y - 4 = -\frac{2}{5}[x + 5]$$

$$y - 4 = -\frac{2}{5}x - 2$$

$$y = -\frac{2}{5}x + 2$$

29. A point and the slope are given; we use point-slope form.

$$y - 5 = 0(x - 5)$$
$$y - 5 = 0$$
$$y = 5$$

31. Two points are given; we first find slope, then use point-slope form.

$$m = \frac{-2 - 6}{5 - 1} = \frac{-8}{4} = -2$$

$$y - 6 = (-2)(x - 1)$$
$$y - 6 = -2x + 2$$
$$y = -2x + 8$$

or

$$y - (-2) = (-2)(x - 5)$$
$$y + 2 = -2x + 10$$
$$y = -2x + 8$$

Thus, it does not matter which point is chosen in substituting into the point-slope form; both points must give rise to the same equation.

33. We proceed as in problem 31.

$$m = \frac{8 - 0}{-4 - 2} = \frac{8}{-6} = -\frac{4}{3}$$

$$y - 0 = -\frac{4}{3}(x - 2)$$

$$y = -\frac{4}{3}x + \frac{8}{3}$$

35. We proceed as in problem 31.

$$m = \frac{4 - 4}{5 - (-3)} = \frac{0}{8} = 0$$

$$y - 4 = 0(x - 5)$$
$$y - 4 = 0$$
$$y = 4$$

37. We proceed as in problem 31.

$$m = \frac{-3 - 6}{4 - 4} = \frac{-9}{0}$$

slope is undefined.
A vertical line through (4, 6) has equation $x = 4$.

39. We proceed as in problem 31, using (6, 0) and (0, 2) as the two given points.

$$m = \frac{0 - 2}{6 - 0} = \frac{-2}{6} = -\frac{1}{3}$$

$$y - 0 = -\frac{1}{3}(x - 6)$$

$$y = -\frac{1}{3}x + 2$$

41. We proceed as in problem 31, using (-4, 0) and (0, 3) as the two given points.

$$m = \frac{3 - 0}{0 - (-4)} = \frac{3}{4}$$

$$y - 0 = \frac{3}{4}[x - (-4)]$$

$$y = \frac{3}{4}x + 3$$

43. A line parallel to $y = 3x - 5$ will have the same slope, namely 3. We now use the point slope form.

$$y - 4 = 3[x - (-3)]$$
$$y - 4 = 3x + 9$$
$$y = 3x + 13$$
$$-3x + y = 13$$
$$3x - y = -13$$

45. A line perpendicular to
$y = -\frac{1}{3}x$ will have slope satisfying

$-\frac{1}{3}m = -1$, or $m = 3$. We use the
point-slope form.
$$y - (-3) = 3(x - 2)$$
$$y + 3 = 3x - 6$$
$$9 = 3x - y$$
$$3x - y = 9$$

47. The equation of the vertical line through (2, 5) is $x = 2$.

49. The equation of the vertical line through (3, -2) is $x = 3$.

51. A line parallel to $3x - 2y = 4$ will have the same slope. The slope of $3x - 2y = 4$, or $2y = 3x - 4$, or $y = \frac{3}{2}x - 2$ is $\frac{3}{2}$. We use the point-slope form,

$$y - 0 = \frac{3}{2}(x - 5)$$
$$y = \frac{3}{2}x - \frac{15}{2}$$
$$\frac{15}{2} = \frac{3}{2}x - y$$
$$15 = 3x - 2y$$
$$3x - 2y = 15$$

(Alternatively, we could notice that a line parallel to $3x - 2y = 4$ will have an equation of the form $3x - 2y = C$. Since the required line must contain $(5, 0)$, $(5, 0)$ must satisfy its equation. Therefore, $3(5) - 2(0) = C$. Since $15 = C$, the equation desired is $3x - 2y = 15$.)

53. A line perpendicular to $x + 3y = 9$, which has slope $-\frac{1}{3}$, will have slope satisfying $-\frac{1}{3}m = -1$, or $m = 3$. We use the point-slope form.

$$y - (-4) = 3(x - 0)$$
$$y + 4 = 3x$$
$$4 = 3x - y$$
$$3x - y = 4$$

55.

x	$y = 2x + 2$	$y = \frac{1}{2}x + 2$	$y = 0x + 2$	$y = -\frac{1}{2}x + 2$	$y = -2x + 2$
0	2	2	2	2	2
2	6	3	2	1	-2
-2	-2	1	2	3	6

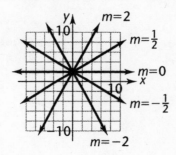

57. slope of $AB = \dfrac{-1 - 2}{4 - 0} = -\dfrac{3}{4}$

slope of $DC = \dfrac{-5 - (-2)}{1 - (-3)} = -\dfrac{3}{4}$

Therefore $AB \parallel DC$.

59. slope of $AB = -\dfrac{3}{4}$

slope of $BC = \dfrac{-5 - (-1)}{1 - 4} = \dfrac{4}{3}$

(slope AB)(slope BC) $= \left(-\dfrac{3}{4}\right)\left(\dfrac{4}{3}\right) = -1$

Therefore $AB \perp BC$.

61. midpoint of $AD = \left(\dfrac{0 + (-3)}{2}, \dfrac{2 + (-2)}{2}\right) = \left(-\dfrac{3}{2}, 0\right)$

slope of $AD = \dfrac{-2 - 2}{-3 - 0} = \dfrac{4}{3}$

We require the equation of a line, through the midpoint of AD, which is perpendicular to AD. Its slope will satisfy $\frac{4}{3}m = -1$, or $m = -\frac{3}{4}$. We use the point-slope form.

$$y - 0 = -\frac{3}{4}\left[x - \left(-\frac{3}{2}\right)\right]$$
$$y = -\frac{3}{4}\left(x + \frac{3}{2}\right)$$
$$y = -\frac{3}{4}x - \frac{9}{8}$$
$$8y = -6x - 9$$
$$6x + 8y = -9$$

63. The circle has center (0, 0). The radius drawn from (0, 0) to the given point (3, 4) has slope given by

$$M_R = \frac{4 - 0}{3 - 0} = \frac{4}{3}$$

Therefore, the slope of the tangent line is given by

$$\left(\frac{4}{3}\right)m = -1 \quad \text{or} \quad m = -\frac{3}{4}$$

We require the equation of a line through (3, 4) with slope $-\frac{3}{4}$. We use the point-slope form.

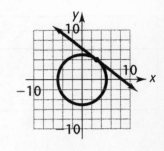

$$y - 4 = -\frac{3}{4}(x - 3)$$

$$y - 4 = -\frac{3}{4}x + \frac{9}{4}$$

$$4y - 16 = -3x + 9$$

$$3x + 4y = 25$$

65. The circle has center (0, 0). The radius drawn from (0, 0) to the given point (5, -5) has slope given by

$$M_R = \frac{-5 - 0}{5 - 0} = -1$$

Therefore, the slope of the tangent line is given by

$$(-1)m = -1 \quad \text{or} \quad m = 1$$

We require the equation of a line through (5, -5) with slope 1. We use the point-slope form.

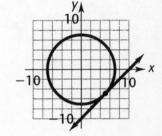

$$y - (-5) = 1(x - 5)$$

$$y + 5 = x - 5$$

$$x - y = 10$$

67. The circle has center (3, -4). The radius drawn from (3, -4) to the given point (8, -16) has slope given by

$$M_R = \frac{-16 - (-4)}{8 - 3} = -\frac{12}{5}$$

Therefore, the slope of the tangent line is given by

$$\left(-\frac{12}{5}\right)m = -1 \quad \text{or} \quad m = \frac{5}{12}$$

We require the equation of a line through (8, -16) with slope $\frac{5}{12}$. We use the point-slope form.

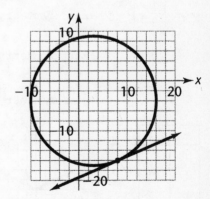

$$y - (-16) = \frac{5}{12}(x - 8)$$

$$y + 16 = \frac{5}{12}x - \frac{40}{12}$$

$$12y + 192 = 5x - 40$$

$$232 = 5x - 12y$$

69. (A) Solve each equation for y in terms of x, then graph.

$$3x + 2y = 6 \qquad 3x + 2y = 6 \qquad 3x + 2y = 3 \qquad 3x + 2y = -3$$
$$2y = 6 - 3x \qquad 2y = -6 - 3x \qquad 2y = 3 - 3x \qquad 2y = -3 - 3x$$
$$y = \frac{6 - 3x}{2} \qquad y = \frac{-6 - 3x}{2} \qquad y = \frac{3 - 3x}{2} \qquad y = \frac{-3 - 3x}{2}$$

x	$y = \dfrac{6 - 3x}{2}$	$y = \dfrac{-6 - 3x}{2}$	$y = \dfrac{3 - 3x}{2}$	$y = \dfrac{-3 - 3x}{2}$
0	3	-3	$\frac{3}{2}$	$-\frac{3}{2}$
2	0	-6	$-\frac{3}{2}$	$-\frac{9}{2}$
-2	6	0	$\frac{9}{2}$	$\frac{3}{2}$

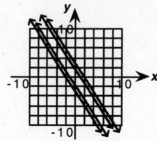

(B) The family of lines all have slope $y = -\dfrac{A}{B}$ and y intercept $\dfrac{C}{B}$. The slopes are therefore equal and the lines are parallel.

(C) Solve $Ax + By = C$ for y in terms of x.

$$By = -Ax + C$$
$$y = \frac{-Ax + C}{B}$$
$$y = -\frac{A}{B}x + \frac{C}{B}$$

For fixed A and B and varying C this is a family of lines as described in part (B).

71. $y = \left| \dfrac{1}{2}x \right|$ y axis symmetry? $y = \left| \dfrac{1}{2}(-x) \right|$ $y = \left| \dfrac{1}{2}x \right|$ Yes, equivalent

x axis symmetry? $-y = \left| \dfrac{1}{2}x \right|$ $y = -\left| \dfrac{1}{2}x \right|$ No, not equivalent

origin symmetry? $-y = \left| \dfrac{1}{2}(-x) \right|$ $y = -\left| \dfrac{1}{2}x \right|$ No, not equivalent

x	y
0	0
2	1
4	2

The graph has y axis symmetry.

We reflect the portion of the graph in quadrant I through the y axis, using the y axis symmetry.

73. $y = 2|x| - 4$ y axis symmetry $y = 2|-x| - 4$ $y = 2|x| - 4$ Yes, equivalent
   x axis symmetry $-y = 2|x| - 4$ $y = -2|x| + 4$ No, not equivalent
   origin symmetry $-y = 2|-x| - 4$ $y = -2|x| + 4$ No, not equivalent

x	y
0	-4
1	-2
2	0
3	2

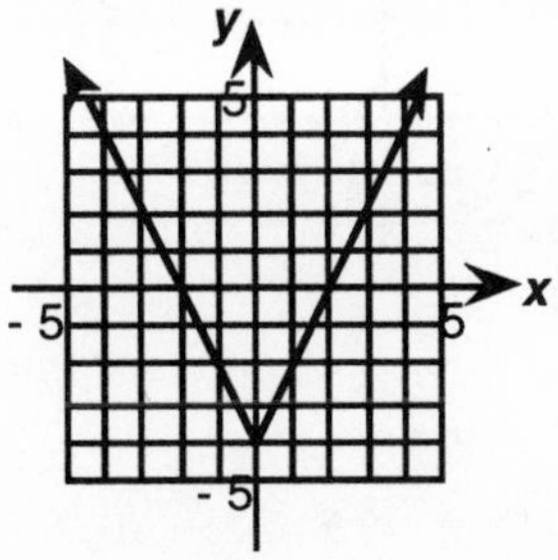

The graph has y axis symmetry. We reflect the portion of the graph to the right of the y axis through the y axis, using the y axis symmetry.

75. $x^2 - y^2 = 0$ y axis symmetry? $(-x)^2 - y^2 = 0$ $x^2 - y^2 = 0$ Yes, equivalent
   x axis symmetry? $x^2 - (-y)^2 = 0$ $x^2 - y^2 = 0$ Yes, equivalent
   origin symmetry follows automatically.

x	y
0	0
4	4

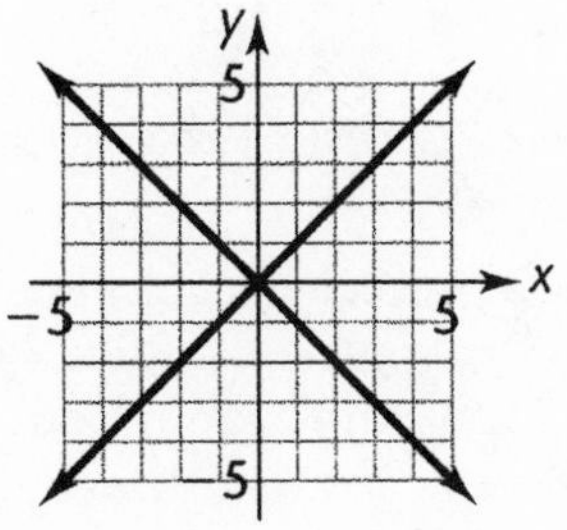

The graph has all three symmetries.

We reflect the portion of the graph in quadrant I through the y axis, and the x axis, and the origin, using all three symmetries.

77. The graph of $y = |mx + b|$ is the same as the graph of $y = mx + b$ for $mx + b$ positive, that is, for $x > -\dfrac{b}{m}$ (m positive) or for $x < -\dfrac{b}{m}$ (m negative). It is however, the reflection of the graph of $y = mx + b$ across the x axis for the remaining portion of the graph. Thus the graph has the shape of a v with vertex at $\left(-\dfrac{b}{m}, 0\right)$.

79. Two points given; we first find slope, then use point-slope form.
$$m = \frac{0 - b}{a - 0} = -\frac{b}{a} \quad a \neq 0$$
$$y - b = -\frac{b}{a}(x - 0)$$
$$y - b = -\frac{bx}{a}$$
Divide both sides by b, then ($b \neq 0$)
$$\frac{y - b}{b} = -\frac{x}{a}$$
$$\frac{y}{b} - 1 = -\frac{x}{a}$$
$$\frac{y}{b} = 1 - \frac{x}{a}$$
$$\frac{x}{a} + \frac{y}{b} = 1$$

81. (A)

x	0	5,000	10,000	15,000	20,000	25,000	30,000
$212 - 0.0018x = B$	212	203	194	185	176	167	158

(B) The boiling point drops 9°F for each 5,000 foot increase in altitude.

83. The rental charges are $25 per day plus $0.25 per mile driven.

85. (A)

x	0	1	2	3	4
Sales	5.9	6.5	7.7	8.6	9.7
$5.74 + 0.97x = y$	5.7	6.7	7.7	8.6	9.6

(B)

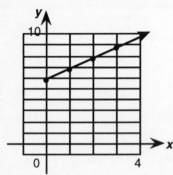

(C) Since 1988 corresponds to $x = 0$, 1993 corresponds to $x = 5$ and 2000 corresponds to $x = 12$.

When $x = 5$, $y = 5.74 + 0.97(5)$
$y \approx 10.6$ (billion dollars)

When $x = 12$, $y = 5.74 + 0.97(12)$
$y \approx 17.4$ (billion dollars)

(D) The sales rose approximately linearly from $5.9 billion in 1988 to $9.7 billion in 1992, a rate of 0.97 billion per year.

87. (A) If F is linearly related to C, then we are looking for an equation whose graph passes through $(C_1, F_1) = (0, 32)$ and $(C_2, F_2) = (100, 212)$. We find the slope, and then use the point-slope form to find the equation.

$$m = \frac{F_2 - F_1}{C_2 - C_1} = \frac{212 - 32}{100 - 0} = \frac{180}{100} = \frac{9}{5}$$

$$F - F_1 = m(C - C_1)$$
$$F - 32 = \frac{9}{5}(C - 0)$$
$$F - 32 = \frac{9}{5}C$$
$$F = \frac{9}{5}C + 32$$

(B) We are asked for F when $C = 20$.

$$F = \frac{9}{5}(20) + 32 = 36 + 32 = 68°$$

We are then asked for C when $F = 86°$

$$86 = \frac{9}{5}C + 32$$
$$430 = 9C + 160$$
$$270 = 9C$$
$$C = 30°$$

(C) The slope is $m = \frac{9}{5}$.

89. (A) If V is linearly related to t, then we are looking for an equation whose graph passes through $(t_1, V_1) = (0, 8,000)$ and $(t_2, V_2) = (5, 0)$. We find the slope, and then we use the point-slope form to find the equation.

$$m = \frac{V_2 - V_1}{t_2 - t_1} = \frac{0 - 8,000}{5 - 0} = \frac{-8,000}{5} = -1,600$$

$$V - V_1 = m(t - t_1)$$
$$V - 8,000 = -1,600(t - 0)$$
$$V - 8,000 = -1,600t$$
$$V = -1,600t + 8,000$$

(B) We are asked for V when $t = 3$

$$V = -1,600(3) + 8,000$$
$$V = -4,800 + 8,000$$
$$V = 3,200$$

(C) The slope is $m = -1,600$

91. (A) If T is linearly related to A, then we are looking for an equation whose graph passes through $(A_1, T_1) = (0, 70)$ and $(A_2, T_2) = (18, -20)$. We find the slope, and then we use the point-slope form to find the equation.

$$m = \frac{T_2 - T_1}{A_2 - A_1} = \frac{(-20) - 70}{18 - 0} = \frac{-90}{18} = -5$$

$$T - T_1 = m(A - A_1)$$
$$T - 70 = -5(A - 0)$$
$$T - 70 = -5A$$
$$T = -5A + 70 \quad A \geq 0$$

(B) We are asked for A when $T = 0$.

$$0 = -5A + 70$$
$$5A = 70$$
$$A = 14 \text{ thousand feet}$$

(C) The slope is $m = -5$; the temperature changes $-5°F$ for each 1,000 foot rise in altitude.

93. (A) If h is linearly related to t, then we are looking for an equation whose graph passes through $(t_1, h_1) = (9, 23)$ and $(t_2, h_2) = (24, 40)$. We find the slope, and then we use the point-slope form to find the equation.

$$m = \frac{h_2 - h_1}{t_2 - t_1} = \frac{40 - 23}{24 - 9} = \frac{17}{15} \approx 1.13$$

$$h - h_1 = m(t - t_1)$$

$$h - 23 = \frac{17}{15}(t - 9)$$

$$h - 23 = \frac{17}{15}t - \frac{51}{5}$$

$$h = \frac{17}{15}t - \frac{51}{5} + 23$$

$$h = 1.13t + 12.8$$

(B) We are asked for t when $h = 50$.

$$50 = 1.13t + 12.8$$
$$37.2 = 1.13t$$
$$t = 32.9 \text{ hours}$$

95. (A) If R is linearly related to C, then we are looking for an equation whose graph passes through $(C_1, R_1) = (210, 0.160)$ and $(C_2, R_2) = (231, 0.192)$. We find the slope, and then we use the point-slope form to find the equation.

$$m = \frac{R_2 - R_1}{C_2 - C_1} = \frac{0.192 - 0.160}{231 - 210} = \frac{0.032}{21} = 0.00152$$

$$R - R_1 = m(C - C_1)$$
$$R - 0.160 = 0.00152(C - 210)$$
$$R - 0.160 = 0.00152C - 0.319$$
$$R = 0.00152C - 0.159, \quad C \geq 210$$

(B) We are asked for R when $C = 260$.

$$R = 0.00152(260) - 0.159$$
$$R = 0.236$$

(C) The slope is $m = 0.00152$; coronary risk increases 0.00152 per unit increase in cholesterol above the 210 cholesterol level.

Exercise 3-3

Key Ideas and Formulas

A function is a rule that produces a correspondence between two sets of elements such that to each element in the first set there corresponds one and only one element in the second set. The first set is called the domain and the second set is called the range.

In an equation in two variables, if to each value of the independent variable (input) there corresponds exactly one value of the dependent variable (output), then the equation defines a function. If there is any value of the independent variable to which there corresponds more than one value of the dependent variable, then the equation does not define a function.

An equation defines a function if each vertical line in the coordinate system passes through at most one point on the graph of the equation. If any vertical line passes through two or more points on the graph of an equation, then the equation does not define a function. (Vertical Line Test)

If a function is defined by an equation and the domain is not indicated, we assume that the domain is the set of all real number replacements of the independent variable (inputs) that produce real values for the dependent variable (outputs).

The symbol $f(x)$ represents the real number in the range of the function f corresponding to the domain value x. Symbolically, $f:x \to f(x)$. The ordered pair $(x, f(x))$ belongs to the function f. If x is a real number that is not in the domain of f, then f is not defined at x and $f(x)$ does not exist.

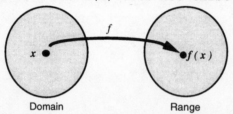

Domain   Range

---

1. A function   **3.** Not a function (two range values correspond to some domain values)

5. A function   **7.** A function; domain = {2, 3, 4, 5}; range = {4, 6, 8, 10}

9. Not a function (two range values correspond to some domain values)

11. A function; domain = {0, 1, 2, 3, 4, 5}; range = {1, 2}

13. A function   **15.** Not a function (fails vertical line test)

17. Not a function (fails vertical line test)   **19.** $f(-1) = 3(-1) - 5 = -8$

21. $G(-2) = (-2) - (-2)^2 = -6$ **23.** $F(-1) + f(3) = 3(-1)^2 + 2(-1) - 4 + 3(3) - 5$
$$= 3 - 2 - 4 + 9 - 5 = 1$$

25. $2F(-2) - G(-1) = 2[3(-2)^2 + 2(-2) - 4] - [(-1) - (-1)^2]$
$$= 2[12 - 4 - 4] - [-1 - 1] = 10$$

27. $\dfrac{f(0) \cdot g(-2)}{F(-3)} = \dfrac{[3(0) - 5] \cdot [4 - (-2)]}{3(-3)^2 + 2(-3) - 4} = \dfrac{(-5)(6)}{27 - 6 - 4} = -\dfrac{30}{17}$

29. When $x = -4$, $y = 7$, so $f(-4) = 7$

31. When $y = 4$, x can equal -5, -1, or 6.

33. Solving for the dependent variable y, we have

$$2x - 5y = 20$$
$$-5y = 20 - 2x$$
$$y = \frac{2x - 20}{5}$$

Since $\frac{2x - 20}{5}$ is a real number for each real number x, the equation defines a function with domain R.

35. Solving for the dependent variable y, we have

$$y^2 - x = 2$$
$$y^2 = x + 2$$
$$y = \pm\sqrt{x + 2}$$

Since each positive number has two real square roots, the equation does not define a function. For example, when $x = 2$, $y = \pm\sqrt{2 + 2} = \pm\sqrt{4} = \pm 2$.

37. Solving for the dependent variable y, we have

$$|x| + y^2 = 5$$
$$y^2 = 5 - |x|$$
$$y = \pm\sqrt{5 - |x|}$$

Since each positive number has two real square roots, the equation does not define a function. For example, when $x = 1$, $y = \pm\sqrt{5 - |1|} = \pm\sqrt{4} = \pm 2$.

39. Solving for the dependent variable y, we have

$$xy - 5y = 10$$
$$y(x - 5) = 10$$
$$y = \frac{10}{x - 5}$$

Since $\frac{10}{x - 5}$ is a real number for each real number x, except when $x - 5 = 0$, that is, $x = 5$, the equation defines a function with domain all real numbers except $x = 5$.

41. Solving for the dependent variable y, we have

$$x^2 + y^2 = 81$$
$$y^2 = 81 - x^2$$
$$y = \pm\sqrt{81 - x^2}$$

Since each positive number has two real square roots, the equation does not define a function. For example, when $x = 0$, $y = \pm\sqrt{81 - 0} = \pm\sqrt{81} = \pm 9$.

43. The domain is all real numbers, since $3x + 8$ represents a real number for all replacements of x by real numbers.

45. The domain is the set of all real numbers x such that $\sqrt{x + 2}$ is a real number, that is, such that $x + 2 \geq 0$, or $x \geq -2$.

47. $\frac{2 + 3x}{4 - x}$ represents a real number for all replacements of x by real numbers except for $x = 4$. All real numbers except 4.

> **Common Error:** $3x + 2 = 0$, $x = -\frac{2}{3}$ is *not* excluded from the domain. Zero *can* be divided by a non-zero number.

49. $\frac{x^2 - 2x + 9}{x^2 - 2x - 8} = \frac{x^2 - 2x + 9}{(x - 4)(x + 2)}$. This represents a real number for all replacements of x by real numbers except -2 and 4 (division by zero is not defined). All real numbers except -2 and 4.

51. The domain is the set of all real numbers x such that $\sqrt{4 - x^2}$ is a real number, that is, such that $4 - x^2 \geq 0$. Solving by the methods of Section 2-8, we obtain the sign chart:

$$\xrightarrow{\quad -- \quad \overset{\bullet}{\underset{-2}{}} \quad ++ \quad \overset{\bullet}{\underset{2}{}} \quad -- \quad} x$$

Thus, $4 - x^2$ is non-negative and $\sqrt{4 - x^2}$ is real for $-2 \leq x \leq 2$.

53. The domain is the set of all real numbers x such that $\sqrt{x^2 - 3x - 4}$ is a real number, that is, such that $x^2 - 3x - 4 = (x - 4)(x + 1) \geq 0$. Solving by the methods of Section 2-8, we obtain the sign chart:

$$\xrightarrow{\quad ++ \quad \overset{\bullet}{\underset{-1}{}} \quad -- \quad \overset{\bullet}{\underset{4}{}} \quad ++ \quad} x$$

Thus, $x^2 - 3x - 4$ is non-negative and $\sqrt{x^2 - 3x - 4}$ is real for $x \leq -1$ or $x \geq 4$.

55. The domain is the set of all real numbers x such that $\sqrt{\dfrac{5 - x}{x - 2}}$ is a real number, that is, such that $\dfrac{5 - x}{x - 2} \geq 0$. Solving by the methods of Section 2-8, we obtain the sign chart:

$$\xrightarrow{\quad -- \quad \overset{\circ}{\underset{2}{}} \quad ++ \quad \overset{\bullet}{\underset{5}{}} \quad -- \quad} x$$

Thus, $\dfrac{5 - x}{x - 2}$ is non-negative and $\sqrt{\dfrac{5 - x}{x - 2}}$ is real for $2 < x \leq 5$.

57. $g(x) = 2x^3 - 5$ **59.** $G(x) = 2\sqrt{x} - x^2$

61. Function f multiplies the domain element by 2 and subtracts 3 from the result.

63. Function F multiplies the cube of the domain element by 3 and subtracts twice the square root of the domain element from the result.

65.
$$F(s) = 3s + 15$$
$$F(2 + h) = 3(2 + h) + 15$$
$$F(2) = 3(2) + 15$$
$$\frac{F(2 + h) - F(2)}{h} = \frac{[3(2 + h) + 15] - [3(2) + 15]}{h} = \frac{[6 + 3h + 15] - [21]}{h}$$
$$= \frac{3h + 21 - 21}{h} = \frac{3h}{h} = 3$$

67.
$$g(x) = 2 - x^2$$
$$g(3 + h) = 2 - (3 + h)^2$$
$$g(3) = 2 - (3)^2$$
$$\frac{g(3 + h) - g(3)}{h} = \frac{[2 - (3 + h)]^2 - [2 - (3)^2]}{h} = \frac{[2 - 9 - 6h - h^2] - [-7]}{h}$$
$$= \frac{-6h - h^2}{h} = \frac{h(-6 - h)}{h} = -6 - h$$

69.
$$L(w) = -2w^2 + 3w - 1$$
$$L(-2 + h) = -2(-2 + h)^2 + 3(-2 + h) - 1$$
$$L(-2) = -2(-2)^2 + 3(-2) - 1$$
$$\frac{L(-2 + h) - L(-2)}{h} = \frac{[-2(-2 + h)^2 + 3(-2 + h) - 1] - [-2(-2)^2 + 3(-2) - 1]}{h}$$
$$= \frac{[-2(4 - 4h + h^2) - 6 + 3h - 1] - [-15]}{h}$$
$$= \frac{-8 + 8h - 2h^2 - 6 + 3h - 1 + 15}{h}$$
$$= \frac{11h - 2h^2}{h} = \frac{h(11 - 2h)}{h} = 11 - 2h$$

71. To find $f(x)$, replace $x + h$ by x wherever $x + h$ occurs. $f(x) = 2x^2 - 4x + 6$.

73. To find $m(x)$, replace $x + h$ by x wherever $x + h$ occurs. $m(x) = 4x - 3\sqrt{x} + 9$.

75. (A)

$$f(x) = 3x - 4$$
$$f(x + h) = 3(x + h) - 4$$

$$\frac{f(x + h) - f(x)}{h} = \frac{[3(x + h) - 4] - [3x - 4]}{h}$$
$$= \frac{3x + 3h - 4 - 3x + 4}{h}$$
$$= \frac{3h}{h}$$
$$= 3$$

(B)

$$f(x) = 3x - 4$$
$$f(a) = 3a - 4$$

$$\frac{f(x) - f(a)}{x - a} = \frac{(3x - 4) - (3a - 4)}{x - a}$$
$$= \frac{3x - 4 - 3a + 4}{x - a}$$
$$= \frac{3x - 3a}{x - a}$$
$$= \frac{3(x - a)}{x - a}$$
$$= 3$$

77. (A)

$$f(x) = x^2 - 1$$
$$f(x + h) = (x + h)^2 - 1$$

$$\frac{f(x + h) - f(x)}{h} = \frac{[(x + h)^2 - 1] - [x^2 - 1]}{h}$$
$$= \frac{x^2 + 2xh + h^2 - 1 - x^2 + 1}{h}$$
$$= \frac{2xh + h^2}{h}$$
$$= \frac{h(2x + h)}{h}$$
$$= 2x + h$$

(B)

$$f(x) = x^2 - 1$$
$$f(a) = a^2 - 1$$

$$\frac{f(x) - f(a)}{x - a} = \frac{(x^2 - 1) - (a^2 - 1)}{x - a}$$
$$= \frac{x^2 - 1 - a^2 + 1}{x - a}$$
$$= \frac{x^2 - a^2}{x - a}$$
$$= \frac{(x - a)(x + a)}{x - a}$$
$$= x + a$$

79. (A)

$$f(x) = -3x^2 + 9x - 12$$
$$f(x + h) = -3(x + h)^2 + 9(x + h) - 12$$

$$\frac{f(x + h) - f(x)}{h} = \frac{[-3(x + h)^2 + 9(x + h) - 12] - [-3x^2 + 9x - 12]}{h}$$
$$= \frac{-3(x^2 + 2xh + h^2) + 9x + 9h - 12 + 3x^2 - 9x + 12}{h}$$
$$= \frac{-3x^2 - 6xh - 3h^2 + 9x + 9h - 12 + 3x^2 - 9x + 12}{h}$$
$$= \frac{-6xh - 3h^2 + 9h}{h}$$
$$= \frac{h(-6x - 3h + 9)}{h}$$
$$= -6x - 3h + 9$$

(B)

$$f(x) = -3x^2 + 9x - 12$$
$$f(a) = -3a^2 + 9a - 12$$

$$\frac{f(x) - f(a)}{x - a} = \frac{(-3x^2 + 9x - 12) - (-3a^2 + 9a - 12)}{x - a}$$
$$= \frac{-3x^2 + 9x - 12 + 3a^2 - 9a + 12}{x - a}$$
$$= \frac{-3x^2 + 3a^2 + 9x - 9a}{x - a}$$
$$= \frac{(x - a)(-3x - 3a) + 9(x - a)}{x - a}$$
$$= \frac{(x - a)(-3x - 3a + 9)}{x - a}$$
$$= -3x - 3a + 9$$

81. (A)
$$f(x) = x^3$$
$$f(x + h) = (x + h)^3$$
$$\frac{f(x + h) - f(x)}{h} = \frac{(x + h)^3 - x^3}{h}$$
$$= \frac{x^3 + 3x^2h + 3xh^2 + h^3 - x^3}{h}$$
$$= \frac{3x^2h + 3xh^2 + h^3}{h}$$
$$= \frac{h(3x^2 + 3xh + h^2)}{h}$$
$$= 3x^2 + 3xh + h^2$$

(B)
$$f(x) = x^3$$
$$f(a) = a^3$$
$$\frac{f(x) - f(a)}{x - a} = \frac{x^3 - a^3}{(x - a)}$$
$$= \frac{(x - a)(x^2 + xa + a^2)}{x - a}$$
$$= x^2 + ax + a^2$$

83. Given w = width and Area = 64, we use $A = \ell w$ to write $\ell = \dfrac{A}{w} = \dfrac{64}{w}$.

Then $P = 2w + 2\ell = 2w + 2\left(\dfrac{64}{w}\right) = 2w + \dfrac{128}{w}$. Since w must be positive, the domain of $P(w)$ is $w > 0$.

85.

Using the given letters, the Pythagorean theorem gives
$$h^2 = b^2 + 5^2$$
$$h^2 = b^2 + 25$$
$$h = \sqrt{b^2 + 25} \text{ (since } h \text{ is positive)}$$
Since b must be positive, the domain of $h(b)$ is $b > 0$.

87. Daily cost = fixed cost + variable cost
$$C(x) = \$300 + (\$1.75 \text{ per dozen doughnuts}) \times (\text{number of dozen doughnuts})$$
$$C(x) = 300 + 1.75x$$

89. (A) $s(t) = 16t^2$
$$s(0) = 16(0)^2 = 0$$
$$s(1) = 16(1)^2 = 16$$
$$s(2) = 16(2)^2 = 64$$
$$s(3) = 16(3)^2 = 144$$

(B)
$$s(2 + h) = 16(2 + h)^2$$
$$s(2) = 64 \text{ from (A)}$$
$$\frac{s(2 + h) - s(2)}{h} = \frac{16(2 + h)^2 - 64}{h} = \frac{16(4 + 4h + h^2) - 64}{h}$$
$$= \frac{64 + 64h + 16h^2 - 64}{h} = \frac{64h + 16h^2}{h} = \frac{h(64 + 16h)}{h} = 64 + 16h$$

(C) As h tends to 0, $16h$ tends to 0, so $64 + 16h$ tends to 64. If we interpret $s(2 + h)$ as the distance fallen after $2 + h$ seconds, and $s(2)$ as the distance fallen after 2 seconds, then we can think of $s(2 + h) - s(2)$ as the distance fallen in h seconds, starting from $t = 2$. Then
$$\frac{s(2 + h) - s(2)}{h} = \frac{\text{distance fallen in } h \text{ seconds}}{h \text{ seconds}}$$
$$= \text{speed during } h \text{ seconds after } t = 2$$

As h tends to zero, this appears to tend to the speed of the object at the end of 2 seconds.

91.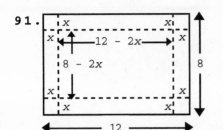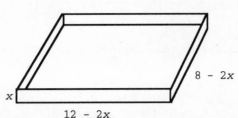

From the above figures it should be clear that
V = length × width × height = $(12 - 2x)(8 - 2x)x$.
Since all distances must be positive, $x > 0$, $8 - 2x > 0$, $12 - 2x > 0$.
Thus, $0 < x$, $4 > x$, $6 > x$, or $0 < x < 4$ (the last condition, $6 > x$,
will be automatically satisfied if $x < 4$.) Domain: $0 < x < 4$.

93. From the text diagram, since each pen must have area 50 square feet, we see
Area = (length)(width) or 50 = (length)x. Thus, the length of each pen is $\dfrac{50}{x}$ feet.
The total amount of fencing = 4(width) + 5(length) + 4(width − gate width)

$$F(x) = 4x + 5\left(\frac{50}{x}\right) + 4(x - 3)$$

$$F(x) = 4x + \frac{250}{x} + 4x - 12$$

$$F(x) = 8x + \frac{250}{x} - 12$$

Then $F(4) = 8(4) + \dfrac{250}{4} - 12 = 82.5$ $\qquad$ $F(5) = 8(5) + \dfrac{250}{5} - 12 = 78$

$F(6) = 8(6) + \dfrac{250}{6} - 12 = 77.7$ $\qquad$ $F(7) = 8(7) + \dfrac{250}{7} - 12 = 79.7$

95.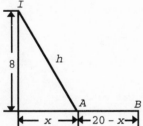

We note that the pipeline consists of the lake section IA,
and shore section AB. The shore section has length
$20 - x$. The lake section has length h, where $h^2 = 8^2 + x^2$,
thus $h = \sqrt{64 + x^2}$.
The cost of all the pipeline =

$$\begin{pmatrix} \text{cost of shore} \\ \text{section per mile} \end{pmatrix}\begin{pmatrix} \text{number of} \\ \text{shore miles} \end{pmatrix} + \begin{pmatrix} \text{cost of lake} \\ \text{section per mile} \end{pmatrix}\begin{pmatrix} \text{number of} \\ \text{lake miles} \end{pmatrix}$$

$C(x) = 10,000(20 - x) + 15,000\sqrt{64 + x^2}$
From the diagram we see that x must be non-negative, but no more than 20.
Domain: $0 \le x \le 20$

97. Cost = (cost per hour) × number of hours
= (Fuel cost per hour + Fixed cost per hour) × number of hours.
For a 500 mile trip, since $D = rt$, $t = D/r = 500/v$ hours.

Hence cost = $\left(\dfrac{v^2}{5} + 400\right)\left(\dfrac{500}{v}\right)$

$$C(v) = \frac{v^2}{5} \cdot \frac{500}{v} + 400 \cdot \frac{500}{v}$$

$$= 100v + \frac{200,000}{v}$$

Exercise 3-4

Key Ideas and Formulas

The graph of a function f is the same as the graph of the equation $y = f(x)$. The abscissa of a point where the graph of a function intersects the x axis is called an **x intercept** or **zero** of the function. The x intercept is also a real solution or **root** of the equation $f(x) = 0$. The ordinate of a point where the graph of a function crosses the y axis is called the **y intercept** of the function. The y intercept is given by $f(0)$, provided 0 is in the domain of f. The domain of a function is the set of all the x coordinates of points on the graph of the function and the range is the set of all the y coordinates.

Let I be an interval in the domain of a function f. Then:

1. f is increasing on I if $f(b) > f(a)$ whenever $b > a$ in I.

2. f is decreasing on I if $f(b) < f(a)$ whenever $b > a$ in I.

3. f is constant on I if $f(a) = f(b)$ for all a and b in I.

A function f is a linear function if $f(x) = mx + b$ $m \neq 0$ where m and b are real numbers.

The graph of a linear function f is a nonvertical and nonhorizontal straight line with slope m and y intercept b.

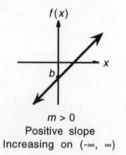

$m > 0$
Positive slope
Increasing on $(-\infty, \infty)$

Domain: All real numbers

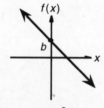

$m < 0$
Negative slope
Decreasing on $(-\infty, \infty)$

Range: All real numbers

A function f is a quadratic function if $f(x) = ax^2 + bx + c$ $a \neq 0$ where a, b, and c are real numbers.

The graph of a quadratic function is a parabola. For $f(x) = ax^2 + bx + c = a(x - h)^2 + k$ $a \neq 0$. The parabola is as shown, with the properties given.

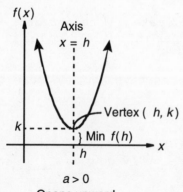

$a > 0$
Opens upward

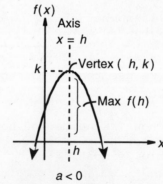

$a < 0$
Opens downward

Vertex: (h, k) (Parabola increases on one side of the vertex and decreases on the other.)

Axis (of symmetry): $x = h$ (Parallel to y axis)

$f(h) = k$ is the minimum if $a > 0$ and the maximum if $a < 0$.

Domain: All real numbers; Range: $(-\infty, k]$ if $a < 0$ or $[k, \infty)$ if $a > 0$

Functions whose definitions involve more than one formula are called piecewise defined functions. Examples include:

1. the absolute value function

$$f(x) = |x| = \begin{cases} -x \text{ if } x < 0 \\ x \text{ if } x \geq 0 \end{cases}$$

2. the greatest integer function
   $f(x) = [\![x]\!]$ = the integer n such that $n \leq x < n + 1$

---

1. (A) [-4, 4) (B) [-3, 3) (C) 0 (D) 0 (E) [-4, 4) (F) None
   (G) None (H) None

3. (A) (-∞, ∞) (B) [-4, ∞) (C) -3, 1 (D) -3 (E) [-1, ∞) (F) (-∞, -1]
   (G) None (H) None

5. (A) (-∞, 2) ∪ (2, ∞) (The function is not defined at $x = 2$.)
   (B) (-∞, -1) ∪ [1, ∞) (C) None (D) 1 (E) None (F) (-∞, -2] ∪ (2, ∞)
   (G) [-2, 2) (H) $x = 2$

7. One possible answer: **9.** One possible answer: **11.** One possible answer:

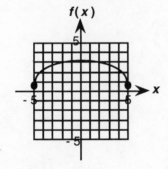   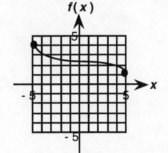   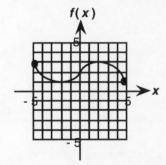

13. $f(x) = \underset{\text{slope}}{2}x + 4$

The y intercept is $f(0) = 4$, and the slope is 2. To find the x intercept, we solve the equation $f(x) = 0$ for x.

$$f(x) = 0$$
$$2x + 4 = 0$$
$$2x = -4$$
$$x = -2$$

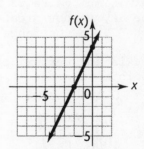

The x intercept is -2.

15. $f(x) = -\underbrace{\dfrac{1}{2}}_{\text{slope}} x - \dfrac{5}{3}$

The y intercept is $f(0) = -\dfrac{5}{3}$, and the slope is $-\dfrac{1}{2}$. To find the x intercept, we solve the equation $f(x) = 0$ for x.

$f(x) = 0$

$-\dfrac{1}{2}x - \dfrac{5}{3} = 0$

$-\dfrac{1}{2}x = \dfrac{5}{3}$

$x = (-2)\dfrac{5}{3}$

$x = \dfrac{-10}{3}$

The x intercept is $\dfrac{-10}{3}$.

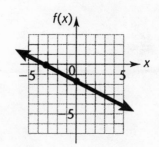

17. A linear function must have the form $f(x) = mx + b$. We are given $f(-2) = 7$, hence $7 = f(-2) = m(-2) + b$. Thus $b = 7 + 2m$. Since $f(4) = -2$, we also have $-2 = f(4) = 4m + b$. Substituting, we have

$b = 7 + 2m$
$-2 = 4m + b$
$-2 = 4m + 7 + 2m$
$-2 = 6m + 7$
$-9 = 6m$
$m = -\dfrac{3}{2}$

$b = 7 + 2m = 7 + 2\left(-\dfrac{3}{2}\right) = 4$

Hence $f(x) = mx + b$ becomes $f(x) = -\dfrac{3}{2}x + 4$.

19. $f(x) = (x - 3)^2 + 2$
Comparing with $f(x) = a(x - h)^2 + k$, $h = 3$ and $k = 2$. Thus, the vertex is $(3, 2)$, the axis of symmetry is $x = 3$, and the minimum value is 2. $y = f(x)$ can be any number greater than or equal to 2, so the range is $[2, \infty)$.
Graph: Locate axis and vertex, then plot several points on either side of the axis.

x	$f(x)$
0	11
1	6
2	3
3	2
4	3
5	6

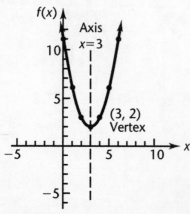

21. $f(x) = -(x + 3)^2 - 2$
Comparing with $f(x) = a(x - h)^2 + k$, $h = -3$ and $k = -2$. Thus, the vertex is $(-3, -2)$, the axis of symmetry is $x = -3$, and the maximum value is -2. $y = f(x)$ can be any number less than or equal to -2, so the range is $(-\infty, -2]$.
Graph: Locate axis and vertex, then plot several points on either side of the axis.

x	$f(x)$
-5	-6
-4	-3
-3	-2
-2	-3
-1	-6
0	-11

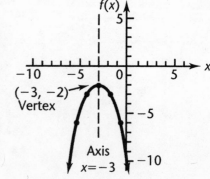

23. $f(x) = x^2 - 4x - 5$. Complete the square.

$$f(x) = (x^2 - 4x + 4) - 4 - 5$$
$$= (x - 2)^2 - 9$$

Comparing with $f(x) = a(x - h)^2 + k$, $h = 2$ and $k = 9$. Thus, the vertex is $(2, 9)$ and the axis of symmetry is $x = 2$.

y intercept: Set $x = 0$, then $f(0) = -5$ is the y intercept.

x intercepts: Set $f(x) = 0$, then
$$x^2 - 4x - 5 = 0$$
$$(x - 5)(x + 1) = 0$$
$x = 5$ or $x = -1$ are the x intercepts

Graph: Locate axis and vertex, then plot several points on either side of the axis.

x	$f(x)$
-1	0
0	-5
1	-8
2	-9
3	-8
4	-5
5	0

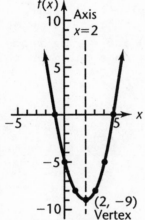

25. $f(x) = -x^2 + 6x$. Complete the square.
$$f(x) = -(x^2 - 6x)$$
$$= -(x^2 - 6x + 9) + 9$$
$$= -(x - 3)^2 + 9$$

Comparing with $f(x) = a(x - h)^2 + k$, $h = 3$ and $k = 9$. Thus, the vertex is $(3, 9)$ and the axis of symmetry is $x = 3$.

y intercept: Set $x = 0$, then $f(0) = 0$ is the y intercept.

x intercepts: Set $f(x) = 0$, then
$$-x^2 + 6x = 0$$
$$x(-x + 6) = 0$$
$x = 0$ or $x = 6$ are the x intercepts.

Graph: Locate axis and vertex, then plot several points on either side of the axis.

x	$f(x)$
0	0
1	5
2	8
3	9
4	8
5	5
6	0

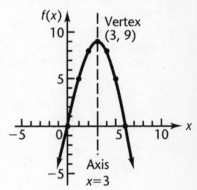

27. $f(x) = x^2 + 6x + 11$. Complete the square.

$$f(x) = (x^2 + 6x + 9) - 9 + 11$$
$$= (x + 3)^2 + 2$$

Comparing with $f(x) = a(x - h)^2 + k$, $h = -3$ and $k = 2$. Thus, the vertex is $(-3, 2)$ and the axis of symmetry is $x = -3$.

Graph: Locate axis and vertex, then plot several points on either side of the axis.

x	$f(x)$
-5	6
-4	3
-3	2
-2	3
-1	6
0	11

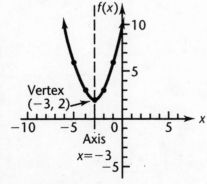

From the graph we see that f is decreasing on $(-\infty, -3]$ and increasing on $[-3, \infty)$.

29. $f(x) = -x^2 + 6x - 6$. Complete the square.

$$f(x) = -(x^2 - 6x + 9) + 9 - 6$$
$$= -(x - 3)^2 + 3$$

Comparing with $f(x) = a(x - h)^2 + k$, $h = 3$ and $k = 3$. Thus, the vertex is $(3, 3)$ and the axis of symmetry is $x = 3$.

Graph: Locate axis and vertex, then plot several points on either side of the axis.

x	$f(x)$
0	-6
1	-1
2	2
3	3
4	2
5	-1

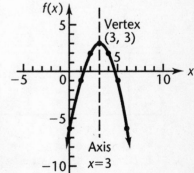

From the graph we see that f is increasing on $(-\infty, 3]$ and decreasing on $[3, \infty)$.

31.

x	$y = x + 1$	x	$y = -x + 1$
-1	0	0	1
$-\frac{1}{2}$	$\frac{1}{2}$	$\frac{1}{2}$	$\frac{1}{2}$
		1	0

Domain: [-1, 1]
Range: [0, 1]

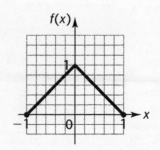

33.

x	$y = -2$	x	$y = 4$
-3	-2	0	4
-2	-2	1	4
		2	4

Domain: [-3, -1) ∪ (-1, 2]
We graph using open dots at (-1, -2) and (-1, 4) to indicate that these points do not belong to the graph of f. In fact f is not defined at -1.

Range: The only possible values of y are -2 and 4. The range consists of these two values, in set notation: {-2, 4}. (This does not mean the interval (-2, 4).)

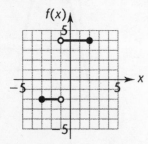

Common Error: Not including the portions of line segments between (-2, -2) and (-2, -1), or (-1, 4) and (0, 4). The domain extends up to -1 on both sides.

Discontinuous: at $x = -1$

35.

x	$y = x + 2$	x	$y = x - 2$
-5	-3	-1	-3
-2	0	5	3

Domain: All real numbers
We graph using an open dot at (-1, 1) and a solid dot at (-1, -3) to show that only the latter belongs to the graph of f.

Range: All real numbers   Discontinuous: at $x = -1$

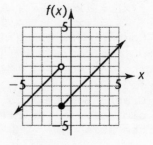

37.

x	$y = x^2 + 1$	x	$y = -x^2 - 1$
-2	5	1	-2
-1	2	2	-5

Domain: (-∞, 0) ∪ (0, ∞)
We graph using open dots at (0, 1), and (0, -1) to indicate that these points do not belong to the graph of g. g is discontinuous at $x = 0$.

Range: (-∞, -1) ∪ (1, ∞)

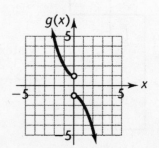

39. $f(x) = \frac{1}{2}x^2 + 2x + 3$. Complete the square.

$f(x) = \frac{1}{2}(x^2 + 4x + 4) - \frac{1}{2}(4) + 3$

$ = \frac{1}{2}(x + 2)^2 + 1$

Comparing with $f(x) = a(x - h)^2 + k$, $h = -2$ and $k = 1$. Thus, the vertex is $(-2, 1)$, the axis of symmetry is $x = -2$, and the minimum value is 1. $y = f(x)$ can be any number greater then or equal to 1, so the range is $[1, \infty)$.

y intercept: Set $x = 0$, then $f(0) = 3$ is the y intercept.

x intercept: Solutions of

$0 = \frac{1}{2}x^2 + 2x + 3$

$0 = x^2 + 4x + 6$

$x = \dfrac{-4 \pm \sqrt{-8}}{2}$

Graph:

x	$f(x)$
-5	5.5
-4	3
-3	1.5
-2	1
-1	1.5
0	3
1	5.5

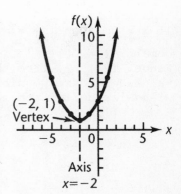

no real solutions, hence no x intercepts

From the graph, we see that f is decreasing on $(-\infty, -2]$ and increasing on $[-2, \infty)$.

41. $f(x) = 4x^2 - 12x + 9$. Complete the square.

$f(x) = 4(x^2 - 3x + \frac{9}{4}) - 4 \cdot \frac{9}{4} + 9$

$ = 4(x - \frac{3}{2})^2$

Comparing with $f(x) = a(x - h)^2 + k$, $h = \frac{3}{2}$ and $k = 0$. Thus, the vertex is $(\frac{3}{2}, 0)$, the axis of symmetry is $x = \frac{3}{2}$, and the minimum value is 0. $y = f(x)$ can be any number greater then or equal to 0, so the range is $[0, \infty)$.

y intercept: Set $x = 0$, then $f(0) = 9$ is the y intercept.

x intercept: Solutions of $\quad 0 = 4x^2 - 12x + 9$

$(2x - 3)^2 = 0$

$2x - 3 = 0$

$x = \frac{3}{2}$

Graph:

x	$f(x)$
0	9
1	1
$\frac{3}{2}$	0
2	1
3	9

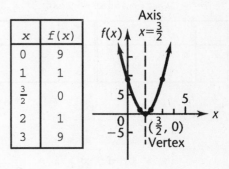

From the graph, we see that f is decreasing on $(-\infty, \frac{3}{2}]$ and increasing on $[\frac{3}{2}, \infty)$.

43. $f(x) = -2x^2 - 8x - 2$. Complete the square.

$f(x) = -2(x^2 + 4x + 4) + 2 \cdot 4 - 2$

$ = -2(x + 2)^2 + 6$

Comparing with $f(x) = a(x - h)^2 + k$, $h = -2$ and $k = 6$. Thus, the vertex is $(-2, 6)$, the axis of symmetry is $x = -2$, and the maximum value is 6. $y = f(x)$ can be any number less then or equal to 6, so the range is $(-\infty, 6]$.

y intercept: Set $x = 0$, then $f(0) = -2$ is the y intercept.

x intercept: Solutions of

$0 = -2x^2 - 8x - 2$

$0 = x^2 + 4x + 1$

$x = \dfrac{-4 \pm \sqrt{12}}{2} = -2 \pm \sqrt{3}$

Graph:

x	$f(x)$
-4	-2
-3	4
-2	6
-1	4
0	-2

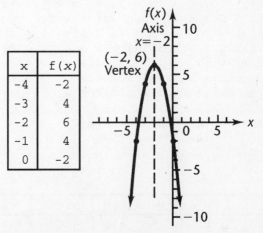

From the graph, we see that f is increasing on $(-\infty, -2]$ and decreasing on $[-2, \infty)$.

45. $f(x) = \dfrac{|x|}{x}$

If $x < 0$ then $|x| = -x$ and $f(x) \dfrac{|x|}{x} = \dfrac{-x}{x} = -1$

If $x = 0$, then f is not defined. f is discontinuous at $x = 0$.

If $x > 0$, then $|x| = x$ and $f(x) = \dfrac{|x|}{x} = \dfrac{x}{x} = 1$

Domain: $(-\infty, 0) \cup (0, \infty)$, that is, $x \neq 0$
Range: $\{-1, 1\}$, from the graph we see that only two
values of f are possible.
Piecewise definition for f:

$$f(x) = \begin{cases} -1 \text{ if } x < 0 \\ 1 \text{ if } x > 0 \end{cases}$$

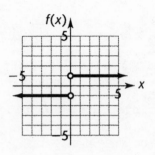

47. $f(x) = x + \dfrac{|x - 1|}{x - 1}$

If $x < 1$, then $|x - 1| = -(x - 1)$ and $f(x) = x + \dfrac{-(x - 1)}{x - 1} = x - 1$

If $x = 1$, then $f(x)$ is not defined. f is discontinuous at $x = 1$.

If $x > 1$, then $|x - 1| = x - 1$ and $f(x) = x + \dfrac{x - 1}{x - 1} = x + 1$

x	$y = x - 1$	x	$y = x + 1$
-1	-2	2	3
0	-1	3	4

Domain: $(-\infty, 1) \cup (1, \infty)$, that is, $x \neq 1$
Piecewise definition for f:

$$f(x) = \begin{cases} x - 1 & \text{if } x < 1 \\ x + 1 & \text{if } x > 1 \end{cases}$$

Range: From the graph, we see that y is less than 0
or y is greater than 2:
$(-\infty, 0) \cup (2, \infty)$

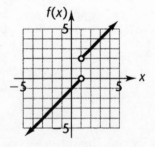

49. $f(x) = |x| + |x - 2|$
If $x < 0$, then $|x| = -x$ and $|x - 2| = -(x - 2) = 2 - x$ and
$f(x) = |x| + |x - 2| = -x + (2 - x) = 2 - 2x$
If $0 \leq x < 2$, then $|x| = x$ but $|x - 2| = -(x - 2) = 2 - x$ and
$f(x) = |x| + |x - 2| = x + (2 - x) = 2$
If $2 \leq x$, then $|x| = x$ and $|x - 2| = x - 2$ and
$f(x) = |x| + |x - 2| = x + (x - 2) = 2x - 2$
Domain: All real numbers
Piecewise definition for f:

$$f(x) = \begin{cases} 2 - 2x & x < 0 \\ 2 & 0 \leq x < 2 \\ 2x - 2 & x \geq 2 \end{cases}$$

There are no discontinuities for f.

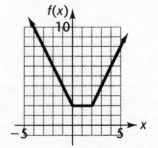

x	$y = 2 - 2x$	x	$y = 2$	x	$y = 2x - 2$
-2	6	0	2	2	2
-1	4	1	2	3	4

Range: From the graph we see that y is never less than 2. $[2, \infty)$

51. We note: $f(0) = [\![0/2]\!] = [\![0]\!] = 0$

$f(1) = [\![1/2]\!] = 0$

$f(2) = [\![2/2]\!] = [\![1]\!] = 1$

$f(3) = [\![3/2]\!] = 1 \qquad f(x) = [\![x/2]\!]$ appears to jump at intervals of 2 units.

Generalizing, we can write: $f(x) = \begin{cases} \vdots & & \vdots \\ -2 & \text{if} & -4 \le x < -2 \\ -1 & \text{if} & -2 \le x < 0 \\ 0 & \text{if} & 0 \le x < 2 \\ 1 & \text{if} & 2 \le x < 4 \\ 2 & \text{if} & 4 \le x < 6 \\ \vdots & & \vdots \end{cases}$

Domain: All real numbers
Range: All integers
Discontinuous at the even integers

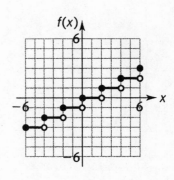

53. Having noted in problem 51 that $[\![\frac{x}{2}]\!]$ jumps at intervals of 2 units, we can investigate $[\![3x]\!] = [\![x \div \frac{1}{3}]\!]$ to see if it jumps at intervals of $\frac{1}{3}$ unit.

$f(0) = [\![3 \cdot 0]\!] = [\![0]\!] = 0$

$f(\frac{1}{6}) = [\![3 \cdot \frac{1}{6}]\!] = [\![\frac{1}{2}]\!] = 0$

$f(\frac{1}{3}) = [\![3 \cdot \frac{1}{3}]\!] = [\![1]\!] = 1$

$f(\frac{1}{2}) = [\![3 \cdot \frac{1}{2}]\!] = [\![\frac{3}{2}]\!] = 1$

$f(\frac{2}{3}) = [\![3 \cdot \frac{2}{3}]\!] = [\![2]\!] = 2$

$f(\frac{5}{6}) = [\![3 \cdot \frac{5}{6}]\!] = [\![\frac{5}{2}]\!] = 2$

Generalizing, we can write: $f(x) = \begin{cases} \vdots & & \vdots \\ -2 & \text{if} & -\frac{2}{3} \le x < -\frac{1}{3} \\ -1 & \text{if} & -\frac{1}{3} \le x < 0 \\ 0 & \text{if} & 0 \le x < \frac{1}{3} \\ 1 & \text{if} & \frac{1}{3} \le x < \frac{2}{3} \\ 2 & \text{if} & \frac{2}{3} \le x < 1 \\ \vdots & & \vdots \end{cases}$

Domain: All real numbers
Range: All integers
Discontinuous at rational numbers of the
form $\frac{k}{3}$ where k is an integer

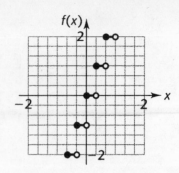

55. This is discontinuous at every integer value of x. If x is an integer, $x - [\![x]\!]$
$= x - x = 0$. If $x = n + \varepsilon$, where $0 < \varepsilon < 1$, then $[\![x]\!] = n$, $-[\![x]\!] = -n$, and
$x - [\![x]\!] = \varepsilon = x - n$. So the graph between $x = n$ and $x = n + 1$ is a line segment
of form $y = x - n$, connecting $(n, 0)$ and $(n + 1, 1)$. However, $f(n + 1) \neq 1$
since $f(n + 1) = 0$. We can write:

$$f(x) = \begin{cases} \vdots & & \vdots \\ x + 2 & \text{if} & -2 \leq x < -1 \\ x + 1 & \text{if} & -1 \leq x < 0 \\ x & \text{if} & 0 \leq x < 1 \\ x - 1 & \text{if} & 1 \leq x < 2 \\ x - 2 & \text{if} & 2 \leq x < 3 \\ \vdots & & \vdots \end{cases}$$

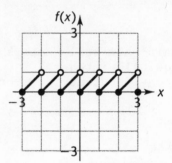

Domain: all real numbers
Range: [0, 1)

57. $f(x) = a(x - h)^2 + k$. Since Min $f(x) = f(2) = 4$, $h = 2$, $k = 4$, and the vertex
is $(2, 4)$. The axis is $x = h$ or $x = 2$. Since y is never less than 4, the range
is $[4, \infty)$ and there are no x intercepts since y cannot be equal to 0.

59. (A) One possible answer:

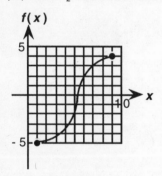

(B) This graph crosses the x axis once.
To meet the conditions specified a graph
must cross the x axis exactly once. If
it crossed more times the function would
have to be decreasing somewhere; if it
did not cross at all the function would
have to be discontinuous somewhere.

61. (A) One possible answer:

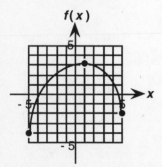

(B) This graph crosses the x axis twice.
To meet the conditions specified a graph
must cross the x axis at least twice. If
it crossed fewer times the function
would have to be discontinuous
somewhere. However, the graph could
cross more times; in fact there is no
upper limit on the number of times it
can cross the x axis.

63. The secant line is the line through $(-1, -3)$ and $(3, 5)$. To find its equation we find its slope, then use the point-slope form of the equation of a line.

Graph:

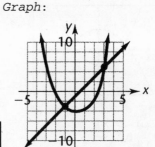

$$m = \frac{f(x_2) - f(x_1)}{x_2 - x_1} = \frac{5 - (-3)}{3 - (-1)} = \frac{8}{4} = 2$$

$y - f(x_1) = m(x - x_1)$
$y - (-3) = 2[x - (-1)]$
$y + 3 = 2[x + 1]$
$y + 3 = 2x + 2$
$y = 2x - 1$

x	$y = 2x - 1$	$y = x^2 - 4$
-2	-5	0
-1	-3	-3
0	-1	-4
1	1	-3
2	3	0
3	5	5

65. (A) The slope of the secant line is given by

$$m = \frac{f(2 + h) - f(2)}{(2 + h) - 2} = \frac{f(2 + h) - f(2)}{h}$$

Since $f(2 + h) = (2 + h)^2 - 3(2 + h) + 5 = 4 + 4h + h^2 - 6 - 3h + 5 = 3 + h + h^2$
and $f(2) = (2)^2 - 3(2) + 5 = 4 - 6 + 5 = 3$
we have

$$m = \frac{f(2 + h) - f(2)}{h} = \frac{(3 + h + h^2) - (3)}{h} = \frac{h + h^2}{h} = \frac{h(1 + h)}{h} = 1 + h$$

(B)

h	$slope = 1 + h$
1	2
0.1	1.1
0.01	1.01
0.001	1.001

The slope seems to be approaching 1.

67. Graphs of f and g

Graph of $m(x) = 0.5[-2x + 0.5x + |-2x - 0.5x|]$
$= 0.5[-1.5x + |-2.5x|]$

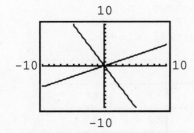

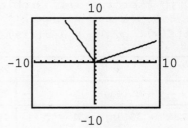

Graph of $n(x) = 0.5[-2x + 0.5x - |-2x - 0.5x|] = 0.5[-1.5x - |-2.5x|]$

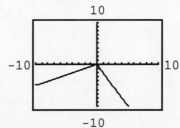

69. Graphs of f and g

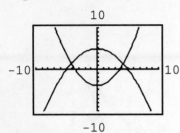

10

-10   10

-10

Graph of $m(x) =$
$0.5[5 - 0.2x^2 + 0.3x^2 - 4 +$
    $|5 - 0.2x^2 - (0.3x^2 - 4)|]$
$= 0.5[1 + 0.1x^2 + |9 - 0.5x^2|]$

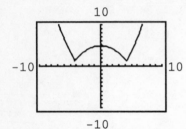

10

-10   10

-10

Graph of $n(x) =$
$0.5[5 - 0.2x^2 + 0.3x^2 - 4 -$
    $|5 - 0.2x^2 - (0.3x^2 - 4)|]$
$= 0.5[1 + 0.1x^2 - |9 - 0.5x^2|]$

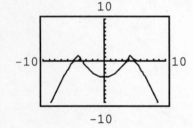

10

-10   10

-10

71. Graphs of f and g

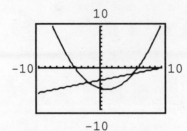

10

-10   10

-10

Graph of $m(x) =$
$0.5[0.2x^2 - 0.4x - 5 + 0.3x - 3 +$
    $|0.2x^2 - 0.4x - 5 - (0.3x - 3)|]$
    $= 0.5[0.2x^2 - 0.1x - 8 + |0.2x^2 - 0.7x - 2|]$

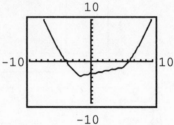

10

-10   10

-10

Graph of $n(x) =$
$0.5[0.2x^2 - 0.4x - 5 + 0.3x - 3 -$
    $|0.2x^2 - 0.4x - 5 - (0.3x - 3)|]$
    $= 0.5[0.2x^2 - 0.1x - 8 - |0.2x^2 - 0.7x - 2|]$

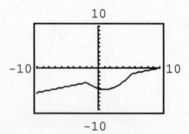

10

-10   10

-10

73. The graphs of $m(x)$ show that the value of $m(x)$ is always the larger of the two values for $f(x)$ and $g(x)$. In other words, $m(x) = \max[f(x), g(x)]$. This is confirmed in Exercise 2.4, Problem 89 where it was shown that
$\max(a, b) = 0.5[a + b + |a - b|]$,
hence $\max[f(x), g(x)] = 0.5[f(x) + g(x) + |f(x) - g(x)|] = m(x)$.

75. (A)

x	28	30	32	34	36
Mileage	45	52	55	51	47
$-0.518x^2 + 33.3x - 481 = f(x)$	45.3	51.8	54.2	52.4	46.5

(B)

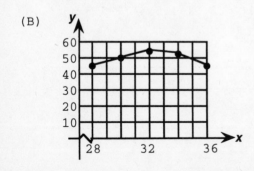

(C) When $x = 31$, $y = -0.518(31)^2 + 33.3(31) - 481$
                $y \approx 53.50$ thousand miles
When $x = 35$, $y = -0.518(35)^2 + 33.3(35) - 481$
                $y \approx 49.95$ thousand miles

(D) The mileage increases with tire pressure to a maximum of 55 thousand miles at tire pressure 32 pounds per square inch and decreases with greater tire pressure after that.

77. (A) Using the hint, we have $f(10) = 1$ and $f(0) = 0$. Since $f(w) = mw + b$, we must have $1 = f(10) = m(10) + b$ and $0 = f(0) = m(0) + b$.

Therefore $b = 0$ and $m = \dfrac{1}{10}$. So $s = f(w) = \dfrac{w}{10}$.

(B) $f(w) = \dfrac{w}{10}$. Therefore $f(15) = \dfrac{15}{10} = 1.5$ inches and $f(30) = \dfrac{30}{10} = 3$ inches.

(C) The slope is the coefficient of w, or $\dfrac{1}{10}$.

(D)

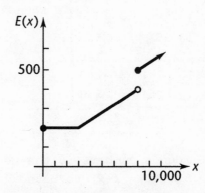

79. If $0 \leq x \leq 3,000$, $E(x) = 200$

	Base Salary	+	Commission on Sales over $3,000
If $\$3,000 < x < 8,000$, $E(x)$	= 200	+	$0.04(x - 3,000)$
	= 200	+	$0.04x - 120$
	= 80	+	$0.04x$

There is a point of discontinuity at $x = 8,000$.

	Salary	+	Bonus
If $x \geq 8,000$, $E(x)$ =	80 + 0.04x	+	100
	= 180	+	0.04x

Summarizing, $E(x) = \begin{cases} 200 & \text{if} \quad 0 \leq x \leq 3,000 \\ 80 + 0.04x & \text{if } 3,000 < x < 8,000 \\ 180 + 0.04x & \text{if } 8,000 \leq x \end{cases}$

$E(5,750) = 80 + 0.04(5,750) = \$310$
$E(9,200) = 180 + 0.04(9,200) = \$548$

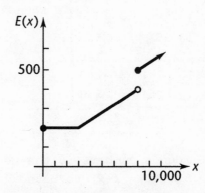

x	$y = 200$	x	$y = 80 + 0.04x$	x	$y = 180 + 0.04x$
0	200	3,000	200	8,000	500
2,000	200	7,000	360	10,000	580

81. (A) Let x = width of pen

Then $2x + 2\ell = 100$

so $2\ell = 100 - 2x$

Therefore $\ell = 50 - x$ = length of pen

$A(x) = x(50 - x) = 50x - x^2$

(B) Since all distances must be positive, $x > 0$ and $50 - x > 0$.
Therefore, $0 < x < 50$ or $(0, 50)$ is the domain of $A(x)$.

(C) Note: 0, 50 were excluded from the domain for geometrical reasons; but can be used to help draw the graph since the *polynomial* $50x - x^2$ has domain including these values.

x	$A(x)$
0	0
10	400
20	600
30	600
40	400
50	0

(D) From the graph, the maximum value of $A(x)$ seems to occur when $x = 25$. Checking, we see that $50x - x^2 = -x^2 + 50x$ is a quadratic function that can be written as $-(x^2 - 50x) = -(x^2 - 50x + 625) + 625 = -(x - 25)^2 + 625$. Thus it has its maximum value at $x = 25$. Then $50 - x = 25$, and the dimensions are 25 feet by 25 feet.

83.

$\begin{aligned}
f(4) &= 10[\![0.5 + 0.4]\!] &= 10(0) &= 0 \\
f(-4) &= 10[\![0.5 - 0.4]\!] &= 10(0) &= 0 \\
f(6) &= 10[\![0.5 + 0.6]\!] &= 10(1) &= 10 \\
f(-6) &= 10[\![0.5 - 0.6]\!] &= 10(-1) &= -10 \\
f(24) &= 10[\![0.5 + 2.4]\!] &= 10(2) &= 20 \\
f(25) &= 10[\![0.5 + 2.5]\!] &= 10(3) &= 30 \\
f(247) &= 10[\![0.5 + 24.7]\!] &= 10(25) &= 250 \\
f(-243) &= 10[\![0.5 - 24.3]\!] &= 10(-24) &= -240 \\
f(-245) &= 10[\![0.5 - 24.5]\!] &= 10(-24) &= -240 \\
f(-246) &= 10[\![0.5 - 24.6]\!] &= 10(-25) &= -250
\end{aligned}$

f rounds numbers to the tens place

85. Since $f(x) = [\![10x + 0.5]\!]/10$ rounds numbers to the nearest tenth, (see text example 6) we try $[\![100x + 0.5]\!]/100 = f(x)$ to round to the nearest hundredth.

$f(3.274) = [\![327.9]\!]/100 = 3.27$

$f(7.846) = [\![785.1]\!]/100 = 7.85$

$f(-2.8783) = [\![-287.33]\!]/100 = -2.88$

A few examples suffice to convince us that this is probably correct.
(A proof would be out of place in this book.)

$f(x) = [\![100x + 0.5]\!]/100$.

87. (A)

$$C(x) = \begin{cases} 15 & 0 < x \le 1 \\ 18 & 1 < x \le 2 \\ 21 & 2 < x \le 3 \\ 24 & 3 < x \le 4 \\ 27 & 4 < x \le 5 \\ 30 & 5 < x \le 6 \end{cases}$$

(B) The two functions appear to coincide, for example

$$C(3.5) = 24 \quad f(3.5) = 15 + 3[\![3.5]\!] = 15 + 3 \cdot 3 = 24$$

However,

$$C(1) = 15 \quad f(1) = 15 + 3[\![1]\!] = 15 \cdot 3 \cdot 1 = 18$$

The functions are not the same, therefore. In fact, $f(x) \ne C(x)$ at $x = 1, 2, 3, 4, 5, 6$.

89. Let x = rate per car per day.
Then (number of cars rented) = 300 - 5(increase of rate over \$40)

$$= 300 - 5(x - 40)$$
$$= 300 - 5x + 200$$
$$= 500 - 5x$$

Income = $I(x) = \begin{pmatrix} \text{Numbers of} \\ \text{Cars Rented} \end{pmatrix} \begin{pmatrix} \text{Rate Per} \\ \text{Car Per Day} \end{pmatrix}$

$$I(x) = (500 - 5x)x$$
$$I(x) = 500x - 5x^2$$

This is a quadratic function. Completing the square yields

$$I(x) = -5x^2 + 500x$$
$$= -5(x^2 - 100x)$$
$$= -5(x^2 - 100x + 2500) + 5(2500)$$
$$= -5(x - 50)^2 + 12,500$$

Thus the maximum value of this function occurs when $x = 50$. Hence the rate should be \$50 per day and the resulting maximum income is $I(50) = \$12,500$.

91. (A) The given function $f(x) = \frac{1}{4}x - \left(\frac{16}{v^2}\right)x^2$ is a quadratic function for every positive value of v. In the coordinate system given, this function has x intercepts 0 and 80, and maximum value when $x = 40$. Completing the square yields

$$f(x) = -\frac{16}{v^2}x^2 + \frac{1}{4}x$$

$$= -\frac{16}{v^2}\left(x^2 - \frac{v^2}{64}x\right)$$

$$= -\frac{16}{v^2}\left(x^2 - \frac{v^2}{64}x + \left(\frac{v^2}{128}\right)^2\right) + \frac{16}{v^2}\left(\frac{v^2}{128}\right)^2$$

$$= -\frac{16}{v^2}\left(x - \frac{v^2}{128}\right)^2 + \frac{v^2}{1024}$$

Comparing with $a(x - h)^2 + k$, the maximum value occurs when $x = h = 40$, thus

$$\frac{v^2}{128} = 40$$
$$v^2 = 5120$$
$$v = \sqrt{5120} \text{ (discarding the negative solution)}$$
$$v = 32\sqrt{5} \approx 71.55 \text{ ft/sec}$$

(B) The maximum height of the cycle above the ground is $10 + k = 10 + \dfrac{v^2}{1024}$.

Substituting $v = 32\sqrt{5}$ yields $10 + k = 10 + \dfrac{v^2}{1024} = 10 + \dfrac{(32\sqrt{5})^2}{1024} = 10 + \dfrac{5120}{1024} =$ $10 + 5 = 15$ feet.

Exercise 3-5

Key Ideas and Formulas

The intersection of sets A and B is the set of all elements in A that are also in B. The sum, difference, product and quotient of the function f and g are the functions defined by:

$$(f + g)(x) = f(x) + g(x) \qquad \text{Sum function}$$
$$(f - g)(x) = f(x) - g(x) \qquad \text{Difference function}$$
$$(fg)(x) = f(x)g(x) \qquad \text{Product function}$$
$$\left(\frac{f}{g}\right)(x) = \frac{f(x)}{g(x)} \qquad \text{Quotient function}$$

Each function is defined on the intersection of the domains of f and g, with the exception that the values of x where $g(x) = 0$ must be excluded from the domain of the quotient function.

Given functions f and g, then $f \cdot g$ is called their composite and is defined by the equation

$$(f \cdot g)(x) = f[g(x)]$$

The domain of $f \cdot g$ is the set of all real numbers x in the domain of g where $g(x)$ is in the domain of f.

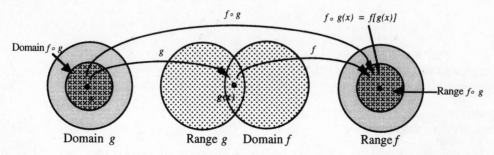

Graphs of Basic Functions:

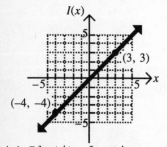

(a) Identity function
$I(x) = x$
Domain: R
Range: R

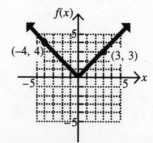

(b) Absolute Value Function
$f(x) = |x|$
Domain: R
Range: $[0, \infty)$

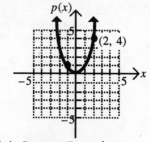

(c) Square Function
$p(x) = x^2$
Domain: R
Range: $[0, \infty)$

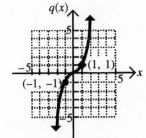

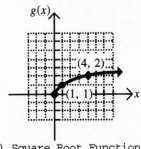

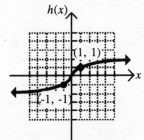

(d) Cube Function

$q(x) = x^3$
Domain: R
Range: R

(e) Square Root Function

$g(x) = \sqrt{x}$
Domain: $[0, \infty)$
Range: $[0, \infty)$

(f) Cube Root Function

$h(x) = \sqrt[3]{x}$
Domain: R
Range: R

Graph Transformations:

Vertical Translation:

$y = f(x) + k$ $\begin{cases} k > 0 & \text{Shift graph of } y = f(x) \text{ up } k \text{ units} \\ k < 0 & \text{Shift graph of } y = f(x) \text{ down } |k| \text{ units} \end{cases}$

Horizontal Translation:

$y = f(x + h)$ $\begin{cases} h > 0 & \text{Shift graph of } y = f(x) \text{ left } h \text{ units} \\ h < 0 & \text{Shift graph of } y = f(x) \text{ right } |h| \text{ units} \end{cases}$

Reflection:

$y = -f(x)$ Reflect the graph of $y = f(x)$ in the x axis

Vertical Expansion and Contraction:

$y = Af(x)$ $\begin{cases} A > 1 & \text{Vertically expand graph of } y = f(x) \\ & \text{by multiplying each ordinate value by } A \\ 0 < A < 1 & \text{Vertically contract graph of } y = f(x) \\ & \text{by multiplying each ordinate value by } A \end{cases}$

---

1. Domain: Since $x \geq 0$, the domain is $[0, \infty)$

Range: Since the range of $f(x) = \sqrt{x}$ is $y \geq 0$, for $h(x) = -\sqrt{x}$, $y \leq 0$. Thus, the range of h is $(-\infty, 0]$.

3. Domain: R

Range: Since the range of $f(x) = x^2$ is $y \geq 0$, for $g(x) = -2x^2$, $y \leq 0$. Thus, the range of g is $(-\infty, 0]$.

5. Domain: R; Range: R

7. $(f + g)(x) = f(x) + g(x) = 4x + x + 1 = 5x + 1$ Domain: $(-\infty, \infty)$
$(f - g)(x) = f(x) - g(x) = 4x - (x + 1) = 3x - 1$ Domain: $(-\infty, \infty)$
$(fg)(x) = f(x)g(x) = 4x(x + 1) = 4x^2 + 4x$ Domain: $(-\infty, \infty)$

$\left(\dfrac{f}{g}\right)(x) = \dfrac{f(x)}{g(x)} = \dfrac{4x}{x + 1}$ Domain: $\{x \mid x \neq -1\}$, *or*

                                                    $(-\infty, -1) \cup (-1, \infty)$

> **Common Error:**
> $f(x) - g(x) \neq 4x - x + 1$. The parentheses are necessary.

9. $(f + g)(x) = f(x) + g(x) = 2x^2 + x^2 + 1 = 3x^2 + 1$ Domain: $(-\infty, \infty)$
$(f - g)(x) = f(x) - g(x) = 2x^2 - (x^2 + 1) = x^2 - 1$ Domain: $(-\infty, \infty)$
$(fg)(x) = f(x)g(x) = 2x^2(x^2 + 1) = 2x^4 + 2x^2$ Domain: $(-\infty, \infty)$

$\left(\dfrac{f}{g}\right)(x) = \dfrac{f(x)}{g(x)} = \dfrac{2x^2}{x^2 + 1}$ Domain: $(-\infty, \infty)$

                                             (since $g(x)$ is never 0.)

11. $(f + g)(x) = f(x) + g(x) = 3x + 5 + x^2 - 1$
$\qquad = x^2 + 3x + 4$ Domain: $(-\infty, \infty)$
$\quad (f - g)(x) = f(x) - g(x) = 3x + 5 - (x^2 - 1)$
$\qquad = 3x + 5 - x^2 + 1 = -x^2 + 3x + 6$ Domain: $(-\infty, \infty)$
$\quad (fg)(x) = f(x)g(x) = (3x + 5)(x^2 - 1)$
$\qquad = 3x^3 - 3x + 5x^2 - 5 = 3x^3 + 5x^2 - 3x - 5$ Domain: $(-\infty, \infty)$
$\left(\dfrac{f}{g}\right)(x) = \dfrac{f(x)}{g(x)} = \dfrac{3x + 5}{x^2 - 1}$ Domain: $\{x \mid x \neq \pm 1\}$, or

$\qquad\qquad\qquad\qquad\qquad\qquad (-\infty, -1) \cup (-1, 1) \cup (1, \infty)$

13. $(f \circ g)(x) = f[g(x)] = f(x^2 - x + 1) = (x^2 - x + 1)^3$ Domain: $(-\infty, \infty)$
$\quad (g \circ f)(x) = g[f(x)] = g(x^3) = (x^3)^2 - x^3 + 1 = x^6 - x^3 + 1$ Domain: $(-\infty, \infty)$

15. $(f \circ g)(x) = f[g(x)] = f(2x + 3) = |2x + 3 + 1| = |2x + 4|$ Domain: $(-\infty, \infty)$
$\quad (g \circ f)(x) = g[f(x)] = g(|x + 1|) = 2|x + 1| + 3$ Domain: $(-\infty, \infty)$

17. $(f \circ g)(x) = f[g(x)] = f(2x^3 + 4) = (2x^3 + 4)^{1/3}$ Domain: $(-\infty, \infty)$
$\quad (g \circ 61$

$f)(x) = g[f(x)] = g(x^{1/3}) = 2(x^{1/3})^3 + 4 = 2x + 4$ Domain: $(-\infty, \infty)$

19. The graph of $f(x)$ is shifted up 2 units

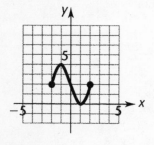

21. The graph of $g(x)$ is shifted up 2 units

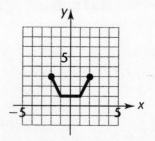

23. The graph of $f(x)$ is shifted right 2 units.

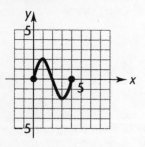

25. The graph of $g(x)$ is shifted left 2 units.

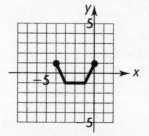

> **Common Errors:**
> Confusing the -2 in $f(x - 2)$
> with a shift left.
> Confusing the positive 2 in
> $g(x + 2)$ with a shift right.

27. The graph of $f(x)$ is reflected in the x axis.

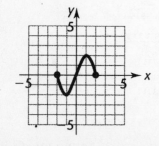

29. The graph of $g(x)$ is vertically expanded by multiplying each ordinate by 2.

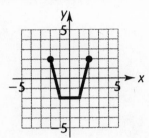

31. The graph of $y = |x|$ is shifted two units to the left and reflected in the x axis.

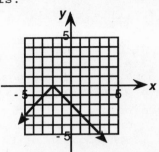

33. The graph of $y = x^2$ is shifted 2 units to the right and 4 units down.

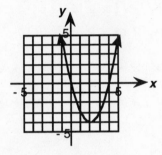

35. The graph of $y = \sqrt{x}$ is vertically expanded by a factor of 2, reflected in the x axis, and shifted 4 units up.

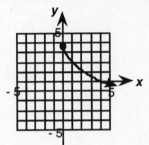

37. $(f + g)(x) = f(x) + g(x) = \sqrt{2 - x} + \sqrt{x + 3}$

$(f - g)(x) = f(x) - g(x) = \sqrt{2 - x} - \sqrt{x + 3}$

$(fg)(x) = f(x)g(x) = \sqrt{2 - x}\sqrt{x + 3} = \sqrt{(2 - x)(3 + x)} = \sqrt{6 - x - x^2}$

$\left(\dfrac{f}{g}\right)(x) = \dfrac{f(x)}{g(x)} = \dfrac{\sqrt{2 - x}}{\sqrt{x + 3}} = \sqrt{\dfrac{2 - x}{x + 3}}$

The domains of f and g are:
Domain of $f = \{x \mid 2 - x \geq 0\} = (-\infty, 2]$

Domain of $g = \{x \mid x + 3 \geq 0\} = [-3, \infty)$
The intersection of these domains is $[-3, 2]$.
This is the domain of the functions of $f + g$, $f - g$, and fg.

Since $g(-3) = 0$, $x = -3$ must be excluded from the domain of $\dfrac{f}{g}$, so its domain is $(-3, 2]$.

39. $(f + g)(x) = f(x) + g(x) = \sqrt{x} + 2 + \sqrt{x} - 4 = 2\sqrt{x} - 2$

$(f - g)(x) = f(x) - g(x) = \sqrt{x} + 2 - (\sqrt{x} - 4) = \sqrt{x} + 2 - \sqrt{x} + 4 = 6$

$(fg)(x) = f(x)g(x) = (\sqrt{x} + 2)(\sqrt{x} - 4) = x - 2\sqrt{x} - 8$

$\left(\dfrac{f}{g}\right)(x) = \dfrac{f(x)}{g(x)} = \dfrac{\sqrt{x} + 2}{\sqrt{x} - 4}$

The domains of f and g are both $\{x \mid x \geq 0\} = [0, \infty)$
This is the domain of $f + g$, $f - g$, and fg. We note that in the domain of $\dfrac{f}{g}$, $g(x) \neq 0$. Thus $\sqrt{x} - 4 \neq 0$. To solve this, we solve

$\sqrt{x} - 4 = 0$

$\sqrt{x} = 4$ | **Common Error:** $x \neq \sqrt{4}$ |

$x = 16$

Hence, 16 must be excluded from $\{x \mid x \geq 0\}$ to find the domain of $\dfrac{f}{g}$.

Domain of $\dfrac{f}{g} = \{x \mid x \geq 0, x \neq 16\} = [0, 16) \cup (16, \infty)$.

41. $(f + g)(x) = f(x) + g(x) = \sqrt{x^2 + x - 6} + \sqrt{7 + 6x - x^2}$

$(f - g)(x) = f(x) - g(x) = \sqrt{x^2 + x - 6} - \sqrt{7 + 6x - x^2}$

$(fg)(x) = f(x)g(x) = \sqrt{x^2 + x - 6} \sqrt{7 + 6x - x^2} = \sqrt{-x^4 + 5x^3 + 19x^2 - 29x - 42}$

$\left(\dfrac{f}{g}\right)(x) = \dfrac{f(x)}{g(x)} = \dfrac{\sqrt{x^2 + x - 6}}{\sqrt{7 + 6x - x^2}} = \sqrt{\dfrac{x^2 + x - 6}{7 + 6x - x^2}}$

The domains of f and g are:

Domain of $f = \{x \mid x^2 + x - 6 \geq 0\} = \{x \mid (x + 3)(x - 2) \geq 0\} = (-\infty, -3] \cup [2, \infty)$

Domain of $g = \{x \mid 7 + 6x - x^2 \geq 0\} = \{x \mid (7 - x)(1 + x) \geq 0\} = [-1, 7]$

The intersection of these domains is $[2, 7]$.

This is the domain of the functions $f + g$, $f - g$, and fg.

Since $g(x) = 7 + 6x - x^2 = (7 - x)(1 + x)$, $g(7) = 0$ and $g(-1) = 0$, hence 7 must be excluded from the domain of $\dfrac{f}{g}$, so its domain is $[2, 7)$.

43. $(f \circ g)(x) = f[g(x)] = f(x - 4) = \sqrt{x - 4}$ Domain: $\{x \mid x \geq 4\}$ or $[4, \infty)$

$(g \circ f)(x) = g[f(x)] = g(\sqrt{x}) = \sqrt{x} - 4$ Domain: $\{x \mid x \geq 0\}$ or $[0, \infty)$

45. $(f \circ g)(x) = f[g(x)] = f\left(\dfrac{1}{x}\right) = \dfrac{1}{x} + 2$ Domain: $\{x \mid x \neq 0\}$ or $(-\infty, 0) \cup (0, \infty)$

$(g \circ f)(x) = g[f(x)] = g(x + 2) = \dfrac{1}{x + 2}$ Domain: $\{x \mid x \neq -2\}$ or $(-\infty, -2) \cup (-2, \infty)$

47. $(f \circ g)(x) = f[g(x)] = f\left(\dfrac{1}{x - 1}\right) = \left|\dfrac{1}{x - 1}\right| = \dfrac{1}{|x - 1|}$

Domain: $\{x \mid x \neq 1\}$ or $(-\infty, 1) \cup (1, \infty)$

$(g \circ f)(x) = g[f(x)] = g(|x|) = \dfrac{1}{|x| - 1}$

To find the domain we must exclude $|x| - 1 = 0$ from the domain of f.

$|x| - 1 = 0$

$|x| = 1$

$x = -1$ or 1

Domain of $g \circ f$: $\{x \mid x \neq -1 \text{ or } 1\} = (-\infty, -1) \cup (-1, 1) \cup (1, \infty)$

49. The graph of the function $f(x) = |x|$ has been shifted one unit to the left and two units down. Equation: $y = |x + 1| - 2$.

51. The graph of the function $f(x) = \sqrt[3]{x}$ has been reflected in the x axis and shifted 3 units up. Equation: $y = -\sqrt[3]{x} + 3$ or $y = 3 - \sqrt[3]{x}$.

53. The graph of the function $y = x^3$ has been reflected in the x axis and shifted two units to the left and one unit up. Equation: $y = -(x + 2)^3 + 1$ or $y = 1 - (x + 2)^3$.

55. $y = \sqrt{x + 2} + 3$ **57.** $y = -|x - 3|$ **59.** $y = -(x + 2)^3 + 1$

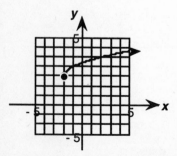   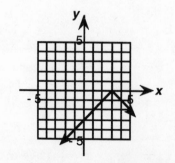   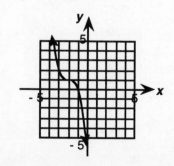

61. $f(x) = 2x^2 - 8x + 4$
$f(x) = 2(x^2 - 4x) + 4$
$f(x) = 2(x^2 - 4x + 4) - 8 + 4$
$f(x) = 2(x - 2)^2 - 4$ $h = -2, k = -4, C = 2.$
It is the same as the graph of $p(x) = x^2$ shifted to
the right 2 units, expanded by a factor of 2, and
shifted down 4 units.

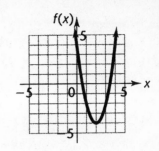

63. $f(x) = -\dfrac{1}{2}x^2 + 2x + 1$

$f(x) = -\dfrac{1}{2}(x^2 - 4x) + 1$

$f(x) = -\dfrac{1}{2}(x^2 - 4x + 4) + 2 + 1$

$f(x) = -\dfrac{1}{2}(x - 2)^2 + 3$ $h = -2, k = 3, C = -\dfrac{1}{2}$

It is the same as the graph of $p(x) = x^2$ shifted to

the right 2 units, contracted by a factor of $\dfrac{1}{2}$,

reflected with respect to the x axis, and shifted up 3 units.

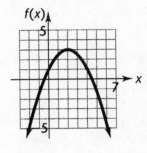

65. If we let $g(x) = 2x - 7$, then
$h(x) = [g(x)]^4$
Now if we let $f(x) = x^4$, we have
$h(x) = [g(x)]^4 = f[g(x)] = (f \circ g)(x)$

67. If we let $g(x) = 4 + 2x$, then
$h(x) = \sqrt{g(x)}$
Now if we let $f(x) = x^{1/2}$, we have
$h(x) = \sqrt{g(x)} = [g(x)]^{1/2}$
$= f[g(x)] = (f \circ g)(x).$

69. If we let $f(x) = x^7$, then
$h(x) = 3f(x) - 5$
Now if we let $g(x) = 3x - 5$, we have
$h(x) = 3f(x) - 5 = g[f(x)] = (g \circ f)(x)$

71. If we let $f(x) = x^{-1/2}$, then
$h(x) = 4f(x) + 3$
Now if we let $g(x) = 4x + 3$, we have
$h(x) = 4f(x) + 3 = g[f(x)] = (g \circ f)(x)$

73. The graph of $y = |x|$ is reflected in the x axis and vertically expanded by a
factor of 3. Equation: $y = -3|x|$.

75. The graph of $y = x^3$ is reflected in the x axis and vertically contracted by a
factor of 0.5. Equation: $y = -0.5x^3$.

77. $(f + g)(x) = f(x) + g(x) = x + \dfrac{1}{x} + x - \dfrac{1}{x} = 2x$

> **Common Error:**
> Domain is not $(-\infty, \infty)$. See below.

$(f - g)(x) = f(x) - g(x) = x + \dfrac{1}{x} - \left(x - \dfrac{1}{x}\right) = \dfrac{2}{x}$

$(fg)(x) = f(x)g(x) = \left(x + \dfrac{1}{x}\right)\left(x - \dfrac{1}{x}\right) = x^2 - \dfrac{1}{x^2}$

$\left(\dfrac{f}{g}\right)(x) = \dfrac{f(x)}{g(x)} = \dfrac{x + \dfrac{1}{x}}{x - \dfrac{1}{x}} = \dfrac{x^2 + 1}{x^2 - 1}$

The domains of f and g are both $\{x \mid x \neq 0\} = (-\infty, 0) \cup (0, \infty)$
This is therefore the domain of $f + g$, $f - g$, and fg. To find the domain of $\dfrac{f}{g}$,
we must exclude from this domain the set of values of x for which $g(x) = 0$.

$x - \dfrac{1}{x} = 0$
$x^2 - 1 = 0$
$x^2 = 1$
$x = -1, 1$

Hence, the domain of $\dfrac{f}{g}$ is $\{x \mid x \neq 0, -1, \text{ or } 1\}$ or

$(-\infty, -1) \cup (-1, 0) \cup (0, 1) \cup (1, \infty).$

79. $(f + g)(x) = f(x) + g(x) = 1 - \dfrac{x}{|x|} + 1 + \dfrac{x}{|x|} = 2$

$(f - g)(x) = f(x) - g(x) = 1 - \dfrac{x}{|x|} - \left(1 + \dfrac{x}{|x|}\right) = 1 - \dfrac{x}{|x|} - 1 - \dfrac{x}{|x|} = \dfrac{-2x}{|x|}$

$(fg)(x) = f(x)g(x) = \left(1 - \dfrac{x}{|x|}\right)\left(1 + \dfrac{x}{|x|}\right) = (1)^2 - \left(\dfrac{x}{|x|}\right)^2 = 1 - \dfrac{x^2}{|x|^2}$

$$= 1 - \dfrac{x^2}{x^2} = 1 - 1 = 0$$

$\left(\dfrac{f}{g}\right)(x) = \dfrac{f(x)}{g(x)} = \dfrac{1 - \frac{x}{|x|}}{1 + \frac{x}{|x|}} = \dfrac{|x| - x}{|x| + x}.$ This can be further simplified

however, when we examine the domain of $\dfrac{f}{g}$ below.

The domains of f and g are both
$\{x \mid x \neq 0\} = (-\infty, 0) \cup (0, \infty)$
This is therefore the domain of $f + g$, $f - g$, and fg. To find the domain of $\dfrac{f}{g}$,
we must exclude from this domain the set of values of x for which $g(x) = 0$.

$1 + \dfrac{x}{|x|} = 0$
$|x| + x = 0$
$\quad |x| = -x$
$\qquad x$ is negative

Thus the domain of $\dfrac{f}{g}$ is the positive numbers, $(0, \infty)$. On this domain,

$|x| = x$, hence $\left(\dfrac{f}{g}\right)(x) = \dfrac{|x| - x}{|x| + x} = \dfrac{x - x}{x + x} = \dfrac{0}{2x} = 0$

81. $(f \circ g)(x) = f[g(x)] = f(x^2) = \sqrt{4 - x^2}$
The domain of f is $(-\infty, 4]$. The domain of g is all real numbers. Hence the
domain of $f \circ g$ is the set of those real numbers x for which $g(x)$ is in $(-\infty, 4]$,
that is, for which $x^2 \leq 4$, or $-2 \leq x \leq 2$.
Domain of $f \circ g = \{x \mid -2 \leq x \leq 2\} = [-2, 2]$

$(g \circ f)(x) = g[f(x)] = g(\sqrt{4 - x}) = (\sqrt{4 - x})^2 = 4 - x$
The domain of $g \circ f$ is the set of those numbers x in $(-\infty, 4]$ for which $f(x)$ is
in $(-\infty, \infty)$, that is, $(-\infty, 4]$.

83. $(f \circ g)(x) = f[g(x)] = f\left(\dfrac{x}{x - 2}\right) = \dfrac{\frac{x}{x - 2} + 5}{\frac{x}{x - 2}} = \dfrac{x + 5(x - 2)}{x} = \dfrac{x + 5x - 10}{x} = \dfrac{6x - 10}{x}$
The domain of f is $\{x \mid x \neq 0\}$. The domain of g is $\{x \mid x \neq 2\}$. Hence the domain of
$f \circ g$ is the set of those numbers in $\{x \mid x \neq 2\}$ for which $g(x)$ is in $\{x \mid x \neq 0\}$.

Thus we must exclude from $\{x \mid x \neq 2\}$ those numbers x for which $\dfrac{x}{x - 2} = 0$, or
$x = 0$. Hence the domain of $f \circ g$ is $\{x \mid x \neq 0, x \neq 2\}$, or $(-\infty, 0) \cup (0, 2) \cup (2, \infty)$.

$(g \circ f)(x) = g[f(x)] = g\left(\dfrac{x + 5}{x}\right) = \dfrac{\frac{x + 5}{x}}{\frac{x + 5}{x} - 2} = \dfrac{x + 5}{x + 5 - 2x} = \dfrac{x + 5}{5 - x}$
The domain of $g \circ f$ is the set of those numbers in $\{x \mid x \neq 0\}$ for which $f(x)$ is in
$\{x \mid x \neq 2\}$. Thus we must exclude from $\{x \mid x \neq 0\}$ those numbers x for which
$\dfrac{x + 5}{x} = 2$, or $x + 5 = 2x$, or $x = 5$. Hence the domain of $g \circ f$ is $\{x \mid x \neq 0, x \neq 5\}$
or $(-\infty, 0) \cup (0, 5) \cup (5, \infty)$.

85. $(f \circ g)(x) = f[g(x)] = f(\sqrt{9 + x^2}) = \sqrt{25 - (\sqrt{9 + x^2})^2} = \sqrt{25 - (9 + x^2)} = \sqrt{16 - x^2}$
The domain of f is $[-5, 5]$. The domain of g is $(-\infty, \infty)$. Hence the domain of $f \circ g$ is the set of those real numbers x for which $g(x)$ is in $[-5, 5]$, that is, $\sqrt{9 + x^2} \le 5$, or $9 + x^2 \le 25$, or $x^2 \le 16$, or $-4 \le x \le 4$. Hence the domain of $f \circ g$ is $\{x \mid -4 \le x \le 4\}$ or $[-4, 4]$.

$(g \circ f)(x) = g[f(x)] = g(\sqrt{25 - x^2}) = \sqrt{9 + (\sqrt{25 - x^2})^2} = \sqrt{9 + 25 - x^2} = \sqrt{34 - x^2}$
The domain of $g \circ f$ is the set of those numbers x in $[-5, 5]$ for which $g(x)$ is real. Since $g(x)$ is real for all x, the domain of $g \circ f$ is $[-5, 5]$.

Common Error: The domain of $g \circ f$ is not evident from the final form $\sqrt{34 - x^2}$. It is not $[-\sqrt{34}, \sqrt{34}]$.

87. Graph of $f(x)$

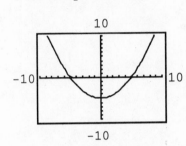

Graph of
$|f(x)| = |0.2x^2 - 5|$

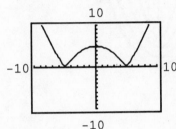

Graph of
$-|f(x)| = -|0.2x^2 - 5|$

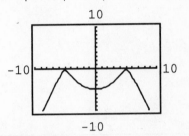

89. Graph of $f(x)$

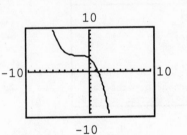

Graph of
$|f(x)| = |4 - 0.1(x + 2)^3|$

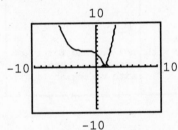

Graph of
$-|f(x)| = -|4 - 0.1(x + 2)^3|$

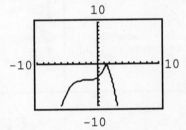

91. The graph of $y = |f(x)|$ is the same as the graph of $y = f(x)$ whenever $f(x) \ge 0$ and is the reflection of the graph of $y = f(x)$ with respect to the x axis whenever $f(x) < 0$.

93. The profit function is the difference of the revenue and the cost functions.
$P = R - C$
Hence $P(x) = R(x) - C(x)$

$$= \left(20x - \frac{1}{200}x^2\right) - (10x + 30,000)$$

$$= 20x - \frac{1}{200}x^2 - 10x - 30,000$$

$$= 10x - \frac{1}{200}x^2 - 30,000$$

Next we use composition to express P as a function of the price p.
$(P \circ f)(p) = P[f(p)]$

$$= P(4,000 - 200p)$$

$$= 10(4,000 - 200p) - \frac{1}{200}(4,000 - 200p)^2 - 30,000$$

$$= 40,000 - 2,000p - \frac{1}{200}(16,000,000 - 1,600,000p + 40,000p^2) - 30,000$$

$$= 40,000 - 2,000p - 80,000 + 8,000p - 200p^2 - 30,000$$

$$= -70,000 + 6,000p - 200p^2$$

95. $y = \frac{1}{C}x^2 - C$

$C = 1$: $y = x^2 - 1$ We sketch the graph of $y = x^2$ shifted down 1 unit

$C = 2$: $y = \frac{1}{2}x^2 - 2$ We sketch the graph of $y = x^2$ contracted by a factor of $\frac{1}{2}$, and shifted down 2 units.

$C = 3$: $y = \frac{1}{3}x^2 - 3$ We sketch the graph of $y = x^2$ contracted by a factor of $\frac{1}{3}$, and shifted down 3 units.

$C = 4$: $y = \frac{1}{4}x^2 - 4$ We sketch the graph of $y = x^2$ contracted by a factor of $\frac{1}{4}$, and shifted down 4 units.

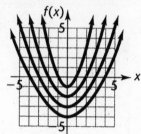

97. $V(t) = \frac{64}{c^2}(C - t)^2 \quad 0 \le t \le C$

t	$C = 1$ $V = 64(1 - t)^2$	$C = 2$ $V = 16(2 - t)^2$	$C = 4$ $V = 4(4 - t)^2$	$C = 8$ $V = (8 - t)^2$
0	64	64	64	64
1	0	16	36	49
2	not defined for $t > 1$	0	16	36
4	arc of a parabola	not defined for $t > 2$	0	16
6	with vertex at (1,0)	arc of a parabola with	not defined for $t > 4$	4
8		vertex at (2,0)	arc of a parabola	0
			with vertex at (4,0)	arc of a parabola with vertex at (8,0)

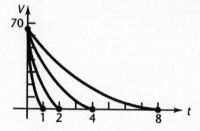

99.

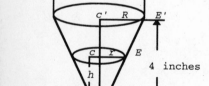

(A) We note: In the figure, triangles VCE and $VC'E'$ are similar. Moreover

R = radius of cup = $\frac{1}{2}$ diameter of cup = $\frac{1}{2}(4)$ = 2 inches.

Hence $\frac{r}{2} = \frac{h}{4}$ or $r = \frac{1}{2}h$. We write $r(h) = \frac{1}{2}h$.

(B) Since $V = \frac{1}{3}\pi r^2 h$ and $r = \frac{1}{2}h$, $V = \frac{1}{3}\pi\left(\frac{1}{2}h\right)^2 h = \frac{1}{3}\pi\frac{1}{4}h^2 h = \frac{1}{12}\pi h^3$.
We write $V(h) = \frac{1}{12}\pi h^3$.

(C) Since $V(h) = \frac{1}{12}\pi h^3$ and $h(t) = 0.5\sqrt{t}$, we use composition to express V as a function of t.

$$(V \circ h)(t) = V[h(t)]$$

$$= \frac{1}{12}\pi[h(t)]^3$$

$$= \frac{1}{12}\pi[0.5\sqrt{t}]^3$$

$$= \frac{1}{12}\pi(0.125)(t^{1/2})^3$$

$$= \frac{0.125}{12}\pi t^{3/2}$$

We write $V(t) = \frac{0.125}{12}\pi t^{3/2}$.

Exercise 3-6

Key Ideas and Formulas

A function is one-to-one if no two ordered pairs in the function have the same second component and different first components.

If $f(a) = f(b)$ for at least one pair of domain values a and b, $a \neq b$, then f is not one-to-one.

If the assumption of $f(a) = f(b)$ always implies that the domain values a and b are equal, then f is one-to-one.

A function is one-to-one if and only if each horizontal line intersects the graph of the function in at most one point. (Horizontal line test.)

If a function f is increasing throughout its domain or decreasing throughout its domain, then f is a one-to-one function.

If f is a one-to-one function, then the inverse of f, denoted f^{-1}, is the function formed by reversing all the ordered pairs in f. $f^{-1} = \{(y, x) \mid (x, y) \text{ is in } f\}$.

If f is not one-to-one, then f does not have an inverse and f^{-1} does not exist.

If f^{-1} exists, then

1. f^{-1} is a one-to-one function.
2. Domain of f^{-1} = Range of f.
3. Range of f^{-1} = Domain of f

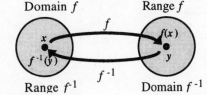

If f^{-1} exists, then

1. $x = f^{-1}(y)$ if and only if $y = f(x)$
2. $f^{-1}[f(x)] = x$ for all x in the domain of f
3. $f[f^{-1}(y)] = y$ for all y in the domain of f^{-1}

To find the inverse of a function f

1. Find the domain of f and verify that f is one-to-one. If f is not one-to-one, then stop—f^{-1} does not exist.

2. Solve the equation $y = f(x)$ for x. The result is an equation of the form $x = f^{-1}(y)$.

3. Interchange x and y in the equation found in Step 2. The result is an equation of the form $y = f^{-1}(x)$.

4. Find the domain of f^{-1}. (This is the same as the range of f.)

5. Check by showing that
   $f^{-1}[f(x)] = x$ for all x in domain of f
   $f[f^{-1}(x)] = x$ for all x in the domain of f^{-1}

The graphs of $y = f(x)$ and $y = f^{-1}(x)$ are symmetric with respect to the line $y = x$.

1. One-to-one

3. The range element 3 corresponds to more than one domain element. Not one-to-one.

5. One-to-one

7. The range element 7 corresponds to more than one domain element. Not one-to-one.

9. One-to-one

11. Some range elements (0, for example) correspond to more than one domain element. Not one-to-one.

13. One-to-one   **15.** One-to-one   **17.** Assume $F(a) = F(b)$

$$\frac{1}{2}a + 1 = \frac{1}{2}b + 1$$

Then $\frac{1}{2}a = \frac{1}{2}b$

$a = b$

Therefore F is one-to-one.

19. $H(x) = 4 - x^2$
Since $H(1) = 4 - 1^2 = 3$ and $H(-1) = 4 - (-1)^2 = 3$, both $(1, 3)$ and $(-1, 3)$ belong to H.
H is not one-to-one.

21. Assume $M(a) = M(b)$

$$\sqrt{a + 1} = \sqrt{b + 1}$$

Then $a + 1 = b + 1$

$a = b$

M is one-to-one.

23. $f(x) = \dfrac{x^2 + |x|}{x}$

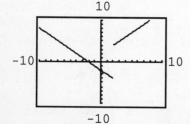

This graph passes the horizontal line test. f is one-to-one.

25. $f(x) = \dfrac{x^3 + |x|}{x}$

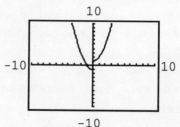

This graph does not pass the horizontal line test. f is not one-to-one.

27. $f(x) = \dfrac{x^2 - 4}{|x - 2|}$

This graph does not pass the horizontal line test. f is not one-to-one.

29. $f(x) = \dfrac{x^3 - 9x}{|x^2 - 9|}$

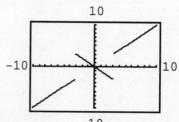

This graph passes the horizontal line test. f is one-to-one.

31. From the given graph of f, we see
Domain of $f = [-4, 4]$ Range of $f = [1, 5]$
Hence
Range of $f^{-1} = [-4, 4]$ Domain of $f^{-1} = [1, 5]$
The graph of f^{-1} is drawn by reflecting the
given graph of f in the line $y = x$.

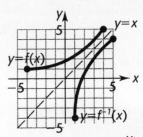

33. From the given graph of f, we see
Domain of $f = [-5, 3]$ Range of $f = [-3, 5]$
Hence
Range of $f^{-1} = [-5, 3]$ Domain of $f^{-1} = [-3, 5]$
The graph of f^{-1} is drawn by reflecting the given graph
of f in the line $y = x$.

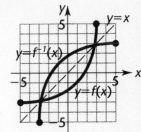

35. $f(x) = 3x + 6$ $g(x) = \dfrac{1}{3}x - 2$

$$g[f(x)] = g(3x + 6) = \frac{1}{3}(3x + 6) - 2 = x + 2 - 2 = x$$

$$f[g(x)] = f\left(\frac{1}{3}x - 2\right) = 3\left(\frac{1}{3}x - 2\right) + 6 = x - 6 + 6 = x$$

37. $f(x) = 4 + x^2$ $x \geq 0$ $g(x) = \sqrt{x - 4}$

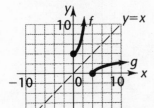

$$g[f(x)] = g(4 + x^2) = \sqrt{4 + x^2 - 4} = \sqrt{x^2} = x \text{ since } x \geq 0$$

$$f[g(x)] = f(\sqrt{x - 4}) = 4 + (\sqrt{x - 4})^2 = 4 + x - 4 = x$$

39. $f(x) = -\sqrt{x - 2}$ $g(x) = x^2 + 2, \quad x \leq 0$

$$f[g(x)] = f(x^2 + 2) = -\sqrt{x^2 + 2 - 2} = -\sqrt{x^2} = -(-x) = x$$

(Note: $\sqrt{x^2} = -x$ since $x \leq 0$)

$$g[f(x)] = g(-\sqrt{x - 2}) = (-\sqrt{x - 2})^2 + 2 = x - 2 + 2 = x$$

Common Error: $(-\sqrt{x - 2})^2 \neq -(x - 2) \text{ nor } -x - 2$

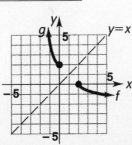

41. $f(x) = 3x$

Solve $y = f(x)$ for x:

$$y = 3x$$

$$x = \frac{1}{3}y = f^{-1}(y)$$

Interchange x and y:

$$y = f^{-1}(x) = \frac{1}{3}x$$

Domain of f^{-1} = Range of $f = (-\infty, \infty)$

Check: $f^{-1}[f(x)] = f^{-1}(3x) = \frac{1}{3}(3x) = x$

$$f[f^{-1}(x)] = f\left(\frac{1}{3}x\right) = 3\left(\frac{1}{3}x\right) = x$$

$$f^{-1}(x) = \frac{1}{3}x$$

43. $f(x) = 4x - 3$

Solve $y = f(x)$ for x:

$$y = 4x - 3$$

$$y + 3 = 4x$$

$$x = \frac{y + 3}{4} = f^{-1}(y)$$

Interchange x and y:

$$y = f^{-1}(x) = \frac{x + 3}{4}$$

Domain of f^{-1} = Range of $f = (-\infty, \infty)$

Check: $f^{-1}[f(x)] = f^{-1}(4x - 3)$

$$= \frac{4x - 3 + 3}{4}$$

$$= \frac{4x}{4}$$

$$= x$$

$$f[f^{-1}(x)] = f\left(\frac{x + 3}{4}\right)$$

$$= 4\left(\frac{x + 3}{4}\right) - 3$$

$$= x + 3 - 3$$

$$= x$$

$$f^{-1}(x) = \frac{x + 3}{4}$$

45. $f(x) = \frac{1}{10}x + \frac{3}{5}$

Solve $y = f(x)$ for x:

$$y = \frac{1}{10}x + \frac{3}{5}$$

$$10y = x + 6$$

$$x = 10y - 6 = f^{-1}(y)$$

Interchange x and y:

$$y = f^{-1}(x) = 10x - 6$$

Domain of f^{-1} = Range of $f = (-\infty, \infty)$

Check: $f^{-1}[f(x)] = f^{-1}\left(\frac{1}{10}x + \frac{3}{5}\right)$

$$= 10\left(\frac{1}{10}x + \frac{3}{5}\right) - 6$$

$$= x + 6 - 6$$

$$= x$$

$$f[f^{-1}(x)] = f(10x - 6)$$

$$= \frac{1}{10}(10x - 6) + \frac{3}{5}$$

$$= x - \frac{6}{10} + \frac{3}{5}$$

$$= x$$

$$f^{-1}(x) = 10x - 6$$

47. $f(x) = \dfrac{2}{x - 1}$

Solve $y = f(x)$ for x:

$$y = \frac{2}{x - 1} \quad x \neq 1$$

$$y(x - 1) = 2$$
$$xy - y = 2$$
$$xy = y + 2$$
$$x = \frac{y + 2}{y} = f^{-1}(y)$$

Interchange x and y:

$$y = f^{-1}(x) = \frac{x + 2}{x}$$

Domain of $f^{-1} = (-\infty, 0) \cup (0, \infty)$

Check: $f^{-1}[f(x)] = f^{-1}\left(\dfrac{2}{x - 1}\right)$

$$= \frac{\frac{2}{x - 1} + 2}{\frac{2}{x - 1}}$$

$$= \frac{2 + 2(x - 1)}{2}$$

$$= \frac{2 + 2x - 2}{2}$$

$$= \frac{2x}{2}$$

$$= x$$

$$f^{-1}(x) = \frac{x + 2}{x}$$

Graph of $y = f(x)$:

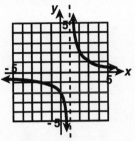

Range of $f = (-\infty, 0) \cup (0, \infty)$

$$f[f^{-1}(x)] = \frac{2}{\frac{x + 2}{x} - 1}$$

$$= \frac{2x}{x + 2 - x}$$

$$= \frac{2x}{2}$$

$$= x$$

49. $f(x) = \dfrac{x}{x + 2}$

Solve $y = f(x)$ for x:

$$y = \frac{x}{x + 2} \quad x \neq -2$$

$$y(x + 2) = x$$
$$xy + 2y = x$$
$$2y = x - xy$$
$$2y = x(1 - y)$$
$$x = \frac{2y}{1 - y} = f^{-1}(y)$$

Interchange x and y:

$$y = f^{-1}(x) = \frac{2x}{1 - x}$$

Domain of $f^{-1} = (-\infty, 1) \cup (1, \infty)$

Check: $f^{-1}[f(x)] = \dfrac{2\frac{x}{x + 2}}{1 - \frac{x}{x + 2}}$

$$= \frac{2x}{x + 2 - x}$$

$$= \frac{2x}{2}$$

$$= x$$

$$f^{-1}(x) = \frac{2x}{1 - x}$$

Graph of $y = f(x)$:

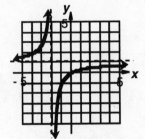

Range of $f = (-\infty, 1) \cup (1, \infty)$

$$f[f^{-1}(x)] = \frac{\frac{2x}{1 - x}}{\frac{2x}{1 - x} + 2}$$

$$= \frac{2x}{2x + 2(1 - x)}$$

$$= \frac{2x}{2x + 2 - 2x}$$

$$= \frac{2x}{2}$$

$$= x$$

51. $f(x) = \dfrac{2x + 5}{3x - 4}$

Solve $y = f(x)$ for x:

$$y = \frac{2x + 5}{3x - 4}$$

$$(3x - 4)y = 2x + 5$$

$$3xy - 4y = 2x + 5$$

$$3xy - 2x = 4y + 5$$

$$x(3y - 2) = 4y + 5$$

$$x = \frac{4y + 5}{3y - 2} = f^{-1}(y)$$

Interchange x and y:

$$y = f^{-1}(x) = \frac{4x + 5}{3x - 2}$$

Domain of $f^{-1} = \left(-\infty, \dfrac{2}{3}\right) \cup \left(\dfrac{2}{3}, \infty\right)$

Check: $f^{-1}[f(x)] = \dfrac{4\frac{2x + 5}{3x - 4} + 5}{3\frac{2x + 5}{3x - 4} - 2}$

$= \dfrac{4(2x + 5) + 5(3x - 4)}{3(2x + 5) - 2(3x - 4)}$

$= \dfrac{8x + 20 + 15x - 20}{6x + 15 - 6x + 8}$

$= \dfrac{23x}{23}$

$= x$

$f^{-1}(x) = \dfrac{4x + 5}{3x - 2}$

Graph of $y = f(x)$:

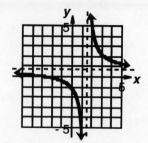

Range of $f = \left(-\infty, \dfrac{2}{3}\right) \cup \left(\dfrac{2}{3}, \infty\right)$

$f[f^{-1}(x)] = \dfrac{2\frac{4x + 5}{3x - 2} + 5}{3\frac{4x + 5}{3x - 2} - 4}$

$= \dfrac{2(4x + 5) + 5(3x - 2)}{3(4x + 5) - 4(3x - 2)}$

$= \dfrac{8x + 10 + 15x - 10}{12x + 15 - 12x + 8}$

$= \dfrac{23x}{23}$

$= x$

53. $f(x) = x^3 + 1$

Solve $y = f(x)$ for x:

$y = x^3 + 1$

$y - 1 = x^3$

$x = \sqrt[3]{y - 1} = f^{-1}(y)$

Interchange x and y:

$y = f^{-1}(x) = \sqrt[3]{x - 1}$ Domain: $(-\infty, \infty)$

Check: $f^{-1}[f(x)] = \sqrt[3]{x^3 + 1 - 1} = \sqrt[3]{x^3} = x$

$f[f^{-1}(x)] = (\sqrt[3]{x - 1})^3 + 1 = x - 1 + 1 = x$

$f^{-1}(x) = \sqrt[3]{x - 1}$

55. $f(x) = 4 - \sqrt[5]{x + 2}$

Solve $y = f(x)$ for x:

$y = 4 - \sqrt[5]{x + 2}$

$y - 4 = -\sqrt[5]{x + 2}$

$4 - y = \sqrt[5]{x + 2}$

$(4 - y)^5 = x + 2$

$x = (4 - y)^5 - 2 = f^{-1}(y)$

Interchange x and y:

$y = f^{-1}(x) = (4 - x)^5 - 2$ Domain: $(-\infty, \infty)$

Check:

$f^{-1}[f(x)] = [4 - (4 - \sqrt[5]{x + 2})]^5 - 2$ $f[f^{-1}(x)] = 4 - \sqrt[5]{(4 - x)^5 - 2 + 2}$

$\qquad\qquad = [4 - 4 + \sqrt[5]{x + 2}]^5 - 2$ $\qquad\qquad = 4 - \sqrt[5]{(4 - x)^5}$

$\qquad\qquad = [\sqrt[5]{x + 2}]^5 - 2$ $\qquad\qquad = 4 - (4 - x)$

$\qquad\qquad = x + 2 - 2$ $\qquad\qquad = 4 - 4 + x$

$\qquad\qquad = x$ $\qquad\qquad = x$

$f^{-1}(x) = (4 - x)^5 - 2$

57. $f(x) = \frac{1}{2}\sqrt{16 - x}$

Solve $y = f(x)$ for x:

$y = \frac{1}{2}\sqrt{16 - x}$

$\left.\begin{array}{l} 2y = \sqrt{16 - x} \\ 4y^2 = 16 - x \end{array}\right\}$ these are equivalent only if $y \geq 0$

$x = 16 - 4y^2 \quad y \geq 0 \quad f^{-1}(y) = 16 - 4y^2$

Interchange x and y:

$y = f^{-1}(x) = 16 - 4x^2, \ x \geq 0 \quad$ Domain: $[0, \infty)$

Check: $f^{-1}[f(x)] = 16 - 4\left[\frac{1}{2}\sqrt{16 - x}\right]^2$ $f[f^{-1}(x)] = \frac{1}{2}\sqrt{16 - (16 - 4x^2)}$

$\qquad\qquad\qquad = 16 - 4\left[\frac{1}{4}(16 - x)\right]$ $\qquad\qquad\quad = \frac{1}{2}\sqrt{16 - 16 + 4x^2}$

$\qquad\qquad\qquad = 16 - (16 - x)$ $\qquad\qquad\quad = \frac{1}{2}\sqrt{4x^2} \quad \sqrt{4x^2} = 2x$ since $x \geq 0$

$\qquad\qquad\qquad = 16 - 16 + x$ $\qquad\qquad\quad = \frac{1}{2}(2x)$

$\qquad\qquad\qquad = x$ $\qquad\qquad\quad = x$

$f^{-1}(x) = 16 - 4x^2, \ x \geq 0$

59. $f(x) = 3 - \sqrt{x - 2}$

Solve $y = f(x)$ for x:

$y = 3 - \sqrt{x - 2}$

$y - 3 = -\sqrt{x - 2}$

$\left.\begin{array}{l} 3 - y = \sqrt{x - 2} \\ (3 - y)^2 = x - 2 \end{array}\right\}$ these are equivalent only if $y \leq 3$

$x = (3 - y)^2 + 2 \quad y \leq 3 \quad f^{-1}(y) = (3 - y)^2 + 2$

Interchange x and y

$y = f^{-1}(x) = (3 - x)^2 + 2 \quad x \leq 3 \quad$ Domain: $(-\infty, 3]$

Check:

$f^{-1}[f(x)] = [3 - (3 - \sqrt{x - 2})]^2 + 2$ $f[f^{-1}(x)] = 3 - \sqrt{(3 - x)^2 + 2 - 2}$

$\qquad\qquad = (\sqrt{x - 2})^2 + 2$ $\qquad\qquad = 3 - \sqrt{(3 - x)^2}$

$\qquad\qquad\qquad\qquad\qquad\qquad\qquad\qquad\qquad\quad \sqrt{(3 - x)^2} = 3 - x$ since $x \leq 3$

$\qquad\qquad = x - 2 + 2$ $\qquad\qquad = 3 - (3 - x)$

$\qquad\qquad = x$ $\qquad\qquad = 3 - 3 + x$

$\qquad\qquad\qquad\qquad\qquad\qquad\qquad\qquad = x$

$f^{-1}(x) = (3 - x)^2 + 2, \ x \leq 3$

61. Since in passing from a function to its inverse, x and y are interchanged, the x intercept of f is the y intercept of f^{-1} and the y intercept of f is the x intercept of f^{-1}.

63. $f(x) = (x - 1)^2 + 2 \quad x \geq 1$
Solve $y = f(x)$ for x:
$$y = (x - 1)^2 + 2 \quad x \geq 1$$
$$y - 2 = (x - 1)^2 \quad x \geq 1$$
$$\sqrt{y - 2} = \sqrt{(x - 1)^2} \quad \text{Since } x \geq 1, \ \sqrt{(x - 1)^2} = x - 1$$
$$\sqrt{y - 2} = x - 1$$
$$x = 1 + \sqrt{y - 2} = f^{-1}(y)$$
Interchange x and y:

$y = f^{-1}(x) = 1 + \sqrt{x - 2}$ Domain: $x \geq 2$
Check:

$$f^{-1}[f(x)] = 1 + \sqrt{(x - 1)^2 + 2 - 2} \qquad\qquad f[f^{-1}(x)] = (1 + \sqrt{x - 2} - 1)^2 + 2$$
$$= 1 + \sqrt{(x - 1)^2} \qquad\qquad\qquad\qquad = (\sqrt{x - 2})^2 + 2$$
$$\sqrt{(x - 1)^2} = x - 1 \text{ since } x \geq 1 \qquad\qquad = x - 2 + 2$$
$$= 1 + x - 1 \qquad\qquad\qquad\qquad\qquad = x$$
$$= x$$

$f^{-1}(x) = 1 + \sqrt{x - 2}, \ x \geq 2$

65. $f(x) = x^2 + 2x - 2 \quad x \leq -1$
Solve $y = f(x)$ for x:
$$y = x^2 + 2x - 2 \qquad\qquad x \leq -1$$
$$y + 3 = x^2 + 2x + 1 \qquad\quad x \leq -1$$
$$y + 3 = (x + 1)^2 \qquad\qquad x \leq -1$$
$$\sqrt{y + 3} = \sqrt{(x + 1)^2} \quad \text{Since } x \leq -1 \quad \sqrt{(x + 1)^2} = -(x + 1)$$
$$\sqrt{y + 3} = -(x + 1)$$
$$-\sqrt{y + 3} = x + 1$$
$$x = -1 - \sqrt{y + 3} = f^{-1}(y)$$
Interchange x and y:

$y = f^{-1}(x) = -1 - \sqrt{x + 3} \quad x \geq -3$ Domain: $[-3, \infty)$

Check: $f^{-1}[f(x)] = -1 - \sqrt{x^2 + 2x - 2 + 3} \qquad\qquad x \leq -1$
$$= -1 - \sqrt{x^2 + 2x + 1} \qquad\qquad x \leq -1$$
$$= -1 - \sqrt{(x + 1)^2} \qquad\qquad\quad x \leq -1$$
$$\sqrt{(x + 1)^2} = -(x + 1) \text{ since } x \leq -1$$
$$= -1 - [-(x + 1)]$$
$$= -1 + x + 1$$
$$= x$$

$$f[f^{-1}(x)] = (-1 - \sqrt{x + 3})^2 + 2(-1 - \sqrt{x + 3}) - 2$$
$$= 1 + 2\sqrt{x + 3} + x + 3 - 2 - 2\sqrt{x + 3} - 2$$
$$= 4 + x - 4$$
$$= x$$

$f^{-1}(x) = -1 - \sqrt{x + 3}, \ x \geq -3$

67. $f(x) = -\sqrt{9 - x^2}$ $0 \le x \le 3$ f is one-to-one. See graph below.

Solve $y = f(x)$ for x:

$$y = -\sqrt{9 - x^2} \quad 0 \le x \le 3$$
$$y^2 = 9 - x^2 \qquad y \le 0$$
$$y^2 - 9 = -x^2 \qquad y \le 0$$
$$9 - y^2 = x^2 \qquad y \le 0$$
$$x = \sqrt{9 - y^2} \quad y \le 0 \quad f^{-1}(y) = \sqrt{9 - y^2}$$

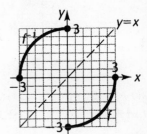

positive square root only because $0 \le x$

Interchange x and y:

$$y = f^{-1}(x) = \sqrt{9 - x^2} \quad \text{Domain: } -3 \le x \le 0$$

Check: $f^{-1}[f(x)] = \sqrt{9 - (-\sqrt{9 - x^2})^2}$

$$= \sqrt{9 - (9 - x^2)}$$
$$= \sqrt{9 - 9 + x^2}$$
$$= \sqrt{x^2}$$
$$\sqrt{x^2} = x \text{ since } x \ge 0 \text{ in the domain of } f$$
$$= x$$

$$f[f^{-1}(x)] = -\sqrt{9 - (\sqrt{9 - x^2})^2}$$
$$= -\sqrt{9 - (9 - x^2)}$$
$$= -\sqrt{9 - 9 + x^2}$$
$$= -\sqrt{x^2}$$
$$\sqrt{x^2} = -x \text{ since } x \le 0 \text{ in the domain of } f^{-1}$$
$$= -(-x)$$
$$= x$$

$f^{-1}(x) = \sqrt{9 - x^2}$ Domain of $f^{-1} = [-3, 0]$
Range of $f^{-1} = $ Domain of $f = [0, 3]$

69. $f(x) = \sqrt{9 - x^2}$ $-3 \le x \le 0$ f is one-to-one. See graph below.

Solve $y = f(x)$ for x

$$y = \sqrt{9 - x^2} \quad -3 \le x \le 0$$
$$y^2 = 9 - x^2 \qquad y \ge 0$$
$$y^2 - 9 = -x^2 \qquad y \ge 0$$
$$9 - y^2 = x^2 \qquad y \ge 0$$
$$x = -\sqrt{9 - y^2} \quad y \ge 0 \quad f^{-1}(y) = -\sqrt{9 - y^2}$$

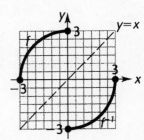

negative square root only because $x \le 0$

Interchange x and y:

$$y = f^{-1}(x) = -\sqrt{9 - x^2} \quad \text{Domain: } 0 \le x \le 3$$

Check: $f^{-1}[f(x)] = -\sqrt{9 - (\sqrt{9 - x^2})^2}$

$$= -\sqrt{9 - (9 - x^2)}$$
$$= -\sqrt{9 - 9 + x^2}$$
$$= -\sqrt{x^2}$$
$$\sqrt{x^2} = -x \text{ since } x \le 0 \text{ in the domain of } f$$
$$= -(-x)$$
$$= x$$

$$f[f^{-1}(x)] = \sqrt{9 - (-\sqrt{9 - x^2})^2}$$
$$= \sqrt{9 - (9 - x^2)}$$
$$= \sqrt{9 - 9 + x^2}$$
$$= \sqrt{x^2}$$
$$\sqrt{x^2} = x \text{ since } x \geq 0 \text{ in the domain of } f^{-1}$$
$$= x$$

$f^{-1}(x) = -\sqrt{9 - x^2}$ Domain: $f^{-1} = [0, 3]$
Range of f^{-1} = Domain of f = $[-3, 0]$

71. $f(x) = 1 + \sqrt{1 - x^2}$ $0 \leq x \leq 1$ f is one-to-one. See graph below.

Solve $y = f(x)$ for x:

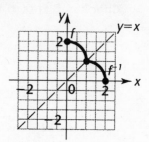

$$y = 1 + \sqrt{1 - x^2} \quad 0 \leq x \leq 1$$
$$y - 1 = \sqrt{1 - x^2}$$
$$(y - 1)^2 = 1 - x^2 \qquad y \geq 1$$
$$y^2 - 2y + 1 = 1 - x^2$$
$$y^2 - 2y = -x^2$$
$$2y - y^2 = x^2$$
$$x = \sqrt{\underbrace{2y - y^2}} \qquad y \geq 1 \quad f^{-1}(y) = \sqrt{2y - y^2}$$

positive square root only because $0 \leq x$

Interchange x and y:

$y = f^{-1}(x) = \sqrt{2x - x^2}$ Domain: $1 \leq x \leq 2$

Check: $f[f^{-1}(x)] = 1 + \sqrt{1 - (\sqrt{2x - x^2})^2}$
$$= 1 + \sqrt{1 - (2x - x^2)}$$
$$= 1 + \sqrt{1 - 2x + x^2}$$
$$= 1 + \sqrt{(1 - x)^2}$$
$$= 1 + [-(1 - x)] \text{ because } 1 \leq x \text{ in the domain of } f^{-1}$$
$$= 1 - 1 + x$$
$$= x$$

$$f^{-1}[f(x)] = \sqrt{2(1 + \sqrt{1 - x^2}) - (1 + \sqrt{1 - x^2})^2}$$
$$= \sqrt{2 + 2\sqrt{1 - x^2} - (1 + 2\sqrt{1 - x^2} + 1 - x^2)}$$
$$= \sqrt{2 + 2\sqrt{1 - x^2} - 1 - 2\sqrt{1 - x^2} - 1 + x^2}$$
$$= \sqrt{x^2}$$
$$= x$$
$$\text{because } x \geq 0 \text{ in the domain of } f$$

$f^{-1}(x) = \sqrt{2x - x^2}$ Domain of f^{-1} = $[1, 2]$
Range of f^{-1} = Domain of f = $[0, 1]$

73. $f(x) = 1 - \sqrt{1 - x^2}$ $-1 \leq x \leq 0$ f is one-to-one. See graph below.

Solve $y = f(x)$ for x:

$$y = 1 - \sqrt{1 - x^2}   -1 \leq x \leq 0$$
$$y - 1 = -\sqrt{1 - x^2}$$
$$1 - y = \sqrt{1 - x^2}$$
$$(1 - y)^2 = 1 - x^2   y \leq 1$$
$$1 - 2y + y^2 = 1 - x^2$$
$$-2y + y^2 = -x^2$$
$$2y - y^2 = x^2$$

$$x = -\sqrt{\underbrace{2y - y^2}}   y \leq 1   f^{-1}(y) = -\sqrt{2y - y^2}$$

negative square root only because $x \leq 0$

Interchange x and y:

$$y = f^{-1}(x) = -\sqrt{2x - x^2}   \text{Domain: } 0 \leq x \leq 1$$

Check: $f[f^{-1}(x)] = 1 - \sqrt{1 - (-\sqrt{2x - x^2})^2}$

$$= 1 - \sqrt{1 - (2x - x^2)}$$
$$= 1 - \sqrt{1 - 2x + x^2}$$
$$= 1 - \sqrt{(1 - x)^2}$$
$$= 1 - (1 - x) \text{ because } x \leq 1 \text{ in the domain of } f^{-1}$$
$$= 1 - 1 + x$$
$$= x$$

$$f^{-1}[f(x)] = -\sqrt{2(1 - \sqrt{1 - x^2}) - (1 - \sqrt{1 - x^2})^2}$$
$$= -\sqrt{2 - 2\sqrt{1 - x^2} - (1 - 2\sqrt{1 - x^2} + 1 - x^2)}$$
$$= -\sqrt{2 - 2\sqrt{1 - x^2} - 1 + 2\sqrt{1 - x^2} - 1 + x^2}$$
$$= -\sqrt{x^2}$$
$$= -(-x) \text{ because } x \leq 0 \text{ in the domain of } f$$
$$= x$$

$f^{-1}(x) = -\sqrt{2x - x^2}$ Domain of $f^{-1} = [0, 1]$
Range of $f^{-1} = $ Domain of $f = [-1, 0]$

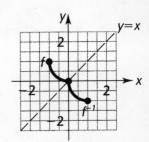

75. $f(x) = ax + b$ $a \neq 0$ f is one-to-one.
Solve $y = f(x)$ for x:

$$y = ax + b$$
$$y - b = ax$$
$$x = \frac{y - b}{a} = f^{-1}(y)$$

Interchange x and y:

$$y = f^{-1}(x) = \frac{x - b}{a}   \text{Domain: } (-\infty, \infty)$$

Check: $f^{-1}[f(x)] = \dfrac{ax + b - b}{a} = \dfrac{ax}{a} = x$

$$f[f^{-1}(x)] = a\frac{x - b}{a} + b = x - b + b = x$$

$$f^{-1}(x) = \frac{x - b}{a}$$

Proof: Assume $f(u) = f(v)$
$$au + b = av + b$$
Then $au = av$
$$u = v$$

77. For f to be its own inverse, $f(x)$ must equal $f^{-1}(x)$ for all x.

Thus $ax + b = \dfrac{x - b}{a}$ or

$a^2x + ab = x - b$ or

$a^2x - x = -ab - b$ or

$(a^2 - 1)x = -ab - b$

This can be true for all x only if $a^2 - 1 = 0$ and $-ab - b = 0$.

If $a^2 - 1 = 0$, then $a = -1$ or $a = 1$.

If $a = -1$, then $-ab - b = -(-1)b - b = 0$ for any b.

If $a = 1$, then $-ab - b = -b - b = -2b = 0$ only if $b = 0$.

Summarizing, f will be its own inverse if either $a = -1$ (b arbitrary), or $a = 1$ and $b = 0$.

79. The slope of the line through (a, b) and (b, a) is

$m_1 = \dfrac{a - b}{b - a} = \dfrac{-(b - a)}{b - a} = -1$

The slope of the line $y = x$ is $1 = m_2$.

Thus, $m_1 m_2 = (-1)(1) = -1$. Hence the lines are perpendicular.

81. $f(x) = (2 - x)^2$

(A) $x \leq 2$

Solve $y = f(x)$ for x:

$y = (2 - x)^2 \qquad x \leq 2$

$\sqrt{y} = 2 - x \quad$ Since $2 - x \geq 0$

$\sqrt{y} - 2 = -x$

$x = 2 - \sqrt{y} = f^{-1}(y)$

Interchange x and y:

$y = f^{-1}(x) = 2 - \sqrt{x} \quad$ Domain: $x \geq 0$

Check: $f^{-1}[f(x)] = 2 - \sqrt{(2 - x)^2}$

$\qquad\qquad\qquad = 2 - (2 - x)$ since $2 - x \geq 0$ *in the domain of f*

$\qquad\qquad\qquad = 2 - 2 + x$

$\qquad\qquad\qquad = x$

$\qquad f[f^{-1}(x)] = [2 - (2 - \sqrt{x})]^2$

$\qquad\qquad\qquad = [2 - 2 + \sqrt{x}]^2$

$\qquad\qquad\qquad = [\sqrt{x}]^2$

$\qquad\qquad\qquad = x$

$f^{-1}(x) = 2 - \sqrt{x}$

(B) $x \geq 2$

Solve $y = f(x)$ for x:

$y = (2 - x)^2 \qquad\qquad x \geq 2$

$\sqrt{y} = -(2 - x) \quad$ Since $2 - x \leq 0$

$\sqrt{y} = -2 + x$

$x = 2 + \sqrt{y} = f^{-1}(y)$

Interchange x and y:

$y = f^{-1}(x) = 2 + \sqrt{x} \quad$ Domain: $x \geq 0$

Check: $f^{-1}[f(x)] = 2 + \sqrt{(2 - x)^2}$

$\qquad\qquad\qquad = 2 + [-(2 - x)]$ since $2 - x \leq 0$ *in the domain of f*

$\qquad\qquad\qquad = 2 - 2 + x$

$\qquad\qquad\qquad = x$

$\qquad\qquad\qquad\qquad f[f^{-1}(x)] = [2 - (2 + \sqrt{x})]^2$

$\qquad\qquad\qquad\qquad\qquad\qquad = [2 - 2 - \sqrt{x}]^2$

$\qquad\qquad\qquad\qquad\qquad\qquad = [-\sqrt{x}]^2$

$\qquad\qquad\qquad\qquad\qquad\qquad = x$

$f^{-1}(x) = 2 + \sqrt{x}$

83. $f(x) = \sqrt{4x - x^2}$

(A) $0 \le x \le 2$

Solve $y = f(x)$ for x:

$$y = \sqrt{4x - x^2}$$
$$y^2 = 4x - x^2 \qquad y \ge 0$$
$$-y^2 = x^2 - 4x$$
$$4 - y^2 = x^2 - 4x + 4$$
$$4 - y^2 = (x - 2)^2$$
$$-\underbrace{\sqrt{4 - y^2}}_{} = x - 2$$

negative square root only because $x \le 2$

$x = 2 - \sqrt{4 - y^2}$ $y \ge 0$ $f^{-1}(y) = 2 - \sqrt{4 - y^2}$

Interchange x and y:

$y = f^{-1}(x) = 2 - \sqrt{4 - x^2}$ Domain: $0 \le x \le 2$

Check: $f^{-1}[f(x)] = 2 - \sqrt{4 - (\sqrt{4x - x^2})^2}$

$\qquad\qquad\quad = 2 - \sqrt{4 - (4x - x^2)}$

$\qquad\qquad\quad = 2 - \sqrt{4 - 4x + x^2}$

$\qquad\qquad\quad = 2 - \sqrt{(2 - x)^2}$

$\qquad\qquad\quad = 2 - (2 - x)$ since $2 - x \ge 0$

$\qquad\qquad\quad = x$

$f[f^{-1}(x)] = \sqrt{4(2 - \sqrt{4 - x^2}) - (2 - \sqrt{4 - x^2})^2}$

$\qquad\qquad\quad = \sqrt{8 - 4\sqrt{4 - x^2} - (4 - 4\sqrt{4 - x^2} + 4 - x^2)}$

$\qquad\qquad\quad = \sqrt{8 - 4\sqrt{4 - x^2} - 4 + 4\sqrt{4 - x^2} - 4 + x^2}$

$\qquad\qquad\quad = \sqrt{x^2}$

$\qquad\qquad\quad = x$ since $0 \le x$

$f^{-1}(x) = 2 - \sqrt{4 - x^2}$, $0 \le x \le 2$

(B) $2 \le x \le 4$

Solve $y = f(x)$ for x:

$$y = \sqrt{4x - x^2}$$
$$y^2 = 4x - x^2 \quad y \ge 0$$
$$-y^2 = x^2 - 4x$$
$$4 - y^2 = x^2 - 4x + 4$$
$$4 - y^2 = (x - 2)^2$$
$$\underbrace{\sqrt{4 - y^2}}_{} = x - 2$$

positive square root only because $x \ge 2$

$x = 2 + \sqrt{4 - y^2}$ $y \ge 0$ $f^{-1}(y) = 2 + \sqrt{4 - y^2}$

Interchange x and y:

$y = f^{-1}(x) = 2 + \sqrt{4 - x^2}$ Domain: $0 \le x \le 2$

Check: $f^{-1}[f(x)] = 2 + \sqrt{4 - (\sqrt{4x - x^2})^2}$

$\qquad\qquad\quad = 2 + \sqrt{4 - (4x - x^2)}$

$\qquad\qquad\quad = 2 + \sqrt{4 - 4x + x^2}$

$\qquad\qquad\quad = 2 + \sqrt{(2 - x)^2}$

$\qquad\qquad\quad = 2 + [-(2 - x)]$ since $2 \le x$ *in the domain of* f

$\qquad\qquad\quad = 2 - 2 + x$

$\qquad\qquad\quad = x$

$$f[f^{-1}(x)] = \sqrt{4(2 + \sqrt{4 - x^2}) - (2 + \sqrt{4 - x^2})^2}$$
$$= \sqrt{8 + 4\sqrt{4 - x^2} - (4 + 4\sqrt{4 - x^2} + 4 - x^2)}$$
$$= \sqrt{8 + 4\sqrt{4 - x^2} - 4 - 4\sqrt{4 - x^2} - 4 + x^2}$$
$$= \sqrt{x^2}$$
$$= x \text{ since } 0 \le x \text{ in the domain of } f^{-1}$$
$$f^{-1}(x) = 2 + \sqrt{4 - x^2}, \; 0 \le x \le 2$$

CHAPTER 3 REVIEW

1. (A) $d(A, B) = \sqrt{[4 - (-2)]^2 + (0 - 3)^2}$
$\qquad\qquad = \sqrt{36 + 9} = \sqrt{45}$

(B) $m = \dfrac{0 - 3}{4 - (-2)} = \dfrac{-3}{6} = -\dfrac{1}{2}$

(C) The slope m_1 of a line perpendicular to AB must satisfy

$$m_1\left(-\dfrac{1}{2}\right) = -1.$$

Therefore, $m_1 = 2$.   *(3-1, 3-2)*

2. (A) Center at $(0, 0)$ and radius $\sqrt{7}$
$x^2 + y^2 = r^2$
$x^2 + y^2 = (\sqrt{7})^2$
$x^2 + y^2 = 7$

(B) Center at $(3, -2)$ and radius $\sqrt{7}$
$\qquad\qquad (h, k) = (3, -2)$
$\qquad\qquad\quad r = \sqrt{7}$
$\qquad (x - h)^2 + (y - k)^2 = r^2$
$(x - 3)^2 + [y - (-2)]^2 = (\sqrt{7})^2$
$\quad (x - 3)^2 + (y + 2)^2 = 7$ *(3-1)*

3. $(x + 3)^2 + (y - 2)^2 = 5$
$[x - (-3)]^2 + (y - 2)^2 = (\sqrt{5})^2$
$\qquad$ Center: $C(h, k) = (-3, 2)$
$\qquad\qquad$ Radius: $r = \sqrt{5}$ *(3-1)*

4. $3x + 2y = 9$
$\qquad 2y = -3x + 9$
$\qquad\quad y = -\dfrac{3}{2}x + \dfrac{9}{2}$

slope: $-\dfrac{3}{2}$

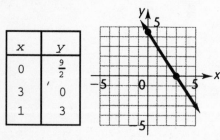

x	y
0	$\frac{9}{2}$
3	0
1	3

     (3-2)

5. The line passes through the two given points, $(6, 0)$ and $(0, 4)$. Thus, its slope is given by

$$m = \dfrac{0 - 4}{6 - 0} = \dfrac{-4}{6} = -\dfrac{2}{3}$$

The equation of the line is, therefore, using the point-slope form,

$$y - 0 = -\dfrac{2}{3}(x - 6)$$
$$\text{or} \quad 3y = -2(x - 6)$$
$$\text{or} \quad 3y = -2x + 12.$$
$$2x + 3y = 12 \qquad\qquad (3-2)$$

6. $y = mx + b \quad m = -\dfrac{2}{3} \quad b = 2$

$y = -\dfrac{2}{3}x + 2$ *(3-2)*

7. *vertical:* $x = -3$, slope not defined; *horizontal:* $y = 4$, slope $= 0$ *(3-2)*

8. (A) A function; domain $= \{1, 2, 3\}$; range $= \{1, 4, 9\}$
(B) Not a function (two range values correspond to some domain values)
(C) A function; domain $= \{-2, -1, 0, 1, 2\}$; range $= \{2\}$ *(3-3)*

9. (A) Not a function (fails vertical line test)
(B) A function
(C) A function
(D) Not a function (fails vertical line test) *(3-3)*

10. (A) Function

(B) Not a function—two range elements correspond to some domain elements; for example 2 and -2 correspond to 4.

(C) Function

(D) Not a function—two range elements correspond to some domain elements; for example 2 and -2 correspond to 2. *(3-3)*

11. $f(2) = 3(2) + 5 = 11$
$g(-2) = 4 - (-2)^2 = 0$
$k(0) = 5$
Therefore
$f(2) + g(-2) + k(0)$
$\quad = 11 + 0 + 5 = 16$ *(3-3)*

12. $m(-2) = 2|-2| - 1 = 3$
$g(2) = 4 - (2)^2 = 0$
Therefore $\dfrac{m(-2) + 1}{g(2) + 4} = \dfrac{3 + 1}{0 + 4} = 1$ *(3-3)*

13. $\dfrac{f(2 + h) - f(2)}{h} = \dfrac{[3(2 + h) + 5] - [3(2) + 5]}{h}$

$\qquad\qquad = \dfrac{6 + 3h + 5 - 11}{h}$

$\qquad\qquad = \dfrac{3h}{h}$

$\qquad\qquad = 3$ *(3-3)*

14. $\dfrac{g(a + h) - g(a)}{h} = \dfrac{[4 - (a + h)^2] - [4 - a^2]}{h}$

$\qquad\qquad = \dfrac{4 - a^2 - 2ah - h^2 - 4 + a^2}{h}$

$\qquad\qquad = \dfrac{-2ah - h^2}{h}$

$\qquad\qquad = \dfrac{h(-2a - h)}{h}$

$\qquad\qquad = -2a - h$ *(3-3)*

15. $(f + g)(x) = f(x) + g(x)$
$\qquad\qquad = 3x + 5 + 4 - x^2 = 9 + 3x - x^2$ *(3-5)*

16. $(f - g)(x) = f(x) - g(x) = 3x + 5 - (4 - x^2) = 3x + 5 - 4 + x^2 = x^2 + 3x + 1$ *(3-5)*

17. $(fg)(x) = f(x)g(x) = (3x + 5)(4 - x^2)$
$\qquad\qquad = 12x - 3x^3 + 20 - 5x^2 = 20 + 12x - 5x^2 - 3x^3$ *(3-5)*

18. $\left(\dfrac{f}{g}\right)(x) = \dfrac{f(x)}{g(x)} = \dfrac{3x + 5}{4 - x^2}$
Domain: $\{x \mid 4 - x^2 \neq 0\}$ or $\{x \mid x \neq \pm 2\}$ *(3-5)*

19. $(f \circ g)(x) = f[g(x)] = f(4 - x^2) = 3(4 - x^2) + 5 = 12 - 3x^2 + 5 = 17 - 3x^2$ *(3-5)*

20. $(g \circ f)(x) = g[f(x)] = g(3x + 5) = 4 - (3x + 5)^2 = 4 - (9x^2 + 30x + 25)$
$\qquad\qquad = 4 - 9x^2 - 30x - 25 = -21 - 30x - 9x^2$ *(3-5)*

21. (A) The graph of $f(x)$ is reflected across the x axis. (B) The graph of $f(x)$ is shifted up 4 units. (C) The graph of $f(x)$ is shifted right 2 units. (D) The graph of $f(x)$ is shifted left 3 units, reflected across the x axis and shifted down 3 units.

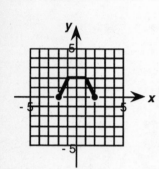
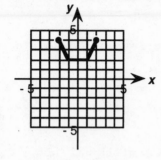
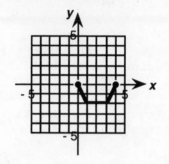
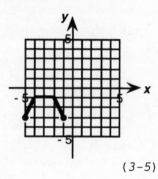

(3-5)

22. (A) The graph that opens up and has a vertex at $(2, -4)$ is g.
(B) The graph that opens down and has vertex at $(-2, 4)$ is m.
(C) The graph that opens down and has vertex at $(2, 4)$ is n.
(D) The graph that opens up and has vertex at $(-2, -4)$ is f. *(3-4, 3-5)*

23. The equation corresponding to graph f is $y = (x + 2)^2 - 4$.

(A) y intercept: Set $x = 0$, then $y = (0 + 2)^2 - 4 = 0$
x intercepts: Set $y = 0$, then $0 = (x + 2)^2 - 4$
$$(x + 2)^2 = 4$$
$$x + 2 = \pm 2$$
$$x = 0, -4$$

(B) $(-2, -4)$

(C) The minimum of -4 occurs at the vertex.

(D) Since y is never less than -4, the range is $[-4, \infty)$.

(E) y is increasing on $[-2, \infty)$.

(F) y is decreasing on $(-\infty, -2]$. *(3-4)*

24. $f(x) = x^2 - 6x + 11$. Complete the square:
$f(x) = (x^2 - 6x + 9) - 9 + 11$
$ = (x - 3)^2 + 2$

Comparing with $f(x) = a(x - h)^2 + k$; $h = 3$ and $k = 2$. Thus, the minimum value is 2 and the vertex is $(3, 2)$. *(3-4)*

25. (A) Reflected across x axis (B) Shifted down 3 units
(C) Shifted left 3 units *(3-5)*

26. (A) 0 (B) 1 (C) 2 (D) 0 *(3-4)*

27. (A) $f(x) = 0$ when $x = -2$ or $x = 0$
(B) $f(x) = 1$ when $x = -1$ or $x = 1$
(C) There is no value of x for which $f(x) = -3$. No solution.
(D) $f(x) = 3$ when $x = 3$ and also for any value of $x < -2$. *(3-4)*

28. Domain: $(-\infty, \infty)$ **29.** $[-2, -1]$, $[1, \infty)$ **30.** $[-1, 1)$ **31.** $(-\infty, -2)$
Range $= (-3, \infty)$ *(3-4)* *(3-4)* *(3-4)*
(3-4)

32. $x = -2$, $x = 1$ *(3-4)* **33.** $f(x) = 4x^3 - \sqrt{x}$ *(3-3)*

34. The function f multiplies the square of the domain element by 3, adds 4 times the domain element, and then subtracts 6. *(3-3)*

35. (A) Since two points are given, we find the slope, then apply the point-slope form.

$$m = \frac{-3 - 3}{0 - (-4)} = \frac{-6}{4} = -\frac{3}{2}$$

$$y - 3 = -\frac{3}{2}[x - (-4)]$$

$$2(y - 3) = -3(x + 4)$$
$$2y - 6 = -3x - 12$$
$$3x + 2y = -6$$

(B) $d(P, Q) = \sqrt{(-3 - 3)^2 + [0 - (-4)]^2} = \sqrt{36 + 16} = \sqrt{52} = 2\sqrt{13}$ $\qquad$ *(3-1, 3-2)*

36. The line $6x + 3y = 5$, or
$3y = -6x + 5$,

or $y = -2x + \frac{5}{3}$, has slope -2.

(A) We require a line through $(-2, 1)$, with slope -2. Applying the point-slope form, we have
$$y - 1 = -2[x - (-2)]$$
$$y - 1 = -2x - 4$$
$$y = -2x - 3$$

(B) We require a line with slope m satisfying $-2m = -1$, or $m = \frac{1}{2}$. Again applying the point-slope form, we have

$$y - 1 = \frac{1}{2}[x - (-2)]$$

$$y - 1 = \frac{1}{2}x + 1$$

$$y = \frac{1}{2}x + 2$$ $\qquad$ *(3-2)*

37. Since the equation is unchanged by any substitution of $-x$ for x or $-y$ for y, or both, the graph must be symmetric with respect to all three. $\qquad$ *(3-1)*

38. The domain is the set of all real numbers x such that $\dfrac{1}{\sqrt{3 - x}}$ is a real number— that is, such that $3 - x > 0$, or $x < 3$. Domain: $(-\infty, 3)$ $\qquad$ *(3-3)*

39. $f(x) = x^2 - 6x + 5$. Complete the square.
$f(x) = (x^2 - 6x + 9) - 9 + 5$
$\quad\; = (x - 3)^2 - 4$

Comparing with $f(x) = a(x - h)^2 + k$, $h = 3$ and $k = -4$. Thus, the vertex is $(3, -4)$, the axis of symmetry is $x = 3$, and the minimum value is -4. $y = f(x)$ can be any number greater then or equal to -4, so the range is $[-4, \infty)$. y intercept: Set $x = 0$, then $f(0) = 5$ is the y intercept.

x intercepts: Set $f(x) = 0$, then
$$x^2 - 6x + 5 = 0$$
$$(x - 5)(x - 1) = 0$$
$$x = 5 \text{ or } x = 1 \text{ are the } x \text{ intercepts.}$$

x	$f(x)$
0	5
1	0
2	-3
3	-4
4	-3
5	0

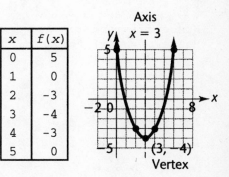

Graph: Locate axis and vertex, then plot several points on either side of the axis. $\qquad$ *(3-4)*

40. The domain is the set of all real numbers x such that $\dfrac{1}{4 - \sqrt{x}}$ is a real number, that is, such that $x \geq 0$ and $4 - \sqrt{x} \neq 0$. The latter condition is equivalent to $\sqrt{x} \neq 4$ or $x \neq 16$. Thus, the domain is all x such that $x \geq 0$ except $x \neq 16$. $[0, 16) \cup (16, \infty)$. *(3-3)*

41. $f(x) = \sqrt{x} - 8 \qquad g(x) = |x|$

(A) $(f \circ g)(x) = f[g(x)] = f(|x|) = \sqrt{|x|} - 8$

$\quad (g \circ f)(x) = g[f(x)] = g(\sqrt{x} - 8) = |\sqrt{x} - 8|$

(B) The domain of f is $\{x \mid x \geq 0\}$. The domain of g is all real numbers. Hence the domain of $f \circ g$ is the set of those real numbers x for which $g(x)$ is non-negative, that is, all real numbers.

The domain of $(g \circ f)$ is the set of all those non-negative numbers x for which $f(x)$ is real, that is all $\{x \mid x \geq 0\}$ or $[0, \infty)$ *(3-5)*

42. (A) $f(x) = x^3$. The graph passes the horizontal line test, so f is one-to-one. Also, assume $\quad f(a) = f(b)$

$$a^3 = b^3$$
$$a^3 - b^3 = 0$$
$$(a - b)(a^2 + ab + b^2) = 0$$

The only real solutions of this equation are those for which $a - b = 0$, hence $a = b$. Thus $f(x)$ is one-to-one.

(B) $g(x) = (x - 2)^2$. Since $g(3) = g(1) = 1$, g is not one-to-one.

(C) $h(x) = 2x - 3$
Assume $h(a) = h(b)$
$$2a - 3 = 2b - 3$$
Then $\qquad 2a = 2b$
$$a = b$$
Thus h is one-to-one.

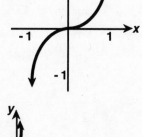

(D) $F(x) = (x + 3)^2 \quad x \geq -3$
The graph passes the horizontal line test, so F is one-to-one.

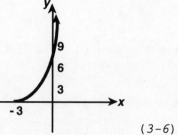

(3-6)

43. (A) $f(x) = 3x - 7$
Assume $f(a) = f(b)$
$$3a - 7 = 3b - 7$$
$$3a = 3b$$
$$a = b$$
Hence f is one-to-one.
Solve $y = f(x)$ for x:
$$y = 3x - 7$$
$$y + 7 = 3x$$
$$x = \frac{1}{3}y + \frac{7}{3} = f^{-1}(y)$$

Interchange x and y:
$$y = f^{-1}(x) = \frac{1}{3}x + \frac{7}{3}$$
Domain: $(-\infty, \infty)$

Check: $\quad f^{-1}[f(x)] = \frac{1}{3}(3x - 7) + \frac{7}{3}$

$$= x - \frac{7}{3} + \frac{7}{3} = x$$

$$f[f^{-1}(x)] = 3\left(\frac{1}{3}x + \frac{7}{3}\right) - 7$$

$$= x + 7 - 7 = x$$

$$f^{-1}(x) = \frac{1}{3}x + \frac{7}{3} = \frac{x + 7}{3}$$

(B) $f^{-1}(5) = \dfrac{5 + 7}{3} = 4$

(C) $f^{-1}[f(x)] = x$ (See part A).

(D) Since $a < b$ implies $3a < 3b$, which implies $3a - 7 < 3b - 7$, or $f(a) < f(b)$, f is increasing.   (3-6)

44.

x	$y = 2 - x$	x	$y = x^2$
-1	3	0	0
		$\frac{1}{2}$	$\frac{1}{4}$
$-\frac{1}{2}$	$2\frac{1}{2}$	1	1

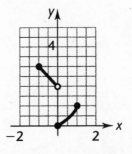

Domain: $[-1, 1]$
Range: $[0, 1] \cup (2, 3]$ (see graph)
Discontinuous at $x = 0$.   (3-4)

45. The graph of $y = x^2$ is vertically expanded by a factor of 2, reflected in the x axis and shifted to the left 3 units. Equation: $y = -2(x + 3)^2$.   (3-5)

46. $g(x) = 5 - 3|x - 2|$

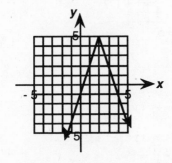

(3-5)

47. The graph of $y = x^2$ has been reflected across the x axis, shifted right 4 units and up 3 units so that the parabola has vertex $(4, 3)$.
Equation: $y = -(x - 4)^2 + 3$.   (3-4, 3-5)

48. (A) This is the same as the graph of $y = |x|$ shifted down 2 units.

(B) This is the same as the graph of $y = |x|$ shifted left 1 unit.

(C) This is the same as the graph of $y = |x|$ contracted by a factor of $\frac{1}{2}$.

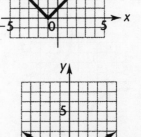

(3-5)

49. $f(x) = \sqrt{x - 1}$.
Assume $f(a) = f(b)$
$$\sqrt{a - 1} = \sqrt{b - 1}$$
$$a - 1 = b - 1$$
$$a = b$$
Thus f is one-to-one.

(A) Solve $y = f(x)$ for x

$y = \sqrt{x - 1}$
$y^2 = x - 1 \quad y \geq 0$
$x = 1 + y^2 \quad y \geq 0 \quad f^{-1}(y) = 1 + y^2$

Interchange x and y:
$y = f^{-1}(x) = 1 + x^2 \quad$ Domain: $x \geq 0$

Check:

$f^{-1}[f(x)] = 1 + (\sqrt{x - 1})^2 = 1 + x - 1 = x$
$f[f^{-1}(x)] = \sqrt{1 + x^2 - 1} = \sqrt{x^2} = x$
since $x \geq 0$ in the domain of f^{-1}

(B) Domain of $f = [1, \infty) = $ Range of f^{-1}
Range of $f = [0, \infty) = $ Domain of f^{-1}

(C)

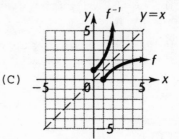

$(3-6)$

50. We are given $C(h, k) = (3, 0)$. To find r we use the distance formula. $r = $ distance from the center to $(-1, 4)$

$$= \sqrt{[(-1) - 3]^2 + (0 - 4)^2}$$
$$= \sqrt{16 + 16}$$
$$= \sqrt{32}$$

Then the equation of the circle is
$(x - h)^2 + (y - k)^2 = r^2$

$(x - 3)^2 + (y - 0)^2 = (\sqrt{32})^2$
$$(x - 3)^2 + y^2 = 32 \qquad (3-1)$$

51.
$$x^2 + y^2 + 4x - 6y = 3$$
$$(x^2 + 4x + ?) + (y^2 - 6y + ?)$$
$$= 3$$
$$(x^2 + 4x + 4) + (y^2 - 6y + 9)$$
$$= 3 + 4 + 9$$
$$(x + 2)^2 + (y - 3)^2 = 16$$
$$[x - (-2)]^2 + (y - 3)^2 = 4^2$$
Center: $C(h, k) = C(-2, 3)$
Radius $r = \sqrt{16} = 4 \qquad (3-1)$

52. $xy = 4$ 　y axis symmetry? $(-x)y = 4$ 　$-xy = 4$ 　No, not equivalent
　　　　　　　x axis symmetry? $x(-y) = 4$ 　$-xy = 4$ 　No, not equivalent
　　　　　　　origin symmetry? $(-x)(-y) = 4$ 　$xy = 4$ 　Yes, equivalent

The graph has symmetry with respect to the origin.

We reflect the portion of the graph in the first quadrant through the origin, using origin symmetry.

x	y
1	4
2	2
3	$\frac{4}{3}$
4	1

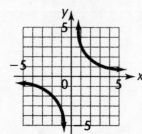

$(3-1)$

53. decreasing 　　　　$(3-2, 3-3)$

54. (A) Domain of $f = [0, \infty)$ = Range of f^{-1}
Since $x^2 \geq 0$, $x^2 - 1 \geq -1$, so Range of $f = [-1, \infty)$ = Domain of f^{-1}

(B) $f(x) = x^2 - 1 \quad x \geq 0$
f is one-to-one on its domain (steps omitted)

Solve $y = f(x)$ for x:
$y = x^2 - 1 \quad x \geq 0$
$x^2 = y + 1 \quad x \geq 0$

$x = \sqrt{y + 1} = f^{-1}(y)$

$\underbrace{\qquad\qquad}$
positive square root since $x \geq 0$

Interchange x and y:

$y = f^{-1}(x) = \sqrt{x + 1} \quad$ Domain: $[-1, \infty)$

Check:
$f^{-1}[f(x)] = \sqrt{x^2 - 1 + 1} \quad x \geq 0$
$= \sqrt{x^2} \quad x \geq 0$
$= x$ since $x \geq 0$
$f[f^{-1}(x)] = (\sqrt{x + 1})^2 - 1$
$= x + 1 - 1$
$= x$

(C) $f^{-1}(3) = \sqrt{3 + 1} = 2$ (D) $f^{-1}[f(4)] = 4$ (E) $f^{-1}[f(x)] = x$ *(3-6)*

55. The graph of $y = \sqrt[3]{x}$ is vertically expanded by a factor of 2, reflected in the
x axis, shifted 1 unit left and 1 unit down. Equation: $y = -2\sqrt[3]{x + 1} - 1$. *(3-4)*

56. It is the same as the graph of g shifted to the right 2 units and down 1 unit,
then reflected in the x axis.

57. This is the same as the graph of $y = |x|$,
reflected across the x axis, and shifted to the
left 1 unit and down 1 unit.

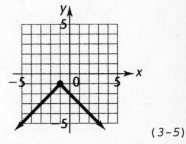

(3-5)

58. The domain of the set of all real numbers x such that $\sqrt{25 - x^2}$ is a real
number—that is, such that $25 - x^2 \geq 0$. Solving by the methods of Section 2-8,
we obtain $-5 \leq x \leq 5$ or $[-5, 5]$ *(2-8, 3-3)*

59. (A) $(fg)(x) = f(x)g(x) = x^2\sqrt{1 - x}$
The domain of f is $(-\infty, \infty)$. The domain of g is $(-\infty, 1]$. Hence the domain of fg
is the intersection of these sets, that is, $(-\infty, 1]$.

(B) $\left(\dfrac{f}{g}\right)(x) = \dfrac{f(x)}{g(x)} = \dfrac{x^2}{\sqrt{1 - x}}$

To find the domain of $\dfrac{f}{g}$, we exclude from $(-\infty, 1]$ the set of values of x for
which $g(x) = 0$
$\sqrt{1 - x} = 0$
$1 - x = 0$
$x = 1$

Thus the domain of $\dfrac{f}{g}$ is $(-\infty, 1)$

(C) $(f \circ g)(x) = f[g(x)] = f(\sqrt{1 - x}) = [\sqrt{1 - x}]^2 = 1 - x$.
The doman of $f \circ g$ is the set of those numbers in $(-\infty, 1]$ for which $g(x)$ is
real, that is $(-\infty, 1]$.

(D) $(g \circ f)(x) = g[f(x)] = g(x^2) = \sqrt{1 - x^2}$
The domain of $g \circ f$ is the set of those real numbers for which $f(x)$ is in
$(-\infty, 1]$, that is, $x^2 \leq 1$, or $-1 \leq x \leq 1$. $[-1, 1]$.   *(3-5)*

60. (A) $f(x) = \dfrac{x + 2}{x - 3}$

Solve $y = f(x)$ for x:

$$y = \frac{x + 2}{x - 3}$$
$$y(x - 3) = x + 2$$
$$xy - 3y = x + 2$$
$$xy - x = 3y + 2$$
$$x(y - 1) = 3y + 2$$
$$x = \frac{3y + 2}{y - 1} = f^{-1}(y)$$

Interchange x and y:

$y = f^{-1}(x) = \dfrac{3x + 2}{x - 1}$ Domain: $x \neq 1$

Check: $f^{-1}[f(x)] = \dfrac{3\frac{x + 2}{x - 3} + 2}{\frac{x + 2}{x - 3} - 1}$ $\qquad$ $f[f^{-1}(x)] = \dfrac{3\frac{3x + 2}{x - 1} + 2}{\frac{3x + 2}{x - 1} - 3}$

$\qquad\qquad\quad = \dfrac{3(x + 2) + 2(x - 3)}{(x + 2) - (x - 3)}$ $\qquad\qquad = \dfrac{3x + 2 + 2(x - 1)}{3x + 2 - 3(x - 1)}$

$\qquad\qquad\quad = \dfrac{3x + 6 + 2x - 6}{x + 2 - x + 3}$ $\qquad\qquad = \dfrac{3x + 2 + 2x - 2}{3x + 2 - 3x + 3}$

$\qquad\qquad\quad = \dfrac{5x}{5}$ $\qquad\qquad\qquad\qquad = \dfrac{5x}{5}$

$\qquad\qquad\quad = x$ $\qquad\qquad\qquad\qquad\quad = x$

(B) $f^{-1}(3) = \dfrac{3(3) + 2}{3 - 1} = \dfrac{11}{2}$ $\qquad$ (C) $f^{-1}[f(x)] = x$ (See part A) $\qquad$ (3-6)

61. $f(x) = |x + 1| - |x - 1|$
If $x < -1$, then $|x + 1| = -(x + 1)$ and $|x - 1| = -(x - 1)$, hence
$f(x) = -(x + 1) - [-(x - 1)]$
$\qquad = -x - 1 + x - 1$
$\qquad = -2$
If $-1 \leq x < 1$, then $|x + 1| = x + 1$ but $|x - 1| = -(x - 1)$, hence
$f(x) = x + 1 - [-(x - 1)]$
$\qquad = x + 1 + x - 1$
$\qquad = 2x$
If $x \geq 1$, then $|x + 1| = x + 1$ and $|x - 1| = x - 1$, hence
$f(x) = x + 1 - (x - 1)$
$\qquad = x + 1 - x + 1$
$\qquad = 2$
Domain: $(-\infty, \infty)$

Piecewise definition for f: $f(x) = \begin{cases} -2 & \text{if } x < -1 \\ 2x & \text{if } -1 \leq x < 1 \\ 2 & \text{if } x \geq 1 \end{cases}$

Range: Since if $-1 \leq x < 1$, then $-2 \leq 2x < 2$, $f(x)$ is always between -2 and 2.
The range is $[-2, 2]$. $\qquad$ (3-4)

62. Let (x, y) be a point equidistant from $(3, 3)$ and $(6, 0)$. Then
$$\sqrt{(x - 3)^2 + (y - 3)^2} = \sqrt{(x - 6)^2 + (y - 0)^2}$$
$$(x - 3)^2 + (y - 3)^2 = (x - 6)^2 + y^2$$
$$x^2 - 6x + 9 + y^2 - 6y + 9 = x^2 - 12x + 36 + y^2$$
$$-6x - 6y + 18 = -12x + 36$$
$$6x - 6y = 18$$
$$x - y = 3$$
This is the equation of a line. $\qquad$ (3-1, 3-2)

63. We will show separately:
(A) If two nonvertical lines are parallel, then they have the same slope.
(B) If two lines have the same slope they are nonvertical and parallel.

(A) Let $y = m_1x + b_1$ and $y = m_2x + b_2$ be parallel. Then there is no point with coordinates that satisfy both equations. But for any y, $m_1x + b_1 = m_2x + b_2$. Then (x, y) will satisfy both equations unless this equation has no solution. But this equation will have a solution
$$m_1x - m_2x = b_2 - b_1$$
$(m_1 - m_2)x = b_2 - b_1$,   that is, $x = \dfrac{b_2 - b_1}{m_1 - m_2}$

unless $m_1 - m_2 = 0$. So $m_1 = m_2$. So the lines have the same slope.

(B) Assume the two lines have equations $y = mx + b_1$, and $y = mx + b_2$. Then both have slopes, hence are nonvertical. For any y a point that lies on both lines must have coordinates that satisfy $mx + b_1 = mx + b_2$ or $b_1 = b_2$. So unless the lines are the same line there is no point which lies on both of them. Hence the lines do not intersect, that is, they are parallel.   *(3-2)*

64. If $m = 0$, the two lines have equations $-y = b$ and $x = c$. The first line is horizontal and the second is vertical, hence they are perpendicular. Otherwise, $m \neq 0$, and neither line is horizontal or vertical. The two equations can be written as
$$\ell_1: y = mx - b \text{ (slope } m\text{)}$$
$$\ell_2: y = -\frac{1}{m}x + \frac{c}{m} \quad \text{(slope} = -\frac{1}{m}\text{)}$$

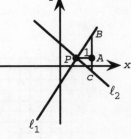

Note that $m = -\dfrac{1}{m}$ has no real solutions, hence the lines have unequal slopes, are not parallel, and must intersect.

Sketch a figure.

The case $m > 0$ is illustrated; if $m < 0$ the argument is similar (and left to the student).

In the figure, PA is constructed horizontal with length 1 unit. BC is constructed vertical, perpendicular to PA at A. We will show that BPC is a right triangle.

1. Since ℓ_1 has slope $m = \dfrac{\text{rise}}{\text{run}} = \dfrac{AB}{AP} = \dfrac{AB}{1}$, AB has length m.

2. Since ℓ_2 has slope $-\dfrac{1}{m} = \dfrac{\text{rise}}{\text{run}} = -\dfrac{AC}{AP} = -\dfrac{AC}{1}$, AC has length $\dfrac{1}{m}$.

3. PAB is constructed as a right triangle, hence by the Pythagorean theorem
$$PB = \sqrt{PA^2 + AB^2} = \sqrt{1 + m^2}.$$

4. PAC is constructed as a right triangle, hence by the Pythagorean theorem,
$$PC = \sqrt{PA^2 + AC^2} = \sqrt{1 + \frac{1}{m^2}}$$

5. $BC = BA + AC = m + \dfrac{1}{m}$.

6. Therefore $PB^2 + PC^2 = BC^2$ since
$$(\sqrt{1 + m^2})^2 + \left(\sqrt{1 + \frac{1}{m^2}}\right)^2 = (m + \frac{1}{m})^2 \quad \text{(check!)}$$

7. Hence, by the converse of the Pythagorean theorem, BPC is a right triangle and the two lines are perpendicular.   *(3-2)*

65. (A) The portion of the graph in the first quadrant is the same as the graph of $[\![x]\!]$, since $|x| = x$ for $x \geq 0$. We draw this portion and reflect it in the y axis.

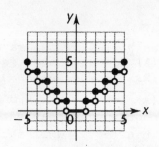

(B) If $x \geq 0$, $[\![x]\!] \geq 0$, $\left| [\![x]\!] \right| = [\![x]\!]$.
If $x < 0$, $[\![x]\!] \leq 0$, $\left| [\![x]\!] \right| = -[\![x]\!]$.
Hence for non-negative x we draw the graph of $[\![x]\!]$, and for negative x we draw the graph of $-[\![x]\!]$.

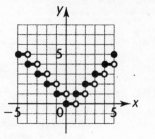

(3-5)

66. Domain: All real numbers except $x = 2$; Range: $y > -3$ or $(-3, \infty)$

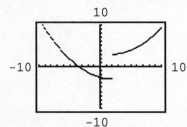

The function is discontinuous at $x = 2$.   *(3-4)*

67. (A) Reflect the given graph across the y axis:

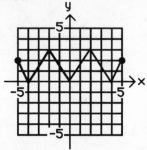

(B) Reflect the given graph across the origin:

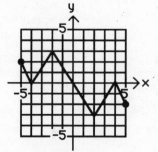

(3-1, 3-5)

68. (A) The graph must cross the x axis exactly once. Some possible graphs are shown:

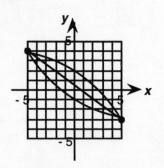

(B) The graph may cross the x axis once, but it may fail to cross the x axis at all. A possible graph of the latter type is shown:

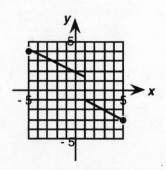

(3-4)

69. (A) If V is linearly related to t, then we are looking for an equation whose graph passes through $(t_1, V_1) = (0, 12{,}000)$ and $(t_2, V_2) = (8, 2{,}000)$. We find the slope, and then we use the point-slope form to find the equation.

$$m = \frac{V_2 - V_1}{t_2 - t_1} = \frac{2{,}000 - 12{,}000}{8 - 0} = \frac{-10{,}000}{8} = -1{,}250$$

$$V - V_1 = m(t - t_1)$$
$$V - 12{,}000 = -1{,}250(t - 0)$$
$$V - 12{,}000 = -1{,}250t$$
$$V = -1{,}250t + 12{,}000$$

(B) We are asked for V when $t = 5$
$$v = -1{,}250(5) + 12{,}000$$
$$V = -6{,}250 + 12{,}000$$
$$V = \$5{,}750$$

(3-2)

70. (A) If R is linearly related to C, then we are looking for an equation whose graph passes through $(C_1, R_1) = (30, 48)$ and $(C_2, R_2) = (20, 32)$. We find the slope, and then we use the point-slope form to find the equation.

$$m = \frac{R_2 - R_1}{C_2 - C_1} = \frac{32 - 48}{20 - 30} = \frac{-16}{-10} = 1.6$$

$$R - R_1 = m(C - C_1)$$
$$R - 48 = 1.6(C - 30)$$
$$R - 48 = 1.6C - 48$$
$$R = 1.6C$$

(B) We are asked for R when $C = 105$.
$$R = 1.6(105)$$
$$= \$168$$

(3-2)

71. If $0 \le x \le 3{,}000$, $E(x) = 200$

$$\begin{pmatrix} \text{Base} \\ \text{Salary} \end{pmatrix} \quad + \quad \begin{pmatrix} \text{Commission on} \\ \text{sales over \$3{,}000} \end{pmatrix}$$

If $x > 3{,}000$, $E(x) = 200 \quad + \quad 0.1(x - 3{,}000)$
$$= 200 \quad + \quad 0.1x - 300$$
$$= 0.1x - 100$$

Summarizing,

$$E(x) = \begin{cases} 200 & \text{if } 0 \le x \le 3{,}000 \\ 0.1x - 100 & \text{if } x > 3{,}000 \end{cases}$$

$$E(2{,}000) = 200$$
$$E(5{,}000) = 0.1(5{,}000) - 100 = 500 - 100 = 400$$

(3-4)

72. (A)

x	0	5	10	15	20
Consumption	309	276	271	255	233
$303.4 - 3.46x = f(x)$	303	286	269	252	234

(B)

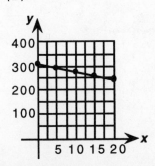

(C) In 1995, $x = 25$, hence $y = 303.4 - 3.46(25) \approx 217$
In 2000, $x = 30$, hence $y = 303.4 - 3.46(30) \approx 200$

(D) Per capita egg consumption is dropping about 17 eggs every five years. *(3-4)*

73. (A) If $0 \leq x < 36$, $C(x) = 0.49x$
If $36 \leq x < 72$, $C(x) = 0.44x$
If $72 \leq x$, $\quad C(x) = 0.39x$

Summarizing,
$$C(x) = \begin{cases} 0.49x & \text{for } 0 \leq x < 36 \\ 0.44x & \text{for } 36 \leq x < 72 \\ 0.39x & \text{for } 72 \leq x \end{cases}$$

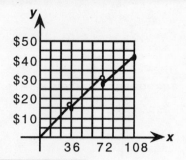

(B) There are points of discontinuity at $x = 36$ and $x = 72$.

(3-4)

x	$y = 0.49x$	x	$y = 0.44x$	x	$y = 0.39x$
0	0	36	15.84	72	28.08
18	8.82	54	23.76	108	42.12

74. (A) Let x = number of units sold. Then
$$C = \begin{pmatrix} \text{cost of shooting} \\ \text{video} \end{pmatrix} + \begin{pmatrix} \text{number of} \\ \text{units} \end{pmatrix} \times \begin{pmatrix} \text{cost per} \\ \text{unit} \end{pmatrix}$$
$$= \quad 84{,}000 \quad + \quad x \quad \cdot \quad 15$$
$$= \quad 84{,}000 + 15x$$

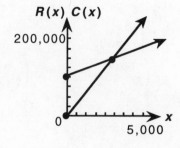

$$R = \begin{pmatrix} \text{number of} \\ \text{units} \end{pmatrix} \times \begin{pmatrix} \text{price per} \\ \text{unit} \end{pmatrix}$$
$$= \quad x \quad \cdot \quad 50$$
$$= \quad 50x$$

(B) $R = C$ when $50x = 84{,}000 + 15x$.
Solving, we obtain
$$35x = 84{,}000$$
$$x = 2{,}400 \text{ units.}$$

From the graph, $R < C$ when $x < 2{,}400$ and $R > C$ when $x > 2{,}400$. *(3-2)*

75. The profit function is the difference of the revenue and the cost functions
$P = R - C$
Hence
$$P(x) = R(x) - C(x)$$
$$= \left(50x - \frac{1}{10}x^2\right) - (20x + 4{,}000)$$
$$= 50x - \frac{1}{10}x^2 - 20x - 4{,}000$$
$$= 30x - \frac{1}{10}x^2 - 4{,}000$$

Next we use composition to express P as a function of the price p.
$$(P \circ f)(p) = P[f(p)]$$
$$= P(500 - 10p)$$
$$= 30(500 - 10p) - \frac{1}{10}(500 - 10p)^2 - 4{,}000$$
$$= 15{,}000 - 300p - \frac{1}{10}(250{,}000 - 10{,}000p + 100p^2) - 4{,}000$$
$$= 15{,}000 - 300p - 25{,}000 + 1{,}000p - 10p^2 - 4{,}000$$
$$= -14{,}000 + 700p - 10p^2$$

(3-5)

76. In the sketch, we note that the point $(4, r - 2)$
is on the circle with equation $x^2 + y^2 = r^2$, hence
$(4, r - 2)$ must satisfy this equation.

$$4^2 + (r - 2)^2 = r^2$$
$$16 + r^2 - 4r + 4 = r^2$$
$$-4r + 20 = 0$$
$$r = 5 \text{ feet}$$

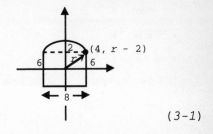

(3-1)

77. (A) From the figure, we see that $A = x(y + y) = 2xy$. Since the fence consists
of four pieces of length y and three pieces of length x, we have $3x + 4y = 120$.
Hence $4y = 120 - 3x$

$$y = 30 - \frac{3}{4}x$$

$$A = 2x\left(30 - \frac{3}{4}x\right)$$

$$A(x) = 60x - \frac{3}{2}x^2$$

(B) Since both x and y must be positive, we have $x > 0$

$30 - \frac{3}{4}x > 0$ or $-\frac{3}{4}x > -30$ or $x < 40$

Hence $0 < x < 40$ is the domain of A.

(C) The function A is a quadratic function. Completing the square yields:

$$A(x) = -\frac{3}{2}x^2 + 60x$$

$$= -\frac{3}{2}(x^2 - 40x)$$

$$= -\frac{3}{2}(x^2 - 40x + 400) + \frac{3}{2} \cdot 400$$

$$= -\frac{3}{2}(x - 20)^2 + 600$$

Comparing with $f(x) = a(x - h)^2 + k$, the total area will be maximum when
$x = 20$. Then

$$y = 30 - \frac{3}{4}x = 30 - \frac{3}{4}(20) = 15$$

(3-4)

78. (A) $f(1) = 1 - (\llbracket\sqrt{1}\rrbracket)^2 = 1 - 1 = 0$

(B) $f(2) = 2 - (\llbracket\sqrt{2}\rrbracket)^2 = 2 - 1 = 1$

(C) $f(3) = 3 - (\llbracket\sqrt{3}\rrbracket)^2 = 3 - 1 = 2$

(D) $f(4) = 4 - (\llbracket\sqrt{4}\rrbracket)^2 = 4 - 4 = 0$

(E) $f(5) = 5 - (\llbracket\sqrt{5}\rrbracket)^2 = 5 - 4 = 1$

(F) $f(n^2) = n^2 - (\llbracket\sqrt{n^2}\rrbracket)^2$

$\qquad = n^2 - (\llbracket n\rrbracket)^2$ since $\sqrt{n^2} = n$ if n is positive

$\qquad = n^2 - (n)^2 \quad$ since $\llbracket n\rrbracket = n$ if n is a (positive) integer.

$\qquad = 0$

(3-4)

CHAPTER 4

Exercise 4-1

Key Ideas and Formulas

For the nth degree polynomial function P given by

$$P(x) = a_n x^n + a_{n-1} x^{n-1} + \cdots + a_1 x + a_0 \qquad a_n \neq 0$$

The number r is said to be a zero of the function P, or a zero of the polynomial $P(x)$, or a solution or root of the equation $P(x) = 0$, if $P(r) = 0$.

If the coefficients of a polynomial $P(x)$ are real, then the x intercepts of the graph of $y = P(x)$ are real zeros of P and $P(x)$ and real solutions or roots for the equation $P(x) = 0$.

Synthetic Division: To divide $P(x)$ by $x - r$ arrange the coefficients of $P(x)$ in order of descending powers of x (write 0 as the coefficient for each missing power).

Use the bring down, multiply, add scheme below:

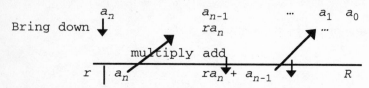

The last number to the right in the third row of numbers is the remainder R; the other numbers in the third row are the coefficients of the quotient $Q(x)$, which is of degree one less than $P(x)$.

$$\frac{P(x)}{x - r} = Q(x) + \frac{R}{x - r}$$

Division Algorithm:

For each polynomial $P(x)$ of degree greater than 0 and each number r, there exists a unique polynomial $Q(x)$ of degree 1 less than $P(x)$ and a unique number R such that

$$P(x) = (x - r)Q(x) + R$$

$Q(x)$ is called the quotient, $x - r$ the divisor, and R the remainder. Note: R may be zero.

Remainder Theorem:

If R is the remainder after dividing $P(x)$ by $x - r$, then $P(r) = R$.

Left and Right Behavior of a Polynomial:

$$P(x) = a_n x^n + a_{n-1} x^{n-1} + \cdots + a_1 x + a_0, \qquad a_n \neq 0$$

1. $a_n > 0$ and n even

Graph of $P(x)$ increases without bound as x decreases to the left and as x increases to the right.

$$P(x) \to \infty \text{ as } x \to -\infty$$
$$P(x) \to \infty \text{ as } x \to \infty$$

2. $a_n > 0$ and n odd

Graph of $P(x)$ decreases without bound as x decreases to the left and increases without bound as x increases to the right.

$$P(x) \to -\infty \text{ as } x \to -\infty$$
$$P(x) \to \infty \text{ as } x \to \infty$$

3. $a_n < 0$ and n even

Graph of $P(x)$ decreases without bound as x decreases to the left and as x increases to the right.

$$P(x) \to -\infty \text{ as } x \to -\infty$$
$$P(x) \to -\infty \text{ as } x \to \infty$$

4. $a_n < 0$ and n odd

Graph of $P(x)$ increases without bound as x decreases to the left and decreases without bound as x increases to the right.

$$P(x) \to \infty \text{ as } x \to -\infty$$
$$P(x) \to -\infty \text{ as } x \to \infty$$

A **turning point** on a continuous graph is a point that separates an increasing portion from a decreasing portion.

Graph Properties of Polynomial Functions:

Let P be an nth degree polynomial function with real coefficients.

1. P is continuous for all real numbers.
2. The graph of P is a smooth curve.
3. The graph of P has at most n x intercepts.
4. P has at most $n - 1$ turning points.

---

1. Since $P(x) \to \infty$ as $x \to \infty$ and $P(x) \to -\infty$ as $x \to -\infty$, this matches graph c.

3. Since $P(x) \to \infty$ as $x \to \infty$ and $P(x) \to \infty$ as $x \to -\infty$, this matches graph d.

5. Since the graph of a second-degree polynomial has at most 1 turning point, only graph h could be an answer. h.

7. Since the graph of a fourth-degree polynomial can have 3 turning points, graph k could be an answer. But graph k is not the only graph which decreases without bound as x decreases to the left and as x increases to the right; graph h could also be an answer. h, k.

9.

$$
\begin{array}{r}
2m + 1 \\
2m - 1 \overline{\smash{)}\, 4m^2 + 0m - 1} \\
\underline{4m^2 - 2m} \\
2m - 1 \\
\underline{2m - 1} \\
0
\end{array}
$$

$2m + 1,\ R = 0$

11.

$$
\begin{array}{r}
4x - 5 \\
2x + 1 \overline{\smash{)}\, 8x^2 - 6x + 6} \\
\underline{8x^2 + 4x} \\
-10x + 6 \\
\underline{-10x - 5} \\
11
\end{array}
$$

$4x - 5,\ R = 11$

13.

$$
\begin{array}{r}
x^2 + x + 1 \\
x - 1 \overline{\smash{)}\, x^3 - 0x^2 + 0x - 1} \\
\underline{x^3 - x^2} \\
x^2 + 0x \\
\underline{x^2 - x} \\
x - 1 \\
\underline{x - 1} \\
0
\end{array}
$$

$x^2 + x + 1,\ R = 0$

15.

$$
\begin{array}{r}
2y^2 - 5y + 13 \\
y + 2 \overline{\smash{)}\, 2y^3 - y^2 + 3y - 1} \\
\underline{2y^3 + 4y^2} \\
-5y^2 + 3y \\
\underline{-5y^2 - 10y} \\
13y - 1 \\
\underline{13y + 26} \\
-27
\end{array}
$$

$2y^2 - 5y + 13,\ R = -27$

17.

$$
\begin{array}{r}
\ 1 \quad 3 \quad -7 \\
\ 2 \quad 10 \\
\hline
2\,\big|\ 1 \quad 5 \quad 3
\end{array}
$$

$$\frac{x^2 + 3x - 7}{x - 2} = x + 5 + \frac{3}{x - 2}$$

19.

$$
\begin{array}{r}
\ 4 \quad 10 \quad -9 \\
\ -12 \quad 6 \\
\hline
-3\,\big|\ 4 \quad -2 \quad -3
\end{array}
$$

$$\frac{4x^2 + 10x - 9}{x + 3} = 4x - 2 - \frac{3}{x + 3}$$

21.

$$
\begin{array}{r}
\ 2 \quad 0 \quad -3 \quad 1 \\
\ 4 \quad 8 \quad 10 \\
\hline
2\,\big|\ 2 \quad 4 \quad 5 \quad 11
\end{array}
$$

$$\frac{2x^3 - 3x + 1}{x - 2} = 2x^2 + 4x + 5 + \frac{11}{x - 2}$$

> **Common Error:**
> The first row is *not*
> 2 -3 1
> The 0 must be inserted for the missing power.

23.

$$
\begin{array}{r}
\ 3 \quad -1 \quad -10 \\
\ -6 \quad 14 \\
\hline
-2\,\big|\ 3 \quad -7 \quad 4
\end{array}
$$

$P(-2) = 4$

25.

$$
\begin{array}{r}
\ 2 \quad -5 \quad 7 \quad -7 \\
\ 4 \quad -2 \quad 10 \\
\hline
2\,\big|\ 2 \quad -1 \quad 5 \quad 3
\end{array}
$$

$P(2) = 3$

27.

```
      1    0   -10   25   -2
          -4    16  -24   -4
  -4│ 1   -4    6    1   -6
```

$P(-4) = -6$

29.

```
      3    0    0   -1   -4
          -3    3   -3    4
  -1│ 3   -3    3   -4    0
```

$3x^3 - 3x^2 + 3x - 4, \; R = 0$

31.

```
      1    0    0    0    0    1
          -1    1   -1    1   -1
  -1│ 1   -1    1   -1    1    0
```

$x^4 - x^3 + x^2 - x + 1, \; R = 0$

33.

```
      3    2    0   -4   -1
          -9   21  -63  201
  -3│ 3   -7   21  -67  200
```

$3x^3 - 7x^2 + 21x - 67, \; R = 200$

35.

```
      2  -13    0   75    2    0   -50
          10  -15  -75    0   10   50
  5│ 2   -3  -15    0    2   10    0
```

$2x^5 - 3x^4 - 15x^3 + 2x + 10, \; R = 0$

37.

```
         4    2   -6   -5    1
             -2    0    3    1
  -½│ 4    0   -6   -2    2
```

$4x^3 - 6x - 2, \; R = 2$

39.

```
         4    4   -7   -6
             -6    3    6
  -³⁄₂│ 4   -2   -4    0
```

$4x^2 - 2x - 4, \; R = 0$

41.

```
       3   -2     2      -3       1
           1.2  -0.32   0.672  -0.9312
  0.4│ 3  -0.8  1.68  -2.328   0.0688
```

$3x^3 - 0.8x^2 + 1.68x - 2.328, \; R = 0.0688$

43.

```
        3    2       5      0       -7       -3
           -2.4    0.32  -4.256   3.4048   2.87616
  -0.8│ 3  -0.4    5.32  -4.256  -3.5952  -0.12384
```

$3x^4 - 0.4x^3 + 5.32x^2 - 4.256x - 3.5952, \; R = -0.12384$

45. We form a synthetic division table:

	1	-5	2	8	
-2	1	-7	16	-24	= $P(-2)$
-1	1	-6	8	0	= $P(-1)$
0	1	-5	2	8	= $P(0)$
1	1	-4	-2	6	= $P(1)$
2	1	-3	-4	0	= $P(2)$
3	1	-2	-4	-4	= $P(3)$
4	1	-1	-2	0	= $P(4)$
5	1	0	2	18	= $P(5)$

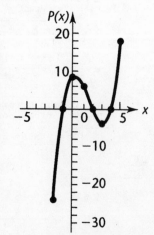

The graph has three x intercepts and two turning points.

$P(x) \to \infty$ as $x \to \infty$ and $P(x) \to -\infty$ as $x \to -\infty$.

47. We form a synthetic division table:

	1	4	-1	-4	
-5	1	-1	4	-24	= P(-5)
-4	1	0	-1	0	= P(-4)
-3	1	1	-4	8	= P(-3)
-2	1	2	-5	6	= P(-2)
-1	1	3	-4	0	= P(-1)
0	1	4	-1	-4	= P(0)
1	1	5	4	0	= P(1)
2	1	6	11	18	= P(2)

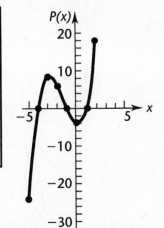

The graph has three x intercepts and two turning points.

$P(x) \to \infty$ as $x \to \infty$ and $P(x) \to -\infty$ as $x \to -\infty$.

49. We form a synthetic division table:

	-1	2	0	-3	
-2	-1	4	-8	13	= P(-2)
-1	-1	3	-3	0	= P(-1)
0	-1	2	0	-3	= P(0)
1	-1	1	-1	-2	= P(1)
2	-1	0	0	-3	= P(2)
3	-1	-1	-3	-12	= P(3)

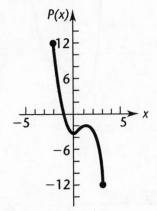

The graph has one x intercept and two turning points.

$P(x) \to -\infty$ as $x \to \infty$ and $P(x) \to \infty$ as $x \to -\infty$.

51. We form a synthetic division table:

	-1	3	-3	2	
-1	-1	4	-7	9	= P(-1)
-0.5	-1	3.5	-4.75	4.375	= P(-0.5)
0	-1	3	0	2	= P(0)
0.5	-1	2.5	-1.75	1.125	= P(0.5)
1	-1	2	-1	1	= P(1)
1.5	-1	1.5	-0.75	0.875	= P(1.5)
2	-1	1	-1	0	= P(2)
2.5	-1	0.5	-1.75	-2.375	= P(2.5)
3	-1	0	-3	-7	= P(3)

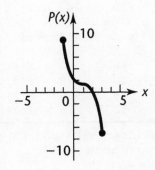

53. $P(x) = x^3$ is an example of a third-degree polynomial with one x intercept ($x = 0$).

55. No such polynomial exists; the graph of a third-degree polynomial must cross the x axis at least once.

57.

$$
\begin{array}{r|rrrr}
 & 2.14 & -5.23 & -8.71 & 6.85 \\
 &   & 7.21 & 6.68 & -6.85 \\
\hline
3.37 & 2.14 & 1.98 & -2.03 & 0.00 \\
\end{array}
$$

$2.14x^2 + 1.98x - 2.03$, $R = 0.00$

59.

$$
\begin{array}{r|rrrrr}
 & 0.96 & 0 & 4.09 & 9.44 & -1.87 \\
 &   & -1.32 & 1.80 & -8.07 & -1.87 \\
\hline
-1.37 & 0.96 & -1.32 & 5.89 & 1.37 & -3.74 \\
\end{array}
$$

$0.96x^3 - 1.32x^2 + 5.89x + 1.37$, $R = -3.74$

61.

$$
3x^2 + 2x - 4 \overline{\smash{\big)} 6x^4 - 5x^3 - 8x^2 + 16x - 8}
$$
$$
\underline{6x^4 + 4x^3 - 8x^2}
$$
$$
-9x^3 \qquad\quad + 16x
$$
$$
\underline{-9x^3 - 6x^2 + 12x}
$$
$$
6x^2 + 4x - 8
$$
$$
\underline{6x^2 + 4x - 8}
$$
$$
0
$$

quotient: $2x^2 - 3x + 2$

$2x^2 - 3x + 2, \ R = 0$

63.

i	1	-3	1	-3
		i	$-3i - 1$	3
	1	$-3 + i$	$-3i$	0

$x^2 + (-3 + i)x - 3i, \ R = 0$

65. We form a synthetic division table:

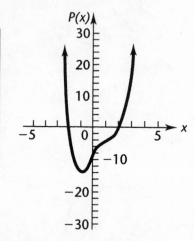

	1	-2	-2	8	-8	
-3	1	-5	13	-31	85	$= P(-3)$
-2	1	-4	6	-4	0	$= P(-2)$
-1	1	-3	1	7	-15	$= P(-1)$
0	1	-2	-2	8	-8	$= P(0)$
1	1	-1	-3	5	-3	$= P(1)$
2	1	0	-2	4	0	$= P(2)$
3	1	1	1	11	25	$= P(3)$

The graph has two x intercepts and one turning point.

$P(x) \to \infty$ as $x \to \infty$ and as $x \to -\infty$.

67. (A) We form a synthetic division table:

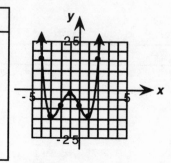

	1	4	-1	-10	-8	
-5	1	-1	4	-30	142	$= P(-5)$
-4	1	0	-1	-6	16	$= P(-4)$
-3	1	1	-4	2	-14	$= P(-3)$
-2	1	2	-5	0	-8	$= P(-2)$
-1	1	3	-4	-6	-2	$= P(-1)$
0	1	4	-1	-10	-8	$= P(0)$
1	1	5	4	-6	-14	$= P(1)$
2	1	6	11	12	16	$= P(2)$

The graph has two x intercepts and three turning points.
$P(x) \to \infty$ as $x \to \infty$ and as $x \to -\infty$.

69. (A) We form a synthetic division table:

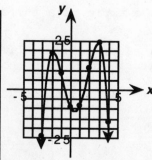

	-1	2	10	-10	-9	
-3	-1	5	-5	5	-24	$= P(-3)$
-2	-1	4	2	-14	19	$= P(-2)$
-1	-1	3	7	-17	8	$= P(-1)$
0	-1	2	10	-10	-9	$= P(0)$
1	-1	1	11	1	-8	$= P(1)$
2	-1	0	10	10	11	$= P(2)$
3	-1	-1	7	11	24	$= P(3)$
4	-1	-2	2	-2	-17	$= P(4)$

The graph has four x intercepts and three turning points.
$P(x) \to -\infty$ as $x \to \infty$ and as $x \to -\infty$.

71. (A) We form a synthetic division table:

	1	-6	4	17	-5	-7
-2	1	-8	20	-23	41	-89 = P(-2)
-1	1	-7	11	6	-11	4 = P(-1)
0	1	-6	4	17	-5	-7 = P(0)
1	1	-5	-1	16	11	4 = P(1)
2	1	-4	-4	9	13	19 = P(2)
3	1	-3	-5	2	1	-4 = P(3)
4	1	-2	-4	1	-1	-11 = P(4)
5	1	-1	-1	12	55	268 = P(5)

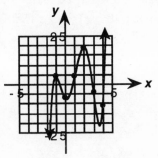

The graph has five x intercepts and four turning points.
$P(x) \to \infty$ as $x \to \infty$ and $P(x) \to -\infty$ as $x \to -\infty$.

73. (A)

$$a_2(x) + (a_1 + a_2r)$$
$$x - r \overline{\smash{)}\, a_2x^2 + a_1x \qquad\qquad\qquad + a_0}$$
$$\underline{a_2x^2 - a_2rx}$$
$$(a_1 + a_2r)x \qquad\qquad + a_0$$
$$\underline{(a_1 + a_2r)x - r(a_1 + a_2r)}$$
$$a_0 + r(a_1 + a_2r)$$

$$
\begin{array}{c|ccc}
 & a_2 & a_1 & a_0 \\
 & & a_2r & (a_1 + a_2r)r \\
\hline
r & a_2 & a_1 + a_2r & a_0 + (a_1 + a_2r)r
\end{array}
$$

In both cases the coefficient of x is a_2, the constant term $a_2r + a_1$, and the remainder is $(a_2r + a_1)r + a_0$.

(B) The remainder expanded is $a_2r^2 + a_1r + a_0 = P(r)$.

75.
$$P(x) = \{[(2x - 3)x + 2]x - 5\}x + 7$$
$$P(-2) = \{[\{2(-2) - 3\}(-2) + 2](-2) - 5\}(-2) + 7$$
$$= \{[\{-7\}(-2) + 2](-2) - 5\}(-2) + 7$$
$$= \{[16](-2) - 5\}(-2) + 7$$
$$= \{-37\}(-2) + 7$$
$$= 81$$
$$P(1.7) = \{[\{2(1.7) - 3\}(1.7) + 2](1.7) - 5\}(1.7) + 7$$
$$= \{[\{0.4\}(1.7) + 2](1.7) - 5\}(1.7) + 7$$
$$= \{[2.68](1.7) - 5\}(1.7) + 7$$
$$= \{-0.444\}(1.7) + 7$$
$$= 6.2452 \text{ or } 6.2 \text{ to two significant digits.}$$

Exercise 4-2

Key Ideas and Formulas

Factor Theorem:
If r is a zero of the polynomial $P(x)$, then $x - r$ is a factor of $P(x)$; conversely, if $x - r$ is a factor of $P(x)$, then r is a zero of $P(x)$.

Fundamental Theorem of Algebra:
Every polynomial $P(x)$ of degree $n > 0$ has at least one zero.

n Zeros Theorem:
Every polynomial $P(x)$ of degree $n > 0$ can be expressed as the product of n linear factors. Hence, $P(x)$ has exactly n zeros—not necessarily distinct.

If $P(x)$ is represented as the product of linear factors and $x - r$ occurs m times, then r is called a **zero of multiplicity m**.

Imaginary Zeros Theorem:

Imaginary zeros of polynomials with real coefficients, if they exist, occur in conjugate pairs.

A polynomial of odd degree with real coefficients always has at least one real zero.

Rational Zero Theorem:

If the rational number b/c, in lowest terms, is a zero of the polynomial

$$P(x) = a_n x^n + a_{n-1} x^{n-1} + \cdots + a_1 x + a_0 \qquad a_n \neq 0$$

with integer coefficients, then b must be an integer factor of a_0 and c must be an integer factor of a_n.

Strategy for Finding Rational Zeros:

Assume that $P(x)$ is a polynomial with integer coefficients and is of degree greater than 2.

Step 1.   List the possible rational zeros of $P(x)$ using the rational zero theorem.

Step 2.   Construct a synthetic division table. If a rational zero r is found, stop, write

$$P(x) = (x - r)Q(x)$$

and immediately proceed to find the rational zeros for $Q(x)$, the reduced polynomial relative to $P(x)$. If the degree of $Q(x)$ is greater than 2, return to step 1 using $Q(x)$ in place of $P(x)$. If $Q(x)$ is quadratic, find all its zeros using standard methods for solving quadratic equations.

---

1. -8 (multiplicity 3), 6 (multiplicity 2); degree of $P(x)$ is 5

3. -4 (multiplicity 3), 3 (multiplicity 2); -1; degree of $P(x)$ is 6.

5. $P(x) = (x - 3)^2(x + 4)$; degree 3

7. $P(x) = (x + 7)^3[x - (-3 + \sqrt{2})][x - (-3 - \sqrt{2})]$; degree 5

9. $P(x) = [x - (2 - 3i)][x - (2 + 3i)](x + 4)^2$; degree 4

11. Since -2, 1, and 3 are zeros, $P(x) = (x + 2)(x - 1)(x - 3)$ is the lowest degree polynomial that has this graph. The degree of $P(x)$ is 3.

13. Since -2 and 1 are zeros, each with multiplicity 2, $P(x) = (x + 2)^2(x - 1)^2$ is the lowest degree polynomial that has this graph. The degree of $P(x)$ is 4.

15. Since -3, -2, 0, 1, and 2 are zeros, $P(x) = (x + 3)(x + 2)x(x - 1)(x - 2)$ is the lowest degree polynomial that has this graph. The degree of $P(x)$ is 5.

17. $x - 1$ will be a factor of $P(x)$ if $P(1) = 0$. Since $P(x) = x^{18} - 1$, $P(1) = 1^{18} - 1 = 0$. Therefore $x - 1$ is a factor of $x^{18} - 1$.

19. $x + 1$ will be a factor of $P(x)$ if $P(-1) = 0$. Since $P(x) = 3x^3 - 7x^2 - 8x + 2$, $P(-1) = 3(-1)^3 - 7(-1)^2 - 8(-1) + 2 = -3 - 7 + 8 + 2 = 0$. Therefore $x + 1$ is a factor of $3x^3 - 7x^2 - 8x + 2$.

21. Possible factors of 6 are ±1, ±2, ±3, ±6. Possible factors of 1 are ±1. Therefore the possible rational zeros are ±1, ±2, ±3, ±6.

23. Possible factors of 4 are ±1, ±2, ±4. Possible factors of 3 are ±1, ±3.

Therefore the possible rational zeros are ± 1, ± 2, ± 4, $\pm \dfrac{1}{3}$, $\pm \dfrac{2}{3}$, $\pm \dfrac{4}{3}$.

25. Possible factors of 3 are ±1, ±3. Possible factors of 12 are ±1, ±2, ±3, ±4, ±6, ±12. Therefore the possible rational zeros are ±1, ±3, ±$\frac{1}{2}$, ±$\frac{3}{2}$, ±$\frac{1}{3}$, ±$\frac{1}{4}$, ±$\frac{3}{4}$, ±$\frac{1}{6}$, ±$\frac{1}{12}$.

27. Let $P(x) = 2x^3 - 5x^2 + 1$. Possible rational zeros: ±1, ±$\frac{1}{2}$.

We form a synthetic division table:

	2	-5	0	1	
1	2	-3	-3	-2	
-1	2	-7	7	-6	
$\frac{1}{2}$	2	-4	-2	0	$\frac{1}{2}$ is a zero

So $P(x) = \left(x - \frac{1}{2}\right)(2x^2 - 4x - 2) = (2x - 1)(x^2 - 2x - 1)$

To find the remaining zeros, we solve $x^2 - 2x - 1 = 0$, by completing the square:

$$x^2 - 2x = 1$$
$$x^2 - 2x + 1 = 2$$
$$(x - 1)^2 = 2$$
$$x - 1 = \pm\sqrt{2}$$
$$x = 1 \pm \sqrt{2}$$

Hence the zeros are $\frac{1}{2}$, $1 \pm \sqrt{2}$. These are the roots of the equation.

29. Let $P(x) = x^4 + 4x^3 - x^2 - 20x - 20$. Possible rational zeros: ±1, ±2, ±4, ±5, ±10, ±20. We form a synthetic division table:

	1	4	-1	-20	-20	
1	1	5	4	-16	-36	
2	1	6	11	2	-16	
4	1	8	31	104	396	
-1	1	3	-4	-16	-4	
-2	1	2	-5	-10	0	-2 is a zero

We now examine $x^3 + 2x^2 - 5x - 10$. The only remaining possible rational zeros are -2, 5, -5, 10, -10. We form a synthetic division table for the reduced polynomial:

	1	2	-5	-10	
-2	1	0	-5	0	-2 is a zero

> **Common Error:**
> Students often miss double zeros. It is necessary to test whether -2 is again a zero of the reduced polynomial.

So $P(x) = (x + 2)^2(x^2 - 5) = (x + 2)^2(x - \sqrt{5})(x + \sqrt{5})$. So the zeros of the polynomial are -2 (double), ±$\sqrt{5}$. These are the roots of the equation.

31. Let $P(x) = x^4 - 2x^3 - 5x^2 + 8x + 4$. Possible rational zeros: ±1, ±2, ±4. We form a synthetic division table:

	1	-2	-5	8	4	
1	1	-1	-6	2	6	
2	1	0	-5	-2	0	2 is a zero

We now examine $x^3 - 5x - 2$. The only remaining possible rational zeros are -1, ± 2. We form a synthetic division table for the reduced polynomial:

	1	0	-5	-2	
2	1	2	-1	-4	
-1	1	-1	-4	2	
-2	1	-2	-1	0	-2 is a zero

So $P(x) = (x - 2)(x + 2)(x^2 - 2x - 1)$. The zeros of $x^2 - 2x - 1$ are $1 \pm \sqrt{2}$ (see problem 27). Hence the zeros of the polynomial are ± 2, $1 \pm \sqrt{2}$. These are the roots of the equation.

33. Let $P(x) = 2x^5 - 3x^4 - 2x + 3$. The possible rational zeros are ± 1, ± 3, $\pm\frac{1}{2}$, $\pm\frac{3}{2}$. We form a synthetic division table:

	2	-3	0	0	-2	3	
1	2	-1	-1	-1	-3	0	1 is a zero

We examine $2x^4 - x^3 - x^2 - x - 3$. The possible rational zeros are the same. We form a synthetic division table for the reduced polynomial.

	2	-1	-1	-1	-3	
1	2	1	0	-1	-4	
3	2	5	14	41	120	
-1	2	-3	2	-3	0	-1 is a zero

We examine $Q(x) = 2x^3 - 3x^2 + 2x - 3$. This clearly factors by grouping into $(2x - 3)(x^2 + 1)$, but if we don't notice this we can proceed as before. Then the possible remaining rational zeros are -1, -3, $\pm\frac{1}{2}$, $\pm\frac{3}{2}$.

We form a synthetic division table for $Q(x)$:

	2	-3	2	-3	
$\frac{1}{2}$	2	-2	1	$-\frac{5}{2}$	
$\frac{3}{2}$	2	0	2	0	$\frac{3}{2}$ is a zero

So $P(x) = (x - 1)(x + 1)\left(x - \frac{3}{2}\right)(2x^2 + 2)$

$$= (x - 1)(x + 1)\left(x - \frac{3}{2}\right)2(x^2 + 1)$$
$$= (x - 1)(x + 1)(2x - 3)(x^2 + 1)$$
$$= (x - 1)(x + 1)(2x - 3)(x - i)(x + i)$$

The zeros are ± 1, $\frac{3}{2}$, $\pm i$. These are the roots of the equation.

35. The possible rational zeros are ± 1, ± 2, ± 3, ± 5, ± 6, ± 10, ± 15, ± 30. We form a synthetic division table:

	1	0	-19	30	
1	1	1	-18	12	
2	1	2	-15	0	2 is a zero

So $P(x) = (x - 2)(x^2 + 2x - 15)$
$\qquad\quad = (x - 2)(x - 3)(x + 5)$
So the zeros of $P(x)$ are 2, 3, -5.

37. $P(x) = x^4 - \frac{21}{10}x^3 + \frac{2}{5}x = \frac{1}{10}(10x^4 - 21x^3 + 4x) = \frac{1}{10}x(10x^3 - 21x^2 + 4)$. Clearly, 0 is a zero. We examine $Q(x) = 10x^3 - 21x^2 + 4$. Possible factors of 4 are ± 1, ± 2, ± 4. Possible factors of 10 are ± 1, ± 2, ± 5, ± 10. Hence the possible rational zeros of $Q(x)$ are ± 1, ± 2, ± 4, $\pm\frac{1}{2}$, $\pm\frac{1}{5}$, $\pm\frac{2}{5}$, $\pm\frac{4}{5}$ $\pm\frac{1}{10}$. We form a synthetic division table:

	10	-21	0	4	
1	10	-11	-11	-7	
2	10	-1	-2	0	2 is a zero

So $P(x) = \frac{1}{10}x(x - 2)(10x^2 - x - 2)$

$= \frac{1}{10}x(x - 2)(5x + 2)(2x - 1)$

So the zeros of $P(x)$ are 0, 2, $-\frac{2}{5}$, $\frac{1}{2}$.

39. $P(x) = x^4 - 5x^3 + \frac{15}{2}x^2 - 2x - 2 = \frac{1}{2}(2x^4 - 10x^3 + 15x^2 - 4x - 4)$. Possible factors of -4 are ± 1, ± 2, ± 4. Possible factors of 2 are ± 1, ± 2. Hence the possible rational zeros are ± 1, ± 2, ± 4, $\pm\frac{1}{2}$. We form a synthetic division table:

	2	-10	15	-4	-4	
1	2	-8	7	3	-1	
2	2	-6	3	2	0	2 is a zero

We examine $Q(x) = 2x^3 - 6x^2 + 3x + 2$. The only remaining possible rational zeros are -1, ± 2, $\pm\frac{1}{2}$. We form a synthetic division table for $Q(x)$:

	2	-6	3	2	
2	2	-2	-1	0	2 is a zero

Thus 2 is a double zero of $P(x)$.

$P(x) = \frac{1}{2}(x - 2)^2(2x^2 - 2x - 1)$

To find the remaining zeros, we solve $2x^2 - 2x - 1 = 0$.
Applying the quadratic formula with $a = 2$, $b = -2$, $c = -1$, we obtain

$$x = \frac{-(-2) \pm \sqrt{(-2)^2 - 4(2)(-1)}}{2(2)} = \frac{2 \pm \sqrt{12}}{4} = \frac{1 \pm \sqrt{3}}{2} \text{ or } \frac{1}{2} \pm \frac{1}{2}\sqrt{3}.$$

So the zeros of $P(x)$ are 2 (double), $\frac{1}{2} \pm \frac{1}{2}\sqrt{3}$.

41. Possible factors of 5 are ± 1, ± 5. Possible factors of 3 are ± 1, ± 3. Hence the possible rational zeros are ± 1, ± 5, $\pm\frac{1}{3}$, $\pm\frac{5}{3}$.
We form a synthetic division table:

	3	-5	-8	16	21	5	
1	3	-2	-10	6	27	32	
5	3	10	42	226	1151	5760	
-1	3	-8	0	16	5	0	-1 is a zero

We examine $Q(x) = 3x^4 - 8x^3 + 16x + 5$. The only remaining possible rational zeros are -1, -5, $\pm\frac{1}{3}$, $\pm\frac{5}{3}$. We form a synthetic division table for $Q(x)$:

	3	-8	0	16	5	
-1	3	-11	11	5	0	-1 is a zero

So -1 is a double zero of $P(x)$.

We examine $R(x) = 3x^3 - 11x^2 + 11x + 5$. The possible rational zeros are the same. We form a synthetic division table for $R(x)$:

	3	-11	11	5
-1	3	-14	25	-20
$\frac{1}{3}$	3	-10	$\frac{23}{3}$	$\frac{68}{9}$
$-\frac{1}{3}$	3	-12	15	0 $-\frac{1}{3}$ is a zero

So $P(x) = (x + 1)^2\left(x + \frac{1}{3}\right)(3x^2 - 12x + 15) = (x + 1)^2\left(x + \frac{1}{3}\right)3(x^2 - 4x + 5)$

To find the remaining zeros, we solve $x^2 - 4x + 5 = 0$ by completing the square.

$$x^2 - 4x = -5$$
$$x^2 - 4x + 4 = -1$$
$$(x - 2)^2 = -1$$
$$x - 2 = \pm i$$
$$x = 2 \pm i$$

So the zeros of $P(x)$ are -1 (double), $-\frac{1}{3}$, $2 \pm i$.

43. The possible rational zeros are ± 1, ± 2, ± 4, $\pm\frac{1}{2}$, $\pm\frac{1}{3}$, $\pm\frac{2}{3}$, $\pm\frac{4}{3}$, $\pm\frac{1}{6}$. We form a synthetic division table:

	6	13	0	-4
1	6	19	19	15
-1	6	7	-7	3
-2	6	1	-2	0 -2 is a zero

$P(x) = (x + 2)(6x^2 + x - 2) = (x + 2)(3x + 2)(2x - 1)$.

45. The possible rational zeros are ± 1, ± 2, ± 4. We form a synthetic division table:

	1	2	-9	-4
1	1	3	-6	-10
2	1	4	-1	-6
4	1	6	15	56
-1	1	1	-10	6
-2	1	0	-9	14
-4	1	-2	-1	0 -4 is a zero

So $P(x) = (x + 4)(x^2 - 2x - 1)$. The zeros of $x^2 - 2x - 1$ are $1 \pm \sqrt{2}$ (see problem 27). Hence $P(x) = (x + 4)[x - (1 + \sqrt{2})][x - (1 - \sqrt{2})]$

47. The possible rational zeros are ± 1, ± 2, $\pm\frac{1}{2}$, $\pm\frac{1}{4}$.

We form a synthetic division table:

	4	-4	-9	1	2
1	4	0	-9	-8	-6
2	4	4	-1	-1	0 2 is a zero

So $P(x) = (x - 2)(4x^3 + 4x^2 - x - 1)$. This clearly factors by grouping into $(x - 2)(x + 1)(4x^2 - 1) = (x - 2)(x + 1)(2x - 1)(2x + 1)$, but if we don't notice this we can proceed as before. Then we examine $Q(x) = 4x^3 + 4x^2 - x - 1$.

The possible remaining rational zeros are -1, $\pm\frac{1}{2}$, $\pm\frac{1}{4}$.

We form a synthetic division table for $Q(x)$:

	4	4	-1	-1
-1	4	0	-1	0 -1 is a zero

So $P(x) = (x - 2)(x + 1)(4x^2 - 1) = (x - 2)(x + 1)(2x + 1)(2x - 1)$.

49.
$$x^2 \le 4x - 1$$
$$x^2 - 4x + 1 \le 0$$
We factor $x^2 - 4x + 1$ by solving $x^2 - 4x + 1 = 0$ by completing the square.
$$x^2 - 4x = -1$$
$$x^2 - 4x + 4 = 3$$
$$(x - 2)^2 = 3$$
$$x - 2 = \pm\sqrt{3}$$
$$x = 2 \pm \sqrt{3}$$
Hence $x^2 - 4x + 1 = [x - (2 + \sqrt{3})][x - (2 - \sqrt{3})]$.
To solve $[x - (2 + \sqrt{3})][x - (2 - \sqrt{3})] \le 0$ we form a sign chart, noting $2 - \sqrt{3} < 2 + \sqrt{3}$, hence $2 - \sqrt{3}$ is to the left of $2 + \sqrt{3}$.

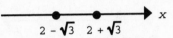

$x^2 - 4x + 1 = [x - (2 + \sqrt{3})][x - (2 - \sqrt{3})]$			
Test Number	0	2	4
Value of Polynomial for Test Number	1	-3	1
Sign of Polynomial in Interval	+	-	+
Interval	$(-\infty, 2 - \sqrt{3})$	$(2 - \sqrt{3}, 2 + \sqrt{3})$	$(2 + \sqrt{3}, \infty)$

$x^2 - 4x + 1 \le 0$ and $x^2 \le 4x - 1$ within the interval $[2 - \sqrt{3}, 2 + \sqrt{3}]$, or $2 - \sqrt{3} \le x \le 2 + \sqrt{3}$

51. $x^3 + 3 \le 3x^2 + x$
$$x^3 - 3x^2 - x + 3 \le 0$$
To factor $x^3 - 3x^2 - x + 3$, we can use factoring by grouping to obtain $(x - 3)(x^2 - 1)$ or $(x - 3)(x - 1)(x + 1)$.
However, if we don't notice this, we can search for zeros by the methods of this section. A synthetic division table immediately gives:

	1	-3	-1	3
1	1	-2	-3	0

Hence $x^3 - 3x^2 - x + 3 = (x - 1)(x^2 - 2x - 3) = (x - 1)(x - 3)(x + 1)$.
We form a sign chart.
Zeros: -1, 1, 3

$x^3 - 3x^2 - x + 3 = (x - 1)(x - 3)(x + 1)$				
Test Number	-2	0	2	4
Value of Polynomial for Test Number	-15	3	-3	15
Sign of Polynomial in Interval	-	+	-	+
Interval	$(-\infty, -1)$	$(-1, 1)$	$(1, 3)$	$(3, \infty)$

$x^3 - 3x^2 - x + 3 \le 0$ and $x^3 + 3 \le 3x^2 + x$ within the intervals $(-\infty, -1]$ and $[1, 3]$, or $x \le -1$ or $1 \le x \le 3$.

53. $2x^3 + 6 \geq 13x - x^2$

$2x^3 + x^2 - 13x + 6 \geq 0$

To factor $2x^3 + x^2 - 13x + 6$, we search for zeros of the polynomial.

Possible rational zeros are ± 1, ± 2, ± 3, ± 6, $\pm\frac{1}{2}$, $\pm\frac{3}{2}$.

We form a synthetic division table:

	2	1	-13	6
1	2	3	-10	-4
2	2	5	-3	0 2 is a zero

$2x^3 + x^2 - 13x + 6 = (x - 2)(2x^2 + 5x - 3) = (x - 2)(2x - 1)(x + 3)$. Hence we must examine the sign behavior of $(x - 2)(2x - 1)(x + 3)$.

We form a sign chart.

Zeros: -3, $\frac{1}{2}$, 2

$2x^3 + x^2 - 13x + 6 = (x - 2)(2x - 1)(x + 3)$				
Test Number	-4	0	1	3
Value of Polynomial for Test Number	-54	6	-4	30
Sign of Polynomial in Interval	-	+	-	+
Interval	$(-\infty, -3)$	$(-3, \frac{1}{2})$	$(\frac{1}{2}, 2)$	$(2, \infty)$

$2x^3 + x^2 - 13x + 6 \geq 0$ and $2x^3 + 6 \geq 13x - x^2$ within the intervals $\left[-3, \frac{1}{2}\right]$ and $[2, \infty)$, or $-3 \leq x \leq \frac{1}{2}$ or $x \geq 2$.

55. $[x - (4 - 5i)][x - (4 + 5i)] = [x - 4 + 5i][x - 4 - 5i]$

$= [(x - 4) + 5i][(x - 4) - 5i]$

$= (x - 4)^2 - 25i^2$

$= x^2 - 8x + 16 + 25$

$= x^2 - 8x + 41$

57. $[x - (3 + 4i)][x - (3 - 4i)] = [x - 3 - 4i][x - 3 + 4i]$

$= [(x - 3) - 4i][(x - 3) + 4i]$

$= (x - 3)^2 - 16i^2$

$= x^2 - 6x + 9 + 16$

$= x^2 - 6x + 25$

59. $[x - (a + bi)][x - (a - bi)] = [x - a - bi][x - a + bi]$

$= [(x - a) - bi][(x - a) + bi]$

$= (x - a)^2 - b^2i^2$

$= (x - a)^2 + b^2$

$= x^2 - 2ax + a^2 + b^2$

61. If $3 - i$ is a zero, then $3 + i$ is a zero. So $Q(x) = [x - (3 - i)][x - (3 + i)]$ divides $P(x)$ evenly. Applying problem 59, $Q(x) = x^2 - 6x + 9 + 1 = x^2 - 6x + 10$. Dividing, we see

$$
\begin{array}{r}
x + 1 \\
x^2 - 6x + 10 \overline{\smash)\ x^3 - 5x^2 + 4x + 10} \\
\underline{x^3 - 6x^2 + 10x} \\
x^2 - 6x + 10 \\
\underline{x^2 - 6x + 10} \\
0
\end{array}
$$

So $P(x) = (x + 1)Q(x)$ and the two other zeros are -1 and $3 + i$.

63. If $-5i$ is a zero, then so is $5i$. So $Q(x) = (x - 5i)(x + 5i)$ divides $P(x)$ evenly. $Q(x) = x^2 - 25i^2 = x^2 + 25$. Dividing, we see

$$
\begin{array}{r}
x - 3 \\
x^2 + 25 \overline{\smash{\big)}\, x^3 - 3x^2 + 25x - 75} \\
\underline{x^3 \qquad\quad + 25x} \\
-3x^2 \qquad\quad - 75 \\
\underline{-3x^2 \qquad\quad - 75} \\
0
\end{array}
$$

So $P(x) = (x - 3)Q(x)$ and the two other zeros are $5i$ and 3.

65. If $2 + i$ is a zero, then $2 - i$ is a zero. So $Q(x) = [x - (2 + i)][x - (2 - i)]$ divides $P(x)$ evenly. Applying problem 59, $Q(x) = x^2 - 4x + 4 + 1 = x^2 - 4x + 5$. Dividing, we see

$$
\begin{array}{r}
x^2 - 2 \\
x^2 - 4x + 5 \overline{\smash{\big)}\, x^4 - 4x^3 + 3x^2 + 8x - 10} \\
\underline{x^4 - 4x^3 + 5x^2} \\
-2x^2 + 8x - 10 \\
\underline{-2x^2 + 8x - 10} \\
0
\end{array}
$$

So $P(x) = Q(x)(x^2 - 2)$. $x^2 - 2$ has two zeros: $\sqrt{2}$ and $-\sqrt{2}$. Summarizing, $P(x)$ has 4 zeros: $2 + i$, $2 - i$, $\sqrt{2}$, $-\sqrt{2}$.

67. $\dfrac{4}{2x^3 + 5x^2 - 2x - 5} \geq 0$

We need to form a sign chart for $\dfrac{P}{Q} = \dfrac{4}{2x^3 + 5x^2 - 2x - 5}$. We first locate the zeros of P and Q. $P = 4$ has no zeros. To find the zeros of $Q = 2x^3 + 5x^2 - 2x - 5$ we can factor by grouping into $(2x + 5)(x^2 - 1) = (2x + 5)(x - 1)(x + 1)$. However, if we don't notice this, we can search for zeros by the methods of this section. A synthetic division table immediately gives

	2	5	-2	-5
1	2	7	5	0

Hence $2x^3 + 5x^2 - 2x - 5 = (x - 1)(2x^2 + 7x + 5) = (x - 1)(x + 1)(2x + 5)$.
We form a sign chart.

Zeros of Q: $-\dfrac{5}{2}$, -1, 1

$(-\infty, -5/2)\ (-5/2, -1)\ (-1, 1)\ (1, \infty)$

	$\dfrac{P}{Q} = \dfrac{4}{(x - 1)(x + 1)(2x + 5)}$			
Test Number	-3	-2	0	2
Value of $\dfrac{P}{Q}$	$-\dfrac{1}{2}$	$\dfrac{4}{3}$	$-\dfrac{4}{5}$	$\dfrac{4}{27}$
Sign of $\dfrac{P}{Q}$	$-$	$+$	$-$	$+$
Interval	$\left(-\infty, -\dfrac{5}{2}\right)$	$\left(-\dfrac{5}{2}, -1\right)$	$(-1, 1)$	$(1, \infty)$

$\dfrac{4}{2x^3 + 5x^2 - 2x - 5} \geq 0$ within the intervals $\left(-\dfrac{5}{2}, -1\right)$ and $(1, \infty)$, or $-\dfrac{5}{2} < x < -1$ or $x > 1$.

69. $\dfrac{x^2 - 3x - 10}{x^3 - 4x^2 + x + 6} \leq 0$

We need to form a sign chart for $\dfrac{P}{Q} = \dfrac{x^2 - 3x - 10}{x^3 - 4x^2 + x + 6}$. We first locate the
zeros of P and Q. $x^2 - 3x - 10 = (x - 5)(x + 2)$, hence P has zeros at 5 and -2.
To find the zeros of Q, we note that the possible rational zeros are ±1, ±2, ±3
and ±6. We form a synthetic division table:

	1	-4	1	6	
1	1	-3	-2	4	
2	1	-2	-3	0	2 is a zero

Hence $Q = (x - 2)(x^2 - 2x - 3) = (x - 2)(x - 3)(x + 1)$.

We form a sign chart.
Zeros of P, Q: -2, -1, 2, 3, 5
Open dots at zeros of Q
Solid dots at zeros of P

	$\dfrac{P}{Q} = \dfrac{(x - 5)(x + 2)}{(x - 2)(x - 3)(x + 1)}$					
Test Number	-3	$-\dfrac{3}{2}$	0	$\dfrac{5}{2}$	4	6
Value of $\dfrac{P}{Q}$	$-\dfrac{2}{15}$	$\dfrac{26}{63}$	$-\dfrac{5}{3}$	$\dfrac{90}{7}$	$-\dfrac{3}{5}$	$\dfrac{2}{21}$
Sign of $\dfrac{P}{Q}$	$-$	$+$	$-$	$+$	$-$	$+$
Interval	$(-\infty, -2)$	$(-2, -1)$	$(-1, 2)$	$(2, 3)$	$(3, 5)$	$(5, \infty)$

$\dfrac{x^2 - 3x - 10}{x^3 - 4x^2 + x + 6} \leq 0$ within the intervals $(-\infty, -2]$ or $(-1, 2)$ or $(3, 5]$. $x \leq -2$
or $-1 < x < 2$ or $3 < x \leq 5$.

71. $\sqrt{6}$ is a root of $x^2 = 6$ or $x^2 - 6 = 0$. The possible rational roots of this
equation are ±1, ±2, ±3, ±6. Since none of them satisfies $x^2 = 6$, there are no
rational roots, hence $\sqrt{6}$ is not rational.

73. $\sqrt[3]{5}$ is a root of $x^3 = 5$ or $x^3 - 5 = 0$. The possible rational roots of this
equation are ±1, ±5. Since none of them satisfies $x^3 = 5$, there are no rational
roots, hence $\sqrt[3]{5}$ is not rational.

75. Here is a computer-generated graph
of $P(x) = 3x^3 - 37x^2 + 84x - 24$.

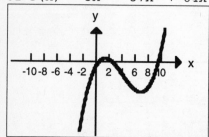

The rational zero theorem gives ±1, ±2,
±3, ±4, ±6, ±8, ±12, ±24, $\pm\dfrac{1}{3}$, $\pm\dfrac{2}{3}$, $\pm\dfrac{4}{3}$,
$\pm\dfrac{8}{3}$ as possible rational zeros. However,
the computer graph suggests that $P(x)$
has no integer zeros; there are no negative zeros, and the positive zeros would
appear to lie between 0 and 1, 2 and 3, and 9 and 10. This suggests that we
consider only the possible zeros $\dfrac{1}{3}$, $\dfrac{2}{3}$, and $\dfrac{8}{3}$.

We form a synthetic division table:

	3	-37	84	-24	
$\frac{1}{3}$	3	-36	72	0	$\frac{1}{3}$ is a zero

So $P(x) = \left(x - \frac{1}{3}\right)(3x^2 - 36x + 72) = \left(x - \frac{1}{3}\right)3(x^2 - 12x + 24)$

To find the remaining zeros, we solve $x^2 - 12x + 24 = 0$ by completing the square:

$$x^2 - 12x = -24$$
$$x^2 - 12x + 36 = 12$$
$$(x - 6)^2 = 12$$
$$x - 6 = \pm\sqrt{12}$$
$$x = 6 \pm \sqrt{12} \text{ or } 6 \pm 2\sqrt{3}$$

So the zeros of $P(x)$ are $\frac{1}{3}$, $6 \pm 2\sqrt{3}$.

77. Here is a computer-generated graph of
$P(x) = 4x^4 + 4x^3 + 49x^2 + 64x - 240.$

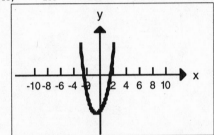

The rational zero theorem gives ±1, ±2, ±3, ±4, ±5, ±6, ±8, ±10, ±12, ±15, ±16, ±20, ±24, ±30, ±40, ±48, ±60, ±80, ±120, ±240, $\pm\frac{1}{2}$, $\pm\frac{3}{2}$, $\pm\frac{5}{2}$, $\pm\frac{15}{2}$, $\pm\frac{1}{4}$, $\pm\frac{3}{4}$, $\pm\frac{5}{4}$, $\pm\frac{15}{4}$ as possible rational zeros. However, the computer graph suggests that $P(x)$ has no integer zeros, and the zeros appear to

lie between -3 and -2, and 1 and 2. This suggests that we consider only the possible zeros $\frac{3}{2}$, $\frac{5}{4}$, and $-\frac{5}{2}$.

We form a synthetic division table:

	4	4	49	64	-240	
$\frac{3}{2}$	4	10	64	160	0	$\frac{3}{2}$ is a zero

So $P(x) = \left(x - \frac{3}{2}\right)(4x^3 + 10x^2 + 64x + 160) = \left(x - \frac{3}{2}\right)2(2x^3 + 5x^2 + 32x + 80)$

We consider the reduced polynomial $Q(x) = 2x^3 + 5x^2 + 32x + 80$. The computer graph suggests that the other real zero is negative, so we try $-\frac{5}{2}$ next.

	2	5	32	80	
$-\frac{5}{2}$	2	0	32	0	$-\frac{5}{2}$ is a zero

So $P(x) = \left(x - \frac{3}{2}\right)2\left(x + \frac{5}{2}\right)(2x^2 + 32) = 4\left(x - \frac{3}{2}\right)\left(x + \frac{5}{2}\right)(x^2 + 16)$

The remaining zeros of $P(x)$ are the zeros of $x^2 + 16$, that is, $\pm 4i$. So the zeros of $P(x)$ are $\frac{3}{2}$, $-\frac{5}{2}$, $\pm 4i$.

79. Here is a computer-generated graph of
$P(x) = 4x^4 - 44x^3 + 145x^2 - 192x + 90.$

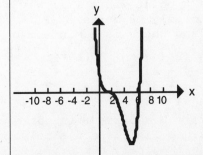

The rational zero theorem gives ± 1, ± 2, ± 3, ± 5, ± 6, ± 9, ± 10, ± 15, ± 18, ± 30, ± 45, ± 90, $\pm\frac{1}{2}$, $\pm\frac{3}{2}$, $\pm\frac{5}{2}$, $\pm\frac{9}{2}$, $\pm\frac{15}{2}$, $\pm\frac{45}{2}$, $\pm\frac{1}{4}$, $\pm\frac{3}{4}$, $\pm\frac{5}{4}$, $\pm\frac{9}{4}$, $\pm\frac{15}{4}$, $\pm\frac{45}{4}$ as possible rational zeros. However, the computer graph suggests that the only possible integer zeros are 1 and 2; there are no negative zeros, and the zeros appear to lie between 0 and 3, and between 6 and 7. This suggests that we consider only the possible zeros 1, 2, $\frac{1}{2}$, $\frac{3}{2}$, $\frac{5}{2}$, $\frac{1}{4}$, $\frac{3}{4}$, $\frac{5}{4}$, $\frac{9}{4}$.

We form a synthetic division table:

	4	-44	145	-192	90	
1	4	-40	105	-87	3	
2	4	-36	73	-46	-2	
$\frac{3}{2}$	4	-38	88	-60	0	$\frac{3}{2}$ is a zero

So $P(x) = \left(x - \frac{3}{2}\right)(4x^3 - 38x^2 + 88x - 60) = \left(x - \frac{3}{2}\right)2(2x^3 - 19x^2 + 44x - 30)$

We consider the reduced polynomial $Q(x) = 2x^3 - 19x^2 + 44x - 30$. The remaining possibilities from the reduced list are $\frac{1}{2}$, $\frac{3}{2}$, $\frac{5}{2}$.

We form a synthetic division table:

	2	-19	44	-30	
$\frac{1}{2}$	2	-18	35	$-\frac{25}{2}$	
$\frac{3}{2}$	2	-16	20	0	$\frac{3}{2}$ is a zero

So $P(x) = \left(x - \frac{3}{2}\right)^2 2(2x^2 - 16x + 20) = \left(x - \frac{3}{2}\right)^2 4(x^2 - 8x + 10)$

To find the remaining zeros, we solve $x^2 - 8x + 10 = 0$, by completing the square.

$$x^2 - 8x = -10$$
$$x^2 - 8x + 16 = 6$$
$$(x - 4)^2 = 6$$
$$x - 4 = \pm\sqrt{6}$$
$$x = 4 \pm \sqrt{6}$$

So the zeros of $P(x)$ are $\frac{3}{2}$ (double), $4 \pm \sqrt{6}$.

81. (A) Since there are 3 zeros of $x^3 - 1$, there are 3 cube roots of 1.

(B) $x^3 - 1 = (x - 1)(x^2 + x + 1)$. The other cube roots of 1 will be solutions to $x^2 + x + 1 = 0$. Applying the quadratic formula with $a = b = c = 1$, we have $x = \dfrac{-1 \pm \sqrt{(1)^2 - 4(1)(1)}}{2(1)} = \dfrac{-1 \pm \sqrt{-3}}{2} = \dfrac{-1 \pm i\sqrt{3}}{2}$. Thus $-\frac{1}{2} + \frac{\sqrt{3}}{2}i$ and $-\frac{1}{2} - \frac{\sqrt{3}}{2}i$ are the other cube roots of 1.

83. $P(x)$ can have at most n and must have at least one real zero. Each zero of $P(x)$ represents a point where $P(x) = y = 0$ so the graph of $P(x)$ will cross the x axis at and only at zeros of $P(x)$. Thus there can be a maximum of n axis crossings and there is a minimum of 1 axis crossing.

85. $P(2 + i) = (2 + i)^2 + 2i(2 + i) - 5$
$\qquad\qquad = 4 + 4i + i^2 + 4i + 2i^2 - 5$
$\qquad\qquad = 4 + 4i - 1 + 4i - 2 - 5$
$\qquad\qquad = -4 + 8i$

So $P(2 + i) \neq 0$ and $2 + i$ is not a zero of $P(x)$. This does not contradict the theorem, since $P(x)$ is not a polynomial with real coefficients (the coefficient of x is the imaginary number $2i$).

87. Let x = the amount of increase.

Then old volume = $1 \times 2 \times 3 = 6$

new volume = $(x + 1)(x + 2)(x + 3) = x^3 + 6x^2 + 11x + 6$

Since (new volume) = 10 (old volume), we must solve

$x^3 + 6x^2 + 11x + 6 = 10(6)$

$x^3 + 6x^2 + 11x + 6 = 60$

$x^3 + 6x^2 + 11x - 54 = 0$

The possible rational zeros are ±1, ±2, ±3, ±6, ±9, ±18, ±27, ±54. We form a synthetic division table:

	1	6	11	-54	
1	1	7	18	-36	
2	1	8	27	0	2 is a zero

2 is the only positive zero. Hence the increase must equal 2 feet.

89.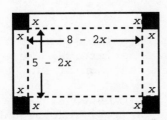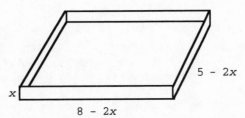

From the figure, it should be clear that

Volume = $x(5 - 2x)(8 - 2x) = 14$

Since x, $5 - 2x$, and $8 - 2x$ must all be positive, the domain of x is $0 < x < \dfrac{5}{2}$ or $(0, 2.5)$. We solve $x(5 - 2x)(8 - 2x) = 14$, or $4x^3 - 26x^2 + 40x = 14$, for x in this domain.

$4x^3 - 26x^2 + 40x = 14$

$4x^3 - 26x^2 + 40x - 14 = 0$

$2x^3 - 13x^2 + 20x - 7 = 0$

Possible rational zeros: ±1, ±7, ±$\dfrac{1}{2}$, ±$\dfrac{7}{2}$.

We form a synthetic division table

	2	-13	20	-7	
1	2	-11	9	2	
$\frac{1}{2}$	2	-12	14	0	$\frac{1}{2}$ is a zero

So $2x^3 - 13x^2 + 20x - 7 = \left(x - \dfrac{1}{2}\right)(2x^2 - 12x + 14) = (2x - 1)(x^2 - 6x + 7)$

To find the remaining zeros, we solve $x^2 - 6x + 7 = 0$, by completing the square:

$x^2 - 6x = -7$

$x^2 - 6x + 9 = 2$

$(x - 3)^2 = 2$

$x - 3 = \pm\sqrt{2}$

$x = 3 \pm \sqrt{2}$

Hence the zeros are $\dfrac{1}{2}$, $3 - \sqrt{2}$, $3 + \sqrt{2}$, or 0.5, 1.59, 4.44 to two significant digits. We discard $3 + \sqrt{2}$ or 4.44, since it is not in the interval $(0, 2.5)$. The square should be 0.5 × 0.5 inches or 1.59 × 1.59 inches.

Exercise 4-3

Key Ideas and Formulas

Location Theorem

If f is continuous on an interval I, a and b are two numbers in I, and $f(a)$ and $f(b)$ are of opposite sign, then there is at least one x intercept for f between a and b.

Upper and Lower Bounds of Real Zeros:

Given an nth degree polynomial $P(x)$ with real coefficients, $n > 0$, $a_n > 0$ and $P(x)$ divided by $x - r$ using synthetic division:

  1. **Upper bound.** If $r > 0$ and all numbers in the quotient row of the synthetic division including the remainder are nonnegative, then r is an upper bound of the real zeros of $P(x)$.

  2. **Lower bound.** If $r < 0$ and all numbers in the quotient row of the synthetic division including the remainder alternate in sign, then r is a lower bound of the real zeros of $P(x)$.

Bisection Method used to approximate real zeros

Let $P(x)$ be a polynomial with real coefficients. If $P(x)$ has opposite signs at the endpoints of the interval (a, b), then a real zero r lies in this interval. We bisect this interval [find the midpoint $m = (a + b)/2$], check the sign of $P(m)$, and choose the interval (a, m) or (m, b) on which $P(x)$ has opposite signs at the endpoints. We repeat this bisecting process (producing a set of "nested" intervals each half the size of the preceding one, and each containing the real zero r) until we get the desired decimal accuracy for the zero approximation. At any point in the process if $P(m) = 0$, we stop, since m is a real zero.

---

1. Since $P(x)$ has opposite signs at -5 and -1, at -1 and -3, and at 5 and 8, there is at least one x intercept in each of the intervals $(-5, -1)$, $(-1, -3)$, and $(5, 8)$.

3. Since $P(x)$ has opposite signs at -6 and -4, at -4 and 0, at 2 and 4, and at 4 and 7, there is at least one x intercept in each of the intervals $(-6, -4)$, $(-4, 0)$, $(2, 4)$, and $(4, 7)$.

5. We form a synthetic division table.

	1	-9	23	-14
0	1	-9	23	-14
1	1	-8	15	1
2	1	-7	9	4
3	1	-6	5	1
4	1	-5	3	-2
5	1	-4	3	1

Since $P(x)$ has opposite signs at 0 and 1, at 3 and 4, and at 4 and 5, it has at least one zero in each of the intervals $(0, 1)$, $(3, 4)$, $(4, 5)$. Since $P(x)$ is a third degree polynomial, there can be only 3 real zeros, and they must each lie in one of these intervals.

7. We form a synthetic division table.

	1	3	-1	-5
0	1	3	-1	-5
1	1	4	3	-2
2	1	5	9	13
-1	1	2	-3	-2
-2	1	1	-3	1
-3	1	0	-1	-2

Since $P(x)$ has opposite signs at 1 and 2, at -2 and -1, and at -3 and -2, it has at least one zero in each of the intervals $(1, 2)$, $(-2, -1)$, $(-3, -2)$. Since $P(x)$ is a third degree polynomial, there can be only 3 real zeros, and they must each lie in one of these intervals.

9. We form a synthetic division table.

	1	0	-3	1	
0	1	0	-3	1	
1	1	1	-2	-1	
2	1	2	1	3	an upper bound
-1	1	-1	-2	3	
-2	1	-2	1	-1	a lower bound

2 is an upper bound; -2 is a lower bound.

11. We form a synthetic division table.

	1	-3	4	2	-9	
0	1	-3	4	2	-9	
1	1	-2	2	4	-5	
2	1	-1	2	6	3	
3	1	0	4	14	33	an upper bound
-1	1	-4	8	-6	-3	
-2	1	-5	14	-26	43	a lower bound

3 is an upper bound; -2 is a lower bound.

13. We form a synthetic division table.

	1	0	-3	3	2	-2	
0	1	0	-3	3	2	-2	
1	1	1	-2	1	3	1	
2	1	2	1	5	12	22	an upper bound
-1	1	-1	-2	5	-3	1	
-2	1	-2	1	1	0	-2	
-3	1	-3	6	-15	47	-143	a lower bound

2 is an upper bound; -3 is a lower bound.

15. (A) We form a synthetic division table.

	1	-2	-5	4	
0	1	-2	-5	4	
1	1	-1	-6	-2	a zero in (0, 1)
2	1	0	-5	-6	
3	1	1	-2	-2	
4	1	2	3	16	an upper bound, also a zero in (3, 4)
-1	1	-3	-2	6	
-2	1	-4	3	-2	a lower bound, also a zero in (-2, -1)

(B) We search for the real zero in (3, 4). We organize our calculations in a table.

Sign Change Interval (a, b)	Midpoint m	$P(a)$	$P(m)$	$P(b)$
(3, 4)	3.5	-	+	+
(3, 3.5)	3.25	-	+	+
(3, 3.25)	3.125	-	-	+
(3.125, 3.25)	3.1875	-	+	+
(3.125, 3.1875)	3.15625	-	-	+
(3.156, 3.188)	We stop here	-		+

Since each endpoint rounds to 3.2, a real zero lies on this last interval and is given by 3.2 to one decimal place accuracy.

17. (A) We form a synthetic division table.

	1	-2	-1	5	
0	1	-2	-1	5	
1	1	-1	-2	3	
2	1	0	-1	3	
3	1	1	2	11	an upper bound
-1	1	-3	2	3	
-2	1	-4	7	-9	a lower bound, also a zero in (-2, -1)

(B) We search for the real zero in (-2, -1). We organize our calculations in a table.

Sign Change Interval (a, b)	Midpoint m	Sign of P $P(a)$	$P(m)$	$P(b)$
(-2, -1)	-1.5	-	-	+
(-1.5, -1)	-1.25	-	+	+
(-1.5, -1.25)	-1.375	-	-	+
(-1.375, -1.25)	-1.3125	-	+	+
(-1.375, -1.3125)	-1.34375	-	+	+
(-1.375, -1.34375)	-1.359375	-	+	+
(-1.375, -1.359375)	We stop here	-		+

Since each endpoint rounds to -1.4, a real zero lies on this last interval and is given by -1.4 to one decimal place accuracy.

19. (A) We form a synthetic division table.

	1	-2	-7	9	7	
0	1	-2	-7	9	7	
1	1	-1	-8	1	8	
2	1	0	-7	-5	-3	a zero in (1, 2)
3	1	1	-4	-3	-2	
4	1	2	1	13	59	an upper bound, also a zero in (3, 4)
-1	1	-3	-4	13	-6	a zero in (-1, 0)
-2	1	-4	1	7	-7	
-3	1	-5	8	-15	52	a lower bound, also a zero in (-3, -2)

(B) We search for the real zero in (3, 4). We organize our calculations in a table.

Sign Change Interval (a, b)	Midpoint m	Sign of P $P(a)$	$P(m)$	$P(b)$
(3, 4)	3.5	-	+	+
(3, 3.5)	3.25	-	+	+
(3, 3.25)	3.125	-	+	+
(3, 3.125)	3.0625	-	-	+
(3.0625, 3.125)	We stop here	-		+

Since each endpoint rounds to 3.1, a real zero lies on this last interval and is given by 3.1 to one decimal place accuracy.

21. (A) We form a synthetic division table.

	1	-1	-4	4	3	
0	1	-1	-4	4	3	
1	1	0	-4	0	3	
2	1	1	-2	0	3	
3	1	2	2	10	33	an upper bound
-1	1	-2	-2	6	-3	a zero in (-1, 0)
-2	1	-3	2	0	3	a lower bound, also a zero in (-2, -1)

(B) We search for the real zero in (-1, 0). We organize our calculations in a table.

Sign Change Interval (a, b)	Midpoint m	Sign of P		
		$P(a)$	$P(m)$	$P(b)$
(-1, 0)	-0.5	-	+	+
(-1, -0.5)	-0.75	-	-	+
(-0.75, -0.5)	-0.625	-	-	+
(-0.625, -0.5)	-0.5625	-	-	+
(-0.5625, -0.5)	-0.53125	-	-	+
(-0.53125, -0.5)	We stop here	-		+

Since each endpoint rounds to -0.5, a real zero lies on this last interval and is given by -0.5 to one decimal place accuracy.

23. (A) We form a synthetic division table.

	1	-2	3	-8
0	1	-2	3	-8
1	1	-1	2	-6
2	1	0	3	-2
3	1	1	6	10
-1	1	-3	6	-14

From the table, 3 is an upper bound and -1 is a lower bound.

(B) The only interval in which a real zero is indicated is (2, 3). We search for this real zero. We organize our calculations in a table.

Sign Change Interval (a, b)	Midpoint m	Sign of P		
		$P(a)$	$P(m)$	$P(b)$
(2, 3)	2.5	-	+	+
(2, 2.5)	2.25	-	+	+
(2, 2.25)	2.125	-	-	+
(2.125, 2.25)	2.1875	-	-	+
(2.1875, 2.25)	2.21875	-	-	+
(2.21875, 2.25)	2.234375	-	-	+
(2.234375, 2.25)	2.2421875	-	-	+
(2.2421875, 2.25)	2.24609375	-	-	+
(2.24609375, 2.25)	We stop here	-		+

Since each endpoint rounds to 2.25, a real zero lies on this last interval and is given by 2.25 to two decimal place accuracy. A glance at a computer-generated graph of $P(x)$ suggests that this is the only real zero.

25. (A) We form a synthetic division table.

	1	1	-5	7	-22
0	1	1	-5	7	-22
1	1	2	-3	4	-18
2	1	3	1	9	-4
3	1	4	7	28	62
-1	1	0	-5	12	-34
-2	1	-1	-3	13	-48
-3	1	-2	1	4	-34
-4	1	-3	7	-21	62

From the table, 3 is an upper bound and -4 is a lower bound.

(B) There are real zeros in the invervals (2, 3) and (-4, -3) indicated in the table. We search for the zero in (2, 3). We organize our calculations in a table.

Sign Change Interval (a, b)	Midpoint m	Sign of P $P(a)$	$P(m)$	$P(b)$
(2, 3)	2.5	−	+	+
(2, 2.5)	2.25	−	+	+
(2, 2.25)	2.125	−	+	+
(2, 2.125)	2.0625	−	−	+
(2.0625, 2.125)	2.09375	−	−	+
(2.09375, 2.125)	2.109375	−	−	+
(2.109375, 2.125)	2.1171875	−	−	+
(2.1171875, 2.125)	2.12109375	−	+	+
(2.1171875, 2.12109375)	We stop here	−		+

Since each endpoint rounds to 2.12, a real zero lies on this last interval and is given by 2.12 to two decimal place accuracy. A similar search (details omitted) leads to -3.51 as the other indicated real zero. A glance at a computer-generated graph of $P(x)$ suggests that these are the only real zeros.

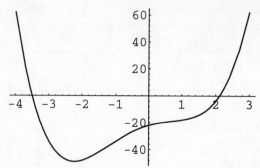

27. (A) We form a synthetic division table.

	1	0	-3	0	-4	4
0	1	0	-3	0	-4	4
1	1	1	-2	-2	-6	-2
2	1	2	1	2	0	4
-1	1	-1	-2	2	-6	10
-2	1	-2	1	-2	0	4
-3	1	-3	6	-18	50	-146

From the table, 2 is an upper bound and -3 is a lower bound.

(B) There are real zeros in the invervals (0, 1), (1, 2), and (-3, -2) indicated in the table. We search for the real zero in (0, 1). We organize our calculations in a table.

Sign Change Interval (a, b)	Midpoint m	Sign of P P(a)	P(m)	P(b)
(0, 1)	0.5	+	+	-
(0.5, 1)	0.75	+	-	-
(0.5, 0.75)	0.625	+	+	-
(0.625, 0.75)	0.6875	+	+	-
(0.6875, 0.75)	0.71875	+	+	-
(0.71875, 0.75)	0.734375	+	+	-
(0.734375, 0.75)	0.7421875	+	+	-
(0.7421875, 0.75)	0.74609875	+	+	-
(0.74609875, 0.75)	We stop here	+		-

Since each endpoint rounds to 0.75, a real zero lies on this last interval and is given by 0.75 to two decimal place accuracy. Similar searches (details omitted) lead to -2.09 and 1.88 as the other indicated real zeros. A glance at a computer-generated graph of $P(x)$ suggests that these are the only real zeros.

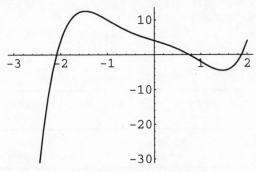

29. (A) We form a synthetic division table.

	1	1	3	1	2	-5
0	1	1	3	1	2	-5
1	1	2	5	6	8	3
-1	1	0	3	-2	4	-9

From the table, 1 is an upper bound and -1 is a lower bound.

(B) There is a real zero in the inverval (0, 1) indicated in the table. We search for this real zero. We organize our calculations in a table.

Sign Change Interval (a, b)	Midpoint m	Sign of P P(a)	P(m)	P(b)
(0, 1)	0.5	-	-	+
(0.5, 1)	0.75	-	-	+
(0.75, 1)	0.875	-	+	+
(0.75, 0.875)	0.8125	-	-	+
(0.8125, 0.875)	0.84375	-	+	+
(0.8125, 0.84375)	0.828125	-	-	+
(0.828125, 0.84375)	0.8359375	-	+	+
(0.828125, 0.8359375)	0.83203125	-	-	+
(0.83203125, 0.8359375)	0.833984375	-	-	+
(0.833984375, 0.8359375)	0.8349609375	-	+	+
(0.833984375, 0.8349609375)	We stop here	-		+

Since each endpoint rounds to 0.83, a real zero lies on this last interval and is given by 0.83 to two decimal place accuracy. A glance at a computer-generated graph of $P(x)$ suggests that this is the only real zero.

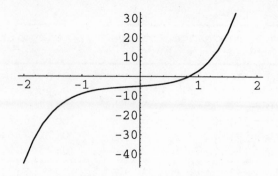

31. (A) We form a synthetic division table.

	1	-5	5	5	-10	5	
0	1	-5	5	5	-10	5	
1	1	-4	1	6	-4	1	
2	1	-3	-1	3	-4	-3	a zero in (1, 2)
3	1	-2	-1	2	-4	-7	
4	1	-1	1	9	26	109	a zero in (3, 4)
5	1	0	5	30	140	705	an upper bound
-1	1	-6	11	-6	-4	9	
-2	1	-7	19	-33	56	-107	a lower bound, also a zero in (-2, -1)

(B) We search for the real zero in (3, 4). We organize our calculations in a table.

Sign Change Interval (a, b)	Midpoint m	P(a)	P(m)	P(b)
(3, 4)	3.5	−	+	+
(3, 3.5)	3.25	−	+	+
(3, 3.25)	3.125	−	−	+
(3.125, 3.25)	3.1875	−	−	+
(3.1875, 3.25)	3.21875	−	+	+
(3.1875, 3.21875)	3.203125	−	−	+
(3.203125, 3.21875)	3.2109375	−	−	+
(3.2109375, 3.21875)	3.21484375	−	−	+
(3.21484375, 3.21875)	3.216796875	−	+	+
(3.21484375, 3.216796875)	3.2158203125	−	+	+
(3.2148375, 3.2158203125)	3.21532890625	−	−	+
(3.21532…, 3.2158…)	We stop here	−		+

Since each endpoint rounds to 3.22, a real zero lies on this last interval and is given by 3.22 to two decimal place accuracy.

33. (A) We form a synthetic division table.

	1	0	-10	0	9	10	
0	1	0	-10	0	9	10	
1	1	1	-9	-9	0	10	
2	1	2	-6	-12	-15	-20	a zero in (1, 2)
3	1	3	-1	-3	0	10	a zero in (2, 3)
4	1	4	6	24	105	430	an upper bound
-1	1	-1	-9	9	0	10	
-2	1	-2	-6	12	-15	40	
-3	1	-3	-1	3	0	10	
-4	1	-4	6	-24	105	-410	a lower bound, also a zero in (-4, -3)

(B) We search for the real zero in (2, 3). We organize our calculations in a table.

Sign Change Interval (a, b)	Midpoint m	Sign of P		
		P(a)	P(m)	P(b)
(2, 3)	2.5	−	−	+
(2.5, 3)	2.75	−	−	+
(2.75, 3)	2.875	−	−	+
(2.875, 3)	2.9375	−	+	+
(2.875, 2.9375)	2.90625	−	−	+
(2.90625, 2.9375)	2.921875	−	−	+
(2.921875, 2.9375)	2.9296875	−	+	+
(2.921875, 2.9296875)	2.92578125	−	+	+
(2.921875, 2.92578125)	2.923828125	−	+	+
(2.921875, 2.923828125)	We stop here	−		+

Since each endpoint rounds to 2.92, a real zero lies on this last interval and is given by 2.92 to two decimal place accuracy.

35. (A) We form a synthetic division table.

	1	−24	−25	10
0	1	−24	−25	10
10	1	−14	−165	−1640
20	1	−4	−105	−2090
30	1	6	155	4660
−10	1	−34	315	−3140

From the table, 30 is an upper bound and −10 is a lower bound.

(B) The real zeros are found by the bisection method to be −1.29, 0.31, and 24.98 to two decimal place accuracy (details omitted).

37. (A) We form a synthetic division table.

	1	12	−900	0	5,000
0	1	12	−900	0	5,000
10	1	22	−680	−680	−63,000
20	1	32	−260	−520	−99,000
30	1	42	360	10,800	329,000
−10	1	2	−920	9,200	−87,000
−20	1	−8	−740	14,800	−291,000
−30	1	−18	−360	10,800	−319,000
−40	1	−28	220	−8,800	357,000

From the table, 30 is an upper bound and −40 is a lower bound.

(B) The real zeros are found by the bisection method to be −36.53, −2.33, 2.40, and 24.46 to two decimal place accuracy (details omitted).

39. (A) We form a synthetic division table.

	1	0	−100	−1,000	−5,000
0	1	0	−100	−1,000	−5,000
10	1	10	0	−1,000	−15,000
20	1	20	300	5,000	95,000
−10	1	−10	0	−1,000	5,000

From the table, 20 is an upper bound and −10 is a lower bound.

(B) The real zeros are found by the bisection method to be −7.47 and 14.03 to two decimal place accuracy (details omitted).

41. (A) We form a synthetic division table.

	4	-40	-1,475	7,875	-10,000
0	4	-40	-1,475	7,875	-10,000
10	4	0	-1,475	-6,875	-78,750
20	4	40	-675	-5,625	-122,500
30	4	80	925	35,625	1,058,750
-10	4	-80	-675	14,625	-156,250
-20	4	-120	925	-10,625	202,500

From the table, 30 is an upper bound and -20 is a lower bound.

(B) Two real zeros can be found by the bisection method to be -17.66 and 22.66 to two decimal place accuracy. In finding the remaining zeros, we note:

	4	-40	-1,475	7,875	-10,000
2.5	4	-30	-1,550	4,000	0

2.5 is a zero. Moreover, $P(x) = (x - 2.5)(4x^3 - 30x^2 - 1,550x + 4,000)$. Testing 2.5 again,

	4	-30	-1,550	4,000
2.5	4	-20	-1,600	0

2.5 is in fact a double zero of $P(x)$.
Hence $P(x) = (x - 2.5)^2(4x^2 - 20x - 1600)$.
Solving $4x^2 - 20x - 1600 = 0$ yields exact solutions

$$x = \frac{20 \pm \sqrt{26,000}}{8}$$

or, again, $x = -17.66$ and 22.66 to two decimal place accuracy.

43. (A) We form a synthetic division table.

	0.01	-0.1	-12	0	0	9,000
0	0.01	-0.1	-12	0	0	9,000
10	0.01	0	-12	-120	-1,200	-3,000
20	0.01	0.1	-10	-200	-4,000	-71,000
30	0.01	0.2	-6	-180	-5,400	-153,000
40	0.01	0.3	0	0	0	9,000
-10	0.01	-0.2	-10	100	-1,000	19,000
-20	0.01	-0.3	-6	120	-2,400	57,000
-30	0.01	-0.4	0	0	0	9,000
-40	0.01	-0.5	8	-320	12,800	-503,000

From the table, 40 is an upper bound and -40 is a lower bound.

(B) The real zeros are found by the bisection method to be -30.45, 9.06, and 39.80 to two decimal place accuracy (details omitted).

45. Let (x, x^2) be a point on the graph of $y = x^2$. Then the distance from $(1, 2)$ to (x, x^2) must equal 1 unit. Applying the distance formula, we have,

$$\sqrt{(x - 1)^2 + (x^2 - 2)^2} = 1$$
$$(x - 1)^2 + (x^2 - 2)^2 = 1$$
$$x^2 - 2x + 1 + x^4 - 4x^2 + 4 = 1$$
$$x^4 - 3x^2 - 2x + 4 = 0$$

Let $P(x) = x^4 - 3x^2 - 2x + 4$
The only rational zero is 1.

	1	0	-3	-2	4
1	1	1	-2	-4	0

We examine $Q(x) = x^3 + x^2 - 2x - 4$.

We form a synthetic division table for $Q(x)$.

	1	1	-2	-4	
0	1	1	-2	-4	
1	1	2	0	-4	
2	1	3	4	4	a zero in (1, 2)
-1	1	0	-2	-2	
-2	1	-1	0	-4	
-3	1	-2	4	-16	a lower bound

The table is inconclusive as to the existence of negative zeros. However, the graph of $Q(x)$ strongly suggests that there are none. We can now apply the bisection method (details omitted) to locate the positive zero of $Q(x)$ to any desired accuracy. To one decimal place, it is 1.7. Therefore, the two real zeros of $P(x)$ are 1 and 1.7. Hence the two required points (x, x^2) are $(1, 1)$ and $(1.7, 2.9)$.

Graph of $y = Q(x) = x^3 + x^2 - 2x - 4$

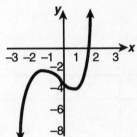

47.

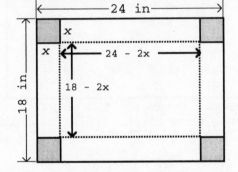

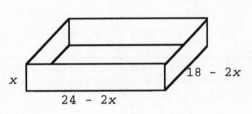

From the above figures it should be clear that
V = length × width × height = $(24 - 2x)(18 - 2x)x$ $0 < x < 9$ (why?)

We solve $(24 - 2x)(18 - 2x)x = 600$
$$432x - 84x^2 + 4x^3 = 600$$
$$4x^3 - 84x^2 + 432x - 600 = 0$$
$$x^3 - 21x^2 + 108x - 150 = 0$$

Let $P(x) = x^3 - 21x^2 + 108x - 150$. We search for positive rational zeros. We form a synthetic division table.

	1	-21	108	-150	
0	1	-21	108	-150	
1	1	-20	88	-62	
2	1	-19	70	-10	
3	1	-18	54	12	a zero in (2,3)
5	1	-16	28	-10	a zero in (3, 5)
6	1	-15	18	-42	
10	1	-11	-2	-170	
15	1	-6	18	120	a zero in (10, 15)
25	1	4	208	5050	an upper bound

There are no rational zeros. Applying the bisection method (details omitted), the positive zeros in (0, 9) are found to be 2.3 and 4.6 to one decimal place accuracy.

$x = 2.3$ inches or 4.6 inches.

49. We note:

$$\begin{pmatrix} \text{Volume} \\ \text{of} \\ \text{tank} \end{pmatrix} = \begin{pmatrix} \text{Volume of} \\ \text{two hemispheres} \\ \text{of radius } x \end{pmatrix} + \begin{pmatrix} \text{Volume of cylinder} \\ \text{with radius } x, \\ \text{height } 10 - 2x \end{pmatrix}$$

$$20\pi = \frac{4}{3}\pi x^3 + \pi x^2 (10 - 2x)$$

$$20 = \frac{4}{3}x^3 + 10x^2 - 2x^3$$

$$60 = 4x^3 + 30x^2 - 6x^3$$

$$2x^3 - 30x^2 + 60 = 0$$

$$x^3 - 15x^2 + 30 = 0 \quad \text{From physical considerations, we are interested only in solutions in } (0, 5).$$

Let $P(x) = x^3 - 15x^2 + 30$

There are no rational zeros.

We form a synthetic division table. We are only interested in the positive real zeros.

	1	-15	0	30	
0	1	-15	0	30	
1	1	-14	-14	16	
2	1	-13	-26	-22	There is a zero in (1, 2)
3	1	-12	-36	-78	
4	1	-11	-44	-146	
5	1	-10	-50	-220	There does not seem to be another zero in (0, 5)
15	1	0	0	30	There is a zero in (5, 15)

From the table, we see that the only zero of physical interest lies in (1, 2). Applying the bisection method (details omitted) the zero in (1, 2) is found to be 1.5 to one decimal place accuracy.

$x = 1.5$ feet.

Exercise 4-4

Key Ideas and Formulas

A function f is a rational function if $f(x) = \dfrac{n(x)}{d(x)}$, $d(x) \neq 0$, where $n(x)$ and $d(x)$ are polynomials. The domain of f is the set of all real numbers such that $d(x) \neq 0$.

If $f(x) = \dfrac{n(x)}{d(x)}$ and $d(a) = 0$, then f is discontinuous at $x = a$ and the graph of f has a hole or break at $x = a$. If $n(c) = 0$ and $d(c) \neq 0$, then $x = c$ is an x intercept for the graph of f.

Asymptotes: the line $x = a$ is a vertical asymptote for the graph of $y = f(x)$ if $f(x)$ either increases or decreases without bound as x approaches a from the right or from the left, $f(x) \to \infty$ or $f(x) \to -\infty$ as $x \to a^+$ or $x \to a^-$.

If $f(x) = \dfrac{n(x)}{d(x)}$, $d(a) = 0$ and $n(a) \neq 0$, then the line $x = a$ is a vertical asymptote.

The line $y = b$ is a horizontal asymptote for the graph of $y = f(x)$ if $f(x)$ approaches b as x increases without bound or as x decreases without bound, $f(x) \to b$ as $x \to \infty$ or $x \to -\infty$.

If $f(x) = \dfrac{a_m x^m + \cdots + a_1 x + a_0}{b_n x^n + \cdots + b_1 x + b_0}$, a_m, $b_n \neq 0$, then

1. for $m < n$ the x axis is a horizontal asymptote

2. for $m = n$ the line $y = \dfrac{a_m}{b_n}$ is a horizontal asymptote

3. for $m > n$ the graph will increase or decrease without bound, depending on m, n, a_m, and b_n, and there are no horizontal asymptotes

If m, the degree of $n(x)$, is one more than n, the degree of $d(x)$, then $f(x)$ can be written in the form $f(x) = mx + b + \dfrac{r(x)}{d(x)}$, where the degree of $f(x)$ is less than the degree of $d(x)$. Then the line $y = mx + b$ is an oblique asymptote for the graph of f.

$$[f(x) - (mx + b)] \to 0 \text{ as } x \to -\infty \text{ or } x \to \infty$$

To graph a rational function, follow the steps
1. Intercepts
2. Vertical asymptotes
3. Sign chart
4. Horizontal asymptotes
5. Complete the sketch. For details, see text.

---

1. This graph has a vertical asymptote $x = 2$, and a horizontal asymptote $y = -2$. This corresponds to $g(x)$.

3. This graph has a vertical asymptote $x = 2$, and a horizontal asymptote $y = 2$. This corresponds to $h(x)$.

5. $\dfrac{2x - 4}{x + 1}$ *Domain:* $d(x) = x + 1$ zero: $x = -1$ domain: $(-\infty, -1) \cup (-1, \infty)$

x *intercepts:* $n(x) = 2x - 4$ zero: $x = 2$ x intercept: 2

7. $\dfrac{x^2 - 1}{x^2 - 16}$ *Domain:* $d(x) = x^2 - 16$ zeros: $x^2 - 16 = 0$

$$x^2 = 16$$
$$x = \pm 4$$

domain: $(-\infty, -4) \cup (-4, 4) \cup (4, \infty)$

x *intercepts:* $n(x) = x^2 - 1$ zeros: $x^2 - 1 = 0$
$$x^2 = 1$$
$$x = \pm 1$$

x intercepts: $-1, 1$

9. $\dfrac{x^2 - x - 6}{x^2 - x - 12}$ *Domain:* $d(x) = x^2 - x - 12$ zeros: $x^2 - x - 12 = 0$

$$(x + 3)(x - 4) = 0$$
$$x = -3, 4$$

domain: $(-\infty, -3) \cup (-3, 4) \cup (4, \infty)$

x *intercepts:* $n(x) = x^2 - x - 6$ zeros: $x^2 - x - 6 = 0$
$$(x + 2)(x - 3) = 0$$
$$x = -2, 3$$

x intercepts: $-2, 3$

11. $\dfrac{x}{x^2 + 4}$ *Domain:* $d(x) = x^2 + 4$ no real zeros

Domain: all real numbers

x *intercepts:* $n(x) = x$ zero: $x = 0$

x intercept: 0

13. $\dfrac{2x}{x - 4}$ *vertical asymptotes:* $d(x) = x - 4$ zero: $x = 4$

vertical asymptote $x = 4$

> **Common Error:**
> $x = 0$ is not a vertical asymptote.

horizontal asymptotes: Since $n(x)$ and $d(x)$ have the same degree, the line $y = 2$ is a horizontal asymptote.

15. $\dfrac{2x^2 + 3x}{3x^2 - 48}$ *vertical asymptotes:* $d(x) = 3x^2 - 48$ zeros: $3x^2 - 48 = 0$

$$3(x + 4)(x - 4) = 0$$
$$x = -4, 4$$

vertical asymptotes: $x = -4$, $x = 4$

horizontal asymptotes: Since $n(x)$ and $d(x)$ have the same degree, the line $y = \dfrac{2}{3}$ is a horizontal asymptote.

17. $\dfrac{2x}{x^4 + 1}$ *vertical asymptote:* $d(x) = x^4 + 1$ No real zeros:
No vertical asymptotes

   horizontal asymptotes: Since the degree of $n(x)$ is less than the degree of
$d(x)$, the x axis is a horizontal asymptote
horizontal asymptote: $y = 0$

19. $\dfrac{6x^4}{3x^2 - 2x - 5}$ *vertical asymptotes:* $d(x) = 3x^2 - 2x - 5$ zeros: $3x^2 - 2x - 5 = 0$

$$(3x - 5)(x + 1) = 0$$

$$x = \frac{5}{3}, -1$$

   vertical asymptotes: $x = -1$, $x = \dfrac{5}{3}$

   horizontal asymptotes: Since the degree of $n(x)$ is greater than the degree
of $d(x)$, there are no horizontal asymptotes.

21. $f(x) = \dfrac{1}{x - 4} = \dfrac{n(x)}{d(x)}$

   Intercepts. There are no real zeros of $n(x) = 1$. No x intercept

   $f(0) = -\dfrac{1}{4}$ $y = -\dfrac{1}{4}$ y intercept

   Vertical asymptotes. $d(x) = x - 4$ zeros: 4 $x = 4$

   Sign Chart.

Test numbers	3	5
Value of f	-1	1
Sign of f	–	+

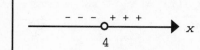

   Since $x = 4$ is a vertical asymptote and $f(x) < 0$ for $x < 4$,
$f(x) \to -\infty$ as $x \to 4^-$
Since $x = 4$ is a vertical asymptote and $f(x) > 0$ for $x > 4$,
$f(x) \to \infty$ as $x \to 4^+$
Horizontal asymptotes. Since the degree of $n(x)$ is less that the degree of
$d(x)$, the x axis is a horizontal asymptote.
Complete the sketch.

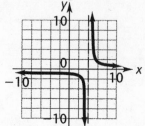

23. $f(x) = \dfrac{x}{x + 1} = \dfrac{n(x)}{d(x)}$

   Intercepts. Real zeros of $n(x) = x$ $x = 0$ x intercept
$f(0) = 0$ $y = 0$ y intercept
The graph crosses the coordinate axes only at the origin.
   Vertical asymptotes. $d(x) = x + 1$ zeros: -1 $x = -1$
   Sign chart.

Test numbers	-2	$-\frac{1}{2}$	1
Value of f	2	-1	$\frac{1}{2}$
Sign of f	+	–	+

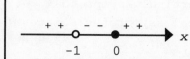

   Since $x = -1$ is a vertical asymptote and $f(x) > 0$ for $x < -1$,
$f(x) \to \infty$ as $x \to -1^-$
Since $x = -1$ is a vertical asymptote and $f(x) < 0$ for $-1 < x < 0$,
$f(x) \to -\infty$ as $x \to -1^+$

Horizontal asymptotes. Since $n(x)$ and $d(x)$ have the same degree, the line $y = 1$ is a horizontal asymptote.

Complete the sketch.

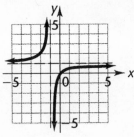

25. $h(x) = \dfrac{x}{2x - 2} = \dfrac{n(x)}{d(x)}$

Intercepts. Real zeros of $n(x) = x$ $x = 0$ x intercept

$h(0) = 0$ $y = 0$ y intercept

The graph crosses the coordinate axes only at the origin.

Vertical asymptotes. $d(x) = 2x - 2$ zeros: 1 $x = 1$

Sign Chart.

Test numbers	-1	$\frac{1}{2}$	2
Value of h	$\frac{1}{4}$	$-\frac{1}{2}$	1
Sign of h	$+$	$-$	$+$

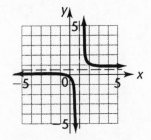

Since $x = 1$ is a vertical asymptote and $h(x) < 0$ for $0 < x < 1$,
$h(x) \to -\infty$ as $x \to 1^-$

Since $x = 1$ is a vertical asymptote and $h(x) > 0$ for $x > 1$
$h(x) \to \infty$ as $x \to 1^+$

Horizontal asymptotes. Since $n(x)$ and $d(x)$ have the same degree, the line $y = \dfrac{1}{2}$ is a horizontal asymptote.

Complete the sketch.

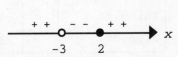

27. $f(x) = \dfrac{2x - 4}{x + 3} = \dfrac{n(x)}{d(x)}$

Intercepts. Real zeros of $n(x) = 2x - 4$ $x = 2$ x intercept

$f(0) = -\dfrac{4}{3}$ y intercept

Vertical asymptotes. $d(x) = x + 3$ zeros: -3 $x = -3$

Sign Chart.

Test numbers	-4	0	3
Value of f	12	$-\frac{4}{3}$	$\frac{1}{3}$
Sign of f	$+$	$-$	$+$

Since $x = -3$ is a vertical asymptote and $f(x) > 0$ for $x < -3$
$f(x) \to \infty$ as $x \to -3^-$

Since $x = -3$ is a vertical asymptote and $f(x) < 0$ for $-3 < x < 2$
$f(x) \to -\infty$ as $x \to -3^+$

Horizontal asymptotes. Since $n(x)$ and $d(x)$ have the same degree, the line $y = 2$ is a horizontal asymptote.

Complete the sketch.

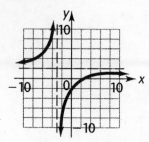

29. $g(x) = \dfrac{1 - x^2}{x^2} = \dfrac{n(x)}{d(x)}$

Intercepts. Real zeros of $n(x) = 1 - x^2$ $1 - x^2 = 0$

$$x^2 = 1$$
$$x = \pm 1 \quad x \text{ intercepts}$$

               $g(0)$ is not defined   no y intercepts

Vertical asymptotes. $d(x) = x^2$ zeros: 0 $x = 0$

Sign Chart.

Test numbers	-2	$-\frac{1}{2}$	$\frac{1}{2}$	2
Value of g	$-\frac{3}{4}$	3	3	$-\frac{3}{4}$
Sign of g	$-$	$+$	$+$	$-$

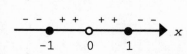

Since $x = 0$ is a vertical asymptote and $g(x) > 0$ for $-1 < x < 0$ and $0 < x < 1$,
     $g(x) \to \infty$ as $x \to 0^+$ and as $x \to 0^-$

Horizontal asymptotes. Since $n(x)$ and $d(x)$ have the same degree, the line $y = -1$ is a horizontal asymptote.

Complete the sketch.

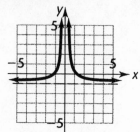

31. $f(x) = \dfrac{9}{x^2 - 9} = \dfrac{n(x)}{d(x)}$

Intercepts. There are no real zeros of $n(x) = 9$. No x intercept

                          $f(0) = -1$ y intercept

Vertical asymptotes. $d(x) = x^2 - 9$ zeros: $x^2 - 9 = 0$

$$x^2 = 9$$
$$x = \pm 3$$

Sign Chart.

Test numbers	-4	0	4
Value of f	$\frac{9}{7}$	-1	$\frac{9}{7}$
Sign of f	$+$	$-$	$+$

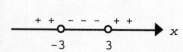

Since $x = -3$ is a vertical asymptote and $f(x) > 0$ for $x < -3$ and $f(x) < 0$ for $-3 < x < 3$,
     $f(x) \to \infty$ as $x \to -3^-$ and $f(x) \to -\infty$ as $x \to -3^+$

Since $x = 3$ is a vertical asymptote and $f(x) < 0$ for $-3 < x < 3$ and $f(x) > 0$ for $x > 3$,
     $f(x) \to -\infty$ as $x \to 3^-$ and $f(x) \to \infty$ as $x \to 3^+$

Horizontal asymptotes. Since the degree of $n(x)$ is less than the degree of $d(x)$, the x axis is a horizontal asymptote.

Complete the sketch.

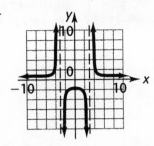

33. $f(x) = \dfrac{x}{x^2 - 1} = \dfrac{n(x)}{d(x)}$

Intercepts. Real zeros of $n(x) = x$ $x = 0$ x intercept

$\qquad\qquad\qquad\qquad\qquad\qquad f(0) = 0$ $y = 0$ y intercept

The graph crosses the coordinate axes only at the origin.

Vertical asymptotes. $d(x) = x^2 - 1$ zeros: $x^2 - 1 = 0$

$$x^2 = 1$$
$$x = \pm 1$$

Sign Chart.

Test numbers	-2	$-\frac{1}{2}$	$\frac{1}{2}$	2
Value of f	$-\frac{2}{3}$	$\frac{2}{3}$	$-\frac{2}{3}$	$\frac{2}{3}$
Sign of f	$-$	$+$	$-$	$+$

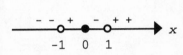

Since $x = -1$ is a vertical asymptote and $f(x) < 0$ for $x < -1$ and $f(x) > 0$ for $-1 < x < 0$,

$\qquad f(x) \to -\infty$ as $x \to -1^-$ and $f(x) \to \infty$ as $x \to -1^+$

Since $x = 1$ is a vertical asymptote and $f(x) < 0$ for $0 < x < 1$ and $f(x) > 0$ for $x > 1$,

$\qquad f(x) \to -\infty$ as $x \to 1^-$ and $f(x) \to \infty$ as $x \to 1^+$

Horizontal asymptotes. Since the degree of $n(x)$ is less than the degree of $d(x)$, the x axis is a horizontal asymptote.

Complete the sketch.

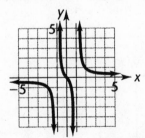

35. $g(x) = \dfrac{2}{x^2 + 1} = \dfrac{n(x)}{d(x)}$

Intercepts. There are no real zeros of $n(x) = 2$.   No x intercept

$\qquad\qquad\qquad\qquad\qquad\qquad g(0) = 2$ y intercept

Vertical asymptotes. There are no real zeros of $d(x) = x^2 + 1$

$\qquad$ No vertical asymptotes

Sign behavior. $g(x)$ is always positive.

Horizontal asymptotes. Since the degree of $n(x)$ is less than the degree of $d(x)$, the x axis is a horizontal asymptote.

Complete the sketch.

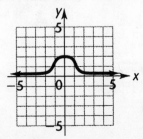

37. $f(x) = \dfrac{12x^2}{(3x+5)^2} = \dfrac{n(x)}{d(x)}$

Intercepts. Real zeros of $n(x) = 12x^2$ $x = 0$ x intercept
$f(0) = 0$ $y = 0$ y intercept
The graph crosses the coordinate axes only at the origin.
Vertical asymptotes. $d(x) = (3x+5)^2$ zeros: $(3x+5)^2 = 0$
$$3x + 5 = 0$$
$$x = -\frac{5}{3}$$

Sign Chart.

Test numbers	-2	-1	1
Value of f	48	3	$\frac{3}{16}$
Sign of f	+	+	+

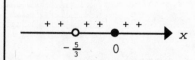

Since $x = -\dfrac{5}{3}$ is a vertical asymptote and $f(x) > 0$ for $x < -\dfrac{5}{3}$ and $-\dfrac{5}{3} < x < 0$,

$f(x) \to \infty$ as $x \to -\dfrac{5}{3}^-$ and as $x \to -\dfrac{5}{3}^+$

Horizontal asymptotes. Since $n(x)$ and $d(x)$ have the same degree, the line
$y = \dfrac{12}{3^2} = \dfrac{4}{3}$ is a horizontal asymptote.

Complete the sketch.

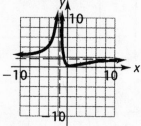

39. $f(x) = \dfrac{x^2 - 1}{x^2 + 7x + 10} = \dfrac{n(x)}{d(x)}$

Intercepts. Real zeros of $n(x) = x^2 - 1$ $x^2 - 1 = 0$
$$x^2 = 1$$
$$x = \pm 1 \quad x \text{ intercepts}$$
$$f(0) = -\frac{1}{10} \quad y \text{ intercept}$$

Vertical asymptotes. Real zeros of $d(x) = x^2 + 7x + 10$ $x^2 + 7x + 10 = 0$
$$(x + 2)(x + 5) = 0$$
$$x = -2, -5$$

Sign Chart.

Test numbers	-6	-3	$-\frac{3}{2}$	0	2
Value of f	$\frac{35}{4}$	-4	$\frac{5}{7}$	$-\frac{1}{10}$	$\frac{3}{28}$
Sign of f	+	-	+	-	+

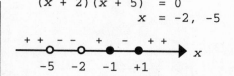

Since $x = -5$ is a vertical asymptote and $f(x) > 0$ for $x < -5$ and $f(x) < 0$ for $-5 < x < -2$,
$f(x) \to \infty$ as $x \to -5^-$ and $f(x) \to -\infty$ as $x \to -5^+$
Since $x = -2$ is a vertical asymptote and $f(x) < 0$ for $-5 < x < -2$ and $f(x) > 0$ for $-2 < x < -1$,
$f(x) \to -\infty$ as $x \to -2^-$ and $f(x) \to \infty$ as $x \to -2^+$

Horizontal asymptotes. Since $n(x)$ and $d(x)$ have the same degree, the line $y = 1$ is a horizontal asymptote.

Complete the sketch.

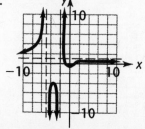

41. The x intercepts of $f(x) = \dfrac{n(x)}{d(x)}$ are the real zeros of $n(x)$. Since $n(x)$ is a quadratic function, the maximum number of real zeros is 2 and the minimum number is 0. Therefore, the maximum number of x intercepts is 2 and the minimum number is 0. For example, $\dfrac{x^2 - 1}{x^2}$ has two intercepts and $\dfrac{x^2 + 1}{x^2}$ has none.

43. $f(x) = \dfrac{2x^2}{x - 1} = \dfrac{n(x)}{d(x)}$

Vertical asymptotes. Real zeroes of $d(x) = x - 1$ $x = 1$

Horizontal asymptote. Since the degree of $n(x)$ is greater than the degree of $d(x)$, there is no horizontal asymptote.

Oblique asymptote.

$$
\begin{array}{r}
2x + 2 \\
x - 1 \overline{)\, 2x^2 } \\
\underline{2x^2 - 2x} \\
2x \\
\underline{2x - 2} \\
2
\end{array}
$$

Thus, $f(x) = 2x + 2 + \dfrac{2}{x - 1}$. Hence, the line $y = 2x + 2$ is an oblique asymptote.

45. $p(x) = \dfrac{x^3}{x^2 + 1} = \dfrac{n(x)}{d(x)}$

Vertical asymptotes. There are no real zeros of $d(x) = x^2 + 1$.
No vertical asymptotes.

Horizontal asymptotes. Since the degree of $n(x)$ is greater than the degree of $d(x)$, there is no horizontal asymptote.

Oblique asymptote:

$$
\begin{array}{r}
x \\
x^2 + 1 \overline{)\, x^3 } \\
\underline{x^3 + x} \\
- x
\end{array}
$$

Thus, $p(x) = x + \dfrac{-x}{x^2 + 1}$. Hence, the line $y = x$ is an oblique asymptote.

47. $r(x) = \dfrac{2x^2 - 3x + 5}{x} = \dfrac{n(x)}{d(x)}$

Vertical asymptotes. Real zeros of $d(x) = x$ $x = 0$

Horizontal asymptote. Since the degree of $n(x)$ is greater than the degree of $d(x)$, there is no horizontal asymptote.

Oblique asymptote. $\dfrac{2x^2 - 3x + 5}{x} = \dfrac{2x^2}{x} - \dfrac{3x}{x} + \dfrac{5}{x}$

$$= 2x - 3 + \dfrac{5}{x}$$

Thus $r(x) = 2x - 3 + \dfrac{5}{x}$. Hence the line $y = 2x - 3$ is an oblique asymptote.

49. Here is a computer-generated graph
of $f(x)$.

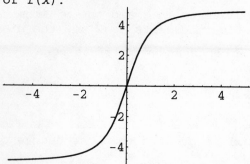

From the graph, we can see that
$f(x) \to 5$ as $x \to \infty$ and $f(x) \to -5$ as
$x \to -\infty$; the lines $y = 5$ and $y = -5$
are horizontal asymptotes.

51. Here is a computer-generated graph
of $f(x)$.

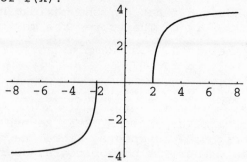

From the graph, we can see that
$f(x) \to 4$ as $x \to \infty$ and $f(x) \to -4$ as
$x \to -\infty$; the lines $y = 4$ and $y = -4$
are horizontal asymptotes.

53. $f(x) = \dfrac{x^2 + 1}{x} = \dfrac{n(x)}{d(x)}$

Intercepts. There are no real zeros of $n(x) = x^2 + 1$.   No x intercept
$\qquad\qquad$ $f(0)$ is not defined $\qquad\qquad\qquad\qquad$ No y intercept

Vertical asymptotes. Real zeros of $d(x) = x$. $x = 0$

Sign Chart.

Test numbers	-1	1
Value of f	-2	2
Sign of f	–	+

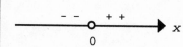

Since $x = 0$ is a vertical asymptote and $f(x) < 0$ for $x < 0$ and $f(x) > 0$
for $x > 0$,

$\qquad f(x) \to -\infty$ as $x \to 0^-$ and $f(x) \to \infty$ as $x \to 0^+$

Horizontal asymptote. Since the degree of $n(x)$ is greater than the degree of
$d(x)$, there is no horizontal asymptote.

Oblique asymptote. $f(x) = \dfrac{x^2 + 1}{x} = \dfrac{x^2}{x} + \dfrac{1}{x} = x + \dfrac{1}{x}$

Hence, the line $y = x$ is an oblique asymptote.

Complete the sketch.

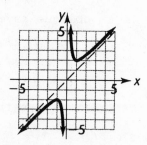

55. $k(x) = \dfrac{x^2 - 4x + 3}{2x - 4} = \dfrac{n(x)}{d(x)}$

Intercepts. Real zeros of $n(x) = x^2 - 4x + 3$ $\qquad x^2 - 4x + 3 = 0$
$\qquad\qquad\qquad\qquad\qquad\qquad\qquad\qquad\qquad (x - 1)(x - 3) = 0$
$\qquad\qquad\qquad\qquad\qquad\qquad\qquad\qquad\qquad\qquad x = 1, 3$ x intercepts

$$k(0) = -\frac{3}{4} \quad y \text{ intercept}$$

Vertical asymptotes. Real zeros of $d(x) = 2x - 4$ $\qquad 2x - 4 = 0$
$\qquad\qquad\qquad\qquad\qquad\qquad\qquad\qquad\qquad\qquad\qquad 2x = 4$
$\qquad\qquad\qquad\qquad\qquad\qquad\qquad\qquad\qquad\qquad\qquad x = 2$

Sign Chart.

Test numbers	0	$1\frac{1}{2}$	$2\frac{1}{2}$	4
Value of k	$-\frac{3}{4}$	$\frac{3}{4}$	$-\frac{3}{4}$	$\frac{3}{4}$
Sign of k	$-$	$+$	$-$	$+$

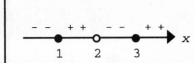

Since $x = 2$ is a vertical asymptote and $k(x) > 0$ for $1 < x < 2$ and $k(x) < 0$ for $2 < x < 3$,

   $k(x) \to \infty$ as $x \to 2^-$ and $k(x) \to -\infty$ as $x \to 2^+$

Horizontal asymptote. Since the degree of $n(x)$ is greater than the degree of $d(x)$, there is no horizontal asymptote.

Oblique asymptote:

$$
\begin{array}{r}
\frac{1}{2}x - 1 \\
2x - 4 \overline{)\, x^2 - 4x + 3} \\
\underline{x^2 - 2x} \\
-2x + 3 \\
\underline{-2x + 4} \\
- 1
\end{array}
$$

Thus, $k(x) = \frac{1}{2}x - 1 + \dfrac{-1}{2x - 4}$. Hence, the line $y = \frac{1}{2}x - 1$ is an oblique asymptote.

Complete the sketch.

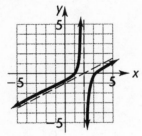

57. $F(x) = \dfrac{8 - x^3}{4x^2} = \dfrac{n(x)}{d(x)}$

Intercepts. Real zeros of $n(x) = 8 - x^3$ $8 - x^3 = 0$

$$(2 - x)(4 + 2x + x^2) = 0$$

$$2 - x = 0 \quad 4 + 2x + x^2 = 0$$

$$x = 2 \quad \text{No real zeros}$$

   $x = 2$ x intercept
   $F(0)$ is not defined. No y intercept.

Vertical asymptotes. Real zeros of $d(x) = 4x^2$. $x = 0$

Sign chart.

Test numbers	-1	1	3
Value of F	$\frac{9}{4}$	$\frac{7}{4}$	$-\frac{19}{36}$
Sign of F	$+$	$+$	$-$

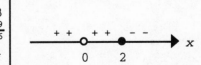

Since $x = 0$ is a vertical asymptote and $F(x) > 0$ for $x < 0$ and $0 < x < 2$,

   $F(x) \to \infty$ as $x \to 0^-$ and $F(x) \to \infty$ as $x \to 0^+$

Horizontal asymptote. Since the degree of $n(x)$ is greater than the degree of $d(x)$, there is no horizontal asymptote.

Oblique asymptote. $F(x) = \dfrac{8 - x^3}{4x^2} = \dfrac{8}{4x^2} - \dfrac{x^3}{4x^2} = -\dfrac{1}{4}x + \dfrac{2}{x^2}$. Hence, the line

$y = -\dfrac{1}{4}x$ is an oblique asymptote.

Complete the sketch.

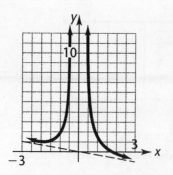

59.
$$x^2 + 1 \overline{\smash{\big)}\, x^4 } \atop x^2 - 1$$

$$\underline{x^4 + x^2}$$
$$-x^2$$
$$\underline{-x^2 - 1}$$
$$1$$

Thus, $f(x) = x^2 - 1 + \dfrac{1}{x^2 + 1}$.

Let $p(x) = x^2 - 1$; then $[f(x) - p(x)] = \dfrac{1}{x^2 + 1} \to 0$ as $x \to \infty$ and as $x \to -\infty$.

Thus, the graph of $f(x)$ approaches the graph of $p(x)$ (a parabola) asymptotically.

61.
$$x^2 - 1 \overline{\smash{\big)}\, x^5 } \atop x^3 + x$$

$$\underline{x^5 - x^3}$$
$$x^3$$
$$\underline{x^3 - x}$$
$$x$$

Thus, $f(x) = x^3 + x + \dfrac{x}{x^2 - 1}$.

Let $p(x) = x^3 + x$; then $[f(x) - p(x)] = \dfrac{x}{x^2 - 1} \to 0$ as $x \to \infty$ and as $x \to -\infty$.

Thus, the graph of $f(x)$ approaches the graph of $p(x)$ asymptotically.

63. $f(x) = \dfrac{x^2 - 4}{x - 2}$. $f(x)$ is not defined if $x - 2 = 0$,

that is, $x = 2$

Domain: $(-\infty, 2) \cup (2, \infty)$

$f(x) = \dfrac{(x - 2)(x + 2)}{(x - 2)}$

$f(x) = x + 2$

The graph is a straight line with slope 1 and y intercept 2, except that the point $(2, 4)$ is not on the graph.

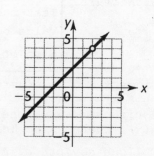

65. $r(x) = \dfrac{x + 2}{x^2 - 4}$. $f(x)$ is not defined if $x^2 - 4 = 0$, that is,

$$x^2 = 4$$
$$x = \pm 2$$

Domain: $(-\infty, -2) \cup (-2, 2) \cup (2, \infty)$

$$r(x) = \frac{x + 2}{(x + 2)(x - 2)}$$

$$r(x) = \frac{1}{x - 2}$$

The graph is the same as the graph of the function $\dfrac{1}{x - 2}$, except that the

point $\left(-2, -\dfrac{1}{4}\right)$ is not on the graph.

Intercepts: $y = -\dfrac{1}{2}$. No x intercept.

Vertical asymptote: $x = 2$

Sign chart.

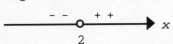

As $x \to 2^-$, $r(x) \to -\infty$. As $x \to 2^+$, $r(x) \to \infty$
Horizontal asymptote: $y = 0$

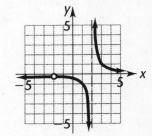

67. $N(t) = \dfrac{50t}{t + 4}$ $t \geq 0$

Intercepts: Real zeros of $50t$: $t = 0$ $N(0) = 0$
Vertical asymptotes: None, since -4, the only zero of $t + 4$, is not in the domain of N.
Sign behavior: $N(t)$ is always positive.
Horizontal asymptote: $N = 50$. As $t \to \infty$, $N \to 50$

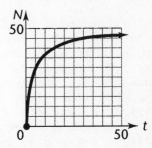

69. $N(t) = \dfrac{5t + 30}{t}$. $t \geq 1$

Intercepts: Real zeros of $5t + 30$, $t \geq 1$. None, since -6, the only zero of $5t + 30$, is not in the domain of N.
Vertical asymptotes: None, since 0, the only zero of t, is not in the domain of N.
Sign behavior: $N(t)$ is always positive.
Horizontal asymptote: $N = 5$. As $t \to \infty$, $N \to 5$

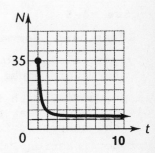

71. (A) $\overline{C}(n) = \dfrac{C(n)}{n} = \dfrac{2,500 + 175n + 25n^2}{n} = 25n + 175 + \dfrac{2,500}{n}$

(B) The minimum value of the function $\overline{C}(n)$ is $\overline{C}\left(\sqrt{\dfrac{c}{a}}\right)$, where $a = 25$ and $c = 2,500$

$\min \overline{C}(n) = \overline{C}\left(\sqrt{\dfrac{2,500}{25}}\right) = \overline{C}(\sqrt{100}) = \overline{C}(10)$

This minimum occurs when $n = 10$, after 10 years.

(C) *Intercepts:* Real zeros of $2{,}500 + 175n + 25n^2$. None. No n intercepts.

0 is not in the domain of n, so there are no $\overline{C}$ intercepts.
Vertical asymptotes: Real zeros of n. The line $n = 0$ is a vertical asymptote.

Sign behavior: $\overline{C}$ is always positive since $n \geq 0$.
Horizontal asymptote: None, since the degree of $C(n)$ is greater than the degree of n.

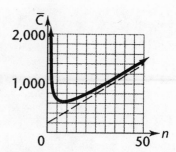

Oblique asymptote: The line $\overline{C} = 25n + 175$ is an oblique asymptote.

73. (A) Since Area = length × width, length = $\dfrac{\text{Area}}{\text{width}} = \dfrac{225}{x}$. Then total length of fence = 2 × width + 2 × length

$$L(x) = 2x + \frac{450}{x} = \frac{2x^2 + 450}{x}$$

(B) x can be any positive number, thus, domain = $(0, \infty)$

(C) The minimum value of the function $L(x)$ is $L\left(\sqrt{\dfrac{c}{a}}\right)$ where $a = 2$ and $c = 450$.

$$\min L(x) = L\left(\sqrt{\frac{450}{2}}\right) = L(\sqrt{225}) = L(15)$$

This minimum occurs when $x = 15$.
Width = 15 feet. Length = $\dfrac{225}{15} = 15$ feet.

(D) *Intercepts:* Real zeros of $2x^2 + 450$. None, hence, no x intercepts. 0 is not in the domain of L, so there are no L intercepts.
Vertical asymptotes: Real zeros of x. The line $x = 0$ is a vertical asymptote.
Sign behavior: $L(x)$ is always positive since $x > 0$.
Horizontal asymptote: None, since the degree of $2x^2 + 450$ is greater than the degree of x.
Oblique asymptote: The line $L = 2x$ is an oblique asymptote.

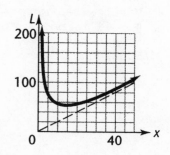

Exercise 4-5

Key Ideas and Formulas

Two polynomials are equal to each other if and only if the coefficients of like-degree terms are equal.

For a polynomial with real coefficients, there always exists a complete factoring involving only linear and/or quadratic factors with real coefficients where the linear and quadratic factors are prime relative to the real numbers.

Any proper fraction $P(x)/D(x)$ reduced to lowest terms can be decomposed into a sum of partial fractions as follows:

If $D(x)$ has:	then the decomposition of $P(x)/D(x)$ has a term of form:
1. a non-repeating linear factor of the form $ax + b$	$\dfrac{A}{ax + b}$ A a constant
2. a k-repeating linear factor of the form $(ax + b)^k$	$\dfrac{A_1}{ax + b} + \dfrac{A_2}{(ax + b)^2} + \cdots + \dfrac{A_k}{(ax + b)^k}$ $A_1, A_2, \cdots, A_k$ constant

3. a non-repeating quadratic factor of the form $ax^2 + bx + c$, prime relative to the real numbers

$$\frac{Ax + B}{ax^2 + bx + c}$$

A and B constants

4. a k-repeating quadratic factor of the form $(ax^2 + bx + c)^k$, where $ax^2 + bx + c$ is prime relative to the real numbers

$$\frac{A_1 x + B_1}{ax^2 + bx + c} + \frac{A_2 x + B_2}{(ax^2 + bx + c)^2} + \cdots + \frac{A_k x + B_k}{(ax^2 + bx + c)^k}$$

$A_1, \cdots, A_k$ and $B_1, \cdots, B_k$ constants

Common Errors:

1. Confusing $(ax + b)^2$ and $ax^2 + bx + c$. The first is a repeated linear factor, and leads to a partial fraction of form

$$\frac{A_1}{ax + b} + \frac{A_2}{(ax + b)^2}$$

The second is a quadratic factor and leads to a partial fraction of form

$$\frac{Ax + B}{ax^2 + bx + c}$$

2. Neglecting the first-degree terms in the numerator in cases 3 and 4.

Always consider $\dfrac{Ax + B}{ax^2 + bx + c}$. Never write $\dfrac{A}{ax^2 + bx + c}$.

---

1. $\dfrac{7x - 14}{(x - 4)(x + 3)} = \dfrac{A}{x - 4} + \dfrac{B}{x + 3}$

$\qquad\qquad\qquad = \dfrac{A(x + 3) + B(x - 4)}{(x - 4)(x + 3)}$

Thus, for all x

$7x - 14 = A(x + 3) + B(x - 4)$

If $x = -3$, then

$-35 = -7B$

$\quad B = 5$

If $x = 4$, then

$14 = 7A$

$\quad A = 2$

$A = 2, \ B = 5$

3. $\dfrac{17x - 1}{(2x - 3)(3x - 1)} = \dfrac{A}{2x - 3} + \dfrac{B}{3x - 1}$

$\qquad\qquad\qquad = \dfrac{A(3x - 1) + B(2x - 3)}{(2x - 3)(3x - 1)}$

Thus, for all x

$17x - 1 = A(3x - 1) + B(2x - 3)$

If $x = \dfrac{1}{3}$, then

$17\left(\dfrac{1}{3}\right) - 1 = B\left[2\left(\dfrac{1}{3}\right) - 3\right]$

$\qquad \dfrac{14}{3} = -\dfrac{7}{3}B$

$\qquad\quad B = -2$

If $x = \dfrac{3}{2}$, then

$17\left(\dfrac{3}{2}\right) - 1 = A\left[3\left(\dfrac{3}{2}\right) - 1\right]$

$\qquad \dfrac{49}{2} = \dfrac{7}{2}A$

$\qquad\quad A = 7$

$A = 7, \ B = -2$

5. $\dfrac{3x^2 + 7x + 1}{x(x+1)^2} = \dfrac{A}{x} + \dfrac{B}{x+1} + \dfrac{C}{(x+1)^2}$

$\qquad = \dfrac{A(x+1)^2 + Bx(x+1) + Cx}{x(x+1)^2}$

Thus, for all x

$3x^2 + 7x + 1$

$\qquad = A(x+1)^2 + Bx(x+1) + Cx$

If $x = -1$, then

$-3 = -C$

$\;C = 3$

If $x = 0$, then

$1 = A$

If $x = 1$, then using $A = 1$ and $C = 3$,

we have

$11 = 4 + 2B + 3$

$\;B = 2$

$A = 1$, $B = 2$, $C = 3$

7. $\dfrac{3x^2 + x}{(x-2)(x^2+3)} = \dfrac{A}{x-2} + \dfrac{Bx+C}{x^2+3}$

$\qquad = \dfrac{A(x^2+3) + (Bx+C)(x-2)}{(x-2)(x^2+3)}$

Thus, for all x

$3x^2 + x = A(x^2+3) + (Bx+C)(x-2)$

If $x = 2$, then

$14 = 7A$

$\;A = 2$

If $x = 0$, then using $A = 2$, we have

$0 = 6 - 2C$

$C = 3$

If $x = 1$, then using $A = 2$, and

$C = 3$, we have

$4 = 8 + (B+3)(-1)$

$-4 = (-1)(B+3)$

$\;B = 1$

$A = 2$, $B = 1$, $C = 3$

9. $\dfrac{2x^2 + 4x - 1}{(x^2+x+1)^2} = \dfrac{Ax+B}{x^2+x+1} + \dfrac{Cx+D}{(x^2+x+1)^2} = \dfrac{(Ax+B)(x^2+x+1) + Cx + D}{(x^2+x+1)^2}$

Thus, for all x

$2x^2 + 4x - 1 = (Ax+B)(x^2+x+1) + Cx + D$

Multiplying out the right side, we have

$2x^2 + 4x - 1 = Ax^3 + Ax^2 + Ax + Bx^2 + Bx + B + Cx + D$

$\qquad\qquad\qquad = Ax^3 + (A+B)x^2 + (A+B+C)x + B + D$

Equating coefficients of like terms we have

$\;0 = A$

$\;2 = A + B$

$\;4 = A + B + C$

$-1 = B + D$

Hence $A = 0$, $B = 2$, $C = 2$, $D = -3$

11. Since $x^2 - 2x - 8 = (x+2)(x-4)$,

we write

$\dfrac{-x+22}{x^2-2x-8} = \dfrac{A}{x+2} + \dfrac{B}{x-4}$

$\qquad = \dfrac{A(x-4) + B(x+2)}{(x+2)(x-4)}$

Thus, for all x

$-x + 22 = A(x-4) + B(x+2)$

If $x = 4$

$18 = 6B$

$\;B = 3$

If $x = -2$

$24 = -6A$

$\;A = -4$

So $\dfrac{-x+22}{x^2-2x-8} = \dfrac{-4}{x+2} + \dfrac{3}{x-4}$

13. Since $6x^2 - x - 12 = (3x+4)(2x-3)$,

we write

$\dfrac{3x-13}{6x^2-x-12} = \dfrac{A}{3x+4} + \dfrac{B}{2x-3}$

$\qquad = \dfrac{A(2x-3) + B(3x+4)}{(3x+4)(2x-3)}$

Thus, for all x

$3x - 13 = A(2x-3) + B(3x+4)$

If $x = \dfrac{3}{2}$

$3\left(\dfrac{3}{2}\right) - 13 = B\left[3\left(\dfrac{3}{2}\right) + 4\right]$

$-\dfrac{17}{2} = \dfrac{17}{2}B$

$B = -1$

If $x = -\dfrac{4}{3}$

$3\left(-\dfrac{4}{3}\right) - 13 = A\left[2\left(-\dfrac{4}{3}\right) - 3\right]$

$-17 = -\dfrac{17}{3}A$

$A = 3$

So $\dfrac{3x-13}{6x^2-x-12} = \dfrac{3}{3x+4} - \dfrac{1}{2x-3}$

15. Since $x^3 - 6x^2 + 9x = x(x^2 - 6x + 9) = x(x - 3)^2$, we write

$$\frac{x^2 - 12x + 18}{x^3 - 6x^2 + 9x} = \frac{A}{x} + \frac{B}{x - 3} + \frac{C}{(x - 3)^2} = \frac{A(x - 3)^2 + Bx(x - 3) + Cx}{x(x - 3)^2}$$

Thus, for all x

$$x^2 - 12x + 18 = A(x - 3)^2 + Bx(x - 3) + Cx$$

If $x = 3$

$-9 = 3C$

$C = -3$

If $x = 0$

$18 = 9A$

$A = 2$

If $x = 2$

$-2 = A - 2B + 2C$

$-2 = 2 - 2B - 6$ using $A = 2$ and $C = -3$

$2 = -2B$

$B = -1$

So $\dfrac{x^2 - 12x + 18}{x^3 - 6x^2 + 9x} = \dfrac{2}{x} - \dfrac{1}{x - 3} - \dfrac{3}{(x - 3)^2}$

17. Since $x^3 + 2x^2 + 3x = x(x^2 + 2x + 3)$, we write

$$\frac{5x^2 + 3x + 6}{x^3 + 2x^2 + 3x} = \frac{A}{x} + \frac{Bx + C}{x^2 + 2x + 3} = \frac{A(x^2 + 2x + 3) + (Bx + C)x}{x(x^2 + 2x + 3)}$$

> **Common Error:**
> Writing $\dfrac{B}{x^2 + 2x + 3}$. Since $x^2 + 2x + 3$ is quadratic, the numerator must be first degree.

Thus, for all x

$$5x^2 + 3x + 6 = A(x^2 + 2x + 3) + (Bx + C)x = (A + B)x^2 + (2A + C)x + 3A$$

Equating coefficients of like terms, we have

$5 = A + B$

$3 = 2A + C$

$3A = 6$

Hence $A = 2$, $C = -1$, $B = 3$

So $\dfrac{5x^2 + 3x + 6}{x^3 + 2x^2 + 3x} = \dfrac{2}{x} + \dfrac{3x - 1}{x^2 + 2x + 3}$

19. Since $x^4 + 4x^2 + 4 = (x^2 + 2)^2$, we write

$$\frac{2x^3 + 7x + 5}{x^4 + 4x^2 + 4} = \frac{Ax + B}{x^2 + 2} + \frac{Cx + D}{(x^2 + 2)^2} = \frac{(Ax + B)(x^2 + 2) + Cx + D}{(x^2 + 2)^2}$$

Thus, for all x

$$2x^3 + 7x + 5 = (Ax + B)(x^2 + 2) + Cx + D = Ax^3 + Bx^2 + (2A + C)x + 2B + D$$

Equating coefficients of like terms, we have

$2 = A$

$0 = B$

$7 = 2A + C$

$5 = 2B + D$

Hence $A = 2$, $B = 0$, $C = 3$, $D = 5$

$$\frac{2x^3 + 7x + 5}{x^4 + 4x^2 + 4} = \frac{2x}{x^2 + 2} + \frac{3x + 5}{(x^2 + 2)^2}$$

21. First we divide to obtain a polynomial plus a proper fraction.

$$
\begin{array}{r}
x - 2 \\
x^2 - 5x + 6 \overline{\smash{\big)}\, x^3 - 7x^2 + 17x - 17} \\
\underline{x^3 - 5x^2 + 6x} \\
-2x^2 + 11x - 17 \\
\underline{-2x^2 + 10x - 12} \\
x - 5
\end{array}
$$

So, $\dfrac{x^3 - 7x^2 + 17x - 17}{x^2 - 5x + 6} = x - 2 + \dfrac{x - 5}{x^2 - 5x + 6}$

To decompose the proper fraction, we note $x^2 - 5x + 6 = (x - 2)(x - 3)$ and we write:

$$\frac{x - 5}{x^2 - 5x + 6} = \frac{A}{x - 2} + \frac{B}{x - 3} = \frac{A(x - 3) + B(x - 2)}{(x - 2)(x - 3)}$$

Thus for all x

$x - 5 = A(x - 3) + B(x - 2)$

If $x = 3$

$-2 = B$

If $x = 2$

$-3 = -A$

$A = 3$

So $\dfrac{x^3 - 7x^2 + 17x - 17}{x^2 - 5x + 6} = x - 2 + \dfrac{3}{x - 2} - \dfrac{2}{x - 3}$

23. First, we must factor $x^3 - 6x - 9$. Possible rational zeros of this polynomial are ± 1, ± 3, ± 9. Forming a synthetic division table, we see

	1	0	-6	-9
1	1	1	-5	-14
3	1	3	3	0

Thus $x^3 - 6x - 9 = (x - 3)(x^2 + 3x + 3)$

$x^2 + 3x + 3$ cannot be factored further in the real numbers.

So $\dfrac{4x^2 + 5x - 9}{x^3 - 6x - 9} = \dfrac{A}{x - 3} + \dfrac{Bx + C}{x^2 + 3x + 3} = \dfrac{A(x^2 + 3x + 3) + (Bx + C)(x - 3)}{(x - 3)(x^2 + 3x + 3)}$

Thus, for all x

$$\begin{aligned}
4x^2 + 5x - 9 &= A(x^2 + 3x + 3) + (Bx + C)(x - 3) \\
&= Ax^2 + 3Ax + 3A + Bx^2 - 3Bx + Cx - 3C \\
&= (A + B)x^2 + (3A - 3B + C)x + 3A - 3C
\end{aligned}$$

Before equating coefficients of like terms, we note that if $x = 3$

$4(3)^2 + 5(3) - 9 = A(3^2 + 3\cdot 3 + 3)$

So $42 = 21A$

$A = 2$

Since $4 = A + B$

$5 = 3A - 3B + C$

$-9 = 3A - 3C$

we have $A = 2$, $B = 2$, $C = 5$

So $\dfrac{4x^2 + 5x - 9}{x^3 - 6x - 9} = \dfrac{2}{x - 3} + \dfrac{2x + 5}{x^2 + 3x + 3}$

25. First, we must factor $x^3 + 2x^2 - 15x - 36$. The possible rational zeros are ± 1, ± 2, ± 3, ± 4, ± 6, ± 9, ± 12, ± 18, ± 36. Forming a synthetic division table, we see

	1	2	-15	-36
1	1	3	-12	-48
2	1	4	-7	-50
3	1	5	0	-36
4	1	6	9	0

Hence $x^3 + 2x^2 - 15x - 36 = (x - 4)(x^2 + 6x + 9) = (x - 4)(x + 3)^2$.

So $\dfrac{x^2 + 16x + 18}{x^3 + 2x^2 - 15x - 36} = \dfrac{A}{x - 4} + \dfrac{B}{x + 3} + \dfrac{C}{(x + 3)^2}$

$= \dfrac{A(x + 3)^2 + B(x - 4)(x + 3) + C(x - 4)}{(x - 4)(x + 3)^2}$

Thus, for all x

$x^2 + 16x + 18 = A(x + 3)^2 + B(x - 4)(x + 3) + C(x - 4)$

If $x = -3$

$-21 = -7C$

$C = 3$

If $x = 4$

$98 = 49A$

$A = 2$

If $x = 5$

$123 = 64A + 8B + C$

$= 128 + 8B + 3$ using $A = 2$ and $C = 3$

$-8 = 8B$

$B = -1$

So $\dfrac{x^2 + 16x + 18}{x^3 + 2x^2 - 15x - 36} = \dfrac{2}{x - 4} - \dfrac{1}{x + 3} + \dfrac{3}{(x + 3)^2}$

27. First, we must factor $x^4 - 5x^3 + 9x^2 - 8x + 4$. The possible rational zeros are ± 1, ± 2, ± 4. We form a synthetic division table:

	1	-5	9	-8	4
1	1	-4	5	-3	1
2	1	-3	3	-2	0

We examine $x^3 - 3x^2 + 3x - 2$; possibly 2 is a double zero.

	1	-3	3	-2
2	1	-1	1	0

Thus $x^4 - 5x^3 + 9x^2 - 8x + 4 = (x - 2)^2(x^2 - x + 1)$. $x^2 - x + 1$ cannot be factored further in the real numbers, so

$$\frac{-x^2 + x - 7}{x^4 - 5x^3 + 9x^2 - 8x + 4} = \frac{A}{x - 2} + \frac{B}{(x - 2)^2} + \frac{Cx + D}{x^2 - x + 1}$$

$$= \frac{A(x - 2)(x^2 - x + 1) + B(x^2 - x + 1) + (Cx + D)(x - 2)^2}{(x - 2)^2(x^2 - x + 1)}$$

Thus, for all x

$-x^2 + x - 7 = A(x - 2)(x^2 - x + 1) + B(x^2 - x + 1) + (Cx + D)(x - 2)^2$

If $x = 2$

$-9 = 3B$

$B = -3$

$-x^2 + x - 7 = A(x^3 - 3x^2 + 3x - 2) + B(x^2 - x + 1) + (Cx + D)(x^2 - 4x + 4)$

$\qquad = (A + C)x^3 + (-3A + B - 4C + D)x^2 + (3A - B + 4C - 4D)x - 2A + B + 4D$

We have already $B = -3$, so equating coefficients of like terms,

$0 = A + C$

$-1 = -3A - 3 - 4C + D$

$1 = 3A + 3 + 4C - 4D$

$-7 = -2A - 3 + 4D$

Since $C = -A$, we can write

$-1 = -3A - 3 + 4A + D$

$1 = 3A + 3 - 4A - 4D$

$-7 = -2A - 3 + 4D$

So $D = 0$ (adding the first equations), $A = 2$, $C = -2$.

Hence $\dfrac{-x^2 + x - 7}{x^4 - 5x^3 + 9x^2 - 8x + 4} = \dfrac{2}{x - 2} - \dfrac{3}{(x - 2)^2} - \dfrac{2x}{x^2 - x + 1}$

29. First we divide to obtain a polynomial plus a proper fraction.

$$
\begin{array}{r}
x + 2 \\
4x^4 + 4x^3 - 5x^2 + 5x - 2 \overline{\smash{\big)}\, 4x^5 + 12x^4 - x^3 + 7x^2 - 4x + 2} \\
\underline{4x^5 + 4x^4 - 5x^3 + 5x^2 - 2x} \\
8x^4 + 4x^3 + 2x^2 - 2x + 2 \\
\underline{8x^4 + 8x^3 - 10x^2 + 10x - 4} \\
- 4x^3 + 12x^2 - 12x + 6
\end{array}
$$

Now we must decompose $\dfrac{-4x^3 + 12x^2 - 12x + 6}{4x^4 + 4x^3 - 5x^2 + 5x - 2}$, starting by factoring $4x^4 + 4x^3 - 5x^2 + 5x - 2$.

Possible rational zeros are ± 1, ± 2, $\pm\frac{1}{2}$, $\pm\frac{1}{4}$. We form a synthetic division table:

	4	4	-5	5	-2	
1	4	8	3	8	6	1 is an upper bound, eliminating 2
-1	4	0	-5	10	-12	
-2	4	-4	3	-1	0	-2 is a zero

We investigate $4x^3 - 4x^2 + 3x - 1$; the only remaining rational zeros are $\pm\frac{1}{2}$, $\pm\frac{1}{4}$. We form a synthetic division table:

	4	-4	3	-1
$\frac{1}{2}$	4	-2	2	0

So $4x^4 + 4x^3 - 5x^2 + 5x - 2 = (x + 2)\left(x - \dfrac{1}{2}\right)(4x^2 - 2x + 2)$

$$= (x + 2)\left(x - \dfrac{1}{2}\right)2(2x^2 - x + 1)$$

$$= (x + 2)(2x - 1)(2x^2 - x + 1)$$

$2x^2 - x + 1$ cannot be factored further in the real numbers, so we write:

$$\dfrac{-4x^3 + 12x^2 - 12x + 6}{4x^4 + 4x^3 - 5x^2 + 5x - 2}$$

$$= \dfrac{A}{x + 2} + \dfrac{B}{2x - 1} + \dfrac{Cx + D}{2x^2 - x + 1}$$

$$= \dfrac{A(2x - 1)(2x^2 - x + 1) + B(x + 2)(2x^2 - x + 1) + (Cx + D)(x + 2)(2x - 1)}{(x + 2)(2x - 1)(2x^2 - x + 1)}$$

Thus, for all x

$-4x^3 + 12x^2 - 12x + 6$

$= A(2x - 1)(2x^2 - x + 1) + B(x + 2)(2x^2 - x + 1) + (Cx + D)(x + 2)(2x - 1)$

If $x = -2$

$-4(-2)^3 + 12(-2)^2 - 12(-2) + 6 = A(-5)[2(-2)^2 - (-2) + 1]$

$$32 + 48 + 24 + 6 = -55A$$

$$110 = -55A$$

$$A = -2$$

If $x = \dfrac{1}{2}$

$$-4\left(\dfrac{1}{2}\right)^3 + 12\left(\dfrac{1}{2}\right)^2 - 12\left(\dfrac{1}{2}\right) + 6 = B\left(\dfrac{5}{2}\right)\left[2\left(\dfrac{1}{2}\right)^2 - \dfrac{1}{2} + 1\right]$$

$$-\dfrac{1}{2} + 3 - 6 + 6 = B\left(\dfrac{5}{2}\right)(1)$$

$$B = 1$$

If $x = 0$

$6 = A(-1)(1) + B(2)(1) + D(2)(-1)$

$6 = -A + 2B - 2D$

$6 = 2 + 2 - 2D$

$D = -1$

If $x = 1$

$-4 + 12 - 12 + 6 = A(1)(2) + B(3)(2) + (C + D)(3)(1)$

$$2 = 2A + 6B + 3C + 3D$$

$$= -4 + 6 + 3C - 3$$

$$C = 1$$

So $\dfrac{4x^5 + 12x^4 - x^3 + 7x^2 - 4x + 2}{4x^4 + 4x^3 - 5x^2 + 5x - 2} = x + 2 - \dfrac{2}{x + 2} + \dfrac{1}{2x - 1} + \dfrac{x - 1}{2x^2 - x + 1}$

CHAPTER 4 REVIEW

1.
```
      2   3   0  -1
         -4   2  -4
  -2│ 2  -1   2  -5
```

$2x^3 + 3x^2 - 1$
$= (x + 2)(2x^2 - x + 2) - 5$ $(4\text{-}1)$

2.
```
      1  -4   0   9   0  -8
          3  -3  -9   0   0
   3│ 1  -1  -3   0   0  -8
```
$P(3) = -8$ $(4\text{-}1, 4\text{-}2)$

3. 2, -4, -1 $(4\text{-}2)$

4. Since complex zeros come in conjugate pairs, $1 - i$ is a zero. $(4\text{-}3)$

5. (A) Since the graph has x intercepts -2, 0, and 2, these are zeros of $P(x)$.
Therefore, $P(x) = (x + 2)x(x - 2) = x^3 - 4x$.

(B) $P(x) \to \infty$ as $x \to \infty$ and $P(x) \to -\infty$ as $x \to -\infty$. $(4\text{-}1)$

6. We form a synthetic division table:

	1	-4	0	2	
-2	1	-6	12	-22	both are lower bounds, since
-1	1	-5	5	-3	both rows alternate in sign
3	1	-1	-3	-7	
4	1	0	0	2	upper bound

$(4\text{-}3)$

7. We investigate $P(1)$ and $P(2)$ by forming a synthetic division table.

	2	-3	1	-5
1	2	-1	0	-5
2	2	1	3	1

Since $P(1)$ and $P(2)$ have opposite signs, there is at least one real zero between 1 and 2. $(4\text{-}3)$

8. The factors of 6 are ±1, ±2, ±3, ±6. $(4\text{-}2)$

9. Using the possibilities found in problem 8, we form a synthetic division table:

	1	-4	1	6	
1	1	-3	-2	4	
2	1	-2	-3	0	2 is a zero

Thus $x^3 - 4x^2 + x + 6 = (x - 2)(x^2 - 2x - 3) = (x - 2)(x - 3)(x + 1)$.
The rational zeros are 2, 3, -1. $(4\text{-}2)$

10. (A) $f(x) = \dfrac{2x - 3}{x + 4} = \dfrac{n(x)}{d(x)}$
The domain of f is the set of all real numbers x such that $d(x) = x + 4 \neq 0$, that is, $(-\infty, -4) \cup (-4, \infty)$. f has an x intercept where $n(x) = 2x - 3 = 0$, that is, $x = \dfrac{3}{2}$

(B) $g(x) = \dfrac{3x}{x^2 - x - 6} = \dfrac{n(x)}{d(x)}$
The domain of g is the set of all real numbers x such that $d(x) = x^2 - x - 6 \neq 0$, that is, $(x + 2)(x - 3) \neq 0$, that is, $x \neq -2, 3$, or $(-\infty, -2) \cup (-2, 3) \cup (3, \infty)$. g has an x intercept where $n(x) = 3x = 0$, that is, $x = 0$ $(4\text{-}4)$

11. (A) Horizontal asymptote: since $n(x)$ and $d(x)$ have the same degree, the line $y = 2$ is a horizontal asymptote. Vertical asymptotes: zeros of $d(x)$
$x + 4 = 0$
$x = -4$

(B) Horizontal asymptote: since the degree of $n(x)$ is less than the degree of $d(x)$, the line $y = 0$ is a horizontal asymptote.
Vertical asymptotes: zeros of $d(x)$

$$x^2 - x - 6 = 0$$
$$(x + 2)(x - 3) = 0$$
$$x = -2, \quad x = 3$$

(4-4)

12. We write

$$\frac{7x - 11}{(x - 3)(x + 2)} = \frac{A}{x - 3} + \frac{B}{x + 2} = \frac{A(x + 2) + B(x - 3)}{(x - 3)(x + 2)}$$

Thus for all x
$7x - 11 = A(x + 2) + B(x - 3)$

If $x = -2$
$-25 = -5B$
$B = 5$

If $x = 3$
$10 = 5A$
$A = 2$

So

$$\frac{7x - 11}{(x - 3)(x + 2)} = \frac{2}{x - 3} + \frac{5}{x + 2}$$

(4-5)

13. (A) We form a synthetic division table:

	1	-3	-3	4
-2	1	-5	7	$-10 = P(-2)$
-1	1	-4	1	$3 = P(-1)$
0	1	-3	-3	$4 = P(0)$
1	1	-2	-5	$-1 = P(1)$
2	1	-1	-5	$-6 = P(2)$
3	1	0	-3	$-5 = P(3)$
4	1	1	1	$8 = P(4)$

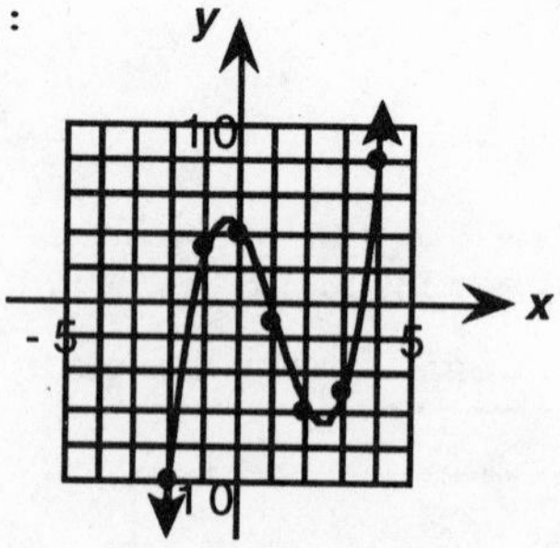

The graph of $P(x)$ has three x intercepts and two turning points; $P(x) \to \infty$ as $x \to \infty$ and $P(x) \to -\infty$ as $x \to -\infty$

(B) We search for the real zero in (3, 4). We organize our calculations in a table.

Sign Change Interval (a, b)	Midpoint m	Sign of P: $P(a)$	$P(m)$	$P(b)$
(3, 4)	3.5	−	−	+
(3.5, 4)	3.75	−	+	+
(3.5, 3.75)	3.625	−	+	+
(3.5, 3.625)	3.5625	−	+	+
(3.5, 3.5625)	3.53125	−	+	+
(3.5, 3.53125)	We stop here	−		+

Since each endpoint rounds to 3.5, a real zero lies on this last interval and is given by 3.5 to one decimal place accuracy.

(4-1, 4-3)

14. We use synthetic divison:

```
     8   -14   -13    -4    7
          2     -3    -4   -2
  _____
1/4│ 8   -12   -16    -8    5
```

Thus,

$$P(x) = \left(x - \frac{1}{4}\right)(8x^3 - 12x^2 - 16x - 8) + 5$$

$$P\left(\frac{1}{4}\right) = 5 \qquad\qquad (4\text{-}1)$$

15.

```
              4    -8    -3    -3
                   -2     5    -1
        _____
  -1/2│  4   -10     2    -4
```

$$P\left(-\frac{1}{2}\right) = -4 \qquad\qquad (4\text{-}1)$$

16. The quadratic formula tells us that $x^2 - 2x - 1 = 0$ if

$$x = \frac{-b \pm \sqrt{b^2 - 4ac}}{2a} \qquad a = 1, \ b = -2, \ c = -1$$

$$x = \frac{-(-2) \pm \sqrt{(-2)^2 - 4(1)(-1)}}{2(1)}$$

$$= \frac{2 \pm \sqrt{8}}{2}$$

$$= 1 \pm \sqrt{2}$$

Since $1 \pm \sqrt{2}$ are zeros of $x^2 - 2x - 1$, its factors are $x - (1 + \sqrt{2})$ and $x - (1 - \sqrt{2})$, that is, $x^2 - 2x - 1 = [x - (1 + \sqrt{2})][x - (1 - \sqrt{2})]$ $\qquad (4\text{-}2)$

17. $x + 1$ will be a factor of $P(x)$ if $P(-1) = 0$.
$P(-1) = 9(-1)^{26} - 11(-1)^{17} + 8(-1)^{11} - 5(-1)^4 - 7 = 9 + 11 - 8 - 5 - 7 = 0$, so the answer is yes, $x + 1$ is a factor. $\qquad (4\text{-}2)$

18. The possible rational zeros are ± 1, ± 2, ± 4, ± 8, $\pm\frac{1}{2}$. We form a synthetic division table:

	2	-3	-18	-8
1	2	-1	-19	-27
2	2	1	-16	-40
4	2	5	2	0

So $2x^3 - 3x^2 - 18x - 8 = (x - 4)(2x^2 + 5x + 2)$
$\qquad\qquad\qquad\qquad\qquad\quad = (x - 4)(2x + 1)(x + 2)$

Zeros: 4, $-\frac{1}{2}$, -2 $\qquad\qquad\qquad\qquad (4\text{-}2)$

19. $(x - 4)(2x + 1)(x + 2)$ $\qquad\qquad\qquad\qquad (4\text{-}2)$

20. The possible rational zeros are ± 1, ± 5. We form a synthetic division table:

	1	-3	0	5
1	1	-2	-2	3
5	1	2	10	55
-1	1	-4	4	1
-5	1	-8	40	-195

There are no rational zeros, since all possibilities fail. $\qquad (4\text{-}2)$

21. $P(x) = 2x^4 - x^3 + 2x - 1$

We can factor $P(x)$ by grouping into $(2x - 1)(x^3 + 1) = (2x - 1)(x + 1)(x^2 - x + 1)$. However, if we don't notice this, we find the possible rational zeros to be ± 1, $\pm \frac{1}{2}$. We form a synthetic division table:

	2	-1	0	2	-1	
1	2	1	1	3	2	
-1	2	-3	3	-1	0	-1 is a zero

We now examine $2x^3 - 3x^2 + 3x - 1$. Only -1, $\frac{1}{2}$, and $-\frac{1}{2}$ remain as possible rational zeros; We form a synthetic division table:

	2	-3	3	-1	
-1	2	-5	8	-9	Not a double zero
$\frac{1}{2}$	2	-2	2	0	$\frac{1}{2}$ is a zero

So $P(x) = (x + 1)\left(x - \dfrac{1}{2}\right)(2x^2 - 2x + 2) = (x + 1)(2x - 1)(x^2 - x + 1)$

To find the remaining zeros, we solve $x^2 - x + 1 = 0$, by the quadratic formula.
$x^2 - x + 1 = 0$

$$x = \frac{-b \pm \sqrt{b^2 - 4ac}}{2a} \quad a = 1, \ b = -1, \ c = 1$$

$$x = \frac{-(-1) \pm \sqrt{(-1)^2 - 4(1)(1)}}{2(1)}$$

$$x = \frac{1 \pm \sqrt{-3}}{2}$$

$$x = \frac{1 \pm i\sqrt{3}}{2}$$

The four zeros are -1, $\dfrac{1}{2}$, and $\dfrac{1 \pm i\sqrt{3}}{2}$

$(4-2)$

22. $(x + 1)\left(x - \dfrac{1}{2}\right)2\left(x - \dfrac{1 + i\sqrt{3}}{2}\right)\left(x - \dfrac{1 - i\sqrt{3}}{2}\right) = (x + 1)(2x - 1)\left(x - \dfrac{1 + i\sqrt{3}}{2}\right)\left(x - \dfrac{1 - i\sqrt{3}}{2}\right)$

$(4-2)$

23. $2x^3 + 3x^2 \le 11x + 6$
$2x^3 + 3x^2 - 11x - 6 \le 0$
To factor $2x^3 + 3x^2 - 11x - 6$, we search for zeros of the polynomial. Possible rational zeros are ± 1, ± 2, ± 3, ± 6, $\pm \frac{1}{2}$, and $\pm \frac{3}{2}$. We form a synthetic division table:

	2	3	-11	-6	
1	2	5	-6	-12	
2	2	7	3	0	2 is a zero

$2x^3 + 3x^2 - 11x - 6 = (x - 2)(2x^2 + 7x + 3) = (x - 2)(x + 3)(2x + 1)$
Hence we must examine the sign behavior of $(x - 2)(x + 3)(2x + 1)$.
We form a sign chart.
Zeros: -3, $-\dfrac{1}{2}$, 2

$(-\infty,-3)\ (-3,-1/2)\ (-1/2,2)\ (2,\infty)$

$-3 \quad -\frac{1}{2} \quad\quad 2 \quad\quad x$

$2x^3 + 3x^2 - 11x - 6 = (x - 2)(x + 3)(2x + 1)$				
Test Number	-4	-1	0	3
Value of Polynomial for Test Number	-42	6	-6	42
Sign of Polynomial in Interval	$-$	$+$	$-$	$+$
Interval	$(-\infty, -3)$	$(-3, -\frac{1}{2})$	$(-\frac{1}{2}, 2)$	$(2, \infty)$

$2x^3 + 3x^2 - 11x - 6 \leq 0$ and $2x^3 + 3x^2 \leq 11x + 6$ within the intervals $(-\infty, -3]$,

and $\left[-\dfrac{1}{2}, 2\right]$, or $x \leq -3$ or $-\dfrac{1}{2} \leq x \leq 2$.   *(4-2, 2-8)*

24. (A) We form a synthetic division table.

	1	-2	-30	0	-25	
0	1	-2	-30	0	-25	
1	1	-1	-31	-31	-56	
2	1	0	-30	-60	-145	
3	1	1	-27	-81	-268	
4	1	2	-22	-88	-377	
5	1	3	-15	-75	-400	
6	1	4	-6	-36	-241	
7	1	5	5	35	220	7 is an upper bound
-1	1	-3	-27	27	-52	
-2	1	-4	-22	44	-113	
-3	1	-5	-15	45	-160	
-4	1	-6	-6	24	-121	
-5	1	-7	5	-25	100	-5 is a lower bound

(B) We search for the real zero in $(6, 7)$ indicated in the table. We organize our calculations in a table.

Sign Change Interval (a, b)	Midpoint m	Sign of P		
		$P(a)$	$P(m)$	$P(b)$
(6, 7)	6.5	$-$	$-$	$+$
(6.5, 7)	6.75	$-$	$+$	$+$
(6.5, 6.75)	6.625	$-$	$+$	$+$
(6.5, 6.625)	6.5625	$-$	$-$	$+$
(6.5625, 6.625)	6.59375	$-$	$-$	$+$
(6.59375, 6.625)	6.609375	$-$	$-$	$+$
(6.609375, 6.625)	6.6171875	$-$	$-$	$+$
(6.6171875, 6.625)	6.62109375	$-$	$+$	$+$
(6.6171875, 6.62109375)	We stop here	$-$		$+$

Since each endpoint rounds to 6.62, a real zero lies on this last interval and is given by 6.62 to two decimal place accuracy.

(C) Here is a computer-generated graph of P.

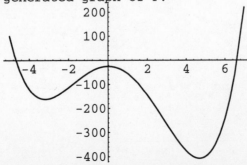

The x intercepts are on the intervals $(-4.8, -4.4)$ and $(6.4, 6.8)$. To estimate more accurately, we zoom in. Here are graphs of P on these intervals.

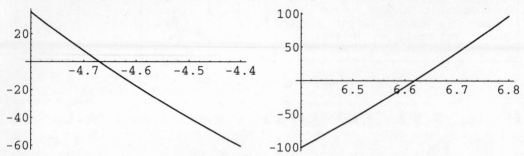

To two decimal place accuracy, the zeros are -4.67 and 6.62. (4-3)

25. $f(x) = \dfrac{x - 1}{2x + 2} = \dfrac{n(x)}{d(x)}$

(A) The domain of f is the set of all real numbers x such that $d(x) = 2x + 2 \neq 0$, that is $(-\infty, -1) \cup (-1, \infty)$.

f has an x intercept where $n(x) = x - 1 = 0$, that is, $x = 1$. $f(0) = -\dfrac{1}{2}$, hence f has a y intercept at $y = -\dfrac{1}{2}$

(B) Vertical asymptote: $x = -1$. Horizontal asymptote: since $n(x)$ and $d(x)$ have the same degree, the line $y = \dfrac{1}{2}$ is a horizontal asymptote.

(C) *Sign Chart.*

Test numbers	-2	0	2
Value of f	$\frac{3}{2}$	$-\frac{1}{2}$	$\frac{1}{6}$
Sign of f	+	–	+

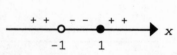

Since $x = -1$ is a vertical asymptote and $f(x) > 0$ for $x < -1$, and $f(x) < 0$ for $-1 < x < 1$
$\qquad f(x) \to \infty$ as $x \to -1^-$
and $f(x) \to -\infty$ as $x \to -1^+$
Complete the sketch.

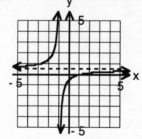

(4-4)

26. $\dfrac{-x^2 + 3x + 4}{x(x - 2)^2} = \dfrac{A}{x} + \dfrac{B}{x - 2} + \dfrac{C}{(x - 2)^2} = \dfrac{A(x - 2)^2 + Bx(x - 2) + Cx}{x(x - 2)^2}$
Thus for all x
$-x^2 + 3x + 4 = A(x - 2)^2 + Bx(x - 2) + Cx$
If $x = 0$
$4 = 4A$
$A = 1$
If $x = 2$
$6 = 2C$
$C = 3$
If $x = 1$, using $A = 1$ and $C = 3$
$6 = 1 - B + 3$
$B = -2$
So $\dfrac{-x^2 + 3x + 4}{x(x - 2)^2} = \dfrac{1}{x} - \dfrac{2}{x - 2} + \dfrac{3}{(x - 2)^2}$

(4-5)

27. First we factor $2x^3 - 3x^2 + 3x = x(2x^2 - 3x + 3)$. $2x^2 - 3x + 3$ cannot be factored further in the real numbers, so we write

$$\frac{8x^2 - 10x + 9}{2x^3 - 3x^2 + 3x} = \frac{A}{x} + \frac{Bx + C}{2x^2 - 3x + 3} = \frac{A(2x^2 - 3x + 3) + (Bx + C)x}{x(2x^2 - 3x + 3)}$$

Thus, for all x

$$8x^2 - 10x + 9 = A(2x^2 - 3x + 3) + (Bx + C)x$$

If $x = 0$

$\quad 9 = 3A$

$\quad A = 3$

$8x^2 - 10x + 9 = 3(2x^2 - 3x + 3) + Bx^2 + Cx = (6 + B)x^2 + (-9 + C)x + 9$

Equating coefficients of like terms, we have

$\quad 8 = 6 + B$

$-10 = -9 + C$

$\quad B = 2$

$\quad C = -1$

So $\dfrac{8x^2 - 10x + 9}{2x^3 - 3x^2 + 3x} = \dfrac{3}{x} + \dfrac{2x - 1}{2x^2 - 3x + 3}$ *(4-5)*

28.
$$\begin{array}{r} \quad 1 \quad\quad 0 \quad\quad\quad 3 \quad\quad\quad 2 \\ \quad\quad\quad 1 + i \quad\quad 2i \quad\quad 1 + 5i \\ \hline 1 + i \,|\, 1 \quad 1 + i \quad 3 + 2i \quad 3 + 5i \end{array}$$

$\quad\quad\quad (1 + i)^2 = (1 + i)(1 + i) = 1 + 2i + i^2 = 1 + 2i - 1 = 2i$

$(1 + i)(3 + 2i) = 3 + 5i + 2i^2 = 3 + 5i - 2 = 1 + 5i$

$P(x) = [x^2 + (1 + i)x + (3 + 2i)][x - (1 + i)] + 3 + 5i$ *(4-1)*

29. $P(x) = \left(x + \dfrac{1}{2}\right)^2 (x + 3)(x - 1)^3$. The degree is 6. *(4-2)*

30. $P(x) = (x + 5)[x - (2 - 3i)][x - (2 + 3i)]$. The degree is 3. *(4-2)*

31. The possible rational zeros are ± 1, ± 2, ± 4, $\pm\dfrac{1}{2}$. We form a synthetic division table:

	2	−5	−8	21	0	−4	
1	2	−3	−11	10	10	6	
2	2	−1	−10	1	2	0	2 is a zero

We continue to examine $2x^4 - x^3 - 10x^2 + x + 2$. The possible rational zeros are -1, ± 2, and $\pm\dfrac{1}{2}$. We form a synthetic division table:

	2	−1	−10	1	2	
2	2	3	−4	−7	−12	Not a double zero
−1	2	−3	−7	8	−6	
−2	2	−5	0	1	0	−2 is a zero

Hence $P(x) = (x - 2)(x + 2)(2x^3 - 5x^2 + 1)$. $2x^3 - 5x^2 + 1$ has been shown (see Exercise 4-3, problem 27 for details) to have zeros $\dfrac{1}{2}$, $1 \pm \sqrt{2}$.

Hence $P(x)$ has zeros $\dfrac{1}{2}$, ± 2, $1 \pm \sqrt{2}$. *(4-2)*

32. $(x - 2)(x + 2)\left(x - \dfrac{1}{2}\right)2[x - (1 - \sqrt{2})][x - (1 + \sqrt{2})]$

$= (x - 2)(x + 2)(2x - 1)[x - (1 - \sqrt{2})][x - (1 + \sqrt{2})]$ *(4-2)*

33. $\dfrac{4x^2 + 4x - 3}{2x^3 + 3x^2 - 11x - 6} \geq 0$

We need to form a sign chart for $\dfrac{P}{Q} = \dfrac{4x^2 + 4x - 3}{2x^3 + 3x^2 - 11x - 6}$.

We first locate the zeros of P and Q.

$4x^2 + 4x - 3 = (2x - 1)(2x + 3)$, hence P has zeros at $\dfrac{1}{2}$ and $-\dfrac{3}{2}$. The zeros of Q were found in problem 23 to be -3, $-\dfrac{1}{2}$, and 2. We now form the sign chart, with zeros -3, $-\dfrac{3}{2}$, $-\dfrac{1}{2}$, $\dfrac{1}{2}$, and 2.

Open dots at zeros of Q.
Solid dots at zeros of P.

$\dfrac{P}{Q} = \dfrac{(2x - 1)(2x + 3)}{(x - 2)(2x + 1)(x + 3)}$						
Test Number	-4	-2	-1	0	1	3
Value of $\dfrac{P}{Q}$	$-\dfrac{15}{14}$	$\dfrac{5}{12}$	$-\dfrac{1}{2}$	$\dfrac{1}{2}$	$-\dfrac{5}{12}$	$\dfrac{15}{14}$
Sign of $\dfrac{P}{Q}$	$-$	$+$	$-$	$+$	$-$	$+$
Interval	$(-\infty, -3)$	$(-3, -\dfrac{3}{2})$	$(-\dfrac{3}{2}, -\dfrac{1}{2})$	$(-\dfrac{1}{2}, \dfrac{1}{2})$	$(\dfrac{1}{2}, 2)$	$(2, \infty)$

$\dfrac{4x^2 + 4x - 3}{2x^3 + 3x^2 - 11x - 6} \geq 0$ within the intervals $\left(-3, -\dfrac{3}{2}\right] \cup \left(-\dfrac{1}{2}, \dfrac{1}{2}\right] \cup (2, \infty)$, or

$-3 < x \leq -\dfrac{3}{2}$ or $-\dfrac{1}{2} < x \leq \dfrac{1}{2}$ or $x > 2$. *(4-2, 2-8)*

34. $P(x)$ changes sign three times. Therefore, it has three zeros and its minimal degree is 3. *(4-3)*

35. Since $1 + 2i$ is a zero, $1 - 2i$ is also a zero. Hence
$[x - (1 - 2i)][x - (1 + 2i)] = [(x - 1) + 2i][(x - 1) - 2i] = (x - 1)^2 - 4i^2$
$= x^2 - 2x + 5$ is a factor. Since $P(x)$ is a cubic polynomial, it must be of the form $a(x - r)(x^2 - 2x + 5)$. Since the constant term of this polynomial, $-5ar$, must be an integer, r must be a rational number. Thus there can be no irrational zeros. *(4-2)*

36. (A) Since $x^3 - 27$ is a cubic polynomial, it has 3 zeros and there are 3 cube roots of 27.

(B) $x^3 - 27 = (x - 3)(x^2 + 3x + 3^2) = (x - 3)(x^2 + 3x + 9)$. We solve $x^2 + 3x + 9 = 0$ by applying the quadratic formula with $a = 1$, $b = 3$, $c = 9$,

to obtain $x = \dfrac{-3 \pm \sqrt{3^2 - 4(1)(9)}}{2(1)} = \dfrac{-3 \pm \sqrt{-27}}{2} = \dfrac{-3 \pm 3i\sqrt{3}}{2}$ or $-\dfrac{3}{2} \pm \dfrac{3i}{2}\sqrt{3}$ *(4-2)*

37. (A) We form a synthetic division table.

	1	2	-500	0	-4,000
0	1	2	-500	0	-4,000
10	1	12	-380	-3,800	-42,000
20	1	22	-60	-1,200	-28,000
30	1	32	460	13,800	410,000
-10	1	-8	-420	4,200	-46,000
-20	1	-18	-140	2,800	-60,000
-30	1	-28	340	-10,200	302,000

From the table, 30 is an upper bound and -30 is a lower bound.

(B) Here is a computer-generated graph of P.

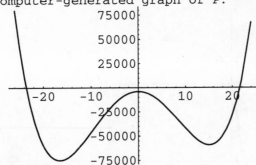

The x intercepts are on the intervals $(-24, -23)$ and $(21, 22)$. To estimate more accurately, we zoom in. Here are graphs of P on the intervals.

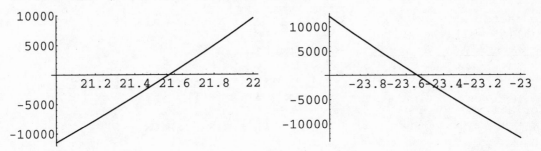

To two decimal place accuracy, the zeros are -23.54 and 21.57. $\qquad$ $(4\text{-}3)$

38. $f(x) = \dfrac{x^2 + 2x + 3}{x + 1} = \dfrac{n(x)}{d(x)}$

Intercepts. There are no real zeros of $n(x) = x^2 + 2x + 3$. No x intercept
$f(0) = 3$ y intercept
Vertical asymptotes. Real zeros of $d(x) = x + 1$ $x = -1$
Sign Chart.

Test numbers	-2	0
Value of f	-3	3
Sign of f	–	+

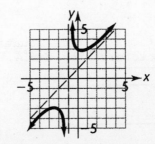

Since $x = -1$ is a vertical asymptote and $f(x) < 0$ for $x < -1$, and $f(x) > 0$ for $x > -1$

$\qquad f(x) \to -\infty$ as $x \to -1^-$ and $f(x) \to \infty$ as $x \to -1^+$

Horizontal asymptotes. Since the degree of $n(x)$ is greater than the degree of $d(x)$, there is no horizontal asymptote.

Oblique asymptote:

$$
\begin{array}{r}
x + 1 \\
x + 1 \overline{)\ x^2 + 2x + 3} \\
\underline{x^2 + x} \\
x + 3 \\
\underline{x + 1} \\
2
\end{array}
$$

Thus, $f(x) = x + 1 + \dfrac{2}{x + 1}$.

Hence, the line $y = x + 1$ is an oblique asymptote.

Complete the sketch

$(4\text{-}4)$

39. Here is a computer-generated graph of $f(x)$.

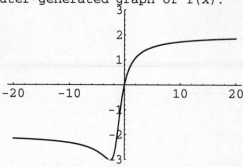

From the graph, we can see that $f(x) \to 2$ as $x \to \infty$ and $f(x) \to -2$ as $x \to -\infty$; the lines $y = 2$ and $y = -2$ are horizontal asymptotes.   (4-4)

40. First we factor $x^4 - 3x^3 + x^2 - 3x = x(x^3 - 3x^2 + x - 3) = x[x^2(x - 3) + 1(x - 3)]$
$= x(x - 3)(x^2 + 1)$. We write

$$\frac{5x^2 + 2x + 9}{x^4 - 3x^3 + x^2 - 3x} = \frac{A}{x} + \frac{B}{x - 3} + \frac{Cx + D}{x^2 + 1}$$

$$= \frac{A(x - 3)(x^2 + 1) + Bx(x^2 + 1) + x(x - 3)(Cx + D)}{x(x - 3)(x^2 + 1)}$$

Thus, for all x

$5x^2 + 2x + 9 = A(x - 3)(x^2 + 1) + Bx(x^2 + 1) + x(x - 3)(Cx + D)$

If $x = 0$
$\quad 9 = -3A$
$\quad A = -3$

If $x = 3$
$\qquad 60 = 30B$
$\qquad B = 2$

$$\begin{aligned}
5x^2 + 2x + 9 &= -3(x - 3)(x^2 + 1) + 2x(x^2 + 1) + x(x - 3)(Cx + D)\\
&= -3(x^3 - 3x^2 + x - 3) + 2x^3 + 2x + (x^2 - 3x)(Cx + D)\\
&= -3x^3 + 9x^2 - 3x + 9 + 2x^3 + 2x + (x^2 - 3x)(Cx + D)\\
&= -x^3 + 9x^2 - x + 9 + Cx^3 - 3Cx^2 + Dx^2 - 3Dx\\
&= x^3(-1 + C) + x^2(9 - 3C + D) + x(-1 - 3D) + 9
\end{aligned}$$

Equating coefficients of like terms, we have
$\quad 0 = -1 + C$
$\quad 5 = 9 - 3C + D$
$\quad 2 = -1 - 3D$

So $C = 1$, $D = -1$.

So $\dfrac{5x^2 + 2x + 9}{x^4 - 3x^3 + x^2 - 3x} = \dfrac{-3}{x} + \dfrac{2}{x - 3} + \dfrac{x - 1}{x^2 + 1}$ (4-5)

41. In the given figure, let y = height of door
$\qquad\qquad\qquad\qquad\quad 2x$ = width of door
$\qquad$ Then Area of door = 48 = $2xy$

Since (x, y) is a point on the parabola $y = 16 - x^2$, its coordinates satisfy the equation of the parabola.

Hence $\qquad 48 = 2x(16 - x^2)$
$\qquad\qquad\quad 48 = 32x - 2x^3$
$2x^3 - 32x + 48 = 0$
$\;x^3 - 16x + 24 = 0$

The possible rational solutions of this equation are ± 1, ± 2, ± 3, ± 4, ± 6, ± 8, ± 12, ± 24. We form a synthetic division table.

	1	0	-16	24	
1	1	1	-15	9	
2	1	2	-12	0	2 is a zero

Thus the equation can be factored
$(x - 2)(x^2 + 2x - 12) = 0$.

To find the remaining zeros, we solve $x^2 + 2x - 12 = 0$, by completing the square.

$$x^2 + 2x = 12$$
$$x^2 + 2x + 1 = 13$$
$$(x + 1)^2 = 13$$
$$x + 1 = \pm\sqrt{13}$$
$$x = -1 + \sqrt{13} \text{ (discarding the negative solution)} \approx 2.61$$

Thus the positive zeros are $x = 2, 2.61$.
Thus the dimensions of the door are either $2x = 4$ feet by $16 - x^2 = 12$ feet, or $2x = 5.2$ feet by $16 - x^2 = 9.2$ feet. $\hfill (4\text{-}2)$

42. We note:

$$\begin{pmatrix} \text{Volume} \\ \text{of} \\ \text{silo} \end{pmatrix} = \begin{pmatrix} \text{Volume of} \\ \text{hemisphere} \\ \text{of radius } x \end{pmatrix} + \begin{pmatrix} \text{Volume of} \\ \text{cylinder with} \\ \text{radius } x, \text{ height } 18 \end{pmatrix}$$

$$486\pi = \frac{2}{3}\pi x^3 + \pi x^2 \cdot 18$$

$$486 = \frac{2}{3}x^3 + 18x^2$$

$$0 = x^3 + 27x^2 - 729$$

There are no rational zeros of $P(x) = x^3 + 27x^2 - 729$. To search for irrational zeros, we form a synthetic division table.

	1	27	0	-729	
0	1	27	0	-729	
1	1	28	28	-701	
2	1	29	58	-613	
3	1	30	90	-459	
4	1	31	124	-233	
5	1	32	160	71	There is a zero between 4 and 5

We have located a zero between successive integers. We now apply the bisection method to locate the zero to one decimal place.

We organize our calculations in a table.

Sign Change Interval (a, b)	Midpoint m	Sign of P $P(a)$	$P(m)$	$P(b)$
(4, 5)	4.5	-	-	+
(4.5, 5)	4.75	-	-	+
(4.75, 5)	4.875	-	+	+
(4.75, 4.875)	4.8125	-	+	+
(4.75, 4.8125)	4.78125	-	-	+
(4.78125, 4.8125)	We stop here	-		+

Since each endpoint rounds to 4.8, a real zero lies on this last interval and is given by 4.8 to one decimal place accuracy. To one decimal place accuracy, the radius is 4.8 feet. $\hfill (4\text{-}3)$

43.

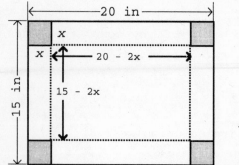 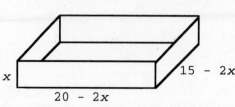

From the above figures it should be clear that

V = length × width × height = $(20 - 2x)(15 - 2x)x$ $0 < x < 7.5$

We solve $(20 - 2x)(15 - 2x)x = 300$

$$300x - 70x^2 + 4x^3 = 300$$
$$4x^3 - 70x^2 + 300x - 300 = 0$$

There are no rational zeros of $P(x) = 4x^3 - 70x^2 + 300x - 300$. To search for irrational zeros, we form a synthetic division table.

	4	-70	300	-300	
0	4	-70	300	-300	
1	4	-66	234	-66	
2	4	-62	176	52	There is a zero between 1 and 2
3	4	-58	126	78	
4	4	-54	84	36	
5	4	-50	50	-50	There is a zero between 4 and 5
6	4	-46	24	-156	
7	4	-42	6	-258	
8	4	-38	-4	-332	

Applying the bisection method (details omitted) the positive zeros in (0, 7.5) are found to be 1.4 and 4.5 to one decimal place accuracy.

x = 1.4 inches or 4.5 inches

(4-3)

44. Let (x, x^2) be a point on the graph of $y = x^2$. Then the distance from $(1, 4)$ to (x, x^2) must equal 3 units. Applying the distance formula, we have,

$$\sqrt{(x - 1)^2 + (x^2 - 4)^2} = 3$$
$$(x - 1)^2 + (x^2 - 4)^2 = 9$$
$$x^2 - 2x + 1 + x^4 - 8x^2 + 16 = 9$$
$$x^4 - 7x^2 - 2x + 8 = 0$$

Let $P(x) = x^4 - 7x^2 - 2x + 8$. We form a synthetic division table.

	1	0	-7	-2	8
1	1	1	-6	-8	0

1 is a zero. We examine $Q(x) = x^3 + x^2 - 6x - 8$. We form a synthetic division table for $Q(x)$.

	1	1	-6	-8	
1	1	2	-4	-12	
2	1	3	0	-8	
4	1	5	14	48	an upper bound
-1	1	0	-6	-2	
-2	1	-1	-4	0	

-2 is a zero. The remaining zeros are found by solving $x^2 - x - 4 = 0$. Applying the quadratic formula with $a = 1$, $b = -1$, $c = -4$, we obtain

$$x = \frac{-(-1) \pm \sqrt{(-1)^2 - 4(1)(-4)}}{2(1)} = \frac{1 \pm \sqrt{17}}{2}$$

To one decimal place accuracy, $x = -1.6$ or 2.6. Thus there are four real zeros of $P(x)$, 1, -2, -1.6, and 2.6. There are four points on the graph of $y = x^2$ that are 3 units from $(1, 4)$ and their coordinates are $(1, 1)$, $(-2, 4)$, $(-1.6, 2.6)$, and $(2.6, 6.8)$

(4-2)

CHAPTER 5

Exercise 5-1

Key Ideas and Formulas

The equation $f(x) = b^x$ $b > 0$, $b \neq 1$ defines an exponential function for each different constant b, called the base. Domain: $(-\infty, \infty)$; Range: $(0, \infty)$

Basic Properties of the graph of $f(x) = b^x$ $b > 0$, $b \neq 1$
1. All graphs pass through $(0, 1)$
2. All graphs are continuous with no holes or jumps.
3. The x-axis is a horizontal asymptote.
4. If $b > 1$, then b^x increases as x increases.
5. If $0 < b < 1$, then b^x decreases as x increases.
6. The function f is one-to-one.

Additional Exponential Function Properties.

For a and b positive, $a \neq 1$, $b \neq 1$, and x and y real:
1. Exponent laws
   $a^x a^y = a^{x+y}$ $(a^x)^y = a^{xy}$ $(ab)^x = a^x b^x$ $(a/b)^x = a^x/b^x$ $a^x/a^y = a^{x-y}$
2. $a^x = a^y$ if and only if $x = y$
3. For $x \neq 0$, then $a^x = b^x$ if and only if $a = b$.

Doubling Time Growth Model:
$P = P_0 2^{t/d}$ where

P = Population at time t
P_0 = Population at time $t = 0$
d = doubling time

Half-Life Decay Model:

$$A = A_0 \left(\frac{1}{2}\right)^{t/h}$$
$$= A_0 2^{-t/h}$$

where
A = amount at time t
A_0 = amount at time $t = 0$
h = half-life

Compound Interest:

If a principal P is invested at an annual rate r compounded n times a year, then the amount A in the account at the end of t years is given by

$$A = P\left(1 + \frac{r}{n}\right)^{nt}$$

1.

x	y
-3	0.04
-2	0.11
-1	0.33
0	1.00
1	3.00
2	9.00
3	27.00

3.

x	y
-3	27.00
-2	9.00
-1	3.00
0	1.00
1	0.33
2	0.11
3	0.04

5.

x	$y = g(x)$
-3	-27.00
-2	-9.00
-1	-3.00
0	-1.00
1	-0.33
2	-0.11
3	-0.04

7.

x	$y = h(x)$
-3	0.19
-2	0.56
-1	1.67
0	5.00
1	15.00
2	45.00
3	135.00

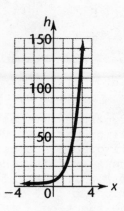

9.

x	y
-6	-4.96
-5	-4.89
-4	-4.67
-3	-4.00
-2	-2.00
-1	4.00
0	22.00

11. $10^{3x-1}10^{4-x} = 10^{3x-1+4-x} = 10^{2x+3}$

13. $\dfrac{3x}{3^{1-x}} = 3^{x-(1-x)} = 3^{x-1+x} = 3^{2x-1}$

15. $\left(\dfrac{4^x}{5^y}\right)^{3z} = \dfrac{4^{3xz}}{5^{3yz}}$

17. $5^{3x} = 5^{4x-2}$ if and only if
$3x = 4x - 2$
$-x = -2$
$x = 2$

19. $7^{x^2} = 7^{2x+3}$ if and only if
$x^2 = 2x + 3$
$x^2 - 2x - 3 = 0$
$(x - 3)(x + 1) = 0$
$x = -1, 3$

21. $(1 - x)^5 = (2x - 1)^5$ if and only if
$1 - x = 2x - 1$
$-3x = -2$
$x = \dfrac{2}{3}$

23. $2^x = 4^{x+1}$
$2^x = (2^2)^{x+1}$
$2^x = 2^{2(x+1)}$ if and only if
$x = 2(x + 1)$
$x = 2x + 2$
$-x = 2$
$x = -2$

25. $25^{x+1} = 125^{2x}$
$(5^2)^{x+1} = (5^3)^{2x}$
$5^{2(x+1)} = 5^{3(2x)}$ if and only if
$2(x + 1) = 3(2x)$
$2x + 2 = 6x$
$2 = 4x$
$x = \dfrac{1}{2}$

27. $9^{x^2} = 3^{3x-1}$
$(3^2)^{x^2} = 3^{3x-1}$
$3^{2x^2} = 3^{3x-1}$ if and only if
$2x^2 = 3x - 1$
$2x^2 - 3x + 1 = 0$
$(2x - 1)(x - 1) = 0$
$x = \dfrac{1}{2}, 1$

29.

$$a^2 = a^{-2}$$

$$a^2 = \frac{1}{a^2}$$

$$a^4 = 1 \quad (a \neq 0)$$

$$a^4 - 1 = 0$$

$$(a - 1)(a + 1)(a^2 + 1) = 0$$

$$a = 1 \text{ or } a = -1$$

This does not violate the exponential property mentioned because $a = 1$ and a negative are excluded from consideration in the statement of the property.

31.

t	$G(t)$
-200	0.11
-150	0.19
-100	0.33
-50	0.58
0	1.00
50	1.73
100	3.00
150	5.20
200	9.00

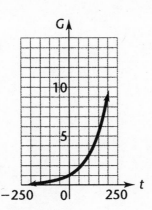

33.

x	y
-9	1543.3
-8	891.0
-7	514.4
-6	297.0
-5	171.5
-4	99
-3	57.2
-2	33
0	11
3	2.1
6	0.4
9	0.1

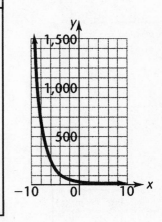

35.

x	$y = g(x)$
-3	0.13
-2	0.25
-1	0.5
0	1.0
1	0.5
2	0.25
3	0.13

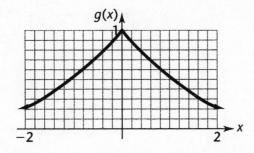

37.

x	y
-10	463
-8	540
-6	630
-4	735
-2	857
0	1,000
2	1,166
4	1,360
6	1,587
8	1,851
10	2,159

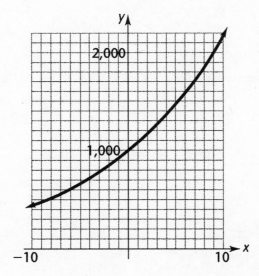

39.

x	y
-2	0.06
-1.5	0.21
-1	0.50
-0.5	0.84
0	1.00
0.5	0.84
1	0.50
1.5	0.21
2	0.06

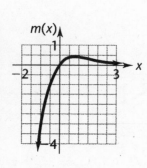

41. 16.24

43. 5.047

45. 4.469

47. $(6^x + 6^{-x})(6^x - 6^{-x}) = (6^x)^2 - (6^{-x})^2$
$$= 6^{2x} - 6^{-2x}$$
(think: $(a + b)(a - b) = a^2 - b^2$)

> **Common Errors:**
> $(6^x)^2 \neq 6^{x^2}$
> $6^{2x} - 6^{-2x} \neq 6^{4x}$

49. $(6^x + 6^{-x})^2 - (6^x - 6^{-x})^2 = (6^x)^2 + 2(6^x)(6^{-x}) + (6^{-x})^2 - [(6^x)^2 - 2(6^x)(6^{-x}) + (6^{-x})^2]$
$$= 6^{2x} + 2 + 6^{-2x} - [6^{2x} - 2 + 6^{-2x}]$$
$$= 6^{2x} + 2 + 6^{-2x} - 6^{2x} + 2 - 6^{-2x}$$
$$= 4$$

51.

x	y = m(x)
-3	-81
-2.5	-39
-2	-18
-1.5	-7.8
-1	-3
-0.5	-0.87
0	0
0.5	0.29
1	0.33
1.5	0.29
2	0.22
2.5	0.16
3	0.11

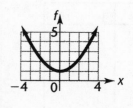

53.

x	y = f(x)
-3	4.06
-2	2.13
-1	1.25
0	1.00
1	1.25
2	2.13
3	4.06

55. (A) Here is a computer-generated graph of $f(x) = 3^x - 5$.

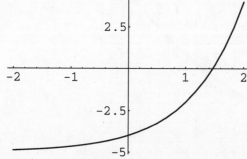

The only zero apparent is on the interval (1.4, 1.5). To estimate more accurately, we zoom in. Here is a graph of f on this interval.

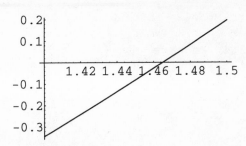

To two decimal place accuracy, the zero is 1.46.

(B) As $x \to \infty$, $f(x) \to \infty$. As $x \to -\infty$, it appears that $f(x) \to -5$. The line $y = -5$ is a horizontal asymptote.

57. (A) Here is a computer-generated graph of $f(x) = 1 + x + 10^x$.

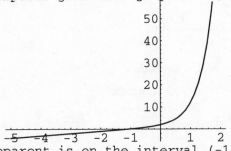

The only zero apparent is on the interval $(-1.1, -1)$. To estimate more accurately, we zoom in. Here is a graph of f on this interval.

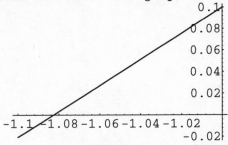

To two decimal place accuracy, the zero is -1.08.

(B) As $x \to \infty$, $f(x) \to \infty$. As $x \to -\infty$, $f(x) \to -\infty$. There is no horizontal asymptote. It appears that the line $y = x + 1$ may be an oblique asymptote as $x \to -\infty$.

59.

n	L
1	2
2	4
3	8
4	16
5	32
6	64
7	128
8	256
9	512
10	1,024

61. We use the Doubling Time Growth Model:
$$P = P_0 2^{t/d}$$
Substituting $P_0 = 10$ and $d = 2.4$, we have
$$P = 10(2^{t/2.4})$$

(A) $t = 7$, hence $P = 10(2^{7/2.4})$
$$= 75.5$$
76 flies

(B) $t = 14$, hence $P = 10(2^{14/2.4})$
$$= 570.2$$
570 flies

63. We use the Half-Life Decay Model
$$A = A_0\left(\frac{1}{2}\right)^{t/h} = A_0 2^{-t/h}$$
Substituting $A_0 = 25$ and $h = 12$, we have
$$A = 25(2^{-t/12})$$

(A) $t = 5$, hence $A = 25(2^{-5/12}) = 19$ pounds

(B) $t = 20$, hence $A = 25(2^{-20/12}) = 7.9$ pounds

65. We use the compound interest formula
$$A = P\left(1 + \frac{r}{n}\right)^{nt} \quad n = \frac{365}{7}$$
$$P = 4,000 \quad r = 0.11$$
$$A = 4,000\left(1 + \frac{0.11}{365/7}\right)^{365t/7}$$
$$= 4,000\left(1 + \frac{0.77}{365}\right)^{365t/7}$$

(A) $t = 0.5$, hence
$$A = 4,000\left(1 + \frac{0.77}{365}\right)^{365(0.5)/7}$$
$$= \$4,225.92$$

(B) $t = 10$, hence
$$A = 4,000\left(1 + \frac{0.77}{365}\right)^{365(10)/7}$$
$$= \$12,002.75$$

67. We use the compound interest formula
$$A = P\left(1 + \frac{r}{n}\right)^{nt} \text{ to find } P: P = \frac{A}{(1 + \frac{r}{n})^{nt}}$$
$$n = 365 \quad r = 0.0825 \quad A = 40,000 \quad t = 17$$
$$P\frac{40,000}{(1 + \frac{0.0825}{365})^{365 \cdot 17}} = \$9,841$$

69. Using the compound interest formula with $P = 10,000$, $r = 0.09$, and $n = 4$, the value of the 9% investment is $A_1(t) = 10,000\left(1 + \frac{.09}{4}\right)^{4t}$.

Using the compound interest formula with $P = 10,000$, $r = 0.089$ and $n = 365$, the value of the 8.9% investment is $A_2(t) = 10,000\left(1 + \frac{.089}{365}\right)^{365t}$.

The two investments are equal at $t = 0$, but after $t = 0$, the values can be calculated to be as follows:

t (in years)	$\frac{1}{4}$	$\frac{2}{4}$	$\frac{3}{4}$	$\frac{4}{4} = 1$	2	5	10
$A_1(t)$	10225	10455.06	10690.30	10930.83	11948.31	15605.90	24351.89
$A_2(t)$	10224.97	10454.97	10690.19	10930.69	11947.39	15604.06	24348.65

Thus, the second investment is less than the first by an ever-increasing amount, and can never be worth more.

Exercise 5-2
Key Ideas and Formulas

As m increases without bound, the value of $\left(1 + \dfrac{1}{m}\right)^m$ approaches an irrational number called e. To twelve decimal places, $e = 2.718\ 281\ 828\ 459$.

For x a real number, the equation $f(x) = e^x$ defines the exponential function with base e.

Exponential Growth and Decay

Description	Equation	Graph	Uses
Unlimited Growth	$y = ce^{kt}$ $c,\ k > 0$		Short term population growth (people, bacteria, etc.) Growth of money at continuous compound interest. ($A = Pe^{rt}$)
Exponential Decay	$y = ce^{-kt}$ $c,\ k > 0$		Radioactive decay. Light absorption in water, glass, etc. Atmosphere pressure. Electric circuits.
Limited Growth	$y = c(1 - e^{-kt})$ $c,\ k > 0$		Learning skills. Sales fads. Company growth. Electric Circuits.
Logistic Growth	$y = \dfrac{M}{1 + ce^{-kt}}$ $c,\ k,\ M > 0$		Long-term population growth. Epidemics. Sales of new products. Company growth.

1.

x	y
-3	-0.05
-2	-0.14
-1	-0.37
0	-1.00
1	-2.72
2	-7.39
3	-20.09

3.

x	y
-5	3.68
-4	4.49
-3	5.49
-2	6.7
-1	8.19
0	10
1	12.21
2	14.92
3	18.22
4	22.26
5	27.18

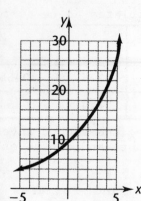

5.

t	f(t)
-5	164.87
-4	149.18
-3	134.99
-2	122.14
-1	110.52
0	100
1	90.48
2	81.87
3	74.08
4	67.03
5	60.65

7. $e^{2x}e^{-3x} = e^{2x-3x} = e^{-x}$

9. e^{3x}

11. $\dfrac{e^{5x}}{e^{2x+1}} = e^{5x-(2x+1)} = e^{5x-2x-1} = e^{3x-1}$

13. (A) Although $1 + \dfrac{1}{m}$ approaches 1, $1 + \dfrac{1}{m}$ is not equal to 1 for any m, hence reasoning as if it were 1 is incorrect.

(B) As $m \to \infty$, $\left(1 + \dfrac{1}{m}\right)^m \to e$.

15.

x	y
-1	2.05
0	2.14
1	2.37
2	3.00
3	4.72
4	9.39
5	22.09

17.

x	y
-3	0.05
-2	0.14
-1	0.37
0	1.00
1	0.37
2	0.14
3	0.05

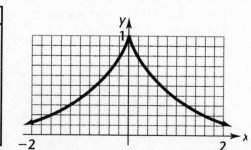

19.

x	$y = M(x)$
-5	12.3
-4	7.5
-3	4.7
-2	3.1
-1	2.3
0	2.0
1	2.3
2	3.1
3	4.7
4	7.5
5	12.3

21.

t	N
-3	3.3
-2	8.6
-1	22
0	50
1	95
2	142
3	174
4	190
5	196

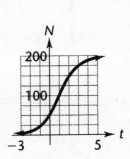

23. $\dfrac{-2x^3e^{-2x} - 3x^2e^{-2x}}{x^6} = \dfrac{x^2e^{-2x}(-2x - 3)}{x^6} = \dfrac{e^{-2x}(-2x - 3)}{x^4}$

25. $(e^x + e^{-x})^2 + (e^x - e^{-x})^2 = (e^x)^2 + 2(e^x)(e^{-x}) + (e^{-x})^2 + (e^x)^2 - 2(e^x)(e^{-x}) + (e^{-x})^2$

$= e^{2x} + 2 + e^{-2x} + e^{2x} - 2 + e^{-2x}$

$= 2e^{2x} + 2e^{-2x}$

> **Common Errors:**
> $(e^x)^2 \neq e^{x^2}$
> $e^{2x} + e^{2x} \neq e^{4x}$

27. $\dfrac{e^{-x}(e^x - e^{-x}) + e^{-x}(e^x + e^{-x})}{e^{-2x}} = \dfrac{e^{-x}e^x - e^{-x}e^{-x} + e^{-x}e^x + e^{-x}e^{-x}}{e^{-2x}}$

$= \dfrac{1 - e^{-2x} + 1 + e^{-2x}}{e^{-2x}}$

$= \dfrac{2}{e^{-2x}}$

$= 2e^{2x}$

29. $2xe^{-x} = 0$ if $2x = 0$ or $e^{-x} = 0$. Since e^{-x} is never 0, the only solution is $x = 0$.

31. $x^2e^x - 5xe^x = 0$

$xe^x(x - 5) = 0$

$x = 0$ or $e^x = 0$ or $x - 5 = 0$

$\phantom{x = 0 \text{ or } e^x = 0 \text{ or }}$ never

$x = 0, 5 \qquad\qquad\qquad x = 5$

33.

x	$f(x)$
-2	0.02
-1.5	0.11
-1	0.37
-0.5	0.78
0	1.00
0.5	0.78
1	0.37
1.5	0.11
2	0.02

35. (A)

s	$f(s)$	s	$f(s)$
-0.5	4.0000	0.5	2.2500
-0.2	3.0518	0.2	2.4883
-0.1	2.8680	0.1	2.5937
-0.01	2.7320	0.01	2.7048
-0.001	2.7196	0.001	2.7169
-0.0001	2.7184	0.0001	2.7181

(B) Both tables are "closing in" on 2.7182... or e.

37.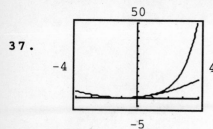

39.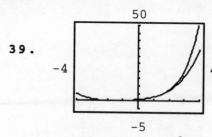

41. Here are computer-generated graphs of $f_1(x) = \dfrac{x}{e^x}$, $f_2(x) = \dfrac{x^2}{e^x}$, and $f_3(x) = \dfrac{x^3}{e^x}$.
In each case as $x \to \infty$, $f_n(x) \to 0$. The line $y = 0$ is a horizontal asymptote.
As $x \to -\infty$, $f_1(x) \to -\infty$ and $f_3(x) \to -\infty$, while $f_2(x) \to \infty$. It appears that as $x \to -\infty$,
$f_n(x) \to \infty$ if n is even and $f_n(x) \to -\infty$ if n is odd.

f_1: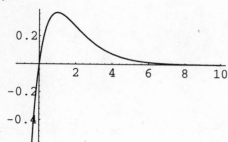

f_2: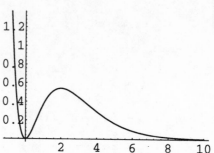

f_3: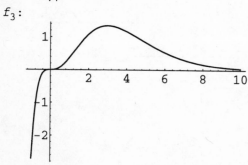

43. We use the continuous compounding formula.
$A = Pe^{rt}$
$P = 6$ billion $r = 0.017$ $t = 10$
$A = (6 \text{ billion})e^{(0.017)(10)}$
$A = 7.1$ billion

45. We use the Continuous Compounding Formula
$A = Pe^{rt}$
For Russia, $P = 148$ million, $r = -0.0062$, $A_1 = (148 \text{ million})e^{-0.0062t}$
For Nigeria, $P = 104$ million, $r = 0.03$, $A_2 = (104 \text{ million})e^{0.03t}$
Here is a computer-generated graph of A_1 and A_2 (vertical axis understood in
millions).

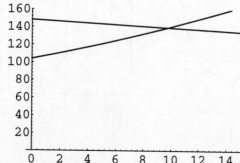

From the graph, assuming $t = 0$ in 1996, it appears that the two populations
will be equal when t is approximately 10, in 2006. After that the population
of Nigeria will be greater than the population of Russia.

47.

t	P
0	75
10	72
20	70
30	68
40	65
50	63
60	61
70	59
80	57
90	55
100	53

49. $I = I_0 e^{-0.00942d}$

(A) $d = 50$ $I = I_0 e^{-0.00942(50)} = 0.62 I_0$ 62%

(B) $d = 100$ $I = I_0 e^{-0.00942(100)} = 0.39 I_0$ 39%

51. We use the Continuous Compound Interest Formula

$A = Pe^{rt}$

$P = 5,250$ $r = 0.1138$ $A = 5,250 e^{0.1138t}$

(A) $t = 6.25$ $A = 5,250 e^{0.1138(6.25)} = \$10,691.81$

(B) $t = 17$ $A = 5,250 e^{0.1138(17)} = \$36,336.69$

53. Gill Savings: Use the Continuous Compound Interest Formula

$A = Pe^{rt}$ $P = 1,000$ $r = 0.083$ $t = 2.5$

$A = 1,000 e^{(0.083)(2.5)}$

$A = \$1,230.60$

Richardson S & L: Use the Compound Interest Formula

$A = P\left(1 + \dfrac{r}{n}\right)^{nt}$ $P = 1,000$ $r = 0.084$ $n = 4$ $t = 2.5$

$A = 1,000\left(1 + \dfrac{0.084}{4}\right)^{(4)(2.5)}$

$A = \$1,231.00$

U.S.A. Savings: Use the Compound Interest Formula

$A = P\left(1 + \dfrac{r}{n}\right)^{nt}$ $P = 1,000$ $r = 0.0825$ $n = 365$ $t = 2.5$

$A = 1,000\left(1 + \dfrac{0.0825}{365}\right)^{(365)(2.5)}$

$A = \$1,229.03$

55. We use the Continuous Compound Interest Formula

$A = Pe^{rt}$

$P = \dfrac{A}{e^{rt}}$ or $P = Ae^{-rt}$

$A = 30,000$ $r = 0.09$ $t = 10$

$P = 30,000 e^{(-0.09)(10)}$

$P = \$12,197.09$

57. We use the Continuous Compounding Formula

$A = Pe^{rt}$

(A) $P = 7.7$ million $r = 0.17$ $t = 4$

$A = (7.7 \text{ million}) e^{0.17(4)}$

$A = 15$ million

(B) $B = 7.7$ million $r = 0.17$ $t = 8$

$A = (7.7 \text{ million}) e^{0.17(8)}$

$A = 30$ million

59.

t	N
0	0
5	18
10	28
15	33
20	36
25	38
30	39

As t increases without bound, $e^{-0.12t}$ approaches 0, hence $N = 40(1 - e^{-0.12t})$ approaches 40. Hence 40 boards is the maximum number of boards an average person could be expected to produce in one day.

61. $T = T_m + (T_0 - T_m)e^{-kt}$

$T_m = 40°$ $T_0 = 72°$ $k = 0.4$ $t = 3$

$T = 40 + (72 - 40)e^{-0.4(3)}$

$T = 50°$

63.

t	q
0	0
1	0.00016
2	0.00030
3	0.00041
4	0.00050
5	0.00057
6	0.00063
7	0.00068
8	0.00072
9	0.00075
10	0.00078

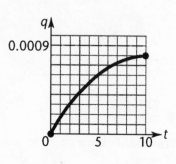

As t increases without bound, $e^{-0.2t}$ approaches 0, hence $q = 0.0009(1 - e^{-0.2t})$ approaches 0.0009. Hence 0.0009 coulomb is the maximum charge on the capacitor.

65.

t	N
0	20
5	33
10	50
15	67
20	80
25	89
30	94

As t increases without bound, $e^{-0.14t}$ approaches 0, hence $N = \dfrac{100}{1 + 4e^{-0.14t}}$ approaches 100. Hence 100 is the number of deer the island can support.

67. $y = \dfrac{e^{0.25x} + e^{-0.25x}}{2(0.25)} = \dfrac{e^{0.25x} + e^{-0.25x}}{0.5}$

x	y
-5	7.6
-4	6.2
-3	5.2
-2	4.5
-1	4.1
0	4.0
1	4.1
2	4.5
3	5.2
4	6.2
5	7.6

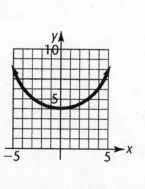

Exercise 5-3

Key Ideas and Formulas

Definition of Logarithmic Function:

For $b > 0$ and $b \neq 1$

logarithmic form   exponential form

$y = \log_b x$ is equivalent to $x = b^y$

The log to the base b of x is the exponent to which b must be raised to obtain x. A logarithm is, therefore, an exponent.

Typical Logarithmic Curves

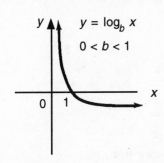

Domain = $(0, \infty)$
Range = $(-\infty, \infty)$

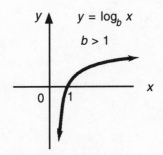

Domain = $(0, \infty)$
Range = $(-\infty, \infty)$

Properties of Logarithmic Functions:

If b, M, and N are positive real numbers, $b \neq 1$, and p and x are real numbers, then:

1. $\log_b 1 = 0$

2. $\log_b b = 1$

3. $\log_b b^x = x$

4. $b^{\log_b x} = x$

5. $\log_b MN = \log_b M + \log_b N$

6. $\log_b \dfrac{M}{N} = \log_b M - \log_b N$

7. $\log_b M^p = p \log_b M$

8. $\log_b M = \log_b N$ if and only if $M = N$

---

1. $81 = 3^4$ **3.** $0.001 = 10^{-3}$ **5.** $3 = 81^{1/4}$ **7.** $16 = \left(\dfrac{1}{2}\right)^{-4}$

9. $\log_{10} 0.0001 = -4$ **11.** $\log_4 8 = \dfrac{3}{2}$ **13.** $\log_{32} \dfrac{1}{2} = -\dfrac{1}{5}$

15. $7 = \sqrt{49}$ is rewritten $7 = 49^{1/2}$. In equivalent logarithmic form this becomes $\log_{49} 7 = \dfrac{1}{2}$.

17. 0 **19.** 1 **21.** 4 **23.** $\log_{10} 0.01 = \log_{10} 10^{-2} = -2$

25. $\log_5 \sqrt[3]{5} = \log_5 5^{1/3} = \dfrac{1}{3}$ **27.** $\sqrt{x}$

29. $e^{2 \log_e x} = e^{\log_e x^2} = x^2$ $\boxed{\text{Common Error: } e^{2 \log_e x} \neq 2x}$

31. Write $\log_2 x = 2$ in equivalent exponential form. $x = 2^2 = 4$

33. $\log_4 16 = \log_4 4^2 = 2$
$ y = 2$

35. Write $\log_b 16 = 2$ in equivalent exponential form.

$16 = b^2$

$b^2 = 16$

$b = 4$ since bases are required to be positive

37. Write $\log_b 1 = 0$ in equivalent exponential form.

$1 = b^0$

This statement is true if b is any real number except 0. However, bases are required to be positive and 1 is not allowed, so the original statement is true if b is any positive real number except 1.

39. Write $\log_4 x = \dfrac{1}{2}$ in equivalent exponential form.

$x = 4^{1/2} = 2$

41. $\log_{1/3} 9 = \log_{1/3} 3^2 = \log_{1/3} \dfrac{1}{(\frac{1}{3})^2} = \log_{1/3} \left(\dfrac{1}{3}\right)^{-2} = -2$

43. Write $\log_b 1000 = \dfrac{3}{2}$ in equivalent exponential form

$\quad 1000 = b^{3/2}$

$\quad\ \ 10^3 = b^{3/2}$

$(10^3)^{2/3} = (b^{3/2})^{2/3}$ (If two numbers are equal the results are

$\qquad\qquad\qquad\qquad\qquad$ equal if they are raised to the same exponent.)

$10^{3(2/3)} = b^{3/2(2/3)}$

$\quad\ \ 10^2 = b$

$\qquad b = 100$

45. $\log_b u^2 v^7 = \log_b u^2 + \log_b v^7 = 2 \log_b u + 7 \log_b v$

47. $\log_b \dfrac{m^{2/3}}{n^{1/2}} = \log_b m^{2/3} - \log_b n^{1/2} = \dfrac{2}{3} \log_b m - \dfrac{1}{2} \log_b n$

49. $\log_b \dfrac{u}{vw} = \log_b u - \log_b vw$

$\qquad\qquad = \log_b u - (\log_b v + \log_b w)$

$\qquad\qquad = \log_b u - \log_b v - \log_b w$

> **Common Error:**
> Forgetting the parentheses.
> $-\log_b v + \log_b w$ is incorrect

51. $\log_b \dfrac{1}{a^2} = \log_b a^{-2} = -2 \log_b a$

53. $\log_b \sqrt[3]{x^2 - y^2} = \log_b (x^2 - y^2)^{1/3} = \dfrac{1}{3} \log_b (x^2 - y^2)$

> **Common Error:**
> $\log_b (x^2 - y^2)$ is not $\log_b x^2 - \log_b y^2$.

55. $\log_b \dfrac{\sqrt[3]{N}}{p^2 q^3} = \log_b \sqrt[3]{N} - \log_b p^2 q^3$

$\qquad\qquad = \log_b N^{1/3} - (\log_b p^2 + \log_b q^3)$

$\qquad\qquad = \dfrac{1}{3} \log_b N - 2 \log_b p - 3 \log_b q$

57. $\log_b \sqrt[4]{\dfrac{x^2 y^3}{\sqrt{z}}} = \log_b \left(\dfrac{x^2 y^3}{z^{1/2}}\right)^{1/4} = \dfrac{1}{4} \log_b \dfrac{x^2 y^3}{z^{1/2}} = \dfrac{1}{4} (\log_b x^2 + \log_b y^3 - \log_b z^{1/2})$

$\qquad\qquad = \dfrac{1}{4} (2 \log_b x + 3 \log_b y - \dfrac{1}{2} \log_b z)$

59. $2 \log_b x - \log_b y = \log_b x^2 - \log_b y = \log_b \dfrac{x^2}{y}$

61. $\log_b w - \log_b x - \log_b y = \log_b \dfrac{w}{x} - \log_b y$

$$= \log_b\left(\dfrac{w}{x} \div y\right)$$

$$= \log_b\left(\dfrac{w}{x} \cdot \dfrac{1}{y}\right)$$

$$= \log_b \dfrac{w}{xy}$$

63. $3 \log_b x + 2 \log_b y - \dfrac{1}{4} \log_b z = \log_b x^3 + \log_b y^2 - \log_b z^{1/4}$

$$= \log_b x^3 y^2 - \log_b z^{1/4} = \log_b \dfrac{x^3 y^2}{z^{1/4}}$$

65. $5\left(\dfrac{1}{2} \log_b u - 2 \log_b v\right) = 5(\log_b u^{1/2} - \log_b v^2) = 5 \log_b \dfrac{u^{1/2}}{v^2} = \log_b\left(\dfrac{u^{1/2}}{v^2}\right)^5$

67. $\dfrac{1}{5}(2 \log_b x + 3 \log_b y) = \dfrac{1}{5}(\log_b x^2 + \log_b y^3) = \dfrac{1}{5} \log_b x^2 y^3 = \log_b(x^2 y^3)^{1/5}$

$$= \log_b \sqrt[5]{x^2 y^3}$$

69. $\log_b[(x + 3)^5 (2x - 7)^2] = \log_b(x + 3)^5 + \log_b(2x - 7)^2$
$$= 5 \log_b(x + 3) + 2 \log_b(2x - 7)$$

71. $\log_b \dfrac{(x + 10)^7}{(1 + 10x)^2} = \log_b(x + 10)^7 - \log_b(1 + 10x)^2 = 7 \log_b(x + 10) - 2 \log_b(1 + 10x)$

73. $\log_b \dfrac{x^2}{\sqrt{x + 1}} = \log_b x^2 - \log_b \sqrt{x + 1}$

$$= \log_b x^2 - \log_b(x + 1)^{1/2}$$

$$= 2 \log_b x - \dfrac{1}{2} \log_b(x + 1)$$

75. $\log_b(x^4 + x^3 - 20x^2) = \log_b[x^2(x^2 + x - 20)]$
$$= \log_b[x^2(x + 5)(x - 4)]$$
$$= \log_b x^2 + \log_b(x + 5) + \log_b(x - 4)$$
$$= 2 \log_b x + \log_b(x + 5) + \log_b(x - 4)$$

77. $\log_2(x + 5) = 2 \log_2 3$
$\log_2(x + 5) = \log_2 3^2$
$\quad x + 5 = 3^2$
$\quad x + 5 = 9$
$\qquad x = 4$

Check: $\log_2(x + 5) = 2 \log_2 3$

$\log_2(4 + 5) \overset{?}{=} 2 \log_2 3$

$\log_2 9 \overset{\checkmark}{=} \log_2 9$

79. $2 \log_5 x = \log_5(x^2 - 6x + 2)$
$\quad \log_5 x^2 = \log_5(x^2 - 6x + 2)$
$\qquad x^2 = x^2 - 6x + 2$
$\qquad 6x = 2$
$\qquad x = \dfrac{1}{3}$

Check: $2 \log_5 x = \log_5(x^2 - 6x + 2)$

$2 \log_5 \dfrac{1}{3} \overset{?}{=} \log_5\left[\left(\dfrac{1}{3}\right)^2 - 6\left(\dfrac{1}{3}\right) + 2\right]$

$\log_5 \dfrac{1}{9} \overset{\checkmark}{=} \log_5\left(\dfrac{1}{9}\right)$

81. $\log_e(x + 8) - \log_e x = 3 \log_e 2$

$$\log_e \frac{x + 8}{x} = \log_e 2^3$$

$$\frac{x + 8}{x} = 2^3$$

$$x + 8 = 8x$$

$$8 = 7x$$

$$x = \frac{8}{7}$$

Check: $\log_e(x + 8) - \log_e x = 3 \log_e 2$

$$\log_e\left(\frac{8}{7} + 8\right) - \log_e \frac{8}{7} \overset{?}{=} 3 \log_e 2$$

$$\log_e\left(\frac{64}{7}\right) - \log_e \frac{8}{7} \overset{?}{=} \log_e 8$$

$$\log_e\left(\frac{64}{7} \div \frac{8}{7}\right) \overset{?}{=} \log_e 8$$

$$\log_e 8 \overset{\checkmark}{=} \log_e 8$$

83.

$$2 \log_3 x = \log_3 2 + \log_3(4 - x)$$

$$\log_3 x^2 = \log_3 2(4 - x)$$

$$x^2 = 2(4 - x)$$

$$x^2 = 8 - 2x$$

$$x^2 + 2x - 8 = 0$$

$$(x + 4)(x - 2) = 0$$

$$x = -4 \text{ or } x = 2$$

Check: $2 \log_3 x = \log_3 2 + \log_3(4 - x)$

$$x = -4$$

$$2 \log_3(-4) \overset{?}{=} \log_3 2 + \log_3[4 - (-4)]$$

False. -4 is not in the domain of $\log_3 x$.
-4 is not a solution.

$$x = 2$$

$$2 \log_3 2 \overset{?}{=} \log_3 2 + \log_3(4 - 2)$$

$$2 \log_3 2 \overset{\checkmark}{=} 2 \log_3 2$$

Solution: 2

85. $3 \log_b 2 + \frac{1}{2} \log_b 25 - \log_b 20 = \log_b x$

$$\log_b 2^3 + \log_b 25^{1/2} - \log_b 20 = \log_b x$$

$$\log_b \frac{2^3 \cdot 25^{1/2}}{20} = \log_b x$$

$$\frac{2^3 \cdot 25^{1/2}}{20} = x$$

$$x = \frac{8 \cdot 5}{20}$$

$$x = 2$$

Check:

$$3 \log_b 2 + \frac{1}{2} \log_b 25 - \log_b 20 = \log_b x$$

$$3 \log_b 2 + \frac{1}{2} \log_b 25 - \log_b 20 \overset{?}{=} \log_b 2$$

$$\log_b 8 + \log_b 5 - \log_b 20 \overset{?}{=} \log_b 2$$

$$\log_b \frac{40}{20} \overset{?}{=} \log_b 2$$

$$\log_b 2 \overset{\checkmark}{=} \log_b 2$$

87. $\log_b(30) = \log_b(2 \cdot 3 \cdot 5) = \log_b 2 + \log_b 3 + \log_b 5 = 0.69 + 1.10 + 1.61 = 3.40$

89. $\log_b \frac{2}{5} = \log_b 2 - \log_b 5 = 0.69 - 1.61 = -0.92$

91. $\log_b 27 = \log_b 3^3 = 3 \log_b 3 = 3(1.10) = 3.30$

93. $\log_b \sqrt[3]{2} = \log_b 2^{1/3} = \frac{1}{3} \log_b 2 = \frac{1}{3}(0.69) = 0.23$

95. $\log_b \sqrt{0.9} = \log_b\left(\frac{9}{10}\right)^{1/2} = \frac{1}{2} \log_b \frac{3^2}{2 \cdot 5} = \frac{1}{2}[\log_b 3^2 - \log_b 2 - \log_b 5]$

$$= \frac{1}{2}[2 \log_b 3 - \log_b 2 - \log_b 5] = \frac{1}{2}[2.20 - 0.69 - 1.61] = \frac{1}{2}(-0.10)$$

$$= -0.05$$

97.

x	y
10	3
6	2
4	1
3	0
$2\frac{1}{2}$	-1
$2\frac{1}{4}$	-2

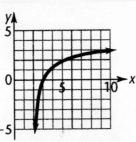

The line $x = 2$ is a vertical asymptote. Note: The graph is the same as the graph of $y = \log_2 x$ shifted 2 units to the right.

99.

x	y
8	1
4	0
2	-1
1	-2
$\frac{1}{2}$	-3
$\frac{1}{4}$	-4

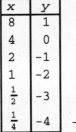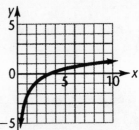

The y axis is a vertical asymptote. Note: The graph is the same as the graph of $y = \log_2 x$ shifted 2 units down.

101.

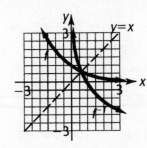

(B) Domain $f = (-\infty, \infty) =$ Range f^{-1}
Range $f = (0, \infty) =$ Domain f^{-1}

(C) $f^{-1}(x) = \log_{1/2} x = -\log_2 x$

103. $f(x) = 5^{3x-1} + 4$
Solve $y = f(x)$ for x
$$y = 5^{3x-1} + 4$$
$$y - 4 = 5^{3x-1}$$
$$\log_5(y - 4) = 3x - 1$$
$$3x - 1 = \log_5(y - 4)$$
$$3x = 1 + \log_5(y - 4)$$
$$x = \frac{1}{3}[1 + \log_5(y - 4)] = f^{-1}(y)$$

Interchange x and y:
$$y = f^{-1}(x) = \frac{1}{3}[1 + \log_5(x - 4)]$$
Domain: $x > 4$
(Check omitted for lack of space)

105. $g(x) = 3\log_e(5x - 2)$
Solve $y = g(x)$ for x
$$y = 3\log_e(5x - 2)$$
$$\frac{y}{3} = \log_e(5x - 2)$$
$$5x - 2 = e^{y/3}$$
$$5x = e^{y/3} + 2$$
$$x = \frac{1}{5}(e^{y/3} + 2) = g^{-1}(y)$$

Interchange x and y:
$$y = g^{-1}(x) = \frac{1}{5}(e^{x/3} + 2)$$
Domain: $(-\infty, \infty)$
(Check omitted for lack of space)

107. $y = 3^{x^2}$ is not a one-to-one function, since for $x = 1$ and $x = -1$, y has the same value ($y = 3^1 = 3$). Therefore the inverse function does not exist and the reflection is not the graph of a function.

109. $\log_e x - \log_e 100 = -0.08t$

$\log_e \frac{x}{100} = -0.08t$

Write this in equivalent exponential form

$\frac{x}{100} = e^{-0.08t}$
$x = 100e^{-0.08t}$

111. Let $u = \log_b M$ and $v = \log_b N$; then
$M = b^u$ and $N = b^v$. Thus,
$\log_b M/N = \log_b b^u/b^v = \log_b b^{u-v}$
$\qquad\qquad = u - v = \log_b M - \log_b N$.

Exercise 5-4

Key Ideas and Formulas

Logarithmic Notation

$\log x = \log_{10} x$ Common logarithm

$\ln x = \log_e x$ Natural logarithm

$\log x = y$ is equivalent to $x = 10^y$

$\ln x = y$ is equivalent to $x = e^y$

Sound Intensity	Earthquake Intensity	Rocket Flight Velocity
$D = 10 \log \dfrac{I}{I_0}$ D = decibel level I = intensity of sound I_0 = threshold of hearing	$M = \dfrac{2}{3} \log \dfrac{E}{E_0}$ M = magnitude on Richter Scale E = Energy released by earthquake E_0 = Energy released by small reference earthquake	$v = c \ln \dfrac{W_t}{W_b}$ v = Velocity at fuel burnout level c = exhaust velocity W_t = take off weight W_b = burnout weight

1. 4.9177   **3.** -2.8419   **5.** 3.7623   **7.** -2.5128   **9.** 200,800

11. $6.648 \times 10^{-4} = 0.0006648$ **13.** 47.73   **15.** 0.6760   **17.** 4.959

19. 7.861   **21.** 3.301   **23.** 4.561

25. $x = \log(5.3147 \times 10^{12}) = \log 5.3147 + \log 10^{12} = 0.725 + 12 = 12.725$ (Round off error is significantly reduced if we enter 5.3147 rather than 5.3147×10^{12})

27. -25.715

29. $\log x = 32.068523 = .068523 + 32 = \log 1.1709 + \log 10^{32} = \log(1.1709 \times 10^{32})$ Hence $x = 1.1709 \times 10^{32}$

31. 4.2672×10^{-7}

33.

x	y
0.5	-0.7
1	0
2	0.7
3	1.1
4	1.4
5	1.6

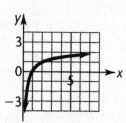

The y axis is a vertical asymptote.

35.

x	y
0.25	1.4
0.5	0.7
0.75	0.3
1	0
2	0.7
3	1.1
4	1.4
5	1.6

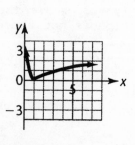

The y axis is a vertical asymptote.

37.

x	y
-1.5	-1.4
-1	0
-0.5	0.8
0	1.4
1	2.2
2	2.8
3	3.2

The line $x = -2$ is a vertical asymptote. Note: The graph is the same of $y = \ln x$ shifted 2 units to the left and expanded by a factor of 2.

39.

x	y
0.5	-5.8
1	-3.0
2	-0.2
3	1.4
4	2.5
5	3.4
6	4.2
7	4.8
8	5.3
9	5.8
10	6.2

The y axis is a vertical asymptote. Note: The graph is the same as the graph of $y = \ln x$ expanded by a factor of 4 and shifted 3 units down.

41. The inequality sign in the last step reverses because $\log \frac{1}{3}$ is negative.

43. (A)

x	log x	log(log x)
10	1	0
100	2	$\log 2 \approx 0.30$
1000	3	$\log 3 \approx 0.48$
10^6	6	$\log 6 \approx 0.78$
10^{100}	100	2
$10^{10^{10}}$	10^{10}	10

(B) Since the range of log x is all real numbers, the domain of log(log x) is those real numbers for which log x is in the domain of log x, that is, log $x > 0$, or $x > 1$. The domain is thus $(1, \infty)$. Since if $y = \log(\log x)$, $10^y = \log x$ or $x = 10^{10^y}$, y can take on any real value and the range of g is $(-\infty, \infty)$.

(C) The graph of f has a vertical asymptote at $x = 0$ and no horizontal asymptote. The graph of g has a vertical asymptote at $x = 1$ and no horizontal asymptote. The graph of f has an x intercept at $x = 1$. The graph of g has an x intercept at $x = 10$. Here are computer-generated graphs of f and g in the same coordinate system.

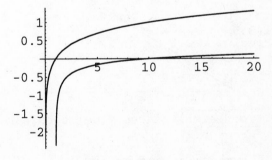

45. Here is a computer-generated graph showing $f(x) = \ln x$ and $g(x) = 0.1x - 0.2$.

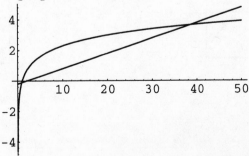

The graphs appear to intersect on the intervals (0, 2) and (38, 40). To estimate the coordinates of the points of intersection, we zoom in. Here are computer-generated graphs of f and g on (0.85, 0.95) and (38.4, 38.6).

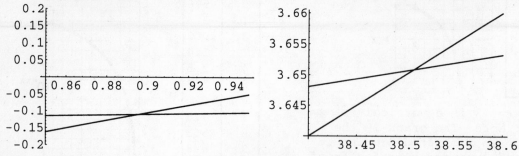

To two decimal places, the coordinates are (0.90, -0.11) and (38.51, 3.65).

47. Here is a computer-generated graph showing $f(x) = \ln x$ and $g(x) = x^{1/3}$.

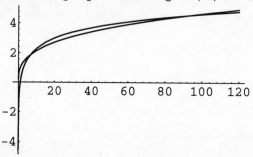

The graphs appear to intersect on the intervals (4, 8) and (92, 96). To estimate the coordinates of the points of intersection, we zoom in. Here are computer-generated graphs of f and g on (6.3, 6.5) and (93.3, 93.5).

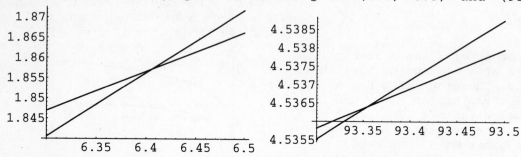

To two decimal places, the coordinates are (6.41, 1.86) and (93.35, 4.54).

49.

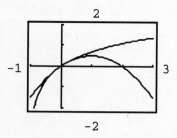

51.

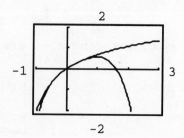

53. We use the decibel formula

$$D = 10 \log \frac{I}{I_0}$$

(A) $I = I_0$

$$D = 10 \log \frac{I_0}{I_0}$$

$$D = 10 \log 1$$
$$D = 0 \text{ decibels}$$

(B) $I_0 = 1.0 \times 10^{-12}$ $I = 1.0$

$$D = 10 \log \frac{1.0}{1.0 \times 10^{-12}}$$

$$D = 120 \text{ decibels}$$

55. We use the decibel formula

$$D = 10 \log \frac{I}{I_0}$$

$$I_2 = 1000 I_1$$

$$D_1 = 10 \log \frac{I_1}{I_0} \quad D_2 = 10 \log \frac{I_2}{I_0}$$

$$D_2 - D_1 = 10 \log \frac{I_2}{I_0} - 10 \log \frac{I_1}{I_0}$$

$$= 10 \log \left(\frac{I_2}{I_0} \div \frac{I_1}{I_0} \right)$$

$$= 10 \log \frac{I_2}{I_1}$$

$$= 10 \log \frac{1000 I_1}{I_1}$$

$$= 10 \log 1000$$
$$= 30 \text{ decibels}$$

57. We use the magnitude formula

$$M = \frac{2}{3} \log \frac{E}{E_0}$$

$$E_0 = 10^{4.40} \quad E = 1.99 \times 10^{17}$$

$$M = \frac{2}{3} \log \frac{1.99 \times 10^{17}}{10^{4.40}}$$

$$= \frac{2}{3} \log(1.99 \times 10^{12.6})$$

$$= \frac{2}{3}(\log 1.99 + \log 10^{12.6})$$

$$= \frac{2}{3}(0.299 + 12.6)$$

$$= 8.6$$

59. We use the magnitude formula

$$M = \frac{2}{3} \log \frac{E}{E_0}$$

For the Long Beach earthquake,

$$6.3 = \frac{2}{3} \log \frac{E_1}{E_0}$$

$$9.45 = \log \frac{E_1}{E_0}$$

$$\frac{E_1}{E_0} = 10^{9.45}$$

$$E_1 = E_0 \cdot 10^{9.45}$$

For the Anchorage earthquake,

$$8.3 = \frac{2}{3} \log \frac{E_2}{E_0}$$

$$12.45 = \log \frac{E_2}{E_0}$$

$$\frac{E_2}{E_0} = 10^{12.45}$$

$$E_2 = E_0 \cdot 10^{12.45}$$

Hence, the Anchorage earthquake compares to the Long Beach earthquake as follows:

$$\frac{E_2}{E_1} = \frac{E_0 \cdot 10^{12.45}}{E_0 \cdot 10^{9.45}} = 10^3$$

$$E_2 = 10^3 E_1, \text{ or } 1000 \text{ times as powerful}$$

61. We use the rocket equation.

$$v = c \ln \frac{W_t}{W_b}$$

$$v = 2.57 \ln (19.8)$$

$$v = 7.67 \text{ km/s}$$

63. (A) $pH = -\log[H^+] = -\log(4.63 \times 10^{-9}) = 8.3$. Since this is greater than 7, the substance is basic.

(B) $pH = -\log[H^+] = -\log(9.32 \times 10^{-4}) = 3.0$ Since this is less than 7, the substance is acidic.

65. Since $pH = -\log[H^+]$, we have

$$5.2 = -\log[H^+], \text{ or}$$

$$[H^+] = 10^{-5.2} = 6.3 \times 10^{-6} \text{ moles per liter}$$

Exercise 5-5

Key Ideas and Formulas

To solve equations in which the variable appears only in the exponent, use logarithms. If $a^x = b$, then $x \log a = \log b$.

To solve equations in which the variable appears only in the argument of a logarithm, change to equivalent exponential form.

If $\log_a x = b$, then $x = a^b$

Change-of-Base Formula: $\log_b N = \dfrac{\log_a N}{\log_a b}$

---

1. $10^{-x} = 0.0347$

$-x = \log_{10} 0.0347$

$x = -\log_{10} 0.0347$

$x = 1.46$

3. $10^{3x+1} = 92$

$3x + 1 = \log_{10} 92$

$3x = \log_{10} 92 - 1$

$x = \dfrac{\log_{10} 92 - 1}{3}$

$x = 0.321$

5. $e^x = 3.65$

$x = \ln 3.65$

$x = 1.29$

7. $e^{2x-1} = 405$

$2x - 1 = \ln 405$

$2x = 1 + \ln 405$

$x = \dfrac{1 + \ln 405}{2}$

$x = 3.50$

9. $5^x = 18$

$x \ln 5 = \ln 18$

$x = \dfrac{\ln 18}{\ln 5}$

$x = 1.80$

11. $2^{-x} = 0.238$

$-x \ln 2 = \ln 0.238$

$-x = \dfrac{\ln 0.238}{\ln 2}$

$x = -\dfrac{\ln 0.238}{\ln 2}$

$x = 2.07$

13. $\log 5 + \log x = 2$

$\log(5x) = 2$

$5x = 10^2$

$5x = 100$

$x = 20$

15. $\log x + \log(x - 3) = 1$

$\log[x(x - 3)] = 1$

$x(x - 3) = 10^1$

$x^2 - 3x = 10$

$x^2 - 3x - 10 = 0$

$(x - 5)(x + 2) = 0$

$x = 5$ or -2

> **Common Error:**
> $\log(x - 3) \neq \log x - \log 3$

Check:

$\log 5 + \log(5 - 3) \overset{\surd}{=} 1$

$\log(-2) + \log(-2 - 3)$ is not defined.

$x = 5$

17. $\log(x + 1) - \log(x - 1) = 1$

$\log \dfrac{x + 1}{x - 1} = 1$

$\dfrac{x + 1}{x - 1} = 10^1$

$\dfrac{x + 1}{x - 1} = 10$

$x + 1 = 10(x - 1)$

$x + 1 = 10x - 10$

$11 = 9x$

$x = \dfrac{11}{9}$

> **Common Error:**
> $\dfrac{x + 1}{x - 1} \neq \log 1$

Check:

$\log\left(\dfrac{11}{9} + 1\right) - \log\left(\dfrac{11}{9} - 1\right) \overset{?}{=} 1$

$\log \dfrac{20}{9} - \log \dfrac{2}{9} \overset{?}{=} 1$

$\log 10 \overset{\surd}{=} 1$

19.
$$2 = 1.05^x$$
$$\ln 2 = x \ln 1.05$$
$$\frac{\ln 2}{\ln 1.05} = x$$
$$x = 14.2$$

21.
$$e^{-1.4x} = 13$$
$$-1.4x = \ln 13$$
$$x = \frac{\ln 13}{-1.4}$$
$$x = -1.83$$

23.
$$123 = 500e^{-0.12x}$$
$$\frac{123}{500} = e^{-0.12x}$$
$$\ln\left(\frac{123}{500}\right) = -0.12x$$
$$\frac{\ln\left(\frac{123}{500}\right)}{-0.12} = x$$
$$x = 11.7$$

25.
$$e^{-x^2} = 0.23$$
$$-x^2 = \ln 0.23$$
$$x^2 = -\ln 0.23$$
$$x = \pm\sqrt{-\ln 0.23}$$
$$x = \pm 1.21$$

27.
$$\log x - \log 5 = \log 2 - \log(x - 3)$$
$$\log \frac{x}{5} = \log \frac{2}{x - 3}$$
$$\frac{x}{5} = \frac{2}{x - 3}$$

Excluded value: $x \neq 3$
$$5(x - 3)\frac{x}{5} = 5(x - 3)\frac{2}{x - 3}$$
$$(x - 3)x = 10$$
$$x^2 - 3x = 10$$
$$x^2 - 3x - 10 = 0$$
$$(x - 5)(x + 2) = 0$$
$$x = 5, -2$$

Check:
$$\log 5 - \log 5 \overset{\surd}{=} \log 2 - \log 2$$
$\log(-2)$ is not defined
Solution: 5

29.
$$\ln x = \ln(2x - 1) - \ln(x - 2)$$
$$\ln x = \ln \frac{2x - 1}{x - 2}$$
$$x = \frac{2x - 1}{x - 2}$$

Excluded value: $x \neq 2$
$$x(x - 2) = (x - 2)\frac{2x - 1}{x - 2}$$
$$x(x - 2) = 2x - 1$$
$$x^2 - 2x = 2x - 1$$
$$x^2 - 4x + 1 = 0$$
$$x = \frac{-b \pm \sqrt{b^2 - 4ac}}{2a}$$
$$a = 1, \ b = -4, \ c = 1$$
$$x = \frac{-(-4) \pm \sqrt{(-4)^2 - 4(1)(1)}}{2(1)}$$
$$x = \frac{4 \pm \sqrt{12}}{2}$$
$$x = 2 \pm \sqrt{3}$$

Check:
$$\ln(2 + \sqrt{3}) \overset{?}{=} \ln[2(2 + \sqrt{3}) - 1]$$
$$- \ln[(2 + \sqrt{3}) - 2]$$
$$\ln(2 + \sqrt{3}) \overset{?}{=} \ln(3 + 2\sqrt{3}) - \ln\sqrt{3}$$
$$\ln(2 + \sqrt{3}) \overset{?}{=} \ln\left(\frac{3 + 2\sqrt{3}}{\sqrt{3}}\right)$$
$$\ln(2 + \sqrt{3}) \overset{\surd}{=} \ln(\sqrt{3} + 2)$$
$\ln(x - 2)$ is not defined if $x = 2 - \sqrt{3}$
Solution: $2 + \sqrt{3}$

31.
$$\log(2x + 1) = 1 - \log(x - 1)$$
$$\log(2x + 1) + \log(x - 1) = 1$$
$$\log[(2x + 1)(x - 1)] = 1$$
$$(2x + 1)(x - 1) = 10$$
$$2x^2 - x - 1 = 10$$
$$2x^2 - x - 11 = 0$$
$$x = \frac{-b \pm \sqrt{b^2 - 4ac}}{2a}$$
$$a = 2, \ b = -1, \ c = -11$$
$$x = \frac{-(-1) \pm \sqrt{(-1)^2 - 4(2)(-11)}}{2(2)}$$
$$x = \frac{1 \pm \sqrt{89}}{4}$$

Check: $\log\left(2\dfrac{1+\sqrt{89}}{4}+1\right) \overset{?}{=} 1 - \log\left(\dfrac{1+\sqrt{89}}{4}-1\right)$

$$\log\left(\dfrac{1+\sqrt{89}+2}{2}\right) \overset{?}{=} 1 - \log\left(\dfrac{1+\sqrt{89}-4}{4}\right)$$

$$\log\left(\dfrac{3+\sqrt{89}}{2}\right) \overset{?}{=} 1 - \log\left(\dfrac{\sqrt{89}-3}{4}\right)$$

$$\log\left(\dfrac{3+\sqrt{89}}{2}\right) \overset{?}{=} \log 10 - \log\left(\dfrac{\sqrt{89}-3}{4}\right)$$

$$\overset{?}{=} \log\left(\dfrac{40}{\sqrt{89}-3}\right)$$

$$\overset{?}{=} \log\left[\dfrac{40(\sqrt{89}+3)}{89-9}\right]$$

$$\overset{\sqrt{}}{=} \log\left(\dfrac{\sqrt{89}+3}{2}\right)$$

$\log(x-1)$ is not defined if $x = \dfrac{1-\sqrt{89}}{4}$

Solution: $\dfrac{1+\sqrt{89}}{4}$

33.
$$(\ln x)^3 = \ln x^4$$
$$(\ln x)^3 = 4 \ln x$$
$$(\ln x)^3 - 4 \ln x = 0$$
$$\ln x[(\ln x)^2 - 4] = 0$$
$$\ln x(\ln x - 2)(\ln x + 2) = 0$$

$$\ln x = 0 \qquad \ln x - 2 = 0 \qquad \ln x + 2 = 0$$
$$x = 1 \qquad \ln x = 2 \qquad \ln x = -2$$
$$x = e^2 \qquad x = e^{-2}$$

Check:
$$(\ln 1)^3 \overset{?}{=} \ln 1^4 \qquad (\ln e^2)^3 \overset{?}{=} \ln(e^2)^4 \qquad (\ln e^{-2})^3 \overset{?}{=} \ln(e^{-2})^4$$
$$0 \overset{\sqrt{}}{=} 0 \qquad\qquad 8 \overset{\sqrt{}}{=} 8 \qquad\qquad -8 \overset{\sqrt{}}{=} -8$$
Solution: $1, \ e^2, \ e^{-2}$

35. $\ln(\ln x) = 1$
$$\ln x = e^1$$
$$\ln x = e$$
$$x = e^e$$

37. $x^{\log x} = 100x$

We start by taking logarithms of both sides.
$$\log(x^{\log x}) = \log 100x$$
$$\log x \log x = \log 100 + \log x$$
$$(\log x)^2 = 2 + \log x$$
$$(\log x)^2 - \log x - 2 = 0$$
$$(\log x - 2)(\log x + 1) = 0$$
$$\log x - 2 = 0 \qquad \log x + 1 = 0$$
$$\log x = 2 \qquad \log x = -1$$
$$x = 10^2 \qquad x = 10^{-1}$$
$$x = 100 \qquad x = 0.1$$

Check: $100^{\log 100} \overset{?}{=} 100 \cdot 100 \qquad (0.1)^{\log(0.1)} \overset{?}{=} 100(0.1)$
$$10^4 \overset{\sqrt{}}{=} 10^4 \qquad\qquad 0.1^{-1} \overset{?}{=} 10$$
$$10 \overset{\sqrt{}}{=} 10$$

39. (A) In solving an exponential equation, we often can get the variable out of the exponent by using logarithms. If we do that in this case, however, we have made no progress:

$$e^{x/2} = 5 \ln x$$
$$\ln e^{x/2} = \ln(5 \ln x)$$
$$\frac{x}{2} = \ln 5 + \ln(\ln x)$$

This equation is no easier than the original to solve. Other algebraic methods seem to be equally useless.

(B) Here is a computer-generated graph of $y = e^{x/2}$ and $y = 5 \ln x$.

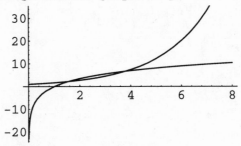

From the graph, it appears that there are two solutions of $e^{x/2} = 5 \ln x$.

41. (A) In solving an exponential equation, we often can get the variable out of the exponent by using logarithms. If we do that in this case, however, we have made no progress:

$$3^x + 2 = 7 + x - e^{-x}$$
$$3^x = -e^{-x} + x + 5$$
$$\log_3 3^x = \log_3(-e^{-x} + x + 5)$$
$$x = \log_3(-e^{-x} + x + 5)$$

This equation is no easier than the original to solve.

(B) Here is a computer-generated graph of $y = 3^x + 2$ and $y = 7 + x - e^{-x}$, $-2 \leq x \leq 2$.

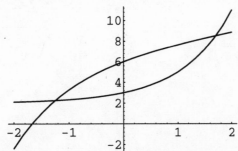

There are zeros between -1.3 and -1.2 and between 1.6 and 1.8. To estimate the zeros to three decimal places, we zoom in. Here are computer-generated graphs of the two functions on (-1.3, -1.2) and (1.7, 1.72).

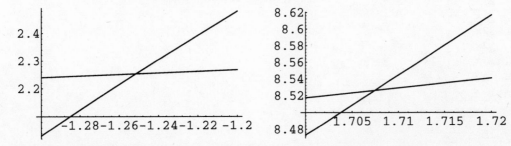

We can now estimate the solutions as -1.252 and 1.707.

43. $\log_5 372 = \dfrac{\ln 372}{\ln 5} = 3.6776$ or $\log_5 372 = \dfrac{\log_{10} 372}{\log_{10} 5} = 3.6776$

45. $\log_8 0.0352 = \dfrac{\ln 0.0352}{\ln 8} = -1.6094$ **47.** $\log_3 0.1483 = \dfrac{\ln 0.1483}{\ln 3} = -1.7372$

49.
$$A = Pe^{rt}$$
$$\frac{A}{P} = e^{rt}$$
$$\ln \frac{A}{P} = rt$$
$$\frac{1}{t} \ln \frac{A}{P} = r$$
$$r = \frac{1}{t} \ln \frac{A}{P}$$

51.
$$D = 10 \log \frac{I}{I_0}$$
$$\frac{D}{10} = \log \frac{I}{I_0}$$
$$\frac{I}{I_0} = 10^{D/10}$$
$$I = I_0(10^{D/10})$$

53.
$$M = 6 - 2.5 \log \frac{I}{I_0}$$
$$6 - M = 2.5 \log \frac{I}{I_0}$$
$$\frac{6 - M}{2.5} = \log \frac{I}{I_0}$$
$$\frac{I}{I_0} = 10^{(6-M)/2.5}$$
$$I = I_0[10^{(6-M)/2.5}]$$

55.
$$I = \frac{E}{R}(1 - e^{-Rt/L})$$
$$RI = E(1 - e^{-Rt/L})$$
$$\frac{RI}{E} = 1 - e^{-Rt/L}$$
$$\frac{RI}{E} - 1 = -e^{-Rt/L}$$
$$-\left(\frac{RI}{E} - 1\right) = e^{-Rt/L}$$
$$-\frac{RI}{E} + 1 = e^{-Rt/L}$$
$$1 - \frac{RI}{E} = e^{-Rt/L}$$
$$\ln\left(1 - \frac{RI}{E}\right) = -\frac{Rt}{L}$$
$$-\frac{L}{R} \ln\left(1 - \frac{RI}{E}\right) = t$$
$$t = -\frac{L}{R} \ln\left(1 - \frac{RI}{E}\right)$$

57.
$$y = \frac{e^x + e^{-x}}{2}$$
$$2y = e^x + e^{-x}$$
$$2y = e^x + \frac{1}{e^x}$$
$$2ye^x = (e^x)^2 + 1$$
$$0 = (e^x)^2 - 2ye^x + 1$$

This equation is quadratic in e^x

$$e^x = \frac{-b \pm \sqrt{b^2 - 4ac}}{2a}$$
$$a = 1, \quad b = -2y, \quad c = 1$$
$$e^x = \frac{-(-2y) \pm \sqrt{(-2y)^2 - 4(1)(1)}}{2(1)}$$
$$e^x = \frac{2y \pm \sqrt{4y^2 - 4}}{2}$$
$$e^x = \frac{2(y \pm \sqrt{y^2 - 1})}{2}$$
$$e^x = y \pm \sqrt{y^2 - 1}$$
$$x = \ln(y \pm \sqrt{y^2 - 1})$$

59.
$$y = \frac{e^x - e^{-x}}{e^x + e^{-x}}$$
$$y = \frac{e^x - \frac{1}{e^x}}{e^x + \frac{1}{e^x}}$$
$$y = \frac{e^x e^x - \frac{1}{e^x} e^x}{e^x e^x + \frac{1}{e^x} e^x}$$
$$y = \frac{e^{2x} - 1}{e^{2x} + 1}$$
$$y(e^{2x} + 1) = e^{2x} - 1$$
$$ye^{2x} + y = e^{2x} - 1$$
$$1 + y = e^{2x} - ye^{2x}$$
$$1 + y = (1 - y)e^{2x}$$
$$e^{2x} = \frac{1 + y}{1 - y}$$
$$2x = \ln \frac{1 + y}{1 - y}$$
$$x = \frac{1}{2} \ln \frac{1 + y}{1 - y}$$

61. $y = 3 + \log_2(2 - x)$

To enter this into the calculator, we use the change-of-base formula to write this as:

$y = 3 + \dfrac{\log(2 - x)}{\log 2}$

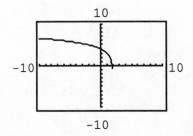

63. $y = \log_3 x - \log_2 x$

To enter this into the calculator, we use the change-of-base formula to write this as:

$Y = \dfrac{\log x}{\log 3} - \dfrac{\log x}{\log 2}$

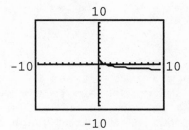

65. Here is a computer-generated graph of $2^{-x} - 2x$, $0 \leq x \leq 1$.

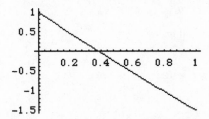

There is a zero between 0.36 and 0.4. To estimate the zero to two decimal places, we "zoom in" on it. Here is the computer-generated graph of $2^{-x} - 2x$, $0.36 \leq x \leq 0.4$

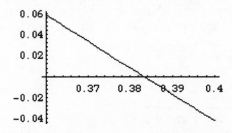

We can now estimate the zero as 0.38.

67. Here is a computer-generated graph of $x3^x - 1$, $0 \leq x \leq 1$.

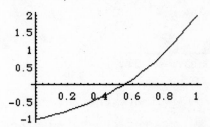

There is a zero between 0.52 and 0.56. To estimate the zero to two decimal places, we zoom in on it. Here is the computer-generated graph of $x3^x - 1$, $0.52 \leq x \leq 0.56$.

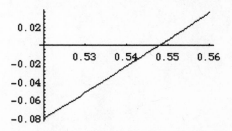

We can now estimate the zero as 0.55.

69. Here is a computer-generated graph of $e^{-x} - x$, $0 \leq x \leq 1$.

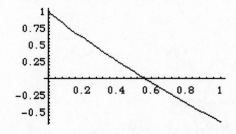

There is a zero between 0.56 and 0.6. To estimate the zero to two decimal places, we zoom in on it. Here is the computer-generated graph of $e^{-x} - x$, $0.56 \leq x \leq 0.6$.

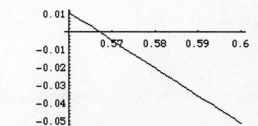

We can now estimate the zero as 0.57.

71. Here is a computer-generated graph of $xe^x - 2$, $0 \le x \le 1$.

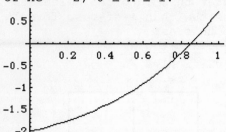

There is a zero between 0.84 and 0.88. To estimate the zero to two decimal places, we zoom in on it. Here is the computer-generated graph of $xe^x - 2$, $0.84 \le x \le 0.88$.

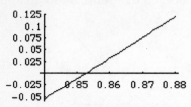

We can now estimate the zero as 0.85.

73. Here is a computer-generated graph of $\ln x + 2x$, $0 < x \le 1$.

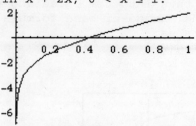

There is a zero between 0.4 and 0.44. To estimate the zero to two decimal places, we zoom in on it. Here is the computer-generated graph of $\ln x + 2x$, $0.4 \le x \le 0.44$.

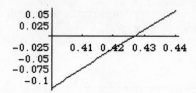

We can now estimate the zero as 0.43.

75. Here is a computer-generated graph of $\ln x + e^x$, $0 < x \le 1$.

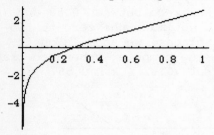

There is a zero between 0.24 and 0.28. To estimate the zero to two decimal places, we zoom in on it. Here is the computer-generated graph of $\ln x + e^x$, $0.24 \le x \le 0.28$.

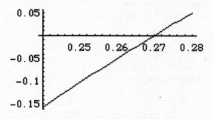

We can now estimate the zero as 0.27.

77. To find the doubling time we replace A in $A = P(1 + 0.15)^n$ with $2P$ and solve for n.

$$2P = P(1.15)^n$$
$$2 = (1.15)^n$$
$$\ln 2 = n \ln 1.15$$
$$n = \frac{\ln 2}{\ln 1.15}$$
$$n = 5 \text{ years to the nearest year}$$

79. We solve $A = Pe^{rt}$ for r, with $A = 2{,}500$, $P = 1{,}000$, $t = 10$

$$2{,}500 = 1{,}000e^{r(10)}$$
$$2.5 = e^{10r}$$
$$10r = \ln (2.5)$$
$$r = \frac{1}{10} \ln 2.5$$
$$r = 0.0916 \text{ or } 9.16\%$$

81. $m = 6 - 2.5 \log \frac{L}{L_0}$

(A) We find m when $L = L_0$

$m = 6 - 2.5 \log \frac{L_0}{L_0}$

$m = 6 - 2.5 \log 1$

$m = 6$

(B) We compare L_1 for $m = 1$ with L_2 for $m = 6$

$1 = 6 - 2.5 \log \frac{L_1}{L_0}$ $\qquad$ $6 = 6 - 2.5 \log \frac{L_2}{L_0}$

$-5 = -2.5 \log \frac{L_1}{L_0}$ $\qquad$ $0 = -2.5 \log \frac{L_2}{L_0}$

$2 = \log \frac{L_1}{L_0}$ $\qquad$ $0 = \log \frac{L_2}{L_0}$

$\frac{L_1}{L_0} = 10^2$ $\qquad$ $\frac{L_2}{L_0} = 1$

$L_1 = 100L_0$ $\qquad$ $L_2 = L_0$

Hence $\frac{L_1}{L_2} = \frac{100L_0}{L_0} = 100$. The star of magnitude 1 is 100 times brighter.

83. We solve $P = P_0 e^{rt}$ for t with $P = 2P_0$, $r = 0.02$.

$2P_0 = P_0 e^{0.02t}$

$2 = e^{0.02t}$

$\ln 2 = 0.02t$

$\frac{\ln 2}{0.02} = t$

$t = 35$ years to the nearest year

85. We solve $A = A_0 e^{-0.000124t}$ for t with $A = 0.1A_0$

$0.1A_0 = A_0 e^{-0.000124t}$

$0.1 = e^{-0.000124t}$

$\ln 0.1 = -0.000124t$

$t = \frac{\ln 0.1}{-0.000124}$

$t = 18,600$ years old

87. We solve $q = 0.0009(1 - e^{-0.2t})$ for t with $q = 0.0007$

$0.0007 = 0.0009(1 - e^{-0.2t})$

$\frac{0.0007}{0.0009} = 1 - e^{-0.2t}$

$\frac{7}{9} = 1 - e^{-0.2t}$

$-\frac{2}{9} = -e^{-0.2t}$

$\frac{2}{9} = e^{-0.2t}$

$\ln \frac{2}{9} = -0.2t$

$t = \frac{\ln \frac{2}{9}}{-0.2}$

$t = 7.52$ seconds

89. First, we solve $T = T_m + (T_0 - T_m)e^{-kt}$ for k, with $T = 61.5°$, $T_m = 40°$, $T_0 = 72°$, $t = 1$

$61.5 = 40 + (72 - 40)e^{-k(1)}$

$21.5 = 32e^{-k}$

$\frac{21.5}{32} = e^{-k}$

$\ln \frac{21.5}{32} = -k$

$k = -\ln \frac{21.5}{32}$

$k = 0.40$

Now we solve $T = T_m + (T_0 - T_m)e^{-0.40t}$ for t, with $T = 50°$, $T_m = 40°$, $T_0 = 72°$

$50 = 40 + (72 - 40)e^{-0.40t}$

$10 = 32e^{-0.40t}$

$\frac{10}{32} = e^{-0.40t}$

$\ln \frac{10}{32} = -0.40t$

$t = \frac{\ln^{10/32}}{-0.40}$

$t = 2.9$ hours

91. We solve $N = \dfrac{100}{1 + 4e^{-0.14t}}$ for t, with $N = 50$

$$50 = \frac{100}{1 + 4e^{-0.14t}}$$

$$\frac{1}{50} = \frac{1 + 4e^{-0.14t}}{100}$$

$$2 = 1 + 4e^{-0.14t}$$

$$1 = 4e^{-0.14t}$$

$$0.25 = e^{-0.14t}$$

$$\ln 0.25 = -0.14t$$

$$t = \frac{\ln 0.25}{-0.14}$$

$$t = 10 \text{ years}$$

CHAPTER 5 REVIEW

1. $\log m = n$ $(5\text{-}3)$ **2.** $\ln x = y$ $(5\text{-}3)$ **3.** $x = 10^y$ $(5\text{-}3)$ **4.** $y = e^x$ $(5\text{-}3)$

5. $\dfrac{7^{x+2}}{7^{2-x}} = 7^{(x+2)-(2-x)}$

$$= 7^{x+2-2+x}$$

$$= 7^{2x} \quad (5\text{-}1)$$

6. $\left(\dfrac{e^x}{e^{-x}}\right)^x = [e^{x-(-x)}]^x$

$$= (e^{2x})^x = e^{2x \cdot x}$$

$$= e^{2x^2} \quad (5\text{-}1)$$

7. $\log_2 x = 3$

$$x = 2^3$$

$$x = 8 \quad (5\text{-}3)$$

8. $\log_x 25 = 2$

$$25 = x^2$$

$$x = 5$$

since bases are
restricted positive
$(5\text{-}3)$

9. $\log_3 27 = x$

$$\log_3 3^3 = x$$

$$3 = x \quad (5\text{-}3)$$

10. $10^x = 17.5$

$$x = \log_{10} 17.5$$

$$x = 1.24 \quad (5\text{-}3)$$

11. $e^x = 143{,}000$

$$x = \ln 143{,}000$$

$$x = 11.9 \quad (5\text{-}3)$$

12. $\ln x = -0.01573$

$$x = e^{-0.01573}$$

$$x = 0.984 \quad (5\text{-}3)$$

13. $\log x = 2.013$

$$x = 10^{2.013}$$

$$x = 103 \quad (5\text{-}3)$$

14. $\ln(2x - 1) = \ln(x + 3)$

$$2x - 1 = x + 3$$

$$x = 4$$

Check:

$\ln(2 \cdot 4 - 1) \overset{?}{=} \ln(4 + 3)$

$\ln 7 \overset{\surd}{=} \ln 7$ $(5\text{-}3)$

15. $\log(x^2 - 3) = 2 \log(x - 1)$

$$\log(x^2 - 3) = \log(x - 1)^2$$

$$x^2 - 3 = (x - 1)^2$$

$$x^2 - 3 = x^2 - 2x + 1$$

$$-3 = -2x + 1$$

$$-4 = -2x$$

$$x = 2$$

Check:

$\log(2^2 - 3) \overset{?}{=} \log(2 - 1)$

$\log 1 \overset{?}{=} 2 \log 1$

$0 \overset{\surd}{=} 0$ $(5\text{-}3)$

16.

$$e^{x^2-3} = e^{2x}$$

$$x^2 - 3 = 2x$$

$$x^2 - 2x - 3 = 0$$

$$(x - 3)(x + 1) = 0$$

$$x = 3, -1$$

$(5\text{-}2)$

17.

$$4^{x-1} = 2^{1-x}$$

$$(2^2)^{x-1} = 2^{1-x}$$

$$2^{2(x-1)} = 2^{1-x}$$

$$2(x - 1) = 1 - x$$

$$2x - 2 = 1 - x$$

$$3x = 3$$

$$x = 1 \quad (5\text{-}1)$$

18.
$$2x^2 e^{-x} = 18e^{-x}$$
$$2x^2 e^{-x} - 18e^{-x} = 0$$
$$2e^{-x}(x^2 - 9) = 0$$
$$2e^{-x}(x - 3)(x + 3) = 0$$
$$2e^{-x} = 0 \quad x - 3 = 0 \quad x + 3 = 0$$
never $\quad x = 3 \quad\quad x = -3$
Solution: 3, -3 $\quad\quad$ (5-2)

19.
$$\log_{1/4} 16 = x$$
$$\log_{1/4} 4^2 = x$$
$$\log_{1/4} \left(\frac{1}{4}\right)^{-2} = x$$
$$-2 = x \quad\quad (5-3)$$

20. $\log_x 9 = -2$
$$x^{-2} = 9$$
$$\frac{1}{x^2} = 9$$
$$1 = 9x^2$$
$$\frac{1}{9} = x^2$$
$$x = \pm\sqrt{\frac{1}{9}}$$
$$x = \frac{1}{3}$$
since bases are
restricted positive
$\quad\quad$ (5-3)

21. $\log_{16} x = \frac{3}{2}$
$$16^{3/2} = x$$
$$64 = x$$
$$x = 64 \quad\quad (5-3)$$

22. $\log_x e^5 = 5$
$$e^5 = x^5$$
$$x = e \quad (5-3)$$

23. $10^{\log_{10} x} = 33$
$$\log_{10} x = \log_{10} 33$$
$$x = 33 \quad\quad (5-3)$$

24. $\ln x = 0$
$$e^0 = x$$
$$x = 1 \quad\quad (5-3)$$

25. 1.145 $\quad$ (5-3)

26. Not defined. (-e is not in the domain of the logarithm function.) $\quad\quad$ (5-3)

27. 2.211 $\quad\quad$ (5-3)

28. 11.59 (5-3)

29. $x = 2(10^{1.32})$
$$x = 41.8 \quad\quad (5-1)$$

30. $x = \log_5 23$
$$x = \frac{\log 23}{\log 5} \text{ or } \frac{\ln 23}{\ln 5}$$
$$x = 1.95 \quad\quad (5-3)$$

31. $\ln x = -3.218$
$$x = e^{-3.218}$$
$$x = 0.0400 \quad (5-3)$$

32. $x = \log(2.156 \times 10^{-7})$
$$x = \log 2.156 + \log 10^{-7}$$
$$x = \log 2.156 - 7$$
$$x = -6.67 \quad\quad (5-3)$$

33. $x = \frac{\ln 4}{\ln 2.31}$
$$x = 1.66 \quad\quad (5-3)$$

34.
$$25 = 5(2)^x$$
$$\frac{25}{5} = 2^x$$
$$5 = 2^x$$
$$\ln 5 = x \ln 2$$
$$\frac{\ln 5}{\ln 2} = x$$
$$x = 2.32 \quad (5-5)$$

35. $4,000 = 2,500e^{0.12x}$
$$\frac{4,000}{2,500} = e^{0.12x}$$
$$0.12x = \ln \frac{4,000}{2,500}$$
$$x = \frac{1}{0.12} \ln \frac{4,000}{2,500}$$
$$x = 3.92 \quad\quad (5-5)$$

36.
$$0.01 = e^{-0.05x}$$
$$-0.05x = \ln 0.01$$
$$x = \frac{\ln 0.01}{-0.05}$$
$$x = 92.1 \quad\quad (5-5)$$

37.
$$5^{2x-3} = 7.08$$
$$(2x - 3)\log 5 = \log 7.08$$
$$2x - 3 = \frac{\log 7.08}{\log 5}$$
$$x = \frac{1}{2}\left[3 + \frac{\log 7.08}{\log 5}\right]$$
$$x = 2.11 \quad\quad (5-5)$$

38.
$$\frac{e^x - e^{-x}}{2} = 1$$
$$e^x - e^{-x} = 2$$
$$e^x - \frac{1}{e^x} = 2$$
$$e^x e^x - e^x\left(\frac{1}{e^x}\right) = 2e^x$$
$$(e^x)^2 - 1 = 2e^x$$
$$(e^x)^2 - 2e^x - 1 = 0$$

This equation is quadratic in e^x
$$e^x = \frac{-b \pm \sqrt{b^2 - 4ac}}{2a}$$
$$a = 1, \ b = -2, \ c = -1$$
$$e^x = \frac{-(-2) \pm \sqrt{(-2)^2 - 4(1)(-1)}}{2}$$
$$e^x = \frac{2 \pm \sqrt{8}}{2}$$
$$e^x = 1 \pm \sqrt{2}$$
$$x = \ln(1 \pm \sqrt{2})$$

$1 - \sqrt{2}$ is negative, hence not in the domain
of the logarithm function.
$$x = \ln(1 + \sqrt{2})$$
$$x = 0.881$$

39. $\log 3x^2 - \log 9x = 2$
$$\log \frac{3x^2}{9x} = 2$$
$$\frac{3x^2}{9x} = 10^2$$
$$\frac{x}{3} = 100$$
$$x = 300$$

Check:
$$\log(3\cdot300^2) - \log(9\cdot300) \overset{?}{=} 2$$
$$\log(270,000) - \log(2,700) \overset{?}{=} 2$$
$$\log \frac{270,000}{2,700} \overset{?}{=} 2$$
$$\log 100 \overset{\checkmark}{=} 2 \qquad (5\text{-}5)$$

$(5\text{-}5)$

40. $\log x - \log 3 = \log 4 - \log(x + 4)$
$$\log \frac{x}{3} = \log \frac{4}{x + 4}$$
$$\frac{x}{3} = \frac{4}{x + 4} \quad \text{excluded value:}$$
$$x \neq -4$$
$$3(x + 4)\frac{x}{3} = 3(x + 4)\frac{4}{x + 4}$$
$$(x + 4)x = 12$$
$$x^2 + 4x = 12$$
$$x^2 + 4x - 12 = 0$$
$$(x + 6)(x - 2) = 0$$
$$x = -6 \quad x = 2$$

Check: $\log(-6)$ is not defined
$$\log 2 - \log 3 \overset{?}{=} \log 4 - \log(2 + 4)$$
$$\log \frac{2}{3} \overset{?}{=} \log \frac{4}{6}$$
$$\log \frac{2}{3} \overset{\checkmark}{=} \log \frac{2}{3}$$

Solution: 2 $\qquad (5\text{-}5)$

41. $\ln(x + 3) - \ln x = 2 \ln 2$
$$\ln \frac{x + 3}{x} = \ln 2^2$$
$$\frac{x + 3}{x} = 2^2$$
$$\frac{x + 3}{x} = 4$$
$$x + 3 = 4x$$
$$3 = 3x$$
$$x = 1$$

Check:
$$\ln(1 + 3) - \ln 1 \overset{?}{=} 2 \ln 2$$
$$\ln 4 - 0 \overset{?}{=} 2 \ln 2$$
$$\ln 4 \overset{\checkmark}{=} \ln 4 \qquad (5\text{-}5)$$

42. $\ln(2x + 1) - \ln(x - 1) = \ln x$
$$\ln \frac{2x + 1}{x - 1} = \ln x$$
$$\frac{2x + 1}{x - 1} = x \quad \text{Excluded value: } x \neq 1$$
$$(x - 1)\frac{2x + 1}{x - 1} = x(x - 1)$$
$$2x + 1 = x^2 - x$$
$$0 = x^2 - 3x - 1$$

$$x = \frac{-b \pm \sqrt{b^2 - 4ac}}{2a} \quad a = 1, \ b = -3, \ c = -1$$

$$x = \frac{-(-3) \pm \sqrt{(-3)^2 - 4(1)(-1)}}{2(1)}$$

$$x = \frac{3 \pm \sqrt{13}}{2}$$

Check: $\ln\left(\dfrac{3 - \sqrt{13}}{2}\right)$ is not defined

$$\ln\left(2 \cdot \frac{3 + \sqrt{13}}{2} + 1\right) - \ln\left(\frac{3 + \sqrt{13}}{2} - 1\right) \overset{?}{=} \ln\left(\frac{3 + \sqrt{13}}{2}\right)$$

$$\ln(3 + \sqrt{13} + 1) - \ln\left(\frac{3 + \sqrt{13} - 2}{2}\right) \overset{?}{=} \ln\left(\frac{3 + \sqrt{13}}{2}\right)$$

$$\ln(4 + \sqrt{13}) - \ln\left(\frac{1 + \sqrt{13}}{2}\right) \overset{?}{=} \ln\left(\frac{3 + \sqrt{13}}{2}\right)$$

$$\ln\left(\frac{4 + \sqrt{13}}{1} \cdot \frac{2}{1 + \sqrt{13}}\right) \overset{?}{=} \ln\left(\frac{3 + \sqrt{13}}{2}\right)$$

$$\ln\left(\frac{(4 + \sqrt{13})2}{1 + \sqrt{13}}\right) \overset{?}{=} \ln\left(\frac{3 + \sqrt{13}}{2}\right)$$

$$\ln\left(\frac{(4 + \sqrt{13})2(1 - \sqrt{13})}{(1 + \sqrt{13})(1 - \sqrt{13})}\right) \overset{?}{=} \ln\left(\frac{3 + \sqrt{13}}{2}\right)$$

$$\ln\left(\frac{2(4 - 3\sqrt{13} - 13)}{1 - 13}\right) \overset{?}{=} \ln\left(\frac{3 + \sqrt{13}}{2}\right)$$

$$\ln\left(\frac{-18 - 6\sqrt{13}}{-12}\right) \overset{?}{=} \ln\left(\frac{3 + \sqrt{13}}{2}\right)$$

$$\ln\left(\frac{3 + \sqrt{13}}{2}\right) \overset{\checkmark}{=} \ln\left(\frac{3 + \sqrt{13}}{2}\right)$$

Solution: $\dfrac{3 + \sqrt{13}}{2}$ $\hspace{6cm}$ (5-5)

43.

$$(\log x)^3 = \log x^9$$
$$(\log x)^3 = 9 \log x$$
$$(\log x)^3 - 9 \log x = 0$$
$$\log x[(\log x)^2 - 9] = 0$$
$$\log x(\log x - 3)(\log x + 3) = 0$$
$$\log x = 0 \quad \log x - 3 = 0 \quad \log x + 3 = 0$$
$$x = 1 \quad \log x = 3 \quad \log x = -3$$
$$x = 10^3 \quad x = 10^{-3}$$

Check:

$$(\log 1)^3 \overset{?}{=} \log 1^9$$
$$0 \overset{\checkmark}{=} 0$$
$$(\log 10^3)^3 \overset{?}{=} \log(10^3)^9$$
$$27 \overset{\checkmark}{=} 27$$
$$(\log 10^{-3})^3 \overset{?}{=} \log(10^{-3})^9$$
$$-27 \overset{\checkmark}{=} -27$$

Solution: $1, \ 10^3, \ 10^{-3}$ (5-5)

44. $\ln(\log x) = 1$
$\log x = e$
$x = 10^e$ (5-5)

45. $(e^x + 1)(e^{-x} - 1) - e^x(e^{-x} - 1) = e^xe^{-x} - e^x + e^{-x} - 1 - e^xe^{-x} + e^x$
$$= 1 - e^x + e^{-x} - 1 - 1 + e^x = e^{-x} - 1 \qquad (5\text{-}2)$$

46. $(e^x + e^{-x})(e^x - e^{-x}) - (e^x - e^{-x})^2 = (e^x)^2 - (e^{-x})^2 - [(e^x)^2 - 2e^xe^{-x} + (e^{-x})^2]$
$$= e^{2x} - e^{-2x} - [e^{2x} - 2 + e^{-2x}]$$
$$= e^{2x} - e^{-2x} - e^{2x} + 2 - e^{-2x}$$
$$= 2 - 2e^{-2x} \qquad (5\text{-}2)$$

47.

x	y
-2	0.13
-1	0.25
0	0.5
1	1
2	2
3	4
4	8

$(5\text{-}1)$

48.

t	f(t)
-25	74
-20	50
-15	33
-10	22
-5	15
0	10
5	6.7
10	4.5
15	3.0

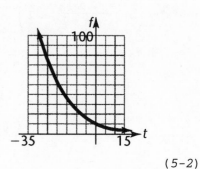

$(5\text{-}2)$

49.

x	y
-0.5	-0.7
0	0
1	0.7
2	1.1
4	1.6
6	1.9
8	2.2
10	2.4

The line $x = -1$ is a vertical asymptote. Note: The graph is the same as the graph of $y = \ln x$ shifted 1 unit to the left $(5\text{-}3)$

50.

t	N
-3	1.6
-2	4.3
-1	11
0	25
1	48
2	71
3	87
4	95
5	98

$(5\text{-}2)$

51. If the graph of $y = e^x$ is reflected in the x axis, y is replaced by $-y$ and the graph becomes the graph of $-y = e^x$ or $y = -e^x$.
If the graph of $y = e^x$ is reflected in the y axis, x is replaced by $-x$ and the graph becomes the graph of $y = e^{-x}$ or $y = \dfrac{1}{e^x}$ or $y = \left(\dfrac{1}{e}\right)^x$. $(5\text{-}3)$

52. (A) For $x > -1$, $y = e^{-x/3}$ decreases from $e^{1/3}$ to 0 while $\ln(x + 1)$ increases from $-\infty$ to ∞. Consequently, the graphs can intersect at exactly one point.

(B) Here is a computer-generated graph of $e^{-x/3}$ and $\ln(x + 1)$, $-1 < x \le 2$.

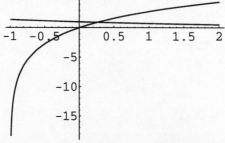

There is a point of intersection between 0.2 and 0.3. To estimate the solution to three decimal places, we zoom in on it. Here is the computer-generated graph of $e^{-x/3}$ and $\ln(x + 1)$, $0.25 \leq x \leq 0.27$.

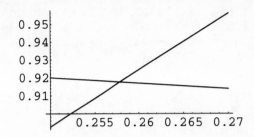

We can now estimate the solution as 0.258. *(5-5)*

53. Here is a computer-generated graph of $4 - x^2 + \ln x$, $0 < x \leq 3$.

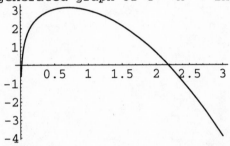

There are zeros between 0 and 0.1 and also between 2.1 and 2.2. To estimate the zeros to three decimal places, we zoom in. Here are computer-generated graphs of $4 - x^2 + \ln x$, $0.01 \leq x \leq 0.02$ and $2.18 \leq x \leq 2.19$.

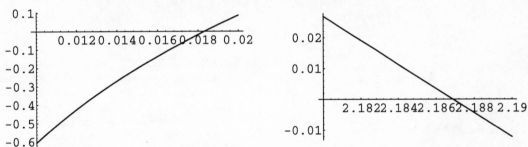

We can now estimate the zeros as 0.018 and 2.187. *(5-3)*

54. Here is a computer-generated graph of 10^{x-3} and $8 \log x$, $0 < x \leq 4$.

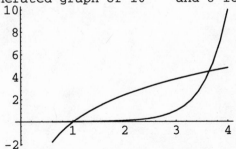

There are points of intersection just to the right of 1, and also between 3.6 and 3.8. To estimate the coordinates to three decimal places, we zoom in. Here are computer-generated graphs of 10^{x-3} and $8 \log x$, $1.00 \le x \le 1.01$ and $3.65 \le x \le 3.66$.

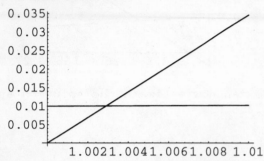 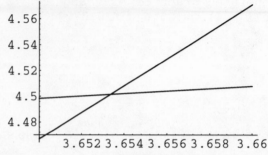

We can now estimate the coordinates of the points of intersection as $(1.003, 0.010)$ and $(3.653, 4.502)$.

$(5-4)$

55.
$$D = 10 \log \frac{I}{I_0}$$
$$\frac{D}{10} = \log \frac{I}{I_0}$$
$$10^{D/10} = \frac{I}{I_0}$$
$$I_0 10^{D/10} = I$$
$$I = I_0(10^{D/10})$$
$(5-5)$

56.
$$y = \frac{1}{\sqrt{2\pi}} e^{-x^2/2}$$
$$\sqrt{2\pi} y = e^{-x^2/2}$$
$$-\frac{x^2}{2} = \ln(\sqrt{2\pi} y)$$
$$x^2 = -2 \ln(\sqrt{2\pi} y)$$
$$x = \pm\sqrt{-2 \ln(\sqrt{2\pi} y)}$$
$(5-5)$

57.
$$x = -\frac{1}{k} \ln \frac{I}{I_0}$$
$$-kx = \ln \frac{I}{I_0}$$
$$\frac{I}{I_0} = e^{-kx}$$
$$I = I_0(e^{-kx})$$
$(5-5)$

58.
$$r = P \frac{i}{1 - (1 + i)^{-n}}$$
$$\frac{r}{P} = \frac{i}{1 - (1 + i)^{-n}}$$
$$\frac{P}{r} = \frac{1 - (1 + i)^{-n}}{i}$$
$$\frac{Pi}{r} = 1 - (1 + i)^{-n}$$
$$\frac{Pi}{r} - 1 = -(1 + i)^{-n}$$
$$1 - \frac{Pi}{r} = (1 + i)^{-n}$$
$$\ln\left(1 - \frac{Pi}{r}\right) = -n \ln(1 + i)$$
$$\frac{\ln\left(1 - \frac{Pi}{r}\right)}{-\ln(1 + i)} = n$$
$$n = -\frac{\ln\left(1 - \frac{Pi}{r}\right)}{\ln(1 + i)} \qquad (5-5)$$

59. $f(x) = 2 \ln(x - 1)$
Since the graph of f is the same as the graph of $y = \ln x$ shifted right one unit and expanded by a factor of 2, it passes the horizontal line test, and f is one-to-one.
Solve $y = f(x)$ for x
$$y = 2 \ln(x - 1)$$
$$\frac{y}{2} = \ln(x - 1)$$
$$x - 1 = e^{y/2}$$
$$x = e^{y/2} + 1 = f^{-1}(y)$$
Interchange x and y:
$y = f^{-1}(x) = e^{x/2} + 1$ Domain: $(-\infty, \infty)$
Check: $f^{-1}[f(x)] = e^{2\ln(x-1)/2} + 1$
$$= e^{\ln(x-1)} + 1$$
$$= x - 1 + 1$$
$$= x$$
$$f[f^{-1}(x)] = 2 \ln(e^{x/2} + 1 - 1)$$
$$= 2 \ln e^{x/2}$$
$$= 2\left(\frac{x}{2}\right)$$
$$= x \qquad (5-5, 3-6)$$

60. $f(x) = \dfrac{e^x - e^{-x}}{2}$ f is one-to-one (proof omitted)

Solve $y = f(x)$ for x

$$y = \frac{e^x - e^{-x}}{2}$$

$$2y = e^x - e^{-x}$$

$$2y = e^x - \frac{1}{e^x}$$

$$2ye^x = e^x e^x - e^x\left(\frac{1}{e^x}\right)$$

$$2ye^x = (e^x)^2 - 1$$

$$0 = (e^x)^2 - 2ye^x - 1$$

This equation is quadratic in e^x

$$e^x = \frac{-b \pm \sqrt{b^2 - 4ac}}{2a} \quad a = 1, \ b = -2y, \ c = -1$$

$$e^x = \frac{-(-2y) \pm \sqrt{(-2y)^2 - 4(1)(-1)}}{2(1)}$$

$$e^x = \frac{2y \pm \sqrt{4y^2 + 4}}{2}$$

$$e^x = y \pm \sqrt{y^2 + 1}$$

Note: Since $0 < 1$, $y^2 < y^2 + 1$, $\sqrt{y^2} < \sqrt{y^2 + 1}$ and $y < \sqrt{y^2 + 1}$ for all real y. Hence $y - \sqrt{y^2 + 1}$ is always negative. Also, $y + \sqrt{y^2 + 1}$ is always positive.

$x = \ln(y + \sqrt{y^2 + 1}) = f^{-1}(y)$

Interchange x and y:

$y = f^{-1}(x) = \ln(x + \sqrt{x^2 + 1})$ Domain: $(-\infty, \infty)$ (since $x + \sqrt{x^2 + 1}$ is always positive)

(Check omitted for lack of space)   *(5-5, 3-6)*

61. $\ln y = -5t + \ln c$

$\ln y - \ln c = -5t$

$$\ln\left(\frac{y}{c}\right) = -5t$$

$$\frac{y}{c} = e^{-5t}$$

$$y = ce^{-5t} \quad\quad (5\text{-}3, \ 5\text{-}5)$$

62.

x	$y = \log_2 x$	$x = \log_2 y$	y
1	0	0	1
2	1	1	2
4	2	2	4
8	3	3	8

Domain $f = (0, \infty)$ = Range f^{-1}
Range $f = (-\infty, \infty)$ = Domain f^{-1}

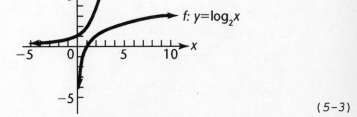

(5-3)

63. If $\log_1 x = y$, then we would have to have $1^y = x$; that is, $1 = x$ for arbitrary positive x, which is impossible.   *(5-3)*

64. Let $u = \log_b M$ and $v = \log_b N$; then $M = b^u$ and $N = b^v$. Thus,
$\log(M/N) = \log_b(b^u/b^v) = \log_b b^{u-v} = u - v = \log_b M - \log_b N.$ *(5-3)*

65. We solve $P = P_0(1.03)^t$ for t, using $P = 2P_0$.

$$2P_0 = P_0(1.03)^t$$
$$2 = (1.03)^t$$
$$\ln 2 = t \ln 1.03$$
$$\frac{\ln 2}{\ln 1.03} = t$$
$$t = 23.4 \text{ years} \qquad (5\text{-}5)$$

66. We solve $P = P_0 e^{0.03t}$ for t using $P = 2P_0$.

$$2P_0 = P_0 e^{0.03t}$$
$$2 = e^{0.03t}$$
$$\ln 2 = 0.03t$$
$$\frac{\ln 2}{0.03} = t$$
$$t = 23.1 \text{ years} \qquad (5\text{-}5)$$

67. A_0 = original amount
$0.01A_0$ = 1 percent of original amount

We solve $A = A_0 e^{-0.000124t}$ for t, using $A = 0.01A_0$.

$$0.01A_0 = A_0 e^{-0.000124t}$$
$$0.01 = e^{-0.000124t}$$
$$\ln 0.01 = -0.000124t$$
$$\frac{\ln 0.01}{-0.000124} = t$$
$$t = 37{,}100 \text{ years} \qquad (5\text{-}5)$$

68. (A) When $t = 0$, $N = 1$. As t increases by $1/2$, N doubles.
Hence $N = 1 \cdot (2)^{t+1/2}$
$$N = 2^{2t} \text{ (or } N = 4^t)$$

(B) We solve $N = 4^t$ for t, using $N = 10^9$
$$10^9 = 4^t$$
$$9 = t \log 4$$
$$t = \frac{9}{\log 4}$$
$$t = 15 \text{ days} \qquad (5\text{-}5)$$

69. We use $A = Pe^{rt}$ with $P = 1$, $r = 0.03$, and $t = 2000$.
$A = 1e^{0.03(2000)}$
$A = 1.1 \times 10^{26}$ dollars *(5-2)*

70. (A)

t	P
0	1,000
5	670
10	450
15	301
20	202
25	135
30	91

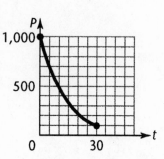

(B) As t tends to infinity, P appears to tend to 0. *(5-2)*

71. $M = \frac{2}{3} \log \frac{E}{E_0}$ $E_0 = 10^{4.40}$

We use $E = 1.99 \times 10^{14}$
$$M = \frac{2}{3} \log \frac{1.99 \times 10^{14}}{10^{4.40}}$$
$$M = \frac{2}{3} \log(1.99 \times 10^{9.6})$$
$$M = \frac{2}{3}(\log 1.99 + 9.6)$$
$$M = \frac{2}{3}(0.299 + 9.6)$$
$$M = 6.6 \qquad (5\text{-}4)$$

72. We solve $M = \frac{2}{3} \log \frac{I}{I_0}$ for I, using $I_0 = 10^{4.40}$, $M = 8.3$

$$8.3 = \frac{2}{3} \log \frac{I}{10^{4.40}}$$
$$\frac{3}{2}(8.3) = \log \frac{I}{10^{4.40}}$$
$$12.45 = \log \frac{I}{10^{4.40}}$$
$$\frac{I}{10^{4.40}} = 10^{12.45}$$
$$I = 10^{4.40} \cdot 10^{12.45}$$
$$I = 10^{16.85} \text{ or } 7.08 \times 10^{16} \text{ joules}$$
$$(5\text{-}4)$$

73. We use the given formula twice, with

$$I_2 = 100,000I_1$$

$$D_1 = 10 \log \frac{I_1}{I_0} \qquad D_2 = 10 \log \frac{I_2}{I_0}$$

$$D_2 - D_1 = 10 \log \frac{I_2}{I_0} - 10 \log \frac{I_1}{I_0}$$

$$= 10 \log \left(\frac{I_2}{I_0} \div \frac{I_1}{I_0} \right)$$

$$= 10 \log \frac{I_2}{I_1}$$

$$= 10 \log \frac{100,000I_1}{I_1}$$

$$= 10 \log 100,000$$

$$= 50 \text{ decibels}$$

The level of the louder sound is 50 decibels more.   (5-4)

74. $I = I_0 e^{-kd}$

To find k, we solve for k using

$I = \frac{1}{2}I_0$ and $d = 73.6$

$$\frac{1}{2}I_0 = I_0 e^{-k(73.6)}$$

$$\frac{1}{2} = e^{-73.6k}$$

$$-73.6k = \ln \frac{1}{2}$$

$$k = \frac{\ln \frac{1}{2}}{-73.6}$$

$$k = 0.00942$$

We now find the depth at which 1% of the surface light remains. We solve $I = I_0 e^{-0.00942d}$ for d with $I = 0.01I_0$

$$0.01I_0 = I_0 e^{-0.00942d}$$

$$0.01 = e^{-0.00942d}$$

$$-0.00942d = \ln 0.01$$

$$d = \frac{\ln 0.01}{-0.00942}$$

$$d = 489 \text{ feet} \qquad (5-2)$$

75. We solve $N = \dfrac{30}{1 + 29e^{-1.35t}}$ for t with $N = 20$.

$$20 = \frac{30}{1 + 29e^{-1.35t}}$$

$$\frac{1}{20} = \frac{1 + 29e^{-1.35t}}{30}$$

$$1.5 = 1 + 29e^{-1.35t}$$

$$0.5 = 29e^{-1.35t}$$

$$\frac{0.5}{29} = e^{-1.35t}$$

$$-1.35t = \ln \frac{0.5}{29}$$

$$t = \frac{\ln \frac{0.5}{29}}{-1.35}$$

$$t = 3 \text{ years} \qquad (5-5)$$

CUMULATIVE REVIEW EXERCISE (Chapters 3, 4 and 5)

1. (A) $d(A, B) = \sqrt{(5 - 3)^2 + (6 - 2)^2} = \sqrt{4 + 16} = \sqrt{20} = 2\sqrt{5}$

(B) $m = \dfrac{6 - 2}{5 - 3} = \dfrac{4}{2} = 2$

(C) The slope m_1 of a line perpendicular to AB must satisfy $m_1(2) = -1$.

Therefore $m_1 = -\dfrac{1}{2}$ *(3-1, 3-2)*

2. (A) Center at $(0, 0)$ and radius $\sqrt{2}$
$x^2 + y^2 = r^2$
$x^2 + y^2 = (\sqrt{2})^2$
$x^2 + y^2 = 2$

(B) Center at $(-3, 1)$ and radius $\sqrt{2}$
$(h, k) = (-3, 1)$
$r = \sqrt{2}$
$(x - h)^2 + (y - k)^2 = r^2$
$[x - (-3)]^2 + (y - 1)^2 = (\sqrt{2})^2$
$(x + 3)^2 + (y - 1)^2 = 2$ *(3-1)*

3. $2x - 3y = 6$
$-3y = -2x + 6$
$y = \dfrac{2}{3}x - 2$

slope: $\dfrac{2}{3}$ y intercept: -2 x intercept: 3 (if $y = 0$, $2x = 6$, hence $x = 3$)

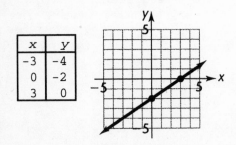

x	y
-3	-4
0	-2
3	0

(3-2)

4. (A) A function; domain = $\{1, 2, 3\}$; range = $\{1\}$
(B) Not a function (three range values correspond to the only domain value)
(C) A function; domain = $\{-2, -1, 0, 1, 2\}$; range = $\{-1, 0, 2\}$ *(3-3)*

5. (A) $f(-2) = (-2)^2 - 2(-2) + 5 = 13$
$g(3) = 3 \cdot 3 - 2 = 7$
Therefore
$f(-2) + g(3) = 13 + 7 = 20$

(B) $(f + g)(x) = f(x) + g(x)$
$= x^2 - 2x + 5 + 3x - 2$
$= x^2 + x + 3$

(C) $(f \circ g)(x) = f[g(x)] = f(3x - 2)$
$= (3x - 2)^2 - 2(3x - 2) + 5$
$= 9x^2 - 12x + 4 - 6x + 4 + 5$
$= 9x^2 - 18x + 13$

(D) $\dfrac{f(a + h) - f(a)}{h} = \dfrac{[(a + h)^2 - 2(a + h) + 5] - (a^2 - 2a + 5)}{h}$

$= \dfrac{a^2 + 2ah + h^2 - 2a - 2h + 5 - a^2 + 2a - 5}{h}$

$= \dfrac{2ah + h^2 - 2h}{h}$

$= \dfrac{h(2a + h - 2)}{h}$

$= 2a + h - 2$ *(3-2, 3-5)*

6. (A) Since the graph has x intercepts -1, 1, and 2, and -1 is at least a double zero (the graph is tangent to the x axis at $x = -1$), the lowest degree equation would be $P(x) = (x + 1)^2(x - 1)(x - 2)$.

(B) $P(x) \to \infty$ as $x \to \infty$ and as $x \to -\infty$. *(4-1)*

7. (A) Expanded by a factor of 2 (B) Shifted right 2 units
(C) Shifted down 2 units *(3-5)*

8. (A) The graph of $f(x)$ is shifted left one unit and reflected across the x axis.

(B) The graph of $f(x)$ is stretched by a factor of two with respect to the y axis and shifted down two units.

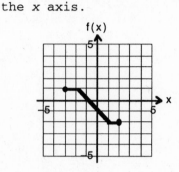

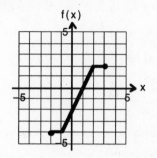

(3-5)

9.
$$
\begin{array}{r|rrrr}
 & 3 & 5 & -18 & -3 \\
 & & -9 & 12 & 18 \\
\hline
-3 & 3 & -4 & -6 & 15
\end{array}
$$

$3x^3 + 5x^2 - 18x - 3 = (x + 3)(3x^2 - 4x - 6) + 15$ *(4-1)*

10. -2, 3, 5   *(4-2)*

11. We investigate $P(1)$ and $P(2)$ by forming a synthetic division table.

	4	-5	-3	-1
1	4	-1	-4	-5
2	4	3	3	5

Since $P(1)$ and $P(2)$ have opposite signs, there is at least one real zero between 1 and 2. *(4-3)*

12. The possible rational zeros are ±1, ±2, ±4, ±8. We form a synthetic division table.

	1	1	-10	8	
1	1	2	-8	0	1 is a zero

Thus $x^3 + x^2 - 10x + 8 = (x - 1)(x^2 + 2x - 8) = (x - 1)(x - 2)(x + 4)$.
The rational zeros are 1, 2, -4. *(4-2)*

13. We write
$$\frac{5x - 4}{(x - 2)(x + 1)} = \frac{A}{x - 2} + \frac{B}{(x + 1)} = \frac{A(x + 1) + B(x - 2)}{(x - 2)(x + 1)}$$

Thus for all x
$5x - 4 = A(x + 1) + B(x - 2)$
If $x = -1$
$-9 = -3B$
$B = 3$
If $x = 2$
$6 = 3A$
$A = 2$

So $\dfrac{5x - 4}{(x - 2)(x + 1)} = \dfrac{2}{x - 2} + \dfrac{3}{x + 1}$ *(4-5)*

$= \log y$ (B) $x = e^y$ (5-4)

$2^3(e^x)^3 = 8e^{3x}$ (B) $\dfrac{e^{3x}}{e^{-2x}} = e^{3x-(-2x)} = e^{5x}$ (5-2)

16
$$\log_3 x = 2$$
$$x = 3^2$$
$$x = 9$$

     (B) $\log_3 81 = x$
               $\log_3 3^4 = x$
                  $x = 4$

(C) $\log_x 4 = -2$
$$x^{-2} = 4$$
$$\frac{1}{x^2} = 4$$
$$1 = 4x^2$$
$$x^2 = \frac{1}{4}$$
$$x = \frac{1}{2}$$

since bases are restricted positive   (5-3)

17. (A) $10^x = 2.35$
$$x = \log 2.35$$
$$x = 0.371$$

     (B) $e^x = 87{,}500$
$$x = \ln 87{,}500$$
$$x = 11.4$$

(C) $\log x = -1.25$
$$x = 10^{-1.25}$$
$$x = 0.0562$$

     (D) $\ln x = 2.75$
$$x = e^{2.75}$$
$$x = 15.6$$ (5-4)

18. (A) All real numbers $(-\infty, \infty)$

(B) From the graph, the possible function values include -2 (only) and all numbers greater than or equal to 1. In set and interval notation: $\{-2\} \cup [1, \infty)$.

(C) $f(-3) = 1$ $f(-2) = 2$ (not -2) $f(2) = -2$ (not 2).
Thus, $f(-3) + f(-2) + f(2) = 1 + 2 + (-2) = 1$

(D) $[-3, -2]$ and $[2, \infty)$

(E) f is discontinuous at $x = -2$ and at $x = 2$.   (3-3, 3-4)

19. The line $3x + 2y = 12$, or
$$2y = -3x + 12,$$
or $y = -\dfrac{3}{2}x + 6$, has slope $-\dfrac{3}{2}$.

(A) We require a line through $(-6, 1)$ with slope $-\dfrac{3}{2}$. Applying the point-slope form, we have
$$y - 1 = -\frac{3}{2}(x + 6)$$
$$y - 1 = -\frac{3}{2}x - 9$$
$$y = -\frac{3}{2}x - 8$$

(B) We require a line with slope m satisfying $-\dfrac{3}{2}m = -1$, or $m = \dfrac{2}{3}$. Again applying the point-slope form, we have
$$y - 1 = \frac{2}{3}(x + 6)$$
$$y - 1 = \frac{2}{3}x + 4$$
$$y = \frac{2}{3}x + 5$$ (3-2)

20. $x + 4 \geq 0$ is equivalent to $x \geq -4$. Domain of $g(x) = \sqrt{x + 4}$ is $[-4, \infty)$ *(3-3)*

21. $f(x) = x^2 - 2x - 8$. Complete the square.
$f(x) = (x^2 - 2x + 1) - 1 - 8$
$\quad = (x - 1)^2 - 9$

Graph: Locate axis and vertex, then plot several points on either side of the axis.

Comparing with $f(x) = a(x - h)^2 + k$, $h = 1$ and $k = -9$. Thus, the vertex is $(1, -9)$, the axis of symmetry is $x = 1$, and the minimum value is -9. $y = f(x)$ can be any number greater than or equal to -9, so the range is $[-9, \infty)$.
y intercept: Set $x = 0$, then $f(0) = -8$ is the y intercept.
x intercepts: Set $f(x) = 0$, then
$\quad 0 = x^2 - 2x - 8$
$\quad 0 = (x - 4)(x + 2)$
$\quad x = 4$ or $x = -2$ are the x intercepts.

x	$f(x)$
-3	7
-2	0
-1	-5
0	-8
1	-9
2	-8
3	-5
4	0
5	7

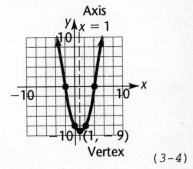

(3-4)

22. $(f \circ g)(x) = f[g(x)] = f\left(\dfrac{x + 3}{x}\right) = \dfrac{1}{\frac{x + 3}{x} - 2} = \dfrac{x(1)}{x(\frac{x + 3}{x} - 2)} = \dfrac{x}{x + 3 - 2x} = \dfrac{x}{3 - x}$

The domain of f is all real numbers except 2.
The domain of g is all non-zero real numbers.
The domain of $f \circ g$ is the set of all non-zero real numbers for which $g(x) \neq 2$.

$g(x) = 2$ only if $\dfrac{x + 3}{x} = 2$, that is $x + 3 = 2x$, or $x = 3$.

Thus the domain of $f \circ g$ is the set of all non-zero real numbers except 3, that is, $(-\infty, 0) \cup (0, 3) \cup (3, \infty)$.

> Common Error:
> The domain of $f \circ g$ cannot be found by looking at the final form $\dfrac{x}{3 - x}$. It is not $\{x \mid x \neq 3\}$.

(3-5)

23. $f(x) = 2x + 5$
Assume $f(a) = f(b)$
$\quad 2a + 5 = 2b + 5$
$\quad\quad 2a = 2b$
$\quad\quad\ a = b$
Thus, f is one-to-one.
Solve $y = f(x)$ for x:
$\quad y = 2x + 5$
$\quad y - 5 = 2x$
$\quad x = \dfrac{y - 5}{2} = f^{-1}(y)$

Interchange x and y
$y = f^{-1}(x) = \dfrac{x - 5}{2}$ or $\dfrac{1}{2}x - \dfrac{5}{2}$ Domain: $(-\infty, \infty)$ *(3-6)*

24. $f(x) = 3 \ln x - \sqrt{x}$ *(3-3, 5-3)*

25. The function f multiplies the base e raised to the power of one-half of the domain element by 100 and then subtracts 50. *(3-3, 5-2)*

26.

x	y = x - 1		x	y = x² + 1
-1	-2		0	1
-2	-3		1	2
			2	5

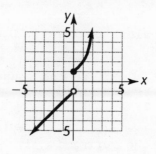

Domain: all real numbers
Range: $(-\infty, -1) \cup [1, \infty)$
Discontinuous at: $x = 0$

(3-4)

27. (A) This is the same as the graph of $y = \sqrt{x}$ expanded by a factor of 2 and shifted up one unit.

(B) This is the same as the graph of $y = \sqrt{x}$ shifted left 1 unit and reflected with respect to the x axis.

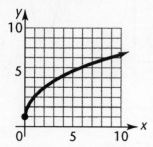

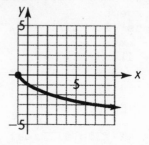

(3-5)

28. The graph of $y = |x|$ is contracted by $\frac{1}{2}$, reflected in the x axis, shifted two units to the right and three units up; $y = -\frac{1}{2}|x - 2| + 3$. (3-5)

29. $f(x) = \dfrac{2x + 8}{x + 2} = \dfrac{n(x)}{d(x)}$

(A) The domain of f is the set of all real numbers x such that $d(x) = x + 2 \neq 0$, that is $(-\infty, -2) \cup (-2, \infty)$ or $x \neq -2$. f has an x intercept where $n(x) = 2x + 8 = 0$, that is, $x = -4$. $f(0) = 4$, hence f has a y intercept at $y = 4$.

(B) *Vertical asymptote:* $x = -2$
Horizontal asymptote: Since $n(x)$ and $d(x)$ have the same degree, the line $y = 2$ is a horizontal asymptote.

(C) *Sign Chart:*

Test numbers	-5	-3	0
Value of f	$\frac{2}{3}$	-2	4
Sign of f	+	-	+

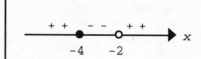

Since $x = -2$ is a vertical asymptote and $f(x) < 0$ for $-4 < x < -2$, and $f(x) > 0$ for $x > -2$, $f(x) \to -\infty$ as $x \to -2^-$ and $f(x) \to \infty$ as $x \to -2^+$.

Complete the sketch:

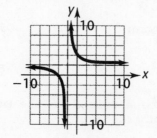

(4-4)

30. $f(x) = \sqrt{x + 4}$

Assume $f(a) = f(b)$
$$\sqrt{a + 4} = \sqrt{b + 4}$$
$$a + 4 = b + 4$$
$$a = b$$

Thus f is one-to-one.

(A) Solve $y = f(x)$ for x

$y = \sqrt{x + 4}$
$y^2 = x + 4 \quad y \geq 0$
$x = y^2 - 4 \quad y \geq 0 \quad f^{-1}(y) = y^2 - 4$

Interchange x and y:
$y = f^{-1}(x) = x^2 - 4 \quad$ Domain: $x \geq 0$

Check: $f^{-1}[f(x)] = (\sqrt{x + 4})^2 - 4 = x + 4 - 4 = x$
$f[f^{-1}(x)] = \sqrt{x^2 - 4 + 4} = \sqrt{x^2} = x \quad$ since $x \geq 0$ in the domain of f^{-1}.

(B) Domain of $f = [-4, \infty) =$ Range of f^{-1}
Range of $f = [0, \infty) =$ Domain of f^{-1}

(C)

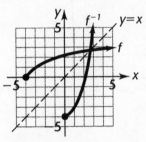

$(3-6)$

31.
$$x^2 - 6x + y^2 + 2y = 0$$
$$(x^2 - 6x + ?) + (y^2 + 2y + ?) = 0$$
$$(x^2 - 6x + 9) + (y^2 + 2y + 1) = 9 + 1$$
$$(x - 3)^2 + (y + 1)^2 = 10$$
$$(x - 3)^2 + [y - (-1)]^2 = (\sqrt{10})^2$$

Center: $C(h, k) = (3, -1) \quad$ Radius $r = \sqrt{10}$

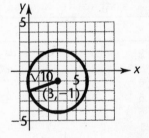

$(3-1)$

32. $xy + |xy| = 5 \quad y$ axis symmetry? $(-x)y + |(-x)y| = 5 \qquad -xy + |xy| = 5$
No, not equivalent

x axis symmetry? $x(-y) + |x(-y)| = 5 \qquad -xy + |xy| = 5$
No, not equivalent

origin symmetry? $(-x)(-y) + |(-x)(-y)| = 5 \quad xy + |xy| = 5$
Yes, equivalent

The graph has symmetry with respect to the origin. $(3-1)$

33. Since the graph is a parabola opening up, $a = +1$. Since the vertex is at $(-2, -3) = (h, k)$, $h = -2$ and $k = -3$. Thus, the equation is $y = (x + 2)^2 - 3$.
$(3-4)$

34.

	2	-5	3	2
		1	-2	$\frac{1}{2}$
$\frac{1}{2}$	2	-4	1	$\frac{5}{2}$

$P\left(\dfrac{1}{2}\right) = \dfrac{5}{2} \qquad (4-2)$

35. $x - 1$ will be a factor of $P(x)$ if $P(1) = 0$
$P(1) = 1^{25} - 1^{20} + 1^{15} + 1^{10} - 1^5 + 1$
$= 1 - 1 + 1 + 1 - 1 + 1 = 2 \neq 0$,
so $x - 1$ is not a factor. $x + 1$ will be a factor of $P(x)$ if $P(-1) = 0$
$P(-1) = (-1)^{25} - (-1)^{20} + (-1)^{15} + (-1)^{10} - (-1)^5 + 1$
$= -1 - 1 - 1 + 1 + 1 + 1 = 0$,
so $x + 1$ is a factor of $P(x)$.
$(4-2)$

36. (A) We form a synthetic division table:

	1	0	-8	0	3
-3	1	-3	1	-3	12 = $P(-3)$
-2	1	-2	-4	8	-13 = $P(-2)$
-1	1	-1	-7	7	-4 = $P(-1)$
0	1	0	-8	0	3 = $P(0)$
1	1	1	-7	-7	-4 = $P(1)$
2	1	2	-4	-8	-13 = $P(2)$
3	1	3	1	3	12 = $P(3)$

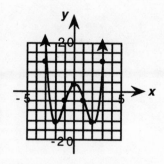

The graph of $P(x)$ has four x intercepts and three turning points; $P(x) \to \infty$ as $x \to \infty$ and as $x \to -\infty$

(B) There are real zeros in the intervals (-3, -2), (-1, 0), (0, 1), and (2, 3) indicated in the table. We search for the real zero in (2, 3). We organize our calculations in a table.

Sign Change Interval (a, b)	Midpoint m	Sign of P $P(a)$	$P(m)$	$P(b)$
(2, 3)	2.5	-	-	+
(2.5, 3)	2.75	-	-	+
(2.75, 3)	2.875	-	+	+
(2.75, 2.875)	2.8125	-	+	+
(2.75, 2.8125)	We stop here	-		+

Since each endpoint rounds to 2.8, a real zero lies on this last interval and is given by 2.8 to one decimal place accuracy.   (4-1, 4-3)

37. (A) We form a synthetic division table:

	1	2	-20	0	-30
0	1	2	-20	0	-30
1	1	3	-17	-17	-47
2	1	4	-12	-24	-78
3	1	5	-5	-15	-75
4	1	6	4	16	34
-1	1	1	-21	21	-51
-2	1	0	-20	40	-110
-3	1	-1	-17	51	-183
-4	1	-2	-12	48	-222
-5	1	-3	-5	25	-155
-6	1	-4	4	-24	114

From the table, 4 is an upper bound and -6 is a lower bound.

(B) There are real zeros in the intervals (-6, -5) and (3, 4) indicated in the table. We search for the real zero in (3, 4). We organize our calculations in a table.

Sign Change Interval (a, b)	Midpoint m	Sign of P $P(a)$	$P(m)$	$P(b)$
(3, 4)	3.5	-	-	+
(3.5, 4)	3.75	-	-	+
(3.75, 4)	3.875	-	+	+
(3.75, 3.875)	3.8125	-	+	+
(3.75, 3.8125)	3.78125	-	-	+
(3.78125, 3.8125)	3.796875	-	-	+
(3.796875, 3.8125)	3.8046875	-	+	+
(3.796875, 3.8046875)	We stop here	-		+

Since each endpoint rounds to 3.80, a real zero lies on this last interval and is given by 3.80 to two decimal place accuracy.

(C) Here is a computer-generated graph of *P*.

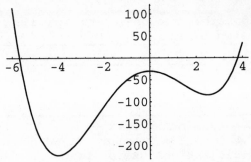

The *x* intercepts are on the invervals (-6, -5.6) and (3.6, 4). To estimate more accurately we zoom in. Here are graphs of *P* on the intervals (-5.7, -5.6) and (3.7, 3.9).

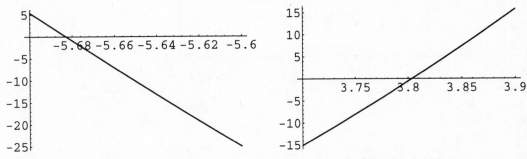

To two decimal place accuracy, the zeros are -5.68 and 3.80.   (4-3)

38. The possible rational zeros are ±1, ±3, ±5, ±15, $\pm\frac{1}{2}$, $\pm\frac{3}{2}$, $\pm\frac{5}{2}$, $\pm\frac{15}{2}$, $\pm\frac{1}{4}$, $\pm\frac{3}{4}$,

$\pm\frac{5}{4}$, $\pm\frac{15}{4}$. There are no negative real zeros. We form a synthetic division table.

	4	-20	29	-15	
1	4	-16	13	-2	
3	4	-8	5	0	3 is a zero

So $P(x) = (x - 3)(4x^2 - 8x + 5)$

To find the remaining zeros, we solve $4x^2 - 8x + 5 = 0$ by the quadratic formula.

$4x^2 - 8x + 5 = 0$

$$x = \frac{-b \pm \sqrt{b^2 - 4ac}}{2a} \quad a = 4, \ b = -8, \ c = 5$$

$$x = \frac{-(-8) \pm \sqrt{(-8)^2 - 4(4)(5)}}{2(4)}$$

$$x = \frac{8 \pm \sqrt{64 - 80}}{8}$$

$$x = \frac{8 \pm \sqrt{-16}}{8}$$

$$x = \frac{8 \pm 4i}{8}$$

$$x = 1 \pm \frac{1}{2}i$$

The zeros are 3, $1 \pm \frac{1}{2}i$ (4-2)

39. The possible rational zeros are ±1, ±2, ±3, ±4, ±6, ±12. We form a synthetic division table.

	1	5	1	-15	-12	
1	1	6	7	-8	-20	
2	1	7	15	15	18	a zero between 1 and 2; 2 is an upperbound
-1	1	4	-3	-12	0	-1 is a zero

We now examine $x^3 + 4x^2 - 3x - 12$. This factors by grouping into $(x + 4)(x^2 - 3)$, however, if we don't notice this, we find the remaining possible rational zeros to be -1, -2, -3, -4, -6, -12. We form a synthetic division table.

	1	4	-3	-12	
-1	1	3	-6	-6	not a double zero
-2	1	2	-7	2	a zero between -1 and -2
-3	1	1	-6	6	
-4	1	0	-3	0	-4 is a zero

So $P(x) = (x + 1)(x + 4)(x^2 - 3) = (x + 1)(x + 4)(x - \sqrt{3})(x + \sqrt{3})$.

The four zeros are -1, -4, $\pm\sqrt{3}$.

(4-2)

40.
$$x^3 + 36 \leq 7x^2$$
$$x^3 - 7x^2 + 36 \leq 0$$
To factor $x^3 - 7x^2 + 36$, we search for zeros of the polynomial. Possible rational zeros are ±1, ±2, ±3, ±4, ±6, ±9, ±12, ±18, ±36. We form a synthetic division table.

	1	-7	0	36	
1	1	-6	-6	30	
2	1	-5	-10	16	
3	1	-4	-12	0	3 is a zero

$$x^3 - 7x^2 + 36 = (x - 3)(x^2 - 4x - 12) = (x - 3)(x - 6)(x + 2).$$

Hence we must examine the sign behavior of $(x - 3)(x - 6)(x + 2)$.

We form a sign chart.
Zeros: 3, 6, -2

$x^3 - 7x^2 + 36 = (x - 3)(x - 6)(x + 2)$				
Test Number	-3	0	4	7
Value of Polynomial for Test Number	-54	36	-12	36
Sign of Polynomial in Interval	−	+	−	+
Interval	$(-\infty, -2)$	$(-2, 3)$	$(3, 6)$	$(6, \infty)$

$x^3 - 7x^2 + 36 \leq 0$ and $x^3 + 36 \leq 7x^2$ within the intervals $(-\infty, -2]$ and $[3, 6]$ or $x \leq -2$ or $3 \leq x \leq 6$.

(4-2, 2-8)

41. $\dfrac{3x^2 - x + 1}{x(x + 1)^2} = \dfrac{A}{x} + \dfrac{B}{x + 1} + \dfrac{C}{(x + 1)^2} = \dfrac{A(x + 1)^2 + Bx(x + 1) + Cx}{x(x + 1)^2}$

Thus for all x
$$3x^2 - x + 1 = A(x + 1)^2 + Bx(x + 1) + Cx$$

If $x = 0$ If $x = -1$ If $x = 1$, using $A = 1$ and $C = -5$
  $1 = A$ $5 = C(-1)$ $3 = 4 + 2B - 5$
                  $C = -5$ $B = 2$

So $\dfrac{3x^2 - x + 1}{x(x + 1)^2} = \dfrac{1}{x} + \dfrac{2}{x + 1} - \dfrac{5}{(x + 1)^2}$

(4-5)

42. First we factor $x^3 - x^2 + x = x(x^2 - x + 1)$. $x^2 - x + 1$ cannot be factored further in the real numbers, so we write

$$\frac{x^2 + x - 2}{x^3 - x^2 + x} = \frac{A}{x} + \frac{Bx + C}{x^2 - x + 1} = \frac{A(x^2 - x + 1) + (Bx + C)x}{x(x^2 - x + 1)}$$

Thus for all x
$$x^2 + x - 2 = A(x^2 - x + 1) + (Bx + C)x$$

If $x = 0$
$$-2 = A$$
$$x^2 + x - 2 = -2(x^2 - x + 1) + Bx^2 + Cx$$
$$= (B - 2)x^2 + (C + 2)x - 2$$

Equating coefficients of like terms, we have
$$1 = B - 2$$
$$1 = C + 2$$
$$B = 3$$
$$C = -1$$

So $\dfrac{x^2 + x - 2}{x^3 - x^2 + x} = \dfrac{-2}{x} + \dfrac{3x - 1}{x^2 - x + 1}$ $\hspace{2cm}$ (4-5)

43.
$$2^{x^2} = 4^{x+4}$$
$$2^{x^2} = (2^2)^{x+4}$$
$$2^{x^2} = 2^{2(x+4)}$$
$$x^2 = 2(x + 4)$$
$$x^2 = 2x + 8$$
$$x^2 - 2x - 8 = 0$$
$$(x - 4)(x + 2) = 0$$
$$x - 4 = 0 \qquad x + 2 = 0$$
$$x = 4 \qquad x = -2 \qquad (5\text{-}1)$$

44.
$$2x^2 e^{-x} + xe^{-x} = e^{-x}$$
$$2x^2 e^{-x} + xe^{-x} - e^{-x} = 0$$
$$e^{-x}(2x^2 + x - 1) = 0$$
$$e^{-x}(2x - 1)(x + 1) = 0$$
$$e^{-x} = 0 \quad 2x - 1 = 0 \quad x + 1 = 0$$
$$\text{never} \qquad x = \frac{1}{2} \qquad x = -1$$

Solutions: $\frac{1}{2}$, -1 $\hspace{1.5cm}$ (5-2)

45. $e^{\ln x} = 2.5$
$\phantom{e^{\ln x}} x = 2.5$ $\hspace{1cm}$ (5-3)

46. $\log_x 10^4 = 4$
$ x^4 = 10^4$
$ x = 10$ $\hspace{1cm}$ (5-3)

47. $\log_9 x = -\dfrac{3}{2}$
$ 9^{-3/2} = x$
$ x = \dfrac{1}{27}$ (5-3)

48. $\ln(x + 4) - \ln(x - 4) = 2 \ln 3$

$\ln \dfrac{x + 4}{x - 4} = \ln 3^2$

$\dfrac{x + 4}{x - 4} = 3^2$

$\dfrac{x + 4}{x - 4} = 9 \qquad x \neq 4$

$x + 4 = 9(x - 4)$
$x + 4 = 9x - 36$
$-8x = -40$
$x = 5$

Check: $\ln(5 + 4) - \ln(5 - 4) \overset{?}{=} 2 \ln 3$
$\ln 9 - \ln 1 \overset{?}{=} 2 \ln 3$
$\ln 9 - 0 \overset{\checkmark}{=} \ln 9$

Solution: $x = 5$ $\hspace{1.5cm}$ (5-5)

49. $\ln(2x^2 + 2) = 2 \ln(2x - 4)$
$\ln(2x^2 + 2) = \ln(2x - 4)^2$
$2x^2 + 2 = (2x - 4)^2$
$2x^2 + 2 = 4x^2 - 16x + 16$
$0 = 2x^2 - 16x + 14$
$0 = 2(x - 7)(x - 1)$
$x = 7, 1$

Check: $x = 7$

$\ln(2 \cdot 7^2 + 2) \overset{?}{=} 2 \ln(2 \cdot 7 - 4)$

$\ln 100 \overset{?}{=} 2 \ln 10$

$\ln 100 \overset{\surd}{=} \ln 100$

$x = 1$

$\ln(2 \cdot 1^2 + 2) \overset{?}{=} 2 \ln(2 \cdot 1 - 4)$

$\ln(4) \neq 2 \ln(-2)$

Solution: $x = 7$ *(5-5)*

50. $\log x + \log(x + 15) = 2$
$\log[x(x + 15)] = 2$
$x(x + 15) = 10^2$
$x^2 + 15x = 100$
$x^2 + 15x - 100 = 0$
$(x - 5)(x + 20) = 0$
$x = 5, -20$

Check: $x = 5$

$\log 5 + \log(5 + 15) \overset{?}{=} 2$

$\log 5 + \log 20 \overset{?}{=} 2$

$\log 100 \overset{\surd}{=} 2$

$x = -20$

$\log(-20) + \log(-20 + 15) \neq 2$

Solution: $x = 5$ *(5-5)*

51. $\log(\ln x) = -1$
$\ln x = 10^{-1}$
$\ln x = 0.1$
$x = e^{0.1}$ *(5-4)*

52.
$4(\ln x)^2 = \ln x^2$
$4(\ln x)^2 = 2 \ln x$
$4(\ln x)^2 - 2 \ln x = 0$
$2 \ln x(2 \ln x - 1) = 0$
$2 \ln x = 0 \qquad 2 \ln x - 1 = 0$
$\ln x = 0 \qquad\qquad \ln x = \frac{1}{2}$
$x = 1 \qquad\qquad x = e^{0.5}$

Check: $x = 1$

$4(\ln 1)^2 \overset{?}{=} \ln 1^2$

$0 \overset{\surd}{=} 0$

$x = e^{0.5}$

$4(\ln e^{0.5})^2 \overset{?}{=} \ln(e^{0.5})^2$

$4(0.5)^2 \overset{?}{=} \ln(e^{2(0.5)})$

$4(0.25) \overset{?}{=} \ln e$

$1 \overset{\surd}{=} 1$

Solution: $1, e^{0.5}$ *(5-5)*

53. $x = \log_3 41$
We use the change of
base formula
$x = \dfrac{\log 41}{\log 3}$
$x = 3.38$ *(5-5)*

54. $\ln x = 1.45$
$x = e^{1.45}$
$x = 4.26$ *(5-4)*

55. $4(2^x) = 20$
$2^x = 5$
$x \log 2 = \log 5$
$x = \dfrac{\log 5}{\log 2}$
$x = 2.32$ *(5-4)*

56.
$$10e^{-0.5x} = 1.6$$
$$e^{-0.5x} = 0.16$$
$$-0.5x = \ln 0.16$$
$$x = \frac{\ln 0.16}{-0.5}$$
$$x = 3.67 \qquad (5\text{-}5)$$

57.
$$\frac{e^x - e^{-x}}{e^x + e^{-x}} = \frac{1}{2}$$
$$\frac{e^x - \frac{1}{e^x}}{e^x + \frac{1}{e^x}} = \frac{1}{2}$$
$$\frac{e^x\left(e^x - \frac{1}{e^x}\right)}{e^x\left(e^x + \frac{1}{e^x}\right)} = \frac{1}{2}$$
$$\frac{(e^x)^2 - 1}{(e^x)^2 + 1} = \frac{1}{2}$$
$$\frac{e^{2x} - 1}{e^{2x} + 1} = \frac{1}{2}$$
$$2(e^{2x} + 1)\frac{e^{2x} - 1}{e^{2x} + 1} = 2(e^{2x} + 1)\frac{1}{2}$$
$$2(e^{2x} - 1) = e^{2x} + 1$$
$$2e^{2x} - 2 = e^{2x} + 1$$
$$e^{2x} = 3$$
$$2x = \ln 3$$
$$x = \frac{1}{2}\ln 3$$
$$x = 0.549 \qquad (5\text{-}5)$$

58.

x	y
-1	9
0	3
1	1
2	0.33
3	0.11

(5-1)

59.

x	f(x)
1	0
0	0.69
-1	1.10
-2	1.39
-3	1.61
-4	1.79
-5	1.95

(5-4)

60.

t	A(t)
-2	182
0	100
2	55
4	30
6	17
8	9
10	5

(5-2)

61. The graph is the same as the graph of $y = \sqrt{x}$ shifted 1 unit to the left, expanded by a factor of 2, reflected with respect to the x axis, and shifted up 3 units.

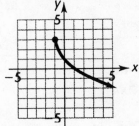

(3-5)

62. The graph is the same as the graph of $y = e^{-x}$ expanded by a factor of 2, reflected with respect to the x axis, and shifted up 3 units. The line $y = 3$ is a horizontal asymptote.

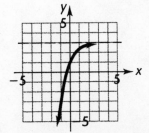

(5-2)

63. If the graph of $y = \ln x$ is reflected in the x axis, y is replaced by $-y$ and the graph becomes the graph of $-y = \ln x$ or $y = -\ln x$.
If the graph of $y = \ln x$ is reflected in the y axis, x is replaced by $-x$ and the graph becomes the graph of $y = \ln(-x)$. *(3-5, 5-3)*

64. (A) For $x > 0$, $y = e^{-x}$ decreases from 1 to 0 while $\ln x$ increases from $-\infty$ to ∞. Consequently, the graphs can intersect at exactly one point.

(B) Here is a computer-generated graph of e^{-x} and $\ln x$, $0 < x \le 4$.

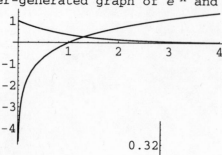

There is a point of intersection between 1.2 and 1.4. To estimate the solution to two decimal places, we zoom in on it. Here is the computer-generated graph of e^{-x} and $\ln x$, $1.3 \le x \le 1.4$.

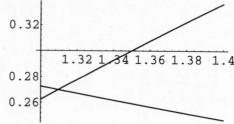

To two decimal places, the solution is 1.31. *(5-3)*

65. $f(x) = |x + 2| + |x - 2|$

If $x < -2$, then $|x + 2| = -(x + 2)$ and $|x - 2| = -(x - 2)$, hence
$$f(x) = -(x + 2) + -(x - 2)$$
$$= -x - 2 - x + 2$$
$$= -2x$$

If $-2 \le x \le 2$ then $|x + 2| = x + 2$ but $|x - 2| = -(x - 2)$, hence
$$f(x) = x + 2 + -(x - 2)$$
$$= x + 2 - x + 2$$
$$= 4$$

If $x > 2$, then $|x + 2| = x + 2$ and $|x - 2| = x - 2$, hence
$$f(x) = x + 2 + x - 2$$
$$= 2x$$

Domain: $(-\infty, \infty)$

x	$f(x) = -2x$	x	$f(x) = 4$	x	$f(x) = 2x$
-5	10	-2	4	3	6
-3	6	0	4	5	10
		2	4		

Piecewise definition for f:
$$f(x) = \begin{cases} -2x & \text{if } x < -2 \\ 4 & \text{if } -2 \le x \le 2 \\ 2x & \text{if } x > 2 \end{cases}$$

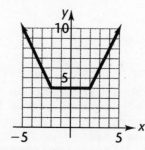

Range: $[4, \infty)$ *(3-4)*

66. (A) $g(x) = \sqrt{4 - x^2}$

The domain of g is those values of x for which $4 - x^2 \geq 0$. Solving by the methods of Section 2-8, we obtain:

Thus, the domain of g is $-2 \leq x \leq 2$, or $[-2, 2]$

(B) $\left(\dfrac{f}{g}\right)(x) = \dfrac{f(x)}{g(x)} = \dfrac{x^2}{\sqrt{4 - x^2}}$. The domain of $\dfrac{f}{g}$ is the intersection of the domains

of f (all real numbers) and g ($[-2, 2]$) with the exclusion of those points (-2 and 2) where $g(x) = 0$. Thus, domain of $f/g = (-2, 2)$.

(C) $(f \circ g)(x) = f[g(x)] = f(\sqrt{4 - x^2}) = (\sqrt{4 - x^2})^2 = 4 - x^2$.

The domain of $f \circ g$ is the set of all real numbers in the domain of g ($[-2, 2]$) for which $f(x)$ is real, that is, all numbers in $[-2, 2]$.   (3-5)

67. (A) $f(x) = x^2 - 2x - 3 \quad x \geq 1$

f passes the horizontal line test (see graph below, in part C), hence f is one-to-one.

Solve $y = f(x)$ for x:
$$y = x^2 - 2x - 3 \quad x \geq 1$$
$$x^2 - 2x = y + 3 \quad x \geq 1$$
$$x^2 - 2x + 1 = y + 4 \quad x \geq 1$$
$$(x - 1)^2 = y + 4 \quad x \geq 1$$
$$x - 1 = \underbrace{\sqrt{y + 4}} \quad x \geq 1$$

positive square root only because $x \geq 1$

$x = 1 + \sqrt{y + 4} = f^{-1}(y)$

Interchange x and y:

$y = f^{-1}(x) = 1 + \sqrt{x + 4}$ Domain: $x \geq -4$ Check omitted for lack of space.

(B) Domain of $f^{-1} = [-4, \infty)$
    Range of f^{-1} = Domain of $f = [1, \infty)$.

(C)

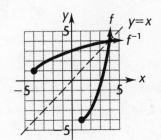

(3-6)

68. $f(x) = \dfrac{x^2 + 4x + 8}{x + 2} = \dfrac{n(x)}{d(x)}$

Intercepts. There are no real zeros of $n(x) = x^2 + 4x + 8$. No x intercept

$\qquad\qquad f(0) = 4$ y intercept

Vertical asymptotes. Real zeros of $d(x) = x + 2$. $x = -2$

Sign Chart.

Test numbers	-3	0
Value of f	-5	4
Sign of f	–	+

Since $x = -2$ is a vertical asymptote and $f(x) < 0$ for $x < -2$ and $f(x) > 0$ for $x > -2$,

$\qquad f(x) \to -\infty$ as $x \to -2^-$ and $f(x) \to \infty$ as $x \to -2^+$

Horizontal asymptote. Since the degree of $n(x)$ is greater than the degree of $d(x)$, there is no horizontal asymptote.

Oblique asymptote.

$$
\begin{array}{r}
x + 2 \\
x + 2 \overline{\smash{)}\; x^2 + 4x + 8} \\
\underline{x^2 + 2x} \\
2x + 8 \\
\underline{2x + 4} \\
4
\end{array}
$$

Thus, $f(x) = x + 2 + \dfrac{4}{x + 2}$

Hence, the line $y = x + 2$ is an oblique asymptote.

Complete the sketch.

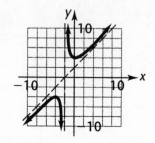

$(4-4)$

69. In problems 51 and 53, Section 3-4, we noted that $[\![x/2]\!]$ jumps at intervals of 2 units, and $[\![3x]\!] = [\![x \div \frac{1}{3}]\!]$ jumps at intervals of $\frac{1}{3}$ unit. We expect $[\![2x]\!]$ to jump at intervals of $\frac{1}{2}$ unit.

$f(x) = 2x - [\![2x]\!]$

$f(0) = 2 \cdot 0 - [\![2 \cdot 0]\!] = 0 - [\![0]\!] = 0 - 0 = 0$

$f(\frac{1}{4}) = 2 \cdot \frac{1}{4} - [\![2 \cdot \frac{1}{4}]\!] = \frac{1}{2} - [\![\frac{1}{2}]\!] = \frac{1}{2} - 0 = \frac{1}{2}$

$f(\frac{1}{2}) = 2 \cdot \frac{1}{2} - [\![2 \cdot \frac{1}{2}]\!] = 1 - [\![1]\!] = 1 - 1 = 0$

$f(\frac{3}{4}) = 2 \cdot \frac{3}{4} - [\![2 \cdot \frac{3}{4}]\!] = \frac{3}{2} - [\![\frac{3}{2}]\!] = \frac{3}{2} - 1 = \frac{1}{2}$

$f(1) = 2 \cdot 1 - [\![2 \cdot 1]\!] = 2 - [\![2]\!] = 2 - 2 = 0$

Thus, $f(x) = 2x$ if $0 \le x \le \frac{1}{2}$, $f(x) = 2x - 1$ if $\frac{1}{2} \le x < 1$

Generalizing, we can write: $f(x) = \begin{cases} \vdots & & \vdots \\ 2x + 2 & \text{if} & -1 \le x < -\frac{1}{2} \\ 2x + 1 & \text{if} & -\frac{1}{2} \le x < 0 \\ 2x & \text{if} & 0 \le x < \frac{1}{2} \\ 2x - 1 & \text{if} & \frac{1}{2} \le x < 1 \\ 2x - 2 & \text{if} & 1 \le x < \frac{3}{2} \\ 2x - 3 & \text{if} & \frac{3}{2} \le x < 2 \\ \vdots & & \vdots \end{cases}$

Domain: All real numbers
Range: [0, 1)
Discontinuous at $x = k/2$, k an integer

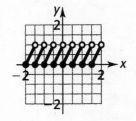

(3-4)

70. $[x - (-1)]^2 (x - 0)^3 [x - (3 + 5i)][x - (3 - 5i)] = (x + 1)^2 x^3 (x - 3 - 5i)(x - 3 + 5i)$
degree 7
(4-3)

71. Yes, for example:
$P(x) = (x + i)(x - i)(x + \sqrt{2})(x - \sqrt{2}) = x^4 - x^2 - 2$ has irrational zeros $\sqrt{2}$
and $-\sqrt{2}$.
(4-2)

72. (A) We form a synthetic division table:

	1	9	-500	0	20,000
0	1	9	-500	0	20,000
10	1	19	-310	-3,100	-11,000
20	1	29	80	1,600	52,000
-10	1	-1	-490	4,900	-29,000
-20	1	-11	-280	5,600	-92,000
-30	1	-21	130	-3,900	137,000
-40	1	-31	740	-29,600	1,204,000

From the table, 20 is an upper bound and -30 is a lower bound.

(B) Here is a computer-generated graph of P, $-30 \le x \le 20$.

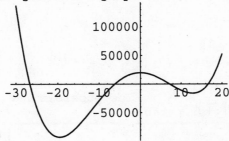

The four real zeros indicated in the table are shown in the graph to lie in $(-30, -20)$, $(-10, 0)$, $(0, 10)$, and $(10, 20)$. To estimate the zero in $(-30, -20)$, we zoom in. Here is a computer-generated graph of P on $(-26.7, -26.6)$

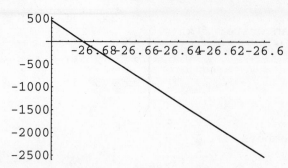

To two decimal places, the zero is -26.68. Similarly, the other zeros can be estimated as -6.22, 7.23, and 16.67 (details omitted). *(4-3)*

73. The possible rational zeros are ± 1, ± 2, ± 3, ± 4, ± 6, ± 12. We form a synthetic division table:

	1	-4	3	10	-10	-12	
1	1	-3	0	10	0	-12	
2	1	-2	-1	8	6	0	2 is a zero

We continue to examine $x^4 - 2x^3 - x^2 + 8x + 6$. The possible rational zeros are -1, ± 2, ± 3, ± 6. We form a synthetic division table:

	1	-2	-1	8	6	
2	1	0	-1	6	18	Not a double zero
3	1	1	2	14	48	an upper bound
-1	1	-3	2	6	0	-1 is a zero

We continue to examine $x^3 - 3x^2 + 2x + 6$. The possible rational zeros are -1, -2, -3, -6. We form a synthetic division table.

	1	-3	2	6	
-1	1	-4	6	0	a double zero

We complete the solution by solving $x^2 - 4x + 6 = 0$ by completing the square.

$$x^2 - 4x = -6$$
$$x^2 - 4x + 4 = -2$$
$$(x - 2)^2 = -2$$
$$x - 2 = \pm i\sqrt{2}$$
$$x = 2 \pm i\sqrt{2}$$

Thus the zeros of $P(x)$ are 2, -1 (double), and $2 \pm i\sqrt{2}$.
$$P(x) = (x - 2)(x + 1)^2[x - (2 + i\sqrt{2})][x - (2 - i\sqrt{2})]$$
$$= (x - 2)(x + 1)^2(x - 2 - i\sqrt{2})(x - 2 + i\sqrt{2})$$

(4-3)

74. $P(x) = x^5 + 4x^4 + x^3 - 11x^2 - 8x + 4$. The possible rational zeros are ± 1, ± 2, ± 4. We form a synthetic division table:

	1	4	1	-11	-8	4	
1	1	5	6	-5	-13	-9	
2	1	6	13	15	22	48	an upper bound
-1	1	3	-2	-9	1	3	
-2	1	2	-3	-5	2	0	-2 is a zero

We continue to examine $x^4 + 2x^3 - 3x^2 - 5x + 2$. The only possible rational zero remaining is -2. We form a synthetic division table.

	1	2	-3	-5	2	
-2 .	1	0	-3	1	0	-2 is a double zero

We contine to examine $x^3 - 3x + 1$.

There are no further rational zeros. The real zeros can be found by the bisection method (details omitted) to be -1.88, 0.35, and 1.53 to two decimal place accuracy.

(4-3)

75. First we must factor $x^4 - x^3 + x^2 - 3x + 2$. The possible rational zeros are ±1, ±2. We form a synthetic division table:

	1	-1	1	-3	2
1	1	0	1	-2	0

We examine $x^3 + x - 2$.

	1	0	1	-2
1	1	1	2	0

Thus $x^4 - x^3 + x^2 - 3x + 2 = (x - 1)^2(x^2 + x + 2)$. $x^2 + x + 2$ cannot be factored further in the real numbers, so

$$\frac{x^2 - 4x + 11}{x^4 - x^3 + x^2 - 3x + 2} = \frac{A}{x - 1} + \frac{B}{(x - 1)^2} + \frac{Cx + D}{x^2 + x + 2}$$

$$= \frac{A(x - 1)(x^2 + x + 2) + B(x^2 + x + 2) + (Cx + D)(x - 1)^2}{(x - 1)^2(x^2 + x + 2)}$$

Thus for all x

$$x^2 - 4x + 11 = A(x - 1)(x^2 + x + 2) + B(x^2 + x + 2) + (Cx + D)(x - 1)^2$$

If $x = 1$

$$8 = 4B$$
$$B = 2$$

$$x^2 - 4x + 11 = A(x^3 + x - 2) + B(x^2 + x + 2) + (Cx + D)(x^2 - 2x + 1)$$
$$= (A + C)x^3 + (B - 2C + D)x^2 + (A + B + C - 2D)x - 2A + 2B + D$$

We have already $B = 2$, so equating coefficients of like terms,

$$0 = A + C$$
$$1 = 2 - 2C + D$$
$$-4 = A + 2 + C - 2D$$
$$11 = -2A + 4 + D$$

Since $C = -A$, we can write

$$1 = 2 + 2A + D$$
$$-4 = 2 - 2D$$
$$11 = -2A + 4 + D$$

So $D = 3$, $A = -2$, $C = 2$

Hence $\dfrac{x^2 - 4x + 11}{x^4 - x^3 + x^2 - 3x + 2} = \dfrac{-2}{x - 1} + \dfrac{2}{(x - 1)^2} + \dfrac{2x + 3}{x^2 + x + 2}$

(4-5)

76. $f(x) = 3 \ln(x - 2)$

(A) Since the graph of f is the same as the graph of $y = \ln x$ shifted 2 units to the right and expanded by a factor of 3, it passes the horizontal line test, and f is one-to-one.

Solve $y = f(x)$ for x

$$y = 3 \ln(x - 2)$$

$$\frac{y}{3} = \ln(x - 2)$$

$$x - 2 = e^{y/3}$$

$$x = e^{y/3} + 2 = f^{-1}(y)$$

Interchange x and y:

$y = f^{-1}(x) = e^{x/3} + 2$ Domain: $(-\infty, \infty)$

Check: $f^{-1}[f(x)] = e^{3\ln(x-2)/3} + 2$

$$= e^{\ln(x-2)} + 2$$

$$= x - 2 + 2$$

$$= x$$

$$f[f^{-1}(x)] = 3 \ln(e^{x/3} + 2 - 2)$$

$$= 3 \ln e^{x/3}$$

$$= 3\left(\frac{x}{3}\right)$$

$$= x$$

(B) Domain of $f = (2, \infty) =$ Range of f^{-1}
     Range of $f = (-\infty, \infty) =$ Domain of f^{-1}

(C)

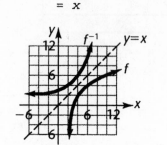

$(3-6, \ 5-5)$

77.

$$A = P\frac{(1 + i)^n - 1}{i}$$

$$Ai = P[(1 + i)^n - 1]$$

$$\frac{Ai}{P} = (1 + i)^n - 1$$

$$1 + \frac{Ai}{P} = (1 + i)^n$$

$$\ln\left(1 + \frac{Ai}{P}\right) = n \ln(1 + i)$$

$$n = \frac{\ln(1 + \frac{Ai}{P})}{\ln(1 + i)}$$ $(5-5)$

78.

$$\ln y = 5x + \ln A$$

$$\ln y - \ln A = 5x$$

$$\ln\left(\frac{y}{A}\right) = 5x$$

$$\frac{y}{A} = e^{5x}$$

$$y = Ae^{5x}$$ $(5-5)$

79. $y = \dfrac{e^x - 2e^{-x}}{2}$

$$2y = e^x - 2e^{-x}$$

$$2y = e^x - \frac{2}{e^x}$$

$$2ye^x = e^x e^x - e^x\left(\frac{2}{e^x}\right)$$

$$2ye^x = (e^x)^2 - 2$$

$$0 = (e^x)^2 - 2ye^x - 2$$

This equation is quadratic in e^x

$$e^x = \frac{-b \pm \sqrt{b^2 - 4ac}}{2a} \quad a = 1, \ b = -2y, \ c = -2$$

$$e^x = \frac{-(-2y) \pm \sqrt{(-2y)^2 - 4(1)(-2)}}{2(1)}$$

$$e^x = \frac{2y \pm \sqrt{4y^2 + 8}}{2}$$

$$e^x = y \pm \sqrt{y^2 + 2}$$

Note: Since $0 < 2$, $y^2 < y^2 + 2$, $\sqrt{y^2} < \sqrt{y^2 + 2}$ and $y < \sqrt{y^2 + 2}$ for all real y. Hence $y - \sqrt{y^2 + 2}$ is always negative. Also, $y + \sqrt{y^2 + 2}$ is always positive.

$$x = \ln(y + \sqrt{y^2 + 2})$$

$(5-5)$

80. If x is linearly related to p, then we are looking for an equation whose graph passes through $(p_1, x_1) = (3.79, 1{,}160)$ and $(p_2, x_2) = (3.59, 1{,}340)$. We find the slope, and then we use the point-slope form to find the equation.

$$m = \frac{x_2 - x_1}{p_2 - p_1} = \frac{1{,}340 - 1{,}160}{3.59 - 3.79} = \frac{180}{-0.2} = -900$$

$$x - x_1 = m(p - p_1)$$
$$x - 1{,}160 = -900(p - 3.79)$$
$$x - 1{,}160 = -900p + 3{,}411$$
$$x = -900p + 4{,}571$$

If the price is lowered to \$3.29, we are asked for x when $p = 3.29$

$$x = -900(3.29) + 4{,}571$$
$$= 1{,}610 \text{ bottles}$$

(3-2)

81. If $0 \le x \le 60$, $C(x) = 0.06x$

If $60 < x \le 60 + 90$,

that is,

$$\begin{pmatrix}\text{Cost of first} \\ 60 \text{ calls}\end{pmatrix} + \begin{pmatrix}\text{Cost of next } x - 60 \\ \text{calls at } 0.05 \text{ per call}\end{pmatrix}$$

$$60 < x \le 150,\ C(x) = 0.06(60) + 0.05(x - 60)$$
$$= 3.60 + 0.05x - 3$$
$$= 0.05x + 0.6$$

If $60 + 90 < x \le 60 + 90 + 150$,

that is,

$$\begin{pmatrix}\text{Cost of first} \\ 150 \text{ calls}\end{pmatrix} + \begin{pmatrix}\text{Cost of next } x - 150 \\ \text{calls at } 0.04 \text{ per call}\end{pmatrix}$$

$$150 < x \le 300,\ C(x) = 0.05(150) + 0.6 + 0.04(x - 150)$$
$$= 7.5 + 0.6 + 0.04x - 6$$
$$= 0.04x + 2.1$$

Finally, if $x > 300$,

$$\begin{pmatrix}\text{Cost of first} \\ 300 \text{ calls}\end{pmatrix} + \begin{pmatrix}\text{Cost of next } x - 300 \\ \text{calls at } 0.03 \text{ per call}\end{pmatrix}$$

$$C(x) = 0.04(300) + 2.1 + 0.03(x - 300)$$
$$= 12 + 2.1 + 0.03x - 9$$
$$= 0.03x + 5.1$$

Summarizing,
$$C(x) = \begin{cases} 0.06x & \text{if } 0 \le x \le 60 \\ 0.05x + 0.6 & \text{if } 60 < x \le 150 \\ 0.04x + 2.1 & \text{if } 150 < x \le 300 \\ 0.03x + 5.1 & \text{if } 300 < x \end{cases}$$

x	$C(x) = 0.06x$	x	$C(x) = 0.05x + 0.6$	x	$C(x) = 0.04x + 2.1$	x	$C(x) = 0.03x + 5.1$
0	0	90	5.1	160	8.5	310	14.4
30	1.8	140	7.6	200	10.1	400	17.1
60	3.6						

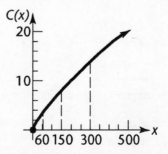

(3-4)

82. (A)   Let x = width of pen
Then $2x + \ell = 80$
So $\ell = 80 - 2x$ = length of pen
$A(x) = x\ell = x(80 - 2x) = 80x - 2x^2$

(B) Since all distances must be positive, $x > 0$ and $80 - 2x > 0$, hence $80 > 2x$ or $x < 40$. Therefore $0 < x < 40$ or $(0, 40)$ is the domain of $A(x)$.

(C) Note: 0, 40 were excluded from the domain for geometrical reasons; but can be used to help draw the graph since the *polynomial* $80x - 2x^2$ has domain including these values.

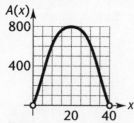

x	$A(x)$
0	0
10	600
20	800
30	600
40	0

From the graph, the maximum value of A seems to occur when $x = 20$. Checking, we see that $80x - 2x^2 = -2x^2 + 80x = -2(x^2 - 40x + 400) + 800 = -2(x - 20)^2 + 800$. Thus the maximum value of A indeed occurs when $x = 20$.
Then $80 - 2x = 40$, and the dimensions are 20 feet by 40 feet.   (3-4)

83. (A) $f(1) = 1 - 2[\![1/2]\!] = 1 - 2(0) = 1$
$f(2) = 2 - 2[\![2/2]\!] = 2 - 2(1) = 0$
$f(3) = 3 - 2[\![3/2]\!] = 3 - 2(1) = 1$
$f(4) = 4 - 2[\![4/2]\!] = 4 - 2(2) = 0$

(B) If n is an integer, n is either odd or even.
If n is even, it can be written as $2k$, where k is an integer.
Then $f(n) = f(2k) = 2k - 2[\![2k/2]\!] = 2k - 2[\![k]\!] = 2k - 2k = 0$.
Otherwise, n is odd, and n can be written as $2k + 1$, where k is an integer.
Then $f(n) = f(2k + 1) = 2k + 1 - 2[\![(2k+1)/2]\!] = 2k + 1 - 2[\![k + \frac{1}{2}]\!]$
$= 2k + 1 - 2k$
$= 1$

Summarizing, if n is an integer, $f(n) = \begin{cases} 1 \text{ if } n \text{ is an odd integer} \\ 0 \text{ if } n \text{ is an even integer} \end{cases}$ (3-4)

84. We are given y = length of container
x = width of one end
Hence $4x$ = girth of container
Length + girth = $y + 4x = 10$
So $y = 10 - 4x$
Since Volume = $8 = x^2y$, we have
$8 = x^2(10 - 4x)$
$8 = 10x^2 - 4x^3$
$4x^3 - 10x^2 + 8 = 0$
$2x^3 - 5x^2 + 4 = 0$

The possible rational solutions of this equation are ± 1, ± 2, ± 4, $\pm \frac{1}{2}$.

We form a synthetic division table.

	2	-5	0	4	
1	2	-3	-3	1	
2	2	-1	-2	0	2 is a zero

Thus the equation can be factored
$(x - 2)(2x^2 - x - 2) = 0$

To find the remaining zeros, we solve $2x^2 - x - 2 = 0$ by the quadratic formula.
$2x^2 - x - 2 = 0$

$$x = \frac{-b \pm \sqrt{b^2 - 4ac}}{2a} \quad a = 2, \ b = -1, \ c = -2$$

$$x = \frac{-(-1) \pm \sqrt{(-1)^2 - 4(2)(-2)}}{2(2)}$$

$$x = \frac{1 + \sqrt{17}}{4} \quad \text{(discarding the negative solution)}$$

Thus, the positive zeros are $x = 2, \ 1.3$.
Thus, the dimensions of the package are $x = 2$ feet and $y = 2$ feet, or $x = 1.3$ feet and $y = 4.8$ feet. *(4-2)*

 85.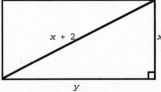

Labelling the rectangle as in the diagram, we have

Area $= xy = 6$

Hence $y = \dfrac{6}{x}$

Applying the Pythagorean theorem, we have

$$(x + 2)^2 = x^2 + y^2$$

$$(x + 2)^2 = x^2 + \left(\frac{6}{x}\right)^2$$

$$x^2 + 4x + 4 = x^2 + \frac{36}{x^2}$$

$$4x + 4 = \frac{36}{x^2}$$

$$4x^3 + 4x^2 = 36$$

$$4x^3 + 4x^2 - 36 = 0$$

$$x^3 + x^2 - 9 = 0$$

There are no rational zeros of $P(x) = x^3 + x^2 - 9$. We form a synthetic division table.

	1	1	0	-9	
0	1	1	0	-9	
1	1	2	2	-7	
2	1	3	6	3	There is a zero between 1 and 2

2 is also an upper bound for the zeros. We apply the bisection method to locate the zero to one decimal place. We organize our calculations in a table.

Sign Change Interval (a, b)	Midpoint m	$P(a)$	$P(m)$	$P(b)$
(1, 2)	1.5	−	−	+
(1.5, 2)	1.75	−	−	+
(1.75, 2)	1.875	−	+	+
(1.75, 1.875)	1.8125	−	+	+
(1.75, 1.8125)	We stop here	−		+

Since each point rounds to 1.8, a real zero lies on this interval and is given by 1.8 to one decimal place accuracy. Hence, $x = 1.8$ feet.

Thus, $y = \dfrac{6}{x} = \dfrac{6}{1.8} = 3.3$ feet to one decimal place. *(4-3)*

86. We use the Doubling Time Growth Model:
$$P = P_0 2^{t/d}$$
    Substituting $P_0 = 40$ million and $d = 22$, we have
$$P = 40(2^{t/22}) \text{ million}$$

(A) $t = 5$, hence $P = 40(2^{5/22})$ million
$$= 46.8 \text{ million}$$
(B) $t = 30$, hence $P = 40(2^{30/22})$ million
$$= 103 \text{ million} \qquad (5-1)$$

87. We solve $P = P_0(1.07)^t$ for t, using $P = 2P_0$
$$2P_0 = P_0(1.07)^t$$
$$2 = (1.07)^t$$
$$\ln 2 = t \ln 1.07$$
$$\frac{\ln 2}{\ln 1.07} = t$$
$$t = 10.2 \text{ years} \qquad (5-5)$$

88. We solve $P = P_0 e^{0.07t}$ for t, using $P = 2P_0$
$$2P_0 = P_0 e^{0.07t}$$
$$2 = e^{0.07t}$$
$$\ln 2 = 0.07t$$
$$t = \frac{\ln 2}{0.07}$$
$$t = 9.90 \text{ years} \qquad (5-5)$$

89. First, we solve $M = \frac{2}{3} \log\left(\frac{E}{E_0}\right)$ for E.

$$M = \frac{2}{3} \log\left(\frac{E}{E_0}\right)$$
$$\frac{3M}{2} = \log \frac{E}{E_0}$$
$$\frac{E}{E_0} = 10^{3m/2}$$
$$E = E_0(10^{3m/2})$$

We now compare E_1 for $M = 8.3$ with E_2 for $M = 7.1$.
$$E_1 = E_0(10^{3 \cdot 8.3/2}) \qquad E_2 = E_0(10^{3 \cdot 7.1/2})$$
$$E_1 = E_0(10^{12.45}) \qquad E_2 = E_0(10^{10.65})$$

Hence $\dfrac{E_1}{E_2} = \dfrac{E_0(10^{12.45})}{E_0(10^{10.65})} = 10^{12.45-10.65} = 10^{1.8}$
$$E_1 = 10^{1.8}E_2 \text{ or } 63.1E_2.$$

The 1906 earthquake was 63.1 times as powerful. $\qquad (5-4)$

90. We solve $D = 10 \log \dfrac{I}{I_0}$ for I, with $D = 88$, $I_0 = 10^{-12}$

$$88 = 10 \log \frac{I}{10^{-12}}$$
$$8.8 = \log \frac{I}{10^{-12}}$$
$$10^{8.8} = \frac{I}{10^{-12}}$$
$$I = 10^{8.8} \cdot 10^{-12}$$
$$I = 10^{-3.2}$$
$$I = 6.31 \times 10^{-4} \ w/m^2 \qquad (5-4)$$

CHAPTER 6

Exercise 6-1

Key Ideas and Formulas

Systems of Linear Equations: Basic Terms

A system of linear equations is **consistent** if it has one or more solutions and **inconsistent** if no solutions exist. Furthermore, a consistent system is said to be **independent** if it has exactly one solution (often referred to as the **unique solution**) and **dependent** if it has more than one solution.

A linear system of equations in two variables

$$ax + by = h$$
$$cx + dy = k$$

must have

1. exactly one solution (consistent and independent) or
2. no solution (inconsistent) or
3. infinitely many solutions (consistent and dependent).

There are no other possibilities.

A linear system in two variables can be solved by graphing, by substitution, or by elimination using addition. (See text for descriptions of each method.)

Equivalent systems of equations are systems with the same solution set.

A system of linear equations is transformed into an equivalent system if:

1. Two equations are interchanged.

2. An equation is multiplied by a non-zero constant.

3. A constant multiple of another equation is added to a given equation.

A **matrix** is a rectangular array of numbers written within brackets. Each number in a matrix is called an **element** of the matrix.

If a matrix has m rows and n columns, it is called an $m \times n$ **matrix** (read "m by n matrix"). The expression $m \times n$ is called the **size** of the matrix and the numbers m and n are called the **dimensions** of the matrix.

A matrix with n rows and n columns is called a **square matrix of order n**. A matrix with only one column is called a **column matrix**, and a matrix with only one row is called a **row matrix**.

Associated with each linear system of the form:

$$a_{11}x_1 + a_{12}x_2 = k_1$$
$$a_{21}x_1 + a_{22}x_2 = k_2$$

where x_1 and x_2 are variables, is the **augmented matrix** of the system:

$$
\begin{array}{l}
\text{Column 1 } (C_1) \\
\text{Column 2 } (C_2) \\
\text{Column 3 } (C_3)
\end{array}
$$

$$
\begin{bmatrix}
a_{11} & a_{12} & k_1 \\
a_{21} & a_{22} & k_2
\end{bmatrix}
\begin{array}{l}
\leftarrow \text{Row 1 } (R_1) \\
\leftarrow \text{Row 2 } (R_2)
\end{array}
$$

We solve a linear system by using row operations to transform it into a row equivalent matrix of one of these types:

$$
\begin{bmatrix}
1 & 0 & m \\
0 & 1 & n
\end{bmatrix}
\quad \text{a unique solution (consistent and independent)}
$$

$$
\begin{bmatrix}
1 & m & n \\
0 & 0 & 0
\end{bmatrix}
\quad \text{infinitely many solutions (consistent and dependent)}
$$

$$\begin{bmatrix} 1 & m & \bigm| & n \\ 0 & 0 & \bigm| & p \end{bmatrix}$$ no solution (inconsistent)

An augmented matrix is transformed into a row-equivalent matrix if any of the following row operations is performed.

1. Two rows are interchanged. ($R_i \leftrightarrow R_j$ means "interchange row i and row j")

2. A row is multiplied by a non-zero constant ($kR_i \rightarrow R_i$ means "multiply row i by the constant k")

3. A constant multiple of another row is added to a given row ($kR_j + R_i \rightarrow R_i$ means "multiply row j by the constant k and add to R_i.")

---

1. Both lines in the given system are different, but they have the same slope $\left(\dfrac{1}{2}\right)$ and are therefore parallel. This system corresponds to (B) and has no solution.

3. In slope-intercept form, these equations are $y = 2x - 5$ and $y = -\dfrac{3}{2}x - \dfrac{3}{2}$. Thus, one has slope 2 and y intercept -5; the other has slope $-\dfrac{3}{2}$ and y intercept $-\dfrac{3}{2}$. This system corresponds to (D) and its solution can be read from the graph as $(1, -3)$. Checking, we see that
$$2x - y = 2 \cdot 1 - (-3) = 5$$
$$3x + 2y = 3 \cdot 1 + 2(-3) = -3$$

5.

$y = x - 3$

$x + y = 7$

$(5, 2)$

7.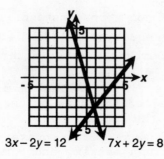

$3x - 2y = 12$ $7x + 2y = 8$

$(2, -3)$

9.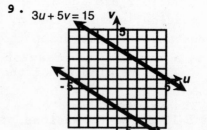

$3u + 5v = 15$

$6u + 10v = -30$

The lines are parallel.
No solution.

11. $2x + 3y = 1$
$3x - y = 7$
If we multiply the bottom equation by 3 and add, we can eliminate y.

$$\begin{array}{r} 2x + 3y = 1 \\ \underline{9x - 3y = 21} \\ 11x = 22 \\ x = 2 \end{array}$$

Now substitute $x = 2$ back into the top equation and solve for y.
$2(2) + 3y = 1$
$ 3y = -3$
$ y = -1$
$(2, -1)$

13. $4x - 3y = 15$
$3x + 4y = 5$
If we multiply the top equation by 4, the bottom by 3, and add, we can eliminate y.

$$\begin{array}{r} 16x - 12y = 60 \\ \underline{9x + 12y = 15} \\ 25x = 75 \\ x = 3 \end{array}$$

Substituting in the bottom equation, we have
$3(3) + 4y = 5$
$9 + 4y = 5$
$ 4y = -4$
$ y = -1$
$(3, -1)$

15. Matrix A has 2 rows and 3 columns, so its size is 2 × 3.
Matrix C has 1 row and 3 columns, so its size is 1 × 3.

17. C is the only row matrix.   **19.** B is the only square matrix.

21. a_{12} in the element is row 1, column 2 of A. This is -2.
a_{23} is the element in row 2, column 3 of A. This is -6.

23. The elements on the principal diagonal are $b_{11} = -2$, $b_{22} = 6$, and $b_{33} = 0$.

25. $R_1 \leftrightarrow R_2$ means interchange Rows 1 and 2.

$$\begin{bmatrix} 4 & -6 & | & -8 \\ 1 & -3 & | & 2 \end{bmatrix}$$

27. $-4R_1 \rightarrow R_1$ means multiply Row 1 by -4.

$$\begin{bmatrix} -4 & 12 & | & -8 \\ 4 & -6 & | & -8 \end{bmatrix}$$

29. $2R_2 \rightarrow R_2$ means multiply Row 2 by 2.

$$\begin{bmatrix} 1 & -3 & | & 2 \\ 8 & -12 & | & -16 \end{bmatrix}$$

31. $(-4)R_1 + R_2 \rightarrow R_2$ means replace Row 2 by itself plus -4 times Row 1.

$$\begin{bmatrix} 1 & -3 & | & 2 \\ 4 & -6 & | & -8 \end{bmatrix} \rightarrow \begin{bmatrix} 1 & -3 & | & 2 \\ 0 & 6 & | & -16 \end{bmatrix}$$
$$-4 \quad 12 \quad -8$$

33. $(-2)R_1 + R_2 \rightarrow R_2$ means replace Row 2 by itself plus -2 times Row 1.

$$\begin{bmatrix} 1 & -3 & | & 2 \\ 4 & -6 & | & -8 \end{bmatrix} \rightarrow \begin{bmatrix} 1 & -3 & | & 2 \\ 2 & 0 & | & -12 \end{bmatrix}$$
$$-2 \quad 6 \quad -4$$

35. $(-1)R_1 + R_2 \rightarrow R_2$ means replace Row 2 by itself plus -1 times Row 1.

$$\begin{bmatrix} 1 & -3 & | & 2 \\ 4 & -6 & | & -8 \end{bmatrix} \rightarrow \begin{bmatrix} 1 & -3 & | & 2 \\ 3 & -3 & | & -10 \end{bmatrix}$$
$$-1 \quad 3 \quad -2$$

37. We write the augmented matrix:

$$\begin{bmatrix} 1 & 1 & | & 7 \\ 1 & -1 & | & 1 \end{bmatrix} \quad (-1)R_1 + R_2 \rightarrow R_2$$
$$-1 \quad -1 \quad -7$$
$$\uparrow$$

Need a 0 here

$$\sim \begin{bmatrix} 1 & 1 & | & 7 \\ 0 & -2 & | & -6 \end{bmatrix} \quad -\tfrac{1}{2}R_2 \rightarrow R_2 \qquad \text{corresponds to the linear system} \quad \begin{matrix} x_1 + x_2 = 7 \\ -2x_2 = -6 \end{matrix}$$
$$\uparrow$$

Need a 1 here
Need a 0 here
$$\downarrow$$

$$\sim \begin{bmatrix} 1 & 1 & | & 7 \\ 0 & 1 & | & 3 \end{bmatrix} \quad (-1)R_2 + R_1 \rightarrow R_1$$
$$0 \quad -1 \quad -3 \qquad \text{corresponds to the linear system} \quad \begin{matrix} x_1 + x_2 = 7 \\ x_2 = 3 \end{matrix}$$

$$\sim \begin{bmatrix} 1 & 0 & | & 4 \\ 0 & 1 & | & 3 \end{bmatrix} \quad \text{corresponds to the linear system} \quad \begin{matrix} x_1 = 4 \\ x_2 = 3 \end{matrix}$$

The solution is $x_1 = 4$, $x_2 = 3$. Each pair of lines graphed below has the same intersection point, (4, 3).

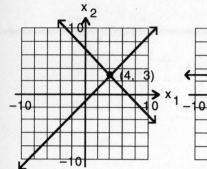

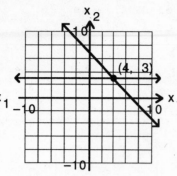

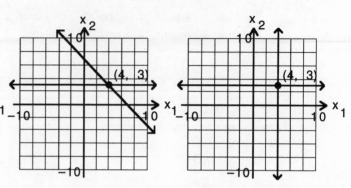

$$x_1 + x_2 = 7$$
$$x_1 - x_2 = 1$$

$$x_1 + x_2 = 7$$
$$-2x_2 = -6$$

$$x_1 + x_2 = 7$$
$$x_2 = 3$$

$$x_1 = 4$$
$$x_2 = 3$$

39. $\begin{bmatrix} 1 & -4 & | & -2 \\ -2 & 1 & | & -3 \end{bmatrix}$ $2R_1 + R_2 \rightarrow R_2$

↑
Need a 0 here

2   -8   -4

$\sim \begin{bmatrix} 1 & -4 & | & -2 \\ 0 & -7 & | & -7 \end{bmatrix}$ $-\frac{1}{7}R_2 \rightarrow R_2$

↑
Need a 1 here
Need a 0 here
↓

$\sim \begin{bmatrix} 1 & -4 & | & -2 \\ 0 & 1 & | & 1 \end{bmatrix}$ $4R_2 + R_1 \rightarrow R_1$

0   4   4

$\sim \begin{bmatrix} 1 & 0 & | & 2 \\ 0 & 1 & | & 1 \end{bmatrix}$

Therefore $x_1 = 2$ and $x_2 = 1$

41. $\begin{bmatrix} 3 & -1 & | & 2 \\ 1 & 2 & | & 10 \end{bmatrix}$ $R_1 \leftrightarrow R_2$

Need a 1 here

$\sim \begin{bmatrix} 1 & 2 & | & 10 \\ 3 & -1 & | & 2 \end{bmatrix}$ $(-3)R_1 + R_2 \rightarrow R_2$

↑
Need a 0 here

-3   -6   -30

$\sim \begin{bmatrix} 1 & 2 & | & 10 \\ 0 & -7 & | & -28 \end{bmatrix}$ $-\frac{1}{7}R_2 \rightarrow R_2$

↑
Need a 1 here
Need a 0 here
↓

$\sim \begin{bmatrix} 1 & 2 & | & 10 \\ 0 & 1 & | & 4 \end{bmatrix}$ $(-2)R_2 + R_1 \rightarrow R_1$

0   -2   -8

$\sim \begin{bmatrix} 1 & 0 & | & 2 \\ 0 & 1 & | & 4 \end{bmatrix}$

Therefore $x_1 = 2$ and $x_2 = 4$

43. $\begin{bmatrix} 1 & 2 & | & 4 \\ 2 & 4 & | & -8 \end{bmatrix}$ $(-2)R_1 + R_2 \rightarrow R_2$

↑
Need a 0 here

-2   -4   -8

$\sim \begin{bmatrix} 1 & 2 & | & 4 \\ 0 & 0 & | & -16 \end{bmatrix}$

This matrix corresponds to the system
$x_1 + 2x_2 = 4$
$0x_1 + 0x_2 = -16$
This system has no solution.

45. $\begin{bmatrix} 2 & 1 & | & 6 \\ 1 & -1 & | & -3 \end{bmatrix}$ $R_1 \leftrightarrow R_2$

Need a 1 here

$\sim \begin{bmatrix} 1 & -1 & | & -3 \\ 2 & 1 & | & 6 \end{bmatrix}$ $(-2)R_1 + R_2 \to R_2$

↑
Need a 0 here
-2   2   6

$\sim \begin{bmatrix} 1 & -1 & | & -3 \\ 0 & 3 & | & 12 \end{bmatrix}$ $\frac{1}{3}R_2 \to R_2$

↑
Need a 1 here
Need a 0 here
↓

$\sim \begin{bmatrix} 1 & -1 & | & -3 \\ 0 & 1 & | & 4 \end{bmatrix}$ $R_2 + R_1 \to R_1$

$\sim \begin{bmatrix} 1 & 0 & | & 1 \\ 0 & 1 & | & 4 \end{bmatrix}$

Therefore $x_1 = 1$ and $x_2 = 4$.

47. $\begin{bmatrix} 3 & -6 & | & -9 \\ -2 & 4 & | & 6 \end{bmatrix}$ $\frac{1}{3}R_1 \to R_1$

Need a 1 here

$\sim \begin{bmatrix} 1 & -2 & | & -3 \\ -2 & 4 & | & 6 \end{bmatrix}$ $2R_1 + R_2 \to R_2$

↑
Need a 0 here
2   -4   -6

$\sim \begin{bmatrix} 1 & -2 & -3 \\ 0 & 0 & 0 \end{bmatrix}$

This matrix corresponds to the system
$x_1 - 2x_2 = -3$
$0x_1 + 0x_2 = 0$
Thus $x_1 = 2x_2 - 3$.
Hence there are infinitely many solutions: for any real number s, $x_2 = s$, $x_1 = 2s - 3$ is a solution.

49. $\begin{bmatrix} 4 & -2 & | & 2 \\ -6 & 3 & | & -3 \end{bmatrix}$ $\frac{1}{4}R_1 \to R_1$

Need a 1 here

$\sim \begin{bmatrix} 1 & -\frac{1}{2} & | & \frac{1}{2} \\ -6 & 3 & | & -3 \end{bmatrix}$ $6R_1 + R_2 \to R_2$

↑
Need a 0 here
6   -3   3

$\sim \begin{bmatrix} 1 & -\frac{1}{2} & | & \frac{1}{2} \\ 0 & 0 & | & 0 \end{bmatrix}$

This matrix corresponds to the system
$x_1 - \frac{1}{2}x_2 = \frac{1}{2}$
$0x_1 + 0x_2 = 0$
Thus $x_1 = \frac{1}{2}x_2 + \frac{1}{2}$.
Hence there are infinitely many solutions: for any real number s, $x_2 = s$, $x_1 = \frac{1}{2}s + \frac{1}{2}$ is a solution.

51. Here is a computer-generated graph of the system , entered as

$y = \dfrac{2x + 5}{3}$

$y = \dfrac{13 - 3x}{4}$

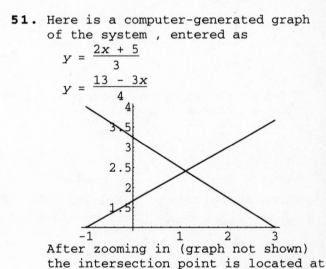

After zooming in (graph not shown) the intersection point is located at (1.12, 2.41) to two decimal places.

53. Here is a computer-generated graph of the system, entered as

$$y = \frac{3.5x - 0.1}{2.4}$$

$$y = \frac{2.6x + 0.2}{1.7}$$

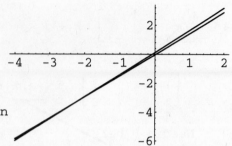

After zooming in (graph not shown) the intersection point is located at (-2.24, -3.31) to two decimal places.

55. Here are computer-generated graphs of the three systems:

(A)

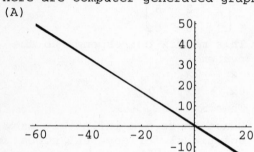

(B)

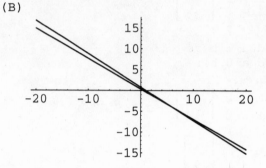

(C)

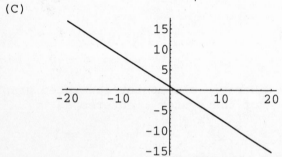

Clearly, the solutions are quite different even though the graph of the second equation changes only slightly from case to case. Solving by elimination yields:

(A)

$$4x + 5y = 4$$
$$9x + 11y = 4$$

Multiply the top equation by $-\frac{9}{4}$ and add to eliminate x.

$$-9x - \frac{45}{4}y = -9$$
$$\underline{9x + 11y = 4}$$
$$-\frac{1}{4}y = -5$$
$$y = 20$$

Now substitute $y = 20$ back into the top equation and solve for x.

$$4x + 5(20) = 4$$
$$4x = -96$$
$$x = -24$$

$(-24, 20)$

(B)

$$4x + 5y = 4$$
$$8x + 11y = 4$$

Multiply the top equation by -2 and add to eliminate x.

$$-8x - 10y = -8$$
$$\underline{8x + 11y = 4}$$
$$y = -4$$

Now substitute $y = -4$ back into the top equation and solve for x.

$$4x + 5(-4) = 4$$
$$4x = 24$$
$$x = 6$$

$(6, -4)$

(C)

$$4x + 5y = 4$$
$$8x + 10y = 4$$

Multiply the top equation by -2 and add.

$$-8x - 10y = -8$$
$$\underline{8x + 10y = 4}$$
$$0 = -4$$

No solution.

In Problems 57 and 59, steps are shown for the TI-82/3 series calculator. The student should consult the manual for the corresponding steps on the other models.

57. Enter the augmented matrix
Matrix [A]   2 × 3
[0.8 2.88   4]
[1.25 4.34   5]

Need a 1 here

Perform $\dfrac{1}{0.8} R_1 \rightarrow R_1$:   Enter   *row(1/.8,[A],1)→[A]
[[1   3.6   5
  1.25   4.34   5]]
↑
Need a 0 here

Perform $(-1.25)R_1 + R_2 \rightarrow R_2$:   Enter   *row+(-1.25,[A],1,2)→[A]
[[1   3.6   5   ]
 [0   -.16 -1.25]]
         ↑
    Need a 1 here

Perform $\dfrac{1}{-0.16} R_2 \rightarrow R_2$:   Enter   *row(-1/0.16,[A],2)→[A]
[[1   3.6   5   ]
 [0   1   7.8125]]

Need a 0 here

Perform $(-3.6)R_2 + R_1 \rightarrow R_1$:   Enter   *row+(-3.6,[A],2,1)→[A]
[[1   0   -23.125]
 [0   1   7.8125]]

This corresponds to the system
x_1 = -23.125
  x_2 = 7.8125
The solution is (-23.125, 7.8125).

59. Enter the augmented matrix
Matrix [A]   2 × 3
[4.8   -40.32   295.2]
[-3.75   28.7   -211.2]

Need a 1 here

Perform $\dfrac{1}{4.8} R_1 \rightarrow R_1$:   Enter   *row(1/4.8,[A],1)→[A]
[[1   -8.4   61.5
 [-3.75   28.7   -211.2]]
 ↑
Need a 0 here

Perform $3.75R_1 + R_2 \rightarrow R_2$:   Enter   *row+(3.75,[A],1,2)→[A]
[[1   -8.4   61.5
 [-1E-13   -2.8   19.425]]
         ↑
    Need a 1 here

Perform $\frac{1}{-2.8} R_2 \rightarrow R_2$:   Enter   *row(1/-2.8,[A],2)→[A]

```
[[1            -8.4      61.5]
 [3.57…E-14     1        -6.9375]]
```

Need a 0 here

Perform $8.4R_2 + R_1 \rightarrow R_1$:   Enter   *row+(8.4,[A],2,1)→[A]

```
[[1            1E-13     3.225]
 [3.57…E-14     1        -6.9375]]
```

This corresponds to the system

$$x_1 \qquad = 3.225$$
$$\quad x_2 = -6.9375$$

The solution is (3.225, -6.9375).

61. Let x = number of 32-cent stamps
       y = number of 23-cent stamps

She bought 75 stamps, hence

$$x + y = 75 \qquad\qquad (1)$$

She spent $19.50, hence

$$0.32x + 0.23y = 19.50 \quad (2)$$

We solve the system of equations (1), (2) using elimination by addition. If we multiply the top equation by -0.23 and add, we can eliminate y.

$$
\begin{array}{rcl}
-0.23x - 0.23y &=& -17.25 \\
0.32x + 0.23y &=& 19.50 \\
\hline
0.09x \qquad\quad &=& 2.25 \\
x &=& 25
\end{array}
$$

Now substitute $x = 25$ back into equation (1) and solve for y

$$25 + y = 75$$
$$\quad\; y = 50$$

25 32-cent stamps, 50 23-cent stamps

63. Let x = amount invested at 6%
       y = amount invested at 9%

Summarize the given information in a table:

	Bond A	Bond B	Total
amount invested	x	y	200,000
interest earned	$0.06x$	$0.09y$	14,775

Form equations from the information:

$$\left(\begin{array}{c}\text{Amount invested} \\ \text{in Bond } A\end{array}\right) + \left(\begin{array}{c}\text{Amount invested} \\ \text{in Bond } B\end{array}\right) = (\text{Total investment})$$
$$\qquad x \qquad\qquad + \qquad\quad y \qquad\quad = \qquad 200,000$$

$$\left(\begin{array}{c}\text{Interest earned} \\ \text{from Bond } A\end{array}\right) + \left(\begin{array}{c}\text{Interest earned} \\ \text{from Bond } B\end{array}\right) = \left(\begin{array}{c}\text{Total} \\ \text{Interest}\end{array}\right)$$
$$\quad 0.06x \qquad\quad + \qquad\quad 0.09y \qquad\quad = \qquad 14,775$$

Solve this system using elimination by addition.

$$
\begin{array}{rcl}
-0.06x - 0.06y &=& -12,000 \\
0.06x + 0.09y &=& 14,775 \\
\hline
0.03y &=& 2,775 \\
y &=& 92,500 \\
x + 92,500 &=& 200,000 \\
x &=& 107,500
\end{array}
$$

$107,500 in bond A and $92,500 in bond B.

65. Let x = amount of 20% solution
y = amount of 80% solution

Summarize the given information in a table:

	20% solution	80% solution	62% solution
amount of solution	x	y	100
amount of acid	$0.2x$	$0.8y$	$0.62(100) = 62$

Form equations from the information:

$$\begin{pmatrix} \text{Amount of} \\ \text{20\% solution} \end{pmatrix} + \begin{pmatrix} \text{Amount of} \\ \text{80\% solution} \end{pmatrix} = \begin{pmatrix} \text{Amount of} \\ \text{62\% solution} \end{pmatrix}$$

$$x \quad + \quad y \quad = \quad 100$$

$$\begin{pmatrix} \text{Amount of acid} \\ \text{in 20\% solution} \end{pmatrix} + \begin{pmatrix} \text{Amount of acid} \\ \text{in 80\% solution} \end{pmatrix} = \begin{pmatrix} \text{Amount of acid} \\ \text{in 62\% solution} \end{pmatrix}$$

$$0.2x \quad + \quad 0.8y \quad = \quad 62$$

Solve this system using elimination by addition.

$$\begin{array}{r} -0.2x - 0.2y = -20 \\ \underline{0.2x + 0.8y = 62} \\ 0.6y = 42 \\ y = 70 \\ x + 70 = 100 \\ x = 30 \end{array}$$

30 liters of 20% solution and 70 liters of 80% solution.

67. Let x = number of grams of mix A
y = number of grams of mix B

Summarize the given information in a table:

	mix A	mix B
protein	0.15	0.3
fat	0.1	0.05

Form equations from the information:

$$\begin{pmatrix} \text{protein in} \\ x \text{ gm of mix } A \end{pmatrix} + \begin{pmatrix} \text{protein in} \\ y \text{ gm of mix } B \end{pmatrix} = \begin{pmatrix} \text{total} \\ \text{protein} \end{pmatrix}$$

$$0.15x \quad + \quad 0.3y \quad = \quad 54$$

$$\begin{pmatrix} \text{fat in } x \text{ gm} \\ \text{of mix } A \end{pmatrix} + \begin{pmatrix} \text{fat in } y \text{ gm} \\ \text{of mix } B \end{pmatrix} = \begin{pmatrix} \text{total} \\ \text{fat} \end{pmatrix}$$

$$0.1x \quad + \quad 0.05y \quad = \quad 24$$

Solve using elimination by addition.

$$\begin{array}{r} 0.15x + 0.3y = 54 \\ \underline{-0.6x - 0.3y = -144} \\ -0.45x = -90 \\ x = 200 \\ 0.1(200) + 0.05y = 24 \\ 0.05y = 4 \\ y = 80 \end{array}$$

200 grams of mix A and 80 grams of mix B.

69. Let x = base price
y = surcharge for each additional pound.

Since a 5-pound package costs the base price plus 4 surcharges,
$x + 4y = 27.75$

Since a 20-pound package costs the base price plus 19 surcharges,
$x + 19y = 64.50$

Solve using elimination by addition.

$$
\begin{array}{rl}
-x - 4y = & -27.75 \\
\underline{x + 19y = } & \underline{64.50} \\
15y = & 36.75 \\
y = & 2.45 \\
x + 4(2.45) = & 27.75 \\
x = & 17.95
\end{array}
$$

The base price is \$17.95 and the surcharge per pound is \$2.45.

71. Let x = number of pounds of robust blend
y = number of pounds of mild blend

Summarize the given information in a table:

	Robust blend	Mild blend
ozs. of Columbian beans	12	6
ozs. of Brazilian beans	4	10

Form equations from the information:

$$
\left(\begin{array}{c} \text{pounds of Columbian} \\ \text{beans needed for} \\ \text{robust blend} \end{array}\right) + \left(\begin{array}{c} \text{pounds of Columbian} \\ \text{beans needed for} \\ \text{mild blend} \end{array}\right) = \left(\begin{array}{c} \text{Total} \\ \text{Columbian beans} \\ \text{available} \end{array}\right)
$$

$$
\frac{12}{16}x \quad + \quad \frac{6}{16}Y \quad = \quad 50(132)
$$

$$
\left(\begin{array}{c} \text{pounds of Brazilian} \\ \text{beans needed for} \\ \text{robust blend} \end{array}\right) + \left(\begin{array}{c} \text{pounds of Brazilian} \\ \text{beans needed for} \\ \text{mild blend} \end{array}\right) = \left(\begin{array}{c} \text{Total} \\ \text{Brazilian beans} \\ \text{available} \end{array}\right)
$$

$$
\frac{4}{16}x \quad + \quad \frac{10}{16}Y \quad = \quad 40(132)
$$

Solve using elimination by addition:

$$
\begin{array}{rl}
\frac{12}{16}x + \frac{6}{16}y = & 6,600 \\[2mm]
-\frac{12}{16}x - \frac{30}{16}y = & -15,840 \\[2mm]
-\frac{24}{16}y = & -9,240 \\[2mm]
y = & 6,160
\end{array}
$$

$$
\frac{4}{16}x + \frac{10}{16}(6,160) = 40(132)
$$

$$
\frac{1}{4}x + 3,850 = 5,280
$$

$$
\frac{1}{4}x = 1,430
$$

$$
x = 5,720
$$

5,720 pounds of the robust blend and 6,160 pounds of the mild blend.

Exercise 6-2

Key Ideas and Formulas

Any linear system must have exactly one solution, no solution, or an infinite number of solutions, regardless of the number of equations or the number of variables in the system. The terms **unique solution**, **consistent**, **inconsistent**, **dependent**, and **independent** are used to describe these solutions, just as they are for systems with two variables.

The method used for solving large systems of linear equations is called Gauss-Jordan elimination. We use a step-by-step procedure to transform the augmented matrix into reduced form, from which the solution to the system can be read by inspection. A matrix is in reduced form if:

1. Each row consisting entirely of 0's is below any row having at least one non-zero element.
2. The leftmost non-zero element in each row is 1.
3. The column containing the leftmost 1 of a given row has 0's above and below the 1.
4. The leftmost 1 in any row is to the right of the leftmost 1 in the preceding row.

Gauss-Jordan Elimination:

Write the augmented matrix of the system. Then

Step 1. Choose the leftmost non-zero column and use appropriate row operations to get a 1 at the top.

Step 2. Use multiples of the row containing the 1 from step 1 to get zeros in all remaining places in the column containing this 1.

Step 3. Repeat step 1 with the **submatrix** formed by (mentally) deleting the row used in step 2 and all rows above this row.

Step 4. Repeat step 2 with the **entire matrix**, including the mentally deleted rows. Continue this process until it is impossible to go further.

Note 1: (inconsistent system)
   If at any point in this process we obtain a row with all zeros to the left of the vertical line and a non-zero number to the right, we can stop, since we will have a contradiction: $0 = n$, $n \neq 0$. We can then conclude that the system has no solution.

Note 2: (dependent system)
   If the number of leftmost 1's in a reduced augmented coefficient matrix is less than the number of variables in the system and there are no contradictions, then the system is dependent and has infinitely many solutions.

1. Yes

3. No. Condition 1 is violated.

5. No. Condition 2 is violated.

7. Yes

9. $x_1 \qquad\quad = -2$
   $\qquad x_2 \qquad = 3$
   $\qquad\quad x_3 = 0$

   The system is already solved.

11. $x_1 \qquad\quad - 2x_3 = 3$
   $\qquad x_2 + \; x_3 = -5$

   Solution:
   $x_3 = t$
   $x_2 = -5 - x_3 = -5 - t$
   $x_1 = 3 + 2x_3 = 3 + 2t$
   Thus $x_1 = 2t + 3$, $x_2 = -t - 5$,
   $x_3 = t$ is the solution for t any
   real number.

13. $x_1 \qquad = 0$
$\qquad x_2 \quad = 0$
$\qquad\qquad 0 = 1$

The system has no solution.

15. $x_1 - 2x_2 \qquad - 3x_4 = -5$
$\qquad\qquad x_3 + 3x_4 = 2$

Solution:
$x_4 = t$
$x_3 = 2 - 3x_4 = 2 - 3t$
$x_2 = s$
$x_1 = -5 + 2x_2 + 3x_4 = -5 + 2s + 3t$
Thus $x_1 = 2s + 3t - 5$, $x_2 = s$,
$x_3 = -3t + 2$, $x_4 = t$ is the
solution, for s and t any real
numbers.

17. $\begin{bmatrix} 1 & 2 & \vline & -1 \\ 0 & 1 & \vline & 3 \end{bmatrix}$ $\quad (-2)R_2 + R_1 \to R_1$

Need a 0 here

$\sim \begin{bmatrix} 1 & 0 & \vline & -7 \\ 0 & 1 & \vline & 3 \end{bmatrix}$

19. $\begin{bmatrix} 1 & 0 & -3 & \vline & 1 \\ 0 & 1 & 2 & \vline & 0 \\ 0 & 0 & 3 & \vline & -6 \end{bmatrix}$ $\quad \frac{1}{3}R_3 \to R_3$

Need a 1 here
Need 0's here

$\sim \begin{bmatrix} 1 & 0 & -3 & \vline & 1 \\ 0 & 1 & 2 & \vline & 0 \\ 0 & 0 & 1 & \vline & -2 \end{bmatrix}$ $\quad \begin{aligned} 3R_3 + R_1 \to R_1 \\ (-2)R_3 + R_2 \to R_2 \end{aligned}$

$\sim \begin{bmatrix} 1 & 0 & 0 & \vline & -5 \\ 0 & 1 & 0 & \vline & 4 \\ 0 & 0 & 1 & \vline & -2 \end{bmatrix}$

21. $\begin{bmatrix} 1 & 2 & -2 & \vline & -1 \\ 0 & 3 & -6 & \vline & 1 \\ 0 & -1 & 2 & \vline & -\frac{1}{3} \end{bmatrix}$ $\quad \frac{1}{3}R_2 \to R_2$

Need a 1 here

$\sim \begin{bmatrix} 1 & 2 & -2 & \vline & -1 \\ 0 & 1 & -2 & \vline & \frac{1}{3} \\ 0 & -1 & 2 & \vline & -\frac{1}{3} \end{bmatrix}$ $\quad \begin{aligned} (-2)R_2 + R_1 \to R_1 \\ \\ R_3 + R_2 \to R_3 \end{aligned}$

Need 0's here

$\sim \begin{bmatrix} 1 & 0 & 2 & \vline & -\frac{5}{3} \\ 0 & 1 & -2 & \vline & \frac{1}{3} \\ 0 & 0 & 0 & \vline & 0 \end{bmatrix}$

23. $\begin{bmatrix} 2 & 4 & -10 & | & -2 \\ 3 & 9 & -21 & | & 0 \\ 1 & 5 & -12 & | & 1 \end{bmatrix}$ $R_1 \leftrightarrow R_3$

Need a 1 here

$\sim \begin{bmatrix} 1 & 5 & -12 & | & 1 \\ 3 & 9 & -21 & | & 0 \\ 2 & 4 & -10 & | & -2 \end{bmatrix}$ $(-3)R_1 + R_2 \to R_2$
$(-2)R_1 + R_3 \to R_3$

Need 0's here

$\sim \begin{bmatrix} 1 & 5 & -12 & | & 1 \\ 0 & -6 & 15 & | & -3 \\ 0 & -6 & 14 & | & -4 \end{bmatrix}$ $-\frac{1}{6}R_2 \to R_2$

Need a 1 here

$\sim \begin{bmatrix} 1 & 5 & -12 & | & 1 \\ 0 & 1 & -\frac{5}{2} & | & \frac{1}{2} \\ 0 & -6 & 14 & | & -4 \end{bmatrix}$ $(-5)R_2 + R_1 \to R_1$
$6R_2 + R_3 \to R_3$

Need 0's here

$\sim \begin{bmatrix} 1 & 0 & \frac{1}{2} & | & -\frac{3}{2} \\ 0 & 1 & -\frac{5}{2} & | & \frac{1}{2} \\ 0 & 0 & -1 & | & -1 \end{bmatrix}$ $-R_3 \to R_3$

↑
Need a 1 here

$\sim \begin{bmatrix} 1 & 0 & \frac{1}{2} & | & -\frac{3}{2} \\ 0 & 1 & -\frac{5}{2} & | & \frac{1}{2} \\ 0 & 0 & 1 & | & 1 \end{bmatrix}$ $(-\frac{1}{2})R_3 + R_1 \to R_1$
$\frac{5}{2}R_3 + R_2 \to R_2$

Need 0's here

$\sim \begin{bmatrix} 1 & 0 & 0 & | & -2 \\ 0 & 1 & 0 & | & 3 \\ 0 & 0 & 1 & | & 1 \end{bmatrix}$

Therefore $x_1 = -2$, $x_2 = 3$, and $x_3 = 1$.

25. $\begin{bmatrix} 3 & 8 & -1 & | & -18 \\ 2 & 1 & 5 & | & 8 \\ 2 & 4 & 2 & | & -4 \end{bmatrix}$ $\frac{1}{2}R_3 \to R_3$
$R_3 \leftrightarrow R_1$

Need a 1 here

$\sim \begin{bmatrix} 1 & 2 & 1 & | & -2 \\ 2 & 1 & 5 & | & 8 \\ 3 & 8 & -1 & | & -18 \end{bmatrix}$ $(-2)R_1 + R_2 \to R_2$
$(-3)R_1 + R_3 \to R_3$

Need 0's here

$\sim \begin{bmatrix} 1 & 2 & 1 & | & -2 \\ 0 & -3 & 3 & | & 12 \\ 0 & 2 & -4 & | & -12 \end{bmatrix}$ $-\frac{1}{3}R_2 \to R_2$

Need a 1 here

$\sim \begin{bmatrix} 1 & 2 & 1 & | & -2 \\ 0 & 1 & -1 & | & -4 \\ 0 & 2 & -4 & | & -12 \end{bmatrix}$ $(-2)R_2 + R_1 \to R_1$
$(-2)R_2 + R_3 \to R_3$

Need 0's here

$\sim \begin{bmatrix} 1 & 0 & 3 & | & 6 \\ 0 & 1 & -1 & | & -4 \\ 0 & 0 & -2 & | & -4 \end{bmatrix}$ $-\frac{1}{2}R_3 \to R_3$

↑
Need a 1 here

$\sim \begin{bmatrix} 1 & 0 & 3 & | & 6 \\ 0 & 1 & -1 & | & -4 \\ 0 & 0 & 1 & | & 2 \end{bmatrix}$ $(-3)R_3 + R_1 \to R_1$
$R_3 + R_2 \to R_2$

Need 0's here

$\begin{bmatrix} 1 & 0 & 0 & | & 0 \\ 0 & 1 & 0 & | & -2 \\ 0 & 0 & 1 & | & 2 \end{bmatrix}$

Therefore $x_1 = 0$, $x_2 = -2$, and $x_3 = 2$.

27. $\begin{bmatrix} 2 & -1 & -3 & | & 8 \\ 1 & -2 & 0 & | & 7 \end{bmatrix}$ $R_1 \leftrightarrow R_2$

$\sim \begin{bmatrix} 1 & -2 & 0 & | & 7 \\ 2 & -1 & -3 & | & 8 \end{bmatrix}$ $(-2)R_1 + R_2 \rightarrow R_2$

$\sim \begin{bmatrix} 1 & -2 & 0 & | & 7 \\ 0 & 3 & -3 & | & -6 \end{bmatrix}$ $\frac{1}{3}R_2 \rightarrow R_2$

$\sim \begin{bmatrix} 1 & -2 & 0 & | & 7 \\ 0 & 1 & -1 & | & -2 \end{bmatrix}$ $2R_2 + R_1 \rightarrow R_1$

$\sim \begin{bmatrix} 1 & 0 & -2 & | & 3 \\ 0 & 1 & -1 & | & -2 \end{bmatrix}$

Let $x_3 = t$. Then

$x_2 - x_3 \quad = -2$

$\quad\quad x_2 = x_3 - 2 = t - 2$

$x_1 - 2x_3 \quad = 3$

$\quad\quad x_1 = 2x_3 + 3 = 2t + 3$

Solution: $x_1 = 2t + 3$, $x_2 = t - 2$,

$x_3 = t$, t any real number.

29. $\begin{bmatrix} 2 & -1 & | & 0 \\ 3 & 2 & | & 7 \\ 1 & -1 & | & -1 \end{bmatrix}$ $R_1 \leftrightarrow R_3$

$\sim \begin{bmatrix} 1 & -1 & | & -1 \\ 3 & 2 & | & 7 \\ 2 & -1 & | & 0 \end{bmatrix}$ $\begin{matrix}(-3)R_1 + R_2 \rightarrow R_2\\ (-2)R_1 + R_3 \rightarrow R_3\end{matrix}$

$\sim \begin{bmatrix} 1 & -1 & | & -1 \\ 0 & 5 & | & 10 \\ 0 & 1 & | & 2 \end{bmatrix}$ $R_2 \leftrightarrow R_3$

$\sim \begin{bmatrix} 1 & -1 & | & -1 \\ 0 & 1 & | & 2 \\ 0 & 5 & | & 10 \end{bmatrix}$ $\begin{matrix}R_2 + R_1 \rightarrow R_1\\ (-5)R_2 + R_3 \rightarrow R_3\end{matrix}$

$\sim \begin{bmatrix} 1 & 0 & | & 1 \\ 0 & 1 & | & 2 \\ 0 & 0 & | & 0 \end{bmatrix}$

Therefore, $x_1 = 1$ and $x_2 = 2$.

31. $\begin{bmatrix} 3 & -4 & -1 & | & 1 \\ 2 & -3 & 1 & | & 1 \\ 1 & -2 & 3 & | & 2 \end{bmatrix}$ $R_1 \leftrightarrow R_3$

$\sim \begin{bmatrix} 1 & -2 & 3 & | & 2 \\ 2 & -3 & 1 & | & 1 \\ 3 & -4 & -1 & | & 1 \end{bmatrix}$ $\begin{matrix}(-2)R_1 + R_2 \rightarrow R_2\\ (-3)R_1 + R_3 \rightarrow R_3\end{matrix}$

$\sim \begin{bmatrix} 1 & -2 & 3 & | & 2 \\ 0 & 1 & -5 & | & -3 \\ 0 & 2 & -10 & | & -5 \end{bmatrix}$ $(-2)R_2 + R_3 \rightarrow R_3$

$\sim \begin{bmatrix} 1 & -2 & 3 & | & 2 \\ 0 & 1 & -5 & | & -3 \\ 0 & 0 & 0 & | & 1 \end{bmatrix}$

Since the last row corresponds to the equation $0x_1 + 0x_2 + 0x_3 = 1$, there is no solution.

33. $\begin{bmatrix} -2 & 1 & 3 & | & -7 \\ 1 & -4 & 2 & | & 0 \\ 1 & -3 & 1 & | & 1 \end{bmatrix}$ $R_1 \leftrightarrow R_2$

$\sim \begin{bmatrix} 1 & -4 & 2 & | & 0 \\ -2 & 1 & 3 & | & -7 \\ 1 & -3 & 1 & | & 1 \end{bmatrix}$ $\begin{matrix}2R_1 + R_2 \rightarrow R_2\\ (-1)R_1 + R_3 \rightarrow R_3\end{matrix}$

$\sim \begin{bmatrix} 1 & -4 & 2 & | & 0 \\ 0 & -7 & 7 & | & -7 \\ 0 & 1 & -1 & | & 1 \end{bmatrix}$ $R_2 \leftrightarrow R_3$

$\sim \begin{bmatrix} 1 & -4 & 2 & | & 0 \\ 0 & 1 & -1 & | & 1 \\ 0 & -7 & 7 & | & -7 \end{bmatrix}$ $\begin{matrix}4R_2 + R_1 \rightarrow R_1\\ 7R_2 + R_3 \rightarrow R_3\end{matrix}$

$\sim \begin{bmatrix} 1 & 0 & -2 & | & 4 \\ 0 & 1 & -1 & | & 1 \\ 0 & 0 & 0 & | & 0 \end{bmatrix}$

Let $x_3 = t$. Then

$\quad\quad x_2 - x_3 = 1$

$\quad\quad\quad\quad x_2 = x_3 + 1 = t + 1$

$\quad\quad x_1 - 2x_3 = 4$

$x_1 = 2x_3 + 4 = 2t + 4$

Solution: $x_1 = 2t + 4$, $x_2 = t + 1$,

$x_3 = t$, t any real number.

35. $\begin{bmatrix} 2 & -2 & -4 & | & -2 \\ -3 & 3 & 6 & | & 3 \end{bmatrix}$ $\begin{array}{l} \frac{1}{2}R_1 \rightarrow R_1 \\ \frac{1}{3}R_2 \rightarrow R_2 \end{array}$

$\sim \begin{bmatrix} 1 & -1 & -2 & | & -1 \\ -1 & 1 & 2 & | & 1 \end{bmatrix}$ $R_1 + R_2 \rightarrow R_2$

$\sim \begin{bmatrix} 1 & -1 & -2 & | & -1 \\ 0 & 0 & 0 & | & 0 \end{bmatrix}$

Let $x_3 = t$, $x_2 = s$. Then
$x_1 - x_2 - 2x_3 = -1$
$\qquad x_1 = x_2 + 2x_3 - 1$
$\qquad\quad = s + 2t - 1$
Solution: $x_1 = s + 2t - 1$, $x_2 = s$, $x_3 = t$, s and t any real numbers.

37. $\begin{bmatrix} 4 & -1 & 2 & | & 3 \\ -4 & 1 & -3 & | & -10 \\ 8 & -2 & 9 & | & -1 \end{bmatrix}$ $\begin{array}{l} R_1 + R_2 \rightarrow R_2 \\ (-2)R_1 + R_3 \rightarrow R_3 \end{array}$

$\sim \begin{bmatrix} 4 & -1 & 2 & | & 3 \\ 0 & 0 & -1 & | & -7 \\ 0 & 0 & 5 & | & -7 \end{bmatrix}$ $5R_2 + R_3 \rightarrow R_3$

$\sim \begin{bmatrix} 4 & -1 & 2 & | & 3 \\ 0 & 0 & -1 & | & -7 \\ 0 & 0 & 0 & | & -42 \end{bmatrix}$ Since the last row corresponds to the equation $0x_1 + 0x_2 + 0x_3 = -42$, there is no solution.

39. $\begin{bmatrix} 2 & -5 & -3 & | & 7 \\ -4 & 10 & 2 & | & 6 \\ 6 & -15 & -1 & | & -19 \end{bmatrix}$ $\begin{array}{l} 2R_1 + R_2 \rightarrow R_2 \\ (-3)R_1 + R_3 \rightarrow R_3 \end{array}$

$\sim \begin{bmatrix} 2 & -5 & -3 & | & 7 \\ 0 & 0 & -4 & | & 20 \\ 0 & 0 & 8 & | & -40 \end{bmatrix}$ $\begin{array}{l} \frac{1}{2}R_1 \rightarrow R_1 \\ -\frac{1}{4}R_2 \rightarrow R_2 \end{array}$

$\sim \begin{bmatrix} 1 & -2.5 & -1.5 & | & 3.5 \\ 0 & 0 & 1 & | & -5 \\ 0 & 0 & 8 & | & -40 \end{bmatrix}$ $\begin{array}{l} 1.5R_2 + R_1 \rightarrow R_1 \\ (-8)R_2 + R_3 \rightarrow R_3 \end{array}$

$\sim \begin{bmatrix} 1 & -2.5 & 0 & | & -4 \\ 0 & 0 & 1 & | & -5 \\ 0 & 0 & 0 & | & 0 \end{bmatrix}$

Let $x_2 = t$. Then $x_3 = -5$ and
$\quad x_1 - 2.5x_2 = -4$
$\qquad\quad x_1 = 2.5x_2 - 4$
$\qquad\quad x_1 = 2.5t - 4$

Solution: $x_1 = 2.5t - 4$, $x_2 = t$, $x_3 = -5$, t any real number.

41. $\begin{bmatrix} 5 & -3 & 2 & | & 13 \\ 2 & -1 & -3 & | & 1 \\ 4 & -2 & 4 & | & 12 \end{bmatrix}$ $\frac{1}{4}R_3 \rightarrow R_3$

$\sim \begin{bmatrix} 5 & -3 & 2 & | & 13 \\ 2 & -1 & -3 & | & 1 \\ 1 & -\frac{1}{2} & 1 & | & 3 \end{bmatrix}$ $R_1 \leftrightarrow R_3$

$$\sim \begin{bmatrix} 1 & -\frac{1}{2} & 1 & \bigm| & 3 \\ 2 & -1 & -3 & \bigm| & 1 \\ 5 & -3 & 2 & \bigm| & 13 \end{bmatrix} \begin{matrix} (-2)R_1 + R_2 \rightarrow R_2 \\ \\ (-5)R_1 + R_3 \rightarrow R_3 \end{matrix}$$

$$\sim \begin{bmatrix} 1 & -\frac{1}{2} & 1 & \bigm| & 3 \\ 0 & 0 & -5 & \bigm| & -5 \\ 0 & -\frac{1}{2} & -3 & \bigm| & -2 \end{bmatrix} \begin{matrix} (-1)R_3 + R_1 \rightarrow R_1 \\ -\frac{1}{5}R_2 \rightarrow R_2 \end{matrix}$$

$$\sim \begin{bmatrix} 1 & 0 & 4 & \bigm| & 5 \\ 0 & 0 & 1 & \bigm| & 1 \\ 0 & -\frac{1}{2} & -3 & \bigm| & -2 \end{bmatrix} \begin{matrix} (-4)R_2 + R_1 \rightarrow R_1 \\ 3R_2 + R_3 \rightarrow R_3 \end{matrix}$$

$$\sim \begin{bmatrix} 1 & 0 & 0 & \bigm| & 1 \\ 0 & 0 & 1 & \bigm| & 1 \\ 0 & -\frac{1}{2} & 0 & \bigm| & 1 \end{bmatrix} (-2)R_3 \rightarrow R_3$$

$$\sim \begin{bmatrix} 1 & 0 & 0 & \bigm| & 1 \\ 0 & 0 & 1 & \bigm| & 1 \\ 0 & 1 & 0 & \bigm| & -2 \end{bmatrix} R_2 \leftrightarrow R_3$$

$$\sim \begin{bmatrix} 1 & 0 & 0 & \bigm| & 1 \\ 0 & 1 & 0 & \bigm| & -2 \\ 0 & 0 & 1 & \bigm| & 1 \end{bmatrix}$$

Solution: $x_1 = 1$, $x_2 = -2$, $x_3 = 1$.

43. (A) The reduced form matrix will have the form

$$\begin{bmatrix} 1 & a & b & \bigm| & c \\ 0 & 0 & 0 & \bigm| & 0 \\ 0 & 0 & 0 & \bigm| & 0 \end{bmatrix}$$

Thus, the system has been shown equivalent to
$$x_1 + ax_2 + bx_3 = c$$
$$0 = 0$$
$$0 = 0$$

The system is dependent, and x_2 and x_3 may assume any real values. Thus, there are two parameters in the solution.

(B) The reduced form matrix will have the form

$$\begin{bmatrix} 1 & 0 & a & \bigm| & b \\ 0 & 1 & c & \bigm| & d \\ 0 & 0 & 0 & \bigm| & 0 \end{bmatrix}$$

Thus, the system has been shown equivalent to
$$x_1 + ax_3 = b$$
$$x_2 + cx_3 = d$$
$$0 = 0$$

The system is dependent, with a solution for any real value of x_3. Thus, there is one parameter in the solution.

(C) The reduced form matrix will have the form

$$\begin{bmatrix} 1 & 0 & 0 & \bigm| & a \\ 0 & 1 & 0 & \bigm| & b \\ 0 & 0 & 1 & \bigm| & c \end{bmatrix}$$

Thus, there is only one solution, $x_1 = a$, $x_2 = b$, $x_3 = c$, and the system is independent.

(D) This is impossible; there are only 3 equations.

45. $\begin{bmatrix} 1 & 2 & -4 & -1 & \bigm| & 7 \\ 2 & 5 & -9 & -4 & \bigm| & 16 \\ 1 & 5 & -7 & -7 & \bigm| & 13 \end{bmatrix}$ $(-2)R_1 + R_2 \rightarrow R_2$
$(-1)R_1 + R_3 \rightarrow R_3$

$\sim \begin{bmatrix} 1 & 2 & -4 & -1 & \bigm| & 7 \\ 0 & 1 & -1 & -2 & \bigm| & 2 \\ 0 & 3 & -3 & -6 & \bigm| & 6 \end{bmatrix}$ $(-2)R_2 + R_1 \rightarrow R_1$

$(-3)R_2 + R_3 \rightarrow R_3$

$\sim \begin{bmatrix} 1 & 0 & -2 & 3 & \bigm| & 3 \\ 0 & 1 & -1 & -2 & \bigm| & 2 \\ 0 & 0 & 0 & 0 & \bigm| & 0 \end{bmatrix}$

Let $x_4 = t$, $x_3 = s$. Then
$$x_2 - x_3 - 2x_4 = 2$$
$$x_2 = s + 2t + 2$$
$$x_1 - 2x_3 + 3x_4 = 3$$
$$x_1 = 2x_3 - 3x_4 + 3$$
$$= 2s - 3t + 3$$

Solution: $x_1 = 2s - 3t + 3$, $x_2 = s + 2t + 2$, $x_3 = s$, $x_4 = t$, s and t any real numbers.

47. $\begin{bmatrix} 1 & -1 & 3 & -2 & \bigm| & 1 \\ -2 & 4 & -3 & 1 & \bigm| & 0.5 \\ 3 & -1 & 10 & -4 & \bigm| & 2.9 \\ 4 & -3 & 8 & -2 & \bigm| & 0.6 \end{bmatrix}$ $2R_1 + R_2 \rightarrow R_2$
$(-3)R_1 + R_3 \rightarrow R_3$
$(-4)R_1 + R_4 \rightarrow R_4$

$\sim \begin{bmatrix} 1 & -1 & 3 & -2 & \bigm| & 1 \\ 0 & 2 & 3 & -3 & \bigm| & 2.5 \\ 0 & 2 & 1 & 2 & \bigm| & -0.1 \\ 0 & 1 & -4 & 6 & \bigm| & -3.4 \end{bmatrix}$ $R_4 \leftrightarrow R_2$

$\sim \begin{bmatrix} 1 & -1 & 3 & -2 & \bigm| & 1 \\ 0 & 1 & -4 & 6 & \bigm| & -3.4 \\ 0 & 2 & 1 & 2 & \bigm| & -0.1 \\ 0 & 2 & 3 & -3 & \bigm| & 2.5 \end{bmatrix}$ $R_2 + R_1 \rightarrow R_1$

$(-2)R_2 + R_3 \rightarrow R_3$
$(-2)R_2 + R_4 \rightarrow R_4$

$\sim \begin{bmatrix} 1 & 0 & -1 & 4 & \bigm| & -2.4 \\ 0 & 1 & -4 & 6 & \bigm| & -3.4 \\ 0 & 0 & 9 & -10 & \bigm| & 6.7 \\ 0 & 0 & 11 & -15 & \bigm| & 9.3 \end{bmatrix}$ $(-1)R_4 + R_3 \rightarrow R_3$

$\sim \begin{bmatrix} 1 & 0 & -1 & 4 & \bigm| & -2.4 \\ 0 & 1 & -4 & 6 & \bigm| & -3.4 \\ 0 & 0 & -2 & 5 & \bigm| & -2.6 \\ 0 & 0 & 11 & -15 & \bigm| & 9.3 \end{bmatrix}$ $-\frac{1}{2}R_3 \leftrightarrow R_3$

$\sim \begin{bmatrix} 1 & 0 & -1 & 4 & \bigm| & -2.4 \\ 0 & 1 & -4 & 6 & \bigm| & -3.4 \\ 0 & 0 & 1 & -2.5 & \bigm| & 1.3 \\ 0 & 0 & 11 & -15 & \bigm| & 9.3 \end{bmatrix}$ $R_3 + R_1 \rightarrow R_1$
$4R_3 + R_2 \rightarrow R_2$

$(-11)R_3 + R_4 \rightarrow R_4$

$\sim \begin{bmatrix} 1 & 0 & 0 & 1.5 & \bigm| & -1.1 \\ 0 & 1 & 0 & -4 & \bigm| & 1.8 \\ 0 & 0 & 1 & -2.5 & \bigm| & 1.3 \\ 0 & 0 & 0 & 12.5 & \bigm| & -5 \end{bmatrix}$ $\frac{1}{12.5}R_4 \rightarrow R_4$

$$\sim \begin{bmatrix} 1 & 0 & 0 & 1.5 & \bigm| & -1.1 \\ 0 & 1 & 0 & -4 & \bigm| & 1.8 \\ 0 & 0 & 1 & -2.5 & \bigm| & 1.3 \\ 0 & 0 & 0 & 1 & \bigm| & -0.4 \end{bmatrix} \quad \begin{array}{l} (-1.5)R_4 + R_1 \rightarrow R_1 \\ 4R_4 + R_2 \rightarrow R_2 \\ 2.5R_4 + R_3 \rightarrow R_3 \end{array}$$

$$\sim \begin{bmatrix} 1 & 0 & 0 & 0 & \bigm| & -0.5 \\ 0 & 1 & 0 & 0 & \bigm| & 0.2 \\ 0 & 0 & 1 & 0 & \bigm| & 0.3 \\ 0 & 0 & 0 & 1 & \bigm| & -0.4 \end{bmatrix}$$

Solution: $x_1 = -0.5$, $x_2 = 0.2$, $x_3 = 0.3$, $x_4 = -0.4$.

49.
$$\begin{bmatrix} 1 & -2 & 1 & 1 & 2 & \bigm| & 2 \\ -2 & 4 & 2 & 2 & -2 & \bigm| & 0 \\ 3 & -6 & 1 & 1 & 5 & \bigm| & 4 \\ -1 & 2 & 3 & 1 & 1 & \bigm| & 3 \end{bmatrix} \quad \begin{array}{l} 2R_1 + R_2 \rightarrow R_2 \\ (-3)R_1 + R_3 \rightarrow R_3 \\ R_1 + R_4 \rightarrow R_4 \end{array}$$

$$\sim \begin{bmatrix} 1 & -2 & 1 & 1 & 2 & \bigm| & 2 \\ 0 & 0 & 4 & 4 & 2 & \bigm| & 4 \\ 0 & 0 & -2 & -2 & -1 & \bigm| & -2 \\ 0 & 0 & 4 & 2 & 3 & \bigm| & 5 \end{bmatrix} \quad \tfrac{1}{4}R_2 \rightarrow R_2$$

$$\sim \begin{bmatrix} 1 & -2 & 1 & 1 & 2 & \bigm| & 2 \\ 0 & 0 & 1 & 1 & 0.5 & \bigm| & 1 \\ 0 & 0 & -2 & -2 & -1 & \bigm| & -2 \\ 0 & 0 & 4 & 2 & 3 & \bigm| & 5 \end{bmatrix} \quad \begin{array}{l} (-1)R_2 + R_1 \rightarrow R_1 \\ \\ 2R_2 + R_3 \rightarrow R_3 \\ (-4)R_2 + R_4 \rightarrow R_4 \end{array}$$

$$\sim \begin{bmatrix} 1 & -2 & 0 & 0 & 1.5 & \bigm| & 1 \\ 0 & 0 & 1 & 1 & 0.5 & \bigm| & 1 \\ 0 & 0 & 0 & 0 & 0 & \bigm| & 0 \\ 0 & 0 & 0 & -2 & 1 & \bigm| & 1 \end{bmatrix} \quad R_3 \leftrightarrow R_4$$

$$\sim \begin{bmatrix} 1 & -2 & 0 & 0 & 1.5 & \bigm| & 1 \\ 0 & 0 & 1 & 1 & 0.5 & \bigm| & 1 \\ 0 & 0 & 0 & -2 & 1 & \bigm| & 1 \\ 0 & 0 & 0 & 0 & 0 & \bigm| & 0 \end{bmatrix} \quad (-\tfrac{1}{2})R_3 \rightarrow R_3$$

$$\sim \begin{bmatrix} 1 & -2 & 0 & 0 & 1.5 & \bigm| & 1 \\ 0 & 0 & 1 & 1 & 0.5 & \bigm| & 1 \\ 0 & 0 & 0 & 1 & -0.5 & \bigm| & -0.5 \\ 0 & 0 & 0 & 0 & 0 & \bigm| & 0 \end{bmatrix} \quad (-1)R_3 + R_2 \rightarrow R_2$$

$$\sim \begin{bmatrix} 1 & -2 & 0 & 0 & 1.5 & \bigm| & 1 \\ 0 & 0 & 1 & 0 & 1 & \bigm| & 1.5 \\ 0 & 0 & 0 & 1 & -0.5 & \bigm| & -0.5 \\ 0 & 0 & 0 & 0 & 0 & \bigm| & 0 \end{bmatrix}$$

Let $x_5 = t$. Then
$$x_4 - 0.5x_5 = -0.5$$
$$x_4 = 0.5x_5 - 0.5 = 0.5t - 0.5$$
$$x_3 + x_5 = 1.5$$
$$x_3 = -x_5 + 1.5 = -t + 1.5$$

Let $x_2 = s$. Then

$$x_1 - 2x_2 + 1.5x_5 = 1$$
$$x_1 = 2x_2 - 1.5x_5 + 1 = 2s - 1.5t + 1$$

Solution: $x_1 = 2s - 1.5t + 1$, $x_2 = s$, $x_3 = -t + 1.5$, $x_4 = 0.5t - 0.5$, $x_5 = t$, s and t any real numbers.

51. Let x_1 = number of 15-cent stamps

x_2 = number of 20-cent stamps

x_3 = number of 35-cent stamps

Then $x_1 + x_2 + x_3 = 45$ (total number of stamps)

$15x_1 + 20x_2 + 35x_3 = 1400$ (total value of stamps)

We write the augmented matrix and solve by Gauss-Jordan elimination.

$$\begin{bmatrix} 1 & 1 & 1 & | & 45 \\ 15 & 20 & 35 & | & 1400 \end{bmatrix} \quad (-15)R_1 + R_2 \to R_2$$

$$\sim \begin{bmatrix} 1 & 1 & 1 & | & 45 \\ 0 & 5 & 20 & | & 725 \end{bmatrix} \quad \tfrac{1}{5}R_2 \to R_2$$

$$\sim \begin{bmatrix} 1 & 1 & 1 & | & 45 \\ 0 & 1 & 4 & | & 145 \end{bmatrix} \quad (-1)R_2 + R_1 \to R_1$$

$$\sim \begin{bmatrix} 1 & 0 & -3 & | & -100 \\ 0 & 1 & 4 & | & 145 \end{bmatrix}$$

This augmented matrix is in reduced form. It corresponds to the system:

$$x_1 - 3x_3 = -100$$
$$x_2 + 4x_3 = 145$$

Let $x_3 = t$. Then

$$x_2 = -4x_3 + 145$$
$$= -4t + 145$$
$$x_1 = 3x_3 - 100$$
$$= 3t - 100$$

A solution is achieved, not for every real value of t, but for integer values of t that give rise to non-negative x_1, x_2, x_3.

$x_1 \geq 0$ means $3t - 100 \geq 0$ or $t \geq 33\tfrac{1}{3}$

$x_2 \geq 0$ means $-4t + 145 \geq 0$ or $t \leq 36\tfrac{1}{4}$

The only integer values of t that satisfy these conditions are 34, 35, 36. Thus we have the solutions:

$x_1 = (3t - 100)$ 15-cent stamps

$x_2 = (145 - 4t)$ 20-cent stamps

$x_3 = t$ 35-cent stamps

where t = 34, 35, or 36.

53. Let x_1 = number of 500-cc containers of 10% solution

x_2 = number of 500-cc containers of 20% solution

x_3 = number of 1,000-cc containers of 50% solution

Then $500x_1 + 500x_2 + 1{,}000x_3 = 12{,}000$ (total number of cc)

$0.10(500x_1) + 0.20(500x_2) + 0.50(1{,}000x_3) = 0.30(12{,}000)$ (total amount of ingredient in solution.)

After simplification, we have

$$x_1 + x_2 + 2x_3 = 24$$
$$x_1 + 2x_2 + 10x_3 = 72$$

We write the augmented matrix and solve by Gauss-Jordan elimination:

$$\begin{bmatrix} 1 & 1 & 2 & | & 24 \\ 1 & 2 & 10 & | & 72 \end{bmatrix} \quad (-1)R_1 + R_2 \to R_2$$

$$\sim \begin{bmatrix} 1 & 1 & 2 & | & 24 \\ 0 & 1 & 8 & | & 48 \end{bmatrix} \quad (-1)R_2 + R_1 \rightarrow R_1$$

$$\sim \begin{bmatrix} 1 & 0 & -6 & | & -24 \\ 0 & 1 & 8 & | & 48 \end{bmatrix}$$

This augmented matrix is in reduced form. It corresponds to the system:

$x_1 - 6x_3 = -24$

$x_2 + 8x_3 = 48$

Let $x_3 = t$. Then

$x_2 = -8x_3 + 48$

$\quad = -8t + 48$

$x_1 = 6x_3 - 24$

$\quad = 6t - 24$

A solution is achieved, not for every real value of t, but for integer values of t that give rise to non-negative x_1, x_2, x_3.

$x_1 \geq 0$ means $6t - 24 \geq 0$ or $t \geq 4$

$x_2 \geq 0$ means $-8t + 48 \geq 0$ or $t \leq 6$

Thus we have the solution:

$x_1 = (6t - 24)$ 500-cc containers of 10% solution

$x_2 = (48 - 8t)$ 500-cc containers of 20% solution

$x_3 = t$ 1000-cc containers of 50% solution

where $t = 4$, 5, or 6.

55. If the curve passes through a point, the coordinates of the point satisfy the equation of the curve. Hence,

$3 = a + b(-2) + c(-2)^2$

$2 = a + b(-1) + c(-1)^2$

$6 = a + b(1) + c(1)^2$

After simplication, we have

$a - 2b + 4c = 3$

$a - b + c = 2$

$a + b + c = 6$

We write the augmented matrix and solve by Gauss-Jordan elimination.

$$\begin{bmatrix} 1 & -2 & 4 & | & 3 \\ 1 & -1 & 1 & | & 2 \\ 1 & 1 & 1 & | & 6 \end{bmatrix} \quad \begin{array}{l} (-1)R_1 + R_2 \rightarrow R_2 \\ (-1)R_1 + R_3 \rightarrow R_3 \end{array}$$

$$\sim \begin{bmatrix} 1 & -2 & 4 & | & 3 \\ 0 & 1 & -3 & | & -1 \\ 0 & 3 & -3 & | & 3 \end{bmatrix} \quad \begin{array}{l} 2R_2 + R_1 \rightarrow R_1 \\ \\ (-3)R_2 + R_3 \rightarrow R_3 \end{array}$$

$$\sim \begin{bmatrix} 1 & 0 & -2 & | & 1 \\ 0 & 1 & -3 & | & -1 \\ 0 & 0 & 6 & | & 6 \end{bmatrix} \quad \frac{1}{6}R_3 \rightarrow R_3$$

$$\sim \begin{bmatrix} 1 & 0 & -2 & | & 1 \\ 0 & 1 & -3 & | & -1 \\ 0 & 0 & 1 & | & 1 \end{bmatrix} \quad \begin{array}{l} 2R_3 + R_1 \rightarrow R_1 \\ 3R_3 + R_2 \rightarrow R_2 \end{array}$$

$$\sim \begin{bmatrix} 1 & 0 & 0 & | & 3 \\ 0 & 1 & 0 & | & 2 \\ 0 & 0 & 1 & | & 1 \end{bmatrix}$$

Thus $a = 3$, $b = 2$, $c = 1$.

57. If the curve passes through a point, the coordinates of the point satisfy the equation of the curve. Hence,

$$6^2 + 2^2 + a(6) + b(2) + c = 0$$
$$4^2 + 6^2 + a(4) + b(6) + c = 0$$
$$(-3)^2 + (-1)^2 + a(-3) + b(-1) + c = 0$$

After simplification, we have

$$6a + 2b + c = -40$$
$$4a + 6b + c = -52$$
$$-3a - b + c = -10$$

We write the augmented matrix and solve by Gauss-Jordan elimination.

$$\begin{bmatrix} 6 & 2 & 1 & | & -40 \\ 4 & 6 & 1 & | & -52 \\ -3 & -1 & 1 & | & -10 \end{bmatrix} \quad \tfrac{1}{6}R_1 \rightarrow R_1$$

$$\sim \begin{bmatrix} 1 & \tfrac{1}{3} & \tfrac{1}{6} & | & -\tfrac{20}{3} \\ 4 & 6 & 1 & | & -52 \\ -3 & -1 & 1 & | & -10 \end{bmatrix} \quad \begin{array}{l} (-4)R_1 + R_2 \rightarrow R_2 \\ 3R_1 + R_3 \rightarrow R_3 \end{array}$$

$$\sim \begin{bmatrix} 1 & \tfrac{1}{3} & \tfrac{1}{6} & | & -\tfrac{20}{3} \\ 0 & \tfrac{14}{3} & \tfrac{1}{3} & | & -\tfrac{76}{3} \\ 0 & 0 & \tfrac{3}{2} & | & -30 \end{bmatrix} \quad \tfrac{2}{3}R_3 \rightarrow R_3$$

$$\sim \begin{bmatrix} 1 & \tfrac{1}{3} & \tfrac{1}{6} & | & -\tfrac{20}{3} \\ 0 & \tfrac{14}{3} & \tfrac{1}{3} & | & -\tfrac{76}{3} \\ 0 & 0 & 1 & | & -20 \end{bmatrix} \quad \begin{array}{l} (-\tfrac{1}{6})R_3 + R_1 \rightarrow R_1 \\ (-\tfrac{1}{3})R_3 + R_2 \rightarrow R_2 \end{array}$$

$$\sim \begin{bmatrix} 1 & \tfrac{1}{3} & 0 & | & -\tfrac{10}{3} \\ 0 & \tfrac{14}{3} & 0 & | & -\tfrac{56}{3} \\ 0 & 0 & 1 & | & -20 \end{bmatrix} \quad \tfrac{3}{14}R_2 \rightarrow R_2$$

$$\sim \begin{bmatrix} 1 & \tfrac{1}{3} & 0 & | & -\tfrac{10}{3} \\ 0 & 1 & 0 & | & -4 \\ 0 & 0 & 1 & | & -20 \end{bmatrix} \quad (-\tfrac{1}{3})R_2 + R_1 \rightarrow R_1$$

$$\begin{bmatrix} 1 & 0 & 0 & | & -2 \\ 0 & 1 & 0 & | & -4 \\ 0 & 0 & 1 & | & -20 \end{bmatrix}$$

Thus $a = -2$, $b = -4$, and $c = -20$.

59. Let x_1 = number of one-person boats
x_2 = number of two-person boats
x_3 = number of four-person boats

We have

$$0.5x_1 + 1.0x_2 + 1.5x_3 = 380 \text{ cutting department}$$
$$0.6x_1 + 0.9x_2 + 1.2x_3 = 330 \text{ assembly department}$$
$$0.2x_1 + 0.3x_2 + 0.5x_3 = 120 \text{ packing department}$$

Clearing of decimals for convenience:

$$x_1 + 2x_2 + 3x_3 = 760$$
$$6x_1 + 9x_2 + 12x_3 = 3300$$
$$2x_1 + 3x_2 + 5x_3 = 1200$$

> **Common Error:**
> The facts in this problem do not justify the equation
> $0.5x_1 + 0.6x_2 + 0.2x_3 = 380$

We write the augmented matrix and solve by Gauss-Jordan elimination:

$$\left[\begin{array}{ccc|c} 1 & 2 & 3 & 760 \\ 6 & 9 & 12 & 3300 \\ 2 & 3 & 5 & 1200 \end{array}\right] \begin{array}{l} \\ (-6)R_1 + R_2 \rightarrow R_2 \\ (-2)R_1 + R_3 \rightarrow R_3 \end{array}$$

$$\sim \left[\begin{array}{ccc|c} 1 & 2 & 3 & 760 \\ 0 & -3 & -6 & -1260 \\ 0 & -1 & -1 & -320 \end{array}\right] \begin{array}{l} \\ -\frac{1}{3}R_2 \rightarrow R_2 \\ \\ \end{array}$$

$$\sim \left[\begin{array}{ccc|c} 1 & 2 & 3 & 760 \\ 0 & 1 & 2 & 420 \\ 0 & -1 & -1 & -320 \end{array}\right] \begin{array}{l} (-2)R_2 + R_1 \rightarrow R_1 \\ \\ R_2 + R_3 \rightarrow R_3 \end{array}$$

$$\sim \left[\begin{array}{ccc|c} 1 & 0 & -1 & -80 \\ 0 & 1 & 2 & 420 \\ 0 & 0 & 1 & 100 \end{array}\right] \begin{array}{l} R_3 + R_1 \rightarrow R_1 \\ (-2)R_3 + R_2 \rightarrow R_2 \\ \\ \end{array}$$

$$\left[\begin{array}{ccc|c} 1 & 0 & 0 & 20 \\ 0 & 1 & 0 & 220 \\ 0 & 0 & 1 & 100 \end{array}\right]$$

Therefore
$x_1 = 20$ one-person boats
$x_2 = 220$ two-person boats
$x_3 = 100$ four-person boats

61. This assumption discards the third equation. The system, cleared of decimals, reads

$$x_1 + 2x_2 + 3x_3 = 760$$
$$6x_1 + 9x_2 + 12x_3 = 3300$$

The augmented matrix becomes

$$\left[\begin{array}{ccc|c} 1 & 2 & 3 & 760 \\ 6 & 9 & 12 & 3300 \end{array}\right]$$

We solve by Gauss-Jordan elimination. We start by introducing a 0 into the lower left corner using $(-6)R_1 + R_2$ as in the previous problem:

$$\sim \left[\begin{array}{ccc|c} 1 & 2 & 3 & 760 \\ 0 & -3 & -6 & -1260 \end{array}\right] \quad -\frac{1}{3}R_2 \rightarrow R_2$$

$$\sim \left[\begin{array}{ccc|c} 1 & 2 & 3 & 760 \\ 0 & 1 & 2 & 420 \end{array}\right] \quad (-2)R_2 + R_1 \rightarrow R_1$$

$$\sim \left[\begin{array}{ccc|c} 1 & 0 & -1 & -80 \\ 0 & 1 & 2 & 420 \end{array}\right]$$

This augmented matrix is in reduced form. It corresponds to the system:
$$x_1 - x_3 = -80$$
$$x_2 + 2x_3 = 420$$
Let $x_3 = t$. Then
$$x_2 = -2x_3 + 420$$
$$= -2t + 420$$
$$x_1 = x_3 - 80$$
$$= t - 80$$
A solution is achieved, not for every real value of t, but for integer values of t that give rise to non-negative x_1, x_2, x_3.
$x_1 \geq 0$ means $t - 80 \geq 0$ or $t \geq 80$
$x_2 \geq 0$ means $-2t + 420 \geq 0$ or $210 \geq t$
Thus we have the solution
$x_1 = (t - 80)$ one-person boats
$x_2 = (-2t + 420)$ two-person boats
$x_3 = t$ four-person boats
$80 \leq t \leq 210$, t an integer

63. In this case we have $x_3 = 0$ from the beginning. The three equations of problem 59, cleared of decimals, read:

$$x_1 + 2x_2 = 760$$
$$6x_1 + 9x_2 = 3300$$
$$2x_1 + 3x_2 = 1200$$

The augmented matrix becomes:

$$\begin{bmatrix} 1 & 2 & | & 760 \\ 6 & 9 & | & 3300 \\ 2 & 3 & | & 1200 \end{bmatrix}$$

Notice that the row operation
$$(-3)R_3 + R_2 \rightarrow R_2$$
transforms this into the equivalent augmented matrix:

$$\begin{bmatrix} 1 & 2 & | & 760 \\ 0 & 0 & | & -300 \\ 2 & 3 & | & 1200 \end{bmatrix}$$

Therefore, since the second row corresponds to the equation
$$0x_1 + 0x_2 = -300$$
there is no solution.

No production schedule will use all the work-hours in all departments.

65. Let x_1 = number of ounces of food A.
x_2 = number of ounces of food B.
x_3 = number of ounces of food C.

Common Error:
The facts in this problem do not justify the equation
$30x_1 + 10x_2 + 10x_3 = 340$

Then
$$30x_1 + 10x_2 + 20x_3 = 340 \text{ (calcium)}$$
$$10x_1 + 10x_2 + 20x_3 = 180 \text{ (iron)}$$
$$10x_1 + 30x_2 + 20x_3 = 220 \text{ (vitamin A)}$$

or
$$3x_1 + x_2 + 2x_3 = 34$$
$$x_1 + x_2 + 2x_3 = 18$$
$$x_1 + 3x_2 + 2x_3 = 22$$

is the system to be solved. We form the augmented matrix and solve by Gauss-Jordan elimination.

$$\begin{bmatrix} 3 & 1 & 2 & | & 34 \\ 1 & 1 & 2 & | & 18 \\ 1 & 3 & 2 & | & 22 \end{bmatrix} R_1 \leftrightarrow R_2$$

$$\sim \begin{bmatrix} 1 & 1 & 2 & | & 18 \\ 3 & 1 & 2 & | & 34 \\ 1 & 3 & 2 & | & 22 \end{bmatrix} \begin{matrix} (-3)R_1 + R_2 \rightarrow R_2 \\ (-1)R_1 + R_3 \rightarrow R_3 \end{matrix}$$

$$\sim \begin{bmatrix} 1 & 1 & 2 & | & 18 \\ 0 & -2 & -4 & | & -20 \\ 0 & 2 & 0 & | & 4 \end{bmatrix} -\tfrac{1}{2}R_2 \rightarrow R_2$$

$$\sim \begin{bmatrix} 1 & 1 & 2 & | & 18 \\ 0 & 1 & 2 & | & 10 \\ 0 & 2 & 0 & | & 4 \end{bmatrix} \begin{matrix} (-1)R_2 + R_1 \rightarrow R_1 \\ (-2)R_2 + R_3 \rightarrow R_3 \end{matrix}$$

$$\sim \begin{bmatrix} 1 & 0 & 0 & | & 8 \\ 0 & 1 & 2 & | & 10 \\ 0 & 0 & -4 & | & -16 \end{bmatrix} -\tfrac{1}{4}R_3 \rightarrow R_3$$

$$\sim \begin{bmatrix} 1 & 0 & 0 & | & 8 \\ 0 & 1 & 2 & | & 10 \\ 0 & 0 & 1 & | & 4 \end{bmatrix} \quad (-2)R_3 + R_2 \to R_2$$

$$\sim \begin{bmatrix} 1 & 0 & 0 & | & 8 \\ 0 & 1 & 0 & | & 2 \\ 0 & 0 & 1 & | & 4 \end{bmatrix}$$

Thus

$x_1 = 8$ ounces food A

$x_2 = 2$ ounces food B

$x_3 = 4$ ounces food C

67. In this case we have $x_3 = 0$ from the beginning. The three equations of problem 65 become

$$30x_1 + 10x_2 = 340$$
$$10x_1 + 10x_2 = 180$$
$$10x_1 + 30x_2 = 220$$

or

$$3x_1 + x_2 = 34$$
$$x_1 + x_2 = 18$$
$$x_1 + 3x_2 = 22$$

The augmented matrix becomes

$$\begin{bmatrix} 3 & 1 & | & 34 \\ 1 & 1 & | & 18 \\ 1 & 3 & | & 22 \end{bmatrix}$$

We solve by Gauss-Jordan elimination, starting by the row operation

$R_1 \leftrightarrow R_2$

$$\begin{bmatrix} 1 & 1 & | & 18 \\ 3 & 1 & | & 34 \\ 1 & 3 & | & 22 \end{bmatrix} \quad \begin{matrix} (-3)R_1 + R_2 \to R_2 \\ (-1)R_1 + R_3 \to R_3 \end{matrix}$$

$$\sim \begin{bmatrix} 1 & 1 & | & 18 \\ 0 & -2 & | & -20 \\ 0 & 2 & | & 4 \end{bmatrix} \quad R_2 + R_3 \to R_3$$

$$\sim \begin{bmatrix} 1 & 1 & | & 18 \\ 0 & -2 & | & -20 \\ 0 & 0 & | & -16 \end{bmatrix}$$

Since the third row corresponds to the equation

$$0x_1 + 0x_2 = -16$$

there is no solution.

69. In this case we discard the third equation. The system becomes

$$30x_1 + 10x_2 + 20x_3 = 340$$
$$10x_1 + 10x_2 + 20x_3 = 180$$

or

$$3x_1 + x_2 + 2x_3 = 34$$
$$x_1 + x_2 + 2x_3 = 18$$

The augmented matrix becomes

$$\begin{bmatrix} 3 & 1 & 2 & | & 34 \\ 1 & 1 & 2 & | & 18 \end{bmatrix}$$

We solve by Gauss-Jordan elimination, starting by the row operation $R_1 \leftrightarrow R_2$.

$$\begin{bmatrix} 1 & 1 & 2 & | & 18 \\ 3 & 1 & 2 & | & 34 \end{bmatrix} \quad (-3)R_1 + R_2 \to R_2$$

$$\sim \begin{bmatrix} 1 & 1 & 2 & | & 18 \\ 0 & -2 & -4 & | & -20 \end{bmatrix} \quad -\tfrac{1}{2} R_2 \to R_2$$

$$\sim \begin{bmatrix} 1 & 1 & 2 & | & 18 \\ 0 & 1 & 2 & | & 10 \end{bmatrix} \quad (-1) R_2 + R_1 \to R_1$$

$$\sim \begin{bmatrix} 1 & 0 & 0 & | & 8 \\ 0 & 1 & 2 & | & 10 \end{bmatrix}$$

This augmented matrix is in reduced form. It corresponds to the system

$$x_1 = 8$$
$$x_2 + 2x_3 = 10$$

Let $x_3 = t$

Then $x_2 = -2x_3 + 10$
$$\qquad = -2t + 10$$

A solution is achieved, not for every real value t, but for values of t that give rise to non-negative x_2, x_3.

$x_3 \geq 0$ means $t \geq 0$

$x_2 \geq 0$ means $-2t + 10 \geq 0$, $5 \geq t$

Thus we have the solution

$x_1 = 8$ ounces food A

$x_2 = -2t + 10$ ounces food B

$x_3 = t$ ounces food C

$0 \leq t \leq 5$

71. Let x_1 = number of hours company A is to be scheduled

     x_2 = number of hours company B is to be scheduled

In x_1 hours, company A can handle $30x_1$ telephone and $10x_1$ house contacts.

In x_2 hours, company B can handle $20x_2$ telephone and $20x_2$ house contacts.

We therefore have:

    $30x_1 + 20x_2 = 600$ telephone contacts

    $10x_1 + 20x_2 = 400$ house contacts

We form the augmented matrix and solve by Gauss-Jordan elimination.

$$\begin{bmatrix} 30 & 20 & | & 600 \\ 10 & 20 & | & 400 \end{bmatrix} \quad \begin{array}{l} \tfrac{1}{10} R_1 \to R_1 \\ \tfrac{1}{10} R_2 \to R_2 \end{array}$$

$$\sim \begin{bmatrix} 3 & 2 & | & 60 \\ 1 & 2 & | & 40 \end{bmatrix} \quad R_1 \leftrightarrow R_2$$

$$\sim \begin{bmatrix} 1 & 2 & | & 40 \\ 3 & 2 & | & 60 \end{bmatrix} \quad (-3) R_1 + R_2 \to R_2$$

$$\sim \begin{bmatrix} 1 & 2 & | & 40 \\ 0 & -4 & | & -60 \end{bmatrix} \quad -\tfrac{1}{4} R_3 \to R_3$$

$$\sim \begin{bmatrix} 1 & 2 & | & 40 \\ 0 & 1 & | & 15 \end{bmatrix} \quad (-2) R_2 + R_1 \to R_1$$

$$\sim \begin{bmatrix} 1 & 0 & | & 10 \\ 0 & 1 & | & 15 \end{bmatrix}$$

Therefore

    $x_1 = 10$ hours company A

    $x_2 = 15$ hours company B

Exercise 6-3

Key Ideas and Formulas

Nonlinear systems are systems that contain at least one nonlinear equation.

Nonlinear systems involving second degree terms can have at most four solutions, some of which may be imaginary.

Nonlinear systems can be solved by the substitution and, in some cases, the elimination methods of Section 6-1.

It is important to check the apparent solutions of any nonlinear system to insure that extraneous roots have not been introduced.

1.
$$x^2 + y^2 = 169$$
$$x = -12$$
$$(-12)^2 + y^2 = 169$$
$$y^2 = 25$$
$$y = \pm 5$$

Solution: $(-12, 5)$, $(-12, -5)$

Check: $-12 \overset{\checkmark}{=} -12$

$(-12)^2 + (\pm 5)^2 \overset{\checkmark}{=} 169$

3.
$$8x^2 - y^2 = 16$$
$$y = 2x$$

Substitute y from the second equation into the first equation.
$$8x^2 - (2x)^2 = 16$$
$$8x^2 - 4x^2 = 16$$
$$4x^2 = 16$$
$$x^2 = 4$$
$$x = \pm 2$$

For $x = 2$ For $x = -2$
   $y = 2(2)$ $y = 2(-2)$
   $y = 4$ $y = -4$

Solutions: $(2, 4)$, $(-2, -4)$

Check:

For $(2, 4)$ For $(-2, -4)$

$4 \overset{\checkmark}{=} 2 \cdot 2$ $-4 \overset{\checkmark}{=} 2(-2)$

$8(2)^2 - 4^2 \overset{\checkmark}{=} 16$ $8(-2)^2 - (-4)^2 \overset{\checkmark}{=} 16$

5. $3x^2 - 2y^2 = 25$
    $x + y = 0$

Solve for y in the first degree equation
$y = -x$
Substitute into the second degree equation.
$$3x^2 - 2(-x)^2 = 25$$
$$x^2 = 25$$
$$x = \pm 5$$

For $x = 5$ For $x = -5$
   $y = -5$ $y = 5$

Solutions: $(5, -5)$, $(-5, 5)$

Check:

For $(5, -5)$ For $(-5, 5)$

$5 + (-5) \overset{\checkmark}{=} 0$ $(-5) + 5 \overset{\checkmark}{=} 0$

$3(5)^2 - 2(-5)^2 \overset{\checkmark}{=} 25$ $3(-5)^2 - 2(5)^2 \overset{\checkmark}{=} 25$

From this point on we will not show the checking steps for lack of space. The student should perform these checking steps, however.

7. $\quad y^2 = x$
$\quad x - 2y = 2$

Solve for x in the first degree equation.
$x = 2y + 2$
Substitute into the second degree equation.
$\quad y^2 = 2y + 2$
$y^2 - 2y - 2 = 0$

$$y = \frac{-b \pm \sqrt{b^2 - 4ac}}{2a} \quad a = 1, \ b = -2, \ c = -2$$

$$y = \frac{-(-2) \pm \sqrt{(-2)^2 - 4(1)(-2)}}{2(1)}$$

$$y = \frac{2 \pm \sqrt{12}}{2}$$

$$y = 1 \pm \sqrt{3}$$

For $y = 1 + \sqrt{3}$ $\qquad$ For $y = 1 - \sqrt{3}$
$\quad x = 2(1 + \sqrt{3}) + 2$ $\qquad x = 2(1 - \sqrt{3}) + 2$
$\quad x = 4 + 2\sqrt{3}$ $\qquad\qquad x = 4 - 2\sqrt{3}$

Solutions: $(4 + 2\sqrt{3}, \ 1 + \sqrt{3}), \ (4 - 2\sqrt{3}, \ 1 - \sqrt{3})$

9. $2x^2 + y^2 = 24$
$\quad x^2 - y^2 = -12$

Solve using elimination by addition. Adding, we obtain:
$3x^2 = 12$
$\quad x^2 = 4$
$\quad x = \pm 2$

For $x = 2$ $\qquad\qquad$ For $x = -2$
$4 - y^2 = -12$ $\qquad 4 - y^2 = -12$
$\quad -y^2 = -16$ $\quad$ Similarly
$\quad y^2 = 16$ $\qquad\qquad y = \pm 4$
$\quad y = \pm 4$

Solutions: $(2, 4), \ (2, -4), \ (-2, 4), \ (-2, -4)$

11. $\quad x^2 + y^2 = 10$
$16x^2 + y^2 = 25$

Solve using elimination by addition. Multiply the top equation by -1 and add.
$\quad -x^2 - y^2 = -10$
$\quad \underline{16x^2 + y^2 = \ \ 25}$
$\quad 15x^2 \qquad = 15$
$\quad\quad x^2 \qquad = 1$
$\quad\quad\quad x \ = \pm 1$

For $x = 1$ $\quad$ For $x = -1$
$1 + y^2 = 10$ $\quad 1 + y^2 = 10$
$\quad y^2 = 9$ $\qquad\quad y = \pm 3$
$\quad y = \pm 3$

Solutions: $(1, 3), \ (1, -3),$
$(-1, 3), \ (-1, -3)$

13. $xy - 4 = 0$
$\quad x - y = 2$

Solve for x in the first degree equation.
$x = y + 2$
Substitute into the second degree equation
$(y + 2)y - 4 = 0$
$\quad y^2 + 2y - 4 = 0$

$$y = \frac{-b \pm \sqrt{b^2 - 4ac}}{2a} \quad \begin{matrix} a = 1, \ b = 2, \\ c = -4 \end{matrix}$$

$$y = \frac{-2 \pm \sqrt{(2)^2 - 4(1)(-4)}}{2(1)}$$

$$y = \frac{-2 \pm \sqrt{20}}{2}$$

$$y = -1 \pm \sqrt{5}$$

For $y = -1 + \sqrt{5}$ $\qquad$ For $y = -1 - \sqrt{5}$
$\quad x = -1 + \sqrt{5} + 2$ $\qquad x = -1 - \sqrt{5} + 2$
$\quad x = 1 + \sqrt{5}$ $\qquad\qquad x = 1 - \sqrt{5}$

Solutions: $(1 + \sqrt{5}, \ -1 + \sqrt{5}),$
$(1 - \sqrt{5}, \ -1 - \sqrt{5})$

15. $x^2 + 2y^2 = 6$
     $xy = 2$

Solve for y in the second equation

$y = \dfrac{2}{x}$

Substitute into the first equation

$$x^2 + 2\left(\dfrac{2}{x}\right)^2 = 6$$

$$x^2 + \dfrac{8}{x^2} = 6 \qquad x \neq 0$$

$$x^2 \cdot x^2 + x^2 \cdot \dfrac{8}{x^2} = 6x^2$$

$$x^4 + 8 = 6x^2$$
$$x^4 - 6x^2 + 8 = 0$$
$$(x^2 - 2)(x^2 - 4) = 0$$
$$(x - \sqrt{2})(x + \sqrt{2})(x - 2)(x + 2) = 0$$
$$x = \sqrt{2},\ -\sqrt{2},\ 2,\ -2$$

For $x = \sqrt{2}$ For $x = -\sqrt{2}$ For $x = 2$ For $x = -2$

$y = \dfrac{2}{\sqrt{2}}$ $y = -\dfrac{2}{\sqrt{2}}$ $y = \dfrac{2}{2}$ $y = \dfrac{2}{-2}$

$y = \sqrt{2}$ $y = -\sqrt{2}$ $y = 1$ $y = -1$

Solutions: $(\sqrt{2}, \sqrt{2})$, $(-\sqrt{2}, -\sqrt{2})$, $(2, 1)$, $(-2, -1)$

17. $2x^2 + 3y^2 = -4$
     $4x^2 + 2y^2 = 8$

Solve using elimination by addition. Multiply the second equation by $-\dfrac{1}{2}$ and add.

$\quad 2x^2 + 3y^2 = -4$
$\underline{-2x^2 - \ y^2 = -4}$
$\qquad\quad 2y^2 = -8$
$\qquad\quad\ y^2 = -4$
$\qquad\qquad y = \pm 2i$

For $y = 2i$ For $y = -2i$
$2x^2 + 3(2i)^2 = -4$ $2x^2 + 3(-2i)^2 = -4$
$\quad 2x^2 - 12 = -4$ $2x^2 - 12 = -4$
$\qquad 2x^2 = 8$ Similarly
$\qquad\ x^2 = 4$ $x = \pm 2$
$\qquad\ x = \pm 2$

Solutions: $(2, 2i)$, $(-2, 2i)$, $(2, -2i)$, $(-2, -2i)$

19. $x^2 - y^2 = 2$
     $y^2 = x$

Substitute y^2 from the second
equation into the first equation.
$\qquad x^2 - x = 2$
$\qquad x^2 - x - 2 = 0$
$\ (x - 2)(x + 1) = 0$
$\qquad\qquad x = 2,\ -1$

For $x = 2$ For $x = -1$
$\quad y^2 = 2$ $y^2 = -1$
$\quad y = \pm\sqrt{2}$ $y = \pm i$

Solutions: $(2, \sqrt{2})$, $(2, -\sqrt{2})$,
$(-1, i)$, $(-1, -i)$

21. $x^2 + y^2 = 9$
     $x^2 = 9 - 2y$

Substitute x^2 from the second
equation into the first equation.
$9 - 2y + y^2 = 9$
$\quad y^2 - 2y = 0$
$\quad y(y - 2) = 0$
$\qquad\quad y = 0,\ 2$

For $y = 0$ For $y = 2$
$\quad x^2 = 9 - 2(0)$ $x^2 = 9 - 2(2)$
$\quad x^2 = 9$ $x^2 = 5$

$\quad x = \pm 3$ $x = \pm\sqrt{5}$

Solutions: $(3, 0)$, $(-3, 0)$,
$(\sqrt{5}, 2)$, $(-\sqrt{5}, 2)$

23. $x^2 - y^2 = 3$
       $xy = 2$

Solve for y in the second equation.

$y = \dfrac{2}{x}$

Substitute into the first equation:

$$x^2 - \left(\dfrac{2}{x}\right)^2 = 3$$

$$x^2 - \dfrac{4}{x^2} = 3 \quad x \neq 0$$

$$x^4 - 4 = 3x^2$$

$$x^4 - 3x^2 - 4 = 0$$

$$(x^2 - 4)(x^2 + 1) = 0$$

$x^2 - 4 = 0 \qquad x^2 + 1 = 0$

$\quad x^2 = 4 \qquad\qquad x^2 = -1$

$\quad x = \pm 2 \qquad\qquad x = \pm i$

For $x = 2$ For $x = -2$ For $x = i$ For $x = -i$

$\quad y = \dfrac{2}{2}$ $y = \dfrac{2}{-2}$ $y = \dfrac{2}{i}$ $y = \dfrac{2}{-i}$

$\quad y = 1$ $y = -1$ $y = -2i$ $y = 2i$

Solutions: $(2, 1)$, $(-2, -1)$, $(i, -2i)$, $(-i, 2i)$

25. $y = 5 - x^2$
       $y = 2 - 2x$

Substitute y from the first equation into the second equation.

$5 - x^2 = 2 - 2x$

$\quad 0 = x^2 - 2x - 3$

$\quad 0 = (x - 3)(x + 1)$

$\quad x = 3, -1$

For $x = 3$ For $x = -1$

$\quad y = 2 - 2(3)$ $y = 2 - 2(-1)$

$\quad y = -4$ $y = 4$

Solutions: $(3, -4)$, $(-1, 4)$

27. $y = x^2 - x$
       $y = 2x$

Substitute y from the first equation into the second equation.

$\quad x^2 - x = 2x$

$\quad x^2 - 3x = 0$

$x(x - 3) = 0$

$\qquad x = 0, 3$

For $x = 0$ For $x = 3$

$\quad y = 2(0)$ $y = 2(3)$

$\quad y = 0$ $y = 6$

Solutions: $(0, 0)$, $(3, 6)$

29. $y = x^2 - 6x + 9$
       $y = 5 - x$

Substitute y from the first equation into the second equation.

$\quad x^2 - 6x + 9 = 5 - x$

$\quad x^2 - 5x + 4 = 0$

$(x - 1)(x - 4) = 0$

$\qquad\qquad x = 1, 4$

For $x = 1$ For $x = 4$

$\quad y = 5 - 1$ $y = 5 - 4$

$\quad y = 4$ $y = 1$

Solutions: $(1, 4)$, $(4, 1)$

31. $y = 8 + 4x - x^2$
       $y = x^2 - 2x$

Substitute y from the first equation into the second equation.

$8 + 4x - x^2 = x^2 - 2x$

$\qquad 0 = 2x^2 - 6x - 8$

$\qquad 0 = x^2 - 3x - 4$

$\qquad 0 = (x - 4)(x + 1)$

$\qquad x = 4, -1$

For $x = 4$ For $x = -1$

$\quad y = 4^2 - 2(4)$ $y = (-1)^2 - 2(-1)$

$\quad y = 8$ $y = 3$

Solutions: $(4, 8)$, $(-1, 3)$

33. (A) The lines are tangent to the circle.

(B) To find values of b such that
$$x^2 + y^2 = 5$$
$$2x - y = b$$
has exactly one solution, we solve the system for arbitrary b. Solve for y in the second equation.
$$y = 2x - b$$

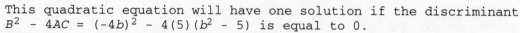

Substitute into the first equation:
$$x^2 + (2x - b)^2 = 5$$
$$x^2 + 4x^2 - 4bx + b^2 = 5$$
$$5x^2 - 4bx + b^2 - 5 = 0$$

This quadratic equation will have one solution if the discriminant $B^2 - 4AC = (-4b)^2 - 4(5)(b^2 - 5)$ is equal to 0.

This will occur when
$$16b^2 - 20b^2 + 100 = 0$$
$$-4b^2 + 100 = 0$$
$$b^2 = 25$$
$$b = \pm 5$$

Consider $b = 5$
Then the solution of the system
$$x^2 + y^2 = 5$$
$$2x - y = 5$$
will be given by solving
$$5x^2 - 4bx + b^2 - 5 = 0$$
for $b = 5$.
$$5x^2 - 4 \cdot 5x + 5^2 - 5 = 0$$
$$5x^2 - 20x + 20 = 0$$
$$5(x - 2)^2 = 0$$
$$x - 2 = 0$$
$$x = 2$$
Since $2x - y = 5$
$$2 \cdot 2 - y = 5$$
$$y = -1$$
The intersection point is $(2, -1)$ for $b = 5$.

Consider $b = -5$
Then the solution of the system
$$x^2 + y^2 = 5$$
$$2x - y = -5$$
will be given by solving
$$5x^2 - 4bx + b^2 - 5 = 0$$
for $b = -5$.
$$5x^2 - 4(-5)x + (-5)^2 - 5 = 0$$
$$5x^2 + 20x + 20 = 0$$
$$5(x + 2)^2 = 0$$
$$x + 2 = 0$$
$$x = -2$$
Since $2x - y = -5$
$$2(-2) - y = -5$$
$$y = 1$$
The intersection point is $(-2, 1)$ for $b = -5$.

(C) The line $x + 2y = 0$ is perpendicular to all the lines in the family and intersects the circle at the intersection points found in part B, since this line passes through the center of the circle and thus includes a diameter of the circle, which is perpendicular to the tangent line at their mutual point of intersection with the circle. Solving the system $x^2 + y^2 = 5$, $x + 2y = 0$ would determine the intersection points.

35. $2x + 5y + 7xy = 8$
$\qquad xy - 3 = 0$

Solve for y in the second equation.
$xy = 3$
$y = \dfrac{3}{x}$

Substitute into the first equation.

$2x + 5\left(\dfrac{3}{x}\right) + 7x\left(\dfrac{3}{x}\right) = 8$

$2x + \dfrac{15}{x} + 21 = 8 \quad x \neq 0$

$2x^2 + 15 + 21x = 8x$
$2x^2 + 13x + 15 = 0$
$(2x + 3)(x + 5) = 0$
$\qquad\qquad x = -\dfrac{3}{2}, -5$

For $x = -\dfrac{3}{2}$ $\qquad$ For $x = -5$

$y = 3 \div \left(-\dfrac{3}{2}\right) \qquad y = \dfrac{3}{-5}$

$y = -2 \qquad\qquad y = -\dfrac{3}{5}$

Solutions: $\left(-\dfrac{3}{2}, -2\right), \left(-5, -\dfrac{3}{5}\right)$

37. $x^2 - 2xy + y^2 = 1$
$\qquad x - 2y = 2$

Solve for x in terms of y in the first-degree equation.
$x = 2y + 2$
Substitute into the second-degree equation.
$(2y + 2)^2 - 2(2y + 2)y + y^2 = 1$
$4y^2 + 8y + 4 - 4y^2 - 4y + y^2 = 1$
$\qquad\qquad\quad y^2 + 4y + 3 = 0$
$\qquad\qquad (y + 1)(y + 3) = 0$
$\qquad\qquad\qquad\qquad\quad y = -1, -3$

For $y = -1$ $\qquad$ For $y = -3$
$\quad x = 2(-1) + 2 \qquad x = 2(-3) + 2$
$\qquad = 0 \qquad\qquad\qquad = -4$
Solutions: $(0, -1), (-4, -3)$

39. $2x^2 - xy + y^2 = 8$
$\qquad x^2 - y^2 = 0$

Factor the left side of the equation that has a zero constant term.

$(x - y)(x + y) = 0$
$\quad x = y \text{ or } x = -y$

> **Common Error:**
> It is incorrect to replace
> $x^2 - y^2 = 0$ or $x^2 = y^2$ by $x = y$.
> This neglects the possibility $x = -y$.

Thus, the original system is equivalent to the two systems
$2x^2 - xy + y^2 = 8 \qquad 2x^2 - xy + y^2 = 8$
$\qquad\quad x = y \qquad\qquad\qquad\quad x = -y$

These systems are solved by substitution.

First system: $\qquad\qquad$ Second system:
$2x^2 - xy + y^2 = 8 \qquad\qquad 2x^2 - xy + y^2 = 8$
$\qquad\quad x = y \qquad\qquad\qquad\qquad\qquad x = -y$
$2y^2 - yy + y^2 = 8 \qquad\quad 2(-y)^2 - (-y)y + y^2 = 8$
$\qquad\quad 2y^2 = 8 \qquad\qquad\qquad 2y^2 + y^2 + y^2 = 8$
$\qquad\quad y^2 = 4 \qquad\qquad\qquad\qquad\quad 4y^2 = 8$
$\qquad\quad y = \pm 2 \qquad\qquad\qquad\qquad\quad y^2 = 2$
$\qquad\qquad\qquad\qquad\qquad\qquad\qquad y = \pm\sqrt{2}$

For $y = 2$ $\quad$ For $y = -2$ $\quad$ For $y = \sqrt{2}$ $\quad$ For $y = -\sqrt{2}$
$\quad x = 2 \qquad\quad x = -2 \qquad\quad x = -\sqrt{2} \qquad\quad x = \sqrt{2}$
Solutions: $(2, 2), (-2, -2), (-\sqrt{2}, \sqrt{2}), (\sqrt{2}, -\sqrt{2})$

41. $x^2 + xy - 3y^2 = 3$
$x^2 + 4xy + 3y^2 = 0$

Factor the left side of the equation that has a zero constant term.
$(x + y)(x + 3y) = 0$

$$x = -y \text{ or } x = -3y$$

Thus the original system is equivalent to the two systems

$x^2 + xy - 3y^2 = 3$ $x^2 + xy - 3y^2 = 3$
$\quad\quad x = -y$ $\quad\quad x = -3y$

These systems are solved by substitution.

First system: Second system:

$x^2 + xy - 3y^2 = 3$ $x^2 + xy - 3y^2 = 3$
$\quad\quad\quad x = -y$ $\quad\quad\quad x = -3y$
$(-y)^2 + (-y)y - 3y^2 = 3$ $(-3y)^2 + (-3y)y - 3y^2 = 3$
$\quad y^2 - y^2 - 3y^2 = 3$ $\quad 9y^2 - 3y^2 - 3y^2 = 3$
$\quad\quad\quad -3y^2 = 3$ $\quad\quad\quad 3y^2 = 3$
$\quad\quad\quad\quad y^2 = -1$ $\quad\quad\quad\quad y^2 = 1$
$\quad\quad\quad\quad y = \pm i$ $\quad\quad\quad\quad y = \pm 1$

For $y = i$ For $y = -i$ For $y = 1$ For $y = -1$
$\quad x = -i$ $x = i$ $x = -3$ $x = 3$

Solutions: $(-i, i)$, $(i, -i)$, $(-3, 1)$, $(3, -1)$

43. Before we can enter these equations in our graphing utility, we must solve for y:

$\quad -x^2 + 2xy + y^2 = 1$ $3x^2 - 4xy + y^2 = 2$
$y^2 + 2xy - 1 - x^2 = 0$ $y^2 - 4xy + 3x^2 - 2 = 0$

Applying the quadratic formula to each equation, we have

$$y = \frac{-2x \pm \sqrt{4x^2 - 4(-1 - x^2)}}{2} \qquad y = \frac{4x \pm \sqrt{16x^2 - 4(3x^2 - 2)}}{2}$$

$$y = \frac{-2x \pm \sqrt{8x^2 + 4}}{2} \qquad\qquad y = \frac{4x \pm \sqrt{4x^2 + 8}}{2}$$

$$y = -x \pm \sqrt{2x^2 + 1} \qquad\qquad y = 2x \pm \sqrt{x^2 + 2}$$

Entering each of these four equations into a graphing utility produces the graph shown at the right.

Zooming in on the four intersection points, or using a built-in intersection routine (details omitted), yields $(-1.41, -0.82)$, $(-0.13, 1.15)$, $(0.13, -1.15)$, and $(1.41, 0.82)$ to two decimal places.

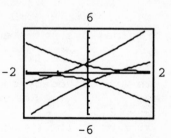

45. Before we can enter these equations in our graphing utility, we must solve for y:

$\quad 3x^2 - 4xy - y^2 = 2$ $2x^2 + 2xy + y^2 = 9$
$y^2 + 4xy + 2 - 3x^2 = 0$ $y^2 + 2xy + 2x^2 - 9 = 0$

Applying the quadratic formula to each equation, we have

$$y = \frac{-4x \pm \sqrt{16x^2 - 4(2 - 3x^2)}}{2} \qquad y = \frac{-2x \pm \sqrt{4x^2 - 4(2x^2 - 9)}}{2}$$

$$y = \frac{-4x \pm \sqrt{28x^2 - 8}}{2} \qquad\qquad y = \frac{-2x \pm \sqrt{36 - 4x^2}}{2}$$

$$y = -2x \pm \sqrt{7x^2 - 2} \qquad\qquad y = -x \pm \sqrt{9 - x^2}$$

Entering each of these four equations into a graphing utility produces the graph shown at the right.

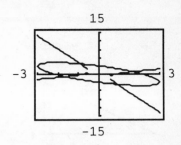

Zooming in on the four intersection points, or using a built-in intersection routine (details omitted), yields (-1.66, -0.84), (-0.91, 3.77), (0.91, -3.77), and (1.66, 0.84) to two decimal places.

47. Before we can enter these equations in our graphing utility, we must solve for y:

$$2x^2 - 2xy + y^2 = 9 \qquad\qquad 4x^2 - 4xy + y^2 + x = 3$$
$$y^2 - 2xy + 2x^2 - 9 = 0 \qquad y^2 - 4xy + 4x^2 + x - 3 = 0$$

Applying the quadratic formula to each equation, we have

$$y = \frac{2x \pm \sqrt{4x^2 - 4(2x^2 - 9)}}{2} \qquad y = \frac{4x \pm \sqrt{16x^2 - 4(4x^2 + x - 3)}}{2}$$

$$y = \frac{2x \pm \sqrt{36 - 4x^2}}{2} \qquad\qquad y = \frac{4x \pm \sqrt{12 - 4x}}{2}$$

$$y = x \pm \sqrt{9 - x^2} \qquad\qquad\quad y = 2x \pm \sqrt{3 - x}$$

Entering each of these four equations into a graphing utility produces the graph shown at the right.

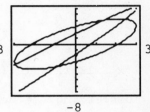

Zooming in on the four intersection points, or using a built-in intersection routine (details omitted), yields (-2.96, -3.47), (-0.89, -3.76), (1.39, 4.05), and (2.46, 4.18) to two decimal places.

49. Let x and y equal the two numbers. We have the system

$$x + y = 3$$
$$xy = 1$$

Solve the first equation for y in terms of x, then substitute into the second degree equation.

$$y = 3 - x$$
$$x(3 - x) = 1$$
$$3x - x^2 = 1$$
$$-x^2 + 3x - 1 = 0$$
$$x^2 - 3x + 1 = 0$$

$$x = \frac{-b \pm \sqrt{b^2 - 4ac}}{2a} \qquad a = 1,\ b = -3,\ c = 1$$

$$x = \frac{-(-3) \pm \sqrt{(-3)^2 - 4(1)(1)}}{2(1)}$$

$$x = \frac{3 \pm \sqrt{5}}{2}$$

For $x = \dfrac{3 + \sqrt{5}}{2}$ For $x = \dfrac{3 - \sqrt{5}}{2}$

$$y = 3 - x \qquad\qquad\qquad y = 3 - x$$

$$= 3 - \frac{3 + \sqrt{5}}{2} \qquad\qquad = 3 - \frac{3 - \sqrt{5}}{2}$$

$$= \frac{6 - 3 - \sqrt{5}}{2} \qquad\qquad = \frac{6 - 3 + \sqrt{5}}{2}$$

$$= \frac{3 - \sqrt{5}}{2} \qquad\qquad\quad = \frac{3 + \sqrt{5}}{2}$$

Thus the two numbers are $\frac{1}{2}(3 - \sqrt{5})$ and $\frac{1}{2}(3 + \sqrt{5})$.

51. Sketch a figure.
Let x and y represent the lengths of the two legs.

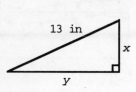

From the Pythagorean Theorem we have
$x^2 + y^2 = 13^2$
From the formula for the area of a triangle we have
$\frac{1}{2}xy = 30$
Thus the system of equations is
$x^2 + y^2 = 169$
$\frac{1}{2}xy = 30$

Solve the second equation for y in terms of x, then substitute into the first equation.

$$xy = 60$$
$$y = \frac{60}{x}$$
$$x^2 + \left(\frac{60}{x}\right)^2 = 169$$
$$x^2 + \frac{3600}{x^2} = 169 \quad x \neq 0$$
$$x^4 + 3600 = 169x^2$$
$$x^4 - 169x^2 + 3600 = 0$$
$$(x^2 - 144)(x^2 - 25) = 0$$
$$(x - 12)(x + 12)(x - 5)(x + 5) = 0$$
$$x = \pm12, \pm5$$

Discarding the negative solutions, we have
$x = 12$ or $x = 5$

For $x = 12$ For $x = 5$
$y = \dfrac{60}{x}$ $y = \dfrac{60}{x}$
$y = 5$ $y = 12$

The lengths of the legs are 5 inches and 12 inches.

53. Let x = width of screen.
    y = height of screen.
From the Pythagorean Theorem, we have
$x^2 + y^2 = (7.5)^2$
From the formula for the area of a rectangle we have
$xy = 27$
Thus the system of equations is:
$x^2 + y^2 = 56.25$
$xy = 27$

Solve the second equation for y in terms of x, then substitute into the first equation.

$$y = \frac{27}{x}$$
$$x^2 + \left(\frac{27}{x}\right)^2 = 56.25 \quad x \neq 0$$
$$x^2 + \frac{729}{x^2} = 56.25 \quad x \neq 0$$
$$x^4 + 729 = 56.25x^2$$
$$x^4 - 56.25x^2 + 729 = 0 \quad \text{quadratic in } x^2$$

$$x^2 = \frac{-b \pm \sqrt{b^2 - 4ac}}{2a} \quad a = 1, \ b = -56.25, \ c = 729$$

$$x^2 = \frac{-(-56.25) \pm \sqrt{(-56.25)^2 - 4(1)(729)}}{2(1)}$$

$$x^2 = \frac{56.25 \pm 15.75}{2}$$

$$x^2 = 36, \ 20.25$$

$$x = 6, \ 4.5 \ \text{(discarding the negative solutions)}$$

For $x = 6$ For $x = 4.5$

$\quad y = \dfrac{27}{6} = 4.5$ $\quad y = \dfrac{27}{4.5} = 6$

The dimensions of the screen must be 6 inches by 4.5 inches.

55.

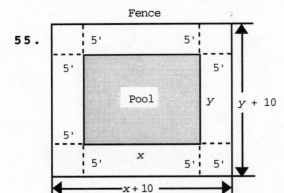

Redrawing and labeling the figure as shown, we have

Area of pool = 572
$$xy = 572$$
Area enclosed by fence = 1,152
$$(x + 10)(y + 10) = 1,152$$

We solve this system by solving for y in terms of x in the first equation, then substituting into the second equation.

$$y = \frac{572}{x}$$

$$(x + 10)\left(\frac{572}{x} + 10\right) = 1,152$$

$$572 + 10x + \frac{5,720}{x} + 100 = 1,152$$

$$10x + \frac{5,720}{x} - 480 = 0 \quad x \neq 0$$

$$10x^2 + 5,720 - 480x = 0$$

$$x^2 - 48x + 572 = 0$$

$$(x - 26)(x - 22) = 0$$

$$x = 26, \ 22$$

For $x = 26$ For $x = 22$

$\quad y = \dfrac{572}{26}$ $\quad y = \dfrac{572}{22}$

$\quad y = 22$ $\quad y = 26$

The dimensions of the pool are 22 feet by 26 feet.

57.   Let x = average speed of Boat B
Then $x + 5$ = average speed of Boat A

Let y = time of Boat B, then $y - \dfrac{1}{2}$ = time of Boat A

Using Distance = rate × time, we have
$$75 = xy$$
$$75 = (x + 5)\left(y - \frac{1}{2}\right)$$

Note: The *faster* boat, A, has the *shorter* time. It is a common error to confuse the signs here. Another common error: if rates are expressed in miles per hour, then $y - 30$ is not the correct time for boat A. Times must be expressed in hours.

Solve the first equation for y in terms of x, then substitute into the second equation.

$$y = \frac{75}{x}$$

$$75 = (x + 5)\left(\frac{75}{x} - \frac{1}{2}\right)$$

$$75 = 75 - \frac{1}{2}x + \frac{375}{x} - \frac{5}{2}$$

$$0 = -\frac{1}{2}x + \frac{375}{x} - \frac{5}{2} \quad x \neq 0$$

$$2x(0) = 2x\left(-\frac{1}{2}x\right) + 2x\left(\frac{375}{x}\right) - 2x\left(\frac{5}{2}\right)$$

$$0 = -x^2 + 750 - 5x$$

$$x^2 + 5x - 750 = 0$$

$$(x - 25)(x + 30) = 0$$

$$x = 25, -30$$

Discarding the negative solution, we have

$x = 25$ mph = average speed of Boat B

$x + 5 = 30$ mph = average speed of Boat A

Exercise 6-4

Key Ideas and Formulas

A line divides a plane into two halves called **half-planes**. A vertical line divides a plane into **left** and **right half-planes**; a nonvertical line divides a plane into **upper** and **lower half-planes**.

The graph of a linear inequality $Ax + By < C$ or $Ax + By > C$ (with $B \neq 0$) is either the upper half-plane or the lower half-plane (but not both) determined by the line $Ax + By = C$. If $B = 0$, then the graph of $Ax < C$ or $Ax > C$ is either the left half-plane or the right half-plane (but not both) determined by the line $Ax = C$. The line $Ax + By = C$ is called the **boundary line** for the half-plane.

Linear inequalities are graphed by the procedure given in the text (not repeated here for reasons of space.)

To graph systems of linear inequalities:

1. Graph each inequality, shading the solution half-plane lightly.

2. The multiply shaded region is the graph of the system, called the solution region or the feasible region.

3. Determine the corner points from the graph, and, as necessary, solving the systems of equations formed by pairs of equations of the appropriate lines.

(A **corner point** is a point in the solution region that is the intersection of two boundary lines.)
A solution region of a system of linear inequalities is bounded if it can be enclosed in a circle; if it cannot be enclosed within a circle, then it is unbounded.

1. Graph $2x - 3y = 6$ as a dashed line, since equality is not included in the original statement. The origin is a suitable test point.
$2x - 3y < 6$
$2(0) - 3(0) = 0 < 6$
Hence (0, 0) is in the solution set. The graph is the half-plane containing (0, 0).

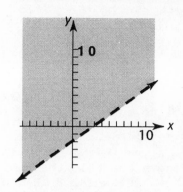

3. Graph $3x + 2y = 18$ as a solid line, since equality is included in the original statement. The origin is a suitable test point.
$3(0) + 2(0) = 0 \not\geq 18$
Hence (0, 0) is not in the solution set. The graph is the line $3x + 2y = 18$ and the half-plane not containing the origin.

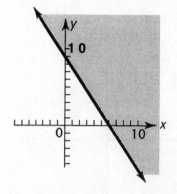

5. Graph $y = \frac{2}{3}x + 5$ as a solid line, since equality is included in the original statement. The origin is a suitable test point.
$0 \overset{?}{\leq} (0) + 5$
$0 \leq 5$
Hence (0, 0) is in the solution set. The graph is the line $y = \frac{2}{3}x + 5$ and the half-plane containing the origin.

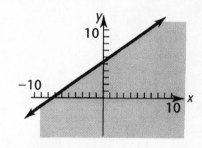

7. Graph $y = 8$ as a dashed line, since equality is not included in the original statement. Clearly the graph consists of all points whose y-coordinates are less than 8, that is, the lower half-plane.

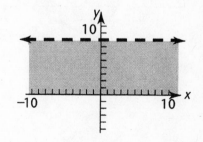

9. This system is equivalent to the system
$y \geq -3$
$y < 2$
and its graph is the intersection of the graphs of these inequalities.

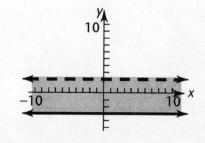

11. $x + 2y \le 8$
$3x - 2y \ge 0$
Choose a suitable test point that lies on neither line, for example, (2, 0).
$2 + 2(0) = 2 \le 8$ Hence, the solution region is *below* the graph
of $x + 2y = 8$.
$3(2) - 2(0) = 6 \ge 0$ Hence, the solution region is *below* the graph
of $3x - 2y = 0$.
Thus the solution region is region IV in the diagram.

13. $x + 2y \ge 8$
$3x - 2y \ge 0$
Choose a suitable test point that lies on neither line, for example, (2, 0).
$2 + 2(0) = 2 \not\ge 8$ Hence, the solution region is *above* the graph
of $x + 2y = 8$.
$3(2) - 2(0) = 6 \ge 0$ Hence, the solution region is *below* the graph
of $3x - 2y = 0$.
Thus the solution region is region I in the diagram.

15.

17.

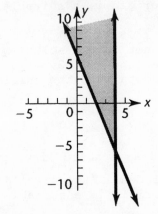

19.

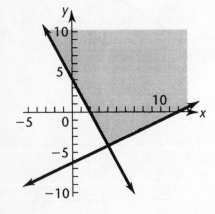

21. Choose a suitable test point that lies on
none of the lines, say (5, 1).
$5 + 3(1) = 8 \le 18$
Hence, the solution region is *below* the
graph of $x + 3y = 18$.

$2(5) + 1 = 11 \not\le 16$
Hence, the solution region is *above* the
graph of $2x + y = 16$.
$5 \ge 0$
$1 \ge 0$
Thus the solution region is region IV in
the diagram. The corner points are the
labelled points (6, 4), (8, 0), and (18,
0).

23. Choose a suitable test point that lies on none of the lines, say (5, 1).
$5 + 3(1) = 8 \not\ge 18$ Hence, the solution region is *above* the graph
of $x + 3y = 18$.
$2(5) + 1 = 11 \not\ge 16$ Hence, the solution region is *above* the graph
of $2x + y = 16$.
$5 \ge 0$
$1 \ge 0$
Thus the solution region is region I in the diagram. The corner points are the
labelled points (0, 16), (6, 4), and (18, 0).

25. The solution region is bounded (contained in, for example, the circle $x^2 + y^2 = 16$). The corner points are obvious from the graph: $(0, 0)$, $(0, 2)$, $(3, 0)$.

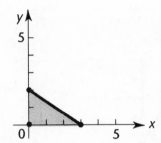

27. The solution region is unbounded. The corner points are obvious from the graph: $(0, 4)$ and $(5, 0)$.

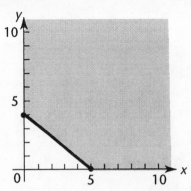

29. The solution region is bounded. Three corner points are obvious from the graph: $(0, 4)$, $(0, 0)$, $(4, 0)$. The fourth corner point is obtained by solving the system
$x + 3y = 12$

$2x + y = 8$ to obtain $\left(\dfrac{12}{5}, \dfrac{16}{5}\right)$.

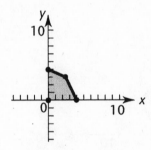

31. The solution region is unbounded. The corner points are obvious from the graph: $(9, 0)$ and $(0, 8)$. The third corner point is obtained by solving the system
$4x + 3y = 24$
$2x + 3y = 18$ to obtain $(3, 4)$.

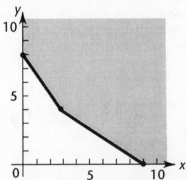

33. The solution region is bounded. Three corner points are obvious from the graph: $(6, 0)$, $(0, 0)$, and $(0, 5)$. The other corner points are obtained by solving:

$2x + y = 12$ and $x + y = 7$
 $x + y = 7$ $x + 2y = 10$
to obtain   to obtain
  $(5, 2)$ $(4, 3)$

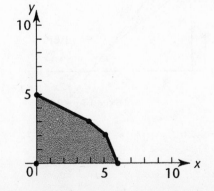

35. The solution region is unbounded. Two of the corner points are obvious from the graph: $(16, 0)$ and $(0, 14)$. The other corner points are obtained by solving:

$x + 2y = 16$ and $x + y = 12$
 $x + y = 12$ $2x + y = 14$
to obtain   to obtain
  $(8, 4)$ $(2, 10)$

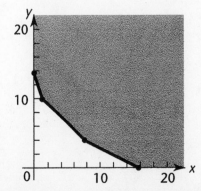

37. The solution region is bounded. The corner points are obtained by solving:

$$\begin{cases} x + y = 11 \\ 5x + y = 15 \end{cases}$$ to obtain (1, 10)

$$\begin{cases} 5x + y = 15 \\ x + 2y = 12 \end{cases}$$ to obtain (2, 5), and

$$\begin{cases} x + y = 11 \\ x + 2y = 12 \end{cases}$$ to obtain (10, 1)

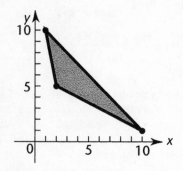

39. From the graph it should be clear that there is no point with x coordinate greater than 4 which satisfies both $3x + 2y \leq 24$ (arrows pointing, roughly, northeast) and $3x + y \leq 15$ (arrows pointing, roughly, southwest). The feasible region is empty.

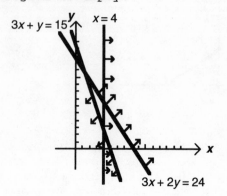

41. The feasible region is bounded. The corner points are obtained by solving:

$$\begin{cases} x + y = 10 \\ 3x - 2y = 15 \end{cases}$$ to obtain (7, 3)

$$\begin{cases} 3x - 2y = 15 \\ 3x + 5y = 15 \end{cases}$$ to obtain (5, 0),

$$\begin{cases} 3x + 5y = 15 \\ -5x + 2y = 6 \end{cases}$$ to obtain (0, 3), and

$$\begin{cases} -5x + 2y = 6 \\ x + y = 10 \end{cases}$$ to obtain (2, 8)

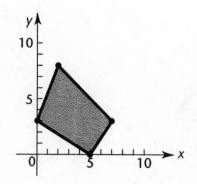

43. The feasible region is bounded. The corner points are obtained by solving:

$$\begin{cases} 16x + 13y = 119 \\ 12x + 16y = 101 \end{cases}$$ to obtain (5.91, 1.88)

$$\begin{cases} 16x + 13y = 119 \\ -4x + 3y = 11 \end{cases}$$ to obtain (2.14, 6.53) and

$$\begin{cases} 12x + 16y = 101 \\ -4x + 3y = 11 \end{cases}$$ to obtain (1.27, 5.36)

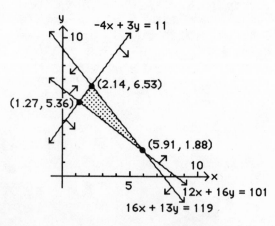

45. Let x = number of trick skis produced per day.
y = number of slalom skis produced per day

Clearly x and y must be non-negative.
Hence $x \geq 0$ (1)
     $y \geq 0$ (2)
To fabricate x trick skis requires $6x$ hours.
To fabricate y slalom skis requires $4y$ hours.
108 hours are available for fabricating; hence
$6x + 4y \leq 108$ (3)
To finish x trick skis requires $1x$ hours.
To finish y slalom skis requires $1y$ hours.
24 hours are available for finishing, hence
$x + y \leq 24$ (4)

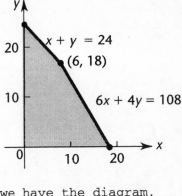

Graphing the inequality system (1), (2), (3), (4), we have the diagram.

47. (A) All production schedules in the feasible region that are on the graph of $50x + 60y = 1{,}100$ will result in a profit of \$1,100.

(B) There are many possible choices. For example, producing 5 trick and 15 slalom skis will produce a profit of \$1,150. The graph of the line $50x + 60y = 1{,}150$ includes all the production schedules in the feasible region that result in a profit of \$1,150.

(C) A graphical approach would involve drawing other lines of the type $50x + 60y = A$. The graphs of these lines include all production schedules that will result in a profit of A. Increase A until the line either intersects the feasible region only in 1 corner point or contains an edge of the feasible region. This value of A will be the maximum profit possible. For more details, see Section 6-5 of the text.

49. Clearly x and y must be non-negative.
Hence $x \geq 0$ (1)
     $y \geq 0$ (2)
x cubic yards of mix A contains $20x$ pounds of phosphoric acid.
y cubic yards of mix B contains $10y$ pounds of phosphoric acid.

At least 460 pounds of phosphoric acid are required, hence
$20x + 10y \geq 460$ (3)
x cubic yards of mix A contains $30x$ pounds of nitrogen.
y cubic yards of mix B contains $30y$ pounds of nitrogen.

At least 960 pounds of nitrogen are required, hence
$30x + 30y \geq 960$ (4)
x cubic yards of mix A contains $5x$ pounds of potash.
y cubic yards of mix B contains $10y$ pounds of potash.

At least 220 pounds of potash are required, hence
$5x + 10y \geq 220$ (5)

Graphing the inequality system (1), (2), (3), (4), (5), we have the diagram:

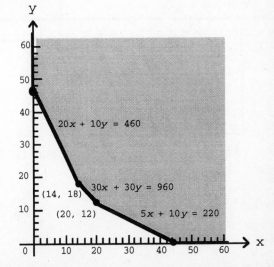

51. Clearly x and y must be non-negative.
Hence $x \geq 0$ (1)
     $y \geq 0$ (2)
Each sociologist will spend 10 hours collecting data: $10x$ hours.
Each research assistant will spend 30 hours collecting data: $30y$ hours.
At least 280 hours must be spent collecting data; hence
$10x + 30y \geq 280$ (3)
Each sociologist will spend 30 hours analyzing data: $30x$ hours.
Each research assistant will spend 10 hours analyzing data: $10y$ hours.
At least 360 hours must be spent analyzing data; hence
$30x + 10y \geq 360$ (4)
Graphing the inequality system (1), (2), (3), (4), we have the diagram.

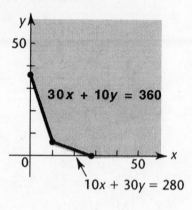

Exercise 6-5

Key Ideas and Formulas

A linear programming problem involves maximizing or minimizing an objective function subject to linear inequalities known as problem constraints.

Fundamental Theorem of Linear Programming:

Let S be the feasible region for a linear programming problem and let $z = ax + by$ be the objective function. If S is bounded, then z has both a maximum and a minimum value on S and each of these occurs at a corner point of S. If S is unbounded, then a maximum or minimum value of z on S may not exist. However, if either does exist, then it must occur at a corner point of S.

Solution of Linear Programming Problems:

1. Form a mathematical model for the problem:
   (A) Introduce decision variables and write a linear objective function.
   (B) Write problem constraints in the form of linear inequalities.
   (C) Write non-negative constraints.

2. Graph the feasible region and find the corner points.

3. Evaluate the objective function at each corner point to determine the optimal solution.

1.

Corner Point (x, y)	Objective Function $z = x + y$	
(0, 12)	12	
(7, 9)	16	Maximum value
(10, 0)	10	
(0, 0)	0	

The maximum value of z on S is 16 at (7, 9).

3.

Corner Point (x, y)	Objective Function z = 3x + 7y	
(0, 12)	84	Maximum value ⎫ Multiple optimal solutions
(7, 9)	84	Maximum value ⎭
(10, 0)	30	
(0, 0)	0	

The maximum value of z on S is 84 at both (0, 12) and (7, 9).

5.

Corner Point (x, y)	Objective Function z = 7x + 4y	
(0, 12)	48	
(12, 0)	84	
(4, 3)	40	
(0, 8)	32	Minimum value

The minimum value of z on S is 32 at (0, 8).

7.

Corner Point (x, y)	Objective Function 3x + 8y	
(0, 12)	96	
(12, 0)	36	Minimum value ⎫ Multiple optimal solutions
(4, 3)	36	Minimum value ⎭
(0, 8)	64	

The minimum value of z on S is 36 at both (12, 0) and (4, 3).

9. The feasible region is graphed as follows:
The corner points (0, 5), (5, 0) and (0, 0) are
obvious from the graph. The corner point (4, 3) is
obtained by solving the system
$x + 2y = 10$
$3x + y = 15$
We now evaluate the objective function at each
corner point.

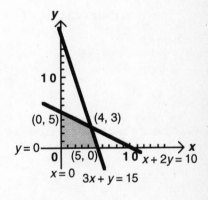

Corner Point (x, y)	Objective Function z = 3x + 2y	
(0, 5)	10	
(0, 0)	0	
(5, 0)	15	
(4, 3)	18	Maximum value

The maximum value of z on S is 18 at (4, 3).

11. The feasible region is graphed as follows:
The corner points (4, 0) and (10, 0) are obvious
from the graph. The corner point (2, 4) is obtained
by solving the system
$x + 2y = 10$
$2x + y = 8$
We now evaluate the objective function at each
corner point.

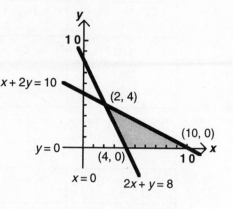

Corner Point (x, y)	Objective Function $z = 3x + 4y$	
(4, 0)	12	Minimum value
(10, 0)	30	
(2, 4)	22	

The minimum value of z on S is 12 at (4, 0).

13. The feasible region is graphed below. The corner points (0, 12), (0, 0), and
(12, 0) are obvious from the graph. The other corner points are obtained by
solving:
$x + 2y = 24$ and $x + y = 14$
 $x + y = 14$ to obtain (4, 10) $2x + y = 24$ to obtain (10, 4)
We now evaluate the objective function at each corner point.

Corner Point (x, y)	Objective Function $z = 3x + 4y$	
(0, 12)	48	
(0, 0)	0	
(12, 0)	36	
(10, 4)	46	
(4, 10)	52	Maximum value

The maximum value of z on S is 52 at (4, 10).

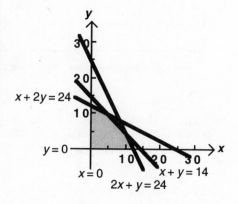

15. The feasible region is graphed as follows:
The corner points (0, 20) and (20, 0) are obvious
from the graph. The third corner point is obtained
by solving:
$x + 4y = 20$
$4x + y = 20$ to obtain (4, 4)
We now evaluate the objective function at each
corner point.

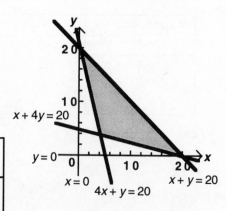

Corner Point (x, y)	Objective Function $z = 5x + 6y$	
(0, 20)	120	
(20, 0)	100	
(4, 4)	44	Minimum value

The minimum value of z on S is 44 at (4, 4).

17. The feasible region is graphed as follows:
The corner points (60, 0) and (120, 0) are
obvious from the graph. The other corner
points are obtained by solving:

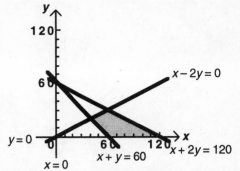

$x + y = 60$ and $x + 2y = 120$
$x - 2y = 0$ $x - 2y = 0$
to obtain (40, 20)   to obtain (60, 30)

We now evaluate the objective function at
each corner point.

Corner Point (x, y)	Objective Function $25x + 50y$	
(60, 0)	1,500	Minimum value
(40, 20)	2,000	
(60, 30)	3,000	Maximum value ⎫
(120, 0)	3,000	Maximum value ⎬ Multiple optimal solutions

The minimum value of z on S is 1,500 at (60, 0). The maximum value of z on S is
3,000 at (60, 30) and (120, 0).

19. The feasible region is graphed as follows:
The corner points (0, 45), (0, 20), (25, 0), and (60, 0) are obvious from the
graph shown below. The other corner points are obtained by solving:

$3x + 4y = 240$ and $3x + 4y = 240$
      $y = 45$ to obtain (20, 45) $x = 60$ to obtain (60, 15)
We now evaluate the objective function at each corner point.

Corner Point (x, y)	Objective Function $25x + 15y$	
(0, 45)	675	
(0, 20)	300	Minimum value
(25, 0)	625	
(60, 0)	1,500	
(60, 15)	1,725	Maximum value
(20, 45)	1,175	

The minimum value of z on S is 300 at (0, 20). The maximum value of z on S is
1,725 at (60, 15).

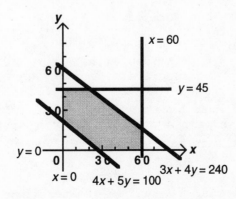

21. The feasible region is graphed as follows:

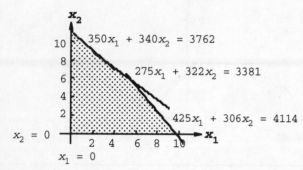

The corner point $(0, 0)$ is obvious from the graph. The other corner points are obtained by solving:

$x_1 = 0$
$275x_1 + 322x_2 = 3381$
to obtain $(0, 10.5)$

$350x_1 + 340x_2 = 3762$
$275x_1 + 322x_2 = 3381$
to obtain $(3.22, 7.75)$

$350x_1 + 340x_2 = 3762$
$425x_1 + 306x_2 = 4114$
to obtain $(6.62, 4.25)$

$425x_1 + 306x_2 = 4114$
$x_2 = 0$
to obtain $(9.68, 0)$

We now evaluate the objective function at each corner point.

Corner Point (x_1, x_2)	Objective Function $525x_1 + 478x_2$	
$(0, 0)$	0	Minimum value
$(0, 10.5)$	5019	
$(3.22, 7.75)$	5395	
$(6.62, 4.25)$	5507	Maximum value
$(9.68, 0)$	5082	

The maximum value of P is 5507 at the corner point $(6.62, 4.25)$.

23. The feasible region is graphed as follows: (heavily outlined for clarity)
Consider the objective function $x + y$. It should be clear that it takes on the value 2 along $x + y = 2$, the value 4 along $x + y = 4$, the value 7 along $x + y = 7$, and so on. The maximum value of the objective function, then, on S, is 7, which occurs at B. Graphically this occurs when the line $x + y = c$ coincides with the boundary of S. Thus to answer questions (A)-(E) we must determine values of a and b such that the appropriate line $ax + by = c$ coincides with the boundary of S only at the specified points.

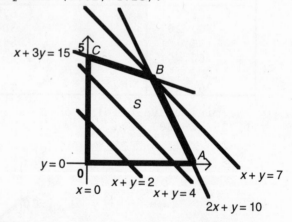

(A) The line $ax + by = c$ must have slope negative, but greater in absolute value that that of line segment AB, $2x + y = 10$. Therefore $a > 2b$.

(B) The line $ax + by = c$ must have slope negative but between that of $x + 3y = 15$ and $2x + y = 10$. Therefore $\frac{1}{3}b < a < 2b$.

(C) The line $ax + by = c$ must have slope greater than that of line segment BC, $x + 3y = 15$. Therefore $a < \frac{1}{3}b$ or $b > 3a$

(D) The line $ax + by = c$ must be parallel to line segment AB, therefore $a = 2b$.

(E) The line $ax + by = c$ must be parallel to line segment BC, therefore $b = 3a$.

25. Much of the mathematical model for this
problem was formed in Section 6-4 problem 45.
We let x = the number of trick skis
        y = the number of slalom skis
The problem constraints were
$6x + 4y \leq 108$
  $x + y \leq 24$
The non-negative constraints were
$x \geq 0$
$y \geq 0$
The feasible region was graphed there.

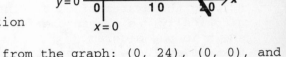

(A) We note now: the linear objective function
    $P = 40x + 30y$ represents the profit.

Three of the corner points are obvious from the graph: (0, 24), (0, 0), and
(18, 0). The fourth corner point is obtained by solving:
    $6x + 4y = 108$
    $x + y = 24$ to obtain (6, 18).
Summarizing: the mathematical model for this problem is:
Maximize $P = 40x + 30y$
subject to: $6x + 4y \leq 108$
             $x + y \leq 24$
              $x, y \geq 0$
We now evaluate the objective function $40x + 30y$ at each corner point.

Corner Point (x, y)	Objective Function 40x + 30y	
(0, 0)	0	
(18, 0)	720	
(6, 18)	780	Maximum value
(0, 24)	720	

The optimal value is 780 at the corner point (6, 18). Thus, 6 trick skis
and 18 slalom skis should be manufactured to obtain the maximum profit of
$780.

(B) The objective function now becomes $40x + 25y$. We evaluate this at each
corner point.

Corner Point (x, y)	Objective Function 40x + 25y	
(0, 0)	0	
(18, 0)	720	Maximum value
(6, 18)	690	
(0, 24)	600	

The optimal value is now 720 at the corner point (18, 0). Thus, 18 trick
skis and no slalom skis should be produced to obtain a maximum profit of
$720.

(C) The objective function now becomes $40x + 45y$. We evaluate this at each
corner point.

Corner Point (x, y)	Objective Function 40x + 45y	
(0, 0)	0	
(18, 0)	720	
(6, 18)	1,050	
(0, 24)	1,080	Maximum value

The optimal value is now 1,080 at the corner point (0, 24). Thus, no trick
skis and 24 slalom skis should be produced to obtain a maximum profit of
$1,080.

27. Let x = number of model A trucks
y = number of model B trucks
We form the linear objective function
$C = 15,000x + 24,000y$
We wish to minimize C, the cost of buying
x trucks @ \$15,000 and y trucks @ \$24,000,
subject to the constraints.
  $x + y \leq 15$ maximum number of trucks constraint
$2x + 3y \geq 36$ capacity constraint
  $x, y \geq 0$ non-negative constraints.

Solving the system of constraint inequalities
graphically, we obtain the feasible region S
shown in the diagram.
Next we evaluate the objective function at each corner point.

Corner Point (x, y)	Objective Function $C = 15,000x + 24,000y$	
(0, 12)	288,000	
(0, 15)	360,000	
(9, 6)	279,000	Minimum value

The optimal value is \$279,000 at the corner point (9, 6). Thus, the company
should purchase 9 model A trucks and 6 model B trucks to realize the minimum
cost of \$279,000.

29. (A) Let x = number of tables
y = number of chairs
We form the linear objective function
$P = 90x + 25y$
We wish to maximize P, the profit from x
tables @ \$90 and y chairs @ \$25, subject to
the constraints
$8x + 2y \leq 400$ assembly department constraint
  $2x + y \leq 120$ finishing department constraint
  $x, y \geq 0$ non-negative constraints

Solving the system of constraint inequalities
graphically, we obtain the feasible region S
shown in the diagram.

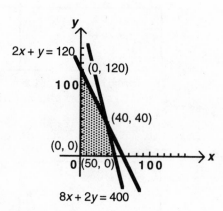

Next we evaluate the objective function at each corner point.

Corner Point (x, y)	Objective Function $P = 90x + 25y$	
(0, 0)	0	
(50, 0)	4,500	
(40, 40)	4,600	Maximum value
(0, 120)	3,000	

The optimal value is 4,600 at the corner point (40, 40). Thus, the company
should manufacture 40 tables and 40 chairs for a maximum profit of \$4,600.

(B) We are faced with the further condition
that $y \geq 4x$. We wish, then, to maximize
$P = 90x + 25y$ under the constraints
$8x + 2y \leq 400$
$2x + y \leq 120$
$\quad\quad y \geq 4x$
$\quad x,\ y \geq 0$
The feasible region is now S' as graphed.

Note that the new condition has the effect
of excluding (40, 40) from the feasible
region.

We now evaluate the objective function at
the new corner points.

Corner Point (x, y)	Objective Function $P = 90x + 25y$	
(0, 120)	3,000	
(0, 0)	0	
(20, 80)	3,800	Maximum value

The optimal value is now 3,800 at the corner point (20, 80). Thus the company
should manufacture 20 tables and 80 chairs for a maximum profit of $3,800.

31. Let x = number of gallons produced using the old process
$\quad\quad y$ = number of gallons produced using the new process

We form the linear objective function
$\quad P = 0.6x + 0.2y$

(A) We wish to maximize P, the profit from x gallons using the old process and
y gallons using the new process, subject to the contraints
$\quad 20x + \ \ 5y \leq 16{,}000 \quad\quad$ sulfur dioxide constraint
$\quad 40x + 20y \leq 30{,}000 \quad\quad$ particulate matter contraint
$\quad\quad\quad\quad x,\ y \geq 0 \quad\quad\quad\quad$ non-negative contraints

Solving the system of constraint
inequalities graphically, we obtain the
feasible region S shown in the diagram.
Note that no corner points are
determined by this (very weak) sulfur
dioxide constraint.

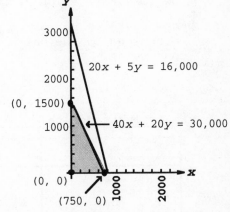

We evaluate the objective function at each corner point.

Corner Point (x, y)	Objective Function $P = 0.6x + 0.2y$	
(0, 0)	0	
(0, 1500)	300	
(750, 0)	450	Maximum value

The optimal value is 450 at the corner point (750, 0). Thus, the company
should manufacture 750 gallons by the old process exclusively, for a profit
of $450.

(B) The sulfur dioxide constraint is now
$20x + 5y \leq 11,500$. We now wish to
maximize P subject to the constraints

$$20x + 5y \leq 11,500$$
$$40x + 20y \leq 30,000$$
$$x, \ y \geq 0$$

The feasible region is now S_1 as shown.

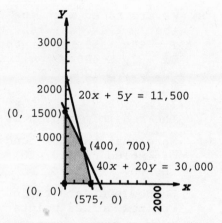

We evaluate the objective function at the new corner points.

Corner Point $(x, \ y)$	Objective Function $P = 0.6x + 0.2y$	
(0, 0)	0	
(575, 0)	345	
(400, 700)	380	Maximum value
(0, 1500)	300	

The optimal value is now 380 at the corner point (400, 700). Thus, the
company should manufacture 400 gallons by the old process and 700 gallons
by the new process, for a profit of $380.

(C) The sulfur dioxide constraint is now
$20x + 5y \leq 7,200$. We now wish to
maximize P subject to the constraints

$$20x + 5y \leq 7,200$$
$$40x + 20y \leq 30,000$$
$$x, \ y \geq 0$$

The feasible region is now S_2 as shown.

Note that now no corner points are
determined by the particulate matter
constraint.

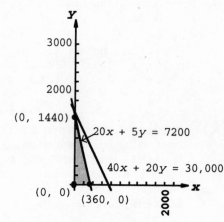

We evaluate the objective function at the new corner points.

Corner Point $(x, \ y)$	Objective Function $P = 0.6x + 0.2y$	
(0, 0)	0	
(360, 0)	216	
(0, 1440)	288	Maximum value

The optimal value is now 288 at the corner point (0, 1440). Thus, the
company should manufacture 1,440 gallons by the new process exclusively,
for a profit of $288.

33. (A) Let x = number of bags of Brand A
           y = number of bags of Brand B
We form the objective function
$N = 6x + 7y$
N represents the amount of nitrogen in x bags @ 6 pounds per bag and y bags @ 7 pounds per bag.

We wish to optimize N subject to the constraints
$2x + 4y \geq 480$ phosphoric acid constraint
$6x + 3y \geq 540$ potash constraint
$3x + 4y \leq 620$ chlorine constraint
   $x, y \geq 0$ non-negative contraints
Solving the system of constraint inequalities graphically, we obtain the feasible region S shown in the diagram.

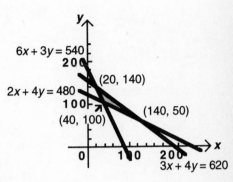

Next we evaluate the objective function at the corner points.

Corner Point (x, y)	Objective Function $N = 6x + 7y$	
(20, 140)	1,100	
(40, 100)	940	Minimum value
(140, 50)	1,190	Maximum value

Hence, the nitrogen will range from a minimum of 940 pounds when 40 bags of brand A and 100 bags of Brand B are used to a maximum of 1,190 pounds when 140 bags of brand A and 50 bags of brand B are used.

CHAPTER 6 REVIEW

1. $2x + y = 7$
     $3x - 2y = 0$
We multiply the top equation by 2 and add.

$4x + 2y = 14$
$\underline{3x - 2y = 0}$
$7x = 14$
     $x = 2$

Substituting $x = 2$ in the top equation, we have
$2(2) + y = 7$
     $y = 3$
Solution: (2, 3) $(6-1)$

2. $3x - 6y = 5$
     $-2x + 4y = 1$
We multiply the top equation by 2, the bottom by 3, and add.

$6x - 12y = 10$
$\underline{-6x + 12y = 3}$
     $0 = 13$

No solution $(6-1)$

3. $4x - 3y = -8$
     $-2x + \dfrac{3}{2}y = 4$

We multiply the bottom equation by 2 and add.

$4x - 3y = -8$
$\underline{-4x + 3y = 8}$
     $0 = 0$

There are infinitely many solutions. For any real number t, $4t - 3y = -8$, hence,
$-3y = -4t - 8$
   $y = \dfrac{4t + 8}{3}$

Thus, $\left(t, \dfrac{4t + 8}{3}\right)$ is a solution for any real number t. $(6-1)$

4. $y = x^2 - 5x - 3$

$y = -x + 2$

We multiply the bottom equation by -1 and add.

$$y = x^2 - 5x - 3$$
$$\underline{-y = \qquad x - 2}$$
$$0 = x^2 - 4x - 5$$
$$0 = (x - 5)(x + 1)$$
$$x = 5, -1$$

For $x = 5$ 　　　　　　　For $x = -1$

$\quad y = -5 + 2$ 　　　　　$\quad y = -(-1) + 2$

$\quad y = -3$ 　　　　　　　$\quad y = 3$

Solutions: $(5, -3)$, $(-1, 3)$

Check: For $(5, -3)$ 　　　For $(-1, 3)$

$\quad y = x^2 - 5x - 3$ 　　　$\quad y = x^2 - 5x - 3$

$-3 \overset{?}{=} 5^2 - 5 \cdot 5 - 3$ 　　$3 \overset{?}{=} (-1)^2 - 5(-1) - 3$

$-3 \overset{\sqrt{}}{=} -3$ 　　　　　　$3 \overset{\sqrt{}}{=} 1 + 5 - 3$

$(6\text{-}3)$

5. $x^2 + y^2 = 2$

$2x - y = 3$

We solve the first-degree equation for y in terms of x, then substitute into the second-degree equation.

$$2x - y = 3$$
$$-y = 3 - 2x$$
$$y = 2x - 3$$
$$x^2 + (2x - 3)^2 = 2$$
$$x^2 + 4x^2 - 12x + 9 = 2$$
$$5x^2 - 12x + 7 = 0$$
$$(5x - 7)(x - 1) = 0$$
$$x = \frac{7}{5}, 1$$

For $x = \dfrac{7}{5}$ 　　　For $x = 1$

$\quad y = 2\left(\dfrac{7}{5}\right) - 3$ 　　$y = 2(1) - 3$

$\quad = -\dfrac{1}{5}$ 　　　　　$= -1$

Solutions: $(1, -1)$, $\left(\dfrac{7}{5}, -\dfrac{1}{5}\right)$ 　　$(6\text{-}3)$

The checking steps are omitted for lack of space.

6. $3x^2 - y^2 = -6$

$2x^2 + 3y^2 = 29$

We use elimination by addition. We multiply the top equation by 3 and add.

$$9x^2 - 3y^2 = -18$$
$$\underline{2x^2 + 3y^2 = \quad 29}$$
$$11x^2 \qquad = 11$$
$$x^2 \qquad = 1$$
$$x = \pm 1$$

For $x = 1$ 　　　　　For $x = -1$

$3(1)^2 - y^2 = -6$ 　　$3(-1)^2 - y^2 = -6$

$\quad -y^2 = -9$ 　　　　$\quad -y^2 = -9$

$\quad y^2 = 9$ 　　　　　　$\quad y = \pm 3$

$\quad y = \pm 3$

Solutions: $(1, 3)$, $(1, -3)$, $(-1, 3)$, $(-1, -3)$ 　　$(6\text{-}3)$

7.

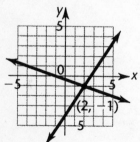

$(6\text{-}1)$

8. Graph $3x - 4y = 24$ as a solid line since equality is included in the original statement. $(0, 0)$ is a suitable test point.
$3(0) - 4(0) = 0 \not\geq 24$
Therefore $(0, 0)$ is not in the solution set. The line $3x - 4y = 24$ and the half-plane not containing $(0, 0)$ form the graph.

9.

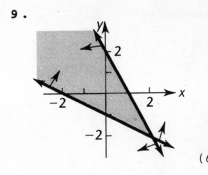

$(6\text{-}4)$

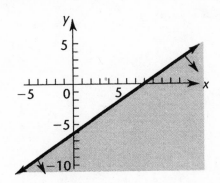

$(6\text{-}4)$

10. $R_1 \leftrightarrow R_2$ means interchange Rows 1 and 2.
$$\begin{bmatrix} 3 & -6 & | & 12 \\ 1 & -4 & | & 5 \end{bmatrix}$$
$(6\text{-}1)$

11. $\frac{1}{3} R_2 \rightarrow R_2$ means multiply Row 2 by $\frac{1}{3}$.
$$\begin{bmatrix} 1 & -4 & | & 5 \\ 1 & -2 & | & 4 \end{bmatrix}$$
$(6\text{-}1)$

12. $(-3)R_1 + R_2 \rightarrow R_2$ means replace Row 2 by itself plus -3 times Row 1.
$$\begin{bmatrix} 1 & -4 & | & 5 \\ 0 & 6 & | & -3 \end{bmatrix}$$
$(6\text{-}1)$

13. $x_1 = 4$
$x_2 = -7$
The solution is $(4, -7)$ $(6\text{-}2)$

14. $x_1 - x_2 = 4$
$\qquad\quad 0 = 1$

No solution $(6\text{-}2)$

15. $x_1 - x_2 = 4$
$\qquad\quad 0 = 0$
Solution:
$x_2 = t$
$x_1 = x_2 + 4 = t + 4$
Thus $x_1 = t + 4$, $x_2 = t$ is the solution, for t any real number.

$(6\text{-}2)$

16.

Corner Point (x, y)	Objective Function $z = 5x + 3y$	
$(0, 10)$	30	
$(0, 6)$	18	Minimum value
$(4, 2)$	26	
$(6, 4)$	42	Maximum value

The maximum value of z on S is 42 at $(6, 4)$. The minimum value of z on S is 18 at $(0, 6)$.
$(6\text{-}5)$

17. We write the augmented matrix:

$$\begin{bmatrix} 1 & -1 & | & 4 \\ 2 & 1 & | & 2 \end{bmatrix} \quad (-2)R_1 + R_2 \rightarrow R_2$$

$$-2 \quad 2 \quad -8$$

Need a 0 here

$$\sim \begin{bmatrix} 1 & -1 & | & 4 \\ 0 & 3 & | & -6 \end{bmatrix} \quad \tfrac{1}{3}R_2 \rightarrow R_2 \qquad \text{corresponds to the linear system} \qquad \begin{aligned} x_1 - x_2 &= 4 \\ 3x_2 &= -6 \end{aligned}$$

↑
Need a 1 here
Need a 0 here
↓

$$\sim \begin{bmatrix} 1 & -1 & | & 4 \\ 0 & 1 & | & -2 \end{bmatrix} \quad R_2 + R_1 \rightarrow R_1 \qquad \text{corresponds to the linear system} \qquad \begin{aligned} x_1 - x_2 &= 4 \\ x_2 &= -2 \end{aligned}$$

$$\sim \begin{bmatrix} 1 & 0 & | & 2 \\ 0 & 1 & | & -2 \end{bmatrix} \qquad\qquad\qquad \text{corresponds to the linear system} \qquad \begin{aligned} x_1 &= 2 \\ x_2 &= -2 \end{aligned}$$

The solution is $x_1 = 2$, $x_2 = -2$. Each pair of lines graphed below has the same intersection point, $(2, -2)$.

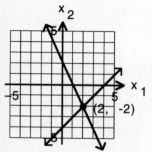

$$x_1 - x_2 = 4$$
$$2x_1 + x_2 = 2$$

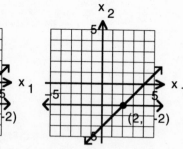

$$x_1 - x_2 = 4$$
$$3x_2 = -6$$

$$x_1 - x_2 = 4$$
$$x_2 = -2$$

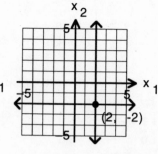

$$x_1 = 2$$
$$x_2 = -2 \quad (6\text{-}1)$$

18. Here is a computer-generated graph of the system , entered as

$$y = \frac{9 - x}{3}$$

$$y = \frac{10 + 2x}{7}$$

After zooming in (graph not shown) the intersection point is located at $(2.54, 2.15)$ to two decimal places.

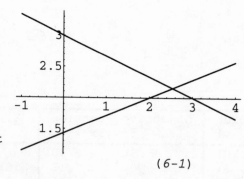

$(6\text{-}1)$

19. $\begin{bmatrix} 3 & 2 & | & 3 \\ 1 & 3 & | & 8 \end{bmatrix}$ $R_1 \leftrightarrow R_2$

$\sim \begin{bmatrix} 1 & 3 & | & 8 \\ 3 & 2 & | & 3 \end{bmatrix}$ $\underbrace{(-3)R_1}_{} + R_2 \rightarrow R_2$

$\quad\quad -3 \quad -9 \quad -24 \blacktriangleleft$

$\sim \begin{bmatrix} 1 & 3 & | & 8 \\ 0 & -7 & | & -21 \end{bmatrix}$ $-\frac{1}{7} R_2 \rightarrow R_2$

$\sim \begin{bmatrix} 1 & 3 & | & 8 \\ 0 & 1 & | & 3 \end{bmatrix}$ $\underbrace{(-3)R_2}_{} + R_1 \rightarrow R_1$

$\quad\quad 0 \quad -3 \quad -9 \blacktriangleleft$

$\sim \begin{bmatrix} 1 & 0 & | & -1 \\ 0 & 1 & | & 3 \end{bmatrix}$

Solution: $x_1 = -1$, $x_2 = 3$ $\quad$ (6-2)

20. $\begin{bmatrix} 1 & 1 & 0 & | & 1 \\ 1 & 0 & -1 & | & -2 \\ 0 & 1 & 2 & | & 4 \end{bmatrix}$ $(-1)R_1 + R_2 \rightarrow R_2$

$\sim \begin{bmatrix} 1 & 1 & 0 & | & 1 \\ 0 & -1 & -1 & | & -3 \\ 0 & 1 & 2 & | & 4 \end{bmatrix}$ $(-1)R_2 \rightarrow R_2$

$\sim \begin{bmatrix} 1 & 1 & 0 & | & 1 \\ 0 & 1 & 1 & | & 3 \\ 0 & 1 & 2 & | & 4 \end{bmatrix}$ $\begin{array}{l}(-1)R_2 + R_1 \rightarrow R_1 \\ (-1)R_2 + R_3 \rightarrow R_3\end{array}$

$\sim \begin{bmatrix} 1 & 0 & -1 & | & -2 \\ 0 & 1 & 1 & | & 3 \\ 0 & 0 & 1 & | & 1 \end{bmatrix}$ $\begin{array}{l}R_3 + R_1 \rightarrow R_1 \\ (-1)R_3 + R_2 \rightarrow R_2\end{array}$

$\sim \begin{bmatrix} 1 & 0 & 0 & | & -1 \\ 0 & 1 & 0 & | & 2 \\ 0 & 0 & 1 & | & 1 \end{bmatrix}$

Solution: $x_1 = -1$, $x_2 = 2$, $x_3 = 1$
$\quad\quad\quad\quad\quad\quad\quad\quad\quad\quad$ (6-2)

21. $\begin{bmatrix} 1 & 2 & 3 & | & 1 \\ 2 & 3 & 4 & | & 3 \\ 1 & 2 & 1 & | & 3 \end{bmatrix}$ $\begin{array}{l}(-2)R_1 + R_2 \rightarrow R_2 \\ (-1)R_1 + R_3 \rightarrow R_3\end{array}$

$\sim \begin{bmatrix} 1 & 2 & 3 & | & 1 \\ 0 & -1 & -2 & | & 1 \\ 0 & 0 & -2 & | & 2 \end{bmatrix}$ $\begin{array}{l}(-1)R_2 \rightarrow R_2 \\ -\frac{1}{2}R_3 \rightarrow R_3\end{array}$

$\sim \begin{bmatrix} 1 & 2 & 3 & | & 1 \\ 0 & 1 & 2 & | & -1 \\ 0 & 0 & 1 & | & -1 \end{bmatrix}$ $(-2)R_2 + R_1 \rightarrow R_1$

$\sim \begin{bmatrix} 1 & 0 & -1 & | & 3 \\ 0 & 1 & 2 & | & -1 \\ 0 & 0 & 1 & | & -1 \end{bmatrix}$ $\begin{array}{l}R_3 + R_1 \rightarrow R_1 \\ (-2)R_3 + R_2 \rightarrow R_2\end{array}$

$\sim \begin{bmatrix} 1 & 0 & 0 & | & 2 \\ 0 & 1 & 0 & | & 1 \\ 0 & 0 & 1 & | & -1 \end{bmatrix}$

Solution: $x_1 = 2$, $x_2 = 1$, $x_3 = -1$
$\quad\quad\quad\quad\quad\quad\quad\quad\quad\quad$ (6-2)

22. $\begin{bmatrix} 1 & 2 & -1 & | & 2 \\ 2 & 3 & 1 & | & -3 \\ 3 & 5 & 0 & | & -1 \end{bmatrix}$ $\begin{array}{l}(-2)R_1 + R_2 \rightarrow R_2 \\ (-3)R_1 + R_3 \rightarrow R_3\end{array}$

$\sim \begin{bmatrix} 1 & 2 & -1 & | & 2 \\ 0 & -1 & 3 & | & -7 \\ 0 & -1 & 3 & | & -7 \end{bmatrix}$ $(-1)R_2 + R_3 \rightarrow R_3$

$\sim \begin{bmatrix} 1 & 2 & -1 & | & 2 \\ 0 & -1 & 3 & | & -7 \\ 0 & 0 & 0 & | & 0 \end{bmatrix}$ $2R_2 + R_1 \rightarrow R_1$

$\sim \begin{bmatrix} 1 & 0 & 5 & | & -12 \\ 0 & -1 & 3 & | & -7 \\ 0 & 0 & 0 & | & 0 \end{bmatrix}$ $(-1)R_2 \rightarrow R_2$

$\sim \begin{bmatrix} 1 & 0 & 5 & | & -12 \\ 0 & 1 & -3 & | & 7 \\ 0 & 0 & 0 & | & 0 \end{bmatrix}$

This corresponds to the system
$x_1 + 5x_3 = -12$
$ x_2 - 3x_3 = 7$

Let $x_3 = t$
Then $x_2 = 3x_3 + 7$
$ = 3t + 7$
$ x_1 = -5x_3 - 12$
$ = -5t - 12$
Hence $x_1 = -5t - 12$, $x_2 = 3t + 7$,
$x_3 = t$ is a solution for every real
number t. There are infinitely many
solutions. $\quad\quad\quad\quad\quad\quad\quad$ (6-2)

23. $\begin{bmatrix} 1 & -2 & | & 1 \\ 2 & -1 & | & 0 \\ 1 & -3 & | & -2 \end{bmatrix}$ $(-2)R_1 + R_2 \rightarrow R_2$
$(-1)R_1 + R_3 \rightarrow R_3$

$\sim \begin{bmatrix} 1 & -2 & | & 1 \\ 0 & 3 & | & -2 \\ 0 & -1 & | & -3 \end{bmatrix}$ $3R_3 + R_2 \rightarrow R_2$

$\sim \begin{bmatrix} 1 & -2 & | & 1 \\ 0 & 0 & | & -11 \\ 0 & -1 & | & -3 \end{bmatrix}$

The second row corresponds to the equation
$0x_1 + 0x_2 = -11$,
hence there is no solution. (6-2)

24. $\begin{bmatrix} 1 & 2 & -1 & | & 2 \\ 3 & -1 & 2 & | & -3 \end{bmatrix}$ $(-3)R_1 + R_2 \rightarrow R_2$

$\sim \begin{bmatrix} 1 & 2 & -1 & | & 2 \\ 0 & -7 & 5 & | & -9 \end{bmatrix}$ $-\frac{1}{7}R_2 \rightarrow R_2$

$\sim \begin{bmatrix} 1 & 2 & -1 & | & 2 \\ 0 & 1 & -\frac{5}{7} & | & \frac{9}{7} \end{bmatrix}$ $(-2)R_2 + R_1 \rightarrow R_1$

$\sim \begin{bmatrix} 1 & 0 & \frac{3}{7} & | & -\frac{4}{7} \\ 0 & 1 & -\frac{5}{7} & | & \frac{9}{7} \end{bmatrix}$

This corresponds to the system
$x_1 + \frac{3}{7}x_3 = -\frac{4}{7}$
$x_2 - \frac{5}{7}x_3 = \frac{9}{7}$

Let $x_3 = t$
Then $x_2 = \frac{5}{7}x_3 + \frac{9}{7}$
$= \frac{5}{7}t + \frac{9}{7}$
$x_1 = -\frac{3}{7}x_3 - \frac{4}{7}$
$= -\frac{3}{7}t - \frac{4}{7}$

Hence $x_1 = -\frac{3}{7}t - \frac{4}{7}$, $x_2 = \frac{5}{7}t + \frac{9}{7}$,
$x_3 = t$ is a solution for every real
number t. There are infinitely many
solutions. (6-2)

The checking steps are omitted in problems 25—27 for lack of space.

25. $x^2 - y^2 = 2$
$y^2 = x$

Substitute y^2 in the second equation into the first equation.
$x^2 - x = 2$
$x^2 - x - 2 = 0$
$(x - 2)(x + 1) = 0$
$x = 2, -1$

For $x = 2$ For $x = -1$
$y^2 = 2$ $y^2 = -1$
$y = \pm\sqrt{2}$ $y = \pm i$

Solutions: $(2, \sqrt{2})$, $(2, -\sqrt{2})$, $(-1, i)$, $(-1, -i)$ (6-3)

26. $x^2 + 2xy + y^2 = 1$
$xy = -2$

Solve for y in the second equation.
$y = \frac{-2}{x}$

Substitute into the first equation
$$x^2 + 2x\left(-\frac{2}{x}\right) + \left(-\frac{2}{x}\right)^2 = 1$$
$$x^2 - 4 + \frac{4}{x^2} = 1 \quad x \neq 0$$
$$x^4 - 4x^2 + 4 = x^2$$
$$x^4 - 5x^2 + 4 = 0$$
$$(x^2 - 4)(x^2 - 1) = 0$$
$$(x - 2)(x + 2)(x - 1)(x + 1) = 0$$
$$x = 2, -2, 1, -1$$

For $x = 2$	For $x = -2$	For $x = 1$	For $x = -1$
$y = \dfrac{-2}{2}$	$y = \dfrac{-2}{-2}$	$y = \dfrac{-2}{1}$	$y = \dfrac{-2}{-1}$
$y = -1$	$y = 1$	$y = -2$	$y = 2$

Solutions: $(2, -1), (-2, 1), (1, -2), (-1, 2)$ $\qquad\qquad$ (6-3)

27. $2x^2 + xy + y^2 = 8$
$\qquad x^2 - y^2 = 0$

We factor the left side of the equation that has a zero constant term.
$(x - y)(x + y) = 0$
$x = y \quad$ or $\quad x = -y$
Thus, the original system is equivalent to the two systems
$2x^2 + xy + y^2 = 8 \qquad 2x^2 + xy + y^2 = 8$
$\qquad\quad x = y \qquad\qquad\qquad\qquad x = -y$
These systems are solved by substitution.

First System: $\qquad\qquad\qquad$ Second System:
$2x^2 + xy + y^2 = 8 \qquad\qquad 2x^2 + xy + y^2 = 8$
$\qquad\qquad x = y \qquad\qquad\qquad\qquad\quad x = -y$
$2y^2 + yy + y^2 = 8 \qquad 2(-y)^2 + (-y)y + y^2 = 8$
$\qquad 4y^2 = 8 \qquad\qquad\qquad\qquad 2y^2 = 8$
$\qquad y^2 = 2 \qquad\qquad\qquad\qquad y^2 = 4$
$\qquad y = \pm\sqrt{2} \qquad\qquad\qquad\qquad y = \pm 2$

For $y = \sqrt{2}$	For $y = -\sqrt{2}$	For $y = 2$	For $y = -2$
$x = \sqrt{2}$	$x = -\sqrt{2}$	$x = -2$	$x = 2$

Solutions: $(\sqrt{2}, \sqrt{2}), (-\sqrt{2}, -\sqrt{2}), (2, -2), (-2, 2)$ $\qquad\quad$ (6-3)

28. The solution region is bounded (contained in,
for example, the circle $x^2 + y^2 = 36$). Three
corner points are obvious from the graph:
$(0, 4), (0, 0)$, and $(4, 0)$. The fourth corner
point is obtained by solving the system
$2x + \ y = 8$
$2x + 3y = 12$ to obtain $(3, 2)$.

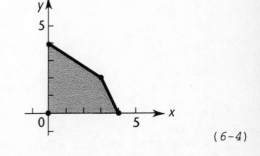

(6-4)

29. The solution region is unbounded. Two corner
points are obvious from the graph: $(0, 8)$ and
$(12, 0)$. The third corner point is obtained by
solving the system
$2x + y = 8$
$x + 3y = 12$

to obtain $\left(\dfrac{12}{5}, \dfrac{16}{5}\right)$.

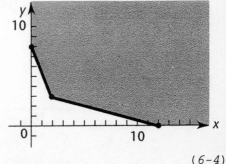

(6-4)

30. The solution region is bounded. The corner point (20, 0) is obvious from the graph. The other corner points are found by solving the system

$$\begin{cases} x + y = 20 \\ x - y = 0 \end{cases} \text{ to obtain (10, 10)}$$

and

$$\begin{cases} x - y = 0 \\ x + 4y = 20 \end{cases} \text{ to obtain (4, 4)}$$

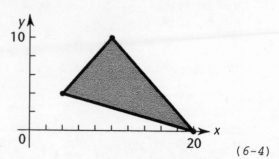

(6-4)

31. The feasible region is graphed as follows: The corner points (0, 4), (0, 0) and (5, 0) are obvious from the graph. The corner point (4, 2) is obtained by solving the system

$x + 2y = 8$
$2x + y = 10$

We now evaluate the objective function at each corner point.

Corner Point (x, y)	Objective Function $z = 7x + 9y$	
(0, 4)	36	
(0, 0)	0	
(5, 0)	35	
(4, 2)	46	Maximum value

The maximum value of z on S is 46 at (4, 2).

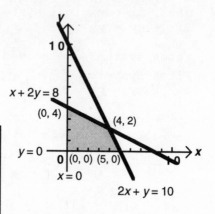

(6-5)

32. The feasible region is graphed as follows: The corner points (0, 20), (0, 15), (15, 0) and (20, 0) are obvious from the graph. The corner point (3, 6) is obtained by solving the system

$3x + y = 15$
$x + 2y = 15$

We now evaluate the objective function at each corner point.

Corner Point (x, y)	Objective Function $z = 5x + 10y$		
(0, 20)	200		
(0, 15)	150		
(3, 6)	75	Minimum value	Multiple optimal solutions
(15, 0)	75	Minimum value	
(20, 0)	100		

The minimum value of z on S is 75 at (3, 6) and (15, 0).

(6-5)

33. The feasible region is graphed below. The corner points (0, 10) and (0, 7) are obvious from the graph. The other corner points are obtained by solving the systems:

$x + 2y = 20$
$3x + y = 15$ to obtain (2, 9)

and $3x + y = 15$
$x + y = 7$ to obtain (4, 3)

We now evaluate the objective function at each corner point.

Corner Point (x, y)	Objective Function $z = 5x + 8y$	
(0, 10)	80	
(0, 7)	56	
(4, 3)	44	Minimum value
(2, 9)	82	Maximum value

The minimum value of z on S is 44 at $(4, 3)$. The maximum value of z on S is 82 at $(2, 9)$.

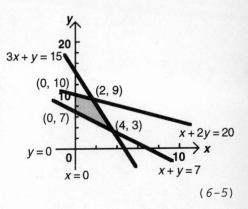

(6-5)

34. $\begin{bmatrix} 1 & 1 & 1 & | & 7000 \\ 0.04 & 0.05 & 0.06 & | & 360 \\ 0.04 & 0.05 & -0.06 & | & 120 \end{bmatrix}$ $\quad (-0.04)R_1 + R_2 \rightarrow R_2$
$\quad\quad\quad\quad\quad\quad\quad\quad\quad\quad\quad\quad\quad (-0.04)R_1 + R_3 \rightarrow R_3$

$\sim \begin{bmatrix} 1 & 1 & 1 & | & 7000 \\ 0 & 0.01 & 0.02 & | & 80 \\ 0 & 0.01 & -0.1 & | & -160 \end{bmatrix}$ $\quad 100R_2 \rightarrow R_2$

$\sim \begin{bmatrix} 1 & 1 & 1 & | & 7000 \\ 0 & 1 & 2 & | & 8000 \\ 0 & 0.01 & -0.1 & | & -160 \end{bmatrix}$ $\quad (-1)R_2 + R_1 \rightarrow R_1$
$\quad\quad\quad\quad\quad\quad\quad\quad\quad\quad\quad\quad\quad\quad (-0.01)R_2 + R_3 \rightarrow R_3$

$\sim \begin{bmatrix} 1 & 0 & -1 & | & -1000 \\ 0 & 1 & 2 & | & 8000 \\ 0 & 0 & -0.12 & | & -240 \end{bmatrix}$ $\quad -\frac{25}{3}R_3 \rightarrow R_3$

$\sim \begin{bmatrix} 1 & 0 & -1 & | & -1000 \\ 0 & 1 & 2 & | & 8000 \\ 0 & 0 & 1 & | & 2000 \end{bmatrix}$ $\quad R_3 + R_1 \rightarrow R_1$
$\quad\quad\quad\quad\quad\quad\quad\quad\quad\quad\quad (-2)R_3 + R_2 \rightarrow R_2$

$\sim \begin{bmatrix} 1 & 0 & 0 & | & 1000 \\ 0 & 1 & 0 & | & 4000 \\ 0 & 0 & 1 & | & 2000 \end{bmatrix}$

Solution: $x_1 = 1000$, $x_2 = 4000$, $x_3 = 2000$

(6-2)

35. $x^2 - xy + y^2 = 4$
$x^2 + xy - 2y^2 = 0$

Factor the left side of the equation that has a zero constant term.
$(x + 2y)(x - y) = 0$
$x = -2y \quad$ or $\quad x = y$
Thus, the original system is equivalent to the two systems
$x^2 - xy + y^2 = 4 \quad\quad\quad x^2 - xy + y^2 = 4$
$\quad\quad x = -2y \quad\quad\quad\quad\quad\quad x = y$
These systems are solved by substitution

First System: | Second System:
$x^2 - xy + y^2 = 4$

$x = -2y$

$(-2y)^2 - (-2y)y + y^2 = 4$
$4y^2 + 2y^2 + y^2 = 4$
$7y^2 = 4$

$y = \pm\sqrt{\dfrac{4}{7}}$

$y = \pm\dfrac{2\sqrt{7}}{7}$

Second System:
$x^2 - xy + y^2 = 4$

$x = y$

$y^2 - yy + y^2 = 4$
$y^2 = 4$
$y = \pm 2$

For $y = \dfrac{2\sqrt{7}}{7}$ For $y = -\dfrac{2\sqrt{7}}{7}$ For $y = -2$ For $y = 2$

$x = -2\left(\dfrac{2\sqrt{7}}{7}\right)$ $x = -2\left(-\dfrac{2\sqrt{7}}{7}\right)$ $x = -2$ $x = 2$

$= -\dfrac{4\sqrt{7}}{7}$ $= \dfrac{4\sqrt{7}}{7}$

Solutions: $\left(\dfrac{4\sqrt{7}}{7}, -\dfrac{2\sqrt{7}}{7}\right)$, $\left(-\dfrac{4\sqrt{7}}{7}, \dfrac{2\sqrt{7}}{7}\right)$, $(2, 2)$, $(-2, -2)$. The checking steps are omitted for lack of space.

(6-3)

36. For convenience we rewrite the constraint conditions:

$1.2x + 0.6y \le 960$ becomes $2x + y \le 1{,}600$
$0.04x + 0.03y \le 36$ becomes $4x + 3y \le 3{,}600$
$0.2x + 0.3y \le 270$ becomes $2x + 3y \le 2{,}700$
       $x, y \ge 0$ are unaltered.

The feasible region is graphed in the diagram. The corner points $(0, 900)$, $(0, 0)$, and $(800, 0)$ are obvious from the graph. The other corner points are obtained by solving the system:

$2x + y = 1{,}600$ and $4x + 3y = 3{,}600$
$4x + 3y = 3{,}600$ to obtain $(600, 400)$ $2x + 3y = 2{,}700$ to obtain $(450, 600)$

We now evaluate the objective function at each corner point.

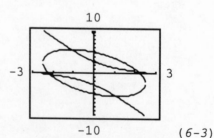

Corner Point (x, y)	Objective Function $z = 30x + 20y$	
$(0, 900)$	$18{,}000$	
$(0, 0)$	0	
$(800, 0)$	$24{,}000$	
$(600, 400)$	$26{,}000$	Maximum value
$(450, 600)$	$25{,}500$	

The maximum value of z on S is $26{,}000$ at $(600, 400)$.

(6-5)

37. Before we can enter these equations in our graphing utility, we must solve for y:

$x^2 + 4xy + y^2 = 8$ $5x^2 + 2xy + y^2 = 25$
$y^2 + 4xy + x^2 - 8 = 0$ $y^2 + 2xy + 5x^2 - 25 = 0$

Applying the quadratic formula to each equation, we have

$y = \dfrac{-4x \pm \sqrt{16x^2 - 4(x^2 - 8)}}{2}$ $y = \dfrac{-2x \pm \sqrt{4x^2 - 4(5x^2 - 25)}}{2}$

$y = \dfrac{-4x \pm \sqrt{12x^2 + 32}}{2}$ $y = \dfrac{-2x \pm \sqrt{100 - 16x^2}}{2}$

$y = -2x \pm \sqrt{3x^2 + 8}$ $y = -x \pm \sqrt{25 - 4x^2}$

Entering each of these four equations into a graphing utility produces the graph shown at the right.

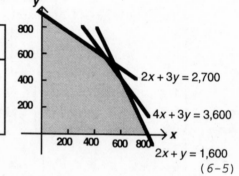

Zooming in on the four intersection points, or using a built-in intersection routine (details omitted), yields $(-2.16, -0.37)$, $(-1.09, 5.59)$, $(1.09, -5.59)$, and $(2.16, 0.37)$ to two decimal places.

(6-3)

38. (A) The system is independent. There is one solution.

(B) The matrix is in fact

$$\begin{bmatrix} 1 & 0 & -3 & | & 4 \\ 0 & 1 & 2 & | & 5 \\ 0 & 0 & 0 & | & n \end{bmatrix}$$

The third row corresponds to the equation $0x_1 + 0x_2 + 0x_3 = n$. This is impossible. The system has no solution.

(C) The matrix is in fact

$$\begin{bmatrix} 1 & 0 & -3 & | & 4 \\ 0 & 1 & 2 & | & 5 \\ 0 & 0 & 0 & | & 0 \end{bmatrix}$$

Thus, there are an infinite number of solutions ($x_3 = t$, $x_2 = 5 - 2t$, $x_1 = 4 + 3t$, for t any real number.)

(6-2)

39. Let x = number of $\frac{1}{2}$ pound packages

y = number of $\frac{1}{3}$ pound packages

There are 120 packages. Hence

$x + y = 120$ (1)

Since x $\frac{1}{2}$-pound packages weigh $\frac{1}{2}x$ pounds and y $\frac{1}{3}$-pound packages weigh $\frac{1}{3}y$ pounds, we have

$\frac{1}{2}x + \frac{1}{3}y = 48$ (2)

We solve the system (1), (2) using elimination by addition. We multiply the second equation by -3 and add.

$$\begin{array}{r} x + y = 120 \\ -\frac{3}{2}x - y = -144 \\ \hline -\frac{1}{2}x = -24 \\ x = 48 \end{array}$$

Substituting into equation (1), we have

$48 + y = 120$

$y = 72$

48 $\frac{1}{2}$-pound packages and 72 $\frac{1}{3}$-pound packages.

(6-1)

40. Using the formulas for perimeter and area of a rectangle, we have

$2a + 2b = 28$

$ab = 48$

Solving the first-degree equation for b and substituting into the second-degree equation, we have

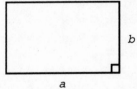

$$\begin{array}{r} 2b = 28 - 2a \\ b = 14 - a \\ a(14 - a) = 48 \\ -a^2 + 14a = 48 \\ a^2 - 14a + 48 = 0 \\ (a - 6)(a - 8) = 0 \\ a = 6, 8 \end{array}$$

For $a = 6$ For $a = 8$

$b = 14 - 6$ $b = 14 - 8$

$= 8$ $= 6$

Dimensions: 6 meters by 8 meters.

(6-3)

41. Let x_1 = number of grams of mix A

x_2 = number of grams of mix B

x_3 = number of grams of mix C

We have

$0.30x_1 + 0.20x_2 + 0.10x_3 = 27$ (protein)

$0.03x_1 + 0.05x_2 + 0.04x_3 = 5.4$ (fat)

$0.10x_1 + 0.20x_2 + 0.10x_3 = 19$ (moisture)

Clearing of decimals for convenience, we have

$3x_1 + 2x_2 + x_3 = 270$

$3x_1 + 5x_2 + 4x_3 = 540$

$x_1 + 2x_2 + x_3 = 190$

Form the augmented matrix and solve by Gauss-Jordan elimination.

$$\left[\begin{array}{ccc|c} 3 & 2 & 1 & 270 \\ 3 & 5 & 4 & 540 \\ 1 & 2 & 1 & 190 \end{array}\right] \quad R_3 \leftrightarrow R_1$$

$$\sim \left[\begin{array}{ccc|c} 1 & 2 & 1 & 190 \\ 3 & 5 & 4 & 540 \\ 3 & 2 & 1 & 270 \end{array}\right] \quad \begin{array}{l} (-3)R_1 + R_2 \rightarrow R_2 \\ (-3)R_1 + R_3 \rightarrow R_3 \end{array}$$

$$\sim \left[\begin{array}{ccc|c} 1 & 2 & 1 & 190 \\ 0 & -1 & 1 & -30 \\ 0 & -4 & -2 & -300 \end{array}\right] \quad (-1)R_2 \rightarrow R_2$$

$$\sim \left[\begin{array}{ccc|c} 1 & 2 & 1 & 190 \\ 0 & 1 & -1 & 30 \\ 0 & -4 & -2 & -300 \end{array}\right] \quad \begin{array}{l} (-2)R_2 + R_1 \rightarrow R_1 \\ \\ 4R_2 + R_3 \rightarrow R_3 \end{array}$$

$$\sim \left[\begin{array}{ccc|c} 1 & 0 & 3 & 130 \\ 0 & 1 & -1 & 30 \\ 0 & 0 & -6 & -180 \end{array}\right] \quad -\tfrac{1}{6}R_3 \rightarrow R_3$$

$$\sim \left[\begin{array}{ccc|c} 1 & 0 & 3 & 130 \\ 0 & 1 & -1 & 30 \\ 0 & 0 & 1 & 30 \end{array}\right] \quad \begin{array}{l} (-3)R_3 + R_1 \rightarrow R_1 \\ R_3 + R_2 \rightarrow R_2 \end{array}$$

$$\sim \left[\begin{array}{ccc|c} 1 & 0 & 0 & 40 \\ 0 & 1 & 0 & 60 \\ 0 & 0 & 1 & 30 \end{array}\right]$$

Therefore

$x_1 = 40$ grams Mix A

$x_2 = 60$ grams Mix B

$x_3 = 30$ grams Mix C

$(6-2)$

42. (A) Let x_1 = number of nickels

x_2 = number of dimes

Then $x_1 + x_2 = 30$ (total number of coins)

$5x_1 + 10x_2 = 190$ (total value of coins)

We form the augmented matrix and solve by Gauss-Jordan elimination.

$$\left[\begin{array}{cc|c} 1 & 1 & 30 \\ 5 & 10 & 190 \end{array}\right] \quad (-5)R_1 + R_2 \rightarrow R_2$$

$$\sim \left[\begin{array}{cc|c} 1 & 1 & 30 \\ 0 & 5 & 40 \end{array}\right] \quad \tfrac{1}{5}R_2 \rightarrow R_2$$

$$\sim \begin{bmatrix} 1 & 1 & | & 30 \\ 0 & 1 & | & 8 \end{bmatrix} \quad (-1)R_2 + R_1 \to R_1$$

$$\sim \begin{bmatrix} 1 & 0 & | & 22 \\ 0 & 1 & | & 8 \end{bmatrix}$$

The augmented matrix is in reduced form. It corresponds to the system
x_1 = 22 nickels
x_2 = 8 dimes

(B) Let x_1 = number of nickels
x_2 = number of dimes
x_3 = number of quarters
Then $x_1 + x_2 + x_3$ = 30 (total number of coins)
$5x_1 + 10x_2 + 25x_3$ = 190 (total value of coins)
We form the augmented matrix and solve by Gauss-Jordan elimination

$$\begin{bmatrix} 1 & 1 & 1 & | & 30 \\ 5 & 10 & 25 & | & 190 \end{bmatrix} \quad (-5)R_1 + R_2 \to R_2$$

$$\sim \begin{bmatrix} 1 & 1 & 1 & | & 30 \\ 0 & 5 & 20 & | & 40 \end{bmatrix} \quad \tfrac{1}{5}R_2 \to R_2$$

$$\sim \begin{bmatrix} 1 & 1 & 1 & | & 30 \\ 0 & 1 & 4 & | & 8 \end{bmatrix} \quad (-1)R_2 + R_1 \to R_1$$

$$\sim \begin{bmatrix} 1 & 0 & -3 & | & 22 \\ 0 & 1 & 4 & | & 8 \end{bmatrix}$$

The augmented matrix is in reduced form. It corresponds to the system:
$x_1 - 3x_3$ = 22
$ x_2 + 4x_3$ = 8

Let $x_3 = t$. Then $x_2 = -4x_3 + 8$
$ = -4t + 8$
$ x_1 = 3x_3 + 22$
$ = 3t + 22$

A solution is achieved, not for every real value of t, but for integer values
of t that give rise to non-negative x_1, x_2, x_3.

$x_1 \geq 0$ means $3t + 22 \geq 0$ or $t \geq -7\tfrac{1}{3}$
$x_2 \geq 0$ means $-4t + 8 \geq 0$ or $t \leq 2$
$x_3 \geq 0$ means $\phantom{-4t + 8 \geq 0 \text{ or }} t \geq 0$
The only integer values of t that satisfy these conditions are 0, 1, 2. Thus we
have the solutions

$x_1 = 3t + 22$ nickels
$x_2 = 8 - 4t$ dimes
$x_3 = t$ quarters
where t = 0, 1, or 2

(6-1, 6-2)

43. Let x = number of regular sails
y = number of competition sails

(A) We form the linear objective function
$P = 60x + 100y$
We wish to maximize P, the profit from x
regular sails @ \$60 and y competition
sails @ \$100, subject to the constraints

$x + 2y \leq 140$ cutting department constraint
$3x + 4y \leq 360$ sewing department constraint
$x,\ y \geq 0$ non-negative constraints

Solving the system of constraint
inequalities graphically, we obtain the
feasible region S shown.

Next we evaluate the objective function at
each corner point.

Corner Point $(x,\ y)$	Objective Function $P = 60x + 100y$	
(0, 0)	0	
(0, 70)	7,000	
(80, 30)	7,800	Maximum value
(120, 0)	7,200	

The optimal value is 7,800 at the corner point (80, 30). Thus, the company
should manufacture 80 regular sails and 30 competition sails for a maximum
profit of \$7,800.

(B) The objective function now becomes $60x + 125y$. We evaluate this at each
corner point.

Corner Point $(x,\ y)$	Objective Function $P = 60x + 125y$	
(0, 0)	0	
(0, 70)	8,750	Maximum value
(80, 30)	8,550	
(120, 0)	7,200	

The optimal value is now 8,750 at the corner point (0, 70). Thus, the
company should manufacture 70 competition sails and no regular sails
for a maximum profit of \$8,750.

(C) The objective function now becomes $60x + 75y$. We evaluate this at each
corner point.

Corner Point $(x,\ y)$	Objective Function $P = 60x + 75y$	
(0, 0)	0	
(0, 70)	5,250	
(80, 30)	7,050	
(120, 0)	7,200	Maximum value

The optimal value is now 7,200 at the corner point (120, 0). Thus, the
company should manufacture 120 regular sails and no competition sails
for a maximum profit of \$7,200.

(6-5)

44. Let x = number of grams of mix A
y = number of grams of mix B

(A) We form the objective function
$C = 0.07x + 0.04y$

C represents the cost of x grams at \$0.07 per gram and y grams at \$0.04 per gram. We wish to minimize C subject to

$5x + 2y \geq 800$	vitamin constraint
$2x + 4y \geq 800$	mineral contraint
$4x + 4y \leq 1300$	calorie constraint
$x, y \geq 0$	non-negative contraints

Solving the system of constraint inequalities graphically, we obtain the feasible region S shown in the diagram.

Next we evaluate the objective function at each corner point.

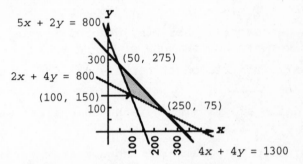

Corner Point (x, y)	Objective Function $C = 0.07x + 0.04y$	
(100, 150)	13	Minimum value
(250, 75)	20.5	
(50, 275)	14.5	

The optimal value is 13 at the corner point (100, 150). Thus, 100 grams of mix A and 150 grams of mix B should be used for a cost of \$13.

(B) The objective function now becomes $0.07x + 0.02y$. We evaluate this at each corner point.

Corner Point (x, y)	Objective Function $C = 0.07x + 0.02y$	
(100, 150)	10	
(250, 75)	19	
(50, 275)	9	Minimum value

The optimal value is now 9 at the corner point (50, 275). Thus, 50 grams of mix A and 275 grams of mix B should be used for a cost of \$9.

(C) The objective function now becomes $0.07x + 0.15y$. We evaluate this at each corner point.

Corner Point (x, y)	Objective Function $C = 0.07x + 0.15y$	
(100, 150)	29.5	
(250, 75)	28.75	Minimum value
(50, 275)	44.75	

The optimal value is now 28.75 at the corner point (250, 75). Thus, 250 grams of mix A and 75 grams of mix B should be used for a cost of \$28.75.

$(6-5)$

CHAPTER 7

Exercise 7-1

Key Ideas and Formulas

Two matrices are equal if they have the same size and their corresponding elements are equal.

The sum of two matrices of the same size, A and B, denoted $A + B$, is a matrix with elements that are the sums of the corresponding elements of A and B.

$A + B = B + A$ (Commutative property of addition).

$(A + B) + C = A + (B + C)$ (Associative property of addition).

The negative of a matrix M, denoted $-M$, is a matrix with elements that are the negatives of the elements of M.

A matrix with elements that are all 0's is called a zero matrix, denoted 0.

$M + (-M) = 0$.

If A and B are matrices of the same size, we define subtraction as follows:

$A - B = A + (-B)$.

The product of a number k and a matrix M, denoted kM, is a matrix formed by multiplying each element of M by k.

The product of a $1 \times n$ row matrix and an $n \times 1$ column matrix is a 1×1 matrix given by

$$[a_1 \quad a_2 \cdots a_n] \begin{bmatrix} b_1 \\ b_2 \\ \vdots \\ b_n \end{bmatrix} = [a_1 b_1 + a_2 b_2 + \cdots + a_n b_n]$$

(A 1×1 matrix is often referred to as a real number.)

The product of two matrices A and B is defined only on the assumption that the number of columns in A is equal to the number of rows in B.

$$\begin{array}{ccc} A & B & = AB \\ m \times p & p \times n & m \times n \end{array}$$

number of columns in A = p = number of rows in B

If A is an $m \times p$ matrix and B is a $p \times n$ matrix, then the **matrix product** of A and B, denoted AB, is an $m \times n$ matrix whose element in the ith row and jth column is the real number obtained from the product of the ith row of A and the jth column of B. If the number of columns in A does not equal the number of rows in B, then the matrix product AB is **not defined**.

Matrix multiplication is not commutative. Depending on A and B, either AB or BA could be not defined, or defined but of different dimensions, or of the same dimension but AB still unequal to BA.

Also, AB could be zero without $A = 0$ or $B = 0$.

---

1. $\begin{bmatrix} -1 & 4 \\ 2 & -6 \end{bmatrix} + \begin{bmatrix} 1 & -2 \\ 0 & 5 \end{bmatrix} = \begin{bmatrix} -1 + 1 & 4 + (-2) \\ 2 + 0 & (-6) + 5 \end{bmatrix} = \begin{bmatrix} 0 & 2 \\ 2 & -1 \end{bmatrix}$

3. $\begin{bmatrix} -3 & 5 \\ 2 & 0 \\ 1 & 4 \end{bmatrix} + \begin{bmatrix} 2 & 1 \\ -6 & 3 \\ 0 & -5 \end{bmatrix} = \begin{bmatrix} (-3) + 2 & 5 + 1 \\ 2 + (-6) & 0 + 3 \\ 1 + 0 & 4 + (-5) \end{bmatrix} = \begin{bmatrix} -1 & 6 \\ -4 & 3 \\ 1 & -1 \end{bmatrix}$

5. These matrices have different sizes, hence the sum is not defined.

7. $\begin{bmatrix} 6 & 2 & -3 \\ 0 & -4 & 5 \end{bmatrix} - \begin{bmatrix} 4 & -1 & 2 \\ -5 & 1 & -2 \end{bmatrix} = \begin{bmatrix} 6-4 & 2-(-1) & (-3)-2 \\ 0-(-5) & (-4)-1 & 5-(-2) \end{bmatrix} = \begin{bmatrix} 2 & 3 & -5 \\ 5 & -5 & 7 \end{bmatrix}$

9. $\begin{bmatrix} 20 & -10 & 30 \\ 0 & -40 & 50 \end{bmatrix}$ **11.** $[2 \quad 4] \begin{bmatrix} 3 \\ 1 \end{bmatrix} = [2 \cdot 3 + 4 \cdot 1] = [10]$

13. $\begin{bmatrix} 3 & 4 \\ -1 & -2 \end{bmatrix} \begin{bmatrix} -1 \\ 2 \end{bmatrix} = \begin{bmatrix} 3(-1) + 4(2) \\ (-1)(-1) + (-2)2 \end{bmatrix} = \begin{bmatrix} 5 \\ -3 \end{bmatrix}$

15. $\begin{bmatrix} 2 & -3 \\ 1 & 2 \end{bmatrix} \begin{bmatrix} 1 & -1 \\ 0 & -2 \end{bmatrix} = \begin{bmatrix} 2 \cdot 1 + (-3)0 & 2(-1) + (-3)(-2) \\ 1 \cdot 1 + 2 \cdot 0 & 1(-1) + 2(-2) \end{bmatrix} = \begin{bmatrix} 2 & 4 \\ 1 & -5 \end{bmatrix}$

17. $\begin{bmatrix} 1 & -1 \\ 0 & -2 \end{bmatrix} \begin{bmatrix} 2 & -3 \\ 1 & 2 \end{bmatrix} = \begin{bmatrix} 1 \cdot 2 + (-1)1 & 1(-3) + (-1)2 \\ 0 \cdot 2 + (-2)1 & 0(-3) + (-2)2 \end{bmatrix} = \begin{bmatrix} 1 & -5 \\ -2 & -4 \end{bmatrix}$

19. $[4 \quad -2] \begin{bmatrix} -5 \\ -3 \end{bmatrix} = [4(-5) + (-2)(-3)] = [-14]$

21. $\begin{bmatrix} -5 \\ -3 \end{bmatrix} [4 \quad -2] = \begin{bmatrix} (-5)4 & (-5)(-2) \\ (-3)4 & (-3)(-2) \end{bmatrix} = \begin{bmatrix} -20 & 10 \\ -12 & 6 \end{bmatrix}$

23. $[3 \quad -2 \quad -4] \begin{bmatrix} 1 \\ 2 \\ -3 \end{bmatrix} = [3 \cdot 1 + (-2)2 + (-4)(-3)] = [11]$

25. $\begin{bmatrix} 1 \\ 2 \\ -3 \end{bmatrix} [3 \quad -2 \quad -4] = \begin{bmatrix} 1 \cdot 3 & 1(-2) & 1(-4) \\ 2 \cdot 3 & 2(-2) & 2(-4) \\ (-3)3 & (-3)(-2) & (-3)(-4) \end{bmatrix} = \begin{bmatrix} 3 & -2 & -4 \\ 6 & -4 & -8 \\ -9 & 6 & 12 \end{bmatrix}$

27. C has 3 columns. A has 2 rows. Therefore, CA is not defined.

29. $BA = \begin{bmatrix} -3 & 1 \\ 2 & 5 \end{bmatrix} \begin{bmatrix} 2 & -1 & 3 \\ 0 & 4 & -2 \end{bmatrix} = \begin{bmatrix} (-3)2 + 1 \cdot 0 & (-3)(-1) + 1 \cdot 4 & (-3)3 + 1(-2) \\ 2 \cdot 2 + 5 \cdot 0 & 2(-1) + 5 \cdot 4 & 2 \cdot 3 + 5(-2) \end{bmatrix}$

$= \begin{bmatrix} -6 & 7 & -11 \\ 4 & 18 & -4 \end{bmatrix}$

31. $C^2 = \begin{bmatrix} -1 & 0 & 2 \\ 4 & -3 & 1 \\ -2 & 3 & 5 \end{bmatrix} \begin{bmatrix} -1 & 0 & 2 \\ 4 & -3 & 1 \\ -2 & 3 & 5 \end{bmatrix}$

$= \begin{bmatrix} (-1)(-1) + 0 \cdot 4 + 2(-2) & (-1)0 + 0(-3) + 2 \cdot 3 & (-1)2 + 0 \cdot 1 + 2 \cdot 5 \\ 4(-1) + (-3)4 + 1(-2) & 4 \cdot 0 + (-3)(-3) + 1 \cdot 3 & 4 \cdot 2 + (-3)1 + 1 \cdot 5 \\ (-2)(-1) + 3 \cdot 4 + 5(-2) & (-2)0 + 3(-3) + 5 \cdot 3 & (-2)2 + 3 \cdot 1 + 5 \cdot 5 \end{bmatrix}$

$= \begin{bmatrix} -3 & 6 & 8 \\ -18 & 12 & 10 \\ 4 & 6 & 24 \end{bmatrix}$

33. $DA = \begin{bmatrix} 3 & -2 \\ 0 & -1 \\ 1 & 2 \end{bmatrix} \begin{bmatrix} 2 & -1 & 3 \\ 0 & 4 & -2 \end{bmatrix}$

$= \begin{bmatrix} 3 \cdot 2 + (-2)0 & 3(-1) + (-2)4 & 3 \cdot 3 + (-2)(-2) \\ 0 \cdot 2 + (-1)0 & 0(-1) + (-1)4 & 0 \cdot 3 + (-1)(-2) \\ 1 \cdot 2 + 2 \cdot 0 & 1(-1) + 2 \cdot 4 & 1 \cdot 3 + 2(-2) \end{bmatrix} = \begin{bmatrix} 6 & -11 & 13 \\ 0 & -4 & 2 \\ 2 & 7 & -1 \end{bmatrix}$

$C + DA = \begin{bmatrix} -1 & 0 & 2 \\ 4 & -3 & 1 \\ -2 & 3 & 5 \end{bmatrix} + \begin{bmatrix} 6 & -11 & 13 \\ 0 & -4 & 2 \\ 2 & 7 & -1 \end{bmatrix} = \begin{bmatrix} 5 & -11 & 15 \\ 4 & -7 & 3 \\ 0 & 10 & 4 \end{bmatrix}$

35. $0.2CD = 0.2 \begin{bmatrix} -1 & 0 & 2 \\ 4 & -3 & 1 \\ -2 & 3 & 5 \end{bmatrix} \begin{bmatrix} 3 & -2 \\ 0 & -1 \\ 1 & 2 \end{bmatrix}$

$= 0.2 \begin{bmatrix} (-1)3 + 0 \cdot 0 + 2 \cdot 1 & (-1)(-2) + 0(-1) + 2 \cdot 2 \\ 4 \cdot 3 + (-3)0 + 1 \cdot 1 & 4(-2) + (-3)(-1) + 1 \cdot 2 \\ (-2)3 + 3 \cdot 0 + 5 \cdot 1 & (-2)(-2) + 3(-1) + 5 \cdot 2 \end{bmatrix}$

$= 0.2 \begin{bmatrix} -1 & 6 \\ 13 & -3 \\ -1 & 11 \end{bmatrix} = \begin{bmatrix} -0.2 & 1.2 \\ 2.6 & -0.6 \\ -0.2 & 2.2 \end{bmatrix}$

37. $DB = \begin{bmatrix} 3 & -2 \\ 0 & -1 \\ 1 & 2 \end{bmatrix} \begin{bmatrix} -3 & 1 \\ 2 & 5 \end{bmatrix} = \begin{bmatrix} 3(-3) + (-2)2 & 3 \cdot 1 + (-2)5 \\ 0(-3) + (-1)2 & 0 \cdot 1 + (-1)5 \\ 1(-3) + 2 \cdot 2 & 1 \cdot 1 + 2 \cdot 5 \end{bmatrix} = \begin{bmatrix} -13 & -7 \\ -2 & -5 \\ 1 & 11 \end{bmatrix}$

$CD = \begin{bmatrix} -1 & 6 \\ 13 & -3 \\ -1 & 11 \end{bmatrix}$ (see problem 35)

Thus, $2DB + 5CD = 2\begin{bmatrix} -13 & -7 \\ -2 & -5 \\ 1 & 11 \end{bmatrix} + 5\begin{bmatrix} -1 & 6 \\ 13 & -3 \\ -1 & 11 \end{bmatrix} = \begin{bmatrix} -26 & -14 \\ -4 & -10 \\ 2 & 22 \end{bmatrix} + \begin{bmatrix} -5 & 30 \\ 65 & -15 \\ -5 & 55 \end{bmatrix} = \begin{bmatrix} -31 & 16 \\ 61 & -25 \\ -3 & 77 \end{bmatrix}$

39. $(-1)AC$ is a matrix of size 2×3. $3DB$ is a matrix of size 3×2. Hence, $(-1)AC + 3DB$ is not defined.

41. $CD = \begin{bmatrix} -1 & 6 \\ 13 & -3 \\ -1 & 11 \end{bmatrix}$ (see problem 35)

Hence $CDA = \begin{bmatrix} -1 & 6 \\ 13 & -3 \\ -1 & 11 \end{bmatrix} \begin{bmatrix} 2 & -1 & 3 \\ 0 & 4 & -2 \end{bmatrix}$

$= \begin{bmatrix} (-1)2 + 6 \cdot 0 & (-1)(-1) + 6 \cdot 4 & (-1)3 + 6(-2) \\ 13 \cdot 2 + (-3)0 & 13(-1) + (-3)4 & 13 \cdot 3 + (-3)(-2) \\ (-1)2 + 11 \cdot 0 & (-1)(-1) + 11 \cdot 4 & (-1)3 + 11(-2) \end{bmatrix} = \begin{bmatrix} -2 & 25 & -15 \\ 26 & -25 & 45 \\ -2 & 45 & -25 \end{bmatrix}$

43. $DB = \begin{bmatrix} -13 & -7 \\ -2 & -5 \\ 1 & 11 \end{bmatrix}$ (see problem 37)

Hence $DBA = \begin{bmatrix} -13 & -7 \\ -2 & -5 \\ 1 & 11 \end{bmatrix} \begin{bmatrix} 2 & -1 & 3 \\ 0 & 4 & -2 \end{bmatrix}$

$= \begin{bmatrix} (-13)2 + (-7)0 & (-13)(-1) + (-7)4 & (-13)3 + (-7)(-2) \\ (-2)2 + (-5)0 & (-2)(-1) + (-5)4 & (-2)3 + (-5)(-2) \\ 1 \cdot 2 + 11 \cdot 0 & 1(-1) + 11 \cdot 4 & 1 \cdot 3 + 11(-2) \end{bmatrix}$

$= \begin{bmatrix} -26 & -15 & -25 \\ -4 & -18 & 4 \\ 2 & 43 & -19 \end{bmatrix}$

45. $\begin{bmatrix} a & b \\ c & d \end{bmatrix} + \begin{bmatrix} 2 & -3 \\ 0 & 1 \end{bmatrix} = \begin{bmatrix} a + 2 & b - 3 \\ c & d + 1 \end{bmatrix} = \begin{bmatrix} 1 & -2 \\ 3 & -4 \end{bmatrix}$

if and only if corresponding elements are equal.

$\begin{array}{cccc} a + 2 = 1 & b - 3 = -2 & c = 3 & d + 1 = -4 \\ a = -1 & b = 1 & c = 3 & d = -5 \end{array}$

47. $\begin{bmatrix} 3x & 5 \\ -1 & 4x \end{bmatrix} + \begin{bmatrix} 2y & -3 \\ -6 & -y \end{bmatrix} = \begin{bmatrix} 3x + 2y & 2 \\ -7 & 4x - y \end{bmatrix} = \begin{bmatrix} 7 & 2 \\ -7 & 2 \end{bmatrix}$

if and only if corresponding elements are equal.

$3x + 2y = 7 \qquad 2 \overset{\lor}{=} 2$ } Two conditions are already met.
$\qquad -7 \overset{\lor}{=} -7 \quad 4x - y = 2$

To find x and y, we solve the system:
$3x + 2y = 7$
$\ 4x - y = 2$ to obtain $x = 1$, $y = 2$.

49. $\begin{bmatrix} 1 & 3 \\ -2 & -2 \end{bmatrix}\begin{bmatrix} x & 1 \\ 3 & 2 \end{bmatrix} = \begin{bmatrix} x + 9 & 7 \\ -2x - 6 & -6 \end{bmatrix} = \begin{bmatrix} y & 7 \\ y & -6 \end{bmatrix}$

if and only if corresponding elements are equal.

$x + 9 = y \qquad 7 \overset{\lor}{=} 7$ } Two conditions are already met.
$-2x - 6 = y \quad -6 \overset{\lor}{=} -6$

To find x and y, we solve the system:
$\ x + 9 = y$
$-2x - 6 = y$ to obtain $x = -5$, $y = 4$.

51. $\begin{bmatrix} 1 & 3 \\ 1 & 4 \end{bmatrix}\begin{bmatrix} a & b \\ c & d \end{bmatrix} = \begin{bmatrix} a + 3c & b + 3d \\ a + 4c & b + 4d \end{bmatrix} = \begin{bmatrix} 6 & -5 \\ 7 & -7 \end{bmatrix}$

if and only if corresponding elements are equal.
$a + 3c = 6 \qquad b + 3d = -5$
$a + 4c = 7 \qquad b + 4d = -7$
Solving these systems we obtain $a = 3$, $b = 1$, $c = 1$, $d = -2$.

53. (A) Since $\begin{bmatrix} a_1 & 0 \\ 0 & d_1 \end{bmatrix} + \begin{bmatrix} a_2 & 0 \\ 0 & d_2 \end{bmatrix} = \begin{bmatrix} a_1 + a_2 & 0 \\ 0 & d_1 + d_2 \end{bmatrix}$, the statement is true.

(B) $A + B = B + A$ is true for any matrices for which $A + B$ is defined, as it is in this case.

(C) Since $\begin{bmatrix} a_1 & 0 \\ 0 & d_1 \end{bmatrix}\begin{bmatrix} a_2 & 0 \\ 0 & d_2 \end{bmatrix} = \begin{bmatrix} a_1 a_2 & 0 \\ 0 & d_1 d_2 \end{bmatrix}$, the statement is true.

(D) Since $\begin{bmatrix} a_1 & 0 \\ 0 & d_1 \end{bmatrix}\begin{bmatrix} a_2 & 0 \\ 0 & d_2 \end{bmatrix} = \begin{bmatrix} a_1 a_2 & 0 \\ 0 & d_1 d_2 \end{bmatrix} = \begin{bmatrix} a_2 a_1 & 0 \\ 0 & d_2 d_1 \end{bmatrix} = \begin{bmatrix} a_2 & 0 \\ 0 & d_2 \end{bmatrix}\begin{bmatrix} a_1 & 0 \\ 0 & d_2 \end{bmatrix}$, the statement is true.

55. $\frac{1}{2}(A + B) = \frac{1}{2}\left(\begin{bmatrix} 30 & 25 \\ 60 & 80 \end{bmatrix} + \begin{bmatrix} 36 & 27 \\ 54 & 74 \end{bmatrix} \right) = \frac{1}{2}\begin{bmatrix} 66 & 52 \\ 114 & 154 \end{bmatrix} = \begin{bmatrix} 33 & 26 \\ 57 & 77 \end{bmatrix} \begin{matrix} \text{Guitar Banjo} \\ \text{Materials} \\ \text{Labor} \end{matrix}$

57. If a quantity is increased b 15%, the result is a multiplication by 1.15. If a quantity is increased by 10%, the result is a multiplication by 1.1. Thus we must calculate $1.1N - 1.15M$. The mark-up matrix is:

$1.1N - 1.15M = 1.1\begin{bmatrix} 13,900 & 783 & 263 & 215 \\ 15,000 & 838 & 395 & 236 \\ 18,300 & 967 & 573 & 248 \end{bmatrix} - 1.15\begin{bmatrix} 10,400 & 682 & 215 & 182 \\ 12,500 & 721 & 295 & 182 \\ 16,400 & 827 & 443 & 192 \end{bmatrix}$

$= \begin{bmatrix} 15,290 & 861.3 & 289.3 & 236.5 \\ 16,500 & 921.8 & 434.5 & 259.6 \\ 20,130 & 1,063.7 & 630.3 & 272.8 \end{bmatrix} - \begin{bmatrix} 11,960 & 784.3 & 247.25 & 209.3 \\ 14,375 & 829.15 & 339.25 & 209.3 \\ 18,860 & 951.05 & 509.45 & 220.8 \end{bmatrix}$

	Basic Car	Air	AM/FM Radio	Cruise Control	
Model A	$3,330	$ 77	$ 42	$27	
= Model B	$2,125	$ 93	$ 95	$50	= Mark up
Model C	$1,270	$113	$121	$52	

59. (A) $[0.6 \quad 0.6 \quad 0.2]\begin{bmatrix} 8 \\ 10 \\ 5 \end{bmatrix} = (0.6)8 + (0.6)10 + (0.2)5 = 11.80$ dollars per boat

(B) $[1.5 \quad 1.2 \quad 0.4]\begin{bmatrix} 9 \\ 12 \\ 6 \end{bmatrix} = (1.5)9 + (1.2)12 + (0.4)6 = 30.30$ dollars per boat

(C) The matrix NM has no obvious meaning, but the matrix MN gives the labor costs per boat at each plant.

(D) $MN = \begin{bmatrix} 0.6 & 0.6 & 0.2 \\ 1.0 & 0.9 & 0.3 \\ 1.5 & 1.2 & 0.4 \end{bmatrix}\begin{bmatrix} 8 & 9 \\ 10 & 12 \\ 5 & 6 \end{bmatrix}$

$= \begin{bmatrix} (0.6)8 + (0.6)10 + (0.2)5 & (0.6)9 + (0.6)12 + (0.2)6 \\ (1.0)8 + (0.9)10 + (0.3)5 & (1.0)9 + (0.9)12 + (0.3)6 \\ (1.5)8 + (1.2)10 + (0.4)5 & (1.5)9 + (1.2)12 + (0.4)6 \end{bmatrix}$

$= \begin{bmatrix} \$11.80 & \$13.80 \\ \$18.50 & \$21.60 \\ \$26.00 & \$30.30 \end{bmatrix}$ One-person boat / Two-person boat / Four-person boat

with column headers Plant I, Plant II

This matrix gives the labor costs for each type of boat at each plant.

61. (A) $A^2 = AA = \begin{bmatrix} 0 & 1 & 0 & 1 & 0 \\ 0 & 0 & 1 & 0 & 0 \\ 1 & 0 & 0 & 0 & 1 \\ 0 & 0 & 1 & 0 & 0 \\ 0 & 0 & 0 & 1 & 0 \end{bmatrix}\begin{bmatrix} 0 & 1 & 0 & 1 & 0 \\ 0 & 0 & 1 & 0 & 0 \\ 1 & 0 & 0 & 0 & 1 \\ 0 & 0 & 1 & 0 & 0 \\ 0 & 0 & 0 & 1 & 0 \end{bmatrix} = \begin{bmatrix} 0 & 0 & 2 & 0 & 0 \\ 1 & 0 & 0 & 0 & 1 \\ 0 & 1 & 0 & 2 & 0 \\ 1 & 0 & 0 & 0 & 1 \\ 0 & 0 & 1 & 0 & 0 \end{bmatrix}$

The 1 in row 2 and column 1 of A^2 indicates that there is one way to travel from Baltimore to Atlanta with one intermediate connection. The 2 in row 1 and column 3 indicates that there are two ways to travel from Atlanta to Chicago with one intermediate connection. In general, the elements in A^2 indicate the number of different ways to travel from the ith city to the jth city with one intermediate connection.

(B) $A^3 = A^2A = \begin{bmatrix} 0 & 0 & 2 & 0 & 0 \\ 1 & 0 & 0 & 0 & 1 \\ 0 & 1 & 0 & 2 & 0 \\ 1 & 0 & 0 & 0 & 1 \\ 0 & 0 & 1 & 0 & 0 \end{bmatrix}\begin{bmatrix} 0 & 1 & 0 & 1 & 0 \\ 0 & 0 & 1 & 0 & 0 \\ 1 & 0 & 0 & 0 & 1 \\ 0 & 0 & 1 & 0 & 0 \\ 0 & 0 & 0 & 1 & 0 \end{bmatrix} = \begin{bmatrix} 2 & 0 & 0 & 0 & 2 \\ 0 & 1 & 0 & 2 & 0 \\ 0 & 0 & 3 & 0 & 0 \\ 0 & 1 & 0 & 2 & 0 \\ 1 & 0 & 0 & 0 & 1 \end{bmatrix}$

The 1 in row 4 and column 2 of A^3 indicates that there is one way to travel from Denver to Baltimore with two intermediate connections. The 2 in row 1 and column 5 indicates that there are two ways to travel from Atlanta to El Paso with two intermediate connections. In general, the elements in A^3 indicate the number of different ways to travel from the ith city to the the jth city with two intermediate connections.

(C) A is given above.

$A + A^2 = \begin{bmatrix} 0 & 1 & 0 & 1 & 0 \\ 0 & 0 & 1 & 0 & 0 \\ 1 & 0 & 0 & 0 & 1 \\ 0 & 0 & 1 & 0 & 0 \\ 0 & 0 & 0 & 1 & 0 \end{bmatrix} + \begin{bmatrix} 0 & 0 & 2 & 0 & 0 \\ 1 & 0 & 0 & 0 & 1 \\ 0 & 1 & 0 & 2 & 0 \\ 1 & 0 & 0 & 0 & 1 \\ 0 & 0 & 1 & 0 & 0 \end{bmatrix} = \begin{bmatrix} 0 & 1 & 2 & 1 & 0 \\ 1 & 0 & 1 & 0 & 1 \\ 1 & 1 & 0 & 2 & 1 \\ 1 & 0 & 1 & 0 & 1 \\ 0 & 0 & 1 & 1 & 0 \end{bmatrix}$

$A + A^2 + A^3 = \begin{bmatrix} 0 & 1 & 2 & 1 & 0 \\ 1 & 0 & 1 & 0 & 1 \\ 1 & 1 & 0 & 2 & 1 \\ 1 & 0 & 1 & 0 & 1 \\ 0 & 0 & 1 & 1 & 0 \end{bmatrix} + \begin{bmatrix} 2 & 0 & 0 & 0 & 2 \\ 0 & 1 & 0 & 2 & 0 \\ 0 & 0 & 3 & 0 & 0 \\ 0 & 1 & 0 & 2 & 0 \\ 1 & 0 & 0 & 0 & 1 \end{bmatrix} = \begin{bmatrix} 2 & 1 & 2 & 1 & 2 \\ 1 & 1 & 1 & 2 & 1 \\ 1 & 1 & 3 & 2 & 1 \\ 1 & 1 & 1 & 2 & 1 \\ 1 & 0 & 1 & 1 & 1 \end{bmatrix}$

A zero element remains, so we must compute A^4.

$$A^4 = A^3 A = \begin{bmatrix} 2 & 0 & 0 & 0 & 2 \\ 0 & 1 & 0 & 2 & 0 \\ 0 & 0 & 3 & 0 & 0 \\ 0 & 1 & 0 & 2 & 0 \\ 1 & 0 & 0 & 0 & 1 \end{bmatrix} \begin{bmatrix} 0 & 1 & 0 & 1 & 0 \\ 0 & 0 & 1 & 0 & 0 \\ 1 & 0 & 0 & 0 & 1 \\ 0 & 0 & 1 & 0 & 0 \\ 0 & 0 & 0 & 1 & 0 \end{bmatrix} = \begin{bmatrix} 0 & 2 & 0 & 4 & 0 \\ 0 & 0 & 3 & 0 & 0 \\ 3 & 0 & 0 & 0 & 3 \\ 0 & 0 & 3 & 0 & 0 \\ 0 & 1 & 0 & 2 & 0 \end{bmatrix}$$

Then $A + A^2 + A^3 + A^4 = \begin{bmatrix} 2 & 1 & 2 & 1 & 2 \\ 1 & 1 & 1 & 2 & 1 \\ 1 & 1 & 3 & 2 & 1 \\ 1 & 1 & 1 & 2 & 1 \\ 1 & 0 & 1 & 1 & 1 \end{bmatrix} + \begin{bmatrix} 0 & 2 & 0 & 4 & 0 \\ 0 & 0 & 3 & 0 & 0 \\ 3 & 0 & 0 & 0 & 3 \\ 0 & 0 & 3 & 0 & 0 \\ 0 & 1 & 0 & 2 & 0 \end{bmatrix} = \begin{bmatrix} 2 & 3 & 2 & 5 & 2 \\ 1 & 1 & 4 & 2 & 1 \\ 4 & 1 & 3 & 2 & 4 \\ 1 & 1 & 4 & 2 & 1 \\ 1 & 1 & 1 & 3 & 1 \end{bmatrix}$

This matrix indicates that it is possible to travel from any origin to any destination with at most 3 intermediate connections.

63. (A) $[1,000 \quad 500 \quad 5,000] \begin{bmatrix} \$0.80 \\ \$1.50 \\ \$0.40 \end{bmatrix} = \begin{aligned} &= 1,000(\$0.80) + 500(\$1.50) + 5,000(\$0.40) \\ &= \$3,550 \end{aligned}$

(B) $[2,000 \quad 800 \quad 8,000] \begin{bmatrix} \$0.80 \\ \$1.50 \\ \$0.40 \end{bmatrix} = \begin{aligned} &= 2,000(\$0.80) + 800(\$1.50) + 8,000(\$0.40) \\ &= \$6,000 \end{aligned}$

(C) The matrix *MN* has no obvious interpretations, but the matrix *NM* represents the total cost of all contacts in each town.

(D) $NM = \begin{bmatrix} 1,000 & 500 & 5,000 \\ 2,000 & 800 & 8,000 \end{bmatrix} \begin{bmatrix} \$0.80 \\ \$1.50 \\ \$0.40 \end{bmatrix} = \begin{bmatrix} 1,000(0.80) + 500(1.50) + 5,000(0.40) \\ 2,000(0.80) + 800(1.50) + 8,000(0.40) \end{bmatrix}$

$= \begin{bmatrix} \$3,550 \\ \$6,000 \end{bmatrix} \begin{matrix} \text{Berkeley} \\ \text{Oakland} \end{matrix} = \text{Cost of all contacts in each town.}$

(E) The matrix $[1 \quad 1]N$ can be used to find the total number of each of the three types of contact:

$[1 \quad 1] \begin{bmatrix} 1,000 & 500 & 5,000 \\ 2,000 & 800 & 8,000 \end{bmatrix} = [1,000 + 2,000 \quad 500 + 800 \quad 5,000 + 8,000]$

[Telephone House Letter] $= [3,000 \quad 1,300 \quad 13,000]$

(F) The matrix $N \begin{bmatrix} 1 \\ 1 \\ 1 \end{bmatrix}$ can be used to find the total number of contacts in each town:

$\begin{bmatrix} 1,000 & 500 & 5,000 \\ 2,000 & 800 & 8,000 \end{bmatrix} \begin{bmatrix} 1 \\ 1 \\ 1 \end{bmatrix}$

$= \begin{bmatrix} 1,000 + 500 + 5,000 \\ 2,000 + 800 + 8,000 \end{bmatrix} = \begin{bmatrix} 6,500 \\ 10,800 \end{bmatrix} = \begin{bmatrix} \text{Berkeley contacts} \\ \text{Oakland contacts} \end{bmatrix}$

Exercise 7-2

Key Ideas and Formulas

The identity element for multiplication for square matrices of order n (size $n \times n$), denoted I, is the square matrix of order n with 1's along the principal diagonal and 0's elsewhere. For A a square matrix of order n,

$$IA = AI = A$$

If M is a square matrix of order n and if there exists a square matrix M^{-1} such that $MM^{-1} = M^{-1}M = I$, M^{-1} is called the (multiplicative) inverse of M.

To find M^{-1} if it exists, form the augmented matrix $[M \mid I]$ and use row operations to transform into $[I \mid B]$ if possible. Then $B = M^{-1}$. If, however, all 0's are obtained at some stage in one or more rows to the left of the vertical line, then M^{-1} will not exist.

---

1. $\begin{bmatrix} 2 & -3 \\ 4 & 5 \end{bmatrix}$

3. $\begin{bmatrix} 2 & -3 \\ 4 & 5 \end{bmatrix}$

5. $\begin{bmatrix} -2 & 1 & 3 \\ 2 & 4 & -2 \\ 5 & 1 & 0 \end{bmatrix}$

7. $\begin{bmatrix} -2 & 1 & 3 \\ 2 & 4 & -2 \\ 5 & 1 & 0 \end{bmatrix}$

9. $\begin{bmatrix} 3 & -4 \\ -2 & 3 \end{bmatrix}\begin{bmatrix} 3 & 4 \\ 2 & 3 \end{bmatrix} = \begin{bmatrix} 3\cdot3 + (-4)2 & 3\cdot4 + (-4)3 \\ (-2)3 + 3\cdot2 & (-2)4 + 3\cdot3 \end{bmatrix} = \begin{bmatrix} 1 & 0 \\ 0 & 1 \end{bmatrix}$

11. $\begin{bmatrix} 3 & -4 \\ 4 & -5 \end{bmatrix}\begin{bmatrix} -5 & 4 \\ -4 & 3 \end{bmatrix} = \begin{bmatrix} 3(-5) + (-4)(-4) & 3\cdot4 + (-4)3 \\ 4(-5) + (-5)(-4) & 4\cdot4 + (-5)3 \end{bmatrix} = \begin{bmatrix} 1 & 0 \\ 0 & 1 \end{bmatrix}$

13. $\begin{bmatrix} 1 & -1 & 1 \\ 0 & 2 & -1 \\ 2 & 3 & 0 \end{bmatrix}\begin{bmatrix} 3 & 3 & -1 \\ -2 & -2 & 1 \\ -4 & -5 & 2 \end{bmatrix}$

$= \begin{bmatrix} 1\cdot3 + (-1)(-2) + 1(-4) & 1\cdot3 + (-1)(-2) + 1(-5) & 1\cdot(-1) + (-1)\cdot1 + 1\cdot2 \\ 0\cdot3 + 2(-2) + (-1)(-4) & 0\cdot3 + 2(-2) + (-1)(-5) & 0(-1) + 2\cdot1 + (-1)\cdot2 \\ 2\cdot3 + 3(-2) + 0(-4) & 2\cdot3 + 3(-2) + 0(-5) & 2(-1) + 3\cdot1 + 0\cdot2 \end{bmatrix}$

$= \begin{bmatrix} 1 & 0 & 0 \\ 0 & 1 & 0 \\ 0 & 0 & 1 \end{bmatrix}$

15. $\begin{bmatrix} 0 & -1 & | & 1 & 0 \\ 1 & 4 & | & 0 & 1 \end{bmatrix} \quad R_1 \leftrightarrow R_2$

$\sim \begin{bmatrix} 1 & 4 & | & 0 & 1 \\ 0 & -1 & | & 1 & 0 \end{bmatrix} \quad 4R_2 + R_1 \rightarrow R_1$

$\sim \begin{bmatrix} 1 & 0 & | & 4 & 1 \\ 0 & -1 & | & 1 & 0 \end{bmatrix} \quad (-1)R_2 \rightarrow R_2$

$\sim \begin{bmatrix} 1 & 0 & | & 4 & 1 \\ 0 & 1 & | & -1 & 0 \end{bmatrix}$

Hence, $M^{-1} = \begin{bmatrix} 4 & 1 \\ -1 & 0 \end{bmatrix}$

Check: $M^{-1}M = \begin{bmatrix} 4 & 1 \\ -1 & 0 \end{bmatrix}\begin{bmatrix} 0 & -1 \\ 1 & 4 \end{bmatrix} = \begin{bmatrix} 4\cdot0 + 1\cdot1 & 4(-1) + 1(4) \\ -1(0) + 0(1) & (-1)(-1) + 0\cdot4 \end{bmatrix} = \begin{bmatrix} 1 & 0 \\ 0 & 1 \end{bmatrix}$

17. $\begin{bmatrix} 1 & 2 & | & 1 & 0 \\ 1 & 3 & | & 0 & 1 \end{bmatrix}$ $(-1)R_1 + R_2 \rightarrow R_2$

$\sim \begin{bmatrix} 1 & 2 & | & 1 & 0 \\ 0 & 1 & | & -1 & 1 \end{bmatrix}$ $(-2)R_2 + R_1 \rightarrow R_1$

$\sim \begin{bmatrix} 1 & 0 & | & 3 & -2 \\ 0 & 1 & | & -1 & 1 \end{bmatrix}$

Hence, $M^{-1} = \begin{bmatrix} 3 & -2 \\ -1 & 1 \end{bmatrix}$

Check: $M^{-1}M = \begin{bmatrix} 3 & -2 \\ -1 & 1 \end{bmatrix}\begin{bmatrix} 1 & 2 \\ 1 & 3 \end{bmatrix} = \begin{bmatrix} 3\cdot1 + (-2)1 & 3\cdot2 + (-2)3 \\ (-1)1 + 1\cdot1 & (-1)2 + 1\cdot3 \end{bmatrix} = \begin{bmatrix} 1 & 0 \\ 0 & 1 \end{bmatrix}$

19. $\begin{bmatrix} 1 & 3 & | & 1 & 0 \\ 2 & 7 & | & 0 & 1 \end{bmatrix}$ $(-2)R_1 + R_2 \rightarrow R_2$

$\sim \begin{bmatrix} 1 & 3 & | & 1 & 0 \\ 0 & 1 & | & -2 & 1 \end{bmatrix}$ $(-3)R_2 + R_1 \rightarrow R_1$

$\sim \begin{bmatrix} 1 & 0 & | & 7 & -3 \\ 0 & 1 & | & -2 & 1 \end{bmatrix}$

Hence, $M^{-1} = \begin{bmatrix} 7 & -3 \\ -2 & 1 \end{bmatrix}$

Check: $M^{-1}M = \begin{bmatrix} 7 & -3 \\ -2 & 1 \end{bmatrix}\begin{bmatrix} 1 & 3 \\ 2 & 7 \end{bmatrix} = \begin{bmatrix} 7\cdot1 + (-3)2 & 7\cdot3 + (-3)7 \\ (-2)1 + 1\cdot2 & (-2)3 + 1\cdot7 \end{bmatrix} = \begin{bmatrix} 1 & 0 \\ 0 & 1 \end{bmatrix}$

21. $\begin{bmatrix} 1 & -2 & 0 & | & 1 & 0 & 0 \\ 0 & 1 & 1 & | & 0 & 1 & 0 \\ 2 & -1 & 2 & | & 0 & 0 & 1 \end{bmatrix}$ $(-2)R_1 + R_3 \rightarrow R_3$

$\sim \begin{bmatrix} 1 & -2 & 0 & | & 1 & 0 & 0 \\ 0 & 1 & 1 & | & 0 & 1 & 0 \\ 0 & 3 & 2 & | & -2 & 0 & 1 \end{bmatrix}$ $\begin{matrix} 2R_2 + R_1 \rightarrow R_1 \\ \\ (-3)R_2 + R_3 \rightarrow R_3 \end{matrix}$

$\sim \begin{bmatrix} 1 & 0 & 2 & | & 1 & 2 & 0 \\ 0 & 1 & 1 & | & 0 & 1 & 0 \\ 0 & 0 & -1 & | & -2 & -3 & 1 \end{bmatrix}$ $\begin{matrix} 2R_3 + R_1 \rightarrow R_1 \\ R_3 + R_2 \rightarrow R_2 \end{matrix}$

$\sim \begin{bmatrix} 1 & 0 & 0 & | & -3 & -4 & 2 \\ 0 & 1 & 0 & | & -2 & -2 & 1 \\ 0 & 0 & -1 & | & -2 & -3 & 1 \end{bmatrix}$ $(-1)R_3 \rightarrow R_3$

$\sim \begin{bmatrix} 1 & 0 & 0 & | & -3 & -4 & 2 \\ 0 & 1 & 0 & | & -2 & -2 & 1 \\ 0 & 0 & 1 & | & 2 & 3 & -1 \end{bmatrix}$

Hence, $M^{-1} = \begin{bmatrix} -3 & -4 & 2 \\ -2 & -2 & 1 \\ 2 & 3 & -1 \end{bmatrix}$

Check $M^{-1}M = \begin{bmatrix} -3 & -4 & 2 \\ -2 & -2 & 1 \\ 2 & 3 & -1 \end{bmatrix}\begin{bmatrix} 1 & -2 & 0 \\ 0 & 1 & 1 \\ 2 & -1 & 2 \end{bmatrix}$

$$= \begin{bmatrix} (-3)1 + (-4)0 + 2\cdot 2 & (-3)(-2) + (-4)1 + 2(-1) & (-3)0 + (-4)1 + 2\cdot 2 \\ (-2)1 + (-2)0 + 1\cdot 2 & (-2)(-2) + (-2)1 + 1(-1) & (-2)0 + (-2)1 + 1\cdot 2 \\ 2\cdot 1 + 3\cdot 0 + (-1)2 & 2(-2) + 3\cdot 1 + (-1)(-1) & 2\cdot 0 + 3\cdot 1 + (-1)2 \end{bmatrix}$$

$$= \begin{bmatrix} 1 & 0 & 0 \\ 0 & 1 & 0 \\ 0 & 0 & 1 \end{bmatrix}$$

23. $\begin{bmatrix} 1 & 1 & 0 & | & 1 & 0 & 0 \\ 0 & 2 & -1 & | & 0 & 1 & 0 \\ 1 & 0 & 1 & | & 0 & 0 & 1 \end{bmatrix}$ $(-1)R_1 + R_3 \to R_1$

$\sim \begin{bmatrix} 1 & 1 & 0 & | & 1 & 0 & 0 \\ 0 & 2 & -1 & | & 0 & 1 & 0 \\ 0 & -1 & 1 & | & -1 & 0 & 1 \end{bmatrix}$ $R_2 \leftrightarrow R_3$

$\sim \begin{bmatrix} 1 & 1 & 0 & | & 1 & 0 & 0 \\ 0 & -1 & 1 & | & -1 & 0 & 1 \\ 0 & 2 & -1 & | & 0 & 1 & 0 \end{bmatrix}$ $\begin{array}{l} R_2 + R_1 \to R_1 \\ 2R_2 + R_3 \to R_3 \end{array}$

$\sim \begin{bmatrix} 1 & 0 & 1 & | & 0 & 0 & 1 \\ 0 & -1 & 1 & | & -1 & 0 & 1 \\ 0 & 0 & 1 & | & -2 & 1 & 2 \end{bmatrix}$ $\begin{array}{l} (-1)R_3 + R_1 \to R_1 \\ (-1)R_3 + R_2 \to R_2 \end{array}$

$\sim \begin{bmatrix} 1 & 0 & 0 & | & 2 & -1 & -1 \\ 0 & -1 & 0 & | & 1 & -1 & -1 \\ 0 & 0 & 1 & | & -2 & 1 & 2 \end{bmatrix}$ $-R_2 \to R_2$

$\sim \begin{bmatrix} 1 & 0 & 0 & | & 2 & -1 & -1 \\ 0 & 1 & 0 & | & -1 & 1 & 1 \\ 0 & 0 & 1 & | & -2 & 1 & 2 \end{bmatrix}$

Hence, $M^{-1} = \begin{bmatrix} 2 & -1 & -1 \\ -1 & 1 & 1 \\ -2 & 1 & 2 \end{bmatrix}$

Check: $M^{-1}M = \begin{bmatrix} 2 & -1 & -1 \\ -1 & 1 & 1 \\ -2 & 1 & 2 \end{bmatrix} \begin{bmatrix} 1 & 1 & 0 \\ 0 & 2 & -1 \\ 1 & 0 & 1 \end{bmatrix}$

$$= \begin{bmatrix} 2\cdot 1 + (-1)0 + (-1)1 & 2\cdot 1 + (-1)2 + (-1)0 & 2\cdot 0 + (-1)(-1) + (-1)1 \\ (-1)1 + 1\cdot 0 + 1\cdot 1 & (-1)1 + 1\cdot 2 + 1\cdot 0 & (-1)0 + 1(-1) + 1\cdot 1 \\ (-2)1 + 1\cdot 0 + 2\cdot 1 & (-2)1 + 1\cdot 2 + 2\cdot 0 & (-2)0 + 1(-1) + 2\cdot 1 \end{bmatrix}$$

$$= \begin{bmatrix} 1 & 0 & 0 \\ 0 & 1 & 0 \\ 0 & 0 & 1 \end{bmatrix}$$

25. If the inverse existed we would find it by row operations on the following matrix:

$= \begin{bmatrix} 3 & 9 & | & 1 & 0 \\ 2 & 6 & | & 0 & 1 \end{bmatrix}$

But consider what happens if we perform $(-\frac{2}{3})R_1 + R_2 \to R_2$

$= \begin{bmatrix} 3 & 9 & | & 1 & 0 \\ 0 & 0 & | & -\frac{2}{3} & 1 \end{bmatrix}$

Since a row of zeros results to the left of the vertical line, no inverse exists.

27. $\begin{bmatrix} 2 & 3 & | & 1 & 0 \\ 3 & 5 & | & 0 & 1 \end{bmatrix} \frac{1}{2} R_1 \to R_1$

$\sim \begin{bmatrix} 1 & 1.5 & | & 0.5 & 0 \\ 3 & 5 & | & 0 & 1 \end{bmatrix} (-3) R_1 + R_2 \to R_2$

$\sim \begin{bmatrix} 1 & 1.5 & | & 0.5 & 0 \\ 0 & 0.5 & | & -1.5 & 1 \end{bmatrix} 2 R_2 \to R_2$

$\sim \begin{bmatrix} 1 & 1.5 & | & 0.5 & 0 \\ 0 & 1 & | & -3 & 2 \end{bmatrix} (-1.5) R_2 + R_1 \to R_1$

$\sim \begin{bmatrix} 1 & 0 & | & 5 & -3 \\ 0 & 1 & | & -3 & 2 \end{bmatrix}$

The inverse is $\begin{bmatrix} 5 & -3 \\ -3 & 2 \end{bmatrix}$

The checking steps are omitted for lack of space in this and subsequent problems.

29. $\begin{bmatrix} 2 & 2 & -1 & | & 1 & 0 & 0 \\ 0 & 4 & -1 & | & 0 & 1 & 0 \\ -1 & -2 & 1 & | & 0 & 0 & 1 \end{bmatrix} R_1 \leftrightarrow R_3$

$\sim \begin{bmatrix} -1 & -2 & 1 & | & 0 & 0 & 1 \\ 0 & 4 & -1 & | & 0 & 1 & 0 \\ 2 & 2 & -1 & | & 1 & 0 & 0 \end{bmatrix} 2R_1 + R_3 \to R_3$

$\sim \begin{bmatrix} -1 & -2 & 1 & | & 0 & 0 & 1 \\ 0 & 4 & -1 & | & 0 & 1 & 0 \\ 0 & -2 & 1 & | & 1 & 0 & 2 \end{bmatrix} \begin{matrix} (-1)R_3 + R_1 \to R_1 \\ 2R_3 + R_2 \to R_2 \end{matrix}$

$\sim \begin{bmatrix} -1 & 0 & 0 & | & -1 & 0 & -1 \\ 0 & 0 & 1 & | & 2 & 1 & 4 \\ 0 & -2 & 1 & | & 1 & 0 & 2 \end{bmatrix} (-1)R_2 + R_3 \to R_3$

$\sim \begin{bmatrix} -1 & 0 & 0 & | & -1 & 0 & -1 \\ 0 & 0 & 1 & | & 2 & 1 & 4 \\ 0 & -2 & 0 & | & -1 & -1 & -2 \end{bmatrix} R_2 \leftrightarrow R_3$

$\sim \begin{bmatrix} -1 & 0 & 0 & | & -1 & 0 & -1 \\ 0 & -2 & 0 & | & -1 & -1 & -2 \\ 0 & 0 & 1 & | & 2 & 1 & 4 \end{bmatrix} \begin{matrix} (-1)R_1 \to R_1 \\ (-\frac{1}{2})R_2 \to R_2 \end{matrix}$

$\sim \begin{bmatrix} 1 & 0 & 0 & | & 1 & 0 & 1 \\ 0 & 1 & 0 & | & \frac{1}{2} & \frac{1}{2} & 1 \\ 0 & 0 & 1 & | & 2 & 1 & 4 \end{bmatrix}$

The inverse is $\begin{bmatrix} 1 & 0 & 1 \\ \frac{1}{2} & \frac{1}{2} & 1 \\ 2 & 1 & 4 \end{bmatrix}$

31. If the inverse existed we would find it by row operations on the following matrix:

$\begin{bmatrix} 2 & 1 & 1 & | & 1 & 0 & 0 \\ 1 & 1 & 0 & | & 0 & 1 & 0 \\ -1 & -1 & 0 & | & 0 & 0 & 1 \end{bmatrix}$

But consider what happens if we perform $R_2 + R_3 \to R_2$

$$\begin{bmatrix} 2 & 1 & 1 & \bigm| & 1 & 0 & 0 \\ 0 & 0 & 0 & \bigm| & 0 & 1 & 1 \\ -1 & -1 & 0 & \bigm| & 0 & 0 & 1 \end{bmatrix}$$

Since a row of zeros results to the left of the vertical line, no inverse exists.

33. $\begin{bmatrix} 1 & 5 & 10 & \bigm| & 1 & 0 & 0 \\ 0 & 1 & 4 & \bigm| & 0 & 1 & 0 \\ 1 & 6 & 15 & \bigm| & 0 & 0 & 1 \end{bmatrix}$ $(-1)R_1 + R_3 \rightarrow R_3$

$\sim \begin{bmatrix} 1 & 5 & 10 & \bigm| & 1 & 0 & 0 \\ 0 & 1 & 4 & \bigm| & 0 & 1 & 0 \\ 0 & 1 & 5 & \bigm| & -1 & 0 & 1 \end{bmatrix}$ $\begin{array}{l}(-5)R_2 + R_1 \rightarrow R_1 \\[4pt] (-1)R_2 + R_3 \rightarrow R_3\end{array}$

$\sim \begin{bmatrix} 1 & 0 & -10 & \bigm| & 1 & -5 & 0 \\ 0 & 1 & 4 & \bigm| & 0 & 1 & 0 \\ 0 & 0 & 1 & \bigm| & -1 & -1 & 1 \end{bmatrix}$ $\begin{array}{l}10R_3 + R_1 \rightarrow R_1 \\[4pt] (-4)R_3 + R_2 \rightarrow R_2\end{array}$

$\sim \begin{bmatrix} 1 & 0 & 0 & \bigm| & -9 & -15 & 10 \\ 0 & 1 & 0 & \bigm| & 4 & 5 & -4 \\ 0 & 0 & 1 & \bigm| & -1 & -1 & 1 \end{bmatrix}$

The inverse is $\begin{bmatrix} -9 & -15 & 10 \\ 4 & 5 & -4 \\ -1 & -1 & 1 \end{bmatrix}$

35. We calculate A^{-1} by row operations on

$\begin{bmatrix} 3 & 4 & \bigm| & 1 & 0 \\ 2 & 3 & \bigm| & 0 & 1 \end{bmatrix}$ $(-1)R_2 + R_1 \rightarrow R_1$

$\sim \begin{bmatrix} 1 & 1 & \bigm| & 1 & -1 \\ 2 & 3 & \bigm| & 0 & 1 \end{bmatrix}$ $(-2)R_1 + R_2 \rightarrow R_2$

$\sim \begin{bmatrix} 1 & 1 & \bigm| & 1 & -1 \\ 0 & 1 & \bigm| & -2 & 3 \end{bmatrix}$ $(-1)R_2 + R_1 \rightarrow R_1$

$\sim \begin{bmatrix} 1 & 0 & \bigm| & 3 & -4 \\ 0 & 1 & \bigm| & -2 & 3 \end{bmatrix}$

Hence $A^{-1} = \begin{bmatrix} 3 & -4 \\ -2 & 3 \end{bmatrix}$

We calculate $(A^{-1})^{-1}$ by row operations on

$\begin{bmatrix} 3 & -4 & \bigm| & 1 & 0 \\ -2 & 3 & \bigm| & 0 & 1 \end{bmatrix}$ $R_2 + R_1 \rightarrow R_1$

$\sim \begin{bmatrix} 1 & -1 & \bigm| & 1 & 1 \\ -2 & 3 & \bigm| & 0 & 1 \end{bmatrix}$ $2R_1 + R_2 \rightarrow R_2$

$\sim \begin{bmatrix} 1 & -1 & \bigm| & 1 & 1 \\ 0 & 1 & \bigm| & 2 & 3 \end{bmatrix}$ $R_2 + R_1 \rightarrow R_1$

$\sim \begin{bmatrix} 1 & 0 & \bigm| & 3 & 4 \\ 0 & 1 & \bigm| & 2 & 3 \end{bmatrix}$

Hence $(A^{-1})^{-1} = \begin{bmatrix} 3 & 4 \\ 2 & 3 \end{bmatrix} = A$

37. Consider the matrix $\begin{bmatrix} a^{-1} & 0 \\ 0 & d^{-1} \end{bmatrix}$.

Since $\begin{bmatrix} a^{-1} & 0 \\ 0 & d^{-1} \end{bmatrix}\begin{bmatrix} a & 0 \\ 0 & d \end{bmatrix} = \begin{bmatrix} 1 & 0 \\ 0 & 1 \end{bmatrix}$, $\begin{bmatrix} a^{-1} & 0 \\ 0 & d^{-1} \end{bmatrix}$ is an inverse for M whenever it exists. This will hold if and only if a and d are non-zero.

Steps are not shown for Problems 39 - 41.

39. $\begin{bmatrix} 0.5 & -0.3 & 0.85 & -0.25 \\ 0 & 0.1 & 0.05 & -0.25 \\ -1 & 0.9 & -1.55 & 0.75 \\ -0.5 & 0.4 & -1.3 & 0.5 \end{bmatrix}$ **41.** $\begin{bmatrix} 1.75 & 5.25 & 8.75 & -1 & -18.75 \\ 1.25 & 3.75 & 6.25 & -1 & -13.25 \\ -4.75 & -13.25 & -22.75 & 3 & 48.75 \\ -1.375 & -4.625 & -7.875 & 1 & 16.375 \\ 3.25 & 8.75 & 15.25 & -2 & -32.25 \end{bmatrix}$

43. Using the assignation numbers 1 to 27 with the letters of the alphabet and a blank as in the text, write

```
C  A  T     I   N      T   H  E     H  A  T
3  1  20 27 9   14 27  20  8  5  27 8  1  20
```
and calculate

$\begin{bmatrix} 3 & 5 \\ 1 & 2 \end{bmatrix}\begin{bmatrix} 3 & 20 & 9 & 27 & 8 & 27 & 1 \\ 1 & 27 & 14 & 20 & 5 & 8 & 20 \end{bmatrix} = \begin{bmatrix} 14 & 195 & 97 & 181 & 49 & 121 & 103 \\ 5 & 74 & 37 & 67 & 18 & 43 & 41 \end{bmatrix}$

The encoded message is thus
```
14  5  195  74  97  37  181  67  49  18  121  43  103  41
```

45. The inverse of matrix A is easily calculated to be

$A^{-1} = \begin{bmatrix} 2 & -5 \\ -1 & 3 \end{bmatrix}$

Putting the coded message into matrix form and multiplying by A^{-1} yields:

$\begin{bmatrix} 2 & -5 \\ -1 & 3 \end{bmatrix}\begin{bmatrix} 111 & 40 & 177 & 50 & 116 & 86 & 62 & 121 & 68 \\ 43 & 15 & 68 & 19 & 45 & 29 & 22 & 43 & 27 \end{bmatrix}$

$= \begin{bmatrix} 7 & 5 & 14 & 5 & 7 & 27 & 14 & 27 & 1 \\ 18 & 5 & 27 & 7 & 19 & 1 & 4 & 8 & 13 \end{bmatrix}$

This decodes to   7 18 5 5 14 27 5 7 7 19 27 1 14 4 27 8 1 13
```
                   G  R  E  E  N     E  G  G  S     A  N  D     H  A  M
```

47. Using the assignation of numbers 1 to 27 with the letters of the alphabet and a blank as in the text, write
```
   D  W  I  G  H  T     D  A  V  I  D     E  I  S  E  N  H  O  W  E  R
   4 23  9  7  8 20 27  4  1 22  9  4 27  5  9 19  5 14  8 15 23  5 18 27 27
```
and calculate

$\begin{bmatrix} 1 & 0 & 1 & 0 & 1 \\ 0 & 1 & 1 & 0 & 3 \\ 2 & 1 & 1 & 1 & 1 \\ 0 & 0 & 1 & 0 & 2 \\ -1 & 1 & 1 & 2 & 1 \end{bmatrix}\begin{bmatrix} 4 & 20 & 9 & 19 & 23 \\ 23 & 27 & 4 & 5 & 5 \\ 9 & 4 & 27 & 14 & 18 \\ 7 & 1 & 5 & 8 & 27 \\ 8 & 22 & 9 & 15 & 27 \end{bmatrix} = \begin{bmatrix} 21 & 46 & 45 & 48 & 68 \\ 56 & 97 & 58 & 64 & 104 \\ 55 & 94 & 63 & 80 & 123 \\ 25 & 48 & 45 & 44 & 72 \\ 58 & 75 & 59 & 69 & 127 \end{bmatrix}$

The encoded message is thus
```
21  56  55  25  58  46  97  94  48  75  45  58  63  45  59  48  64
80  44  69  68  104  123  72  127
```

49. The inverse of B is calculated to be

$$B^{-1} = \begin{bmatrix} -2 & -1 & 2 & 2 & -1 \\ 3 & 2 & -2 & -4 & 1 \\ 6 & 2 & -4 & -5 & 2 \\ -2 & -1 & 1 & 2 & 0 \\ -3 & -1 & 2 & 3 & -1 \end{bmatrix}$$

Putting the coded message into matrix form and multiplying by B^{-1} yields

$$\begin{bmatrix} -2 & -1 & 2 & 2 & -1 \\ 3 & 2 & -2 & -4 & 1 \\ 6 & 2 & -4 & -5 & 2 \\ -2 & -1 & 1 & 2 & 0 \\ -3 & -1 & 2 & 3 & -1 \end{bmatrix} \begin{bmatrix} 41 & 25 & 43 & 44 & 68 \\ 84 & 56 & 54 & 67 & 135 \\ 82 & 67 & 89 & 86 & 136 \\ 44 & 20 & 39 & 44 & 81 \\ 74 & 54 & 102 & 90 & 149 \end{bmatrix} = \begin{bmatrix} 12 & 14 & 14 & 15 & 14 \\ 25 & 27 & 5 & 8 & 27 \\ 14 & 2 & 19 & 14 & 27 \\ 4 & 1 & 27 & 19 & 27 \\ 15 & 9 & 10 & 15 & 27 \end{bmatrix}$$

This decodes to

12 25 14 4 15 14 27 2 1 9 14 5 19 27 10 15 8 14 19 15 14 27 27 27

L Y N D O N   B A I N E S   J O H N S O N

Exercise 7-3

Key Ideas and Formulas

Basic Properties of Matrices

Assuming all products and sums are defined for the indicated matrices A, B, C, I, and 0, then

ADDITION PROPERTIES

ASSOCIATIVE:	$(A + B) + C = A + (B + C)$
COMMUTATIVE:	$A + B = B + A$
ADDITIVE IDENTITY:	$A + 0 = 0 + A = A$
ADDITIVE INVERSE:	$A + (-A) = (-A) + A = 0$

MULTIPLICATION PROPERTIES

ASSOCIATIVE PROPERTY:	$A(BC) = (AB)C$
MULTIPLICATIVE IDENTITY:	$AI = IA = A$
MULTIPLICATIVE INVERSE:	If A is a square matrix and A^{-1} exists, then $AA^{-1} = A^{-1}A = I$.

COMBINED PROPERTIES

LEFT DISTRIBUTIVE:	$A(B + C) = AB + AC$
RIGHT DISTRIBUTIVE:	$(B + C)A = BA + CA$

EQUALITY

ADDITION:	If $A = B$, then $A + C = B + C$
LEFT MULTIPLICATION:	If $A = B$, then $CA = CB$.
RIGHT MULTIPLICATION:	If $A = B$, then $AC = BC$.

A system of n equations in n variables can be written as

$$AX = B$$

where A is the coefficient matrix of the system,

$$X = \begin{bmatrix} x_1 \\ x_2 \\ \vdots \\ x_n \end{bmatrix}$$

and B is the column matrix of constant terms. Then if A^{-1} exists, the solution of the system can be written as $X = A^{-1}B$.

1. $2x_1 - x_2 = 3$
$x_1 + 3x_2 = -2$

3. $-2x_1 + x_3 = 3$
$x_1 + 2x_2 + x_3 = -4$
$x_2 - x_3 = 2$

5. $\begin{bmatrix} 4 & -3 \\ 1 & 2 \end{bmatrix} \begin{bmatrix} x_1 \\ x_2 \end{bmatrix} = \begin{bmatrix} 2 \\ 1 \end{bmatrix}$

7. $\begin{bmatrix} 1 & -2 & 1 \\ -1 & 1 & 0 \\ 2 & 3 & 1 \end{bmatrix} \begin{bmatrix} x_1 \\ x_2 \\ x_3 \end{bmatrix} = \begin{bmatrix} -1 \\ 2 \\ -3 \end{bmatrix}$

9. Since $\begin{bmatrix} 3 & -2 \\ 1 & 4 \end{bmatrix} \begin{bmatrix} -2 \\ 1 \end{bmatrix} = \begin{bmatrix} 3(-2) + (-2)1 \\ 1(-2) + 4 \cdot 1 \end{bmatrix}$

$= \begin{bmatrix} -8 \\ 2 \end{bmatrix}, \begin{bmatrix} x_1 \\ x_2 \end{bmatrix} = \begin{bmatrix} -8 \\ 2 \end{bmatrix}$ if and only if $x_1 = -8$ and $x_2 = 2$.

11. Since $\begin{bmatrix} -2 & 3 \\ 2 & -1 \end{bmatrix} \begin{bmatrix} 3 \\ 2 \end{bmatrix} = \begin{bmatrix} (-2)3 + 3 \cdot 2 \\ 2 \cdot 3 + (-1)2 \end{bmatrix} = \begin{bmatrix} 0 \\ 4 \end{bmatrix}$

$\begin{bmatrix} x_1 \\ x_2 \end{bmatrix} = \begin{bmatrix} 0 \\ 4 \end{bmatrix}$ if and only if $x_1 = 0$ and $x_2 = 4$.

13. $\begin{bmatrix} 1 & 2 \\ 1 & 3 \end{bmatrix} \begin{bmatrix} x_1 \\ x_2 \end{bmatrix} = \begin{bmatrix} k_1 \\ k_2 \end{bmatrix}$

$AX = K$ has solution $X = A^{-1}K$.

We find $A^{-1}K$ for each given K. From problem 17, Exercise 7-2, $A^{-1} = \begin{bmatrix} 3 & -2 \\ -1 & 1 \end{bmatrix}$

(A) $K = \begin{bmatrix} 1 \\ 3 \end{bmatrix}$ $A^{-1}K = \begin{bmatrix} 3 & -2 \\ -1 & 1 \end{bmatrix} \begin{bmatrix} 1 \\ 3 \end{bmatrix} = \begin{bmatrix} -3 \\ 2 \end{bmatrix} = \begin{bmatrix} x_1 \\ x_2 \end{bmatrix}$ $x_1 = -3$, $x_2 = 2$

(B) $K = \begin{bmatrix} 3 \\ 5 \end{bmatrix}$ $A^{-1}K = \begin{bmatrix} 3 & -2 \\ -1 & 1 \end{bmatrix} \begin{bmatrix} 3 \\ 5 \end{bmatrix} = \begin{bmatrix} -1 \\ 2 \end{bmatrix} = \begin{bmatrix} x_1 \\ x_2 \end{bmatrix}$ $x_1 = -1$, $x_2 = 2$

(C) $K = \begin{bmatrix} -2 \\ 1 \end{bmatrix}$ $A^{-1}K = \begin{bmatrix} 3 & -2 \\ -1 & 1 \end{bmatrix} \begin{bmatrix} -2 \\ 1 \end{bmatrix} = \begin{bmatrix} -8 \\ 3 \end{bmatrix} = \begin{bmatrix} x_1 \\ x_2 \end{bmatrix}$ $x_1 = -8$, $x_2 = 3$

15. $\begin{bmatrix} 1 & 3 \\ 2 & 7 \end{bmatrix} \begin{bmatrix} x_1 \\ x_2 \end{bmatrix} = \begin{bmatrix} k_1 \\ k_2 \end{bmatrix}$

$AX = K$ has solution $X = A^{-1}K$.

We find $A^{-1}K$ for each given K. From problem 19, Exercise 7-2, $A^{-1} = \begin{bmatrix} 7 & -3 \\ -2 & 1 \end{bmatrix}$

(A) $K = \begin{bmatrix} 2 \\ -1 \end{bmatrix}$ $A^{-1}K = \begin{bmatrix} 7 & -3 \\ -2 & 1 \end{bmatrix} \begin{bmatrix} 2 \\ -1 \end{bmatrix} = \begin{bmatrix} 17 \\ -5 \end{bmatrix} = \begin{bmatrix} x_1 \\ x_2 \end{bmatrix}$ $x_1 = 17$, $x_2 = -5$

(B) $K = \begin{bmatrix} 1 \\ 0 \end{bmatrix}$ $A^{-1}K = \begin{bmatrix} 7 & -3 \\ -2 & 1 \end{bmatrix} \begin{bmatrix} 1 \\ 0 \end{bmatrix} = \begin{bmatrix} 7 \\ -2 \end{bmatrix} = \begin{bmatrix} x_1 \\ x_2 \end{bmatrix}$ $x_1 = 7$, $x_2 = -2$

(C) $K = \begin{bmatrix} 3 \\ -1 \end{bmatrix}$ $A^{-1}K = \begin{bmatrix} 7 & -3 \\ -2 & 1 \end{bmatrix} \begin{bmatrix} 3 \\ -1 \end{bmatrix} = \begin{bmatrix} 24 \\ -7 \end{bmatrix} = \begin{bmatrix} x_1 \\ x_2 \end{bmatrix}$ $x_1 = 24$, $x_2 = -7$

17. $\begin{bmatrix} 1 & -2 & 0 \\ 0 & 1 & 1 \\ 2 & -1 & 2 \end{bmatrix} \begin{bmatrix} x_1 \\ x_2 \\ x_3 \end{bmatrix} = \begin{bmatrix} k_1 \\ k_2 \\ k_3 \end{bmatrix}$

$AX = K$ has solution $X = A^{-1}K$.

We find $A^{-1}K$ for each given K. From problem 21, Exercise 7-2, $A^{-1} = \begin{bmatrix} -3 & -4 & 2 \\ -2 & -2 & 1 \\ 2 & 3 & -1 \end{bmatrix}$

(A) $K = \begin{bmatrix} 1 \\ 0 \\ 2 \end{bmatrix}$ $A^{-1}K = \begin{bmatrix} -3 & -4 & 2 \\ -2 & -2 & 1 \\ 2 & 3 & -1 \end{bmatrix} \begin{bmatrix} 1 \\ 0 \\ 2 \end{bmatrix} = \begin{bmatrix} 1 \\ 0 \\ 0 \end{bmatrix} = \begin{bmatrix} x_1 \\ x_2 \\ x_3 \end{bmatrix}$ $x_1 = 1, \ x_2 = 0, \ x_3 = 0$

(B) $K = \begin{bmatrix} -1 \\ 1 \\ 0 \end{bmatrix}$ $A^{-1}K = \begin{bmatrix} -3 & -4 & 2 \\ -2 & -2 & 1 \\ 2 & 3 & -1 \end{bmatrix} \begin{bmatrix} -1 \\ 1 \\ 0 \end{bmatrix} = \begin{bmatrix} -1 \\ 0 \\ 1 \end{bmatrix} = \begin{bmatrix} x_1 \\ x_2 \\ x_3 \end{bmatrix}$ $x_1 = -1, \ x_2 = 0, \ x_3 = 1$

(C) $K = \begin{bmatrix} 2 \\ -2 \\ 1 \end{bmatrix}$ $A^{-1}K = \begin{bmatrix} -3 & -4 & 2 \\ -2 & -2 & 1 \\ 2 & 3 & -1 \end{bmatrix} \begin{bmatrix} 2 \\ -2 \\ 1 \end{bmatrix} = \begin{bmatrix} 4 \\ 1 \\ -3 \end{bmatrix} = \begin{bmatrix} x_1 \\ x_2 \\ x_3 \end{bmatrix}$ $x_1 = 4, \ x_2 = 1, \ x_3 = -3$

19. $\begin{bmatrix} 1 & 1 & 0 \\ 0 & 2 & -1 \\ 1 & 0 & 1 \end{bmatrix} \begin{bmatrix} x_1 \\ x_2 \\ x_3 \end{bmatrix} = \begin{bmatrix} k_1 \\ k_2 \\ k_3 \end{bmatrix}$

$AX = K$ has solution $X = A^{-1}K$.

We find $A^{-1}K$ for each given K. From problem 23, Exercise 7-2, $A^{-1} = \begin{bmatrix} 2 & -1 & -1 \\ -1 & 1 & 1 \\ 2 & 1 & 2 \end{bmatrix}$

(A) $K = \begin{bmatrix} 2 \\ 0 \\ 4 \end{bmatrix}$ $A^{-1}K = \begin{bmatrix} 2 & -1 & -1 \\ -1 & 1 & 1 \\ -2 & 1 & 2 \end{bmatrix} \begin{bmatrix} 2 \\ 0 \\ 4 \end{bmatrix} = \begin{bmatrix} 0 \\ 2 \\ 4 \end{bmatrix} = \begin{bmatrix} x_1 \\ x_2 \\ x_3 \end{bmatrix}$ $x_1 = 0, \ x_2 = 2, \ x_3 = 4$

(B) $K = \begin{bmatrix} 0 \\ 4 \\ -2 \end{bmatrix}$ $A^{-1}K = \begin{bmatrix} 2 & -1 & -1 \\ -1 & 1 & 1 \\ -2 & 1 & 2 \end{bmatrix} \begin{bmatrix} 0 \\ 4 \\ -2 \end{bmatrix} = \begin{bmatrix} -2 \\ 2 \\ 0 \end{bmatrix} = \begin{bmatrix} x_1 \\ x_2 \\ x_3 \end{bmatrix}$ $x_1 = -2, \ x_2 = 2, \ x_3 = 0$

(C) $K = \begin{bmatrix} 4 \\ 2 \\ 0 \end{bmatrix}$ $A^{-1}K = \begin{bmatrix} 2 & -1 & -1 \\ -1 & 1 & 1 \\ -2 & 1 & 2 \end{bmatrix} \begin{bmatrix} 4 \\ 2 \\ 0 \end{bmatrix} = \begin{bmatrix} 6 \\ -2 \\ -6 \end{bmatrix} = \begin{bmatrix} x_1 \\ x_2 \\ x_3 \end{bmatrix}$ $x_1 = 6, \ x_2 = -2, \ x_3 = -6$

21. $AX - BX = C$
$(A - B)X = C$ Right distributive property
[To be more careful, we should write
  $AX - BX = AX + -BX$ by the definition of subtraction
            $= [A + (-B)]X$ by the right distributive property
            $= (A - B)X$ by the definition of subtraction
but the distributive properties are generally
understood as applying to subtraction also.]

$(A - B)^{-1}[(A - B)X] = (A - B)^{-1}C$ Left multiplication property

$[(A - B)^{-1}(A - B)]X = (A - B)^{-1}C$ Associative property

$IX = (A - B)^{-1}C$ $A^{-1}A = I$

$X = (A - B)^{-1}C$ $IX = X$

23.
$$AX + X = C$$
$$AX + IX = C \qquad IX = X$$
$$(A + I)X = C \qquad \text{Right distributive property}$$
$$(A + I)^{-1}[(A + I)X] = (A + I)^{-1}C \qquad \text{Left multiplication property}$$
$$[(A + I)^{-1}(A + I)]X = (A + I)^{-1}C \qquad \text{Associative property}$$
$$IX = (A + I)^{-1}C \qquad A^{-1}A = I$$
$$X = (A + I)^{-1}C \qquad IX = X$$

25.
$$AX - C = D - BX$$
$$AX - C + C + BX = D - BX + BX + C \qquad \text{Addition property}$$
$$AX + 0 + BX = D + 0 + C \qquad M + (-M) = 0$$
$$AX + BX = D + C \qquad M + 0 = M$$
$$(A + B)X = D + C \qquad \text{Right distributive property}$$
$$(A + B)^{-1}[(A + B)X] = (A + B)^{-1}(D + C) \qquad \text{Left multiplication property}$$
$$[(A + B)^{-1}(A + B)]X = (A + B)^{-1}(C + D) \qquad \text{Associative and commutative properties}$$
$$IX = (A + B)^{-1}(C + D) \qquad A^{-1}A = I$$
$$X = (A + B)^{-1}(C + D) \qquad IX = X$$

27. $\begin{bmatrix} 1 & 2.001 \\ 1 & 2 \end{bmatrix} \begin{bmatrix} x_1 \\ x_2 \end{bmatrix} = \begin{bmatrix} k_1 \\ k_2 \end{bmatrix}$

$AX = K$ has solution $X = A^{-1}K$

$A^{-1} = \begin{bmatrix} -2000 & 2001 \\ 1000 & -1000 \end{bmatrix}$

We find $A^{-1}K$ for each given K.

(A) $K = \begin{bmatrix} 1 \\ 1 \end{bmatrix}$ $\quad A^{-1}K = \begin{bmatrix} -2000 & 2001 \\ 1000 & -1000 \end{bmatrix} \begin{bmatrix} 1 \\ 1 \end{bmatrix} = \begin{bmatrix} 1 \\ 0 \end{bmatrix} = \begin{bmatrix} x_1 \\ x_2 \end{bmatrix}$ $\quad x_1 = 1, \; x_2 = 0$

(B) $K = \begin{bmatrix} 1 \\ 0 \end{bmatrix}$ $\quad A^{-1}K = \begin{bmatrix} -2000 & 2001 \\ 1000 & -1000 \end{bmatrix} \begin{bmatrix} 1 \\ 0 \end{bmatrix} = \begin{bmatrix} -2000 \\ 1000 \end{bmatrix} = \begin{bmatrix} x_1 \\ x_2 \end{bmatrix}$ $\quad x_1 = -2000, \; x_2 = 1000$

(C) $K = \begin{bmatrix} 0 \\ 1 \end{bmatrix}$ $\quad A^{-1}K = \begin{bmatrix} -2000 & 2001 \\ 1000 & -1000 \end{bmatrix} \begin{bmatrix} 0 \\ 1 \end{bmatrix} = \begin{bmatrix} 2001 \\ -1000 \end{bmatrix} = \begin{bmatrix} x_1 \\ x_2 \end{bmatrix}$ $\quad x_1 = 2001, \; x_2 = -1000$

Because of the size of the entries of A^{-1}, a small change in K has a magnified effect on the solutions of $AX = K$. Geometrically, the lines are very close to being parallel, so a small change in the position (y intercept) of one line displaces their point of intersection a great distance.

29. $\begin{bmatrix} 1 & 8 & 7 \\ 6 & 6 & 8 \\ 3 & 4 & 6 \end{bmatrix} \begin{bmatrix} x_1 \\ x_2 \\ x_3 \end{bmatrix} = \begin{bmatrix} 135 \\ 155 \\ 75 \end{bmatrix}$

$AX = B$ has solution $X = A^{-1}B$.

From a computer mathematics system or a graphing calculator, A^{-1} is found to be

$\begin{bmatrix} -0.08 & 0.4 & -0.44 \\ 0.24 & 0.3 & -0.68 \\ -0.12 & -0.4 & 0.84 \end{bmatrix}$

Hence

$\begin{bmatrix} -0.08 & 0.4 & -0.44 \\ 0.24 & 0.3 & -0.68 \\ -0.12 & -0.4 & 0.84 \end{bmatrix} \begin{bmatrix} 135 \\ 155 \\ 75 \end{bmatrix} = \begin{bmatrix} 18.2 \\ 27.9 \\ -15.2 \end{bmatrix} = \begin{bmatrix} x_1 \\ x_2 \\ x_3 \end{bmatrix}$ $\quad x_1 = 18.2, \; x_2 = 27.9, \; x_3 = -15.2$

31. $\begin{bmatrix} 6 & 9 & 7 & 5 \\ 6 & 4 & 7 & 3 \\ 4 & 5 & 3 & 2 \\ 4 & 3 & 8 & 2 \end{bmatrix} \begin{bmatrix} x_1 \\ x_2 \\ x_3 \\ x_4 \end{bmatrix} = \begin{bmatrix} 250 \\ 195 \\ 145 \\ 125 \end{bmatrix}$

$AX = B$ has solution $X = A^{-1}B$.

From a computer mathematics system or a graphing calculator, A^{-1} is found to be

$$\begin{bmatrix} -0.25 & 0.37 & 0.28 & -0.21 \\ 0 & -0.4 & 0.4 & 0.2 \\ 0 & -0.16 & -0.04 & 0.28 \\ 0.5 & 0.5 & -1 & -0.5 \end{bmatrix}$$

Hence

$$\begin{bmatrix} -0.25 & 0.37 & 0.28 & -0.21 \\ 0 & -0.4 & 0.4 & 0.2 \\ 0 & -0.16 & -0.04 & 0.28 \\ 0.5 & 0.5 & -1 & -0.5 \end{bmatrix}\begin{bmatrix} 250 \\ 195 \\ 145 \\ 125 \end{bmatrix} = \begin{bmatrix} 24 \\ 5 \\ -2 \\ 15 \end{bmatrix} = \begin{bmatrix} x_1 \\ x_2 \\ x_3 \\ x_4 \end{bmatrix}$$

$x_1 = 24$, $x_2 = 5$, $x_3 = -2$, $x_4 = 15$

33. The system to be solved, for an arbitrary return, is derived as follows:
Let x_1 = number of \$4 tickets sold
  x_2 = number of \$8 tickets sold
Then $x_1 + x_2 = 10,000$ number of seats
  $4x_1 + 8x_2 = k_2$ return required
We solve the system by writing it as a matrix equation.

$$\underset{A}{\begin{bmatrix} 1 & 1 \\ 4 & 8 \end{bmatrix}}\underset{X}{\begin{bmatrix} x_1 \\ x_2 \end{bmatrix}} = \underset{B}{\begin{bmatrix} 10,000 \\ k_2 \end{bmatrix}}$$

If A^{-1} exists, then $X = A^{-1}B$. To find A^{-1}, we perform row operations on

$$\begin{bmatrix} 1 & 1 & | & 1 & 0 \\ 4 & 8 & | & 0 & 1 \end{bmatrix} \quad (-4)R_1 + R_2 \rightarrow R_2$$

$$\sim \begin{bmatrix} 1 & 1 & | & 1 & 0 \\ 0 & 4 & | & -4 & 1 \end{bmatrix} \quad 0.25R_2 \rightarrow R_2$$

$$\sim \begin{bmatrix} 1 & 1 & | & 1 & 0 \\ 0 & 1 & | & -1 & 0.25 \end{bmatrix} \quad (-1)R_2 + R_1 \rightarrow R_1$$

$$\sim \begin{bmatrix} 1 & 0 & | & 2 & -0.25 \\ 0 & 1 & | & -1 & 0.25 \end{bmatrix}$$

Hence $A^{-1} = \begin{bmatrix} 2 & -0.25 \\ -1 & 0.25 \end{bmatrix}$

Check: $A^{-1}A = \begin{bmatrix} 2 & -0.25 \\ -1 & 0.25 \end{bmatrix}\begin{bmatrix} 1 & 1 \\ 4 & 8 \end{bmatrix} = \begin{bmatrix} 1 & 0 \\ 0 & 1 \end{bmatrix}$

We can now solve the system as

$$\underset{X}{\begin{bmatrix} x_1 \\ x_2 \end{bmatrix}} = \underset{A^{-1}}{\begin{bmatrix} 2 & -0.25 \\ -1 & 0.25 \end{bmatrix}}\underset{B}{\begin{bmatrix} 10,000 \\ k_2 \end{bmatrix}}$$

If $k_2 = 56,000$ (Concert 1),

$$\begin{bmatrix} x_1 \\ x_2 \end{bmatrix} = \begin{bmatrix} 2 & -0.25 \\ -1 & 0.25 \end{bmatrix}\begin{bmatrix} 10,000 \\ 56,000 \end{bmatrix} = \begin{bmatrix} 6,000 \\ 4,000 \end{bmatrix}$$

Concert 1: 6,000 \$4 tickets and 4,000 \$8 tickets

If $k_2 = 60,000$ (Concert 2),

$$\begin{bmatrix} x_1 \\ x_2 \end{bmatrix} = \begin{bmatrix} 2 & -0.25 \\ -1 & 0.25 \end{bmatrix}\begin{bmatrix} 10,000 \\ 60,000 \end{bmatrix} = \begin{bmatrix} 5,000 \\ 5,000 \end{bmatrix}$$

Concert 2: 5,000 \$4 tickets and 5,000 \$8 tickets

If $k_2 = 68,000$ (Concert 3),

$$\begin{bmatrix} x_1 \\ x_2 \end{bmatrix} = \begin{bmatrix} 2 & -0.25 \\ -1 & 0.25 \end{bmatrix}\begin{bmatrix} 10,000 \\ 68,000 \end{bmatrix} = \begin{bmatrix} 3,000 \\ 7,000 \end{bmatrix}$$

Concert 3: 3,000 \$4 tickets and 7,000 \$8 tickets

35. We solve the system, for arbitrary V_1 and V_2, by writing it as a matrix equation.

$$\begin{array}{ccc} A & J & B \end{array}$$
$$\begin{bmatrix} 1 & -1 & 1 \\ 1 & 1 & 0 \\ 0 & 1 & 2 \end{bmatrix}\begin{bmatrix} I_1 \\ I_2 \\ I_3 \end{bmatrix} = \begin{bmatrix} 0 \\ V_1 \\ V_2 \end{bmatrix}$$

If A^{-1} exists, then $J = A^{-1}B$. To find A^{-1}, we perform row operations on

$$\begin{bmatrix} 1 & -1 & 1 & | & 1 & 0 & 0 \\ 1 & 1 & 0 & | & 0 & 1 & 0 \\ 0 & 1 & 2 & | & 0 & 0 & 1 \end{bmatrix} \quad (-1)R_1 + R_2 \to R_2$$

$$\sim \begin{bmatrix} 1 & -1 & 1 & | & 1 & 0 & 0 \\ 0 & 2 & -1 & | & -1 & 1 & 0 \\ 0 & 1 & 2 & | & 0 & 0 & 1 \end{bmatrix} \quad R_2 \leftrightarrow R_3$$

$$\sim \begin{bmatrix} 1 & -1 & 1 & | & 1 & 0 & 0 \\ 0 & 1 & 2 & | & 0 & 0 & 1 \\ 0 & 2 & -1 & | & -1 & 1 & 0 \end{bmatrix} \quad \begin{array}{l} R_2 + R_1 \to R_1 \\ \\ (-2)R_2 + R_3 \to R_3 \end{array}$$

$$\sim \begin{bmatrix} 1 & 0 & 3 & | & 1 & 0 & 1 \\ 0 & 1 & 2 & | & 0 & 0 & 1 \\ 0 & 0 & -5 & | & -1 & 1 & -2 \end{bmatrix} \quad -\tfrac{1}{5}R_3 \to R_3$$

$$\sim \begin{bmatrix} 1 & 0 & 3 & | & 1 & 0 & 1 \\ 0 & 1 & 2 & | & 0 & 0 & 1 \\ 0 & 0 & 1 & | & \tfrac{1}{5} & -\tfrac{1}{5} & \tfrac{2}{5} \end{bmatrix} \quad \begin{array}{l} (-3)R_3 + R_1 \to R_1 \\ (-2)R_3 + R_2 \to R_2 \end{array}$$

$$\sim \begin{bmatrix} 1 & 0 & 0 & | & \tfrac{2}{5} & \tfrac{3}{5} & -\tfrac{1}{5} \\ 0 & 1 & 0 & | & -\tfrac{2}{5} & \tfrac{2}{5} & \tfrac{1}{5} \\ 0 & 0 & 1 & | & \tfrac{1}{5} & -\tfrac{1}{5} & \tfrac{2}{5} \end{bmatrix}$$

Hence $A^{-1} = \dfrac{1}{5}\begin{bmatrix} 2 & 3 & -1 \\ -2 & 2 & 1 \\ 1 & -1 & 2 \end{bmatrix}$

Check: $A^{-1}A = \dfrac{1}{5}\begin{bmatrix} 2 & 3 & -1 \\ -2 & 2 & 1 \\ 1 & -1 & 2 \end{bmatrix}\begin{bmatrix} 1 & -1 & 1 \\ 1 & 1 & 0 \\ 0 & 1 & 2 \end{bmatrix} = \begin{bmatrix} 1 & 0 & 0 \\ 0 & 1 & 0 \\ 0 & 0 & 1 \end{bmatrix}$

We can now solve the system as

$$\begin{array}{ccc} J & A^{-1} & B \end{array}$$
$$\begin{bmatrix} I_1 \\ I_2 \\ I_3 \end{bmatrix} = \dfrac{1}{5}\begin{bmatrix} 2 & 3 & -1 \\ -2 & 2 & 1 \\ 1 & -1 & 2 \end{bmatrix}\begin{bmatrix} 0 \\ V_1 \\ V_2 \end{bmatrix}$$

(A) $V_1 = 10$ $V_2 = 10$

$$\begin{bmatrix} I_1 \\ I_2 \\ I_3 \end{bmatrix} = \dfrac{1}{5}\begin{bmatrix} 2 & 3 & -1 \\ -2 & 2 & 1 \\ 1 & -1 & 2 \end{bmatrix}\begin{bmatrix} 0 \\ 10 \\ 10 \end{bmatrix} = \begin{bmatrix} 4 \\ 6 \\ 2 \end{bmatrix} \quad I_1 = 4,\ I_2 = 6,\ I_3 = 2 \text{ (amperes)}$$

(B) $V_1 = 10$ $V_2 = 15$

$$\begin{bmatrix} I_1 \\ I_2 \\ I_3 \end{bmatrix} = \frac{1}{5}\begin{bmatrix} 2 & 3 & -1 \\ -2 & 2 & 1 \\ 1 & -1 & 2 \end{bmatrix}\begin{bmatrix} 0 \\ 10 \\ 15 \end{bmatrix} = \begin{bmatrix} 3 \\ 7 \\ 4 \end{bmatrix}$$ $I_1 = 3$, $I_2 = 7$, $I_3 = 4$ (amperes)

(C) $V_1 = 15$ $V_2 = 10$

$$\begin{bmatrix} I_1 \\ I_2 \\ I_3 \end{bmatrix} = \frac{1}{5}\begin{bmatrix} 2 & 3 & -1 \\ -2 & 2 & 1 \\ 1 & -1 & 2 \end{bmatrix}\begin{bmatrix} 0 \\ 15 \\ 10 \end{bmatrix} = \begin{bmatrix} 7 \\ 8 \\ 1 \end{bmatrix}$$ $I_1 = 7$, $I_2 = 8$, $I_3 = 1$ (amperes)

37. If the graph of $f(x) = ax^2 + bx + c$ passes through a point, the coordinates of the point must satisfy the equation of the graph. Hence

$k_1 = a(1)^2 + b(1) + c$
$k_2 = a(2)^2 + b(2) + c$
$k_3 = a(3)^2 + b(3) + c$

After simplification, we obtain:

$a + b + c = k_1$
$4a + 2b + c = k_2$
$9a + 3b + c = k_3$

We solve this system, for arbitrary k_1, k_2, k_3, by writing it as a matrix equation.

$$\overset{A}{\begin{bmatrix} 1 & 1 & 1 \\ 4 & 2 & 1 \\ 9 & 3 & 1 \end{bmatrix}}\overset{X}{\begin{bmatrix} a \\ b \\ c \end{bmatrix}} = \overset{B}{\begin{bmatrix} k_1 \\ k_2 \\ k_3 \end{bmatrix}}$$

If A^{-1} exists, then $X = A^{-1}B$. To find A^{-1} we perform row operations on

$$\left[\begin{array}{ccc|ccc} 1 & 1 & 1 & 1 & 0 & 0 \\ 4 & 2 & 1 & 0 & 1 & 0 \\ 9 & 3 & 1 & 0 & 0 & 1 \end{array}\right] \begin{array}{l} \\ (-4)R_1 + R_2 \to R_2 \\ (-9)R_1 + R_3 \to R_3 \end{array}$$

$$\sim \left[\begin{array}{ccc|ccc} 1 & 1 & 1 & 1 & 0 & 0 \\ 0 & -2 & -3 & -4 & 1 & 0 \\ 0 & -6 & -8 & -9 & 0 & 1 \end{array}\right] \quad -\tfrac{1}{2}R_2 \to R_2$$

$$\sim \left[\begin{array}{ccc|ccc} 1 & 1 & 1 & 1 & 0 & 0 \\ 0 & 1 & \frac{3}{2} & 2 & -\frac{1}{2} & 0 \\ 0 & -6 & -8 & -9 & 0 & 1 \end{array}\right] \begin{array}{l} (-1)R_2 + R_1 \to R_1 \\ \\ 6R_2 + R_3 \to R_3 \end{array}$$

$$\sim \left[\begin{array}{ccc|ccc} 1 & 0 & -\frac{1}{2} & -1 & \frac{1}{2} & 0 \\ 0 & 1 & \frac{3}{2} & 2 & -\frac{1}{2} & 0 \\ 0 & 0 & 1 & 3 & -3 & 1 \end{array}\right] \begin{array}{l} \frac{1}{2}R_3 + R_1 \to R_1 \\ (-\frac{3}{2})R_3 + R_2 \to R_2 \end{array}$$

$$\sim \left[\begin{array}{ccc|ccc} 1 & 0 & 0 & \frac{1}{2} & -1 & \frac{1}{2} \\ 0 & 1 & 0 & -\frac{5}{2} & 4 & -\frac{3}{2} \\ 0 & 0 & 1 & 3 & -3 & 1 \end{array}\right]$$

Hence $A^{-1} = \frac{1}{2}\begin{bmatrix} 1 & -2 & 1 \\ -5 & 8 & -3 \\ 6 & -6 & 2 \end{bmatrix}$

Check: $A^{-1}A = \frac{1}{2}\begin{bmatrix} 1 & -2 & 1 \\ -5 & 8 & -3 \\ 6 & -6 & 2 \end{bmatrix}\begin{bmatrix} 1 & 1 & 1 \\ 4 & 2 & 1 \\ 9 & 3 & 1 \end{bmatrix} = \begin{bmatrix} 1 & 0 & 0 \\ 0 & 1 & 0 \\ 0 & 0 & 1 \end{bmatrix}$

We can now solve the system as

$$\begin{matrix} X & A^{-1} & B \end{matrix}$$

$$\begin{bmatrix} a \\ b \\ c \end{bmatrix} = \frac{1}{2} \begin{bmatrix} 1 & -2 & 1 \\ -5 & 8 & -3 \\ 6 & -6 & 2 \end{bmatrix} \begin{bmatrix} k_1 \\ k_2 \\ k_3 \end{bmatrix}$$

(A) $\begin{bmatrix} a \\ b \\ c \end{bmatrix} = \frac{1}{2} \begin{bmatrix} 1 & -2 & 1 \\ -5 & 8 & -3 \\ 6 & -6 & 2 \end{bmatrix} \begin{bmatrix} -2 \\ 1 \\ 6 \end{bmatrix} = \begin{bmatrix} 1 \\ 0 \\ -3 \end{bmatrix}$ $a = 1$, $b = 0$, $c = -3$

(B) $\begin{bmatrix} a \\ b \\ c \end{bmatrix} = \frac{1}{2} \begin{bmatrix} 1 & -2 & 1 \\ -5 & 8 & -3 \\ 6 & -6 & 2 \end{bmatrix} \begin{bmatrix} 4 \\ 3 \\ -2 \end{bmatrix} = \begin{bmatrix} -2 \\ 5 \\ 1 \end{bmatrix}$ $a = -2$, $b = 5$, $c = 1$

(C) $\begin{bmatrix} a \\ b \\ c \end{bmatrix} = \frac{1}{2} \begin{bmatrix} 1 & -2 & 1 \\ -5 & 8 & -3 \\ 6 & -6 & 2 \end{bmatrix} \begin{bmatrix} 8 \\ -5 \\ 4 \end{bmatrix} = \begin{bmatrix} 11 \\ -46 \\ 43 \end{bmatrix}$ $a = 11$, $b = -46$, $c = 43$

39. The system to be solved, for an arbitrary diet, is derived as follows:

Let x_1 = amount of mix A

$\quad\;\; x_2$ = amount of mix B

Then $0.20x_1 + 0.10x_2 = k_1$ (k_1 = amount of protein)

$\quad\quad\;\; 0.02x_1 + 0.06x_2 = k_2$ (k_2 = amount of fat)

We solve the system by writing it as a matrix equation.

$$\begin{matrix} A & X & B \end{matrix}$$

$$\begin{bmatrix} 0.20 & 0.10 \\ 0.02 & 0.06 \end{bmatrix} \begin{bmatrix} x_1 \\ x_2 \end{bmatrix} = \begin{bmatrix} k_1 \\ k_2 \end{bmatrix}$$

If A^{-1} exists, then $X = A^{-1}B$. To find A^{-1}, we perform row operations on

$$\begin{bmatrix} 0.20 & 0.10 & | & 1 & 0 \\ 0.02 & 0.06 & | & 0 & 1 \end{bmatrix} \quad \begin{matrix} 5R_1 \to R_1 \\ 50R_2 \to R_2 \end{matrix}$$

$$\sim \begin{bmatrix} 1 & 0.5 & | & 5 & 0 \\ 1 & 3 & | & 0 & 50 \end{bmatrix} \quad (-1)R_1 + R_2 \to R_2$$

$$\sim \begin{bmatrix} 1 & 0.5 & | & 5 & 0 \\ 0 & 2.5 & | & -5 & 50 \end{bmatrix} \quad 0.4R_2 \to R_2$$

$$\sim \begin{bmatrix} 1 & 0.5 & | & 5 & 0 \\ 0 & 1 & | & -2 & 20 \end{bmatrix} \quad (-0.5)R_2 + R_1 \to R_1$$

$$\sim \begin{bmatrix} 1 & 0 & | & 6 & -10 \\ 0 & 1 & | & -2 & 20 \end{bmatrix}$$

Hence $A^{-1} = \begin{bmatrix} 6 & -10 \\ -2 & 20 \end{bmatrix}$

Check: $A^{-1}A = \begin{bmatrix} 6 & -10 \\ -2 & 20 \end{bmatrix} \begin{bmatrix} 0.20 & 0.10 \\ 0.02 & 0.06 \end{bmatrix} = \begin{bmatrix} 1 & 0 \\ 0 & 1 \end{bmatrix}$

We can now solve the system as

$$\begin{matrix} X & A^{-1} & B \end{matrix}$$

$$\begin{bmatrix} x_1 \\ x_2 \end{bmatrix} = \begin{bmatrix} 6 & -10 \\ -2 & 20 \end{bmatrix} \begin{bmatrix} k_1 \\ k_2 \end{bmatrix}$$

For Diet 1, $k_1 = 20$ and $k_2 = 6$

$$\begin{bmatrix} x_1 \\ x_2 \end{bmatrix} = \begin{bmatrix} 6 & -10 \\ -2 & 20 \end{bmatrix} \begin{bmatrix} 20 \\ 6 \end{bmatrix} = \begin{bmatrix} 60 \\ 80 \end{bmatrix}$$ Diet 1: 60 ounces Mix A and 80 ounces Mix B

For Diet 2, $k_1 = 10$ and $k_2 = 4$

$$\begin{bmatrix} x_1 \\ x_2 \end{bmatrix} = \begin{bmatrix} 6 & -10 \\ -2 & 20 \end{bmatrix} \begin{bmatrix} 10 \\ 4 \end{bmatrix} = \begin{bmatrix} 20 \\ 60 \end{bmatrix}$$ Diet 2: 20 ounces Mix A and 60 ounces Mix B

For Diet 3, $k_1 = 10$ and $k_2 = 6$

$$\begin{bmatrix} x_1 \\ x_2 \end{bmatrix} = \begin{bmatrix} 6 & -10 \\ -2 & 20 \end{bmatrix} \begin{bmatrix} 10 \\ 6 \end{bmatrix} = \begin{bmatrix} 0 \\ 100 \end{bmatrix}$$ Diet 3: 0 ounces Mix A and 100 ounces Mix B

Exercise 7-4

Key Ideas and Formulas

The determinant of a square matrix A is a number, denoted det A, or by writing the array of elements in A using vertical lines instead of square brackets.

Second-order determinant:

$$\begin{vmatrix} a_{11} & a_{12} \\ a_{21} & a_{22} \end{vmatrix} = a_{11}a_{22} - a_{21}a_{12}$$

Third-order determinant:

$$\begin{vmatrix} a_{11} & a_{12} & a_{13} \\ a_{21} & a_{22} & a_{23} \\ a_{31} & a_{32} & a_{33} \end{vmatrix}$$

The minor of an element a_{ij} is the determinant array obtained by deleting the row and column that contain a_{ij}, that is, row i and column j.

The cofactor of $a_{ij} = (-1)^{i+j}$ (Minor of a_{ij}).

The value of a determinant of order 3 (or $n > 3$) is the sum of the three (or n) products obtained by multiplying each element of any one row (or each element of any one column) by its cofactor.

---

1. $\begin{vmatrix} 2 & 2 \\ -3 & 1 \end{vmatrix} = 2 \cdot 1 - (-3)2 = 8$

3. $\begin{vmatrix} 6 & -2 \\ -1 & -3 \end{vmatrix} = 6(-3) - (-1)(-2) = -20$

5. $\begin{vmatrix} 1.8 & -1.6 \\ -1.9 & 1.2 \end{vmatrix} = (1.8)(1.2) - (-1.9)(-1.6) = -0.88$

7. $\begin{vmatrix} 2 & 3 & 0 \\ 5 & 1 & -2 \\ 7 & -4 & 8 \end{vmatrix} = \begin{vmatrix} 1 & -2 \\ -4 & 8 \end{vmatrix}$

9. $\begin{vmatrix} -2 & 3 & 0 \\ 5 & 1 & -2 \\ 7 & -4 & 8 \end{vmatrix} = \begin{vmatrix} -2 & 0 \\ 5 & -2 \end{vmatrix}$

11. $(-1)^{1+1} \begin{vmatrix} 1 & -2 \\ -4 & 8 \end{vmatrix} = (-1)^2 [1 \cdot 8 - (-2)(-4)] = 0$

13. $(-1)^{3+2} \begin{vmatrix} -2 & 0 \\ 5 & -2 \end{vmatrix} = (-1)^5 [(-2)(-2) - 0(5)] = -4$

15. We expand by row 1

$$\begin{vmatrix} 1 & 0 & 0 \\ -2 & 4 & 3 \\ 5 & -2 & 1 \end{vmatrix} =$$

a_{11} (cofactor of a_{11}) + a_{12} (cofactor of a_{12}) + a_{13} (cofactor of a_{13})

$$= 1(-1)^{1+1} \begin{vmatrix} 4 & 3 \\ -2 & 1 \end{vmatrix} + 0(\diagup) + (\diagdown)$$

It is unnecessary to evaluate these since they are multiplied by 0.

$$= (-1)^2 [4 \cdot 1 - (-2)3]$$
$$= 10$$

17. We expand by column 1

$$\begin{vmatrix} 0 & 1 & 5 \\ 3 & -7 & 6 \\ 0 & -2 & -3 \end{vmatrix} = a_{11} \text{ (cofactor of } a_{11}) + a_{21} \text{ (cofactor of } a_{21}) + a_{31} \text{ (cofactor of } a_{31})$$

$$= 0(\diagup) + 3(-1)^{2+1} \begin{vmatrix} 1 & 5 \\ -2 & -3 \end{vmatrix} + 0(\diagdown)$$

It is unnecessary to evaluate these since they are multiplied by 0.

$$= 3(-1)^3 [1(-3) - (-2)5]$$
$$= -21$$

> **Common Error:** Neglecting the sign of the cofactor. The cofactor is often called the "signed" minor.

19. We expand by column 2

$$\begin{vmatrix} -1 & 2 & -3 \\ -2 & 0 & -6 \\ 4 & -3 & 2 \end{vmatrix} =$$

a_{12} (cofactor of a_{12}) + a_{22} (cofactor of a_{22}) + a_{32} (cofactor of a_{32})

$$= 2(-1)^{1+2} \begin{vmatrix} -2 & -6 \\ 4 & 2 \end{vmatrix} + 0(\quad) + (-3)(-1)^{3+2} \begin{vmatrix} -1 & -3 \\ -2 & -6 \end{vmatrix}$$

$$= 2(-1)^3 [(-2)2 - 4(-6)] + (-3)(-1)^5 [(-1)(-6) - (-2)(-3)]$$
$$= (-2)(20) + 3(0) = -40$$

21. $(-1)^{1+1} \begin{vmatrix} a_{11} & a_{12} & a_{13} & a_{14} \\ a_{21} & a_{22} & a_{23} & a_{24} \\ a_{31} & a_{32} & a_{33} & a_{34} \\ a_{41} & a_{42} & a_{43} & a_{44} \end{vmatrix} = (-1)^{1+1} \begin{vmatrix} a_{22} & a_{23} & a_{24} \\ a_{32} & a_{33} & a_{34} \\ a_{42} & a_{43} & a_{44} \end{vmatrix}$

23. $(-1)^{4+3} \begin{vmatrix} a_{11} & a_{12} & a_{13} & a_{14} \\ a_{21} & a_{22} & a_{23} & a_{24} \\ a_{31} & a_{32} & a_{33} & a_{34} \\ a_{41} & a_{42} & a_{43} & a_{44} \end{vmatrix} = (-1)^{4+3} \begin{vmatrix} a_{11} & a_{12} & a_{14} \\ a_{21} & a_{22} & a_{24} \\ a_{31} & a_{32} & a_{34} \end{vmatrix}$

25. We expand by the second column

$$\begin{vmatrix} 3 & -2 & -8 \\ -2 & 0 & -3 \\ 1 & 0 & -4 \end{vmatrix} =$$

a_{12} (cofactor of a_{12}) + a_{22} (cofactor of a_{22}) + a_{32} (cofactor of a_{32})

$$= (-2)(-1)^{1+2} \begin{vmatrix} -2 & -3 \\ 1 & -4 \end{vmatrix} + 0 + 0$$

$$= (-2)(-1)^3 [(-2)(-4) - 1(-3)]$$
$$= 2(11)$$
$$= 22$$

27. We expand by the first row

$$\begin{vmatrix} 1 & 4 & 1 \\ 1 & 1 & -2 \\ 2 & 1 & -1 \end{vmatrix} = a_{11} \text{ (cofactor of } a_{11}) + a_{12} \text{ (cofactor of } a_{12}) + a_{13} \text{ (cofactor of } a_{13})$$

$$= 1(-1)^{1+1} \begin{vmatrix} 1 & -2 \\ 1 & -1 \end{vmatrix} + 4(-1)^{1+2} \begin{vmatrix} 1 & -2 \\ 2 & -1 \end{vmatrix} + 1(-1)^{1+3} \begin{vmatrix} 1 & 1 \\ 2 & 1 \end{vmatrix}$$

$$= (-1)^2[1(-1) - 1(-2)] + 4(-1)^3[1(-1) - 2(-2)] + (-1)^4[1\cdot1 - 2\cdot1]$$

$$= 1 + (-12) + (-1)$$

$$= -12$$

29. We expand by the first row

$$\begin{vmatrix} 1 & 4 & 3 \\ 2 & 1 & 6 \\ 3 & -2 & 9 \end{vmatrix} = a_{11} \text{ (cofactor of } a_{11}) + a_{12} \text{ (cofactor of } a_{12}) + a_{13} \text{ (cofactor of } a_{13})$$

$$= 1(-1)^{1+1} \begin{vmatrix} 1 & 6 \\ -2 & 9 \end{vmatrix} + 4(-1)^{1+2} \begin{vmatrix} 2 & 6 \\ 3 & 9 \end{vmatrix} + 3(-1)^{1+3} \begin{vmatrix} 2 & 1 \\ 3 & -2 \end{vmatrix}$$

$$= (-1)^2[1\cdot9 - (-2)6] + 4(-1)^3[2\cdot9 - 3\cdot6] + 3(-1)^4[2(-2) - 1\cdot3]$$

$$= 21 + 0 - 21$$

$$= 0$$

31. We expand by the second row. Clearly the only non-zero term will be a_{22} (cofactor of a_{22}), which is

$$3(-1)^{2+2} \begin{vmatrix} 2 & 1 & 7 \\ 3 & 2 & 5 \\ 0 & 0 & 2 \end{vmatrix}$$

The order 3 determinant is expanded by the third row. Again there is only one non-zero term, a_{33} (cofactor of a_{33}). So the original determinant is reduced to

$$3(-1)^{2+2} 2(-1)^{3+3} \begin{vmatrix} 2 & 1 \\ 3 & 2 \end{vmatrix} = 6(-1)^{10}(2\cdot2 - 3\cdot1) = 6$$

33.

$$\begin{vmatrix} -2 & 0 & 0 & 0 & 0 \\ 9 & -1 & 0 & 0 & 0 \\ 2 & 1 & 3 & 0 & 0 \\ -1 & 4 & 2 & 2 & 0 \\ 7 & -2 & 3 & 5 & 5 \end{vmatrix} = (-2)(-1)^{1+1} \begin{vmatrix} -1 & 0 & 0 & 0 \\ 1 & 3 & 0 & 0 \\ 4 & 2 & 2 & 0 \\ -2 & 3 & 5 & 5 \end{vmatrix} + 0 \text{ terms}$$

$$= -2 \begin{vmatrix} -1 & 0 & 0 & 0 \\ 1 & 3 & 0 & 0 \\ 4 & 2 & 2 & 0 \\ -2 & 3 & 5 & 5 \end{vmatrix} = (-2)\left[(-1)(-1)^{1+1} \begin{vmatrix} 3 & 0 & 0 \\ 2 & 2 & 0 \\ 3 & 5 & 5 \end{vmatrix} + 0 \text{ terms} \right]$$

$$= (-2)(-1) \begin{vmatrix} 3 & 0 & 0 \\ 2 & 2 & 0 \\ 3 & 5 & 5 \end{vmatrix} = (-2)(-1)\left[3(-1)^{1+1} \begin{vmatrix} 2 & 0 \\ 5 & 5 \end{vmatrix} + 0 \text{ terms} \right]$$

$$= (-2)(-1)3 \begin{vmatrix} 2 & 0 \\ 5 & 5 \end{vmatrix}$$

$$= (-2)(-1)(3)[2\cdot5 - 5\cdot0]$$

$$= (-2)(-1)(3)(2)(5)$$

$$= 60$$

35.

$$2 \cdot 3 \cdot 1 + 6(-7)(-4) + (-1)(5)(-2) - (-4)(3)(-1) - (-2)(-7)2 - 6 \cdot 5 \cdot 1$$
$$= 6 + 168 + 10 - 12 - 28 - 30$$
$$= 114$$

37. (A) $\begin{vmatrix} a_{11} & a_{12} \\ 0 & a_{22} \end{vmatrix} = a_{11}a_{22} - 0a_{12} = a_{11}a_{22}$

We expand $\begin{vmatrix} a_{11} & a_{12} & a_{13} \\ 0 & a_{22} & a_{23} \\ 0 & 0 & a_{33} \end{vmatrix}$ by the first column.

$\begin{vmatrix} a_{11} & a_{12} & a_{13} \\ 0 & a_{22} & a_{23} \\ 0 & 0 & a_{33} \end{vmatrix} = a_{11}(\text{cofactor of } a_{11}) + 0(\text{cofactor of } a_{21}) + 0(\text{cofactor of } a_{31})$

$$= a_{11}(-1)^{1+1} \begin{vmatrix} a_{22} & a_{23} \\ 0 & a_{33} \end{vmatrix} + 0 \text{ terms}$$

$$= a_{11} \begin{vmatrix} a_{22} & a_{23} \\ 0 & a_{33} \end{vmatrix}$$

$$= a_{11}(a_{22}a_{33} - 0a_{23})$$

$$= a_{11}a_{22}a_{33}$$

(B) In each case, the determinant is the product of the terms on the principal diagonal.

39. $\begin{vmatrix} a & b \\ c & d \end{vmatrix} = ad - bc$ $\begin{vmatrix} c & d \\ a & b \end{vmatrix} = cb - ad = -(ad - bc)$

Hence, $\begin{vmatrix} a & b \\ c & d \end{vmatrix} = -\begin{vmatrix} c & d \\ a & b \end{vmatrix}$; interchanging the rows of this determinant changes its sign.

41. $\begin{vmatrix} a & b \\ c & d \end{vmatrix} = ad - bc$ $\begin{vmatrix} ka & b \\ kc & d \end{vmatrix} = kad - kcb = k(ad - bc)$

Hence, $\begin{vmatrix} ka & b \\ kc & d \end{vmatrix} = k\begin{vmatrix} a & b \\ c & d \end{vmatrix}$; multiplying a column of this determinant by a number k multiplies the value of the determinant by k.

43. $\begin{vmatrix} a & b \\ c & d \end{vmatrix} = ad - bc$ $\begin{vmatrix} kc + a & kd + b \\ c & d \end{vmatrix} = (kc + a)d - (kd + b)c$

$$= kcd + ad - kdc - bc$$
$$= ad - bc$$

Hence, $\begin{vmatrix} kc + a & kd + b \\ c & d \end{vmatrix} = \begin{vmatrix} a & b \\ c & d \end{vmatrix}$; adding a multiple of one row to the other row does not change the value of this determinant.

45. Expanding by the first column

$\begin{vmatrix} a_{11} & a_{12} & a_{13} \\ a_{21} & a_{22} & a_{23} \\ a_{31} & a_{32} & a_{33} \end{vmatrix}$

$$= a_{11}(-1)^{1+1} \begin{vmatrix} a_{22} & a_{23} \\ a_{32} & a_{33} \end{vmatrix} + a_{21}(-1)^{2+1} \begin{vmatrix} a_{12} & a_{13} \\ a_{32} & a_{33} \end{vmatrix} + a_{31}(-1)^{3+1} \begin{vmatrix} a_{12} & a_{13} \\ a_{22} & a_{23} \end{vmatrix}$$

$$= a_{11} \begin{vmatrix} a_{22} & a_{23} \\ a_{32} & a_{33} \end{vmatrix} - a_{21} \begin{vmatrix} a_{12} & a_{13} \\ a_{32} & a_{33} \end{vmatrix} + a_{31} \begin{vmatrix} a_{12} & a_{13} \\ a_{22} & a_{23} \end{vmatrix}$$

$$= a_{11}(a_{22}a_{33} - a_{32}a_{23}) - a_{21}(a_{12}a_{33} - a_{32}a_{13}) + a_{31}(a_{12}a_{23} - a_{22}a_{13})$$
$$\quad\;\; ① \qquad\qquad ② \qquad\qquad ③ \qquad\qquad ④ \qquad\qquad ⑤ \qquad\qquad ⑥$$

$$= a_{11}a_{22}a_{33} - a_{11}a_{32}a_{23} - a_{21}a_{12}a_{33} + a_{21}a_{32}a_{13} + a_{31}a_{12}a_{23} - a_{31}a_{22}a_{13}$$

Expanding by the third row

$$\begin{vmatrix} a_{11} & a_{12} & a_{13} \\ a_{21} & a_{22} & a_{23} \\ a_{31} & a_{32} & a_{33} \end{vmatrix}$$

$$= a_{31}(-1)^{3+1} \begin{vmatrix} a_{12} & a_{13} \\ a_{22} & a_{23} \end{vmatrix} + a_{32}(-1)^{3+2} \begin{vmatrix} a_{11} & a_{13} \\ a_{21} & a_{23} \end{vmatrix} + a_{33}(-1)^{3+3} \begin{vmatrix} a_{11} & a_{12} \\ a_{21} & a_{22} \end{vmatrix}$$

$$= a_{31} \begin{vmatrix} a_{12} & a_{13} \\ a_{22} & a_{23} \end{vmatrix} - a_{32} \begin{vmatrix} a_{11} & a_{13} \\ a_{21} & a_{23} \end{vmatrix} + a_{33} \begin{vmatrix} a_{11} & a_{12} \\ a_{21} & a_{22} \end{vmatrix}$$

$$= a_{31}(a_{12}a_{23} - a_{13}a_{22}) - a_{32}(a_{11}a_{23} - a_{13}a_{21}) + a_{33}(a_{11}a_{22} - a_{12}a_{21})$$
$$\quad\;\; ⑤ \qquad\qquad ⑥ \qquad\qquad ② \qquad\qquad ④ \qquad\qquad ① \qquad\qquad ③$$

$$= a_{31}a_{12}a_{23} - a_{31}a_{13}a_{22} - a_{32}a_{11}a_{23} + a_{32}a_{13}a_{21} + a_{33}a_{11}a_{22} - a_{33}a_{12}a_{21}$$

Comparing the two expressions, with the aid of the numbers over the terms, shows that the expressions are the same.

47. $A = \begin{bmatrix} 2 & 3 \\ 1 & -2 \end{bmatrix}$ $B = \begin{bmatrix} -1 & 3 \\ 2 & 1 \end{bmatrix}$

We calculate $AB = \begin{bmatrix} 2 & 3 \\ 1 & -2 \end{bmatrix}\begin{bmatrix} -1 & 3 \\ 2 & 1 \end{bmatrix} = \begin{bmatrix} 2(-1) + 3\cdot2 & 2\cdot3 + 3\cdot1 \\ 1(-1) + (-2)2 & 1\cdot3 + (-2)\cdot1 \end{bmatrix} = \begin{bmatrix} 4 & 9 \\ -5 & 1 \end{bmatrix}$

$$\det(AB) = \begin{vmatrix} 4 & 9 \\ -5 & 1 \end{vmatrix} = 4\cdot1 - (-5)9 = 49$$

$$\det A = \begin{vmatrix} 2 & 3 \\ 1 & -2 \end{vmatrix} = 2(-2) - 1\cdot3 = -7$$

$$\det B = \begin{vmatrix} -1 & 3 \\ 2 & 1 \end{vmatrix} = (-1)1 - 2\cdot3 = -7$$

Therefore $\det(AB) = 49 = (-7)(-7) = \det A \cdot \det B$

49. The matrix $xI - A$ is calculated as

$$x\begin{bmatrix} 1 & 0 \\ 0 & 1 \end{bmatrix} - \begin{bmatrix} 5 & -4 \\ 2 & -1 \end{bmatrix} = \begin{bmatrix} x & 0 \\ 0 & x \end{bmatrix} - \begin{bmatrix} 5 & -4 \\ 2 & -1 \end{bmatrix} = \begin{bmatrix} x-5 & 4 \\ -2 & x+1 \end{bmatrix}$$

The characteristic polynomial is the determinant of this matrix:

$$\begin{vmatrix} x-5 & 4 \\ -2 & x+1 \end{vmatrix} = (x-5)(x+1) - (4)(-2) = x^2 - 4x - 5 + 8 = x^2 - 4x + 3$$

The zeros of this polynomial are the solutions of $x^2 - 4x + 3 = 0$

$$x^2 - 4x + 3 = 0$$
$$(x-1)(x-3) = 0$$
$$x = 1, 3$$

The eigenvalues of this matrix are 1 and 3.

51. The matrix $xI - A$ is calculated as

$$x\begin{bmatrix} 1 & 0 & 0 \\ 0 & 1 & 0 \\ 0 & 0 & 1 \end{bmatrix} - \begin{bmatrix} 4 & -4 & 0 \\ 2 & -2 & 0 \\ 4 & -8 & -4 \end{bmatrix} = \begin{bmatrix} x & 0 & 0 \\ 0 & x & 0 \\ 0 & 0 & x \end{bmatrix} - \begin{bmatrix} 4 & -4 & 0 \\ 2 & -2 & 0 \\ 4 & -8 & -4 \end{bmatrix} = \begin{bmatrix} x - 4 & 4 & 0 \\ -2 & x + 2 & 0 \\ -4 & 8 & x + 4 \end{bmatrix}$$

The characteristic polynomial is the determinant of this matrix.

$$\begin{vmatrix} x - 4 & 4 & 0 \\ -2 & x + 2 & 0 \\ -4 & 8 & x + 4 \end{vmatrix} = (x + 4)(-1)^{3+3}\begin{vmatrix} x - 4 & 4 \\ -2 & x + 2 \end{vmatrix} \text{ expanding by the third column}$$

$$= (x + 4)(1)[(x - 4)(x + 2) - 4(-2)]$$
$$= (x + 4)(x^2 - 2x - 8 + 8)$$
$$= (x + 4)(x^2 - 2x)$$
$$= x^3 + 2x^2 - 8x$$

The zeros of this polynomial are the solutions of
$$x^3 + 2x^2 - 8x = 0$$
$$x(x^2 + 2x - 8) = 0$$
$$x(x + 4)(x - 2) = 0$$
$$x = 0, -4, 2$$
The eigenvalues of this matrix are 0, -4, and 2.

Exercise 7-5

Key Ideas and Formulas

If each element of any row (or column) of a determinant is multiplied by a constant k, the new determinant is k times the original. (Theorem 1) Caution: This operation and its result is very different from the operation of multiplying a matrix by a number k.

If every element in a row (or column) is 0, the value of the determinant is 0. (Theorem 2)

If two rows (or two columns) of a determinant are interchanged, the new determinant is the negative of the original. (Theorem 3)

If the corresponding elements are equal in two rows (or columns) the value of the determinant is 0. (Theorem 4)

If a multiple of any row (or column) of a determinant is added to any other row (or column) the value of the determinant is unchanged. (Theorem 5)

---

1. Theorem 1   **3.** Theorem 1   **5.** Theorem 2   **7.** Theorem 3   **9.** Theorem 5

11. $3C_1 + C_2 \rightarrow C_2$ has been used. Hence $x = 3(-1) + 3 = 0$

13. $3C_1 + C_3 \rightarrow C_3$ has been used. Hence $x = 3(1) + 2 = 5$

15. Interchanging two rows of a determinant changes the sign of the determinant (Theorem 3). Hence,
$$\begin{vmatrix} c & d \\ a & b \end{vmatrix} = -10$$

17. Adding a multiple (in this case a multiple by 1) of a row to another row does not change the value of the determinant (Theorem 5). Hence,
$$\begin{vmatrix} a + c & b + d \\ c & d \end{vmatrix} = 10$$

19. Adding a multiple of a column to another column does not change the value of the determinant (Theorem 5). Hence,

$$\begin{vmatrix} a & a-b \\ c & c-d \end{vmatrix} = \begin{vmatrix} a & -b \\ c & -d \end{vmatrix}$$

If every element of a column of a determinant is multiplied by -1, the value of the determinant is -1 times the original. Hence,

$$\begin{vmatrix} a & a-b \\ c & c-d \end{vmatrix} = \begin{vmatrix} a & -b \\ c & -d \end{vmatrix} = (-1)\begin{vmatrix} a & b \\ c & d \end{vmatrix} = -1(10) = -10$$

21. $\begin{vmatrix} -1 & 0 & 3 \\ 2 & 5 & 4 \\ 1 & 5 & 2 \end{vmatrix} = \begin{vmatrix} -1 & 0 & 3 \\ 1 & 0 & 2 \\ 1 & 5 & 2 \end{vmatrix}$ $(-1)R_3 + R_2 \rightarrow R_2$

$$= 5(-1)^{3+2}\begin{vmatrix} -1 & 3 \\ 1 & 2 \end{vmatrix} = -5[(-1)2 - 1(3)] = 25$$

23. $\begin{vmatrix} 3 & 5 & 0 \\ 1 & 1 & -2 \\ 2 & 1 & -1 \end{vmatrix} = \begin{vmatrix} 3 & 5 & 0 \\ -3 & -1 & 0 \\ 2 & 1 & -1 \end{vmatrix}$ $(-2)R_3 + R_2 \rightarrow R_2$

$$= (-1)(-1)^{3+3}\begin{vmatrix} 3 & 5 \\ -3 & -1 \end{vmatrix} = (-1)[3(-1) - (-3)5] = -12$$

25. Theorem 1   **27.** Theorem 2   **29.** Theorem 5

31. $2C_3 + C_1 \rightarrow C_1$ has been used. Hence $x = 2 \cdot 1 + 3 = 5$
$C_3 + C_2 \rightarrow C_2$ has been used. Hence $y = (-2) + 2 = 0$

33. $(-4)R_2 + R_1 \rightarrow R_1$ has been used. Hence $x = (-4)3 + 9 = -3$
$2R_2 + R_3 \rightarrow R_3$ has been used. Hence $y = 2 \cdot 3 + 4 = 10$

35. We will generate zeros in the first column by row operations.

$\begin{vmatrix} 1 & 5 & 3 \\ 4 & 2 & 1 \\ 3 & 1 & 2 \end{vmatrix} = \begin{vmatrix} 1 & 5 & 3 \\ 0 & -18 & -11 \\ 0 & -14 & -7 \end{vmatrix}$ $\begin{array}{l} (-4)R_1 + R_2 \rightarrow R_2 \\ (-3)R_1 + R_3 \rightarrow R_3 \end{array}$

$$= 1(-1)^{1+1}\begin{vmatrix} -18 & -11 \\ -14 & -7 \end{vmatrix} + 0 + 0$$

$$= (-1)^2[(-18)(-7) - (-14)(-11)] = -28$$

37. We will generate zeros in the second row by column operations.

$\begin{vmatrix} 5 & 2 & -3 \\ -2 & 4 & 4 \\ 1 & -1 & 3 \end{vmatrix} = \begin{vmatrix} 5 & 12 & 7 \\ -2 & 0 & 0 \\ 1 & 1 & 5 \end{vmatrix}$ $\begin{array}{l} 2C_1 + C_2 \rightarrow C_2 \\ 2C_1 + C_3 \rightarrow C_3 \end{array}$

$$= (-2)(-1)^{2+1}\begin{vmatrix} 12 & 7 \\ 1 & 5 \end{vmatrix} + 0 + 0$$

$$= (-2)(-1)^3[12 \cdot 5 - 1 \cdot 7] = 106$$

39. The column operation $(-3)C_3 + C_1 \rightarrow C_1$ transforms this determinant into

$$\begin{vmatrix} 0 & -4 & 1 \\ 0 & -1 & 2 \\ 0 & 2 & 3 \end{vmatrix}$$

By Theorem 2, the value of this determinant is 0.

41. We start by generating one more zero in the first row.

$$\begin{vmatrix} 0 & 1 & 0 & 1 \\ 1 & -2 & 4 & 3 \\ 2 & 1 & 5 & 4 \\ 1 & 2 & 1 & 2 \end{vmatrix} = \begin{vmatrix} 0 & 0 & 0 & 1 \\ 1 & -5 & 4 & 3 \\ 2 & -3 & 5 & 4 \\ 1 & 0 & 1 & 2 \end{vmatrix} \qquad (-1)C_4 + C_2 \rightarrow C_2$$

$$= 1(-1)^{1+4} \begin{vmatrix} 1 & -5 & 4 \\ 2 & -3 & 5 \\ 1 & 0 & 1 \end{vmatrix} + 0 + 0 + 0$$

$$= (-1) \begin{vmatrix} 1 & -5 & 4 \\ 2 & -3 & 5 \\ 1 & 0 & 1 \end{vmatrix}$$

$$= \begin{vmatrix} 1 & 5 & 4 \\ 2 & 3 & 5 \\ 1 & 0 & 1 \end{vmatrix} \quad \text{by Theorem 1}$$

We now generate one more zero in the third row.

$$\begin{vmatrix} 1 & 5 & 4 \\ 2 & 3 & 5 \\ 1 & 0 & 1 \end{vmatrix} = \begin{vmatrix} 1 & 5 & 3 \\ 2 & 3 & 3 \\ 1 & 0 & 0 \end{vmatrix} \qquad (-1)C_1 + C_3 \rightarrow C_3$$

$$= 1(-1)^{3+1} \begin{vmatrix} 5 & 3 \\ 3 & 3 \end{vmatrix} = (-1)^4 [5 \cdot 3 - 3 \cdot 3] = 6$$

43. We start by generating zeros in the third row.

$$\begin{vmatrix} 3 & 2 & 3 & 1 \\ 3 & -2 & 8 & 5 \\ 2 & 1 & 3 & 1 \\ 4 & 5 & 4 & -3 \end{vmatrix} = \begin{vmatrix} 1 & 1 & 0 & 1 \\ -7 & -7 & -7 & 5 \\ 0 & 0 & 0 & 1 \\ 10 & 8 & 13 & -3 \end{vmatrix} \qquad \begin{matrix} (-2)C_4 + C_1 \rightarrow C_1 \\ (-1)C_4 + C_2 \rightarrow C_2 \\ (-3)C_4 + C_3 \rightarrow C_3 \end{matrix}$$

$$= 1(-1)^{3+4} \begin{vmatrix} 1 & 1 & 0 \\ -7 & -7 & -7 \\ 10 & 8 & 13 \end{vmatrix} + 0 + 0 + 0$$

$$= (-1) \begin{vmatrix} 1 & 1 & 0 \\ -7 & -7 & -7 \\ 10 & 8 & 13 \end{vmatrix}$$

$$= \begin{vmatrix} 1 & 1 & 0 \\ 7 & 7 & 7 \\ 10 & 8 & 13 \end{vmatrix} \quad \text{by Theorem 1}$$

We now generate one more zero in the first row.

$$\begin{vmatrix} 1 & 1 & 0 \\ 7 & 7 & 7 \\ 10 & 8 & 13 \end{vmatrix} = \begin{vmatrix} 1 & 0 & 0 \\ 7 & 0 & 7 \\ 10 & -2 & 13 \end{vmatrix} \qquad (-1)C_1 + C_2 \rightarrow C_2$$

$$= (-2)(-1)^{3+2} \begin{vmatrix} 1 & 0 \\ 7 & 7 \end{vmatrix} + 0 + 0$$

$$= (-2)(-1)^5 [1 \cdot 7 - 0 \cdot 7] = 14$$

45. Expand, for example, by the first column.

$$\begin{vmatrix} a & b & a \\ d & e & d \\ g & h & g \end{vmatrix} = a(-1)^{1+1}\begin{vmatrix} e & d \\ h & g \end{vmatrix} + d(-1)^{2+1}\begin{vmatrix} b & a \\ h & g \end{vmatrix} + g(-1)^{3+1}\begin{vmatrix} b & a \\ e & d \end{vmatrix}$$

$$= a\begin{vmatrix} e & d \\ h & g \end{vmatrix} - d\begin{vmatrix} b & a \\ h & g \end{vmatrix} + g\begin{vmatrix} b & a \\ e & d \end{vmatrix}$$

$$= a(eg - hd) - d(bg - ha) + g(bd - ea)$$

$$= aeg - adh - bdg + adh + bdg - aeg$$

$$= 0$$

47. We expand the left side by the first column, the right side by the second column.

$$\begin{vmatrix} a_1 & b_1 & c_1 \\ a_2 & b_2 & c_2 \\ a_3 & b_3 & c_3 \end{vmatrix} = a_1(-1)^{1+1}\begin{vmatrix} b_2 & c_2 \\ b_3 & c_3 \end{vmatrix} + a_2(-1)^{2+1}\begin{vmatrix} b_1 & c_1 \\ b_3 & c_3 \end{vmatrix} + a_3(-1)^{3+1}\begin{vmatrix} b_1 & c_1 \\ b_2 & c_2 \end{vmatrix}$$

$$= a_1\begin{vmatrix} b_2 & c_2 \\ b_3 & c_3 \end{vmatrix} - a_2\begin{vmatrix} b_1 & c_1 \\ b_3 & c_3 \end{vmatrix} + a_3\begin{vmatrix} b_1 & c_1 \\ b_2 & c_2 \end{vmatrix}$$

$$-\begin{vmatrix} b_1 & a_1 & c_1 \\ b_2 & a_2 & c_2 \\ b_3 & a_3 & c_3 \end{vmatrix} = -\left[a_1(-1)^{1+2}\begin{vmatrix} b_2 & c_2 \\ b_3 & c_3 \end{vmatrix} + a_2(-1)^{2+2}\begin{vmatrix} b_1 & c_1 \\ b_3 & c_3 \end{vmatrix} + a_3(-1)^{3+2}\begin{vmatrix} b_1 & c_1 \\ b_2 & c_2 \end{vmatrix} \right]$$

$$= -\left[-a_1\begin{vmatrix} b_2 & c_2 \\ b_3 & c_3 \end{vmatrix} + a_2\begin{vmatrix} b_1 & c_1 \\ b_3 & c_3 \end{vmatrix} - a_3\begin{vmatrix} b_1 & c_1 \\ b_2 & c_2 \end{vmatrix} \right]$$

$$= a_1\begin{vmatrix} b_2 & c_2 \\ b_3 & c_3 \end{vmatrix} - a_2\begin{vmatrix} b_1 & c_1 \\ b_3 & c_3 \end{vmatrix} + a_3\begin{vmatrix} b_1 & c_1 \\ b_2 & c_2 \end{vmatrix}$$

Hence the two original expressions are equal.

49. The statements: $(2, 5)$ satisfies the equation $\begin{vmatrix} x & y & 1 \\ 2 & 5 & 1 \\ -3 & 4 & 1 \end{vmatrix} = 0$

and $(-3, 4)$ satisfies the equation $\begin{vmatrix} x & y & 1 \\ 2 & 5 & 1 \\ -3 & 4 & 1 \end{vmatrix} = 0$

are equivalent to the statements $\begin{vmatrix} 2 & 5 & 1 \\ 2 & 5 & 1 \\ -3 & 4 & 1 \end{vmatrix} = 0$ and $\begin{vmatrix} -3 & 4 & 1 \\ 2 & 5 & 1 \\ -3 & 4 & 1 \end{vmatrix} = 0.$

The latter statements are true by Theorem 4.

51. The statement $\begin{vmatrix} x & y & 1 \\ x_1 & y_1 & 1 \\ x_2 & y_2 & 1 \end{vmatrix} = 0$ is the equation of a line because, expanding by the

first row, we have $x(-1)^{1+1}\begin{vmatrix} y_1 & 1 \\ y_2 & 1 \end{vmatrix} + y(-1)^{1+2}\begin{vmatrix} x_1 & 1 \\ x_2 & 1 \end{vmatrix} + 1(-1)^{1+3}\begin{vmatrix} x_1 & y_1 \\ x_2 & y_2 \end{vmatrix} = 0$

This is in the standard form for the equation of a line
$Ax + By + C = 0$

To show that the line passes through (x_1, y_1) and (x_2, y_2), we note merely that (x_1, y_1) and (x_2, y_2) satisfy the equation, because

$$\begin{vmatrix} x_1 & y_1 & 1 \\ x_1 & y_1 & 1 \\ x_2 & y_2 & 1 \end{vmatrix} = 0 \text{ and } \begin{vmatrix} x_2 & y_2 & 1 \\ x_1 & y_1 & 1 \\ x_2 & y_2 & 1 \end{vmatrix} = 0 \text{ are true by Theorem 4.}$$

53. Using the result stated in problem 52, we have

$$\begin{vmatrix} x_1 & y_1 & 1 \\ x_2 & y_2 & 1 \\ x_3 & y_3 & 1 \end{vmatrix} = 2 \times \text{(area of triangle formed by the three points).}$$

If the determinant is 0, then the area of the triangle formed by the three points is zero. The only way this can happen is if the three points are on the same line; that is, the points are collinear.

Exercise 7-6

Key Ideas and Formulas

Cramer's Rule for Two Equations and Two Variables

Given the system
$$a_{11}x + a_{12}y = k_1$$
$$a_{21}x + a_{22}y = k_2$$
with
$$D = \begin{vmatrix} a_{11} & a_{12} \\ a_{21} & a_{22} \end{vmatrix} \neq 0$$
then
$$x = \frac{\begin{vmatrix} k_1 & a_{12} \\ k_2 & a_{22} \end{vmatrix}}{D} \text{ and } y = \frac{\begin{vmatrix} a_{11} & k_1 \\ a_{21} & k_2 \end{vmatrix}}{D}$$

Cramer's Rule for Three Equations and Three Variables

Given the system
$$a_{11}x + a_{12}y + a_{13}z = k_1$$
$$a_{21}x + a_{22}y + a_{23}z = k_2$$
$$a_{31}x + a_{32}y + a_{33}z = k_3$$
with
$$D = \begin{vmatrix} a_{11} & a_{12} & a_{13} \\ a_{21} & a_{22} & a_{23} \\ a_{31} & a_{32} & a_{33} \end{vmatrix} \neq 0$$
then
$$x = \frac{\begin{vmatrix} k_1 & a_{12} & a_{13} \\ k_2 & a_{22} & a_{23} \\ k_3 & a_{32} & a_{33} \end{vmatrix}}{D} \quad y = \frac{\begin{vmatrix} a_{11} & k_1 & a_{13} \\ a_{21} & k_2 & a_{23} \\ a_{31} & k_3 & a_{33} \end{vmatrix}}{D} \quad z = \frac{\begin{vmatrix} a_{11} & a_{12} & k_1 \\ a_{21} & a_{22} & k_2 \\ a_{31} & a_{32} & k_3 \end{vmatrix}}{D}$$

The determinant D is called the coefficient determinant. If $D \neq 0$, the system has exactly one solution, given by Cramer's rule. If $D = 0$, the system is either inconsistent or dependent. Cramer's rule does not apply. We use methods of Chapter 6 to find solutions (if any) of the system.

1. $D = \begin{vmatrix} 1 & 2 \\ 1 & 3 \end{vmatrix} = 1$ $\quad x = \dfrac{\begin{vmatrix} 1 & 2 \\ -1 & 3 \end{vmatrix}}{D} = \dfrac{5}{1} = 5$ $\quad y = \dfrac{\begin{vmatrix} 1 & 1 \\ 1 & -1 \end{vmatrix}}{D} = \dfrac{-2}{1} = -2$ $\quad x = 5, \ y = -2$

3. $D = \begin{vmatrix} 2 & 1 \\ 5 & 3 \end{vmatrix} = 1$ $\quad x = \dfrac{\begin{vmatrix} 1 & 1 \\ 2 & 3 \end{vmatrix}}{D} = \dfrac{1}{1} = 1$ $\quad y = \dfrac{\begin{vmatrix} 2 & 1 \\ 5 & 2 \end{vmatrix}}{D} = \dfrac{-1}{1} = -1$ $\quad x = 1, \ y = -1$

5. $D = \begin{vmatrix} 2 & -1 \\ -1 & 3 \end{vmatrix} = 5$ $\quad x = \dfrac{\begin{vmatrix} -3 & -1 \\ 3 & 3 \end{vmatrix}}{D} = \dfrac{-6}{5} = -\dfrac{6}{5}$ $\quad y = \dfrac{\begin{vmatrix} 2 & -3 \\ -1 & 3 \end{vmatrix}}{D} = \dfrac{3}{5}$

7. $D = \begin{vmatrix} 4 & -3 \\ 3 & 2 \end{vmatrix} = 17$ $\quad x = \dfrac{\begin{vmatrix} 4 & -3 \\ -2 & 2 \end{vmatrix}}{D} = \dfrac{2}{17}$ $\quad y = \dfrac{\begin{vmatrix} 4 & 4 \\ 3 & -2 \end{vmatrix}}{D} = \dfrac{-20}{17} = -\dfrac{20}{17}$

9. $D = \begin{vmatrix} 0.9925 & -0.9659 \\ 0.1219 & 0.2588 \end{vmatrix} = 0.37460$

$x = \dfrac{\begin{vmatrix} 0 & -0.9659 \\ 2,500 & 0.2588 \end{vmatrix}}{D} = \dfrac{2,414.75}{0.37460} = 6,400$ to two significant digits

$y = \dfrac{\begin{vmatrix} 0.9925 & 0 \\ 0.1219 & 2,500 \end{vmatrix}}{D} = \dfrac{2,481.25}{0.37460} = 6,600$ to two significant digits

11. $D = \begin{vmatrix} 0.9954 & -0.9942 \\ 0.0958 & 0.1080 \end{vmatrix} = 0.20275$

$x = \dfrac{\begin{vmatrix} 0 & -0.9942 \\ 155 & 0.1080 \end{vmatrix}}{D} = \dfrac{154.10}{0.20275} = 760$ to two significant digits

$y = \dfrac{\begin{vmatrix} 0.9954 & 0 \\ 0.0958 & 155 \end{vmatrix}}{D} = \dfrac{154.29}{0.20275} = 760$ to two significant digits

13. $D = \begin{vmatrix} 1 & 1 & 0 \\ 0 & 2 & 1 \\ -1 & 0 & 1 \end{vmatrix} = 1$ $\quad x = \dfrac{\begin{vmatrix} 0 & 1 & 0 \\ -5 & 2 & 1 \\ -3 & 0 & 1 \end{vmatrix}}{D} = \dfrac{2}{1} = 2$ $\quad y = \dfrac{\begin{vmatrix} 1 & 0 & 0 \\ 0 & -5 & 1 \\ -1 & -3 & 1 \end{vmatrix}}{D} = \dfrac{-2}{1} = -2$

$z = \dfrac{\begin{vmatrix} 1 & 1 & 0 \\ 0 & 2 & -5 \\ -1 & 0 & -3 \end{vmatrix}}{D} = \dfrac{-1}{1} = -1$

15. $D = \begin{vmatrix} 1 & 1 & 0 \\ 0 & 2 & 1 \\ 0 & -1 & 1 \end{vmatrix} = 3$ $x = \dfrac{\begin{vmatrix} 1 & 1 & 0 \\ 0 & 2 & 1 \\ 1 & -1 & 1 \end{vmatrix}}{D} = \dfrac{4}{3}$ $y = \dfrac{\begin{vmatrix} 1 & 1 & 0 \\ 0 & 0 & 1 \\ 0 & 1 & 1 \end{vmatrix}}{D} = \dfrac{-1}{3} = -\dfrac{1}{3}$

$$z = \dfrac{\begin{vmatrix} 1 & 1 & 1 \\ 0 & 2 & 0 \\ 0 & -1 & 1 \end{vmatrix}}{D} = \dfrac{2}{3}$$

17. $D = \begin{vmatrix} 0 & 3 & 1 \\ 1 & 0 & 2 \\ 1 & -3 & 0 \end{vmatrix} = 3$ $x = \dfrac{\begin{vmatrix} -1 & 3 & 1 \\ 3 & 0 & 2 \\ -2 & -3 & 0 \end{vmatrix}}{D} = \dfrac{-27}{3} = -9$ $y = \dfrac{\begin{vmatrix} 0 & -1 & 1 \\ 1 & 3 & 2 \\ 1 & -2 & 0 \end{vmatrix}}{D} = \dfrac{-7}{3} = -\dfrac{7}{3}$

$$z = \dfrac{\begin{vmatrix} 0 & 3 & -1 \\ 1 & 0 & 3 \\ 1 & -3 & -2 \end{vmatrix}}{D} = \dfrac{18}{3} = 6$$

19. $D = \begin{vmatrix} 0 & 2 & -1 \\ 1 & -1 & -1 \\ 1 & -1 & 2 \end{vmatrix} = -6$ $x = \dfrac{\begin{vmatrix} -3 & 2 & -1 \\ 2 & -1 & -1 \\ 4 & -1 & 2 \end{vmatrix}}{D} = \dfrac{-9}{-6} = \dfrac{3}{2}$ $y = \dfrac{\begin{vmatrix} 0 & -3 & -1 \\ 1 & 2 & -1 \\ 1 & 4 & 2 \end{vmatrix}}{D} = \dfrac{7}{-6} = -\dfrac{7}{6}$

$$z = \dfrac{\begin{vmatrix} 0 & 2 & -3 \\ 1 & -1 & 2 \\ 1 & -1 & 4 \end{vmatrix}}{-6} = \dfrac{-4}{-6} = \dfrac{2}{3}$$

21. $x = \dfrac{\begin{vmatrix} -3 & -3 & 1 \\ -11 & 3 & 2 \\ 3 & -1 & -1 \end{vmatrix}}{\begin{vmatrix} 2 & -3 & 1 \\ -4 & 3 & 2 \\ 1 & -1 & -1 \end{vmatrix}} = \dfrac{20}{5} = 4$ **23.** $y = \dfrac{\begin{vmatrix} 12 & 5 & 11 \\ 15 & -13 & -9 \\ 5 & 0 & 2 \end{vmatrix}}{\begin{vmatrix} 12 & -14 & 11 \\ 15 & 7 & -9 \\ 5 & -3 & 2 \end{vmatrix}} = \dfrac{28}{14} = 2$

25. $z = \dfrac{\begin{vmatrix} 3 & -4 & 18 \\ -9 & 8 & -13 \\ 5 & -7 & 33 \end{vmatrix}}{\begin{vmatrix} 3 & -4 & 5 \\ -9 & 8 & 7 \\ 5 & -7 & 10 \end{vmatrix}} = \dfrac{5}{2}$

27. $D = \begin{vmatrix} 1 & -4 & 9 \\ 4 & -1 & 6 \\ 1 & -1 & 3 \end{vmatrix} = \begin{vmatrix} -2 & -1 & 0 \\ 2 & 1 & 0 \\ 1 & -1 & 3 \end{vmatrix} \begin{matrix} (-3)R_3 + R_1 \rightarrow R_1 \\ (-2)R_3 + R_2 \rightarrow R_2 \end{matrix} = -\begin{vmatrix} 2 & 1 & 0 \\ 2 & 1 & 0 \\ 1 & -1 & 3 \end{vmatrix}$ by Theorem 1

$$= 0 \text{ by Theorem 4}$$

Since $D = 0$, the system either has no solution or infinitely many. Since $x = 0$, $y = 0$, $z = 0$ is a solution, the second case must hold.

29. We start with

$a_{11}x + a_{12}y = k_1$

$a_{21}x + a_{22}y = k_2$

We wish to eliminate x. We multiply the top equation by $-a_{21}$ and the bottom equation by a_{11}, then add.

$$-a_{11}a_{21}x - a_{21}a_{12}y = -k_1a_{21}$$
$$a_{11}a_{21}x + a_{11}a_{22}y = k_2a_{11}$$
$$\overline{0x + a_{11}a_{22}y - a_{21}a_{12}y = k_2a_{11} - k_1a_{21}}$$
$$(a_{11}a_{22} - a_{21}a_{12})y = a_{11}k_2 - a_{21}k_1$$

$$y = \frac{a_{11}k_2 - a_{21}k_1}{a_{11}a_{22} - a_{21}a_{12}} = \frac{\begin{vmatrix} a_{11} & k_1 \\ a_{21} & k_2 \end{vmatrix}}{\begin{vmatrix} a_{11} & a_{12} \\ a_{21} & a_{22} \end{vmatrix}}$$

31. (A) $R = xp + yq = (200 - 6p + 4q)p + (300 + 2p - 3q)q$

$= 200p - 6p^2 + 4pq + 300q + 2pq - 3q^2 = 200p + 300q - 6p^2 + 6pq - 3q^2$

(B) Rewrite the demand equations as

$6p - 4q = 200 - x$

$-2p + 3q = 300 - y$

Apply Cramer's rule: $D = \begin{vmatrix} 6 & -4 \\ -2 & 3 \end{vmatrix} = 10$

$$p = \frac{\begin{vmatrix} 200 - x & -4 \\ 300 - y & 3 \end{vmatrix}}{D} = \frac{1800 - 3x - 4y}{10} = -0.3x - 0.4y + 180$$

$$q = \frac{\begin{vmatrix} 6 & 200 - x \\ -2 & 300 - y \end{vmatrix}}{D} = \frac{2200 - 2x - 6y}{10} = -0.2x - 0.6y + 220$$

Then

$R = xp + yq = x(-0.3x - 0.4y + 180) + y(-0.2x - 0.6y + 220)$

$= -0.3x^2 - 0.4xy + 180x - 0.2xy - 0.6y^2 + 220y$

$= 180x + 220y - 0.3x^2 - 0.6xy - 0.6y^2$

CHAPTER 7 REVIEW

1. $A + B = \begin{bmatrix} 1 & 2 \\ 3 & 1 \end{bmatrix} + \begin{bmatrix} 2 & 1 \\ 1 & 1 \end{bmatrix} = \begin{bmatrix} 1+2 & 2+1 \\ 3+1 & 1+1 \end{bmatrix} = \begin{bmatrix} 3 & 3 \\ 4 & 2 \end{bmatrix}$ (7-1)

2. $B + D$ is not defined (7-1)

3. $A - 2B = \begin{bmatrix} 1 & 2 \\ 3 & 1 \end{bmatrix} - 2\begin{bmatrix} 2 & 1 \\ 1 & 1 \end{bmatrix} = \begin{bmatrix} 1 & 2 \\ 3 & 1 \end{bmatrix} - \begin{bmatrix} 4 & 2 \\ 2 & 2 \end{bmatrix} = \begin{bmatrix} -3 & 0 \\ 1 & -1 \end{bmatrix}$ (7-1)

4. $AB = \begin{bmatrix} 1 & 2 \\ 3 & 1 \end{bmatrix}\begin{bmatrix} 2 & 1 \\ 1 & 1 \end{bmatrix} = \begin{bmatrix} 1\cdot2 + 2\cdot1 & 1\cdot1 + 2\cdot1 \\ 3\cdot2 + 1\cdot1 & 3\cdot1 + 1\cdot1 \end{bmatrix} = \begin{bmatrix} 4 & 3 \\ 7 & 4 \end{bmatrix}$ (7-1)

5. AC is not defined (7-1)   **6.** $AD = \begin{bmatrix} 1 & 2 \\ 3 & 1 \end{bmatrix}\begin{bmatrix} 1 \\ 2 \end{bmatrix} = \begin{bmatrix} 1\cdot1 + 2\cdot2 \\ 3\cdot1 + 1\cdot2 \end{bmatrix} = \begin{bmatrix} 5 \\ 5 \end{bmatrix}$ (7-1)

7. $DC = \begin{bmatrix} 1 \\ 2 \end{bmatrix}[2 \quad 3] = \begin{bmatrix} 1\cdot2 & 1\cdot3 \\ 2\cdot2 & 2\cdot3 \end{bmatrix} = \begin{bmatrix} 2 & 3 \\ 4 & 6 \end{bmatrix}$ (7-1)

8. $CD = \begin{bmatrix} 2 & 3 \end{bmatrix} \begin{bmatrix} 1 \\ 2 \end{bmatrix} = [2 \cdot 1 + 3 \cdot 2] = [8]$ (7-1) 9. $C + D$ is not defined. (7-1)

10. $\begin{bmatrix} 3 & 2 & | & 1 & 0 \\ 4 & 3 & | & 0 & 1 \end{bmatrix}$ $(-1)R_1 + R_2 \rightarrow R_2$

$\sim \begin{bmatrix} 3 & 2 & | & 1 & 0 \\ 1 & 1 & | & -1 & 1 \end{bmatrix}$ $R_1 \leftrightarrow R_2$

$\sim \begin{bmatrix} 1 & 1 & | & -1 & 1 \\ 3 & 2 & | & 1 & 0 \end{bmatrix}$ $(-3)R_1 + R_2 \rightarrow R_2$

$\sim \begin{bmatrix} 1 & 1 & | & -1 & 1 \\ 0 & -1 & | & 4 & -3 \end{bmatrix}$ $R_2 + R_1 \rightarrow R_1$

$\sim \begin{bmatrix} 1 & 0 & | & 3 & -2 \\ 0 & -1 & | & 4 & -3 \end{bmatrix}$ $(-1)R_2 \rightarrow R_2$

$\sim \begin{bmatrix} 1 & 0 & | & 3 & -2 \\ 0 & 1 & | & -4 & 3 \end{bmatrix}$

Hence, $A^{-1} = \begin{bmatrix} 3 & -2 \\ -4 & 3 \end{bmatrix}$

$A^{-1}A = \begin{bmatrix} 3 & -2 \\ -4 & 3 \end{bmatrix}\begin{bmatrix} 3 & 2 \\ 4 & 3 \end{bmatrix} = \begin{bmatrix} 3 \cdot 3 + (-2)4 & 3 \cdot 2 + (-2)3 \\ (-4)3 + 3 \cdot 4 & (-4)2 + 3 \cdot 3 \end{bmatrix} = \begin{bmatrix} 1 & 0 \\ 0 & 1 \end{bmatrix} = I$ (7-2)

11. As a matrix equation the system becomes

$\begin{array}{ccc} A & X & B \end{array}$

$\begin{bmatrix} 3 & 2 \\ 4 & 3 \end{bmatrix}\begin{bmatrix} x_1 \\ x_2 \end{bmatrix} = \begin{bmatrix} k_1 \\ k_2 \end{bmatrix}$

The solution of $AX = B$ is $X = A^{-1}B$.

Using the result of problem 10, we have

$X = \begin{bmatrix} 3 & -2 \\ -4 & 3 \end{bmatrix}\begin{bmatrix} k_1 \\ k_2 \end{bmatrix}$

(A) $\begin{bmatrix} x_1 \\ x_2 \end{bmatrix} = \begin{bmatrix} 3 & -2 \\ -4 & 3 \end{bmatrix}\begin{bmatrix} 3 \\ 5 \end{bmatrix} = \begin{bmatrix} 3 \cdot 3 + (-2)5 \\ (-4)3 + 3 \cdot 5 \end{bmatrix} = \begin{bmatrix} -1 \\ 3 \end{bmatrix}$ $x_1 = -1$, $x_2 = 3$

(B) $\begin{bmatrix} x_1 \\ x_2 \end{bmatrix} = \begin{bmatrix} 3 & -2 \\ -4 & 3 \end{bmatrix}\begin{bmatrix} 7 \\ 10 \end{bmatrix} = \begin{bmatrix} 3 \cdot 7 + (-2)10 \\ (-4)7 + 3 \cdot 10 \end{bmatrix} = \begin{bmatrix} 1 \\ 2 \end{bmatrix}$ $x_1 = 1$, $x_2 = 2$

(C) $\begin{bmatrix} x_1 \\ x_2 \end{bmatrix} = \begin{bmatrix} 3 & -2 \\ -4 & 3 \end{bmatrix}\begin{bmatrix} 4 \\ 2 \end{bmatrix} = \begin{bmatrix} 3 \cdot 4 + (-2)2 \\ (-4)4 + 3 \cdot 2 \end{bmatrix} = \begin{bmatrix} 8 \\ -10 \end{bmatrix}$ $x_1 = 8$, $x_2 = -10$ (7-3)

12. $\begin{vmatrix} 2 & -3 \\ -5 & -1 \end{vmatrix} = 2(-1) - (-5)(-3) = -17$ (7-4)

13. $\begin{vmatrix} 2 & 3 & -4 \\ 0 & 5 & 0 \\ 1 & -4 & -2 \end{vmatrix} = 0 + 5(-1)^{2+2}\begin{vmatrix} 2 & -4 \\ 1 & -2 \end{vmatrix} + 0 = 5(-1)^4[2(-2) - 1(-4)] = 0$ (7-4, 7-5)

14. $D = \begin{vmatrix} 3 & -2 \\ 1 & 3 \end{vmatrix} = 11$

$x = \dfrac{\begin{vmatrix} 8 & -2 \\ -1 & 3 \end{vmatrix}}{D} = \dfrac{22}{11} = 2 \quad y = \dfrac{\begin{vmatrix} 3 & 8 \\ 1 & -1 \end{vmatrix}}{D} = \dfrac{-11}{11} = -1$ 　　　　　 (7-6)

15. (A) Interchanging two rows of a determinant changes the sign of the determinant (Theorem 3, Section 7-5). Hence,

$\begin{vmatrix} g & h & i \\ d & e & f \\ a & b & c \end{vmatrix} = -2$

(B) If every element of a column of a determinant is multiplied by 3, (Theorem 5, Section 7-5), the value of the determinant is 3 times the original. Hence,

$\begin{vmatrix} a & 3b & c \\ d & 3e & f \\ g & 3h & i \end{vmatrix} = 3 \cdot 2 = 6$

(C) Adding a multiple of a column to another column does not change the value of the determinant (Theorem 5, Section 7-5). Hence,

$\begin{vmatrix} a & b & a+b+c \\ d & e & d+e+f \\ g & h & g+h+i \end{vmatrix} = \begin{vmatrix} a & b & b+c \\ d & e & e+f \\ g & h & h+i \end{vmatrix} = \begin{vmatrix} a & b & c \\ d & e & f \\ g & h & i \end{vmatrix} = 2$ 　　　 (7-5)

16. $A + D$ is not defined. 　　　　　 (7-1)

17. $DA = \begin{bmatrix} 3 & -2 & 1 \\ -1 & 1 & 2 \end{bmatrix}\begin{bmatrix} 2 & -2 \\ 1 & 0 \\ 3 & 2 \end{bmatrix} = \begin{bmatrix} 3 \cdot 2 + (-2)1 + 1 \cdot 3 & 3(-2) + (-2)0 + 1 \cdot 2 \\ (-1)2 + 1 \cdot 1 + 2 \cdot 3 & (-1)(-2) + 1 \cdot 0 + 2 \cdot 2 \end{bmatrix} = \begin{bmatrix} 7 & -4 \\ 5 & 6 \end{bmatrix}$

$E + DA = \begin{bmatrix} 3 & -4 \\ -1 & 0 \end{bmatrix} + \begin{bmatrix} 7 & -4 \\ 5 & 6 \end{bmatrix} = \begin{bmatrix} 10 & -8 \\ 4 & 6 \end{bmatrix}$ 　　　　　 (7-1)

18. From problem 17, $DA = \begin{bmatrix} 7 & -4 \\ 5 & 6 \end{bmatrix}$

$DA - 3E = \begin{bmatrix} 7 & -4 \\ 5 & 6 \end{bmatrix} - 3\begin{bmatrix} 3 & -4 \\ -1 & 0 \end{bmatrix} = \begin{bmatrix} 7 & -4 \\ 5 & 6 \end{bmatrix} - \begin{bmatrix} 9 & -12 \\ -3 & 0 \end{bmatrix} = \begin{bmatrix} -2 & 8 \\ 8 & 6 \end{bmatrix}$ 　 (7-1)

19. CD is not defined. 　　　　　 (7-1)

20. $CB = \begin{bmatrix} 2 & 1 & 3 \end{bmatrix}\begin{bmatrix} -1 \\ 2 \\ 3 \end{bmatrix} = [2(-1) + 1 \cdot 2 + 3 \cdot 3] = [9]$ 　　　　　 (7-1)

21. $AD = \begin{bmatrix} 2 & -2 \\ 1 & 0 \\ 3 & 2 \end{bmatrix}\begin{bmatrix} 3 & -2 & 1 \\ -1 & 1 & 2 \end{bmatrix} = \begin{bmatrix} 2 \cdot 3 + (-2)(-1) & 2(-2) + (-2)1 & 2 \cdot 1 + (-2)2 \\ 1 \cdot 3 + 0(-1) & 1(-2) + 0 \cdot 1 & 1 \cdot 1 + 0 \cdot 2 \\ 3 \cdot 3 + 2(-1) & 3(-2) + 2 \cdot 1 & 3 \cdot 1 + 2 \cdot 2 \end{bmatrix}$

$= \begin{bmatrix} 8 & -6 & -2 \\ 3 & -2 & 1 \\ 7 & -4 & 7 \end{bmatrix}$

$$BC = \begin{bmatrix} -1 \\ 2 \\ 3 \end{bmatrix} [2 \quad 1 \quad 3] = \begin{bmatrix} -1\cdot 2 & -1\cdot 1 & -1\cdot 3 \\ 2\cdot 2 & 2\cdot 1 & 2\cdot 3 \\ 3\cdot 2 & 3\cdot 1 & 3\cdot 3 \end{bmatrix} = \begin{bmatrix} -2 & -1 & -3 \\ 4 & 2 & 6 \\ 6 & 3 & 9 \end{bmatrix}$$

$$AD - BC = \begin{bmatrix} 8 & -6 & -2 \\ 3 & -2 & 1 \\ 7 & -4 & 7 \end{bmatrix} - \begin{bmatrix} -2 & -1 & -3 \\ 4 & 2 & 6 \\ 6 & 3 & 9 \end{bmatrix} = \begin{bmatrix} 10 & -5 & 1 \\ -1 & -4 & -5 \\ 1 & -7 & -2 \end{bmatrix} \qquad (7\text{-}1)$$

22. $\begin{bmatrix} 1 & 2 & 3 \\ 2 & 3 & 4 \\ 1 & 2 & 1 \end{bmatrix} \left|\begin{matrix} 1 & 0 & 0 \\ 0 & 1 & 0 \\ 0 & 0 & 1 \end{matrix}\right.$ $\begin{matrix} (-2)R_1 + R_2 \to R_2 \\ (-1)R_1 + R_3 \to R_3 \end{matrix}$

$\sim \begin{bmatrix} 1 & 2 & 3 \\ 0 & -1 & -2 \\ 0 & 0 & -2 \end{bmatrix} \left|\begin{matrix} 1 & 0 & 0 \\ -2 & 1 & 0 \\ -1 & 0 & 1 \end{matrix}\right.$ $2R_2 + R_1 \to R_1$

$\sim \begin{bmatrix} 1 & 0 & -1 \\ 0 & -1 & -2 \\ 0 & 0 & -2 \end{bmatrix} \left|\begin{matrix} -3 & 2 & 0 \\ -2 & 1 & 0 \\ -1 & 0 & 1 \end{matrix}\right.$ $\begin{matrix} (-\frac{1}{2})R_3 + R_1 \to R_1 \\ (-1)R_3 + R_2 \to R_2 \end{matrix}$

$\sim \begin{bmatrix} 1 & 0 & 0 \\ 0 & -1 & 0 \\ 0 & 0 & -2 \end{bmatrix} \left|\begin{matrix} -\frac{5}{2} & 2 & -\frac{1}{2} \\ -1 & 1 & -1 \\ -1 & 0 & 1 \end{matrix}\right.$ $\begin{matrix} (-1)R_2 \to R_2 \\ (-\frac{1}{2})R_3 \to R_3 \end{matrix}$

$\sim \begin{bmatrix} 1 & 0 & 0 \\ 0 & 1 & 0 \\ 0 & 0 & 1 \end{bmatrix} \left|\begin{matrix} -\frac{5}{2} & 2 & -\frac{1}{2} \\ 1 & -1 & 1 \\ \frac{1}{2} & 0 & -\frac{1}{2} \end{matrix}\right.$

Hence

$$A^{-1} = \begin{bmatrix} -\frac{5}{2} & 2 & -\frac{1}{2} \\ 1 & -1 & 1 \\ \frac{1}{2} & 0 & -\frac{1}{2} \end{bmatrix} \text{ or } \frac{1}{2}\begin{bmatrix} -5 & 4 & -1 \\ 2 & -2 & 2 \\ 1 & 0 & -1 \end{bmatrix}$$

$$A^{-1}A = \begin{bmatrix} -\frac{5}{2} & 2 & -\frac{1}{2} \\ 1 & -1 & 1 \\ \frac{1}{2} & 0 & -\frac{1}{2} \end{bmatrix}\begin{bmatrix} 1 & 2 & 3 \\ 2 & 3 & 4 \\ 1 & 2 & 1 \end{bmatrix}$$

$$\begin{bmatrix} (-\frac{5}{2})1 + 2\cdot 2 + (-\frac{1}{2})1 & (-\frac{5}{2})2 + 2\cdot 3 + (-\frac{1}{2})2 & (-\frac{5}{2})3 + 2\cdot 4 + (-\frac{1}{2})1 \\ 1\cdot 1 + (-1)2 + 1\cdot 1 & 1\cdot 2 + (-1)3 + 1\cdot 2 & 1\cdot 3 + (-1)4 + 1\cdot 1 \\ (\frac{1}{2})1 + 0\cdot 2 + (-\frac{1}{2})1 & (\frac{1}{2})2 + 0\cdot 3 + (-\frac{1}{2})2 & (\frac{1}{2})3 + 0\cdot 4 + (-\frac{1}{2})1 \end{bmatrix}$$

$$= \begin{bmatrix} 1 & 0 & 0 \\ 0 & 1 & 0 \\ 0 & 0 & 1 \end{bmatrix} = I \qquad (7\text{-}2)$$

$$\begin{matrix} \quad A \quad\quad\quad X \quad\quad B \end{matrix}$$

23. $\begin{bmatrix} 1 & 2 & 3 \\ 2 & 3 & 4 \\ 1 & 2 & 1 \end{bmatrix}\begin{bmatrix} x_1 \\ x_2 \\ x_3 \end{bmatrix} = \begin{bmatrix} k_1 \\ k_2 \\ k_3 \end{bmatrix}$

The solution to $AX = B$ is $X = A^{-1}B$.

Applying the A^{-1} found in problem 22, we have

(A) $B = \begin{bmatrix} 1 \\ 3 \\ 3 \end{bmatrix}$ $X = \begin{bmatrix} x_1 \\ x_2 \\ x_3 \end{bmatrix} = \begin{bmatrix} -\frac{5}{2} & 2 & -\frac{1}{2} \\ 1 & -1 & 1 \\ \frac{1}{2} & 0 & -\frac{1}{2} \end{bmatrix} \begin{bmatrix} 1 \\ 3 \\ 3 \end{bmatrix} = \begin{bmatrix} 2 \\ 1 \\ -1 \end{bmatrix}$ $x_1 = 2, \; x_2 = 1, \; x_3 = -1$

(B) $B = \begin{bmatrix} 0 \\ 0 \\ -2 \end{bmatrix}$ $X = \begin{bmatrix} x_1 \\ x_2 \\ x_3 \end{bmatrix} = \begin{bmatrix} -\frac{5}{2} & 2 & -\frac{1}{2} \\ 1 & -1 & 1 \\ \frac{1}{2} & 0 & -\frac{1}{2} \end{bmatrix} \begin{bmatrix} 0 \\ 0 \\ -2 \end{bmatrix} = \begin{bmatrix} 1 \\ -2 \\ 1 \end{bmatrix}$ $x_1 = 1, \; x_2 = -2, \; x_3 = 1$

(C) $B = \begin{bmatrix} -3 \\ -4 \\ 1 \end{bmatrix}$ $X = \begin{bmatrix} x_1 \\ x_2 \\ x_3 \end{bmatrix} = \begin{bmatrix} -\frac{5}{2} & 2 & -\frac{1}{2} \\ 1 & -1 & 1 \\ \frac{1}{2} & 0 & -\frac{1}{2} \end{bmatrix} \begin{bmatrix} -3 \\ -4 \\ 1 \end{bmatrix} = \begin{bmatrix} -1 \\ 2 \\ -2 \end{bmatrix}$ $x_1 = -1, \; x_2 = 2, \; x_3 = -2$

(7-3)

24. $\begin{vmatrix} -\frac{1}{4} & \frac{3}{2} \\ \frac{1}{2} & \frac{2}{3} \end{vmatrix} = (-\frac{1}{4})(\frac{2}{3}) - (\frac{1}{2})(\frac{3}{2}) = -\frac{1}{6} - \frac{3}{4} = -\frac{11}{12}$ (7-4)

25. $\begin{vmatrix} 2 & -1 & 1 \\ -3 & 5 & 2 \\ 1 & -2 & 4 \end{vmatrix} = \begin{vmatrix} 0 & 0 & 1 \\ -7 & 7 & 2 \\ -7 & 2 & 4 \end{vmatrix}$ $\begin{array}{l} (-2)C_3 + C_1 \rightarrow C_1 \\ C_3 + C_2 \rightarrow C_2 \end{array}$

$= 0 + 0 + 1(-1)^{1+3} \begin{vmatrix} -7 & 7 \\ -7 & 2 \end{vmatrix} = (-1)^4[(-7)2 - (-7)7] = 35$ (7-4, 7-5)

26. $y = \dfrac{\begin{vmatrix} 1 & -6 & 1 \\ 0 & 4 & -1 \\ 2 & 2 & 1 \end{vmatrix}}{\begin{vmatrix} 1 & -2 & 1 \\ 0 & 1 & -1 \\ 2 & 2 & 1 \end{vmatrix}} = \dfrac{\begin{vmatrix} 1 & -6 & 1 \\ 0 & 4 & -1 \\ 0 & 14 & -1 \end{vmatrix}}{\begin{vmatrix} 1 & -2 & -1 \\ 0 & 1 & 0 \\ 2 & 2 & 3 \end{vmatrix}} = \dfrac{1(-1)^{1+1}\begin{vmatrix} 4 & -1 \\ 14 & -1 \end{vmatrix}}{1(-1)^{2+2}\begin{vmatrix} 1 & -1 \\ 2 & 3 \end{vmatrix}} = \dfrac{(-1)^2[4(-1) - 14(-1)]}{(-1)^4[1\cdot 3 - 2(-1)]}$

$= \dfrac{10}{5} = 2$ (7-6)

27. (A) If the coefficient matrix has an inverse, then the system can be written as $AX = B$ and its solution can be written $X = A^{-1}B$. Thus the system has one solution.

(B) If the coefficient matrix does not have an inverse, then the system can be solved by Gauss-Jordan elimination, but it will not have exactly one solution. The other possibilities are that the system has no solution or an infinite number of solutions, and either possibility may occur. (7-3)

28. If we assume that A is a non-zero matrix with an inverse A^{-1}, then if $A^2 = 0$ we can write $A^{-1}A^2 = A^{-1}0$ or $A^{-1}AA = 0$ or $IA = 0$ or $A = 0$. But A was assumed non-zero, so there is a contradiction. Hence A^{-1} cannot exist for such a matrix. (7-3)

29.

$$\begin{array}{ll} AX - B = CX & \\ AX - B + B - CX = CX - CX + B & \text{Addition property} \\ AX + 0 - CX = 0 + B & M + (-M) = 0 \\ AX - CX = B & M + 0 = M \\ (A - C)X = B & \text{Right distributive property} \\ & \text{(applied to subtraction of matrices; see problem 21,} \\ & \text{Exercise 7-3)} \\ (A - C)^{-1}[(A - C)X] = (A - C)^{-1}B & \text{Left multiplication property} \\ [(A - C)^{-1}(A - C)]X = (A - C)^{-1}B & \text{Associative property} \\ IX = (A - C)^{-1}B & A^{-1}A = I \\ X = (A - C)^{-1}B & IX = X \end{array}$$

> **Common Errors:** $(A - C)^{-1}B \neq B(A - C)^{-1}$
> $(A - C)^{-1} \neq A^{-1} - C^{-1}$

(7-3)

30. $\begin{bmatrix} 4 & 5 & 6 & | & 1 & 0 & 0 \\ 4 & 5 & -6 & | & 0 & 1 & 0 \\ 1 & 1 & 1 & | & 0 & 0 & 1 \end{bmatrix}$ $\quad (-1)R_1 + R_2 \to R_2$

$\sim \begin{bmatrix} 4 & 5 & 6 & | & 1 & 0 & 0 \\ 0 & 0 & -12 & | & -1 & 1 & 0 \\ 1 & 1 & 1 & | & 0 & 0 & 1 \end{bmatrix}$ $\quad R_1 \leftrightarrow R_3$

$\sim \begin{bmatrix} 1 & 1 & 1 & | & 0 & 0 & 1 \\ 0 & 0 & -12 & | & -1 & 1 & 0 \\ 4 & 5 & 6 & | & 1 & 0 & 0 \end{bmatrix}$ $\quad (-4)R_1 + R_3 \to R_3$

$\sim \begin{bmatrix} 1 & 1 & 1 & | & 0 & 0 & 1 \\ 0 & 0 & -12 & | & -1 & 1 & 0 \\ 0 & 1 & 2 & | & 1 & 0 & -4 \end{bmatrix}$ $\quad R_2 \leftrightarrow R_3$

$\sim \begin{bmatrix} 1 & 1 & 1 & | & 0 & 0 & 1 \\ 0 & 1 & 2 & | & 1 & 0 & -4 \\ 0 & 0 & -12 & | & -1 & 1 & 0 \end{bmatrix}$ $\quad (-1)R_2 + R_1 \to R_1$

$\sim \begin{bmatrix} 1 & 0 & -1 & | & -1 & 0 & 5 \\ 0 & 1 & 2 & | & 1 & 0 & -4 \\ 0 & 0 & -12 & | & -1 & 1 & 0 \end{bmatrix}$ $\quad -\frac{1}{12}R_3 \to R_3$

$\sim \begin{bmatrix} 1 & 0 & -1 & | & -1 & 0 & 5 \\ 0 & 1 & 2 & | & 1 & 0 & -4 \\ 0 & 0 & 1 & | & \frac{1}{12} & -\frac{1}{12} & 0 \end{bmatrix}$ $\quad \begin{matrix} R_3 + R_1 \to R_1 \\ (-2)R_3 + R_2 \to R_2 \end{matrix}$

$\sim \begin{bmatrix} 1 & 0 & 0 & | & -\frac{11}{12} & -\frac{1}{12} & 5 \\ 0 & 1 & 0 & | & \frac{10}{12} & \frac{2}{12} & -4 \\ 0 & 0 & 1 & | & \frac{1}{12} & -\frac{1}{12} & 0 \end{bmatrix}$

Hence

$$A^{-1} = \begin{bmatrix} -\frac{11}{12} & -\frac{1}{12} & 5 \\ \frac{10}{12} & \frac{2}{12} & -4 \\ \frac{1}{12} & -\frac{1}{12} & 0 \end{bmatrix} \text{ or } \frac{1}{12}\begin{bmatrix} -11 & -1 & 60 \\ 10 & 2 & -48 \\ 1 & -1 & 0 \end{bmatrix}$$

$$A^{-1}A = \frac{1}{12}\begin{bmatrix} -11 & -1 & 60 \\ 10 & 2 & -48 \\ 1 & -1 & 0 \end{bmatrix}\begin{bmatrix} 4 & 5 & 6 \\ 4 & 5 & -6 \\ 1 & 1 & 1 \end{bmatrix}$$

$$= \frac{1}{12}\begin{bmatrix} (-11)4 + (-1)4 + 60\cdot 1 & (-11)5 + (-1)5 + 60\cdot 1 & (-11)6 + (-1)(-6) + 60\cdot 1 \\ 10\cdot 4 + 2\cdot 4 + (-48)1 & 10\cdot 5 + 2\cdot 5 + (-48)1 & 10\cdot 6 + 2(-6) + (-48)1 \\ 1\cdot 4 + (-1)4 + 0\cdot 1 & 1\cdot 5 + (-1)5 + 0\cdot 1 & 1\cdot 6 + (-1)(-6) + 0\cdot 1 \end{bmatrix}$$

$$= \frac{1}{12}\begin{bmatrix} 12 & 0 & 0 \\ 0 & 12 & 0 \\ 0 & 0 & 12 \end{bmatrix} = \begin{bmatrix} 1 & 0 & 0 \\ 0 & 1 & 0 \\ 0 & 0 & 1 \end{bmatrix} = I \qquad (7\text{-}2)$$

31. Multiplying the first two equations by 100, the system becomes

$$4x_1 + 5x_2 + 6x_3 = 36{,}000$$
$$4x_1 + 5x_2 - 6x_3 = 12{,}000$$
$$x_1 + x_2 + x_3 = 7{,}000$$

As a matrix equation, we have

$$\overset{A}{\begin{bmatrix} 4 & 5 & 6 \\ 4 & 5 & -6 \\ 1 & 1 & 1 \end{bmatrix}} \overset{X}{\begin{bmatrix} x_1 \\ x_2 \\ x_3 \end{bmatrix}} = \overset{B}{\begin{bmatrix} 36,000 \\ 12,000 \\ 7,000 \end{bmatrix}}$$

The solution to $AX = B$ is $X = A^{-1}B$. Using A^{-1} from problem 30, we have

$$X = \begin{bmatrix} x_1 \\ x_2 \\ x_3 \end{bmatrix} = \frac{1}{12} \begin{bmatrix} -11 & -1 & 60 \\ 10 & 2 & -48 \\ 1 & -1 & 0 \end{bmatrix} \begin{bmatrix} 36,000 \\ 12,000 \\ 7,000 \end{bmatrix}$$

$$= \frac{1}{12} \begin{bmatrix} (-11)(36,000) + (-1)(12,000) + (60)(7,000) \\ (10)(36,000) + (2)(12,000) + (-48)(7,000) \\ 1(36,000) + (-1)(12,000) + (0)(7,000) \end{bmatrix} = \frac{1}{12} \begin{bmatrix} 12,000 \\ 48,000 \\ 24,000 \end{bmatrix} = \begin{bmatrix} 1,000 \\ 4,000 \\ 2,000 \end{bmatrix}$$

Hence, $x_1 = 1,000$, $x_2 = 4,000$, $x_3 = 2,000$ 　　　　　　　　　　　　　　(7-3)

32. $\begin{vmatrix} -1 & 4 & 1 & 1 \\ 5 & -1 & 2 & -1 \\ 2 & -1 & 0 & 3 \\ -3 & 3 & 0 & 3 \end{vmatrix} = \begin{vmatrix} -1 & 4 & 1 & 1 \\ 7 & -9 & 0 & -3 \\ 2 & -1 & 0 & 3 \\ -3 & 3 & 0 & 3 \end{vmatrix}$ 　　$(-2)R_1 + R_2 \to R_2$

$$= 1(-1)^{1+3} \begin{vmatrix} 7 & -9 & -3 \\ 2 & -1 & 3 \\ -3 & 3 & 3 \end{vmatrix} + 0 + 0 + 0$$

$$= \begin{vmatrix} 7 & -9 & -3 \\ 2 & -1 & 3 \\ -3 & 3 & 3 \end{vmatrix} \quad \begin{matrix} R_1 + R_2 \to R_2 \\ R_1 + R_3 \to R_3 \end{matrix}$$

$$= \begin{vmatrix} 7 & -9 & -3 \\ 9 & -10 & 0 \\ 4 & -6 & 0 \end{vmatrix} = (-3)(-1)^{1+3} \begin{vmatrix} 9 & -10 \\ 4 & -6 \end{vmatrix} + 0 + 0$$

$$= (-3)(-1)^4 [9(-6) - 4(-10)]$$
$$= (-3)(-14) = 42$$ 　　　　　　　　　　　　　　　　　　　　(7-5)

33. $\begin{vmatrix} u + kv & v \\ w + kx & x \end{vmatrix} = (u + kv)x - (w + kx)v = ux + kvx - wv - kvx = ux - wv$

$$= \begin{vmatrix} u & v \\ w & x \end{vmatrix}$$ 　　　　　　　　　　　　　　　　　　　　　　　　(7-5)

34. The statements: $(1, 2)$ satisfies the equation $\begin{vmatrix} x & y & 1 \\ 1 & 2 & 1 \\ -1 & 5 & 1 \end{vmatrix} = 0$ and $(-1, 5)$

satisfies the equation $\begin{vmatrix} x & y & 1 \\ 1 & 2 & 1 \\ -1 & 5 & 1 \end{vmatrix} = 0$ are equivalent to the statements

$\begin{vmatrix} 1 & 2 & 1 \\ 1 & 2 & 1 \\ -1 & 5 & 1 \end{vmatrix} = 0$ and $\begin{vmatrix} -1 & 5 & 1 \\ 1 & 2 & 1 \\ -1 & 5 & 1 \end{vmatrix} = 0$. The latter statements are true by

Theorem 4, Section 7-5. All other points on the line through the given points will also satisfy the equation. 　　　　　　　　　　　　　　　　　(7-5)

35. Let x_1 = number of tons at Big Bend

x_2 = number of tons at Saw Pit

Then

$0.05x_1 + 0.03x_2$ = number of tons of nickel at both mines = k_1

$0.07x_1 + 0.04x_2$ = number of tons of copper at both mines = k_2

We solve

$0.05x_1 + 0.03x_2 = k_1$

$0.07x_1 + 0.04x_4 = k_2,$

For arbitrary k_1 and k_2, by writing the system as a matrix equation.

$$\begin{matrix} A & X & B \end{matrix}$$
$$\begin{bmatrix} 0.05 & 0.03 \\ 0.07 & 0.04 \end{bmatrix} \begin{bmatrix} x_1 \\ x_2 \end{bmatrix} = \begin{bmatrix} k_1 \\ k_2 \end{bmatrix}$$

If A^{-1} exists, then $X = A^{-1}B$. To find A^{-1}, we perform row operations on

$$\begin{bmatrix} 0.05 & 0.03 & | & 1 & 0 \\ 0.07 & 0.04 & | & 0 & 1 \end{bmatrix} \quad 20R_1 \rightarrow R_1$$

$$\sim \begin{bmatrix} 1 & 0.6 & | & 20 & 0 \\ 0.07 & 0.04 & | & 0 & 1 \end{bmatrix} \quad (-0.07)R_1 + R_2 \rightarrow R_2$$

$$\sim \begin{bmatrix} 1 & 0.6 & | & 20 & 0 \\ 0 & -0.002 & | & -1.4 & 1 \end{bmatrix} \quad -500R_2 \rightarrow R_2$$

$$\sim \begin{bmatrix} 1 & 0.6 & | & 20 & 0 \\ 0 & 1 & | & 700 & -500 \end{bmatrix} \quad (-0.6)R_2 + R_1 \rightarrow R_1$$

$$\sim \begin{bmatrix} 1 & 0 & | & -400 & 300 \\ 0 & 1 & | & 700 & -500 \end{bmatrix}$$

Hence $A^{-1} = \begin{bmatrix} -400 & 300 \\ 700 & -500 \end{bmatrix}$

Check: $A^{-1}A = \begin{bmatrix} -400 & 300 \\ 700 & -500 \end{bmatrix} \begin{bmatrix} 0.05 & 0.03 \\ 0.07 & 0.04 \end{bmatrix} = \begin{bmatrix} 1 & 0 \\ 0 & 1 \end{bmatrix}$

We can now solve the system as:

$$\begin{matrix} X & A^{-1} & B \end{matrix}$$
$$\begin{bmatrix} x_1 \\ x_2 \end{bmatrix} = \begin{bmatrix} -400 & 300 \\ 700 & -500 \end{bmatrix} \begin{bmatrix} k_1 \\ k_2 \end{bmatrix}$$

(A) If $k_1 = 3.6$, $k_2 = 5$,

$$\begin{bmatrix} x_1 \\ x_2 \end{bmatrix} = \begin{bmatrix} -400 & 300 \\ 700 & -500 \end{bmatrix} \begin{bmatrix} 3.6 \\ 5 \end{bmatrix} = \begin{bmatrix} 60 \\ 20 \end{bmatrix}$$

60 tons of ore must be produced at Big Bend, 20 tons of ore at Saw Pit.

(B) If $k_1 = 3$, $k_2 = 4.1$,

$$\begin{bmatrix} x_1 \\ x_2 \end{bmatrix} = \begin{bmatrix} -400 & 300 \\ 700 & -500 \end{bmatrix} \begin{bmatrix} 3 \\ 4.1 \end{bmatrix} = \begin{bmatrix} 30 \\ 50 \end{bmatrix}$$

30 tons of ore must be produced at Big Bend, 50 tons of ore at Saw Pit.

(C) If $k_1 = 3.2$, $k_2 = 4.4$,

$$\begin{bmatrix} x_1 \\ x_2 \end{bmatrix} = \begin{bmatrix} -400 & 300 \\ 700 & -500 \end{bmatrix} \begin{bmatrix} 3.2 \\ 4.4 \end{bmatrix} = \begin{bmatrix} 40 \\ 40 \end{bmatrix}$$

40 tons of ore must be produced at Big Bend, 40 tons of ore at Saw Pit. (7-3)

36. (A) The labor cost of producing one printer stand at the South Carolina plant is the product of the stand row of L with South Carolina column of H.

$$[0.9 \quad 1.8 \quad 0.6]\begin{bmatrix} 10.00 \\ 8.50 \\ 4.50 \end{bmatrix} = 27 \text{ dollars}$$

(B) The matrix HL has no obvious meaning, but the matrix LH represents the total labor costs for each item at each plant.

(C) $LH = \begin{bmatrix} 1.7 & 2.4 & 0.8 \\ 0.9 & 1.8 & 0.6 \end{bmatrix}\begin{bmatrix} 11.50 & 10.00 \\ 9.50 & 8.50 \\ 5.00 & 4.50 \end{bmatrix}$

$= \begin{bmatrix} (1.7)(11.50) + (2.4)(9.50) + (0.8)(5.00) & (1.7)(10.00) + (2.4)(8.50) + (0.8)(4.50) \\ (0.9)(11.50) + (1.8)(9.50) + (0.6)(5.00) & (0.9)(10.00) + (1.8)(8.50) + (0.6)(4.50) \end{bmatrix}$

$$\begin{array}{cc} \text{N.C.} & \text{S.C.} \end{array}$$
$$\begin{bmatrix} \$46.35 & \$41.00 \\ \$30.45 & \$27.00 \end{bmatrix}\begin{array}{l} \text{Desk} \\ \text{Stands} \end{array}$$

$(7\text{-}1)$

37. (A) The average monthly production for the months of January and February is represented by the matrix $\frac{1}{2}(J + F)$

$$\frac{1}{2}(J + F) = \frac{1}{2}\left(\begin{bmatrix} 1,500 & 1,650 \\ 850 & 700 \end{bmatrix} + \begin{bmatrix} 1,700 & 1,810 \\ 930 & 740 \end{bmatrix}\right) = \frac{1}{2}\begin{bmatrix} 3,200 & 3,460 \\ 1,780 & 1,440 \end{bmatrix}$$

$$\begin{array}{cc} \text{N.C.} & \text{S.C.} \end{array}$$
$$= \begin{bmatrix} 1,600 & 1,730 \\ 890 & 720 \end{bmatrix}\begin{array}{l} \text{Desks} \\ \text{Stands} \end{array}$$

(B) The increase in production from January to February is represented by the matrix $F - J$.

$$F - J = \begin{bmatrix} 1,700 & 1,810 \\ 930 & 740 \end{bmatrix} - \begin{bmatrix} 1,500 & 1,650 \\ 850 & 700 \end{bmatrix}$$

$$\begin{array}{cc} \text{N.C.} & \text{S.C.} \end{array}$$
$$= \begin{bmatrix} 200 & 160 \\ 80 & 40 \end{bmatrix}\begin{array}{l} \text{Desks} \\ \text{Stands} \end{array}$$

(C) $J\begin{bmatrix} 1 \\ 1 \end{bmatrix} = \begin{bmatrix} 1,500 & 1,650 \\ 850 & 700 \end{bmatrix}\begin{bmatrix} 1 \\ 1 \end{bmatrix} = \begin{bmatrix} 3,150 \\ 1,550 \end{bmatrix}\begin{array}{l} \text{Desks} \\ \text{Stands} \end{array}$

This matrix represents the total production of each item in January.

$(7\text{-}1)$

38. The inverse of matrix B is calculated to be

$$B^{-1} = \begin{bmatrix} 1 & 1 & -1 \\ 0 & -1 & 1 \\ -1 & 0 & 1 \end{bmatrix}$$

Putting the coded message into matrix form and multiplying by B^{-1} yields

$$\begin{bmatrix} 1 & 1 & -1 \\ 0 & -1 & 1 \\ -1 & 0 & 1 \end{bmatrix}\begin{bmatrix} 25 & 24 & 21 & 41 & 21 & 52 \\ 8 & 25 & 41 & 30 & 32 & 52 \\ 26 & 33 & 48 & 50 & 41 & 79 \end{bmatrix} = \begin{bmatrix} 7 & 16 & 14 & 21 & 12 & 25 \\ 18 & 8 & 7 & 20 & 9 & 27 \\ 1 & 9 & 27 & 9 & 20 & 27 \end{bmatrix}$$

This decodes to

7 18 1 16 8 9 14 7 27 21 20 9 12 9 20 25 27 27
G R A P H I N G   U T I L I T Y

$(7\text{-}2)$

CUMULATIVE REVIEW EXERCISE (Chapters 6 and 7)

1. We choose elimination by addition. We multiply the top equation by 3, the bottom by 5, and add.

$$9x - 15y = 33$$
$$\underline{10x + 15y = 5}$$
$$19x = 38$$
$$x = 2$$

Substituting $x = 2$ in the bottom equation, we have

$$2(2) + 3y = 1$$
$$4 + 3y = 1$$
$$3y = -3$$
$$y = -1$$

Solution: $(2, -1)$ $(6\text{-}1)$

2.

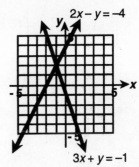

$2x - y = -4$

$3x + y = -1$

$(-1, 2)$ $(6\text{-}1)$

3. $x^2 + y^2 = 2$
$2x - y = 1$

We choose substitution, solving the first-degree equation for y in terms of x, then substituting into the second-degree equation.

$$2x - y = 1$$
$$-y = 1 - 2x$$
$$y = 2x - 1$$
$$x^2 + (2x - 1)^2 = 2$$
$$x^2 + 4x^2 - 4x + 1 = 2$$
$$5x^2 - 4x - 1 = 0$$
$$(5x + 1)(x - 1) = 0$$
$$x = -\frac{1}{5}, \ 1$$

For $x = -\dfrac{1}{5}$ For $x = 1$

$y = 2\left(-\dfrac{1}{5}\right) - 1$ $y = 2(1) - 1$

$= -\dfrac{7}{5}$ $= 1$

Solutions: $\left(-\dfrac{1}{5}, \ -\dfrac{7}{5}\right)$, $(1, 1)$ $(6\text{-}3)$

The checking steps are omitted for lack of space.

4.

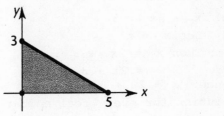

$(6\text{-}4)$

5.

Corner Point (x, y)	Objective Function $z = 2x + 3y$	
$(0, 4)$	12	
$(5, 0)$	10	Minimum value
$(6, 7)$	33	Maximum value
$(0, 10)$	30	

The minimum value of z on S is 10 at $(5, 0)$.
The maximum value of z on S is 33 at $(6, 7)$. $(6\text{-}5)$

6. (A) 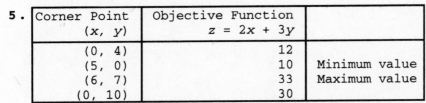 $M - 2N = \begin{bmatrix} 2 & 1 \\ 1 & -3 \end{bmatrix} - 2\begin{bmatrix} 1 & 2 \\ -1 & 3 \end{bmatrix} = \begin{bmatrix} 2 & 1 \\ 1 & -3 \end{bmatrix} - \begin{bmatrix} 2 & 4 \\ -2 & 6 \end{bmatrix} = \begin{bmatrix} 0 & -3 \\ 3 & -9 \end{bmatrix}$

(B) $P + Q$ is not defined

(C) $PQ = [1 \quad 2]\begin{bmatrix} -1 \\ 2 \end{bmatrix} = [1(-1) + 2 \cdot 2] = [3]$

(D) $MN = \begin{bmatrix} 2 & 1 \\ 1 & -3 \end{bmatrix}\begin{bmatrix} 1 & 2 \\ -1 & 3 \end{bmatrix} = \begin{bmatrix} 2 \cdot 1 + 1(-1) & 2 \cdot 2 + 1 \cdot 3 \\ 1 \cdot 1 + (-3)(-1) & 1 \cdot 2 + (-3)3 \end{bmatrix} = \begin{bmatrix} 1 & 7 \\ 4 & -7 \end{bmatrix}$

(E) $PN = [1 \quad 2]\begin{bmatrix} 1 & 2 \\ -1 & 3 \end{bmatrix} = [1 \cdot 1 + 2(-1) \quad 1 \cdot 2 + 2 \cdot 3] = [-1 \quad 8]$

(F) QM is not defined

$(7-1)$

7. $\begin{vmatrix} 0 & 2 & 0 \\ 1 & 3 & 2 \\ -1 & 4 & 3 \end{vmatrix} = 0 + 2(-1)^{1+2}\begin{vmatrix} 1 & 2 \\ -1 & 3 \end{vmatrix} + 0 = 2(-1)^3[1 \cdot 3 - (-1)2] = -10$ $\qquad (7-4)$

8. (A) $x_1 = 3$
$\qquad\quad x_2 = -4$

(B) $x_1 - 2x_2 = 3$
$\qquad$ Let $x_2 = t$
$\qquad$ Then $x_1 = 2x_2 + 3$
$\qquad\qquad\qquad = 2t + 3$
Hence, $x_1 = 2t + 3$, $x_2 = t$ is a
solution for every real number t.

(C) $x_1 - 2x_2 = 3$
$\qquad\quad 0x_1 + 0x_2 = 1$
No solution. $\qquad\qquad (6-1)$

9. (A) $\begin{bmatrix} 1 & 1 & | & 3 \\ -1 & 1 & | & 5 \end{bmatrix}$

(B) $\begin{bmatrix} 1 & 1 & | & 3 \\ -1 & 1 & | & 5 \end{bmatrix}$ $R_1 + R_2 \rightarrow R_2$

$\sim \begin{bmatrix} 1 & 1 & | & 3 \\ 0 & 2 & | & 8 \end{bmatrix}$ $\frac{1}{2}R_2 \rightarrow R_2$

$\sim \begin{bmatrix} 1 & 1 & | & 3 \\ 0 & 1 & | & 4 \end{bmatrix}$ $(-1)R_2 + R_1 \rightarrow R_1$

$\sim \begin{bmatrix} 1 & 0 & | & -1 \\ 0 & 1 & | & 4 \end{bmatrix}$

(C) Solution: $x_1 = -1$, $x_2 = 4$
$\qquad\qquad\qquad\qquad (6-1, \ 6-2)$

10. (A) As a matrix equation the system becomes
$\qquad\quad A \qquad X \qquad B$
$\begin{bmatrix} 1 & -3 \\ 2 & -5 \end{bmatrix}\begin{bmatrix} x_1 \\ x_2 \end{bmatrix} = \begin{bmatrix} k_1 \\ k_2 \end{bmatrix}$

(B) To find A^{-1} we perform row operations on
$\begin{bmatrix} 1 & -3 & | & 1 & 0 \\ 2 & -5 & | & 0 & 1 \end{bmatrix}$ $(-2)R_1 + R_2 \rightarrow R_2$

$\sim \begin{bmatrix} 1 & -3 & | & 1 & 0 \\ 0 & 1 & | & -2 & 1 \end{bmatrix}$ $3R_2 + R_1 \rightarrow R_1$

$\sim \begin{bmatrix} 1 & 0 & | & -5 & 3 \\ 0 & 1 & | & -2 & 1 \end{bmatrix}$

Hence $A^{-1} = \begin{bmatrix} -5 & 3 \\ -2 & 1 \end{bmatrix}$

Check: $A^{-1}A = \begin{bmatrix} -5 & 3 \\ -2 & 1 \end{bmatrix}\begin{bmatrix} 1 & -3 \\ 2 & -5 \end{bmatrix} = \begin{bmatrix} (-5)1 + 3 \cdot 2 & (-5)(-3) + 3(-5) \\ (-2)1 + 1 \cdot 2 & (-2)(-3) + 1(-5) \end{bmatrix} = \begin{bmatrix} 1 & 0 \\ 0 & 1 \end{bmatrix}$

(C) The solution of $AX = B$ is $X = A^{-1}B$. Using the result of (B), we have

$$X = \begin{bmatrix} -5 & 3 \\ -2 & 1 \end{bmatrix}\begin{bmatrix} k_1 \\ k_2 \end{bmatrix}$$

If $k_1 = -2$, $k_2 = 1$, we have

$$\begin{bmatrix} x_1 \\ x_2 \end{bmatrix} = \begin{bmatrix} -5 & 3 \\ -2 & 1 \end{bmatrix}\begin{bmatrix} -2 \\ 1 \end{bmatrix} = \begin{bmatrix} (-5)(-2) + 3\cdot1 \\ (-2)(-2) + 1\cdot1 \end{bmatrix} = \begin{bmatrix} 13 \\ 5 \end{bmatrix} \quad x_1 = 13, \; x_2 = 5$$

(D) If $k_1 = 1$, $k_2 = -2$, reasoning as in (C) we have

$$\begin{bmatrix} x_1 \\ x_2 \end{bmatrix} = \begin{bmatrix} -5 & 3 \\ -2 & 1 \end{bmatrix}\begin{bmatrix} 1 \\ -2 \end{bmatrix} = \begin{bmatrix} (-5)1 + 3(-2) \\ (-2)1 + 1(-2) \end{bmatrix} = \begin{bmatrix} -11 \\ -4 \end{bmatrix} \quad x_1 = -11, \; x_2 = -4 \qquad (7\text{-}3)$$

11. (A) $D = \begin{vmatrix} 2 & -3 \\ 4 & -5 \end{vmatrix} = 2(-5) - 4(-3) = 2$

(B) $x = \dfrac{\begin{vmatrix} 1 & -3 \\ 2 & -5 \end{vmatrix}}{D} = \dfrac{1}{2} \qquad y = \dfrac{\begin{vmatrix} 2 & 1 \\ 4 & 2 \end{vmatrix}}{D} = \dfrac{0}{2} = 0$ \qquad (7\text{-}6)

12. We write the augmented matrix:

$$\begin{bmatrix} 1 & 3 & \big| & 10 \\ 2 & -1 & \big| & -1 \end{bmatrix} \quad (-2)R_1 + R_2 \to R_2$$

$-2 \quad -6 \quad -20$

↑
Need a 0 here

$\sim \begin{bmatrix} 1 & 3 & \big| & 10 \\ 0 & -7 & \big| & -21 \end{bmatrix} \quad -\frac{1}{7}R_2 \to R_2$ \qquad corrresponds to the linear system $\begin{aligned} x_1 + 3x_2 &= 10 \\ -7x_2 &= -21 \end{aligned}$

↑
Need a 1 here
Need a 0 here
↓

$\begin{bmatrix} 1 & 3 & \big| & 10 \\ 0 & 1 & \big| & 3 \end{bmatrix} \quad (-3)R_2 + R_1 \to R_1$ \qquad corresponds to the linear system $\begin{aligned} x_1 + 3x_2 &= 10 \\ x_2 &= 3 \end{aligned}$

$0 \quad -3 \quad -9$

$\sim \begin{bmatrix} 1 & 0 & \big| & 1 \\ 0 & 1 & \big| & 3 \end{bmatrix}$ \qquad corresponds to the linear system $\begin{aligned} x_1 &= 1 \\ x_2 &= 3 \end{aligned}$

The solution is $x_1 = 1$, $x_2 = 3$. Each pair of lines graphed below has the same intersection point, $(1, 3)$.

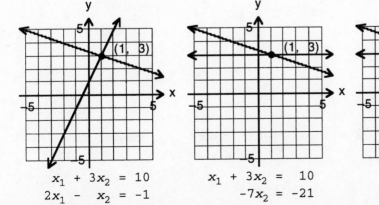

$\begin{aligned} x_1 + 3x_2 &= 10 \\ 2x_1 - x_2 &= -1 \end{aligned}$
$\begin{aligned} x_1 + 3x_2 &= 10 \\ -7x_2 &= -21 \end{aligned}$
$\begin{aligned} x_1 + 3x_2 &= 10 \\ x_2 &= 3 \end{aligned}$

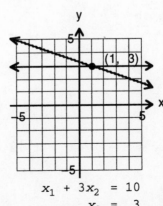
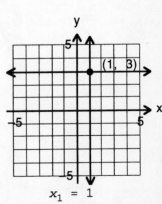

$\begin{aligned} x_1 &= 1 \\ x_2 &= 3 \end{aligned} \qquad (6\text{-}1)$

13. Here is a computer-generated graph of the system, entered as

$$y = \frac{7 + 2x}{3}$$

$$y = \frac{18 - 3x}{4}$$

After zooming in (graph not shown) the intersection point is located at (1.53, 3.35) to two decimal places.

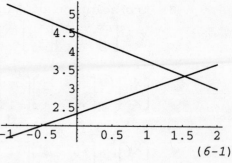

(6-1)

14. $\begin{bmatrix} 1 & 2 & -1 & | & 3 \\ 0 & 1 & 1 & | & -2 \\ 2 & 3 & 1 & | & 0 \end{bmatrix}$ $(-2)R_1 + R_3 \rightarrow R_3$

$\sim \begin{bmatrix} 1 & 2 & -1 & | & 3 \\ 0 & 1 & 1 & | & -2 \\ 0 & -1 & 3 & | & -6 \end{bmatrix}$ $(-2)R_2 + R_1 \rightarrow R_1$

$R_2 + R_3 \rightarrow R_3$

$\sim \begin{bmatrix} 1 & 0 & -3 & | & 7 \\ 0 & 1 & 1 & | & -2 \\ 0 & 0 & 4 & | & -8 \end{bmatrix}$ $\frac{1}{4}R_3 \rightarrow R_3$

$\sim \begin{bmatrix} 1 & 0 & -3 & | & 7 \\ 0 & 1 & 1 & | & -2 \\ 0 & 0 & 1 & | & -2 \end{bmatrix}$ $3R_3 + R_1 \rightarrow R_1$

$(-1)R_3 + R_2 \rightarrow R_2$

$\sim \begin{bmatrix} 1 & 0 & 0 & | & 1 \\ 0 & 1 & 0 & | & 0 \\ 0 & 0 & 1 & | & -2 \end{bmatrix}$

Solution: $x_1 = 1$, $x_2 = 0$, $x_3 = -2$

(6-2)

15. $\begin{bmatrix} 1 & 1 & -2 & | & 2 \\ 0 & 4 & 6 & | & -1 \\ 0 & 6 & 9 & | & 0 \end{bmatrix}$ $(-\frac{2}{3})R_2 + R_3 \rightarrow R_3$

$\sim \begin{bmatrix} 1 & 1 & -2 & | & 2 \\ 0 & 4 & 6 & | & -1 \\ 0 & 0 & 0 & | & \frac{2}{3} \end{bmatrix}$

The last row corresponds to the equation

$$0x_1 + 0x_2 + 0x_3 = \frac{2}{3},$$

hence there is no solution.   (6-2)

16. $\begin{bmatrix} 1 & -2 & 1 & | & 1 \\ 3 & -2 & -1 & | & -5 \end{bmatrix}$ $(-3)R_1 + R_2 \rightarrow R_2$

$\sim \begin{bmatrix} 1 & -2 & 1 & | & 1 \\ 0 & 4 & -4 & | & -8 \end{bmatrix}$ $\frac{1}{4}R_2 \rightarrow R_2$

$\sim \begin{bmatrix} 1 & -2 & 1 & | & 1 \\ 0 & 1 & -1 & | & -2 \end{bmatrix}$ $2R_2 + R_1 \rightarrow R_1$

$\sim \begin{bmatrix} 1 & 0 & -1 & | & -3 \\ 0 & 1 & -1 & | & -2 \end{bmatrix}$

This corresponds to the system

$x_1 \quad - x_3 = -3$

$\quad x_2 - x_3 = -2$

Let $x_3 = t$

Then $x_2 = x_3 - 2$

$\quad = t - 2$

$\quad x_1 = x_3 - 3$

$\quad = t - 3$

Hence $x_1 = t - 3$, $x_2 = t - 2$, $x_3 = t$ is a solution for every real number t.

(6-2)

17. $x^2 - 3xy + 3y^2 = 1$
$xy = 1$

Solve for y in the second equation.

$$y = \frac{1}{x}$$

$$x^2 - 3x\left(\frac{1}{x}\right) + 3\left(\frac{1}{x}\right)^2 = 1$$

$$x^2 - 3 + \frac{3}{x^2} = 1 \quad x \neq 0$$

$$x^4 - 3x^2 + 3 = x^2$$

$$x^4 - 4x^2 + 3 = 0$$

$$(x^2 - 1)(x^2 - 3) = 0$$

$$x^2 - 1 = 0 \qquad x^2 - 3 = 0$$

$$x^2 = 1 \qquad x^2 = 3$$

$$x = \pm 1 \qquad x = \pm\sqrt{3}$$

For $x = 1$ For $x = -1$ For $x = \sqrt{3}$ For $x = -\sqrt{3}$

$$y = \frac{1}{1} \qquad\quad y = \frac{1}{-1} \qquad\quad y = \frac{1}{\sqrt{3}} \qquad\quad y = \frac{1}{-\sqrt{3}}$$

$$y = 1 \qquad\qquad y = -1 \qquad\qquad y = \frac{\sqrt{3}}{3} \qquad\qquad y = -\frac{\sqrt{3}}{3}$$

Solutions: $(1, 1)$, $(-1, -1)$, $\left(\sqrt{3}, \dfrac{\sqrt{3}}{3}\right)$, $\left(-\sqrt{3}, -\dfrac{\sqrt{3}}{3}\right)$ *(6-3)*

18. $x^2 - 3xy + y^2 = -1$
$x^2 - xy = 0$

Factor the left side of the equation that has a zero constant term.
$x(x - y) = 0$
$x = 0 \quad$ or $\quad x = y$
Thus the original system is equivalent to the two systems

$$x^2 - 3xy + y^2 = -1 \qquad\qquad x^2 - 3xy + y^2 = -1$$
$$x = 0 \qquad\qquad\qquad\qquad x = y$$

These systems are solved by substitution.
First system: Second system:

$$x^2 - 3xy + y^2 = -1 \qquad\qquad x^2 - 3xy + y^2 = -1$$
$$x = 0 \qquad\qquad\qquad\qquad x = y$$
$$0^2 - 3(0)y + y^2 = -1 \qquad\qquad y^2 - 3yy + y^2 = -1$$
$$y^2 = -1 \qquad\qquad\qquad -y^2 = -1$$
$$y = \pm i \qquad\qquad\qquad y^2 = 1$$
$$x = 0 \qquad\qquad\qquad y = \pm 1$$

For $y = 1$ For $y = -1$
$x = 1$ $x = -1$

Solutions: $(1, 1)$, $(-1, -1)$, $(0, i)$, $(0, -i)$ *(6-3)*

19. (A) $MN = \begin{bmatrix} 1 & 2 & -1 \end{bmatrix} \begin{bmatrix} 1 \\ -1 \\ 2 \end{bmatrix} = [1\cdot 1 + 2(-1) + (-1)2] = [-3]$

(B) $NM = \begin{bmatrix} 1 \\ -1 \\ 2 \end{bmatrix} \begin{bmatrix} 1 & 2 & -1 \end{bmatrix} = \begin{bmatrix} 1\cdot 1 & 1\cdot 2 & 1(-1) \\ (-1)1 & (-1)2 & (-1)(-1) \\ 2\cdot 1 & 2\cdot 2 & 2(-1) \end{bmatrix} = \begin{bmatrix} 1 & 2 & -1 \\ -1 & -2 & 1 \\ 2 & 4 & -2 \end{bmatrix}$ *(7-1)*

20. (A) $LM = \begin{bmatrix} 2 & -1 & 0 \\ 1 & 2 & 1 \end{bmatrix} \begin{bmatrix} 1 & 2 \\ -1 & 0 \\ 1 & 1 \end{bmatrix} = \begin{bmatrix} 2 \cdot 1 + (-1)(-1) + 0 \cdot 1 & 2 \cdot 2 + (-1)0 + 0 \cdot 1 \\ 1 \cdot 1 + 2(-1) + 1 \cdot 1 & 1 \cdot 2 + 2 \cdot 0 + 1 \cdot 1 \end{bmatrix}$

$= \begin{bmatrix} 3 & 4 \\ 0 & 3 \end{bmatrix}$

$LM - 2N = \begin{bmatrix} 3 & 4 \\ 0 & 3 \end{bmatrix} - 2 \begin{bmatrix} 2 & 1 \\ -1 & 0 \end{bmatrix} = \begin{bmatrix} 3 & 4 \\ 0 & 3 \end{bmatrix} - \begin{bmatrix} 4 & 2 \\ -2 & 0 \end{bmatrix} = \begin{bmatrix} -1 & 2 \\ 2 & 3 \end{bmatrix}$

(B) Since ML is a 3 × 3 matrix and N is a 2 × 2 matrix, $ML + N$ is not defined.

$(7-1)$

21. The solution region is unbounded. Two corner points are obvious from the graph: (0, 6) and (8, 0). The third corner point is obtained by solving the system
$3x + 2y = 12$
$x + 2y = 8$ to obtain (2, 3).

$(6-4)$

22. The feasible region is graphed as follows: The corner points (0, 7), (0, 0) and (8, 0) are obvious from the graph. The corner point (6, 4) is obtained by solving the system
$x + 2y = 14$
$2x + y = 16$
We now evaluate the objective function at each corner point.

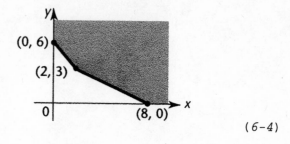

Corner Point (x, y)	Objective Function $z = 4x + 9y$	
(0, 7)	63	Maximum value
(0, 0)	0	
(8, 0)	32	
(6, 4)	60	

$(6-5)$

23. (A) As a matrix equation the system becomes

$\quad\quad A \quad\quad\quad X \quad\quad B$
$\begin{bmatrix} 1 & 4 & 2 \\ 2 & 6 & 3 \\ 2 & 5 & 2 \end{bmatrix} \begin{bmatrix} x_1 \\ x_2 \\ x_3 \end{bmatrix} = \begin{bmatrix} k_1 \\ k_2 \\ k_3 \end{bmatrix}$

(B) To find A^{-1}, we perform row operations on

$\begin{bmatrix} 1 & 4 & 2 & | & 1 & 0 & 0 \\ 2 & 6 & 3 & | & 0 & 1 & 0 \\ 2 & 5 & 2 & | & 0 & 0 & 1 \end{bmatrix}$ $\quad (-2)R_1 + R_2 \rightarrow R_2$
$\quad (-2)R_1 + R_3 \rightarrow R_3$

$$\sim \begin{bmatrix} 1 & 4 & 2 & | & 1 & 0 & 0 \\ 0 & -2 & -1 & | & -2 & 1 & 0 \\ 0 & -3 & -2 & | & -2 & 0 & 1 \end{bmatrix} \quad -\tfrac{1}{2}R_2 \rightarrow R_2$$

$$\sim \begin{bmatrix} 1 & 4 & 2 & | & 1 & 0 & 0 \\ 0 & 1 & \tfrac{1}{2} & | & 1 & -\tfrac{1}{2} & 0 \\ 0 & -3 & -2 & | & -2 & 0 & 1 \end{bmatrix} \quad \begin{array}{l} (-4)R_2 + R_1 \rightarrow R_1 \\ \\ 3R_2 + R_3 \rightarrow R_3 \end{array}$$

$$\sim \begin{bmatrix} 1 & 0 & 0 & | & -3 & 2 & 0 \\ 0 & 1 & \tfrac{1}{2} & | & 1 & -\tfrac{1}{2} & 0 \\ 0 & 0 & -\tfrac{1}{2} & | & 1 & -\tfrac{3}{2} & 1 \end{bmatrix} \quad R_3 + R_2 \rightarrow R_2$$

$$\sim \begin{bmatrix} 1 & 0 & 0 & | & -3 & 2 & 0 \\ 0 & 1 & 0 & | & 2 & -2 & 1 \\ 0 & 0 & -\tfrac{1}{2} & | & 1 & -\tfrac{3}{2} & 1 \end{bmatrix} \quad -2R_3 \rightarrow R_3$$

$$\sim \begin{bmatrix} 1 & 0 & 0 & | & -3 & 2 & 0 \\ 0 & 1 & 0 & | & 2 & -2 & 1 \\ 0 & 0 & 1 & | & -2 & 3 & -2 \end{bmatrix}$$

Hence $A^{-1} = \begin{bmatrix} -3 & 2 & 0 \\ 2 & -2 & 1 \\ -2 & 3 & -2 \end{bmatrix}$

Check: $A^{-1}A = \begin{bmatrix} -3 & 2 & 0 \\ 2 & -2 & 1 \\ -2 & 3 & -2 \end{bmatrix}\begin{bmatrix} 1 & 4 & 2 \\ 2 & 6 & 3 \\ 2 & 5 & 2 \end{bmatrix}$

$$= \begin{bmatrix} (-3)1 + 2\cdot2 + 0\cdot2 & (-3)4 + 2\cdot6 + 0\cdot5 & -3\cdot2 + 2\cdot3 + 0\cdot2 \\ 2\cdot1 + (-2)2 + 1\cdot2 & 2\cdot4 + (-2)6 + 1\cdot5 & 2\cdot2 + (-2)3 + 1\cdot2 \\ (-2)1 + 3\cdot2 + (-2)2 & (-2)4 + 3\cdot6 + (-2)5 & (-2)2 + 3\cdot3 + (-2)2 \end{bmatrix}$$

$$= \begin{bmatrix} 1 & 0 & 0 \\ 0 & 1 & 0 \\ 0 & 0 & 1 \end{bmatrix}$$

(C) The solution of $AX = B$ is $X = A^{-1}B$. Using the result of (B), we have

$$X = \begin{bmatrix} -3 & 2 & 0 \\ 2 & -2 & 1 \\ -2 & 3 & -2 \end{bmatrix}\begin{bmatrix} k_1 \\ k_2 \\ k_3 \end{bmatrix}$$

If $k_1 = -1$, $k_2 = 2$, $k_3 = 1$, we have

$$\begin{bmatrix} x_1 \\ x_2 \\ x_3 \end{bmatrix} = \begin{bmatrix} -3 & 2 & 0 \\ 2 & -2 & 1 \\ -2 & 3 & -2 \end{bmatrix}\begin{bmatrix} -1 \\ 2 \\ 1 \end{bmatrix} = \begin{bmatrix} 7 \\ -5 \\ 6 \end{bmatrix} \quad x_1 = 7,\ x_2 = -5,\ x_3 = 6$$

(D) If $k_1 = 2$, $k_2 = 0$, $k_3 = -1$, reasoning as in (C) we have

$$\begin{bmatrix} x_1 \\ x_2 \\ x_3 \end{bmatrix} = \begin{bmatrix} -3 & 2 & 0 \\ 2 & -2 & 1 \\ -2 & 3 & -2 \end{bmatrix}\begin{bmatrix} 2 \\ 0 \\ -1 \end{bmatrix} = \begin{bmatrix} -6 \\ 3 \\ -2 \end{bmatrix} \quad x_1 = -6,\ x_2 = 3,\ x_3 = -2 \qquad (7\text{-}3)$$

24. (A) $D = \begin{vmatrix} 1 & 2 & -1 \\ 2 & 8 & 1 \\ -1 & 3 & 5 \end{vmatrix} = 1(-1)^{1+1}\begin{vmatrix} 8 & 1 \\ 3 & 5 \end{vmatrix} + 2(-1)^{1+2}\begin{vmatrix} 2 & 1 \\ -1 & 5 \end{vmatrix} + (-1)(-1)^{1+3}\begin{vmatrix} 2 & 8 \\ -1 & 3 \end{vmatrix}$

$$= (-1)^2(8\cdot5 - 3\cdot1) + 2(-1)^3[2\cdot5 - (-1)1] + (-1)^5[2\cdot3 - (-1)8]$$

$$= 37 - 22 - 14 = 1$$

(B) $z = \dfrac{\begin{vmatrix} 1 & 2 & 1 \\ 2 & 8 & -2 \\ -1 & 3 & 2 \end{vmatrix}}{D} = \dfrac{1(-1)^{1+1}\begin{vmatrix} 8 & -2 \\ 3 & 2 \end{vmatrix} + 2(-1)^{1+2}\begin{vmatrix} 2 & -2 \\ -1 & 2 \end{vmatrix} + 1(-1)^{1+3}\begin{vmatrix} 2 & 8 \\ -1 & 3 \end{vmatrix}}{1}$

$= 32$

$(7-5, \ 7-6)$

25. Before we can enter these equations in our graphing utility, we must solve for y:

$$x^2 + 2xy - y^2 = 1 \qquad\qquad 9x^2 + 4xy + y^2 = 15$$
$$y^2 - 2xy - x^2 + 1 = 0 \qquad\qquad y^2 + 4xy + 9x^2 - 15 = 0$$

Applying the quadratic formula to each equation, we have

$$y = \frac{2x \pm \sqrt{4x^2 - 4(-x^2 + 1)}}{2} \qquad\qquad y = \frac{-4x \pm \sqrt{16x^2 - 4(9x^2 - 15)}}{2}$$

$$y = \frac{2x \pm \sqrt{8x^2 - 4}}{2} \qquad\qquad y = \frac{-4x \pm \sqrt{60 - 20x^2}}{2}$$

$$y = x \pm \sqrt{2x^2 - 1} \qquad\qquad y = -2x \pm \sqrt{15 - 5x^2}$$

Entering each of these four equations into a graphing utility produces the graph shown at the right.

Zooming in on the four intersection points, or using a built-in intersection routine (details omitted), yields $(-1.35, 0.28)$, $(-0.87, -1.60)$, $(0.87, 1.60)$, and $(1.35, -0.28)$ to two decimal places.

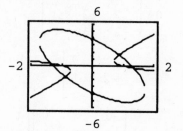

$(6-3)$

26. (A) The matrix is in fact

$$\begin{bmatrix} 1 & 0 & -5 & | & 2 \\ 0 & 1 & 3 & | & 6 \\ 0 & 0 & 0 & | & 0 \end{bmatrix}$$

Thus there are an infinite number of solutions ($x_3 = t$, $x_2 = 6 - 3t$, $x_1 = 2 + 5t$, for t any real number.)

(B) The matrix is in fact

$$\begin{bmatrix} 1 & 0 & -5 & | & 2 \\ 0 & 1 & 3 & | & 6 \\ 0 & 0 & 0 & | & n \end{bmatrix}$$

The third row corresponds to the equation $0x_1 + 0x_2 + 0x_3 = n$. This is impossible. The system has no solution.

(C) The system is independent. There is one solution.

$(6-2)$

27. If $A^2 = A$ and A^{-1} exists, then $A^{-1}A^2 = A^{-1}A$ or $A^{-1}AA = I$ or $IA = I$. Thus, $A = I$, the $n \times n$ identity matrix.

$(7-3)$

28. L, M, and P are in reduced form.
N is not in reduced form; it violates condition 1 of the definition of reduced matrix, Section 6-2: there is a row of 0's above rows having non-zero elements.

$(6-2)$

29. $k\begin{vmatrix} a & b \\ c & d \end{vmatrix} = k(ad - bc) = kad - kbc = kad - kcb = \begin{vmatrix} ka & b \\ kc & d \end{vmatrix}$

$(7-5)$

30. $\begin{vmatrix} a & b \\ c & d \end{vmatrix} = ad - bc = ad + akb - akb - bc = ad + akb - (bc + bka)$

$$= a(d + kb) - b(c + ka) = \begin{vmatrix} a & b \\ c + ka & d + kb \end{vmatrix} \qquad (7\text{-}5)$$

31. We assume that $\det M = ad - bc \neq 0$ and find M^{-1} by applying row operations to

$\begin{bmatrix} a & b & | & 1 & 0 \\ c & d & | & 0 & 1 \end{bmatrix}$ Case I: $a \neq 0$

                             $(-\frac{c}{a})R_1 + R_2 \rightarrow R_2$

$\sim \begin{bmatrix} a & b & | & 1 & 0 \\ 0 & \frac{ad-bc}{a} & | & -\frac{c}{a} & 1 \end{bmatrix}$ $\frac{a}{ad-bc}R_2 \rightarrow R_2$

$\sim \begin{bmatrix} a & b & | & 1 & 0 \\ 0 & 1 & | & \frac{-c}{ad-bc} & \frac{a}{ad-bc} \end{bmatrix}$ $(-b)R_2 + R_1 \rightarrow R_1$

$\sim \begin{bmatrix} a & 0 & | & 1 + \frac{bc}{ad-bc} & \frac{-ab}{ad-bc} \\ 0 & 1 & | & \frac{-c}{ad-bc} & \frac{a}{ad-bc} \end{bmatrix}$

$\sim \begin{bmatrix} a & 0 & | & \frac{ad}{ad-bc} & \frac{-ab}{ad-bc} \\ 0 & 1 & | & \frac{-c}{ad-bc} & \frac{a}{ad-bc} \end{bmatrix}$ $\frac{1}{a}R_1 \rightarrow R_1$

$\sim \begin{bmatrix} 1 & 0 & | & \frac{d}{ad-bc} & \frac{-b}{ad-bc} \\ 0 & 1 & | & \frac{-c}{ad-bc} & \frac{a}{ad-bc} \end{bmatrix}$

Case II: $a = 0$ (b, $c \neq 0$, since $b = 0$ or $c = 0$ would imply $\det M = ad - bc = 0$)

$\begin{bmatrix} 0 & b & | & 1 & 0 \\ c & d & | & 0 & 1 \end{bmatrix}$ $R_1 \leftrightarrow R_2$

$\sim \begin{bmatrix} c & d & | & 0 & 1 \\ 0 & b & | & 1 & 0 \end{bmatrix}$ $\frac{1}{c}R_1 \rightarrow R_1$

$\sim \begin{bmatrix} 1 & \frac{d}{c} & | & 0 & \frac{1}{c} \\ 0 & b & | & 1 & 0 \end{bmatrix}$ $(-\frac{d}{bc})R_2 + R_1 \rightarrow R_1$

$\sim \begin{bmatrix} 1 & 0 & | & -\frac{d}{bc} & \frac{1}{c} \\ 0 & b & | & 1 & 0 \end{bmatrix}$ $\frac{1}{b}R_2 \rightarrow R_2$

$\sim \begin{bmatrix} 1 & 0 & | & -\frac{d}{bc} & \frac{1}{c} \\ 0 & 1 & | & \frac{1}{b} & 0 \end{bmatrix}$

$\sim \begin{bmatrix} 1 & 0 & | & \frac{d}{ad-bc} & \frac{-b}{ad-bc} \\ 0 & 1 & | & \frac{-c}{ad-bc} & \frac{a}{ad-bc} \end{bmatrix}$ since $a = 0$.

In either case,

$$M^{-1} = \begin{bmatrix} \frac{d}{ad-bc} & \frac{-b}{ad-bc} \\ \frac{-c}{ad-bc} & \frac{a}{ad-bc} \end{bmatrix} = \frac{1}{ad-bc}\begin{bmatrix} d & -b \\ -c & a \end{bmatrix} = \frac{1}{\det M}\begin{bmatrix} d & -b \\ -c & a \end{bmatrix} \qquad (7\text{-}2, \ 7\text{-}4)$$

32. Let x = amount invested at 8%
$\quad\quad y$ = amount invested at 14%
Then $x + y = 12{,}000$ (total amount invested) $\quad\quad\quad\quad\quad$ (1)
$0.08x + 0.14y = 0.10(12{,}000)$ (total yield on investment) $\quad\quad$ (2)
We solve the system of equations (1), (2) using elimination by addition.

$\begin{array}{ll} -0.08x - 0.08y = -0.08(12{,}000) & -0.08[\text{equation (1)}] \\ \underline{0.08x + 0.14y = 0.10(12{,}000)} & \text{equation (2)} \end{array}$

$$0.06y = -0.08(12{,}000) + 0.10(12{,}000)$$
$$0.06y = 240$$
$$y = 4{,}000$$
$$x + y = 12{,}000$$
$$x = 8{,}000$$

$8{,}000$ at 8% and $4{,}000$ at 14%. $\hfill (6\text{-}1,\ 6\text{-}2)$

33. Let x_1 = number of ounces of mix A.
$\quad\quad x_2$ = number of ounces of mix B.
$\quad\quad x_3$ = number of ounces of mix C.
Then
$$\begin{array}{llll} 0.2x_1 + & 0.1x_2 + & 0.15x_3 = & 23 \quad \text{(protein)} \\ 0.02x_1 + & 0.06x_2 + & 0.05x_3 = & 6.2 \quad \text{(fat)} \\ 0.15x_1 + & 0.1x_2 + & 0.05x_3 = & 16 \quad \text{(moisture)} \end{array}$$
or
$$\begin{array}{lll} 4x_1 + 2x_2 + 3x_3 = & 460 \\ 2x_1 + 6x_2 + 5x_3 = & 620 \\ 3x_1 + 2x_2 + x_3 = & 320 \end{array}$$
is the system to be solved. We form the augmented matrix and solve by Gauss-Jordan elimination.

$$\left[\begin{array}{ccc|c} 4 & 2 & 3 & 460 \\ 2 & 6 & 5 & 620 \\ 3 & 2 & 1 & 320 \end{array}\right] \quad \tfrac{1}{4}R_1 \to R_1$$

$$\sim \left[\begin{array}{ccc|c} 1 & \tfrac{1}{2} & \tfrac{3}{4} & 115 \\ 2 & 6 & 5 & 620 \\ 3 & 2 & 1 & 320 \end{array}\right] \quad \begin{array}{l} (-2)R_1 + R_2 \to R_2 \\ (-3)R_1 + R_3 \to R_3 \end{array}$$

$$\sim \left[\begin{array}{ccc|c} 1 & \tfrac{1}{2} & \tfrac{3}{4} & 115 \\ 0 & 5 & \tfrac{7}{2} & 390 \\ 0 & \tfrac{1}{2} & -\tfrac{5}{4} & -25 \end{array}\right] \quad R_2 \leftrightarrow R_3$$

$$\sim$$

$$\left[\begin{array}{ccc|c} 1 & \tfrac{1}{2} & \tfrac{3}{4} & 115 \\ 0 & \tfrac{1}{2} & -\tfrac{5}{4} & -25 \\ 0 & 5 & \tfrac{7}{2} & 390 \end{array}\right] \quad \begin{array}{l} (-1)R_2 + R_1 \to R_1 \\[6pt] (-10)R_2 + R_3 \to R_3 \end{array}$$

$$\sim \left[\begin{array}{ccc|c} 1 & 0 & 2 & 140 \\ 0 & \tfrac{1}{2} & -\tfrac{5}{4} & -25 \\ 0 & 0 & 16 & 640 \end{array}\right] \quad \begin{array}{l} 2R_2 \to R_2 \\[6pt] \tfrac{1}{16}R_3 \to R_3 \end{array}$$

$$\sim \left[\begin{array}{ccc|c} 1 & 0 & 2 & 140 \\ 0 & 1 & -\tfrac{5}{2} & -50 \\ 0 & 0 & 1 & 40 \end{array}\right] \quad \begin{array}{l} (-2)R_3 + R_1 \to R_1 \\[6pt] (\tfrac{5}{2})R_3 + R_2 \to R_2 \end{array}$$

$$\sim \left[\begin{array}{ccc|c} 1 & 0 & 0 & 60 \\ 0 & 1 & 0 & 50 \\ 0 & 0 & 1 & 40 \end{array}\right]$$

Thus, $x_1 = 60$g Mix A, $x_2 = 50$g Mix B, $x_3 = 40$g Mix C $\hfill (6\text{-}2)$

34. Let x_1 = number of model A trucks

x_2 = number of model B trucks

x_3 = number of model C trucks

Then $x_1 + x_2 + x_3 = 12$ (total number of trucks)

$18,000x_1 + 22,000x_2 + 30,000x_3 = 300,000$ (total funds needed)

We form the augmented matrix and solve by Gauss-Jordan elimination.

$$\begin{bmatrix} 1 & 1 & 1 & | & 12 \\ 18,000 & 22,000 & 30,000 & | & 300,000 \end{bmatrix} \quad \tfrac{1}{2,000}R_2 \to R_2$$

$$\sim \begin{bmatrix} 1 & 1 & 1 & | & 12 \\ 9 & 11 & 15 & | & 150 \end{bmatrix} \quad (-9)R_1 + R_2 \to R_2$$

$$\sim \begin{bmatrix} 1 & 1 & 1 & | & 12 \\ 0 & 2 & 6 & | & 42 \end{bmatrix} \quad -\tfrac{1}{2}R_2 + R_1 \to R_1$$

$$\sim \begin{bmatrix} 1 & 0 & -2 & | & -9 \\ 0 & 2 & 6 & | & 42 \end{bmatrix} \quad \tfrac{1}{2}R_2 \to R_2$$

$$\sim \begin{bmatrix} 1 & 0 & -2 & | & -9 \\ 0 & 1 & 3 & | & 21 \end{bmatrix}$$

Let $x_3 = t$. Then $x_2 = -3x_3 + 21$

$= -3t + 21$

$x_1 = 2x_3 - 9$

$= 2t - 9$

A solution is achieved, not for every value of t, but for integer values of t that give rise to non-negative x_1, x_2, x_3.

$x_1 \geq 0$ means $2t - 9 \geq 0$ or $t \geq 4\tfrac{1}{2}$

$x_2 \geq 0$ means $-3t + 21 \geq 0$ or $t \leq 7$

The only integer values of t that satisfy these conditions are 5, 6, 7. Thus we have the solutions

$x_1 = 2t - 9$ model A trucks

$x_2 = -3t + 21$ model B trucks

$x_3 = t$ model C trucks

where t = 5, 6, or 7. Thus the distributor can purchase

1 model A truck, 6 model B trucks, and 5 model C trucks, or

3 model A trucks, 3 model B trucks, and 6 model C trucks, or

5 model A trucks and 7 model C trucks.

(6-2)

35. Let a = length of rectangle

b = width of rectangle

Then $ab = 32$ (area)

$2a + 2b = 24$ (perimeter)

We solve by substitution, solving the first-degree equation for b and substituting into the second-degree equation.

$2b = 24 - 2a$

$b = 12 - a$

$a(12 - a) = 32$

$12a - a^2 = 32$

$0 = a^2 - 12a + 32$

$0 = (a - 4)(a - 8)$

$a = 4$ or $a = 8$

$b = 12 - a = 8$ $b = 12 - a = 4$

Dimensions: 8 meters by 4 meters

(6-3)

36. Let x = number of standard day packs
y = number of deluxe day packs

(A) We form the linear objective function
$P = 8x + 12y$
We wish to maximize P, the profit from x standard packs @ \$8 and y deluxe packs @ \$12, subject to the constraints

$$0.5x + 0.5y \leq 300 \quad \text{fabricating constraint}$$
$$0.3x + 0.6y \leq 240 \quad \text{sewing constraint}$$
$$x, y \geq 0 \quad \text{non-negative constraints}$$

Solving the system of constraint inequalities graphically, we obtain the feasible region S shown in the diagram.

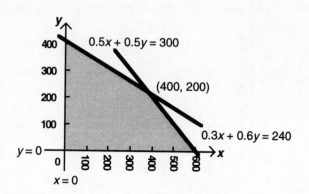

The three corner points (0, 400), (0, 0), and (600, 0) are obvious from the diagram. The corner point (400, 200) is found by solving the system
$0.5x + 0.5y = 300$
$0.3x + 0.6y = 240$

Next we evaluate the objective function at each corner point.

Corner Point (x, y)	Objective Function $P = 8x + 12y$	
(0, 0)	0	
(600, 0)	4,800	
(400, 200)	5,600	Maximum value
(0, 400)	4,800	

The optimal value is 5,600 at the corner point (400, 200). Thus, the company should manufacture 400 standard packs and 200 deluxe packs for a maximum profit of \$5,600.

(B) The objective function now becomes $5x + 15y$. We evaluate this at each corner point.

Corner Point (x, y)	Objective Function $P = 5x + 15y$	
(0, 0)	0	
(600, 0)	3,000	
(400, 200)	5,000	
(0, 400)	6,000	Maximum value

The optimal value is now 6,000 at the corner point (0, 400). Thus, the company should manufacture no standard packs and 400 deluxe packs for a maximum profit of \$6,000.

(C) The objective function now becomes $11x + 9y$. We evaluate this at each corner point.

Corner Point (x, y)	Objective Function $P = 11x + 9y$	
(0, 0)	0	
(600, 0)	6,600	Maximum value
(400, 200)	6,200	
(0, 400)	3,600	

The optimal value is now 6,600 at the corner point (600, 0). Thus, the company should manufacture 600 standard packs and no deluxe packs for a maximum profit of \$6,600.

37. (A) $M\begin{bmatrix} 0.25 \\ 0.25 \\ 0.25 \\ 0.25 \end{bmatrix} = \begin{bmatrix} 78 & 84 & 81 & 86 \\ 91 & 65 & 84 & 92 \\ 95 & 90 & 92 & 91 \\ 75 & 82 & 87 & 91 \\ 83 & 88 & 81 & 76 \end{bmatrix}\begin{bmatrix} 0.25 \\ 0.25 \\ 0.25 \\ 0.25 \end{bmatrix}$

$$= \begin{bmatrix} 0.25(78 + 84 + 81 + 86) \\ 0.25(91 + 65 + 84 + 92) \\ 0.25(95 + 90 + 92 + 91) \\ 0.25(75 + 82 + 87 + 91) \\ 0.25(83 + 88 + 81 + 76) \end{bmatrix} = \begin{bmatrix} 82.25 \\ 83 \\ 92 \\ 83.75 \\ 82 \end{bmatrix} \begin{matrix} \text{Ann} \\ \text{Bob} \\ \text{Carol} \\ \text{Dan} \\ \text{Eric} \end{matrix}$$

(B) $M\begin{bmatrix} 0.2 \\ 0.2 \\ 0.2 \\ 0.4 \end{bmatrix} = \begin{bmatrix} 78 & 84 & 81 & 86 \\ 91 & 65 & 84 & 92 \\ 95 & 90 & 92 & 91 \\ 75 & 82 & 87 & 91 \\ 83 & 88 & 81 & 76 \end{bmatrix}\begin{bmatrix} 0.2 \\ 0.2 \\ 0.2 \\ 0.4 \end{bmatrix}$

$$= \begin{bmatrix} 0.2(78 + 84 + 81) + 0.4 \cdot 86 \\ 0.2(91 + 65 + 84) + 0.4 \cdot 92 \\ 0.2(95 + 90 + 92) + 0.4 \cdot 91 \\ 0.2(75 + 82 + 87) + 0.4 \cdot 91 \\ 0.2(83 + 88 + 81) + 0.4 \cdot 76 \end{bmatrix} = \begin{bmatrix} 83 \\ 84.8 \\ 91.8 \\ 85.2 \\ 80.8 \end{bmatrix} \begin{matrix} \text{Ann} \\ \text{Bob} \\ \text{Carol} \\ \text{Dan} \\ \text{Eric} \end{matrix}$$

(C) $[0.2 \quad 0.2 \quad 0.2 \quad 0.2 \quad 0.2]M = [0.2 \quad 0.2 \quad 0.2 \quad 0.2 \quad 0.2]\begin{bmatrix} 78 & 84 & 81 & 86 \\ 91 & 65 & 84 & 92 \\ 95 & 90 & 92 & 91 \\ 75 & 82 & 87 & 91 \\ 83 & 88 & 81 & 76 \end{bmatrix}$

$= [0.2(78 + 91 + 95 + 75 + 83) \quad 0.2(84 + 65 + 90 + 82 + 88) \quad 0.2(81 + 84 + 92 + 87 + 81) \quad 0.2(86 + 92 + 91 + 91 + 76)]$

Test 1	Test 2	Test 3	Test 4

$= [84.4 \qquad 81.8 \qquad 85 \qquad 87.2]$

$(7\text{-}1)$

CHAPTER 8

Exercise 8-1

Key Ideas and Formulas

A sequence is a function with domain a set of successive integers.

Notation: for $f(n)$ we write a_n. The elements of the range, $f(n)$, are called terms of the sequence, a_n. a_1 is the first term, a_2 is the second term, a_n is the nth term or the general term.

Finite sequence: domain is finite set of successive integers.

Infinite sequence: domain is infinite set of successive integers.

Some sequences are specified by a recursion formula, a formula that defines each term in terms of one or more preceding terms.

If a_1, a_2, a_3, ..., a_n, ... is a sequence, then the expression $a_1 + a_2 + a_3 + \cdots + a_n$ is called a series. If a_1, a_2, ..., a_n are the terms of the sequence, the series is written

$$\sum_{k=1}^{n} a_k = a_1 + a_2 + \cdots + a_n \quad k \text{ is called the summing index.}$$

---

1. $a_1 = 1 - 2 = -1$ $\quad a_2 = 2 - 2 = 0$ $\quad a_3 = 3 - 2 = 1$ $\quad a_4 = 4 - 2 = 2$

3. $a_1 = \dfrac{1-1}{1+1} = 0$ $\quad a_2 = \dfrac{2-1}{2+1} = \dfrac{1}{3}$ $\quad a_3 = \dfrac{3-1}{3+1} = \dfrac{2}{4} = \dfrac{1}{2}$ $\quad a_4 = \dfrac{4-1}{4+1} = \dfrac{3}{5}$

5. $a_1 = (-2)^{1+1} = (-2)^2 = 4$ $\quad a_2 = (-2)^{2+1} = (-2)^3 = -8$
   $a_3 = (-2)^{3+1} = (-2)^4 = 16$ $\quad a_4 = (-2)^{4+1} = (-2)^5 = -32$

7. $a_8 = 8 - 2 = 6$ $\qquad$ 9. $a_{100} = \dfrac{100-1}{100+1} = \dfrac{99}{101}$ $\qquad$ 11. $S_5 = 1 + 2 + 3 + 4 + 5$

13. $S_3 = \dfrac{1}{10^1} + \dfrac{1}{10^2} + \dfrac{1}{10^3} = \dfrac{1}{10} + \dfrac{1}{100} + \dfrac{1}{1000}$

15. $S_4 = (-1)^1 + (-1)^2 + (-1)^3 + (-1)^4 = (-1) + 1 + (-1) + 1 = -1 + 1 - 1 + 1$

17. $a_1 = (-1)^{1+1}1^2 = 1$ $\quad a_2 = (-1)^{2+1}2^2 = -4$ $\quad a_3 = (-1)^{3+1}3^2 = 9$
    $a_4 = (-1)^{4+1}4^2 = -16$ $\quad a_5 = (-1)^{5+1}5^2 = 25$

19.
$$a_1 = \frac{1}{3}\left[1 - \frac{1}{10^1}\right] = \frac{1}{3} \cdot \frac{9}{10} = \frac{3}{10} = 0.3$$

$$a_2 = \frac{1}{3}\left[1 - \frac{1}{10^2}\right] = \frac{1}{3} \cdot \frac{99}{100} = \frac{33}{100} = 0.33$$

$$a_3 = \frac{1}{3}\left[1 - \frac{1}{10^3}\right] = \frac{1}{3} \cdot \frac{999}{1,000} = \frac{333}{1,000} = 0.333$$

$$a_4 = \frac{1}{3}\left[1 - \frac{1}{10^4}\right] = \frac{1}{3} \cdot \frac{9,999}{10,000} = \frac{3,333}{10,000} = 0.3333$$

$$a_5 = \frac{1}{3}\left[1 - \frac{1}{10^5}\right] = \frac{1}{3} \cdot \frac{99,999}{100,000} = \frac{33,333}{100,000} = 0.33333$$

21. $a_1 = \left(-\dfrac{1}{2}\right)^{1-1} = \left(-\dfrac{1}{2}\right)^0 = 1$

$\quad a_2 = \left(-\dfrac{1}{2}\right)^{2-1} = -\dfrac{1}{2}$

$\quad a_3 = \left(-\dfrac{1}{2}\right)^{3-1} = \dfrac{1}{4}$

$\quad a_4 = \left(-\dfrac{1}{2}\right)^{4-1} = -\dfrac{1}{8}$

$\quad a_5 = \left(-\dfrac{1}{2}\right)^{5-1} = \dfrac{1}{16}$

23. $a_1 = 7$

$\quad a_2 = a_{2-1} - 4 = a_1 - 4 = 7 - 4 = 3$

$\quad a_3 = a_{3-1} - 4 = a_2 - 4 = 3 - 4 = -1$

$\quad a_4 = a_{4-1} - 4 = a_3 - 4 = -1 - 4 = -5$

$\quad a_5 = a_{5-1} - 4 = a_4 - 4 = -5 - 4 = -9$

> **Common Error:**
> $a_2 \neq a_2 - 1 - 4$. The 1 is in
> the subscript: a_{2-1}; that is, a_1

25. $a_1 = 4$

$\quad a_2 = \dfrac{1}{4}a_{2-1} = \dfrac{1}{4}a_1 = \dfrac{1}{4} \cdot 4 = 1$

$\quad a_3 = \dfrac{1}{4}a_{3-1} = \dfrac{1}{4}a_2 = \dfrac{1}{4} \cdot 1 = \dfrac{1}{4}$

$\quad a_4 = \dfrac{1}{4}a_{4-1} = \dfrac{1}{4}a_3 = \dfrac{1}{4} \cdot \dfrac{1}{4} = \dfrac{1}{16}$

$\quad a_5 = \dfrac{1}{4}a_{5-1} = \dfrac{1}{4}a_4 = \dfrac{1}{4} \cdot \dfrac{1}{16} = \dfrac{1}{64}$

27. a_n: 4, 5, 6, 7, $\cdots$

$\quad n = 1, 2, 3, 4, \cdots$

Comparing a_n with n, we see that

$a_n = n + 3$

29. a_n: 3, 6, 9, 12, $\cdots$

$\quad n = 1, 2, 3, 4, \cdots$

Comparing a_n with n, we see that

$a_n = 3n$

31. a_n: $\dfrac{1}{2}, \dfrac{2}{3}, \dfrac{3}{4}, \dfrac{4}{5}, \cdots$

$\quad n = 1, 2, 3, 4, \cdots$

Comparing a_n with n, we see that

$a_n = \dfrac{n}{n + 1}$

33. a_n: 1, -1, 1, -1, $\cdots$

$\quad n = 1, 2, 3, 4, \cdots$

Comparing a_n with n, we see that a_n involves (-1) to successively even and odd
powers, hence to a power that depends on n. We could write
$a_n = (-1)^{n-1}$ or $a_n = (-1)^{n+1}$
or other choices. $a_n = (-1)^{n+1}$ is one of many correct answers.

35. a_n: -2, 4, -8, 16, $\cdots$

$\quad n = 1, 2, 3, 4, \cdots$

Comparing a_n with n, we see that a_n involves -1 and 2 to successively higher
powers, hence to powers that depend on n. We write
$a_n = (-1)^n(2)^n$ or $a_n = (-2)^n$

37. a_n: $x, \dfrac{x^2}{2}, \dfrac{x^3}{3}, \dfrac{x^4}{4}, \cdots$

$\quad$ or $\dfrac{x^1}{1}, \dfrac{x^2}{2}, \dfrac{x^3}{3}, \dfrac{x^4}{4}, \cdots$

Comparing a_n with n,
we see that

$a_n = \dfrac{x^n}{n}$

39.

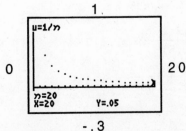

41.

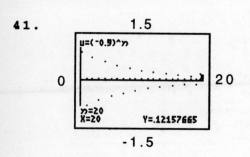

43. $S_4 = \dfrac{(-2)^{1+1}}{1} + \dfrac{(-2)^{2+1}}{2} + \dfrac{(-2)^{3+1}}{3} + \dfrac{(-2)^{4+1}}{4}$

$= \dfrac{4}{1} - \dfrac{8}{2} + \dfrac{16}{3} - \dfrac{32}{4}$

45. $S_3 = \dfrac{1}{1}x^{1+1} + \dfrac{1}{2}x^{2+1} + \dfrac{1}{3}x^{3+1}$

$= x^2 + \dfrac{x^3}{2} + \dfrac{x^4}{3}$

47. $S_5 = \dfrac{(-1)^{1+1}}{1}x^1 + \dfrac{(-1)^{2+1}}{2}x^2 + \dfrac{(-1)^{3+1}}{3}x^3 + \dfrac{(-1)^{4+1}}{4}x^4 + \dfrac{(-1)^{5+1}}{5}x^5$

$= x - \dfrac{x^2}{2} + \dfrac{x^3}{3} - \dfrac{x^4}{4} + \dfrac{x^5}{5}$

49. $S_4 = 1^2 + 2^2 + 3^2 + 4^2$

$k = 1, 2, 3, 4, \cdots$

Clearly, $a_k = k^2$, $k = 1, 2, 3, 4$

$S_4 = \displaystyle\sum_{k=1}^{4} k^2$

51. $S_5 = \dfrac{1}{2^1} + \dfrac{1}{2^2} + \dfrac{1}{2^3} + \dfrac{1}{2^4} + \dfrac{1}{2^5}$

$k = 1, 2, 3, 4, \cdots$

Clearly, $a_k = \dfrac{1}{2^k}$, $k = 1, 2, 3, 4, 5$

$S_5 = \displaystyle\sum_{k=1}^{5} \dfrac{1}{2^k}$

53. $S_n = 1 + \dfrac{1}{2^2} + \dfrac{1}{3^2} + \cdots + \dfrac{1}{n^2}$

$k = 1, 2, 3, \cdots, n$

Clearly, $a_k = \dfrac{1}{k^2}$, $k = 1, 2, 3, \cdots, n$

$S_n = \displaystyle\sum_{k=1}^{n} \dfrac{1}{k^2}$

55. $S_n = 1 - 4 + 9 + \cdots + (-1)^{n+1}n^2$

$k = 1, 2, 3, \cdots, n$

Clearly, $a_k = (-1)^{k+1}k^2$, $k = 1, 2, 3, \cdots, n$

$S_n = \displaystyle\sum_{k=1}^{n} (-1)^{k+1}k^2$

57. (A) $a_1 = 3$

$a_2 = \dfrac{a_{2-1}^2 + 2}{2a_{2-1}} = \dfrac{a_1^2 + 2}{2a_1} = \dfrac{3^2 + 2}{2 \cdot 3} \approx 1.83$

$a_3 = \dfrac{a_{3-1}^2 + 2}{2a_{3-1}} = \dfrac{a_2^2 + 2}{2a_2} = \dfrac{(1.83)^2 + 2}{2(1.83)} \approx 1.46$

$a_4 = \dfrac{a_{4-1}^2 + 2}{2a_{4-1}} = \dfrac{a_3^2 + 2}{2a_3} = \dfrac{(1.46)^2 + 2}{2(1.46)} \approx 1.415$

(B) Calculator $\sqrt{2} = 1.4142135\ldots$

(C) $a_1 = 1$

$a_2 = \dfrac{a_1^2 + 2}{2a_1} = \dfrac{1^2 + 2}{2 \cdot 1} = 1.5$

$a_3 = \dfrac{a_2^2 + 2}{2a_2} = \dfrac{(1.5)^2 + 2}{2(1.5)} \approx 1.417$

$a_4 = \dfrac{a_3^2 + 2}{2a_3} = \dfrac{(1.417)^2 + 2}{2(1.417)} \approx 1.414$

59. The first ten terms of the Fibonacci sequence a_n are

1, 1, 2, 3, 5, 8, 13, 21, 34, 55

The first ten terms of b_n are

1, 3, 4, 7, 11, 18, 29, 47, 76, 123

The first ten terms of $c_n = \dfrac{b_n}{a_n}$ are

$$\frac{1}{1}, \frac{3}{1}, \frac{4}{2}, \frac{7}{3}, \frac{11}{5}, \frac{18}{8}, \frac{29}{13}, \frac{47}{21}, \frac{76}{34}, \frac{123}{55}$$

In decimal notation this becomes

1, 3, 2, 2.33..., 2.2, 2.25, 2.23..., 2.238..., 2.235..., 2.236...

61.
$$e^{0.2} = 1 + \frac{0.2}{1!} + \frac{(0.2)^2}{2!} + \frac{(0.2)^3}{3!} + \frac{(0.2)^4}{4!}$$

$$= 1 + 0.2 + \frac{(0.2)^2}{1 \cdot 2} + \frac{(0.2)^3}{1 \cdot 2 \cdot 3} + \frac{(0.2)^4}{1 \cdot 2 \cdot 3 \cdot 4}$$

$$= 1.2 + 0.2\left[\frac{0.2}{2} + \frac{(0.2)^2}{6} + \frac{(0.2)^3}{24}\right]$$

$$= 1.2 + 0.2\left\{0.2\left[\frac{1}{2} + \frac{0.2}{6} + \frac{(0.2)^2}{24}\right]\right\}$$

$$= 1.2 + 0.2\left\{0.2\left[\frac{1}{2} + 0.2\left(\frac{1}{6} + \frac{0.2}{24}\right)\right]\right\} \quad \text{(In this form this is more easily entered into a scientific calculator.)}$$

$$= 1.2214000$$

$e^{0.2} = 1.2214028$ (calculator—direct evaluation)

63. $\displaystyle\sum_{k=1}^{n} ca_k = ca_1 + ca_2 + ca_3 + \cdots + ca_n = c(a_1 + a_2 + a_3 + \cdots a_n) = c\sum_{k=1}^{n} a_k$

Exercise 8-2

Key Ideas and Formulas

Principle of Mathematical Induction

Let P_n be a statement associated with each positive integer n and suppose the following conditions are satisfied:

1. P_1 is true.

2. For any positive integer k, if P_k is true, then P_{k+1} is also true.

Then the statement P_n is true for all positive integers n.

Extended Principle of Mathematical Induction

Let m be a positive integer, let P_n be a statement associated with each integer $n \geq m$, and suppose the following conditions are satisfied:

1. P_m is true.

2. For any integer $k \geq m$, if P_k is true, then P_{k+1} is also true.

Then the statement P_n is true for all integers $n \geq m$.

Proof by Mathematical Induction: To prove that a statement P_n holds for all integers $n \geq p$: (p is often, but not always, equal to 1.) State the conjecture, P_n.

Part 1. Show that P_p is true.

Part 2. Write out P_k and P_{k+1}. Show that if P_k is true, then P_{k+1} is true.

Then P_n is true for all integers $n \geq p$.

1. $(3 + 5)^1 = 3^1 + 5^1$ True
$(3 + 5)^2 = 3^2 + 5^2$ False
Fails at $n = 2$

3. $1^2 = 3 \cdot 1 - 2$ True
$2^2 = 3 \cdot 2 - 2$ True
$3^2 = 3 \cdot 3 - 2$ False
Fails at $n = 3$

5. $P_1: 2 = 2 \cdot 1^2$ $2 = 2$
      $P_2: 2 + 6 = 2 \cdot 2^2$ $8 = 8$
$P_3: 2 + 6 + 10 = 2 \cdot 3^2$ $18 = 18$

7. $P_1: a^5 a^1 = a^{5+1}$ $a^6 = a^6$
$P_2: a^5 a^2 = a^{5+2}$ $a^5 a^2 = a^5(a^1 a) = (a^5 a)a = a^6 a = a^7 = a^{5+2}$
$P_3: a^5 a^3 = a^{5+3}$ $a^5 a^3 = a^5(a^2 a) = a^5(a^1 a)a = [(a^5 a)a]a = a^8 = a^{5+3}$

9. $P_1: 9^1 - 1 = 8$ is divisible by 4
$P_2: 9^2 - 1 = 80$ is divisible by 4
$P_3: 9^3 - 1 = 728$ is divisible by 4

11. $P_k: 2 + 6 + 10 + \cdots + (4k - 2) = 2k^2$
$P_{k+1}: 2 + 6 + 10 + \cdots + (4k - 2) + [4(k + 1) - 2] = 2(k + 1)^2$
or $2 + 6 + 10 + \cdots + (4k - 2) + (4k + 2) = 2(k + 1)^2$

13. $P_k: a^5 a^k = a^{5+k}$
$P_{k+1}: a^5 a^{k+1} = a^{5+k+1}$

15. $P_k: 9^k - 1 = 4r$ for some integer r
$P_{k+1}: 9^{k+1} - 1 = 4s$ for some integer s

17. Prove: $2 + 6 + 10 + \cdots + (4n - 2) = 2n^2$, $n \in N$
Write: $P_n: 2 + 6 + 10 + \cdots + (4n - 2) = 2n^2$

Part 1: Show that P_1 is true:
$P_1: 2 = 2 \cdot 1^2$
       $= 2$ P_1 is true

Part 2: Show that if P_k is true, then P_{k+1} is true:
Write out P_k and P_{k+1}.
   $P_k: 2 + 6 + 10 + \cdots + (4k - 2) = 2k^2$
   $P_{k+1}: 2 + 6 + 10 + \cdots + (4k - 2) + (4k + 2) = 2(k + 1)^2$
We start with P_k:
   $2 + 6 + 10 + \cdots + (4k - 2) = 2k^2$
Adding $(4k + 2)$ to both sides, we get
   $2 + 6 + 10 + \cdots + (4k - 2) + (4k + 2) = 2k^2 + 4k + 2$
                                        $= 2(k^2 + 2k + 1)$
                                        $= 2(k + 1)^2$
We have shown that if P_k is true, then P_{k+1} is true.

Conclusion: P_n is true for all positive integers n.

> **Common Errors:**
> P_1 is not the nonsensical statement
> $2 + 6 + 10 + \cdots + (4 \cdot 1 - 2) = 2 \cdot 1^2$
> achieved by "substitution" of 1 for n. The left side of P_n has n terms; the left side of P_1 must
> have 1 term. Also, $P_{k+1} \neq P_k + 1$
> Nor is P_{k+1} simply its k + 1st term. $4k + 2 \neq 2(k + 1)^2$

19. Prove: $a^5 a^n = a^{5+n}$, $n \in N$
Write: $P_n: a^5 a^n = a^{5+n}$

Part 1: Show that P_1 is true:
$P_1: a^5 a^1 = a^{5+1}$

      $a^5 a = a^{5+1}$. True by the recursive definition of a^n.

Part 2: Show that if P_k is true, then P_{k+1} is true:
Write out P_k and P_{k+1}.

P_k: $a^5a^k = a^{5+k}$
P_{k+1}: $a^5a^{k+1} = a^{5+k+1}$

We start with P_k:

$a^5a^k = a^{5+k}$

Multiply both sides by a:

$a^5a^ka = a^{5+k}a$
$a^5a^{k+1} = a^{5+k+1}$ by the recursive definition of a^n.

We have shown that if P_k is true, then P_{k+1} is true.

Conclusion: P_n is true for all positive integers n.

21.   Prove: $9^n - 1$ is divisible by 4.
Write: P_n: $9^n - 1$ is divisible by 4

Part 1: Show that P_1 is true:
P_1: $9^1 - 1 = 8 = 4 \cdot 2$ is divisible by 4. Clearly true.

Part 2: Show that if P_k is true, then P_{k+1} is true:
Write out P_k and P_{k+1}.

P_k: $9^k - 1 = 4r$ for some integer r
P_{k+1}: $9^{k+1} - 1 = 4s$ for some integer s

We start with P_k:

$9^k - 1 = 4r$ for some integer r

Now, $9^{k+1} - 1 = 9^{k+1} - 9^k + 9^k - 1$
$= 9^k(9 - 1) + 9^k - 1$
$= 9^k \cdot 8 + 9^k - 1$
$= 4 \cdot 9^k \cdot 2 + 4r$
$= 4(9^k \cdot 2 + r)$

Therefore,
$9^{k+1} - 1 = 4s$ for some integer s $(= 2 \cdot 9^k + r)$
$9^{k+1} - 1$ is divisible by 4.
We have shown that if P_k is true, then P_{k+1} is true.

Conclusion: P_n is true for all positive integers n.

23.   The polynomial $x^4 + 1$ is a counterexample, since it has degree 4, and no real zeros.

25.   23 is a counterexample, since there are no prime numbers p such that $23 < p < 29$.

27. P_n: $2 + 2^2 + 2^3 + \cdots + 2^n = 2^{n+1} - 2$

Part 1: Show that P_1 is true:
P_1: $2 = 2^{1+1} - 2$
$= 2^2 - 2$
$= 4 - 2$
$= 2$ P_1 is true

Part 2: Show that if P_k is true, then P_{k+1} is true:
Write out P_k and P_{k+1}.

P_k: $2 + 2^2 + 2^3 + \cdots + 2^k = 2^{k+1} - 2$
P_{k+1}: $2 + 2^2 + 2^3 + \cdots + 2^k + 2^{k+1} = 2^{k+2} - 2$

We start with P_k:

$2 + 2^2 + 2^3 + \cdots + 2^k = 2^{k+1} - 2$

Adding 2^{k+1} to both sides,

$$
\begin{aligned}
2 + 2^2 + 2^3 + \cdots + 2^k + 2^{k+1} &= 2^{k+1} - 2 + 2^{k+1} \\
&= 2^{k+1} + 2^{k+1} - 2 \\
&= 2 \cdot 2^{k+1} - 2 \\
&= 2^1 \cdot 2^{k+1} - 2 \\
&= 2^{k+2} - 2
\end{aligned}
$$

We have shown that if P_k is true, then P_{k+1} is true.

Conclusion: P_n is true for all positive integers n.

29. P_n: $1^2 + 3^2 + 5^2 + \cdots + (2n - 1)^2 = \frac{1}{3}(4n^3 - n)$

Part 1: Show that P_1 is true:

$$
\begin{aligned}
P_1: 1^2 &= \frac{1}{3}(4 \cdot 1^3 - 1) \\
&= \frac{1}{3}(4 - 1) \\
&= 1 \qquad \text{True.}
\end{aligned}
$$

Part 2: Show that if P_k is true, then P_{k+1} is true:
Write out P_k and P_{k+1}.

$$P_k: 1^2 + 3^2 + 5^2 + \cdots + (2k - 1)^2 = \frac{1}{3}(4k^3 - k)$$

$$P_{k+1}: 1^2 + 3^2 + 5^2 + \cdots + (2k - 1)^2 + (2k + 1)^2 = \frac{1}{3}[4(k + 1)^3 - (k + 1)]$$

We start with P_k:

$$1^2 + 3^2 + 5^2 + \cdots + (2k - 1)^2 = \frac{1}{3}(4k^3 - k)$$

Adding $(2k + 1)^2$ to both sides,

$$
\begin{aligned}
1^2 + 3^2 + 5^2 + \cdots + (2k - 1)^2 + (2k + 1)^2 &= \frac{1}{3}(4k^3 - k) + (2k + 1)^2 \\
&= \frac{1}{3}[4k^3 - k + 3(2k + 1)^2] \\
&= \frac{1}{3}[4k^3 - k + 3(4k^2 + 4k + 1)] \\
&= \frac{1}{3}[4k^3 - k + 12k^2 + 12k + 3] \\
&= \frac{1}{3}[4k^3 + 12k^2 + 12k - k + 3] \\
&= \frac{1}{3}[4k^3 + 12k^2 + 12k + 4 - k - 1] \\
&= \frac{1}{3}[4(k^3 + 3k^2 + 3k + 1) - (k + 1)] \\
&= \frac{1}{3}[4(k + 1)^3 - (k + 1)]
\end{aligned}
$$

We have shown that if P_k is true, then P_{k+1} is true.

Conclusion: P_n is true for all positive integers n.

31. P_n: $1^2 + 2^2 + 3^2 + \cdots + n^2 = \dfrac{n(n + 1)(2n + 1)}{6}$

Part 1: Show that P_1 is true:

$$
\begin{aligned}
P_1: 1^2 &= \frac{1(1 + 1)(2 \cdot 1 + 1)}{6} \\
&= \frac{1 \cdot 2 \cdot 3}{6} \\
&= 1 \qquad P_1 \text{ is true.}
\end{aligned}
$$

Part 2: Show that if P_k is true, then P_{k+1} is true:
Write out P_k and P_{k+1}.

$$P_k: 1^2 + 2^2 + 3^2 + \cdots + k^2 = \frac{k(k + 1)(2k + 1)}{6}$$

$$P_{k+1}: 1^2 + 2^2 + 3^2 + \cdots + k^2 + (k + 1)^2 = \frac{(k + 1)(k + 2)(2k + 3)}{6}$$

We start with P_k:

$$1^2 + 2^2 + 3^2 + \cdots + k^2 = \frac{k(k + 1)(2k + 1)}{6}$$

Adding $(k + 1)^2$ to both sides,

$$1^2 + 2^2 + 3^2 + \cdots + k^2 + (k + 1)^2 = \frac{k(k + 1)(2k + 1)}{6} + (k + 1)^2$$

$$= \frac{k(k + 1)(2k + 1)}{6} + \frac{6(k + 1)^2}{6}$$

$$= \frac{k(k + 1)(2k + 1) + 6(k + 1)^2}{6}$$

$$= \frac{(k + 1)[k(2k + 1) + 6(k + 1)]}{6}$$

$$= \frac{(k + 1)[2k^2 + 7k + 6]}{6}$$

$$= \frac{(k + 1)(k + 2)(2k + 3)}{6}$$

We have shown that if P_k is true, then P_{k+1} is true.

Conclusion: P_n is true for all positive integers n.

33. $P_n: \dfrac{a^n}{a^3} = a^{n-3} \qquad n > 3$

Part 1: Show that P_4 is true:

$P_4: \dfrac{a^4}{a^3} = a^{4-3}$

But $\dfrac{a^4}{a^3} = \dfrac{a^3 a}{a^3} = a = a^1 = a^{4-3}$ So, P_4 is true.

Part 2: Show that if P_k is true, then P_{k+1} is true:
Write out P_k and P_{k+1}.

$$P_k: \frac{a^k}{a^3} = a^{k-3}$$

$$P_{k+1}: \frac{a^{k+1}}{a^3} = a^{k-2}$$

We start with P_k:

$$\frac{a^k}{a^3} = a^{k-3}$$

Multiply both sides by a:

$$\frac{a^k}{a^3} a = a^{k-3} a$$

$$\frac{a^{k+1}}{a^3} = a^{k-3+1} \text{ by the recursive definition of } a^n$$

$$\frac{a^{k+1}}{a^3} = a^{k-2}$$

We have shown that if P_k is true, then P_{k+1} is true.

Conclusion: P_n is true for all integers $n > 3$.

35. Write: P_n: $a^m a^n = a^{m+n}$ m an arbitrary element of N.

Part 1: Show that P_1 is true:

P_1: $a^m a^1 = a^{m+1}$ This is true by the recursive definition of a^n.

Part 2: Show that if P_k is true, then P_{k+1} is true:

Write out P_k and P_{k+1}.

   P_k: $a^m a^k = a^{m+k}$

   P_{k+1}: $a^m a^{k+1} = a^{m+k+1}$

We start with P_k:

   $a^m a^k = a^{m+k}$

Multiplying both sides by a:

   $a^m a^k a = a^{m+k} a$

   $a^m a^{k+1}$ $= a^{m+k+1}$ by the recursive definition of a^n.

We have shown that if P_k is true, then P_{k+1} is true.

Conclusion: $a^m a^n = a^{m+n}$ is true for arbitrary m, n positive integers.

37. Write: P_n: $x^n - 1 = (x - 1)Q_n(x)$ for some polynomial $Q_n(x)$.

Part 1: Show that P_1 is true:

P_1: $x^1 - 1 = (x - 1)Q_1(x)$ for some $Q_1(x)$. This is true since we can choose $Q_1(x) = 1$.

Part 2: Show that if P_k is true, then P_{k+1} is true:

Write out P_k and P_{k+1}.

   P_k: $x^k - 1 = (x - 1)Q_k(x)$ for some $Q_k(x)$

   P_{k+1}: $x^{k+1} - 1 = (x - 1)Q_{k+1}(x)$ for some $Q_{k+1}(x)$

We start with P_k:

$x^k - 1 = (x - 1)Q_k(x)$ for some $Q_k(x)$

Now, $x^{k+1} - 1 = x^{k+1} - x^k + x^k - 1$

$= x^k(x - 1) + x^k - 1$

$= x^k(x - 1) + (x - 1)Q_k(x)$

$= (x - 1)(x^k + Q_k(x))$

$= (x - 1)Q_{k+1}(x)$, where $Q_{k+1}(x) = x^k + Q_k(x)$

Conclusion: $x^n - 1 = (x - 1)Q_n(x)$ for all positive integers n. Thus $x^n - 1$ is divisible by $x - 1$ for all positive integers n.

39. Write: P_n: $x^{2n} - 1 = (x - 1)Q_n(x)$ for some polynomial $Q_n(x)$.

Part 1: Show that P_1 is true:

P_1: $x^2 - 1 = (x - 1)Q_1(x)$ for some polynomial $Q_1(x)$. This is true since we can choose $Q_1(x) = x + 1$.

Part 2: Show that if P_k is true, then P_{k+1} is true:

Write out P_k and P_{k+1}.

   P_k: $x^{2k} - 1 = (x - 1)Q_k(x)$ for some $Q_k(x)$

   P_{k+1}: $x^{2k+2} - 1 = (x - 1)Q_{k+1}(x)$ for some $Q_{k+1}(x)$

Now, $x^{2k+2} - 1$ $= x^{2k+2} - x^{2k} + x^{2k} - 1$

$= x^{2k}(x^2 - 1) + x^{2k} - 1$

$= x^{2k}(x + 1)(x - 1) + (x - 1)Q_k(x)$

$= (x - 1)[x^{2k}(x + 1) + Q_k(x)]$

$= (x - 1)Q_{k+1}(x)$, where $Q_{k+1}(x) = x^{2k}(x + 1) + Q_k(x)$

Conclusion: $x^{2n} - 1 = (x - 1)Q_n(x)$ for all positive integers n. Thus $x^{2n} - 1$ is divisible by $x - 1$ for all positive integers n.

41. P_n: $1^3 + 2^3 + 3^3 + \cdots + n^3 = (1 + 2 + 3 + \cdots + n)^2$

Part 1: Show that P_1 is true:

P_1: $1^3 = 1^2$ True

Part 2: Show that if P_k is true, then P_{k+1} is true:

Write out P_k and P_{k+1}.

   P_k: $1^3 + 2^3 + 3^3 + \cdots + k^3 = (1 + 2 + 3 + \cdots + k)^2$

   P_{k+1}: $1^3 + 2^3 + 3^3 + \cdots + k^3 + (k + 1)^3 = (1 + 2 + 3 + \cdots + k + k + 1)^2$

We take it as proved in text Matched Problem 1 that

   $1 + 2 + 3 + \cdots + n = \dfrac{n(n + 1)}{2}$ for $n \in N$

Thus, $1 + 2 + 3 + \cdots + k = \dfrac{k(k + 1)}{2}$

   $1 + 2 + 3 + \cdots + k + (k + 1) = \dfrac{(k + 1)(k + 2)}{2}$ are known

We start with P_k:

   $1^3 + 2^3 + 3^3 + \cdots + k^3 = (1 + 2 + 3 + \cdots + k)^2$

Adding $(k + 1)^3$ to both sides:

   $1^3 + 2^3 + 3^3 + \cdots + k^3 + (k + 1)^3 = (1 + 2 + 3 + \cdots + k)^2 + (k + 1)^3$

$$= \left[\frac{k(k + 1)}{2}\right]^2 + (k + 1)^3$$

$$= (k + 1)^2\left[\left(\frac{k}{2}\right)^2 + (k + 1)\right]$$

$$= (k + 1)^2\left[\frac{k^2}{4} + \frac{4(k + 1)}{4}\right]$$

$$= (k + 1)^2\left[\frac{k^2 + 4k + 4}{4}\right]$$

$$= \frac{(k + 1)^2(k + 2)^2}{4}$$

$$= \left[\frac{(k + 1)(k + 2)}{2}\right]^2$$

$$= [1 + 2 + 3 + \cdots + k + (k + 1)]^2$$

Conclusion: P_n is true for all positive integers n.

43. We note:

$$2 = \ \ 2 = 1 \cdot 2 \ \ \ n = 1$$
$$2 + 4 = \ \ 6 = 2 \cdot 3 \ \ \ n = 2$$
$$2 + 4 + 6 = 12 = 3 \cdot 4 \ \ \ n = 3$$
$$2 + 4 + 6 + 8 = 20 = 4 \cdot 5 \ \ \ n = 4$$

Hypothesis: P_n: $2 + 4 + 6 + \cdots + 2n = n(n + 1)$

Proof: Part 1: Show P_1 is true

P_1: $2 = 1 \cdot 2$ True

Part 2: Show that if P_k is true, then P_{k+1} is true:

Write out P_k and P_{k+1}.

   P_k: $2 + 4 + 6 + \cdots + 2k = k(k + 1)$

   P_{k+1}: $2 + 4 + 6 + \cdots + 2k + (2k + 2) = (k + 1)(k + 2)$

We start with P_k:

   $2 + 4 + 6 + \cdots + 2k = k(k + 1)$

Adding $2k + 2$ to both sides:

   $2 + 4 + 6 + \cdots + 2k + (2k + 2) = k(k + 1) + 2k + 2$
$$= (k + 1)k + (k + 1)2$$
$$= (k + 1)(k + 2)$$

Conclusion: The hypothesis P_n is true for all positive integers n.

45. $n = 1$: no line is determined.
$n = 2$: one line is determined.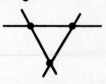
$n = 3$: three lines are determined.

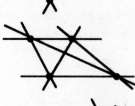

$n = 4$: six lines are determined.

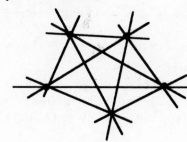

$n = 5$: ten lines are determined.

Hypothesis: P_n: n points (no three collinear) determine

$1 + 2 + 3 + \cdots + (n - 1) = \dfrac{n(n - 1)}{2}$ lines, $n \geq 2$.

Proof: Part 1: Show that P_2 is true.
P_2: 2 points determine one line:

$1 = \dfrac{2(2 - 1)}{2}$

$= \dfrac{2 \cdot 1}{2}$

$= 1$ is true

Part 2: Show that if P_k is true, then P_{k+1} is true:
Write out P_k and P_{k+1}.

P_k: k points determine $1 + 2 + 3 + \cdots + (k - 1) = \dfrac{k(k - 1)}{2}$ lines

P_{k+1}: $k + 1$ points determine $1 + 2 + 3 + \cdots + (k - 1) + k = \dfrac{(k + 1)k}{2}$ lines

We start with P_k:

k points determine $1 + 2 + 3 + \cdots + (k - 1) = \dfrac{k(k - 1)}{2}$ lines

Now, the $k + 1$st point will determine a total of k new lines, one with each of the previously existing k points. These k new lines will be added to the previously existing lines. Hence, $k + 1$ points determine

$1 + 2 + 3 + \cdots + (k - 1) + k = \dfrac{k(k - 1)}{2} + k$ lines

$= k \left[\dfrac{k - 1}{2} + 1 \right]$ lines

$= k \left[\dfrac{k - 1 + 2}{2} \right]$ lines

$= \dfrac{k(k + 1)}{2}$ lines

Conclusion: The hypothesis P_n is true for $n \geq 2$.

47. P_n: $a > 1 \Rightarrow a^n > 1$, $n \in N$.

Part 1: Show that P_1 is true:

P_1: $a > 1 \Rightarrow a^1 > 1$. This is automatically true.

Part 2: Show that if P_k is true, then P_{k+1} is true:

Write out P_k and P_{k+1}.

$\quad P_k$: $a > 1 \Rightarrow a^k > 1$

$\quad P_{k+1}$: $a > 1 \Rightarrow a^{k+1} > 1$

We start with P_k. Further, assume $a > 1$ and try to derive $a^{k+1} > 1$. If this succeeds, we have proved P_{k+1}.

Assume $a > 1$.

From P_k we know that $a^k > 1$, also $1 > 0$, hence $a > 0$. We may therefore multiply both sides of the inequality $a^k > 1$ by a without changing the sense of the inequality. Hence,

$\quad a^k a > 1a$

$\quad a^{k+1} > a$

But, $a > 1$ by assumption. Hence, $a^{k+1} > 1$. We have derived this from P_k and $a > 1$.

Thus, if P_k is true, then P_{k+1} is true.

Conclusion: P_n is true for all $n \in N$.

49. P_n: $n^2 > 2n$, $n \geq 3$.

Part 1: Show that P_3 is true:

P_3: $3^2 > 2 \cdot 3$

$\quad 9 > 6$ True

Part 2: Show that if P_k is true, then P_{k+1} is true:

Write out P_k and P_{k+1}.

$\quad P_k$: $k^2 > 2k$

$\quad P_{k+1}$: $(k + 1)^2 > 2(k + 1)$

We start with P_k:

$\quad k^2 > 2k$

Adding $2k + 1$ to both sides: $k^2 + 2k + 1 > 2k + 2k + 1$

$\quad (k + 1)^2 > 2k + 2 + 2k - 1$

Now, $2k - 1 > 0$, $k \in N$, hence $2k + 2 + 2k - 1 > 2k + 2$.

Therefore, $(k + 1)^2 > 2k + 2$

$\quad\quad\quad (k + 1)^2 > 2(k + 1)$

Thus, if P_k is true, then P_{k+1} is true.

Conclusion: P_n is true for all integers $n \geq 3$.

51. $3^4 + 4^4 + 5^4 + 6^4 = 2,258$

$\quad\quad\quad\quad\quad 7^4 = 2,401$

$3^4 + 4^4 + 5^4 + 6^4 \neq 7^4$

Thus, there is no true obvious generalization of the given facts.

53. To prove $a_n = b_n$, $n \in N$, write:

P_n: $a_n = b_n$.

Proof: Part 1: Show P_1 is true.

P_1: $a_1 = b_1$ $a_1 = 1$ $b_1 = 2 \cdot 1 - 1 = 1$

Thus, $a_1 = b_1$

Part 2: Show that if P_k is true, then P_{k+1} is true:

Write out P_k and P_{k+1}.

$\quad P_k$: $a_k = b_k$

$\quad P_{k+1}$: $a_{k+1} = b_{k+1}$

We start with P_k:

   $a_k = b_k$

Now, $a_{k+1} = a_{k+1-1} + 2 = a_k + 2 = b_k + 2 = 2k - 1 + 2 = 2k + 1 = 2(k + 1) - 1 = b_{k+1}$

Therefore, $a_{k+1} = b_{k+1}$

Thus, if P_k is true, then P_{k+1} is true.

Conclusion: P_n is true for all $n \in N$. Hence, $\{a_n\} = \{b_n\}$.

55. To prove: $a_n = b_n$, $n \in N$, write:

P_n: $a_n = b_n$.

Proof: Part 1: Show P_1 is true.

P_1: $a_1 = b_1$ $a_1 = 2$ $b_1 = 2^{2 \cdot 1 - 1} = 2^1 = 2$

Thus, $a_1 = b_1$

Part 2: Show that if P_k is true, then P_{k+1} is true:

Write out P_k and P_{k+1}.

   P_k: $a_k = b_k$

   P_{k+1}: $a_{k+1} = b_{k+1}$

We start with P_k:

   $a_k = b_k$

Now, $a_{k+1} = 2^2 a_{k+1-1} = 2^2 a_k = 2^2 b_k = 2^2 2^{2k-1} = 2^{2+2k-1} = 2^{2(k+1)-1} = b_{k+1}$

Therefore, $a_{k+1} = b_{k+1}$

Thus, if P_k is true, then P_{k+1} is true.

Conclusion: P_n is true for all $n \in N$. Hence, $\{a_n\} = \{b_n\}$.

Exercise 8-3

Key Ideas and Formulas

A sequence a_1, a_2, a_3 ..., a_n, ... is called an arithmetic sequence or arithmetic progression if there exists a constant d, called the *common difference*, such that $a_n - a_{n-1} = d$, that is, $a_n = a_{n-1} + d$ for every $n > 1$. Then $a_n = a_1 + (n - 1)d$ for every $n > 1$.

A sequence a_1, a_2, a_3, ..., a_n is called a geometric sequence or geometric progression if there exists a constant r, called the common ratio, such that

$$\frac{a_n}{a_{n-1}} = r, \text{ that is, } a_n = ra_{n-1}$$

Then $a_n = a_1 r^{n-1}$ for every $n > 1$

Sum of an arithmetic series: Let $S_n = \sum_{k=1}^{n} a_n$. Then

$$S_n = \frac{n}{2}[2a_1 + (n - 1)d] = \frac{n}{2}(a_1 + a_n)$$

Sum of a geometric series: Let $S_n = \sum_{k=1}^{n} a_k$, then

$$S_n = \frac{a_1 - a_1 r^n}{1 - r} = \frac{a_1 - ra_n}{1 - r} \quad r \neq 1$$

Sum of an infinite geometric series

$S_\infty = \dfrac{a_1}{1 - r}$ $|r| < 1$ *only*. If $r > 1$ or $r < -1$ the infinite geometric series has no sum.

1. (A) Since $(-16) - (-11) = -5$ and $(-21) - (-16) = -5$, the given terms can start an arithmetic sequence with $d = -5$. Then next terms are then $-21 + (-5) = -26$, and $(-26) + (-5) = -31$.

(B) Since $(-4) - 2 = -6$ and $8 - (-4) = 12$, there is no common difference. Since $(-4) \div 2 = -2$ and $8 \div (-4) = -2$, the given terms can start a geometric sequence with $r = -2$. The next terms are then $8 \cdot (-2) = -16$, and $(-16) \cdot (-2) = 32$.

(C) Since $4 - 1 \neq 9 - 4$, there is no common difference, so the sequence is not an arithmetic sequence. Since $4 \div 1 \neq 9 \div 4$, there is no common ratio, so the sequence is not geometric either.

(D) Since $\frac{1}{6} - \frac{1}{2} = -\frac{1}{3}$ and $\frac{1}{18} - \frac{1}{6} = -\frac{1}{9}$, there is no common difference. Since $\frac{1}{6} \div \frac{1}{2} = \frac{1}{3}$ and $\frac{1}{18} \div \frac{1}{6} = \frac{1}{3}$, the given terms can start a geometric sequence with $r = \frac{1}{3}$. The next terms are then $\frac{1}{18} \cdot \frac{1}{3} = \frac{1}{54}$ and $\frac{1}{54} \cdot \frac{1}{3} = \frac{1}{162}$.

3. $a_2 = a_1 + d = -5 + 4 = -1$
$a_3 = a_2 + d = -1 + 4 = 3$
$a_4 = a_3 + d = 3 + 4 = 7$
$a_n = a_1 + (n - 1)d$

5. $a_{15} = a_1 + 14d = -3 + 14 \cdot 5 = 67$
$S_n = \frac{n}{2}[2a_1 + (n - 1)d]$
$S_{11} = \frac{11}{2}[2(-3) + (11 - 1)5]$
$\qquad = \frac{11}{2}(44)$
$\qquad = 242$

7. $a_2 - a_1 = 5 - 1 = d = 4$
$S_n = \frac{n}{2}[2a_1 + (n - 1)d]$
$S_{21} = \frac{21}{2}[2 \cdot 1 + (21 - 1)4]$
$\qquad = \frac{21}{2}(82)$
$\qquad = 861$

9. $a_2 - a_1 = 5 - 7 = -2 = d$
$a_n = a_1 + (n - 1)d$
$a_{15} = 7 + (15 - 1)(-2)$
$\qquad = -21$

11. $a_2 = a_1 r = (-6)\left(-\frac{1}{2}\right) = 3$
$a_3 = a_2 r = 3\left(-\frac{1}{2}\right) = -\frac{3}{2}$
$a_4 = a_3 r = \left(-\frac{3}{2}\right)\left(-\frac{1}{2}\right) = \frac{3}{4}$

13. $a_n = a_1 r^{n-1}$
$a_{10} = 81\left(\frac{1}{3}\right)^{10-1}$
$\qquad = \frac{1}{243}$

15. $S_n = \frac{a_1 - ra_n}{1 - r}$
$S_7 = \frac{3 - 3(2,187)}{1 - 3}$
$\qquad = 3,279$

17. $a_n = a_1 + (n - 1)d$
$a_{20} = a_1 + 19d$
$117 = 3 + 19d$
So, $d = 6$. Therefore
$a_{101} = a_1 + (100)d$
$\qquad = 3 + 100(6)$
$\qquad = 603$

19. $S_n = \frac{n}{2}(a_1 + a_n)$
$S_{40} = \frac{40}{2}(-12 + 22)$
$\qquad = 200$

21. $a_2 - a_1 = d$

$\dfrac{1}{2} - \dfrac{1}{3} = \dfrac{1}{6} = d$

$a_n = a_1 + (n - 1)d$

$a_{11} = \dfrac{1}{3} + (11 - 1)\dfrac{1}{6}$

$= 2$

$S_n = \dfrac{n}{2}(a_1 + a_n)$

$S_{11} = \dfrac{11}{2}\left(\dfrac{1}{3} + 2\right)$

$= \dfrac{77}{6}$

23. $a_n = a_1 + (n - 1)d$

$a_{10} = a_1 + 9d$

$a_3 = a_1 + 2d$

Eliminating d between these two statements by addition, we have

$2a_{10} = 2a_1 + 18d$

$\dfrac{-9a_3 = -9a_1 - 18d}{2a_{10} - 9a_3 = -7a_1}$

$a_1 = \dfrac{2a_{10} - 9a_3}{-7} = \dfrac{2(55) - 9(13)}{-7}$

$= 1$

25. $a_n = a_1 r^{n-1}$

$1 = 100 r^{6-1}$

$\dfrac{1}{100} = r^5$

$r = \sqrt[5]{0.01} = 10^{-2/5}$

$r = 0.398$

27. $S_n = \dfrac{a_1 - a_1 r^n}{1 - r}$

$S_{10} = \dfrac{5 - 5(-2)^{10}}{1 - (-2)}$

$= \dfrac{-5,115}{3}$

$= -1,705$

29. First find r:

$a_n = a_1 r^{n-1}$

$\dfrac{8}{3} = 9 r^{4-1}$

$\dfrac{8}{27} = r^3$

$r = \dfrac{2}{3}$

$a_2 = r a_1 = \dfrac{2}{3}(9) = 6$

$a_3 = r a_2 = \dfrac{2}{3}(6) = 4$

31. $a_n = a_1 + (n - 1)d$

$d = a_2 - a_1$

$= (3 \cdot 2 + 3) - (3 \cdot 1 + 3) = 3$

$a_{51} = a_1 + (51 - 1)d$

$= (3 \cdot 1 + 3) + 50 \cdot 3$

$= 156$

$S_n = \dfrac{n}{2}(a_1 + a_n)$

$S_{51} = \dfrac{51}{2}(3 \cdot 1 + 3 + 156)$

$= \dfrac{51}{2} \cdot 162$

$= 4,131$

33. $S_n = \dfrac{a_1 - a_1 r^n}{1 - r}$

First, note $a_1 = (-3)^{1-1} = 1$

$r = \dfrac{a_2}{a_1} = \dfrac{(-3)^{2-1}}{(-3)^{1-1}} = -3$

$S_7 = \dfrac{1 - 1(-3)^7}{1 - (-3)}$

$= \dfrac{2,188}{4}$

$= 547$

35. $g(t) = 5 - t$

$g(1) = 5 - 1 = 4$

$g(51) = 5 - 51 = -46$

$g(1) + g(2) + g(3) + \cdots + g(51) = S_{51}$

$S_n = \dfrac{n}{2}(a_1 + a_n)$

$S_{51} = \dfrac{51}{2}(g(1) + g(51))$

$= \dfrac{51}{2}[4 + (-46)]$

$= -1,071$

37. $g(1) + g(2) + \cdots + g(10)$ is a geometric series, with

$$g(1) = a_1 = \left(\frac{1}{2}\right)^1 = \frac{1}{2}$$

$$r = \frac{g(2)}{g(1)} = \frac{\left(\frac{1}{2}\right)^2}{\left(\frac{1}{2}\right)^1} = \frac{1}{2}$$

$$S_n = \frac{a_1 - a_1 r^n}{1 - r}$$

$$S_{10} = \frac{\frac{1}{2} - \frac{1}{2}\left(\frac{1}{2}\right)^{10}}{1 - \frac{1}{2}}$$

$$= \frac{1 - \left(\frac{1}{2}\right)^{10}}{2 - 1}$$

$$= \frac{1,023}{1,024}$$

39. First, find n:

$$a_n = a_1 + (n - 1)d$$
$$134 = 22 + (n - 1)2$$
$$n = 57$$

Now, find S_{57}

$$S_n = \frac{n}{2}(a_1 + a_n)$$

$$S_{57} = \frac{57}{2}(22 + 134)$$

$$= 4,446$$

41. To prove:

$$1 + 3 + 5 + \cdots + (2n - 1) = n^2$$

The sequence 1, 3, 5, $\cdots$ is an arithmetic sequence, with $d = 2$. We are to find S_n.

But, $S_n = \frac{n}{2}(a_1 + a_n)$

$$= \frac{n}{2}[1 + (2n - 1)]$$

$$= \frac{n}{2} \cdot 2n$$

So, $S_n = n^2$

43. $\dfrac{a_2}{a_1} = \dfrac{a_3}{a_2} = r$ for $a_1 + a_2 + a_3$ to be a geometric series. Hence,

$$\frac{x}{-2} = \frac{-6}{x}$$
$$x^2 = 12$$
$$x = 2\sqrt{3},$$

since x is specified positive.

45. Note that $a_n - a_{n-1} = 3$. Hence, this is an arithmetic sequence, with $d = 3$. We are to find a_n.

But, $a_n = a_1 + (n - 1)d$

So, $a_n = -3 + (n - 1)3$　or　$3n - 6$

47. Here are computer-generated graphs of $\{a_n\}$ (the dots suggest a straight line) and $\{b_n\}$.

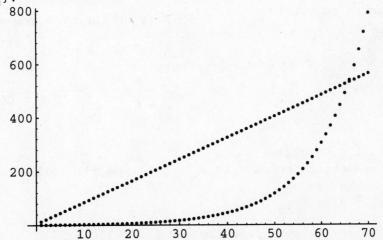

The graph indicates that 66 is the least positive integer n such that $a_n < b_n$. From the table display we note

$$n = 65 \qquad a_n = 525 \qquad b_n = 490.371$$
$$n = 66 \qquad a_n = 533 \qquad b_n = 539.408$$

confirming the graph.

49. Here are computer-generated graphs of $\{a_n\}$ (the dots suggest a straight line) and $\{b_n\}$.

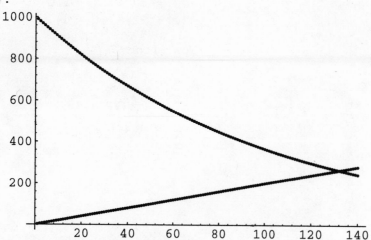

The graph indicates that the least positive integer n such that $a_n < b_n$ is between 132 and 136. From the table display we note

$$n = 132 \qquad a_n = 265.366 \qquad b_n = 265$$
$$n = 133 \qquad a_n = 262.713 \qquad b_n = 267$$

confirming that 133 is the least positive integer n required.

51. $\dfrac{a_2}{a_1} = 1 \div 3 = \dfrac{1}{3} = r. \quad |r| < 1$

Therefore, this infinite geometric series has a sum.

$$S_\infty = \frac{a_1}{1 - r}$$
$$= \frac{3}{1 - \frac{1}{3}} = \frac{9}{2}$$

53. $\dfrac{a_2}{a_1} = \dfrac{4}{2} = 2 = r \geq 1$ Therefore, this infinite geometric series has no sum.

55. $\dfrac{a_2}{a_1} = \left(-\dfrac{1}{2}\right) \div 2 = -\dfrac{1}{4} = r. \quad |r| < 1$

Therefore, this infinite geometric series has a sum.

$$S_\infty = \frac{a_1}{1 - r}$$
$$= \frac{2}{1 - (-\frac{1}{4})}$$
$$= \frac{8}{5}$$

57. $0.\overline{7} = 0.777\ldots = 0.7 + 0.07 + 0.007$
$$+ 0.0007 + \ldots$$

This is an infinite geometric series with

$a_1 = 0.7$ and $r = 0.1$.

Thus,

$$S_\infty = \frac{a_1}{1 - r} = \frac{0.7}{1 - 0.1} = \frac{0.7}{0.9} = \frac{7}{9}$$

59. $0.\overline{54}\ldots = 0.54 + 0.0054 + 0.000054 + \ldots$

This is an infinite geometric series with $a_1 = 0.54$ and $r = 0.01$. Thus,

$$S_\infty = \frac{a_1}{1 - r} = \frac{0.54}{1 - 0.01} = \frac{0.54}{0.99} = \frac{6}{11}$$

61. $3.\overline{216} = 3.216216216\ldots = 3 + 0.216 + 0.000216 + 0.000000216 + \ldots$

Therefore, we note: $0.216 + 0.000216 + 0.000000216 + \ldots$ is an infinite geometric series with $a_1 = 0.216$ and $r = 0.001$. Thus,

$$3.\overline{216} = 3 + S_\infty = 3 + \frac{a_1}{1 - r} = 3 + \frac{0.216}{1 - 0.001} = 3 + \frac{0.216}{0.999} = 3\frac{8}{37} \text{ or } \frac{119}{37}$$

63. Write: $P_n: a_n = a_1 + (n - 1)d$

Proof: Part 1: Show that P_1 is true:
$P_1: a_1 = a_1 + (1 - 1)d = a_1$ P_1 is true.

Part 2: Show that if P_k is true, then P_{k+1} is true:
Write out P_k and P_{k+1}.

$P_k: a_k = a_1 + (k - 1)d$
$P_{k+1}: a_{k+1} = a_1 + kd$
We start with P_k:

$a_k = a_1 + (k - 1)d$
Adding d to both sides:

$a_k + d = a_1 + (k - 1)d + d$
$a_{k+1} = a_1 + d[(k - 1) + 1]$
$a_{k+1} = a_1 + kd$
Thus, if P_k is true, then P_{k+1} is true.

Conclusion: P_n is true for all positive integers n.

65. This is a geometric sequence with ratio $\dfrac{a_n}{a_{n-1}} = -3$. Hence,

$a_n = a_1 r^{n-1}$
$= (-2)(-3)^{n-1}$

67. Assume x, y, z are consecutive terms of an arithmetic progression.
Then, $y - x = d$, $z - y = d$
To show $a = x^2 + xy + y^2$, $b = z^2 + xz + x^2$, $c = y^2 + yz + z^2$, are consecutive
terms of an arithmetic progression, we need only show that
$b - a = c - b$
$b - a = z^2 + xz + x^2 - (x^2 + xy + y^2)$
$= z^2 + xz - xy - y^2$
$= z^2 + x(z - y) - y^2$
$= z^2 - y^2 + x(z - y)$
$= (z - y)(z + y + x)$
$= d(x + y + z)$
$c - b = y^2 + yz + z^2 - (z^2 + xz + x^2)$
$= y^2 + yz - xz - x^2$
$= y^2 - x^2 + yz - xz$
$= y^2 - x^2 + z(y - x)$
$= (y - x)[y + x + z]$
$= d(x + y + z)$
Hence, a, b, c, that is, $x^2 + xy + y^2$, $z^2 + xz + x^2$, $y^2 + yz + z^2$ are
consecutive terms of an arithmetic progression.

69. Write: $P_n: a_n = a_1 r^{n-1}$, $n \in N$

Proof: Part 1: Show that P_1 is true:
$P_1: a_1 = a_1 r^{1-1}$
$= a_1 r^0$
$= a_1$ True.

Part 2: Show that if P_k is true, then P_{k+1} is true:
Write out P_k and P_{k+1}.

$P_k: a_k = a_1 r^{k-1}$
$P_{k+1}: a_{k+1} = a_1 r^k$
We start with P_k:

$a_k = a_1 r^{k-1}$

Multiply both sides by r:
$$a_k r = a_1 r^{k-1} r$$
$$a_k r = a_1 r^k$$

But, we are given that $a_{k+1} = r a_k = a_k r$.

Hence, $a_{k+1} = a_1 r^k$

Thus, if P_k is true, then P_{k+1} is true.

Conclusion: P_n is true for all $n \in N$.

71. Given a, b, c, d, e, f is an arithmetic progression, let D be the common difference. Then, assuming $D \neq 0$:

$b = a + D$, $c = a + 2D$, $d = a + 3D$, $e = a + 4D$, $f = a + 5D$

We know from Cramer's Rule that there will be a unique solution if

$$\begin{vmatrix} a & b \\ d & e \end{vmatrix} \neq 0$$

$$\begin{vmatrix} a & b \\ d & e \end{vmatrix} = \begin{vmatrix} a & a + D \\ a + 3D & a + 4D \end{vmatrix} = a(a + 4D) - (a + 3D)(a + D)$$
$$= a^2 + 4aD - (a^2 + 4aD + 3D^2)$$
$$= -3D^2 \neq 0$$

We can now calculate x and y, since

$$x = \frac{\begin{vmatrix} c & b \\ f & e \end{vmatrix}}{-3D^2} = \frac{ce - fb}{-3D^2} = \frac{(a + 2D)(a + 4D) - (a + 5D)(a + D)}{-3D^2}$$

$$= \frac{a^2 + 6aD + 8D^2 - (a^2 + 6aD + 5D^2)}{-3D^2}$$

$$= \frac{3D^2}{-3D^2}$$

$$= -1$$

$$y = \frac{\begin{vmatrix} a & c \\ d & f \end{vmatrix}}{-3D^2} = \frac{af - dc}{-3D^2} = \frac{a(a + 5D) - (a + 3D)(a + 2D)}{-3D^2}$$

$$= \frac{a^2 + 5aD - (a^2 + 5aD + 6D^2)}{-3D^2}$$

$$= \frac{-6D^2}{-3D^2}$$

$$= 2$$

73. With each firm, the salaries form an arithmetic sequence. We are asked for the sum of fifteen terms, or S_{15}, given a_1 and d.

$$S_n = \frac{n}{2}[2a_1 + (n - 1)d]$$

$$S_{15} = \frac{15}{2}[2a_1 + 14d] = 15(a_1 + 7d)$$

Firm A: $a_1 = 25,000$ $d = 1,200$ $S_{15} = 15(25,000 + 7 \cdot 1,200) = \$501,000$

Firm B: $a_1 = 28,000$ $d = 800$ $S_{15} = 15(28,000 + 7 \cdot 800) = \$504,000$

75. We are asked for the sum of an infinite geometric series.

$a_1 = \$800,000$

$r = 0.8 \quad |r| \leq 1$,

so the series has a sum,

$$S_\infty = \frac{a_1}{1 - r}$$

$$= \frac{\$800,000}{1 - 0.8}$$

$$= \frac{\$800,000}{0.2}$$

$$= \$4,000,000$$

77. After one year, $P(1 + r)$ is present. Hence, the geometric sequence has $a_1 = P(1 + r)$. The ratio is given as $(1 + r)$, hence

$$a_n = a_1(1 + r)^{n-1}$$

$$= P(1 + r)(1 + r)^{n-1}$$

$$= P(1 + r)^n \text{ is the amount present}$$

after n years.

The time taken for P to double is represented by n years. We set $A = 2P$, $r = 0.06$, then solve

$$2P = P(1 + 0.06)^n$$

$$2 = (1.06)^n$$

$$\log 2 = n \log 1.06$$

$$n = \frac{\log 2}{\log 1.06}$$

$$\approx 12 \text{ years}$$

79. This involves an arithmetic sequence. Let d = increase in earnings each year.

$$a_1 = 7,000$$

$$a_{11} = 14,000$$

$$a_n = a_1 + (n - 1)d$$

$$14,000 = 7,000 + (11 - 1)d$$

$$d = \$700$$

The amount of money received over the 11 years is

$$S_{11} = a_1 + a_2 + \cdots + a_{11}$$

$$S_n = \frac{n}{2}(a_1 + a_n)$$

$$S_{11} = \frac{11}{2}(7,000 + 14,000) = \$115,500$$

81. We are asked for the sum of an infinite geometric series.

a_1 = number of revolutions in the first minute = 300

$r = \dfrac{2}{3} \quad |r| < 1$, so the series has a sum,

$$S_\infty = \frac{a_1}{1 - r} = \frac{300}{1 - \frac{2}{3}} = 900 \text{ revolutions.}$$

83. This involves a geometric sequence. Let $a_n = 2,000$ calories. There are five stages: $n = 5$. We require a_1 on the assumption that $r = 20\% = \dfrac{1}{5}$.

$$a_n = a_1 r^{n-1}$$

$$2000 = a_1\left(\frac{1}{5}\right)^{5-1}$$

$$a_1 = 2,000 \cdot 5^4$$

$$= 1,250,000 \text{ calories}$$

85. We have an arithmetic sequence, 16, 48, 80, $\cdots$, with $a_1 = 16$, $d = 32$.

(A) This requires a_{11}: $a_n = a_1 + (n - 1)d$

$$a_{11} = 16 + (11 - 1)32$$

$$= 336 \text{ feet}$$

(B) This requires s_{11}: $s_n = \dfrac{n}{2}(a_1 + a_n)$

$$s_{11} = \frac{11}{2}(16 + 336)$$

$$= 1,936 \text{ feet}$$

(C) This requires s_t: $\quad s_n = \dfrac{n}{2}[2a_1 + (n-1)d]$

$$s_t = \dfrac{t}{2}[2a_1 + (t-1)32]$$

$$= \dfrac{t}{2}[32 + (t-1)32]$$

$$= \dfrac{t}{2} \cdot 32t$$

$$= 16t^2 \text{ feet}$$

87. This involves a geometric sequence. Let $a_1 = 2A_0 = $ number present after 1 half-hour period. In t hours, $2t$ half-hours will have elapsed, hence, $n = 2t$; $r = 2$, since the number of bacteria doubles in each period.

$$a_n = a_1 r^{n-1}$$
$$a_{2t} = 2A_0 2^{2t-1}$$
$$= A_0 2^1 2^{2t-1}$$
$$= A_0 2^{2t}$$

89. If b_n is the brightness of an nth-magnitude star, we find r for the geometric progression b_1, b_2, b_3, ..., given $b_1 = 100b_6$.

$$b_n = b_1 r^{n-1}$$
$$b_6 = b_1 r^{6-1}$$
$$b_6 = b_1 r^5$$
$$b_6 = 100b_6 r^5$$
$$\dfrac{1}{100} = r^5$$
$$r = \sqrt[5]{0.01} = 10^{-0.4} = 0.398$$

91. This involves a geometric sequence. $a_1 = $ amount of money on first square; $a_1 = \$0.01$, $r = 2$; $a_n = $ amount of money on nth square.

$$a_n = a_1 r^{n-1}$$
$$a_{64} = 0.01 \cdot 2^{64-1}$$
$$= 9.22 \times 10^{16} \text{ dollars}$$

The amount of money on the whole board $= a_1 + a_2 + a_3 + \cdots + a_{64} = S_{64}$.

$$S_n = \dfrac{a_1 - ra_n}{1 - r}$$
$$S_{64} = \dfrac{0.01 - 2(9.223 \times 10^{16})}{1 - 2}$$
$$= 1.845 \times 10^{17} \text{ dollars}$$

93. This involves a geometric sequence. Let $a_1 = 15 = $ pressure at sea level, $r = \dfrac{1}{10} = $ factor of decrease. We require $a_5 = $ pressure after four 10-mile increases in altitude.

$$a_5 = 15\left(\dfrac{1}{10}\right)^{5-1}$$

$$= 0.0015 \text{ pounds per square inch.}$$

95. This involves an infinite geometric series. $a_1 = $ perimeter of first triangle $= 1$, $r = \dfrac{1}{2}$, $|r| < 1$, so the series has a sum.

$$S_\infty = \dfrac{a_1}{1 - r} = \dfrac{1}{1 - \frac{1}{2}} = 2$$

97.

From the figure, and ordinary induction, it should be clear that the interior angles of an $n + 2$-sided polygon, $n = 1, 2, 3, \cdots$, are those of n triangles, that is, $180°n$. Thus, for the sequence of interior angles $\{a_n\}$,

$a_n - a_{n-1} = 180°n - 180°(n - 1) = 180° = d$.

A detailed proof by mathematical induction is omitted.

For a 21-sided polygon, we use $a_n = a_1 + (n - 1)d$ with $a_1 = 180°$, $d = 180°$, and $n + 2 = 21$, hence, $n = 19$.

$a_{19} = 180° + (19 - 1)180°$

$\quad = 3,420°$

Exercise 8-4

Key Ideas and Formulas

For $n \in N$, $n! = n(n - 1) \cdots 2 \cdot 1$

$\qquad\qquad 1! = 0! = 1$

$\qquad\qquad n! = n \cdot (n - 1)!$

$(a + b)^0 = 1$

$(a + b)^1 = a + b$

$(a + b)^2 = a^2 + 2ab + b^2$

$(a + b)^3 = a^3 + 3a^2b + 3ab^2 + b^3$

$(a + b)^4 = a^4 + 4a^3b + 6a^2b^2 + 4ab^3 + b^4$

$(a + b)^5 = a^5 + 5a^4b + 10a^3b^2 + 10a^2b^3 + 5ab^4 + b^5$

$$\binom{n}{r} = \frac{n!}{r!(n - r)!} = \frac{n(n - 1)(n - 2)\cdots(n - r + 1)}{r(r - 1) \cdots 2 \cdot 1} = {_nC_r}$$

Binomial Formula

$$(a + b)^n = \sum_{k=0}^{n} \binom{n}{k} a^{n-k}b^k \qquad n \in N, \; n \geq 1$$

---

1. $6! = 6 \cdot 5 \cdot 4 \cdot 3 \cdot 2 \cdot 1 = 720$

3. $\dfrac{20!}{19!} = \dfrac{20 \cdot 19!}{19!} = 20$

5. $\dfrac{10!}{7!} = \dfrac{10 \cdot 9!}{7!} = \dfrac{10 \cdot 9 \cdot 8!}{7!} = \dfrac{10 \cdot 9 \cdot 8 \cdot 7!}{7!} = 10 \cdot 9 \cdot 8 = 720$

7. $\dfrac{6!}{4!2!} = \dfrac{6 \cdot 5 \cdot 4 \cdot 3 \cdot 2 \cdot 1}{4 \cdot 3 \cdot 2 \cdot 1 \cdot 2 \cdot 1} = \dfrac{6 \cdot 5}{2 \cdot 1} = 15$

9. $\dfrac{9!}{0!(9 - 0)!} = \dfrac{9!}{1(9!)} = 1$

11. $\dfrac{8!}{2!(8 - 2)!} = \dfrac{8!}{2!6!} = \dfrac{8 \cdot 7 \cdot 6 \cdot 5 \cdot 4 \cdot 3 \cdot 2 \cdot 1}{2 \cdot 1 \cdot 6 \cdot 5 \cdot 4 \cdot 3 \cdot 2 \cdot 1} = \dfrac{8 \cdot 7}{2 \cdot 1} = 28$

13. Since $n! = n(n - 1)!$, $\dfrac{n!}{(n - 1)!} = n$. Hence, $9 = \dfrac{9!}{(9 - 1)!} = \dfrac{9!}{8!}$.

15. $6 \cdot 7 \cdot 8 = \dfrac{1 \cdot 2 \cdot 3 \cdot 4 \cdot 5 \cdot 6 \cdot 7 \cdot 8}{1 \cdot 2 \cdot 3 \cdot 4 \cdot 5} = \dfrac{8!}{5!}$

17. $\dbinom{9}{5} = \dfrac{9!}{5!(9 - 5)!} = \dfrac{9!}{5!4!} = \dfrac{9 \cdot 8 \cdot 7 \cdot 6 \cdot 5!}{5!4 \cdot 3 \cdot 2 \cdot 1} = \dfrac{9 \cdot 8 \cdot 7 \cdot 6}{4 \cdot 3 \cdot 2 \cdot 1} = 126$

19. $\binom{6}{5} = \dfrac{6!}{5!(6-5)!} = \dfrac{6!}{5!1!} = \dfrac{6 \cdot 5!}{5! \cdot 1} = 6$

21. $\binom{9}{9} = \dfrac{9!}{9!(9-9)!} = \dfrac{9!}{9!0!} = \dfrac{1}{0!} = \dfrac{1}{1} = 1$

23. $\binom{17}{13} = \dfrac{17!}{13!(17-13)!} = \dfrac{17!}{13!4!} = \dfrac{17 \cdot 16 \cdot 15 \cdot 14 \cdot 13!}{13!4 \cdot 3 \cdot 2 \cdot 1} = \dfrac{17 \cdot 16 \cdot 15 \cdot 14}{4 \cdot 3 \cdot 2 \cdot 1} = 2{,}380$

25. The answer to this question depends on the model of calculator. An example: on a TI-82 model calculator, $69! = 1.711 \times 10^{98}$, but $70!$ produces an overflow error.

27. $(m+n)^3 = \displaystyle\sum_{k=0}^{3} \binom{3}{k} m^{3-k} n^k$

$\qquad = \binom{3}{0} m^3 + \binom{3}{1} m^2 n + \binom{3}{2} mn^2 + \binom{3}{3} n^3$

$\qquad = m^3 + 3m^2 n + 3mn^2 + n^3$

29. $(2x - 3y)^3 = [2x + (-3y)]^3 = \displaystyle\sum_{k=0}^{3} \binom{3}{k}(2x)^{3-k}(-3y)^k$

$\qquad = \binom{3}{0}(2x)^3 + \binom{3}{1}(2x)^2(-3y)^1 + \binom{3}{2}(2x)(-3y)^2 + \binom{3}{3}(-3y)^3$

$\qquad = 8x^3 + 3(4x^2)(-3y) + 3(2x)(9y^2) + (-27y^3)$

$\qquad = 8x^3 - 36x^2 y + 54xy^2 - 27y^3$

31. $(x - 2)^4 = [x + (-2)]^4 = \displaystyle\sum_{k=0}^{4} \binom{4}{k} x^{4-k}(-2)^k$

$\qquad = \binom{4}{0}x^4 + \binom{4}{1}x^3(-2)^1 + \binom{4}{2}x^2(-2)^2 + \binom{4}{3}x(-2)^3 + \binom{4}{4}(-2)^4$

$\qquad = x^4 - 8x^3 + 24x^2 - 32x + 16$

33. $(m + 3n)^4 = \displaystyle\sum_{k=0}^{4} \binom{4}{k} m^{4-k}(3n)^k$

$\qquad = \binom{4}{0}m^4 + \binom{4}{1}m^3(3n)^1 + \binom{4}{2}m^2(3n)^2 + \binom{4}{3}m(3n)^3 + \binom{4}{4}(3n)^4$

$\qquad = m^4 + 12m^3 n + 54m^2 n^2 + 108mn^3 + 81n^4$

35. $(2x - y)^5 = [2x + (-y)]^5 = \displaystyle\sum_{k=0}^{5} \binom{5}{k}(2x)^{5-k}(-y)^k$

$\qquad = \binom{5}{0}(2x)^5 + \binom{5}{1}(2x)^4(-y)^1 + \binom{5}{2}(2x)^3(-y)^2 + \binom{5}{3}(2x)^2(-y)^3$

$\qquad\qquad + \binom{5}{4}(2x)(-y)^4 + \binom{5}{5}(-y)^5$

$\qquad = 32x^5 - 80x^4 y + 80x^3 y^2 - 40x^2 y^3 + 10xy^4 - y^5$

37. $(m + 2n)^6 = \sum_{k=0}^{6} \binom{6}{k} m^{6-k} (2n)^k$

$= \binom{6}{0} m^6 + \binom{6}{1} m^5 (2n)^1 + \binom{6}{2} m^4 (2n)^2 + \binom{6}{3} m^3 (2n)^3 + \binom{6}{4} m^2 (2n)^4$

$+ \binom{6}{5} m(2n)^5 + \binom{6}{6} (2n)^6$

$= m^6 + 12m^5 n + 60m^4 n^2 + 160m^3 n^3 + 240m^2 n^4 + 192mn^5 + 64n^6$

39. In the expansion of $(a + b)^n$, the exponent of b in the rth term is $r - 1$ and the exponent of a is $n - (r - 1)$. Here, $r = 7$, $n = 15$.

Seventh term $= \binom{15}{6} u^9 v^6$

$= \frac{15!}{9!6!} u^9 v^6$

$= \frac{15 \cdot 14 \cdot 13 \cdot 12 \cdot 11 \cdot 10}{6 \cdot 5 \cdot 4 \cdot 3 \cdot 2 \cdot 1} u^9 v^6$

$= 5,005 u^9 v^6$

41. In the expansion of $(a + b)^n$, the exponent of b in the rth term is $r - 1$ and the exponent of a is $n - (r - 1)$. Here, $r = 11$, $n = 12$.

Eleventh term $= \binom{12}{10} (2m)^2 n^{10}$

$= \frac{12!}{10!2!} 4m^2 n^{10}$

$= \frac{12 \cdot 11}{2 \cdot 1} 4m^2 n^{10}$

$= 264 m^2 n^{10}$

43. In the expansion of $(a + b)^n$, the exponent of b in the rth term is $r - 1$ and the exponent of a is $n - (r - 1)$. Here, $r = 7$, $n = 12$.

Seventh term $= \binom{12}{6} \left(\frac{w}{2}\right)^6 (-2)^6$

$= \frac{12!}{6!6!} \frac{w^6}{2^6} 2^6$

$= \frac{12 \cdot 11 \cdot 10 \cdot 9 \cdot 8 \cdot 7 \cdot 6!}{6 \cdot 5 \cdot 4 \cdot 3 \cdot 2 \cdot 1 \cdot 6!} w^6$

$= 924 w^6$

45. In the expansion of $(a + b)^n$, the exponent of b in the rth term is $r - 1$ and the exponent of a is $n - (r - 1)$. Here, $r = 6$, $n = 8$.

Sixth term $= \binom{8}{5} (3x)^3 (-2y)^5$

$= \frac{8!}{5!3!} (3x)^3 (-2y)^5$

$= \frac{8 \cdot 7 \cdot 6 \cdot 5!}{5! \cdot 3 \cdot 2 \cdot 1} (27x^3)(-32y^5)$

$= -(56)(27)(32) x^3 y^5$

$= -48,384 x^3 y^5$

47. Here is a computer-generated graph of the sequence

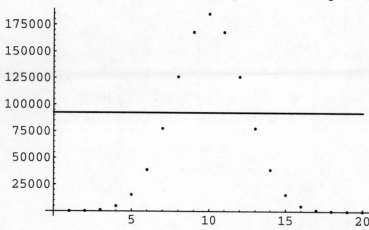

The solid line indicates $\frac{1}{2}(184,756) = 92,378$, that is, half the largest term.

Thus there are 5 terms greater than half the largest term. From the table display we note

$$\binom{20}{8} = 125,970 \qquad \binom{20}{9} = 167,960 \qquad \binom{20}{10} = 184,756$$

$$\binom{20}{11} = 167,960 \qquad \binom{20}{12} = 125,970$$

are all greater than 92,378, but the other terms are less than 92,378.

49. Here is a computer-generated graph of the sequence

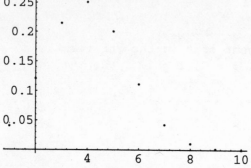

From the table display we find the largest term to be $\binom{10}{4}(.6)^6(.4)^4 = 0.2508$ as displayed in the graph.

(B) According to the binomial formula,

$$\sum_{k=0}^{n} \binom{n}{k} a^{n-k} b^k = (a + b)^n$$

Thus, $a_0 + a_1 + a_2 + \cdots + a_{10} = \sum_{k=0}^{10} \binom{10}{k}(0.6)^{10-k}(0.4)^k = (0.6 + 0.4)^{10} = 1^{10} = 1$

51. $(1.01)^{10} = (1 + 0.01)^{10} = \sum\limits_{k=0}^{10} \binom{10}{k} 1^{10-k}(0.01)^k, \quad 1^{10-k} = 1$

$$= \sum\limits_{k=0}^{10} \binom{10}{k}(0.01)^k$$

$$= \binom{10}{0} + \binom{10}{1}(0.01)^1 + \binom{10}{2}(0.01)^2 + \binom{10}{3}(0.01)^3$$

$$+ \binom{10}{4}(0.01)^4 + \binom{10}{5}(0.01)^5 + \binom{10}{6}(0.01)^6 + \binom{10}{7}(0.01)^7$$

$$+ \binom{10}{8}(0.01)^8 + \binom{10}{9}(0.01)^9 + \binom{10}{10}(0.01)^{10}$$

$$= 1 + 0.1 + 0.0045 + 0.00012 + (0.0000021 + \text{other terms}$$
$$\text{with no effect in fourth decimal place})$$

$$= 1.1046$$

53. $\binom{n}{r} = \dfrac{n!}{r!(n-r)!} = \dfrac{n!}{(n-r)!\,r!} = \dfrac{n!}{(n-r)![n-(n-r)]!} = \binom{n}{n-r}$

55. $\binom{k}{r-1} + \binom{k}{r} = \dfrac{k!}{(r-1)!(k-r+1)!} + \dfrac{k!}{r!(k-r)!}$

$$= \dfrac{rk! + (k-r+1)k!}{r!(k-r+1)!} = \dfrac{(r+k-r+1)k!}{r!(k-r+1)!}$$

$$= \dfrac{(k+1)k!}{r!(k-r+1)!} = \dfrac{(k+1)!}{r!(k-r+1)!}$$

$$= \binom{k+1}{r}$$

57. $\binom{k}{k} = \dfrac{k!}{k!(k-k)!} = 1 = \dfrac{(k+1)!}{(k+1)![(k+1)-(k+1)]!} = \binom{k+1}{k+1}$

59. $2^n = (1+1)^n = \sum\limits_{k=0}^{n} \binom{n}{k} 1^{n-k} 1^k, \quad 1^{n-k} = 1^k = 1$

$$= \sum\limits_{k=0}^{n} \binom{n}{k}$$

$$= \binom{n}{0} + \binom{n}{1} + \binom{n}{2} + \cdots + \binom{n}{n}$$

Exercise 8-5

Key Ideas and Formulas

Multiplication Principle (Fundamental Counting Principle)

1. If two operations O_1 and O_2 are performed in order, with N_1 possible outcomes for the first operation and N_2 possible outcomes for the second operation, then there are

$$N_1 \cdot N_2$$

possible combined operations of the first operation followed by the second.

2. In general, if n operations O_1, O_2, …, O_n are performed in order, with possible number of outcomes N_1, N_2, …, N_n, respectively, then there are

$$N_1 \cdot N_2 \cdot … \cdot N_n$$

possible combined outcomes of the operations performed in the given order.

Number of Permutations of n objects: $P_{n,n} = n!$

Number of Permutations of n objects, taken r at a time:

$$P_{n,r} = n(n-1) … (n-r+1) = \frac{n!}{(n-r)!}$$

Number of Combinations of n objects, taken r at a time:

$$C_{n,r} = \binom{n}{r} = \frac{P_{n,r}}{r!} = \frac{n!}{r!(n-r)!} \quad 0 \le r \le n$$

In a permutation, the order of the objects counts. In a combination, the order of the objects does not count.

---

1. $\dfrac{11!}{8!} = \dfrac{11 \cdot 10 \cdot 9 \cdot 8!}{8!} = 990$

3. $\dfrac{5!}{2!3!} = \dfrac{5 \cdot 4 \cdot 3!}{2 \cdot 1 \cdot 3!} = 10$

5. $\dfrac{7!}{4!(7-4)!} = \dfrac{7!}{4!3!} = \dfrac{7 \cdot 6 \cdot 5 \cdot 4!}{4!3 \cdot 2 \cdot 1} = 35$

7. $\dfrac{7!}{7!(7-7)!} = \dfrac{7!}{7!0!} = \dfrac{7!}{7!(1)} = 1$

9. $P_{5,3} = \dfrac{5!}{(5-3)!} = \dfrac{5!}{2!} = \dfrac{5 \cdot 4 \cdot 3 \cdot 2!}{2!} = 60$

11. $P_{52,4} = \dfrac{52!}{(52-4)!} = \dfrac{52!}{48!}$
$= \dfrac{52 \cdot 51 \cdot 50 \cdot 49 \cdot 48!}{48!} = 6{,}497{,}400$

13. $C_{5,3} = \dfrac{5!}{3!(5-3)!} = \dfrac{5!}{3!2!} = 10$
(problem 3)

15. $C_{52,4} = \dfrac{52!}{4!(52-4)!} = \dfrac{P_{52,4}}{4!}$
$= \dfrac{6{,}497{,}400}{4 \cdot 3 \cdot 2 \cdot 1} = 270{,}725$

17. O_1: Selecting the color N_1: 5 ways
O_2: Selecting the transmission N_2: 3 ways
O_3: Selecting the interior N_3: 4 ways
O_4: Selecting the engine N_4: 2 ways
Applying the multiplication principle, there are $5 \cdot 3 \cdot 4 \cdot 2 = 120$ variations of the car.

19. Order is important here. We use permutations, selecting, in order, three horses out of ten:
$P_{10,3} = 10 \cdot 9 \cdot 8 = 720$ different finishes

21. For the subcommittee, order is not important. We use combinations, selecting three persons out of seven:
$C_{7,3} = \dfrac{7!}{3!(7-3)!} = \dfrac{7!}{3!4!} = \dfrac{7 \cdot 6 \cdot 5 \cdot 4!}{3 \cdot 2 \cdot 1 \cdot 4!} = 35$ subcommittees
In choosing a president, a vice-president, and a secretary, we can use permutations, or apply the multiplication principle.
O_1: Selecting the president N_1: 7 ways
O_2: Selecting the vice-president N_2: 6 ways (the president is not considered)
O_3: Selecting the secretary N_3: 5 ways (the president and vice-president are not considered)
Thus, there are $N_1 \cdot N_2 \cdot N_3 = 7 \cdot 6 \cdot 5 \ (= P_{7,3}) = 210$ ways.

23. For each game, we are selecting two teams out of ten to be opponents. Since the order of the opponents does not matter (this has nothing to do with the order in which the games might be played, which is not under discussion here), we use combinations.

$$C_{10,2} = \frac{10!}{2!(10-2)!} = \frac{10!}{2!8!} = \frac{10 \cdot 9 \cdot 8!}{2 \cdot 1 \cdot 8!} = 45 \text{ games}$$

25.

	No letter can be repeated	Allowing letters to repeat
O_1: Selecting first letter	N_1: 6 ways	N_1: 6 ways
O_2: Selecting second letter	N_2: 5 ways	N_2: 6 ways
O_3: Selecting third letter	N_3: 4 ways	N_3: 6 ways
O_4: Selecting fourth letter	N_4: 3 ways	N_4: 6 ways

$P_{6,4} = 6 \cdot 5 \cdot 4 \cdot 3 = 360$ possible code words

$6 \cdot 6 \cdot 6 \cdot 6 = 1,296$ possible code words

27.

	No digit can be repeated	Allowing digits to repeat
O_1: Selecting first digit	N_1: 10 ways	N_1: 10 ways
O_2: Selecting second digit	N_2: 9 ways	N_2: 10 ways
$\vdots$	$\vdots$	$\vdots$
O_5: Selecting fifth digit	N_5: 6 ways	N_5: 10 ways

$P_{10,5} = 10 \cdot 9 \cdot 8 \cdot 7 \cdot 6 = 30,240$ lock combinations

$10 \cdot 10 \cdot 10 \cdot 10 \cdot 10 = 100,000$ lock combinations

29. We are selecting five cards out of the 13 hearts in the deck. The order is not important, so we use combinations.

$$C_{13,5} = \frac{13!}{5!(13-5)!} = \frac{13!}{5!8!} = \frac{13 \cdot 12 \cdot 11 \cdot 10 \cdot 9 \cdot 8!}{5 \cdot 4 \cdot 3 \cdot 2 \cdot 1 \cdot 8!} = 1,287$$

31.

	Repeats allowed	No repeats allowed
O_1: Selecting first letter	N_1: 26 ways	N_1: 26 ways
O_2: Selecting second letter	N_2: 26 ways	N_2: 25 ways
O_3: Selecting third letter	N_3: 26 ways	N_3: 24 ways
O_4: Selecting first digit	N_4: 10 ways	N_4: 10 ways
O_5: Selecting second digit	N_5: 10 ways	N_5: 9 ways
O_6: Selecting third digit	N_6: 10 ways	N_6: 8 ways

$26 \cdot 26 \cdot 26 \cdot 10 \cdot 10 \cdot 10 = 17,576,000$ license plates

$26 \cdot 25 \cdot 24 \cdot 10 \cdot 9 \cdot 8 = 11,232,000$ license plates

33. O_1: Choosing 5 spades out of 13 possible (order is not important)

N_1: $C_{13,5}$

O_2: Choosing 2 hearts out of 13 possible (order is not important)

N_2: $C_{13,2}$

Using the multiplication principle, we have:

Number of hands $= C_{13,5} \cdot C_{13,2} = \dfrac{13!}{5!(13-5)!} \cdot \dfrac{13!}{2!(13-2)!}$

$= 1,287 \cdot 78$

$= 100,386$

35. O_1: Choosing 3 appetizers out of 8 possible (order is not important)
N_1: $C_{8,3}$
O_2: Choosing 4 main courses out of 10 possible (order is not important)
N_2: $C_{10,4}$
O_3: Choosing 2 desserts out of 7 possible (order is not important)
N_3: $C_{7,2}$
Using the multiplication principle, we have:

Number of banquets = $C_{8,3} \cdot C_{10,4} \cdot C_{7,2} = \dfrac{8!}{3!(8-3)!} \cdot \dfrac{10!}{4!(10-4)!} \cdot \dfrac{7!}{2!(7-2)!}$
$= 56 \cdot 210 \cdot 21$
$= 246,960$

37. (A) $P_{10,0} = 1$ $0! = 1$
      $P_{10,1} = 10$ $1! = 1$
      $P_{10,2} = 90$ $2! = 2$
      $P_{10,3} = 720$ $3! = 6$
      $P_{10,4} = 5,040$ $4! = 24$
      $P_{10,5} = 30,240$ $5! = 120$
      $P_{10,6} = 151,200$ $6! = 720$
      $P_{10,7} = 604,800$ $7! = 5,040$
      $P_{10,8} = 1,814,400$ $8! = 40,320$
      $P_{10,9} = 3,628,800$ $9! = 362,880$
      $P_{10,10} = 3,628,800$ $10! = 3,628,800$

Thus, $P_{10,r} \geq r!$ for $r = 0, 1, \ldots, 10$

(B) If $r = 0$, $P_{10,0} = 0! = 1$. If $r = 10$, $P_{10,10} = 10! = 3,628,800$

(C) $P_{n,r}$ and $r!$ are each the product of r consecutive integers, the largest of which is n for $P_{n,r}$ and r for $r!$. Thus if $r \leq n$, $P_{n,r} \geq r!$

39. O_1: Choosing a left glove out of 12 possible
N_1: 12
O_2: Choosing a right glove out of all the right gloves that do *not* match the left glove already chosen
N_2: 11
Using the multiplication principle, we have:
Number of ways to mismatch gloves = $12 \cdot 11 = 132$.

41. (A) We are choosing 2 points out of the 8 to join by a chord. Order is not important.

$C_{8,2} = \dfrac{8!}{2!(8-2)!} = 28$ chords

(B) No three of the points can be collinear, since no line intersects a circle at more than two points. Thus, we can select any three of the 8 to use as vertices of the triangle. Order is not important.

$C_{8,3} = \dfrac{8!}{3!(8-3)!} = 56$ triangles

(C) We can select any four of the eight points to use as vertices of a quadrilateral. Order is not important.

$C_{8,4} = \dfrac{8!}{4!(8-4)!} = 70$ quadrilaterals

43. To seat two people, we can seat one person, then the second person.

O_1: Seat the first person in any chair

N_1: 5 ways

O_2: Seat the second person in any remaining chair

N_2: 4 ways

Thus, applying the multiplication principle, we can seat two persons in $N_1 \cdot N_2 = 5 \cdot 4 = 20$ ways.

We can continue this reasoning for a third person.

O_3: Seat the third person in any of the three remaining chairs

N_3: 3 ways

Thus we can seat 3 persons in $5 \cdot 4 \cdot 3 = 60$ ways.

For a fourth person:

O_4: Seat the fourth person in any of the two remaining chairs

N_4: 2 ways

Thus we can seat 4 persons in $5 \cdot 4 \cdot 3 \cdot 2 = 120$ ways.

For a fifth person:

O_5: Seat the fifth person. There will be only one chair remaining.

N_5: 1 way

Thus we can seat 5 persons in $5 \cdot 4 \cdot 3 \cdot 2 \cdot 1 = 120$ ways

45. (A) Order is important, so we use permutations, selecting, in order, 5 persons out of 8:

$$P_{8,5} = 8 \cdot 7 \cdot 6 \cdot 5 \cdot 4 = 6{,}720 \text{ teams.}$$

(B) Order is not important, so we use combinations, selecting 5 persons out of 8

$$C_{8,5} = \frac{8!}{5!(8-5)!} = \frac{8!}{5!3!} = \frac{8 \cdot 7 \cdot 6 \cdot 5!}{5! \cdot 3 \cdot 2 \cdot 1} = 56 \text{ teams}$$

(C) O_1: Selecting either Mike or Ken out of {Mike, Ken}

N_1: $C_{2,1}$

O_2: Selecting the 4 remaining players out of the 6 possibilities that do not include either Mike or Ken.

N_2: $C_{6,4}$

Using the multiplication principle, we have

$$
\begin{aligned}
N_1 \cdot N_2 = C_{2,1} \cdot C_{6,4} &= \frac{2!}{1!(2-1)!} \cdot \frac{6!}{4!(6-4)!} \\
&= \frac{2!}{1!1!} \cdot \frac{6!}{4!2!} \\
&= 2 \cdot 15 \\
&= 30 \text{ teams}
\end{aligned}
$$

47. The number of ways to deal a hand containing exactly 1 king is computed as follows:

O_1: Choosing 1 king out of 4 possible (order is not important)

N_1: $C_{4,1}$

O_2: Choosing 4 cards out of 48 possible (order is not important)

N_2: $C_{48,4}$

Using the multiplication principle, we have

$$N_1 \cdot N_2 = C_{4,1} \cdot C_{48,4} = \frac{4!}{1!(4-1)!} \cdot \frac{48!}{4!(48-4)!}$$

$$= \frac{4!}{1!3!} \cdot \frac{48!}{4!44!}$$

$$= 4 \cdot 194,580$$

$$= 778,320 \text{ ways}$$

The number of ways to deal a hand containing no hearts is

$$C_{39,5} = \frac{39!}{5!(39-5)!} = \frac{39!}{5!34!} = 575,757 \text{ ways}$$

Thus the hand that contains exactly one king is more likely.

CHAPTER 8 REVIEW

1. (A) Since $\frac{-8}{16} = \frac{4}{-8} = -\frac{1}{2}$, this could start a geometric sequence.

(B) Since $7 - 5 = 9 - 7 = 2$, this could start an arithmetic sequence.

(C) Since $-5 - (-8) = -2 - (-5) = 3$, this could start an arithmetic sequence.

(D) Since $\frac{3}{2} \neq \frac{5}{3}$ and $3 - 2 \neq 5 - 3$, this could start neither an arithmetic nor a geometric sequence.

(E) Since $\frac{2}{-1} = \frac{-4}{2} = -2$, this could start a geometric sequence. $\qquad$ (8-1, 8-3)

2. $a_n = 2n + 3$

(A) $a_1 = 2 \cdot 1 + 3 = 5$
$a_2 = 2 \cdot 2 + 3 = 7$
$a_3 = 2 \cdot 3 + 3 = 9$
$a_4 = 2 \cdot 4 + 3 = 11$

(B) This is an arithmetic sequence with $d = 2$. Hence

$$a_n = a_1 + (n - 1)d$$
$$a_{10} = 5 + (10 - 1)d$$
$$= 23$$

(C) $S_n = \frac{n}{2}(a_1 + a_n)$

$$S_{10} = \frac{10}{2}(5 + 23)$$
$$= 140$$

$\qquad$ (8-1, 8-3)

3. $a_n = 32\left(\frac{1}{2}\right)^n$

(A) $a_1 = 32\left(\frac{1}{2}\right)^1 = 16$

$a_2 = 32\left(\frac{1}{2}\right)^2 = 8$

$a_3 = 32\left(\frac{1}{2}\right)^3 = 4$

$a_4 = 32\left(\frac{1}{2}\right)^4 = 2$

(B) This is a geometric sequence with $r = \frac{1}{2}$. Hence

$$a_n = a_1 r^{n-1}$$
$$a_{10} = 16\left(\frac{1}{2}\right)^{10-1}$$
$$= \frac{1}{32}$$

(C) $S_n = \frac{a_1 - ra_n}{1 - r}$

$$S_{10} = \frac{16 - \frac{1}{2}\left(\frac{1}{32}\right)}{\frac{1}{2}} = \frac{16 - \frac{1}{64}}{\frac{1}{2}} = 31\frac{31}{32}$$

$\qquad$ (8-1, 8-3)

4. $a_1 = -8$, $a_n = a_{n-1} + 3$, $n \geq 2$

(A) $a_1 = -8$
$a_2 = a_1 + 3 = -8 + 3 = -5$
$a_3 = a_2 + 3 = -5 + 3 = -2$
$a_4 = a_3 + 3 = -2 + 3 = 1$

(B) This is an arithmetic sequence with $d = 3$. Hence
$a_n = a_1 + (n - 1)d$
$a_{10} = -8 + (10 - 1)3$
$= 19$

(C) $S_n = \dfrac{n}{2}(a_1 + a_n)$

$S_{10} = \dfrac{10}{2}(-8 + 19)$

$= 55$

$(8\text{-}1,\ 8\text{-}3)$

5. $a_1 = -1$, $a_n = (-2)a_{n-1}$, $n \geq 2$

(A) $a_1 = -1$
$a_2 = (-2)a_1 = (-2)(-1) = 2$
$a_3 = (-2)a_2 = (-2)2 = -4$
$a_4 = (-2)a_3 = (-2)(-4) = 8$

(B) This is a geometric sequence with $r = -2$. Hence
$a_n = a_1 r^{n-1}$
$a_{10} = (-1)(-2)^{10-1}$
$= 512$

(C) $S_n = \dfrac{a_1 - ra_n}{1 - r}$

$S_{10} = \dfrac{-1 - (-2)(512)}{1 - (-2)}$

$= 341$

$(8\text{-}1,\ 8\text{-}3)$

6. This is a geometric sequence with $a_1 = 16$ and $r = \dfrac{1}{2} < 1$, so the sum exists:

$S_\infty = \dfrac{a_1}{1 - r}$

$= \dfrac{16}{1 - \frac{1}{2}}$

$= 32$

$(8\text{-}3)$

7. $6! = 6 \cdot 5 \cdot 4 \cdot 3 \cdot 2 \cdot 1 = 720$ $\qquad (8\text{-}4)$

8. $\dfrac{22!}{19!} = \dfrac{22 \cdot 21 \cdot 20 \cdot 19!}{19!} = 9,240$ $\qquad (8\text{-}4)$

9. $\dfrac{7!}{2!(7 - 2)!} = \dfrac{7!}{2!5!} = \dfrac{7 \cdot 6 \cdot 5!}{2 \cdot 1 \cdot 5!} = 21$ $\qquad (8\text{-}4)$

10. $C_{6,2} = \dfrac{6!}{2!(6 - 2)!} = \dfrac{6!}{2!4!} = \dfrac{6 \cdot 5 \cdot 4!}{2 \cdot 1 \cdot 4!} = 15$ $\quad P_{6,2} = 6 \cdot 5 = 30$ $\qquad (8\text{-}5)$

11. (A) The outcomes can be displayed in a tree diagram as follows:

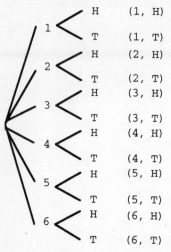

    H   (1, H)

    T   (1, T)

    H   (2, H)

    T   (2, T)

    H   (3, H)

    T   (3, T)

    H   (4, H)

    T   (4, T)

    H   (5, H)

    T   (5, T)

    H   (6, H)

    T   (6, T)

(B) O_1: Rolling the die
N_1: 6 outcomes
O_2: Flipping the coin
N_2: 2 outcomes

Applying the multiplication principle, there are $6 \cdot 2 = 12$ combined outcomes.

(8-5)

12. O_1: Seating the first person N_1: 6 ways
O_2: Seating the second person N_2: 5 ways
O_3: Seating the third person N_3: 4 ways
O_4: Seating the fourth person N_4: 3 ways
O_5: Seating the fifth person N_5: 2 ways
O_6: Seating the sixth person N_6: 1 way

Applying the multiplication principle, there are $6 \cdot 5 \cdot 4 \cdot 3 \cdot 2 \cdot 1 = 720$ arrangements.

(8-5)

13. Order is important here. We use permutations to determine the number of arrangements of 6 objects. $P_{6,6} = 6! = 720$

(8-5)

14. P_1: $5 = 1^2 + 4 \cdot 1 = 5$
P_2: $5 + 7 = 2^2 + 4 \cdot 2$
       $12 = 12$
P_3: $5 + 7 + 9 = 3^2 + 4 \cdot 3$
       $21 = 21$ *(8-2)*

15. P_1: $2 = 2^{1+1} - 2 = 4 - 2 = 2$
P_2: $2 + 4 = 2^{2+1} - 2$
       $6 = 6$
P_3: $2 + 4 + 8 = 2^{3+1} - 2$
       $14 = 14$ *(8-2)*

16. P_1: $49^1 - 1$ is divisible by 6
$48 = 6 \cdot 8$ true
P_2: $49^2 - 1$ is divisible by 6
$2,400 = 6 \cdot 400$ true
P_3: $49^3 - 1$ is divisible by 6
$117,648 = 19,608 \cdot 6$ true

(8-2)

17. P_k: $5 + 7 + 9 + \cdots + (2k + 3) = k^2 + 4k$
P_{k+1}: $5 + 7 + 9 + \cdots + (2k + 3) + (2k + 5) = (k + 1)^2 + 4(k + 1)$ *(8-2)*

18. P_k: $2 + 4 + 8 + \cdots + 2^k = 2^{k+1} - 2$
P_{k+1}: $2 + 4 + 8 + \cdots + 2^k + 2^{k+1} = 2^{k+2} - 2$ *(8-2)*

19. P_k: $49^k - 1 = 6r$ for some integer r
P_{k+1}: $49^{k+1} - 1 = 6s$ for some integer s *(8-2)*

20. Although 1 is less than 4, $1 + \frac{1}{2}$ is less than 4, $1 + \frac{1}{2} + \frac{1}{3}$ is less than 4, and soon, the statement is false. In fact,
$$1 + \frac{1}{2} + \frac{1}{3} + \ldots + \frac{1}{31} \approx 4.027245$$
hence $n = 31$ is a counterexample.　　　　　　　　　　　　　　　　　　　　　*(8-2)*

21. $S_{10} = (2 \cdot 1 - 8) + (2 \cdot 2 - 8) + (2 \cdot 3 - 8) + (2 \cdot 4 - 8) + (2 \cdot 5 - 8) + (2 \cdot 6 - 8)$
$$+ (2 \cdot 7 - 8) + (2 \cdot 8 - 8) + (2 \cdot 9 - 8) + (2 \cdot 10 - 8)$$
$$= (-6) + (-4) + (-2) + 0 + 2 + 4 + 6 + 8 + 10 + 12$$
$$= 30$$
　　　　　　　　　　　　　　　　　　　　　　　　　　　　　　　　　　(8-3)

22. $S_7 = \frac{16}{2^1} + \frac{16}{2^2} + \frac{16}{2^3} + \frac{16}{2^4} + \frac{16}{2^5} + \frac{16}{2^6} + \frac{16}{2^7}$
$$= 8 + 4 + 2 + 1 + \frac{1}{2} + \frac{1}{4} + \frac{1}{8}$$
$$= 15\frac{7}{8}$$
　　　　　　　　　　　　　　　　　　　　　　　　　　　　　　　　　　(8-3)

23. This is an infinite geometric sequence with $a_1 = 27$.

$r = \frac{-18}{27} = -\frac{2}{3}$　$\left| -\frac{2}{3} \right| < 1$, hence the sum exists

$S_\infty = \frac{a_1}{1 - r}$

$= \frac{27}{1 - (-\frac{2}{3})}$

$= \frac{81}{5}$　　　　　　　　*(8-3)*

24. $S_n = \sum_{k=1}^{n} \frac{(-1)^{k+1}}{3^k}$

This geometric sequence has $a_1 = \frac{1}{3}$.

$r = \left(-\frac{1}{9}\right) + \frac{1}{3} = -\frac{1}{3}$　$\left| -\frac{1}{3} \right| < 1$, hence the sum exists.

$S_\infty = \frac{a_1}{1 - r}$

$= \frac{\frac{1}{3}}{1 - (-\frac{1}{3})}$

$= \frac{1}{4}$　　　　　　　*(8-3)*

25. We can select any three of the six points to use as vertices of the triangle. Order is not important. $C_{6,3} = \frac{6!}{3!(6-3)!} = 20$ triangles　　　　*(8-5)*

26. First, find d:
$a_n = a_1 + (n - 1)d$
$31 = 13 + (7 - 1)d$
$31 = 13 + 6d$
$d = 3$
Hence, $a_5 = 13 + (5 - 1)3$
$= 25$　　　　　*(8-3)*

27.

	Case 1	Case 2	Case 3
O_1: select the first letter	N_1: 8 ways	8 ways	8 ways
O_2: select the second letter	N_2: 7 ways	8 ways	7 ways
O_3: select the third letter	N_3: 6 ways	8 ways	7 ways
	(exclude first and second letter)		(exclude second letter.)
	$8 \cdot 7 \cdot 6 = 336$ words	$8 \cdot 8 \cdot 8 = 512$ words	$8 \cdot 7 \cdot 7 = 392$ words

　　　　　　　　　　　　　　　　　　　　　　　　　　　　　　　　　　(8-5)

28. $0.\overline{72} = 0.72 + 0.0072 + 0.000072 + \cdots$
This is an infinite geometric sequence with $a_1 = 0.72$ and $r = 0.01$.

$$0.\overline{72} = S_\infty = \frac{a_1}{1 - r}$$
$$= \frac{0.72}{1 - 0.01}$$
$$= \frac{0.72}{0.99}$$
$$= \frac{8}{11} \tag{8-3}$$

29. (A) Order is important here. We are selecting 3 digits out of 6 possible.
$P_{6,3} = 6 \cdot 5 \cdot 4 = 120$ lock combinations

(B) Order is not important here. We are selecting 2 players out of 5.
$$C_{5,2} = \frac{5!}{2!(5 - 2)!} = 10 \text{ games} \tag{8-5}$$

30. $\dfrac{20!}{18!(20 - 18)!} = \dfrac{20!}{18!2!}$
$$= \frac{20 \cdot 19 \cdot 18!}{18!2 \cdot 1} = 190 \tag{8-4}$$

31. $\dbinom{16}{12} = \dfrac{16!}{12!(16 - 12)!} = \dfrac{16!}{12!4!}$
$$= \frac{16 \cdot 15 \cdot 14 \cdot 13 \cdot 12!}{12! \cdot 4 \cdot 3 \cdot 2 \cdot 1} = 1{,}820 \tag{8-4}$$

32. $\dbinom{11}{11} = \dfrac{11!}{11!(11 - 11)!} = \dfrac{11!}{11!0!} = 1 \tag{8-4}$

33. $(x - y)^5 = [x + (-y)]^5 = \displaystyle\sum_{k=0}^{5} \binom{5}{k}(x)^{5-k}(-y)^k$

$$= \binom{5}{0}x^5 + \binom{5}{1}x^4(-y)^1 + \binom{5}{2}x^3(-y)^2 + \binom{5}{3}x^2(-y)^3$$
$$+ \binom{5}{4}x(-y)^4 + \binom{5}{5}(-y)^5$$
$$= x^5 - 5x^4y + 10x^3y^2 - 10x^2y^3 + 5xy^4 - y^5 \tag{8-4}$$

34. In the expansion of $(a + b)^n$, the exponent of b in the rth term is $r - 1$ and the exponent of a is $n - (r - 1)$. Here, $r = 10$, $n = 12$.

Tenth term $= \dbinom{12}{9}(2x)^3(-y)^9$

$$= \frac{12!}{9!3!}(8x^3)(-y^9)$$
$$= \frac{12 \cdot 11 \cdot 10 \cdot 9!}{9! \cdot 3 \cdot 2 \cdot 1}(-8x^3y^9)$$
$$= -1760x^3y^9 \tag{8-4}$$

35. Write: P_n: $5 + 7 + 9 + \cdots + (2n + 3) = n^2 + 4n$
Proof: Part 1: Show that P_1 is true:
P_1: $5 = 1^2 + 4 \cdot 1$

$= 1 + 4$ Clearly true.

Part 2: Show that if P_k is true, then P_{k+1} is true:
Write out P_k and P_{k+1}.

$\quad P_k$: $5 + 7 + 9 + \cdots + (2k + 3) = k^2 + 4k$

$\quad P_{k+1}$: $5 + 7 + 9 + \cdots + (2k + 3) + (2k + 5) = (k + 1)^2 + 4(k + 1)$

We start with P_k:

$\quad 5 + 7 + 9 + \cdots + (2k + 3) = k^2 + 4k$

Adding $2k + 5$ to both sides:

$$
\begin{aligned}
5 + 7 + 9 + \cdots + (2k + 3) + (2k + 5) &= k^2 + 4k + 2k + 5 \\
&= k^2 + 6k + 5 \\
&= k^2 + 2k + 1 + 4k + 4 \\
&= (k + 1)^2 + 4(k + 1)
\end{aligned}
$$

We have shown that if P_k is true, then P_{k+1} is true.

Conclusion: P_n is true for all positive integers n. $\qquad$ (8-2)

36. Write: P_n: $2 + 4 + 8 + \cdots + 2^n = 2^{n+1} - 2$
Proof: Part 1: Show that P_1 is true:
P_1: $2 = 2^{1+1} - 2$

$\qquad = 4 - 2$ Clearly true.

Part 2: Show that if P_k is true, then P_{k+1} is true:
Write out P_k and P_{k+1}.

$\quad P_k$: $2 + 4 + 8 + \cdots + 2^k = 2^{k+1} - 2$

$\quad P_{k+1}$: $2 + 4 + 8 + \cdots + 2^k + 2^{k+1} = 2^{k+2} - 2$

We start with P_k:

$\quad 2 + 4 + 8 + \cdots + 2^k = 2^{k+1} - 2$

Adding 2^{k+1} to both sides:

$$
\begin{aligned}
2 + 4 + 8 + \cdots + 2^k + 2^{k+1} &= 2^{k+1} - 2 + 2^{k+1} \\
&= 2^{k+1} + 2^{k+1} - 2 \\
&= 2 \cdot 2^{k+1} - 2 \\
&= 2^{k+2} - 2
\end{aligned}
$$

We have shown that if P_k is true, then P_{k+1} is true.

Conclusion: P_n is true for all positive integers n. $\qquad$ (8-2)

37. Write: P_n: $49^n - 1 = 6r$ for some r in N.
Part 1: Show that P_1 is true:

$\quad P_1$: $49^1 - 1 = 6r$ is true $(r = 8)$.

Part 2: Show that if P_k is true, then P_{k+1} is true:
Write out P_k and P_{k+1}.

$\quad P_k$: $49^k - 1 = 6r$, for some integer r

P_{k+1}: $49^{k+1} - 1 = 6s$ for some integer s

We start with P_k:

$\quad 49^k - 1 = 6r$ for some integer r

$$
\begin{aligned}
\text{Now, } 49^{k+1} - 1 &= 49^{k+1} - 49^k + 49^k - 1 \\
&= 49^k(49 - 1) + 49^k - 1 \\
&= 49^k \cdot 8 \cdot 6 + 6r \\
&= 6(49^k \cdot 8 + r) \\
&= 6s \text{ with } s = 49^k \cdot 8 + r
\end{aligned}
$$

We have shown that if P_k is true, then P_{k+1} is true.

Conclusion: P_n is true for all positive integers n. $\qquad$ (8-2)

38. Here are computer-generated graphs of $\{a_n\}$ and $\{b_n\}$.

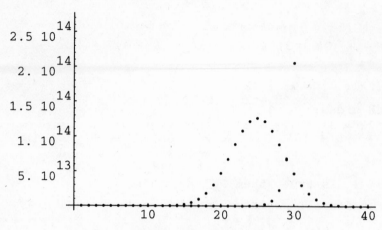

The graph indicates that the least positive integer n such that $a_n < b_n$ is between 28 and 30. From the table display we note

$n = 28$ $a_n = 8.875 \times 10^{13}$ $b_n = 2.287 \times 10^{13}$

$n = 29$ $a_n = 6.733 \times 10^{13}$ $b_n = 6.863 \times 10^{13}$

confirming that 29 is the least positive integer n required.   *(8-4)*

39. Here are computer-generated graphs of $\{a_n\}$ and $\{b_n\}$ (the dots suggest a straight line).

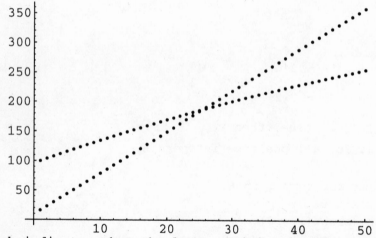

The graph indicates that the least positive integer n such that $a_n < b_n$ is between 25 and 27. From the table display we note

$n = 25$ $a_n = 188.87$ $b_n = 184$

$n = 26$ $a_n = 191.98$ $b_n = 193$

confirming that 26 is the least positive integer n required.   *(8-1)*

40. In the first case, the order matters. We have five successive events, each of which can happen two ways (girl or boy). Applying the multiplication principle, there are $2 \cdot 2 \cdot 2 \cdot 2 \cdot 2 = 32$ possible families. In the second case, the possible families can be listed as {0 girls, 1 girl, 2 girls, 3 girls, 4 girls, 5 girls}, thus there are 6 possibilities.   *(8-5)*

41. An arithmetic sequence is involved, with $a_1 = \dfrac{g}{2}$, $d = \dfrac{3g}{2} - \dfrac{g}{2} = g$. Distance fallen during the twenty-fifth second $= a_{25}$.

$$a_n = a_1 + (n - 1)d$$

$$a_{25} = \frac{g}{2} + (25 - 1)g$$

$$= \frac{49g}{2} \text{ feet}$$

Total distance fallen after twenty-five seconds $= a_1 + a_2 + a_3 + \cdots + a_{25} = S_{25}$

$$S_n = \frac{n}{2}(a_1 + a_n)$$

$$S_{25} = \frac{25}{2}\left(\frac{g}{2} + \frac{49g}{2}\right)$$

$$= \frac{625g}{2} \text{ feet} \tag{8-3}$$

42. To seat two people, we can seat one person, then the second person.

O_1: Seat the first person in any chair.

N_1: 4 ways

O_2: Seat the second person in any remaining chair.

N_2: 3 ways

Thus, applying the multiplication principle, there are $4 \cdot 3 = 12$ ways to seat two persons. (8-5)

43. $(x + i)^6 = \displaystyle\sum_{k=0}^{6} \binom{6}{k} x^{6-k} i^k$

$$= \binom{6}{0}x^6 + \binom{6}{1}x^5 i^1 + \binom{6}{2}x^4 i^2 + \binom{6}{3}x^3 i^3 + \binom{6}{4}x^2 i^4 + \binom{6}{5}x i^5 + \binom{6}{6}i^6$$

$$= x^6 + 6ix^5 - 15x^4 - 20ix^3 + 15x^2 + 6ix - 1 \tag{8-4}$$

44. A route plan can be regarded as a series of choices of stores, thus an arrangement of the 5 stores. Since the order matters, we use permutations: $P_{5,5} = 5! = 120$ route plans. (8-5)

45. Write: P_n: $\displaystyle\sum_{k=1}^{n} k^3 = \left(\sum_{k=1}^{n} k\right)^2$

Proof: Part 1: Show that P_1 is true.

$$P_1: \sum_{k=1}^{1} k^3 = 1^3 = 1^2 = \left(\sum_{k=1}^{1} k\right)^2$$

Part 2: Show that if P_j is true, then P_{j+1} is true.

Write out P_j and P_{j+1}.

$$P_j: \sum_{k=1}^{j} k^3 = \left(\sum_{k=1}^{j} k\right)^2$$

$$P_{j+1}: \sum_{k=1}^{j+1} k^3 = \left(\sum_{k=1}^{j+1} k\right)^2$$

We start with P_j:

$$\sum_{k=1}^{j} k^3 = \left(\sum_{k=1}^{j} k\right)^2$$

Adding $(j + 1)^3$ to both sides:

$$\sum_{k=1}^{j} k^3 + (j + 1)^3 = \left(\sum_{k=1}^{j} k\right)^2 + (j + 1)^3$$

$$\sum_{k=1}^{j+1} k^3 = (1 + 2 + 3 + \ldots + j)^2 + (j + 1)^3$$

$$= \left[\frac{j(j + 1)}{2}\right]^2 + (j + 1)^3 \text{ using Matched Problem 1, Section 8-2}$$

$$= (j + 1)^2 \frac{j^2}{4} + (j + 1)^2(j + 1)$$

$$= (j + 1)^2\left[\frac{j^2}{4} + j + 1\right]$$

$$= (j + 1)^2\left[\frac{j^2 + 4j + 4}{4}\right]$$

$$= (j + 1)^2 \frac{(j + 2)^2}{2^2}$$

$$= \left[\frac{(j + 1)(j + 2)}{2}\right]^2$$

$$= [1 + 2 + 3 + \cdots + (j + 1)]^2 \text{ using Matched Problem 1, Section 8-2}$$

$$= \left(\sum_{k=1}^{j+1} k\right)^2$$

We have shown that if P_j is true, then P_{j+1} is true.

Conclusion: P_n is true for all positive integers n.　　　　　　　　　　(8-2)

46. Write: P_n: $x^{2n} - y^{2n} = (x - y)Q_n(x, y)$, where $Q_n(x, y)$ denotes some polynomial in x and y.

Proof: Part 1: Show that P_1 is true.

P_1: $x^{2 \cdot 1} - y^{2 \cdot 1} = (x - y)(x + y) = (x - y)Q_1(x, y)$　　P_1 is true.

Part 2: Show that if P_k is true, then P_{k+1} is true.
Write out P_k and P_{k+1}.

P_k: $x^{2k} - y^{2k} = (x - y)Q_k(x, y)$

P_{k+1}: $x^{2k+2} - y^{2k+2} = (x - y)Q_{k+1}(x, y)$

We start with P_k:

$x^{2k} - y^{2k} = (x - y)Q_k(x, y)$

Now, $x^{2k+2} - y^{2k+2} = x^{2k+2} - x^{2k}y^2 + x^{2k}y^2 - y^{2k+2}$

$$= x^{2k}(x^2 - y^2) + y^2(x^{2k} - y^{2k})$$

$$= (x - y)x^{2k}(x + y) + y^2(x - y)Q_k(x, y) \text{ by } P_k$$

$$= (x - y)[x^{2k}(x + y) + y^2Q_k(x, y)]$$

$$= (x - y)Q_{k+1}(x, y)$$

We have shown that if P_k is true, then P_{k+1} is true.

Conclusion: P_n is true for all positive integers n.　　　　　　　　　　(8-2)

47. Write: P_n: $\dfrac{a^n}{a^m} = a^{n-m}$, m an arbitrary positive integer, $n > m$.

Proof: Part 1: Show P_{m+1} is true.

P_{m+1}: $\dfrac{a^{m+1}}{a^m} = a^{m+1-m}$

$\qquad \dfrac{a^m a}{a^m} = a^1$ by the recursive definition of a^n

$\qquad\quad a = a \quad P_{m+1}$ is true.

Part 2: Show that if P_k is true, then P_{k+1} is true.
Write out P_k and P_{k+1}.

$\qquad P_k$: $\dfrac{a^k}{a^m} = a^{k-m}$

$\qquad P_{k+1}$: $\dfrac{a^{k+1}}{a^m} = a^{k-m+1}$

We start with P_k:

$\qquad \dfrac{a^k}{a^m} = a^{k-m}$

Multiplying both sides by a:

$\qquad \dfrac{a^k}{a^m}a = a^{k-m}a$

$\qquad \dfrac{a^k a}{a^m} = a^{k-m+1}$

$\qquad \dfrac{a^{k+1}}{a^m} = a^{k-m+1}$

We have show that if P_k is true, then P_{k+1} is true, for m an arbitrary positive integer.

Conclusion: P_n is true for all positive integers m, n. $\hspace{2cm}$ (8-2)

48. To prove $a_n = b_n$, n a positive integer, write:

P_n: $a_n = b_n$

Proof: Part 1: Show P_1 is true.

$\quad a_1 = -3 \qquad b_1 = -5 + 2\cdot 1 = -3$.

Thus, $a_1 = b_1$

Part 2: Show that if P_k is true, then P_{k+1} is true.
Write out P_k and P_{k+1}.

$\qquad P_k$: $a_k = b_k$

$\qquad P_{k+1}$: $a_{k+1} = b_{k+1}$

We start with P_k:

$\qquad a_k = b_k$

Now, $a_{k+1} = a_k + 2 = b_k + 2$

$\qquad\qquad\qquad\quad = -5 + 2k + 2$

$\qquad\qquad\qquad\quad = -5 + 2(k+1)$

$\qquad\qquad\qquad\quad = b_{k+1}$

Therefore, $a_{k+1} = b_{k+1}$.

Thus, if P_k is true, then P_{k+1} is true.

Conclusion: P_n is true for all $n \in N$. Hence, $\{a_n\} = \{b_n\}$ $\hspace{1.5cm}$ (8-2)

49. Write: P_n: $(1!)1 + (2!)2 + (3!)3 + \cdots + (n!)n = (n + 1)! - 1$.

Proof: Part 1: Show that P_1 is true.

P_1: $(1!)1 = 1 = 2 - 1 = 2! - 1$ is true.

Part 2: Show that if P_k is true, then P_{k+1} is true.

Write out P_k and P_{k+1}.

P_k: $(1!)1 + (2!)2 + (3!)3 + \cdots + (k!)k = (k + 1)! - 1$

P_{k+1}: $(1!)1 + (2!)2 + (3!)3 + \cdots + (k!)k + (k + 1)!(k + 1) = (k + 2)! - 1$

We start with P_k:

$(1!)1 + (2!)2 + (3!)3 + \cdots + (k!)k = (k + 1)! - 1$

Adding $(k + 1)!(k + 1)$ to both sides:

$$(1!)1 + (2!)2 + (3!)3 + \cdots + (k!)k + (k + 1)!(k + 1)$$
$$= (k + 1)! - 1 + (k + 1)!(k + 1)$$
$$= (k + 1)!(1 + k + 1) - 1$$
$$= (k + 2)(k + 1)! - 1$$
$$= (k + 2)! - 1$$

Thus, if P_k is true, then P_{k+1} is true.

Conclusion: P_n is true for all positive integers n. $(8-2)$

CHAPTER 9

Exercise 9-1

Key Ideas and Formulas

The graphs of a second degree equation in two variables $Ax^2 + Bxy + Cy^2 + Dx + Ey + F = 0$ (for different values of the coefficients: A, B, C not all zero) are plane curves called conic sections. They include the circle, parabola, ellipse, and hyperbola. A parabola is the set of all points in a plane equidistant from a fixed point F (the focus) and a fixed line L (the directrix) in the plane. A line through the focus perpendicular to the plane is called the **axis** and the point on the axis halfway between the focus and directrix is called the **vertex**.

Standard Equations of a Parabola with Vertex at (0, 0)

1. $y^2 = 4ax$
   Vertex: $(0, 0)$
   Focus: $(a, 0)$
   Directrix: $x = -a$
   Axis: the x axis

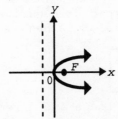

$a < 0$ (opens left) $a > 0$ (opens right)

Symmetric with respect to the x axis.

2. $x^2 = 4ay$
   Vertex: $(0, 0)$
   Focus: $(0, a)$
   Directrix: $y = -a$
   Axis: the y axis

$a < 0$ (opens down) $a > 0$ (opens up)

Symmetric with respect to the y axis.

---

1. To graph $y^2 = 4x$, assign x values that make the right side a perfect square (x must be non-negative for y to be real) and solve for y. Since the coefficient of x is positive, a must be positive, and the parabola opens right.

x	0	1	4
y	0	±2	±4

To find the focus and directrix, solve
$4a = 4$
$a = 1$
Focus: $(1, 0)$ Directrix: $x = -1$

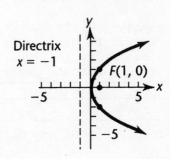

3. To graph $x^2 = 8y$, assign y values that make the right side a perfect square (y must be non-negative for x to be real) and solve for x. Since the coefficient of y is positive, a must be positive, and the parabola opens up.

x	0	±4	±2
y	0	2	$\frac{1}{2}$

To find the focus and directrix, solve
$4a = 8$
 $a = 2$
Focus: $(0, 2)$ Directrix: $y = -2$

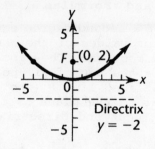

5. To graph $y^2 = -12x$, assign x values that make the right side a perfect square (x must be zero or negative for y to be real) and solve for y. Since the coefficient of x is negative, a must be negative, and the parabola opens left.

x	0	-3	$-\frac{1}{3}$
y	0	±6	±2

To find the focus and directrix, solve
$4a = -12$
 $a = -3$
Focus: $(-3, 0)$ Directrix: $x = -(-3) = 3$

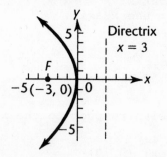

7. To graph $x^2 = -4y$, assign y values that make the right side a perfect square (y must be zero or negative for x to be real) and solve for x. Since the coefficient of y is negative, a must be negative, and the parabola opens down.

x	0	±2	±4
y	0	-1	-4

To find the focus and directrix, solve
$4a = -4$
 $a = -1$
Focus: $(0, -1)$ Directrix: $y = -(-1) = 1$

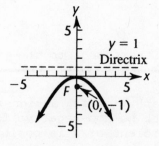

9. To graph $y^2 = -20x$, we may proceed as in problem 5. Alternatively, after noting that x must be zero or negative for y to be real, we may pick convenient values for x and solve for y using a calculator. Since the coefficient of x is negative, a must be negative, and the parabola opens left.

x	0	-1	-2
y	0	$±\sqrt{20} \approx ±4.5$	$±\sqrt{40} \approx ±6.3$

To find the focus and directrix, solve
$4a = -20$
 $a = -5$
Focus: $(-5, 0)$ Directrix: $x = -(-5) = 5$

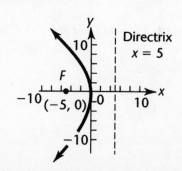

11. To graph $x^2 = 10y$, we may proceed as in Problem 3. Alternatively, after noting that y must be non-negative for x to be real, we may pick convenient values for y and solve for x using a calculator. Since the coefficient of y is positive, a must be positive, and the parabola opens up.

x	0	$\pm\sqrt{10} \approx \pm3.2$	$\pm\sqrt{20} \approx \pm4.5$
y	0	1	2

To find the focus and directrix, solve
$4a = 10$
$\ a = 2.5$
Focus: (0, 2.5) Directrix: $y = -2.5$

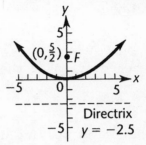

13. Comparing $y^2 = 39x$ with $y^2 = 4ax$, the standard equation of a parabola symmetric with respect to the x axis, we have
$4a = 39$ Focus on x axis
$\ a = 9.75$
Focus: (9.75, 0)

15. Comparing $x^2 = -105y$ with $x^2 = 4ay$, the standard equation of a parabola symmetric with respect to the y axis, we have
$4a = -105$ Focus on y axis
$\ a = -26.25$
Focus: (0, -26.25)

17. Comparing $y^2 = -77x$ with $y^2 = 4ax$, the standard equation of a parabola symmetric with respect to the x axis, we have
$4a = -77$ Focus on x axis
$\ a = -19.25$
Focus: (-19.25, 0)

19. Comparing directrix $y = -3$ with the information in the chart of standard equations for a parabola, we see:
$a = 3$. Axis: the y axis. Equation: $x^2 = 4ay$.
Thus the equation of the parabola must be $x^2 = 4\cdot3y$, or $x^2 = 12y$.

21. Comparing focus (0, -7) with the information in the chart of standard equations for a parabola, we see:
$a = -7$ Axis: the y-axis. Equation: $x^2 = 4ay$.
Thus the equation of the parabola must be $x^2 = 4(-7)y$, or $x^2 = -28y$.

23. Comparing directrix $x = 6$ with the information in the chart of standard equations for a parabola, we see:
$6 = -a$. $a = -6$. Axis: the x-axis. Equation: $y^2 = 4ax$.
Thus the equation of the parabola must be $y^2 = 4(-6)x$, or $y^2 = -24x$.

25. Comparing focus (2, 0) with the information in the chart of standard equations for a parabola, we see:
$a = 2$ Axis: the x-axis. Equation: $y^2 = 4ax$.
Thus the equation of the parabola must be $y^2 = 4(2)x$, or $y^2 = 8x$.

27. The parabola is opening up and has an equation of the form $x^2 = 4ay$. Since $(4, 2)$ is on the graph, we have:

$$x^2 = 4ay$$
$$(4)^2 = 4a(2)$$
$$16 = 8a$$
$$2 = a$$

Thus, the equation of the parabola is

$$x^2 = 4(2)y$$
$$x^2 = 8y$$

29. The parabola is opening left and has an equation of the form $y^2 = 4ax$. Since $(-3, 6)$ is on the graph, we have:

$$y^2 = 4ax$$
$$(6)^2 = 4a(-3)$$
$$-3 = a$$

Thus, the equation of the parabola is

$$y^2 = 4(-3)x$$
$$y^2 = -12x$$

31. The parabola is opening down and has an equation of the form $x^2 = 4ay$. Since $(-6, -9)$ is on the graph, we have:

$$x^2 = 4ay$$
$$(-6)^2 = 4a(-9)$$
$$36 = -36a$$
$$-1 = a$$

Thus, the equation of the parabola is

$$x^2 = 4(-1)y$$
$$x^2 = -4y$$

33. $x^2 = 4y$
$y^2 = 4x$

Solve for y in the first equation, then substitute into the second equation.

$$y = \frac{x^2}{4}$$
$$\left(\frac{x^2}{4}\right)^2 = 4x$$
$$\frac{x^4}{16} = 4x$$
$$x^4 = 64x$$
$$x^4 - 64x = 0$$
$$x(x^3 - 64) = 0$$
$$x(x - 4)(x^2 + 4x + 8) = 0$$
$$x = 0 \quad x - 4 = 0 \quad x^2 + 4x + 8 = 0$$
$$x = 4 \quad \text{No real solutions}$$

For $x = 0$ For $x = 4$

$$y = 0 \qquad\qquad y = \frac{4^2}{4}$$
$$y = 4$$

Solutions: $(0, 0)$, $(4, 4)$

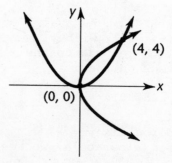

The exact solutions yield more than 3-digit accuracy.

35. $y^2 = 6x$
$x^2 = 5y$

Solve for x in the first equation, then substitute into the second equation.

$$x = \frac{y^2}{6}$$
$$\left(\frac{y^2}{6}\right)^2 = 5y$$
$$\frac{y^4}{36} = 5y$$
$$y^4 = 180y$$
$$y^4 - 180y = 0$$
$$y(y^3 - 180) = 0$$
$$y(y - \sqrt[3]{180})(y^2 + y\sqrt[3]{180} + (\sqrt[3]{180})^2) = 0$$

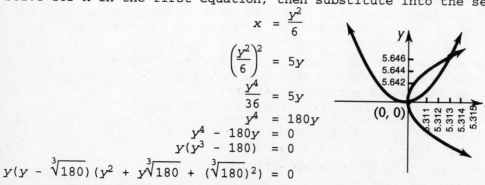

$$y = 0 \quad y - \sqrt[3]{180} = 0 \qquad y^2 + y\sqrt[3]{180} + (\sqrt[3]{180})^2 = 0$$

$$y = \sqrt[3]{180} \quad \text{No real solutions}$$
$$\approx 5.646$$

For $y = 0$ For $y = \sqrt[3]{180}$

$$x = 0 \qquad\qquad x = \frac{(\sqrt[3]{180})^2}{6}$$
$$\approx 5.313$$

Solutions: $(0, 0)$, $(5.313, 5.646)$

37. (A) The line $x = 0$ intersects the parabola $x^2 = 4ay$ only at $(0, 0)$.
The line $y = 0$ intersects the parabola $x^2 = 4ay$ only at $(0, 0)$.
Only these 2 lines intersect the parabola at exactly one point (see part (B)).
(B) A line through $(0, 0)$ with slope $m \neq 0$ has equation $y = mx$.
Solve the system:
$$y = mx$$
$$x^2 = 4ay$$
by substituting y from the first equation into the second equation.
$$x^2 = 4amx$$
$$x^2 - 4amx = 0$$
$$x(x - 4am) = 0$$
$$x = 0 \quad x = 4am$$
For $x = 0$ For $x = 4am$
$$y = 0 \qquad\qquad y = m(4am)$$
$$= 4am^2$$

Solutions: $(0, 0)$, $(4am, 4am^2)$ are the required coordinates.

39. Since A and B lie on the curve $x^2 = 4ay$, their coordinates must satisfy the equation of the curve. Clearly, the y coordinate of A, F, and B is a. Substituting a for y, we have
$$x^2 = 4aa$$
$$x^2 = 4a^2$$
$$x = \pm 2a$$
Therefore, A has coordinates $(-2a, a)$ and B has coordinates $(2a, a)$.

41. Let $P(x, y)$ be a point on the parabola.

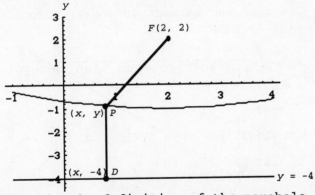

Then, by the definition of the parabola, the distance from $P(x, y)$ to the focus $F(2, 2)$ must equal the perpendicular distance from P to the directrix at $D(x, -4)$. Applying the distance formula, we have
$$d(P, F) = d(P, D)$$
$$\sqrt{(x - 2)^2 + (y - 2)^2} = \sqrt{(x - x)^2 + [y - (-4)]^2}$$
$$(x - 2)^2 + (y - 2)^2 = (x - x)^2 + (y + 4)^2$$
$$x^2 - 4x + 4 + y^2 - 4y + 4 = 0 + y^2 + 8y + 16$$
$$x^2 - 4x - 12y - 8 = 0$$

43. Let $P(x, y)$ be a point on the parabola.

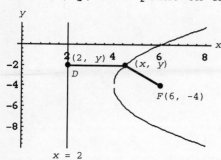

Then, by the definition of the parabola, the distance from $P(x, y)$ to the focus $F(6, -4)$ must equal the perpendicular distance from P to the directrix at $D(2, y)$. Applying the distance formula, we have

$$d(P, F) = d(P, D)$$

$$\sqrt{(x - 6)^2 + [y - (-4)]^2} = \sqrt{(x - 2)^2 + (y - y)^2}$$

$$(x - 6)^2 + (y + 4)^2 = (x - 2)^2 + (y - y)^2$$

$$x^2 - 12x + 36 + y^2 + 8y + 16 = x^2 - 4x + 4 + 0$$

$$y^2 + 8y - 8x + 48 = 0$$

45. Here are computer-generated graphs of $x^2 = 8y$ and $y = 5x + 4$, $-15 \leq x \leq 45$.

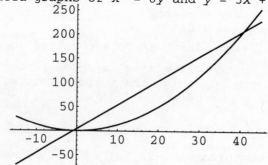

The curves intersect for x between -2 and 0, and between 40 and 42. Zooming in on these intervals, we obtain the following graphs.

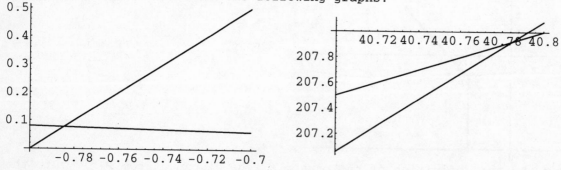

To two decimal places, the coordinates of the points of intersection are (-0.78, 0.08) and (40.78, 207.92).

47. Here are computer-generated
graphs of $x^2 = -8y$ and $y^2 = -5x$
(entered as $y = \pm\sqrt{-5x}$).

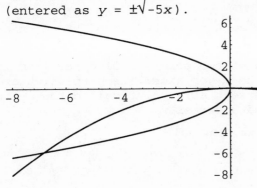

The curves intersect at (0, 0), as can be
easily checked by substitution. The other
point of intersection occurs for x between
-7.2 and -6.8. Zooming in on this interval,
we obtain the following graph.

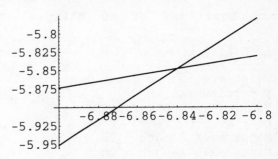

To two decimal places, the coordinates of the
other point of intersection are (-6.84, -5.85).

49. From the figure, we see that the coordinates
of P must be (-100, -50). The parabola is
opening down with axis the y axis, hence it
has an equation of the form $x^2 = 4ay$. Since
(-100, -50) is on the graph, we have
$(-100)^2 = 4a(-50)$
$10{,}000 = -200a$
$a = -50$
Thus, the equation of the parabola is
$x^2 = 4(-50)y$
$x^2 = -200y$

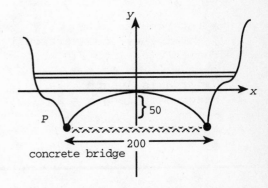

51. (A) From the figure, we see that the parabola is
opening up with axis the y axis, hence it has an
equation of the form $x^2 = 4ay$. Since the focus is
at $(0, a) = (0, 100)$, $a = 100$ and the equation of
the parabola is
$x^2 = 400y$ or $y = 0.0025x^2$

(B) Since the depth represents the y coordinate y_1
of a point on the parabola with $x = 100$, we have

$y_1 = 0.0025(100)^2$
depth $= 25$ feet

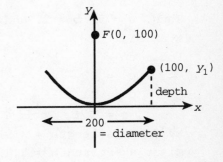

Exercise 9-2

Key Ideas and Formulas

An ellipse is the set of all points P in a plane such that the sum of the distances of P from two fixed points (the foci) in the plane is constant.

The line through the foci intersects the ellipse at two points called vertices; the line segment connecting the vertices is called the major axis; its perpendicular bisector is called the minor axis. The axes intersect at a point called the center of the ellipse. Each end of the major axis is called a vertex.

Standard Equations of an Ellipse—Center at (0, 0)

1. $\dfrac{x^2}{a^2} + \dfrac{y^2}{b^2} = 1 \quad a > b > 0$

   x intercepts: $\pm a$ (vertices)
   y intercepts: $\pm b$
   Foci: $F'(-c, 0)$, $F(c, 0)$
   $c^2 = a^2 - b^2$
   Major axis length = $2a$
   Minor axis length = $2b$

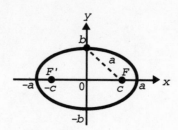

2. $\dfrac{x^2}{b^2} + \dfrac{y^2}{a^2} = 1 \quad a > b > 0$

   x intercepts: $\pm b$
   y intercepts: $\pm a$ (vertices)
   Foci: $F'(0, -c)$, $F(0, c)$
   $c^2 = a^2 - b^2$
   Major axis length = $2a$
   Minor axis length = $2b$

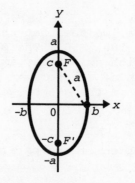

---

1. When $y = 0$, $\dfrac{x^2}{25} = 1$. x intercepts: ± 5

   When $x = 0$, $\dfrac{y^2}{4} = 1$. y intercepts: ± 2

   Thus, $a = 5$, $b = 2$, and the major axis is on the x axis.

   Foci: $c^2 = a^2 - b^2$
   $\qquad c^2 = 25 - 4$
   $\qquad c^2 = 21$
   $\qquad\quad c = \sqrt{21}$ Foci: $F'(-\sqrt{21}, 0)$, $F(\sqrt{21}, 0)$
   Major axis length = $2(5) = 10$
   Minor axis length = $2(2) = 4$

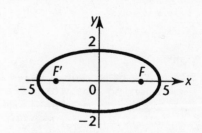

> **Common Error:**
> The relationship $c^2 = a^2 + b^2$ does not apply to a, b, c as defined for ellipses.

3. When $y = 0$, $\frac{x^2}{4} = 1$. x intercepts: ±2

When $x = 0$, $\frac{y^2}{25} = 1$. y intercepts: ±5

Thus, $a = 5$, $b = 2$, and the major axis is on the y axis.
Foci: $c^2 = a^2 - b^2$
$c^2 = 25 - 4$
$c^2 = 21$
$c = \sqrt{21}$ Foci: $F'(0, -\sqrt{21})$, $F(0, \sqrt{21})$
Major axis length = 2(5) = 10 Minor axis length = 2(2) = 4

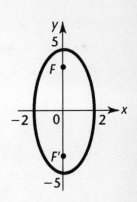

5. First, write the equation in standard form by dividing both sides by 9.
$x^2 + 9y^2 = 9$
$\frac{x^2}{9} + \frac{y^2}{1} = 1$
Locate the intercepts.
When $y = 0$, $\frac{x^2}{9} = 1$. x intercepts: ±3

When $x = 0$, $\frac{y^2}{1} = 1$. y intercepts: ±1

Thus $a = 3$, $b = 1$, and the major axis is on the
x axis.
Foci: $c^2 = a^2 - b^2$
$c^2 = 9 - 1$
$c^2 = 8$
$c = \sqrt{8}$ Foci: $F'(-\sqrt{8}, 0)$, $F(\sqrt{8}, 0)$
Major axis length = 2(3) = 6
Minor axis length = 2(1) = 2

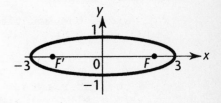

7. First, write the equation in standard form by dividing both sides by 225.
$25x^2 + 9y^2 = 225$
$\frac{x^2}{9} + \frac{y^2}{25} = 1$
Locate the intercepts.
When $y = 0$, $\frac{x^2}{9} = 1$. x intercepts: ±3

When $x = 0$, $\frac{y^2}{25} = 1$. y intercepts: ±5

Thus $a = 5$, $b = 3$, and the major axis is on the y axis.
Foci: $c^2 = a^2 - b^2$
$c^2 = 25 - 9$
$c^2 = 16$
$c = 4$ Foci: $F'(0, -4)$, $F(0, 4)$
Major axis length = 2(5) = 10
Minor axis length = 2(3) = 6

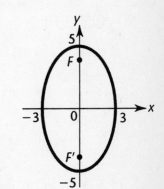

9. First, write the equation in standard form by dividing both sides by 12.

$2x^2 + y^2 = 12$

$\dfrac{x^2}{6} + \dfrac{y^2}{12} = 1$

Locate the intercepts.

When $y = 0$, $\dfrac{x^2}{6} = 1$. x intercepts: $\pm\sqrt{6}$

When $x = 0$, $\dfrac{y^2}{12} = 1$. y intercepts: $\pm\sqrt{12}$

Thus $a = \sqrt{12}$, $b = \sqrt{6}$, and the major axis is on the y axis.

Foci: $c^2 = a^2 - b^2$

$\qquad c^2 = 12 - 6$

$\qquad c^2 = 6$

$\qquad c = \sqrt{6}$ Foci: $F'(0, -\sqrt{6})$, $F(0, \sqrt{6})$

Major axis length $= 2\sqrt{12} \approx 6.93$

Minor axis length $= 2\sqrt{6} \approx 4.90$

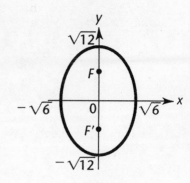

11. First, write the equation in standard form by dividing both sides by 28.

$4x^2 + 7y^2 = 28$

$\dfrac{x^2}{7} + \dfrac{y^2}{4} = 1$

Locate the intercepts.

When $y = 0$, $\dfrac{x^2}{7} = 1$. x intercepts: $\pm\sqrt{7}$

When $x = 0$, $\dfrac{y^2}{4} = 1$. y intercepts: ± 2

Thus $a = \sqrt{7}$, $b = 2$, and the major axis is on the x axis.

Foci: $c^2 = a^2 - b^2$

$\qquad c^2 = 7 - 4$

$\qquad c^2 = 3$

$\qquad c = \sqrt{3}$ Foci: $F'(-\sqrt{3}, 0)$, $F(\sqrt{3}, 0)$

Major axis length $= 2\sqrt{7} \approx 5.29$

Minor axis length $= 2(2) = 4$

13. Make a rough sketch of the ellipse and compute x and y intercepts.

$\dfrac{x^2}{a^2} + \dfrac{y^2}{b^2} = 1$

$a = \dfrac{8}{2} = 4$, $b = \dfrac{6}{2} = 3$

$\dfrac{x^2}{16} + \dfrac{y^2}{9} = 1$

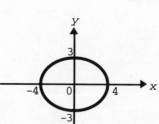

15. Make a rough sketch of the ellipse and compute x and y intercepts.

$\dfrac{x^2}{b^2} + \dfrac{y^2}{a^2} = 1$

$a = \dfrac{22}{2} = 11$, $b = \dfrac{16}{2} = 8$

$\dfrac{x^2}{64} + \dfrac{y^2}{121} = 1$

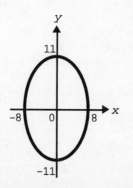

17. Make a rough sketch of the ellipse, locate focus and x intercepts, then determine y intercepts using the special triangle relationship.

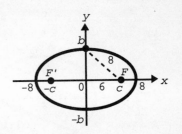

$$\frac{x^2}{a^2} + \frac{y^2}{b^2} = 1$$

$$a = \frac{16}{2} = 8$$

$$b^2 = 8^2 - 6^2 = 64 - 36 = 28$$

$$b = \sqrt{28}$$

$$\frac{x^2}{64} + \frac{y^2}{28} = 1$$

19. Make a rough sketch of the ellipse, locate focus and x intercepts, then determine y intercepts using the special triangle relationship.

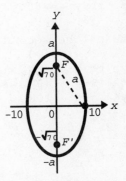

$$\frac{x^2}{b^2} + \frac{y^2}{a^2} = 1$$

$$b = \frac{20}{2} = 10$$

$$a^2 = 10^2 + (\sqrt{70})^2 = 100 + 70 = 170$$

$$a = \sqrt{170}$$

$$\frac{x^2}{100} + \frac{y^2}{170} = 1$$

21. The graph does not pass the vertical line test; most vertical ines that intersect an ellipse do so in two places; hence, the equation does not define a function.

23. $16x^2 + 25y^2 = 400$
　　　$2x - 5y = 10$

We solve for x in terms of y in the second equation, then substitute the expression for x into the first equation.

$$2x = 5y + 10$$

$$x = \frac{5y + 10}{2}$$

$$16\left(\frac{5y + 10}{2}\right)^2 + 25y^2 = 400$$

$$\frac{16(25y^2 + 100y + 100)}{4} + 25y^2 = 400$$

$$100y^2 + 400y + 400 + 25y^2 = 400$$

$$125y^2 + 400y = 0$$

$$25y(5y + 16) = 0$$

$$y = 0 \qquad\qquad y = -\frac{16}{5} \text{ or } -3.2$$

For $y = 0$　　　　　　　For $y = -\frac{16}{5}$

$$x = \frac{5 \cdot 0 + 10}{2} \qquad\qquad x = \frac{5\left(-\frac{16}{5}\right) + 10}{2}$$

$$x = 5 \qquad\qquad\qquad x = \frac{-16 + 10}{2}$$

$$x = -3$$

Solutions: $(5, 0)$, $(-3, -3.2)$.

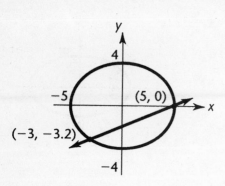

Graphs:
$16x^2 + 25y^2 = 400$

$$\frac{x^2}{25} + \frac{y^2}{16} = 1 \quad \text{Ellipse}$$

$a = 5$, $b = 4$, major axis on the x axis.
x intercepts: ± 5
y intercepts: ± 4
$2x - 5y = 10$
Straight line, intercepts $(0, -2)$ and $(5, 0)$

25. $25x^2 + 16y^2 = 400$
$25x^2 - 36y = 0$

Subtract the second equation from the first.
$16y^2 + 36y = 400$
Solve this equation for y.
$16y^2 + 36y - 400 = 0$
$4y^2 + 9y - 100 = 0$
$(4y + 25)(y - 4) = 0$
$4y + 25 = 0 \qquad y - 4 = 0$

$$y = -\frac{25}{4} \qquad y = 4$$

For $y = -\dfrac{25}{4}$ For $y = 4$

$$25x^2 - 36\left(-\frac{25}{4}\right) = 0 \qquad 25x^2 - 36(4) = 0$$

$$25x^2 + 225 = 0 \qquad 25x^2 - 144 = 0$$

$$x^2 = -9 \qquad x^2 = \frac{144}{25}$$

$$x = \pm 3i \qquad x = \pm\frac{12}{5} \text{ or } \pm 2.4$$

No real solution   Solutions: $(2.4, 4)$ and $(-2.4, 4)$

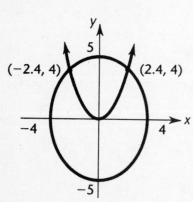

Graphs:
$25x^2 + 16y^2 = 400$

$$\frac{x^2}{16} + \frac{y^2}{25} = 1$$

$a = 5$, $b = 4$, major axis on the y axis.
x intercepts: ± 4
y intercepts: ± 5
$25x^2 - 36y = 0$

$$x^2 = \frac{36}{25}y$$

Parabola, opens up.

27. $5x^2 + 2y^2 = 63$
$2x - y = 0$

Solve for y in the second equation, then substitute into the first equation.

$$y = 2x$$
$$5x^2 + 2(2x)^2 = 63$$
$$5x^2 + 8x^2 = 63$$
$$13x^2 = 63$$
$$x^2 = \frac{63}{13}$$

$$x = \pm\sqrt{\frac{63}{13}}$$

$\approx \pm 2.201$ We discard the negative solution, since we are asked for first quadrant solutions only.

For $x = 2.201$
$$y = 2(2.201)$$
$$= 4.402$$
Solution: $(2.201, 4.402)$

29. $2x^2 + 3y^2 = 33$
$x^2 - 8y = 0$

Solve for y in the second equation, then substitute into the first equation.

$$x^2 = 8y$$
$$y = \frac{x^2}{8}$$

$$2x^2 + 3\left(\frac{x^2}{8}\right)^2 = 33$$

$$2x^2 + \frac{3x^4}{64} = 33$$

$$3x^4 + 128x^2 = 2112$$

$3x^4 + 128x^2 - 2112 = 0$ Quadratic in x^2.

$$x^2 = \frac{-b \pm \sqrt{b^2 - 4ac}}{2a} \quad a = 3, \ b = 128, \ c = -2112$$

$$x^2 = \frac{-128 \pm \sqrt{128^2 - 4(3)(-2112)}}{2(3)}$$

$$x^2 = \frac{-128 + \sqrt{41728}}{6} \text{ discarding the negative solution}$$

$$x = \sqrt{\frac{-128 + \sqrt{41728}}{6}} \text{ discarding the negative solution}$$

$$x \approx 3.565$$
$$y = \frac{x^2}{8}$$
$$= \frac{(3.565)^2}{8}$$
$$\approx 1.589$$
Solution: $(3.565, 1.589)$

31. From the figure, we see that the point $P(x, y)$ is a point on the curve if and only if

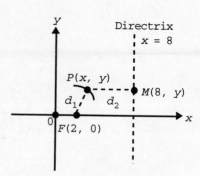

$$d_1 = \frac{1}{2} d_2$$

$$d(P, F) = \frac{1}{2} d(P, M)$$

$$\sqrt{(x - 2)^2 + (y - 0)^2} = \frac{1}{2} \sqrt{(x - 8)^2 + (y - y)^2}$$

$$(x - 2)^2 + y^2 = \frac{1}{4} (x - 8)^2$$

$$x^2 - 4x + 4 + y^2 = \frac{1}{4} (x^2 - 16x + 64)$$

$$4x^2 - 16x + 16 + 4y^2 = x^2 - 16x + 64$$

$$3x^2 + 4y^2 = 48$$

$$\frac{x^2}{16} + \frac{y^2}{12} = 1$$

The curve must be an ellipse since its equation can be written in standard form for an ellipse.

33. Here are computer-generated graphs of $x^2 + 3y^2 = 20$ $\left(\text{entered as } y = \pm\sqrt{\dfrac{20 - x^2}{3}}\right)$

and $4x + 5y = 11$ $\left(\text{entered as } y = \dfrac{11 - 4x}{5}\right)$.

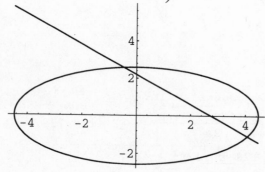

The curves intersect for x between -0.8 and -0.4 and between 4 and 4.4. Zooming in on these intervals, we obtain the following graphs:

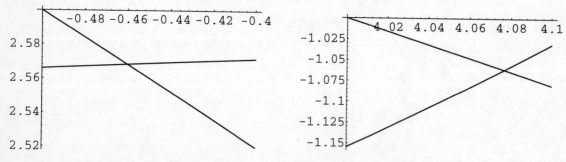

To two decimal places, the coordinates of the points of intersection are $(-0.46, 2.57)$ and $(4.08, -1.06)$.

35. Here are computer-generated graphs of
$50x^2 + 4y^2 = 1025$

$$\left(\text{entered as } y = \pm\sqrt{\frac{1025 - 50x^2}{4}}\right)$$

and $9x^2 + 2y^2 = 300$

$$\left(\text{entered as } y = \pm\sqrt{\frac{300 - 9x^2}{2}}\right).$$

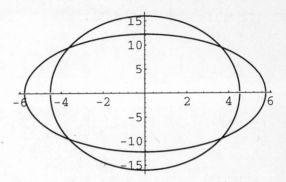

Note that the graphs are symmetric with
respect to both axes and the origin. The
curves intersect in the first quadrant
for x between 3.6 and 4. Zooming in on
this interval, we obtain the graph shown
at the right.

To two decimal places, the coordinates of
the point of intersection are
(3.64, 9.50). From symmetry, we obtain
the other three coordinate pairs:
(-3.64, 9.50), (-3.64, -9.50), and
(3.64, -9.50).

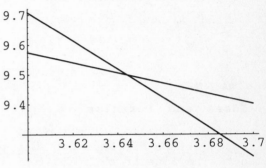

37. From the figure we see that the x and y intercepts of the ellipse must be 20
and 12 respectively. Hence the equation of the ellipse must be

$$\frac{x^2}{(20)^2} + \frac{y^2}{(12)^2} = 1$$
or
$$\frac{x^2}{400} + \frac{y^2}{144} = 1$$

To find the clearance above the water 5 feet from the bank, we need the y
coordinate y_1 of the point P whose x coordinate is $a - 5 = 15$. Since P is on
the ellipse, we have

$$\frac{15^2}{400} + \frac{y_1^2}{144} = 1$$

$$\frac{225}{400} + \frac{y_1^2}{144} = 1$$

$$0.5625 + \frac{y_1^2}{144} = 1$$

$$\frac{y_1^2}{144} = 0.4375$$

$$y_1^2 = 144(0.4375)$$

$$y_1^2 = 63$$

$$y_1 \approx \pm 7.94$$

Therefore, the clearance is 7.94 feet, approximately.

39. (A) From the figure we see that the x intercept of the ellipse must be 24.0.
Hence the equation of the ellipse must have the form

$$\frac{x^2}{(24.0)^2} + \frac{y^2}{b^2} = 1$$

Since the point (23.0, 1.14) is on the ellipse, its coordinates must satisfy

the equation of the ellipse. Hence

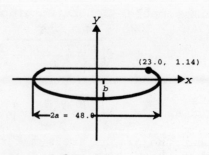

$$\frac{(23.0)^2}{(24.0)^2} + \frac{(1.14)^2}{b^2} = 1$$

$$\frac{(1.14)^2}{b^2} = 1 - \frac{(23.0)^2}{(24.0)^2}$$

$$\frac{(1.14)^2}{b^2} = \frac{47}{576}$$

$$b^2 = \frac{576(1.14)^2}{47}$$

$$b^2 = 15.9$$

Thus, the equation of the ellipse must be $\dfrac{x^2}{576} + \dfrac{y^2}{15.9} = 1$

(B) From the figure, we can see that the width of the wing must equal
$1.14 + b = 1.14 + \sqrt{15.9} = 5.13$ feet.

Exercise 9-3

Key Ideas and Formulas

A hyperbola is the set of all points P in a plane such that the absolute value of the difference of the distances of P to two-fixed points (the foci) is a positive constant. The line through the foci intersects the hyperbola at two points called vertices; the line segment connecting the vertices is called the transverse axis. The midpoint of the transverse axis is called the center of the hyperbola.

Standard Equations of a Hyperbola — Center at (0, 0)

1. $\dfrac{x^2}{a^2} - \dfrac{y^2}{b^2} = 1$

   x intercepts: $\pm a$ (vertices)
   y intercepts: none
   Foci: $F'(-c, 0)$, $F(c, 0)$
   $c^2 = a^2 + b^2$
   Transverse axis length = $2a$
   Conjugate axis length = $2b$

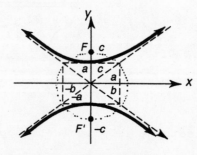

Asymptotes: $y = \pm\dfrac{b}{a}x$

2. $\dfrac{y^2}{a^2} - \dfrac{x^2}{b^2} = 1$

   x intercepts: none
   y intercepts: $\pm a$ (vertices)
   Foci: $F'(0, -c)$, $F(0, c)$
   $c^2 = a^2 + b^2$
   Transverse axis length = $2a$
   Conjugate axis length = $2b$

Asymptotes: $y = \pm\dfrac{a}{b}x$

Note: Both graphs are symmetric with respect to the x axis, y axis, and origin.

1. When $y = 0$, $\dfrac{x^2}{9} = 1$. x intercepts: ± 3 $a = 3$

When $x = 0$, $-\dfrac{y^2}{4} = 1$. There are no y intercepts, but $b = 2$.

Sketch the asymptotes using the asymptote rectangle, then sketch in the hyperbola.
Foci: $c^2 = 3^2 + 2^2$
$c^2 = 13$
$c = \sqrt{13}$

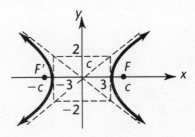

$F'(-\sqrt{13}, 0)$, $F(\sqrt{13}, 0)$
Transverse axis length = 2(3) = 6
Conjugate axis length = 2(2) = 4

3. When $y = 0$, $-\dfrac{x^2}{9} = 1$. There are no x intercepts, but $b = 3$.

When $x = 0$, $\dfrac{y^2}{4} = 1$. y intercepts: ± 2, $a = 2$

Sketch the asymptotes using the asymptote rectangle, then sketch in the hyperbola.
Foci: $c^2 = 2^2 + 3^2$
$c^2 = 13$
$c = \sqrt{13}$

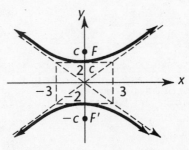

$F'(0, -\sqrt{13})$, $F(0, \sqrt{13})$
Transverse axis length = 2(2) = 4
Conjugate axis length = 2(3) = 6

5. First, write the equation in standard form by dividing both sides by 16.
$4x^2 - y^2 = 16$
$\dfrac{x^2}{4} - \dfrac{y^2}{16} = 1$

Locate intercepts:
When $y = 0$, $x = \pm 2$. x intercepts: ± 2 $a = 2$

When $x = 0$, $-\dfrac{y^2}{16} = 1$. There are no y intercepts,

but $b = 4$.

Sketch the asymptotes using the asymptote rectangle, then sketch in the hyperbola.
Foci: $c^2 = 2^2 + 4^2$
$c^2 = 20$
$c = \sqrt{20}$

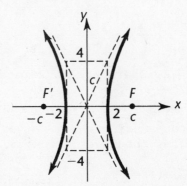

$F'(-\sqrt{20}, 0)$, $F(\sqrt{20}, 0)$
Transverse axis length = 2(2) = 4
Conjugate axis length = 2(4) = 8

7. First, write the equation in standard form by dividing both sides by 144.
$9y^2 - 16x^2 = 144$
$\dfrac{y^2}{16} - \dfrac{x^2}{9} = 1$

Locate intercepts:
When $y = 0$, $-\dfrac{x^2}{9} = 1$. There are no x intercepts, but $b = 3$.

When $x = 0$, $\frac{y^2}{16} = 1$, $y = \pm 4$.

y intercepts: ± 4 $a = 4$

Sketch the asymptotes using the asymptote rectangle, then sketch in the hyperbola.

Foci: $c^2 = 4^2 + 3^2$

$\qquad c^2 = 25$

$\qquad\quad c = 5$

$F'(0, -5)$, $F(0, 5)$

Transverse axis length = $2(4) = 8$

Conjugate axis length = $2(3) = 6$

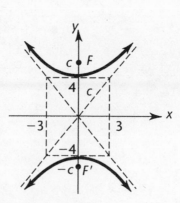

9. First, write the equation in standard form by dividing both sides by 12.

$3x^2 - 2y^2 = 12$

$\dfrac{x^2}{4} - \dfrac{y^2}{6} = 1$

Locate intercepts:

When $y = 0$, $\dfrac{x^2}{4} = 1$. $x = \pm 2$.

x intercepts: ± 2 $a = 2$

When $x = 0$, $-\dfrac{y^2}{6} = 1$. There are no y intercepts, but $b = \sqrt{6}$.

Sketch the asymptotes using the asymptote rectangle, then sketch in the hyperbola.

Foci: $c^2 = 2^2 + (\sqrt{6})^2$

$\qquad c^2 = 10$

$\qquad\quad c = \sqrt{10}$

$F'(-\sqrt{10}, 0)$, $F(\sqrt{10}, 0)$

Transverse axis length = $2(2) = 4$

Conjugate axis length = $2\sqrt{6} \approx 4.90$

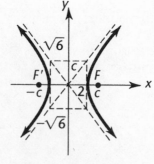

11. First, write the equation in standard form by dividing both sides by 28.

$7y^2 - 4x^2 \qquad = 28$

$\dfrac{y^2}{4} - \dfrac{x^2}{7} \qquad = 1$

Locate intercepts:

When $y = 0$, $-\dfrac{x^2}{7} = 1$. There are no x intercepts, but $b = \sqrt{7}$.

When $x = 0$, $\dfrac{y^2}{4} = 1$, $y = \pm 2$. y intercepts: ± 2 $a = 2$

Sketch the asymptotes using the asymptote rectangle, then sketch in the hyperbola.

Foci: $c^2 = 2^2 + (\sqrt{7})^2$

$\qquad c^2 = 11$

$\qquad\quad c = \sqrt{11}$

$F'(0, -\sqrt{11})$, $F(0, \sqrt{11})$

Transverse axis length = $2(2) = 4$

Conjugate axis length = $2\sqrt{7} \approx 5.29$

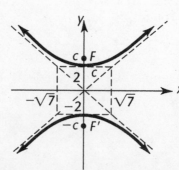

13. Since the transverse axis is on the x axis, start with

$$\frac{x^2}{a^2} - \frac{y^2}{b^2} = 1$$

and find a and b

$$a = \frac{14}{2} = 7 \text{ and } b = \frac{10}{2} = 5$$

Thus, the equation is

$$\frac{x^2}{49} - \frac{y^2}{25} = 1$$

15. Since the transverse axis is on the y axis, start with

$$\frac{y^2}{a^2} - \frac{x^2}{b^2} = 1$$

and find a and b

$$a = \frac{24}{2} = 12 \text{ and } b = \frac{18}{2} = 9$$

Thus, the equation is

$$\frac{y^2}{144} - \frac{x^2}{81} = 1$$

17. Since the transverse axis is on the x axis, start with

$$\frac{x^2}{a^2} - \frac{y^2}{b^2} = 1$$

and find a and b

$$a = \frac{18}{2} = 9$$

To find b, sketch the asymptote rectangle, label known parts, and use the Pythagorean Theorem.

$$b^2 = 11^2 - 9^2$$
$$b^2 = 40$$
$$b = \sqrt{40}$$

Thus, the equation is

$$\frac{x^2}{81} - \frac{y^2}{40} = 1$$

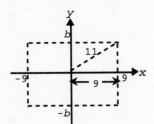

19. Since the conjugate axis is on the x axis, start with

$$\frac{y^2}{a^2} - \frac{x^2}{b^2} = 1$$

and find a and b

$$b = \frac{14}{2} = 7$$

To find a, sketch the asymptote rectangle, label known parts, and use the Pythagorean Theorem.

$$a^2 = (\sqrt{200})^2 - 7^2$$
$$a^2 = 151$$
$$a = \sqrt{151}$$

Thus, the equation is

$$\frac{y^2}{151} - \frac{x^2}{49} = 1$$

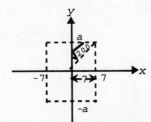

21. (A) If a hyperbola has center at $(0, 0)$ and a focus at $(1, 0)$, its equation must be of form

$$\frac{x^2}{a^2} - \frac{y^2}{b^2} = 1$$

with $c = 1$, thus $a^2 + b^2 = 1$ or $b^2 = 1 - a^2$

Therefore there are an infinite number of such hyperbolas.
Each has an equation of form

$$\frac{x^2}{a^2} - \frac{y^2}{1 - a^2} = 1$$

Note that since $0 < a < c$, we require $0 < a < 1$.

(B) If an ellipse has center at $(0, 0)$ and a focus at $(1, 0)$, its equation must be of form

$$\frac{x^2}{a^2} + \frac{y^2}{b^2} = 1$$

with $c = 1$, thus $a^2 - 1 = b^2$

Therefore there are an infinite number of such ellipses. Each has an equation of form

$$\frac{x^2}{a^2} + \frac{y^2}{a^2 - 1} = 1$$

Note that since $a > c > 0$, we require $a > 1$.

(C) If a parabola has vertex at $(0, 0)$ and focus at $(1, 0)$, its equation must be of form

$$y^2 = 4ax$$

with $a = 1$.

Therefore there is one such parabola; its equation is $y^2 = 4x$.

23. $3y^2 - 4x^2 = 12$
$\quad\quad y^2 + x^2 = 25$

We solve using elimination by addition, multiplying the second equation by 4 and adding to the first equation.

$$3y^2 - 4x^2 = 12$$
$$\underline{4y^2 + 4x^2 = 100}$$
$$7y^2 \quad\quad = 112$$

$$y^2 = 16$$
$$y = \pm 4$$

For $y = 4$ or $y = -4$
$y^2 + x^2 = 25$
$16 + x^2 = 25$
$\quad\quad x^2 = 9$
$\quad\quad x = \pm 3$

Solutions: (3, 4), (-3, 4), (-3, -4), (3, -4)

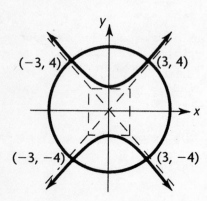

Graphs:
$3y^2 - 4x^2 = 12$ Hyperbola
$$\frac{y^2}{4} - \frac{x^2}{3} = 1$$
y intercepts: ± 2 $a = 2$

x intercepts: none, but $b = \sqrt{3}$
Transverse axis on y axis.
$y^2 + x^2 = 25$ Circle
radius 5, center (0, 0)

25. $2x^2 + y^2 = 24$
$\quad\quad x^2 - y^2 = -12$

We solve by adding the two equations to eliminate y.

$3x^2 = 12$
$\quad x^2 = 4$
$\quad x = \pm 2$

For $x = \pm 2$
$\quad 2x^2 + y^2 = 24$
$\quad 2(4) + y^2 = 24$
$\quad\quad 8 + y^2 = 24$
$\quad\quad\quad y^2 = 16$
$\quad\quad\quad y = \pm 4$

Solutions: (2, 4), (2, -4), (-2, -4), (-2, 4)

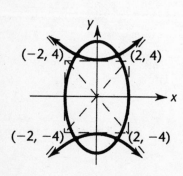

Graphs:
$2x^2 + y^2 = 24$
$$\frac{x^2}{12} + \frac{y^2}{24} = 1$$ Ellipse

$a = \sqrt{24}$, $b = \sqrt{12}$, major axis on the y axis

x intercepts: $\pm\sqrt{12}$

y intercepts: $\pm\sqrt{24}$
$x^2 - y^2 = -12$
$$\frac{y^2}{12} - \frac{x^2}{12} = 1$$ Hyperbola

y intercepts: $\pm\sqrt{12}$ $a = \sqrt{12}$

x intercepts: none, but $b = \sqrt{12}$
Transverse axis on y axis.

27. $y^2 - x^2 = 9$
$2y - x = 8$

Solve for x in the second equation, then substitute into the first equation.

$$2y = 8 + x$$
$$x = 2y - 8$$
$$y^2 - (2y - 8)^2 = 9$$
$$y^2 - (4y^2 - 32y + 64) = 9$$
$$-3y^2 + 32y - 64 = 9$$
$$3y^2 - 32y + 73 = 0$$

$$y = \frac{-b \pm \sqrt{b^2 - 4ac}}{2a} \quad a = 3, \ b = -32, \ c = 73$$

$$y = \frac{-(-32) \pm \sqrt{(-32)^2 - 4(3)(73)}}{2(3)}$$

$$= \frac{32 \pm \sqrt{148}}{6}$$

$$y \approx 3.306, \ 7.361$$

For $y = 3.306$ For $y = 7.361$
   $x = 2(3.306) - 8$ $x = 2(7.361) - 8$
   ≈ -1.389 ≈ 6.722

Solutions: $(-1.389, 3.306)$, $(6.722, 7.361)$

29. $y^2 - x^2 = 4$
$y^2 + 2x^2 = 36$

We solve using elimination by addition, multiplying the first equation by 2 and adding to the second equation.

$$2y^2 - 2x^2 = 8$$
$$\underline{y^2 + 2x^2 = 36}$$
$$3y^2 \qquad\ \ = 44$$

$$y^2 = \frac{44}{3}$$

$$y = \pm\sqrt{\frac{44}{3}} \ \ \text{We discard the negative solution.}$$

$$y \approx 3.830$$
$$y^2 - x^2 = 4$$
$$x^2 = y^2 - 4$$
$$= \frac{44}{3} - 4$$
$$x^2 = \frac{32}{3}$$
$$x = \pm 3.266$$

Solutions: $(3.266, 3.830)$, $(-3.266, 3.830)$ $(y \geq 0)$

31. From the figure, we see that the point $P(x, y)$ is a point on the curve if and only if $\qquad d_1 = \frac{3}{2} d_2$

$$d(P, F) = \frac{3}{2} d(P, M)$$

$$\sqrt{(x - 3)^2 + (y - 0)^2} = \frac{3}{2}\sqrt{\left(x - \frac{4}{3}\right)^2 + (y - y)^2}$$

$$(x - 3)^2 + y^2 = \frac{9}{4}\left(x - \frac{4}{3}\right)^2$$

$$x^2 - 6x + 9 + y^2 = \frac{9}{4}\left(x^2 - \frac{8}{3}x + \frac{16}{9}\right)$$

$$4x^2 - 24x + 36 + 4y^2 = 9x^2 - 24x + 16$$

$$-5x^2 + 4y^2 = -20$$

$$\frac{x^2}{4} - \frac{y^2}{5} = 1$$

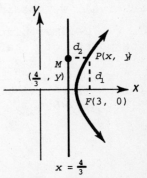

The curve must be a hyperbola since its equation can be written in standard form for a hyperbola.

33. Here are computer-generated graphs of $2x^2 - 3y^2 = 20$

$\left(\text{entered as } y = \pm\sqrt{\dfrac{2x^2 - 20}{3}}\right)$ and $7x + 15y = 10$ $\left(\text{entered as } y = \dfrac{10 - 7x}{15}\right)$.

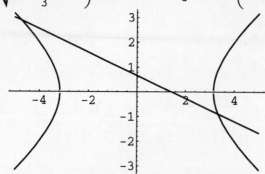

The curves intersect for x between -5.2 and -4.8 and between 3.2 and 3.6. Zooming in on these intervals, we obtain the following graphs.

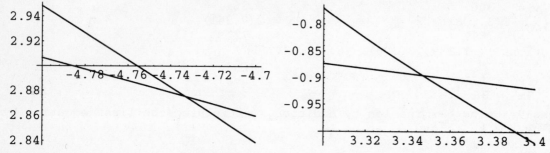

To two decimal places, the coordinates of the points of intersection are $(-4.73, 2.88)$ and $(3.35, -0.90)$.

35. Here are computer-generated graphs of $24y^2 - 18x^2 = 175$

$\left(\text{entered as } y = \pm\sqrt{\dfrac{175 + 18x^2}{24}}\right)$ and $90x^2 + 3y^2 = 200$

$\left(\text{entered as } y = \pm\sqrt{\dfrac{200 - 90x^2}{3}}\right)$.

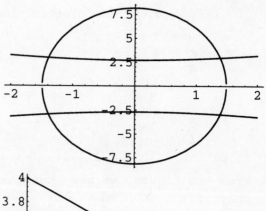

Note that the graphs are symmetric with respect to both axes and the origin. The curves intersect in the first quadrant for x between 1.2 and 1.4. Zooming in on this interval, we obtain the second graph.

To two decimal places, the coordinates of the points of intersection are $(1.39, 2.96)$. From symmetry, we obtain the other three coordinate pairs: $(-1.39, 2.96)$, $(-1.39, -2.96)$, and $(1.39, -2.96)$.

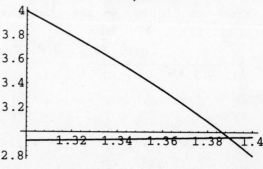

37. From the figure below, we see that the transverse axis of the hyperbola must be on the y axis and $a = 4$. Hence, the equation of the hyperbola must have the form

$$\frac{y^2}{4^2} - \frac{x^2}{b^2} = 1$$

To find b, we note that the point $(8, 12)$ is on the hyperbola, hence its coordinates satisfy the equation. Substituting, we have

$$\frac{12^2}{4^2} - \frac{8^2}{b^2} = 1$$

$$9 - \frac{64}{b^2} = 1$$

$$-\frac{64}{b^2} = -8$$

$$-64 = -8b^2$$

$$b^2 = 8$$

The equation required is

$$\frac{y^2}{16} - \frac{x^2}{8} = 1$$

Using this equation, we can compute y when $x = 6$ to answer the question asked (see figure).

$$\frac{y^2}{16} - \frac{6^2}{8} = 1$$

$$\frac{y^2}{16} - \frac{36}{8} = 1$$

$$y^2 - 72 = 16$$

$$y^2 = 88$$

$$y = 9.38 \text{ to two decimal places}$$

The height above the vertex
$$= y - \text{height of vertex}$$
$$= 9.38 - 4$$
$$= 5.38 \text{ feet}$$

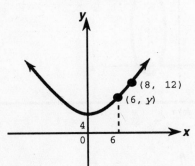

Hyperbola part of dome

39. From the figure below, we can see:
$$FF' = 2c = 120 - 20 = 100$$
Thus $c = 50$
$$FV = c - a = 120 - 110 = 10$$
Thus $a = c - 10 = 50 - 10 = 40$
Since $c^2 = a^2 + b^2$
$$50^2 = 40^2 + b^2$$
$$b = 30$$
Thus the equation of the hyperbola, in standard form, is
$$\frac{y^2}{40^2} - \frac{x^2}{30^2} = 1$$
Expressing y in terms of x, we have
$$\frac{y^2}{40^2} = 1 + \frac{x^2}{30^2}$$
$$\frac{y^2}{40^2} = \frac{1}{30^2}(30^2 + x^2)$$
$$y^2 = \frac{40^2}{30^2}(30^2 + x^2)$$
$$y = \frac{4}{3}\sqrt{x^2 + 30^2}$$

discarding the negative solution, since the reflecting hyperbola is above the x axis.

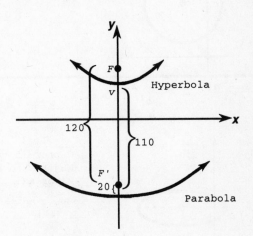

Exercise 9-4
Key Ideas and Formulas

A translation of coordinates occurs when the new coordinate axes have the same direction as and are parallel to the old coordinate axes.

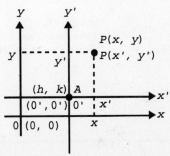

If the coordinates of the origin of the translated system are (h, k) relative to the old system, then (x, y) and (x', y') are related by:

1. $x = x' + h$
   $y = y' + k$

2. $x' = x - h$
   $y' = y - k$

> **Common Error:**
> The signs are very easily confused here, especially if h or k is negative in a problem.

Standard Equations for Translated Conics

PARABOLAS

$$(x - h)^2 = 4a(y - k)$$

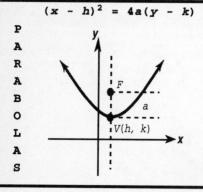

Vertex (h, k)
Focus $(h, k + a)$
$a > 0$ opens up
$a < 0$ opens down

$$(y - k)^2 = 4a(x - h)$$

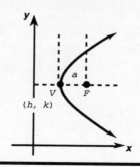

Vertex (h, k)
Focus $(h + a, k)$
$a < 0$ opens left
$a > 0$ opens right

CIRCLES

$$(x - h)^2 + (y - k)^2 = r^2$$

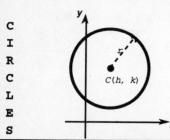

Center (h, k)
Radius r

ELLIPSES

$$\frac{(x - h)^2}{a^2} + \frac{(y - k)^2}{b^2} = 1$$
$$a > b > 0$$

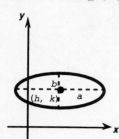

Center (h, k)
Major axis $2a$
Minor axis $2b$

$$\frac{(x - h)^2}{b^2} + \frac{(y - k)^2}{a^2} = 1$$

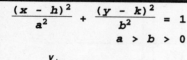

Center (h, k)
Major axis $2a$
Minor axis $2b$

HYPERBOLAS

$$\frac{(x - h)^2}{a^2} - \frac{(y - k)^2}{b^2} = 1$$

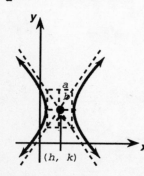

Center (h, k)
Transverse axis $2a$
Conjugate axis $2b$

$$\frac{(y - k)^2}{a^2} - \frac{(x - h)^2}{b^2} = 1$$

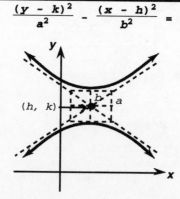

Center (h, k)
Transverse axis $2a$
Conjugate axis $2b$

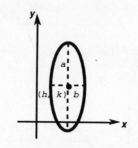

1. (A) Since $(h, k) = (3, 5)$, use translation formulas
$$x' = x - h = x - 3$$
$$y' = y - k = y - 5$$

(B) $x'^2 + y'^2 = 81$

(C) Circle

3. (A) Since $(h, k) = (-7, 4)$, use translation formulas
$$x' = x - h = x + 7$$
$$y' = y - k = y - 4$$

(B) $\dfrac{x'^2}{9} + \dfrac{y'^2}{16} = 1$

(C) Ellipse

5. (A) Since $(h, k) = (4, -9)$, use translation formulas
$$x' = x - h = x - 4$$
$$y' = y - k = y + 9$$

(B) $y'^2 = 16x'$

(C) Parabola

7. (A) Since $(h, k) = (-8, -3)$, use translation formulas
$$x' = x - h = x + 8$$
$$y' = y - k = y + 3$$

(B) $\dfrac{x'^2}{12} + \dfrac{y'^2}{8} = 1$

(C) Ellipse

9. (A) Divide both sides by 144
$$\frac{(x - 3)^2}{9} - \frac{(y + 2)^2}{16} = 1$$

(B) This is the equation of a hyperbola.

11. (A) Divide both sides by 30.
$$\frac{(x + 5)^2}{5} + \frac{(y + 7)^2}{6} = 1$$

(B) This is the equation of an ellipse.

13. (A) Subtract $24(y - 4)$ from both sides.
$$(x + 6)^2 = -24(y - 4)$$

(B) This is the equation of a parabola.

15.
$$4x^2 + 9y^2 - 16x - 36y + 16 = 0$$
$$4x^2 - 16x + 9y^2 - 36y = -16$$
$$4(x^2 - 4x + ?) + 9(y^2 - 4y + ?) = -16$$
$$4(x^2 - 4x + 4) + 9(y^2 - 4y + 4) = -16 + 16 + 36$$
$$4(x - 2)^2 + 9(y - 2)^2 = 36$$
$$\frac{(x - 2)^2}{9} + \frac{(y - 2)^2}{4} = 1$$

This is the equation of an ellipse with center at $(2, 2)$. The equations of translation are $x' = x - 2$, $y' = y - 2$. Making these substitutions, we obtain
$$\frac{x'^2}{9} + \frac{y'^2}{4} = 1$$
We graph this in the $x'y'$ system, following the process discussed in Section 9-2.

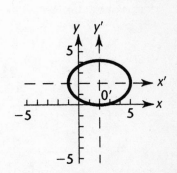

17. $x^2 + 8x + 8y = 0$
$$x^2 + 8x = -8y$$
$$x^2 + 8x + 16 = -8y + 16$$
$$(x + 4)^2 = -8(y - 2)$$

This is the equation of a parabola opening down with vertex at $(h, k) = (-4, 2)$. The equations of translation are $x' = x + 4$, $y' = y - 2$. Making these substitutions, we obtain
$$x'^2 = -8y'$$
We graph this in the $x'y'$ system, following the process discussed in Section 9-1.

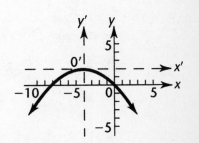

19.

$$x^2 + y^2 + 12x + 10y + 45 = 0$$
$$x^2 + 12x + y^2 + 10y = -45$$
$$x^2 + 12x + 36 + y^2 + 10y + 25 = -45 + 36 + 25$$
$$(x + 6)^2 + (y + 5)^2 = 16$$

This is the equation of a circle with center at $(-6, -5)$ and radius 4. The equations of translation are $x' = x + 6$, $y' = y + 5$. Making these substitutions, we obtain

$$x'^2 + y'^2 = 16$$

We graph this in the $x'y'$ system.

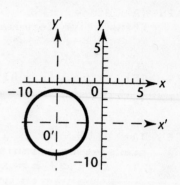

21.

$$-9x^2 + 16y^2 - 72x - 96y - 144 = 0$$
$$-9x^2 - 72x + 16y^2 - 96y = 144$$
$$-9(x^2 + 8x + ?) + 16(y^2 - 6y + ?) = 144$$
$$-9(x^2 + 8x + 16) + 16(y^2 - 6y + 9) = 144 - 144 + 144$$
$$-9(x + 4)^2 + 16(y - 3)^2 = 144$$
$$\frac{(y - 3)^2}{9} - \frac{(x + 4)^2}{16} = 1$$

This is the equation of a hyperbola with center at $(-4, 3)$. The equations of translation are $x' = x + 4$, $y' = y - 3$. Making these substitutions, we obtain

$$\frac{y'^2}{9} - \frac{x'^2}{16} = 1$$

We graph this in the $x'y'$ system, following the process discussed in Section 9-3.

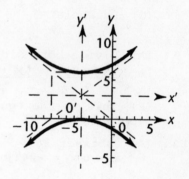

23. If $A \neq 0$, $C = 0$, and $E \neq 0$, write

$$Ax^2 + Dx + Ey + F = 0$$

Complete the square relative to x:

$$A\left(x^2 + \frac{D}{A}x + ?\right) = -Ey - F$$

$$A\left(x^2 + \frac{D}{A}x + \frac{D^2}{4A^2}\right) = -Ey - F + \frac{D^2}{4A}$$

$$\left(x + \frac{D}{2A}\right)^2 = -\frac{E}{A}y - \frac{F}{A} + \frac{D^2}{4A^2}$$

$$\left(x + \frac{D}{2A}\right)^2 = 4 \cdot \left(-\frac{E}{4A}\right)\left[y + \frac{F}{E} - \frac{D^2}{4AE}\right]$$

$$\left[x - \left(-\frac{D}{2A}\right)\right]^2 = 4\left(-\frac{E}{4A}\right)\left[y - \frac{D^2 - 4AF}{4AE}\right]$$

The equations of translation are

$$x' = x - \left(-\frac{D}{2A}\right) \qquad y' = y - \frac{D^2 - 4AF}{4AE}$$

Thus, $h = -\dfrac{D}{2A}$, $k = \dfrac{D^2 - 4AF}{4AE}$

The equation would become $x'^2 = 4\left(-\dfrac{E}{4A}\right)y'$, that is, the equation of a parabola.

25. Locate the vertex and axis in the original coordinate system, then sketch the parabola and translate the origin to the vertex of the parabola. Next write the equation of the parabola in the translated system:

$$x'^2 = 4ay'$$

The origin in the translated system is at $(h, k) = (2, 5)$ and the translation formulas are

$$x' = x - h = x - 2$$
$$y' = y - k = y - 5$$

Thus, the equation of the parabola in the original system is

$$(x - 2)^2 = 4a(y - 5)$$

Since the point $(-2, 1)$ is on the parabola, its coordinates must satisfy the equation of the parabola, hence

$$[(-2) - 2]^2 = 4a(1 - 5)$$
$$16 = -16a$$
$$a = -1$$

Thus, the equation of the parabola is

$$(x - 2)^2 = -4(y - 5)$$
$$x^2 - 4x + 4 = -4y + 20$$
$$x^2 - 4x + 4y - 16 = 0$$

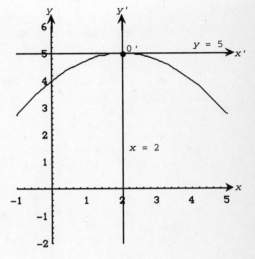

27. Locate the vertices in the original coordinate system, then sketch the ellipse and translate the origin to the center of the ellipse. Next write the equation of the ellipse in the translated system. Since $2a = 8$, and $2b = 4$, we know $a = 4$, $b = 2$, hence the equation is

$$\frac{x'^2}{4^2} + \frac{y'^2}{2^2} = 1$$

The origin in the translated system is at $(h, k) = (-2, -3)$ and the translation formulas are

$$x' = x - h = x - (-2) = x + 2$$
$$y' = y - k = y - (-3) = y + 3$$

Thus, the equation of the ellipse in the original system is

$$\frac{(x + 2)^2}{16} + \frac{(y + 3)^2}{4} = 1$$
$$(x + 2)^2 + 4(y + 3)^2 = 16$$
$$x^2 + 4x + 4 + 4y^2 + 24y + 36 = 16$$
$$x^2 + 4y^2 + 4x + 24y + 24 = 0$$

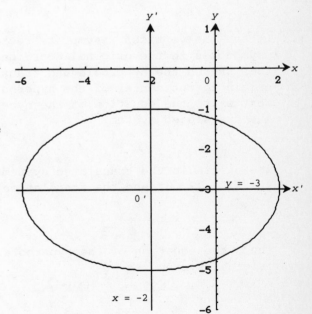

29. Locate the vertices in the original coordinate system, then sketch the ellipse and translate the origin to the center of the ellipse. Since $2a = 3 - (-7) = 10$, $a = 5$. Since $2c = 2 - (-6) = 8$, $c = 4$. Hence, $b = \sqrt{a^2 - c^2} = \sqrt{25 - 16} = 3$. The endpoints of the minor axis are symmetrically placed with respect to the center $(4, -2)$, that is, at $(1, -2)$ and $(7, -2)$.
Next write the equation of the ellipse in the translated system.

$$\frac{x'^2}{9} + \frac{y'^2}{25} = 1$$

The origin in the translated system is at $(h, k) = (4, -2)$ and the translation formulas are
$$x' = x - h = x - 4$$
$$y' = y - k = y - (-2) = y + 2$$
Thus, the equation of the ellipse in the original system is

$$\frac{(x - 4)^2}{9} + \frac{(y + 2)^2}{25} = 1$$
$$25(x - 4)^2 + 9(y + 2)^2 = 225$$
$$25(x^2 - 8x + 16) + 9(y^2 + 4y + 4) = 225$$
$$25x^2 - 200x + 400 + 9y^2 + 36y + 36 = 225$$
$$25x^2 + 9y^2 - 200x + 36y + 211 = 0$$

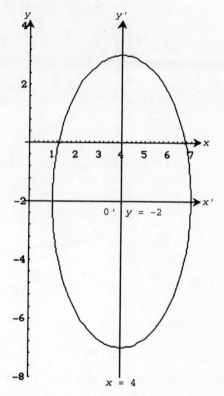

31. Locate the vertices, asymptote rectangle, and asymptotes in the original coordinate system, then sketch the hyperbola and translate the origin to the center of the hyperbola.
Next write the equation of the hyperbola in the translated system.

$$\frac{y'^2}{4} - \frac{x'^2}{1} = 1$$

The origin in the translated system is at $(h, k) = (2, 3)$ and the translation formulas are
$$x' = x - h = x - 2$$
$$y' = y - k = y - 3$$
Thus, the equation of the hyperbola in the original system is
$$\frac{(y - 3)^2}{4} - \frac{(x - 2)^2}{1} = 1$$
$$(y - 3)^2 - 4(x - 2)^2 = 4$$
$$y^2 - 6y + 9 - 4(x^2 - 4x + 4) = 4$$
$$y^2 - 6y + 9 - 4x^2 + 16x - 16 = 4$$
$$-4x^2 + y^2 + 16x - 6y - 11 = 0$$
$$4x^2 - y^2 - 16x + 6y + 11 = 0$$

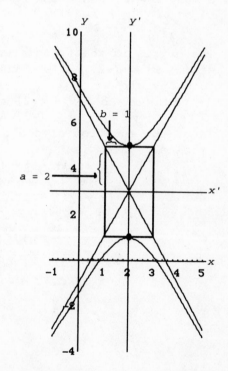

33. First find the coordinates of the foci in the translated system.

$$c'^2 = 3^2 - 2^2 = 5$$

$$c' = \sqrt{5}$$

$$-c' = -\sqrt{5}$$

Thus the coordinates in the translated system are

$F'(-\sqrt{5}, 0)$ and $F(\sqrt{5}, 0)$

Now use

$x = x' + h = x' + 2$

$y = y' + k = y' + 2$

to obtain

$F'(-\sqrt{5} + 2, 2)$ and $F(\sqrt{5} + 2, 2)$

as the coordinates of the foci in the original system.

35. First find the coordinates of the focus in the translated system. Since $a = -2$, and the parabola opens down, the coordinates are $(0, -2)$. Now use

$x = x' + h = x' - 4 = 0 - 4$

$y = y' + k = y' + 2 = -2 + 2$

to obtain $(-4, 0)$ as the coordinates of the focus in the original system.

37. First find the coordinates of the foci in the translated system.

$$c'^2 = 3^2 + 4^2 = 25$$

$$c' = 5$$

$$-c' = -5$$

Thus the coordinates in the translated system are

$F'(0, -5)$ and $F(0, 5)$

Now use

$x = x' + h = x' - 4$

$y = y' + k = y' + 3$

to obtain

$F'(0 - 4, -5 + 3) = F'(-4, -2)$ and $F(0 - 4, 5 + 3) = F(-4, 8)$ as the coordinates of the foci in the original system.

39. Before we can enter these equations in our graphing utility, we must solve for y:

$$3x^2 - 5y^2 + 7x - 2y + 11 = 0 \qquad 6x + 4y = 15$$

$$5y^2 + 2y - (3x^2 + 7x + 11) = 0 \qquad 4y = 15 - 6x$$

$$y = \frac{15 - 6x}{4}$$

Applying the quadratic formula yields

$$y = \frac{-2 \pm \sqrt{4 + 4(5)(3x^2 + 7x + 11)}}{10}$$

$$y = \frac{-1 \pm \sqrt{1 + 5(3x^2 + 7x + 11)}}{5}$$

$$y = \frac{-1 \pm \sqrt{15x^2 + 35x + 56}}{5}$$

Entering these three equations into a graphing utility produces the following graph:

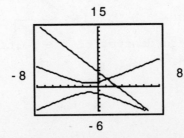

The curves intersect for x between 1 and 1.5 and between 6.5 and 7. Zooming in on these intervals, we obtain the following graphs:

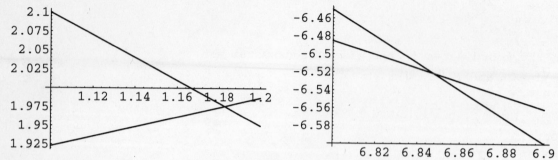

To two decimal places, the coordinates of the points of intersection are (1.18, 1.98) and (6.85, -6.52).

41. Before we can enter these equations in our graphing utility, we must solve for y:

$$7x^2 - 8x + 5y - 25 = 0$$
$$5y = 25 + 8x - 7x^2$$
$$y = \frac{25 + 8x - 7x^2}{5}$$

$$x^2 + 4y^2 + 4x - y - 12 = 0$$
$$4y^2 - y - (12 - 4x - x^2) = 0$$

Applying the quadratic formula yields

$$y = \frac{1 \pm \sqrt{1 + 4(4)(12 - 4x - x^2)}}{8}$$

$$y = \frac{1 \pm \sqrt{193 - 64x - 16x^2}}{8}$$

Entering these three equations into a graphing utility produces the following graph:

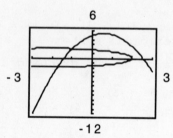

The curves intersect for x between -1.8 and -1.6 and between -1 and -0.8. Zooming in on these intervals, we obtain the following graphs:

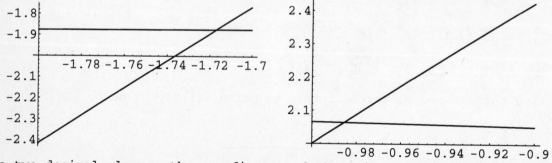

To two decimal places, the coordinates of the points of intersection are (-1.72, -1.87) and (-0.99, 2.06).

CHAPTER 9 REVIEW

1. First write the equation in standard form by dividing both sides by 225.

    $9x^2 + 25y^2 = 225$

    $$\frac{x^2}{25} + \frac{y^2}{9} = 1$$

    In this form the equation is identifiable as that of an ellipse. Locate the intercepts.

    When $y = 0$, $\frac{x^2}{25} = 1$. x intercepts: ± 5

    When $x = 0$, $\frac{x^2}{9} = 1$. y intercepts: ± 3

    Thus, $a = 5$, $b = 3$, and the major axis is on the x axis.

    > **Common Error:**
    > The relationship $c^2 = a^2 + b^2$ applies to a, b, c as defined for hyperbolas but not for ellipses.

    Foci: $c^2 = a^2 - b^2$
      $c^2 = 25 - 9$
      $c^2 = 16$
       $c = 4$
    Foci: $F'(-4, 0)$, $F(4, 0)$
    Major axis length = $2(5) = 10$
    Minor axis length = $2(3) = 6$

    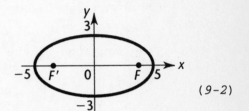

    (9-2)

2. $x^2 = -12y$ is the equation of a parabola. To graph, assign y values that make the right side a perfect square (y must be zero or negative for x to be real) and solve for x. Since the coefficient of y is negative, a must be negative, and the parabola opens down.

x	0	± 6	± 2
y	0	-3	$-\frac{1}{3}$

    To find the focus and directrix, solve
    $4a = -12$
     $a = -3$
    Focus: $(0, -3)$ Directrix: $y = -(-3) = 3$

    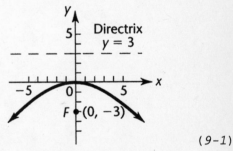

    (9-1)

3. First, write the equation in standard form by dividing both sides by 225.

    $25y^2 - 9x^2 = 225$

    $$\frac{y^2}{9} - \frac{x^2}{25} = 1$$

    In this form the equation is identifiable as that of a hyperbola.

    When $y = 0$, $-\frac{x^2}{25} = 1$. There are no x intercepts, but $b = 5$.

    When $x = 0$, $\frac{y^2}{9} = 1$. y intercepts: ± 3

    Sketch the asymptotes using the asymptote rectangle, then sketch in the hyperbola.
    Foci: $c^2 = 3^2 + 5^2$
      $c^2 = 34$
       $c = \sqrt{34}$
    Foci: $F'(0, -\sqrt{34})$, $F(0, \sqrt{34})$
    Transverse axis length = $2(3) = 6$
    Conjugate axis length = $2(5) = 10$

    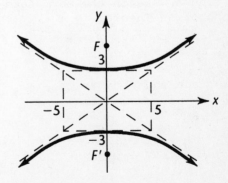

    (9-3)

4. (A) Divide both sides by 100

$$\frac{(y + 2)^2}{25} - \frac{(x - 4)^2}{4} = 1$$

(B) This is the equation of a hyperbola. *(9-4)*

5. (A) Subtract $12(y + 4)$ from both sides.

$$(x + 5)^2 = -12(y + 4)$$

(B) This is the equation of a parabola. *(9-4)*

6. (A) Divide both sides by 144

$$\frac{(x - 6)^2}{9} + \frac{(y - 4)^2}{16} = 1$$

(B) This is the equation of an ellipse. *(9-4)*

7. The parabola is opening either left or right and has an equation of the form $y^2 = 4ax$. Since $(-4, -2)$ is on the graph, we have:

$$(-2)^2 = 4a(-4)$$
$$4 = -16a$$
$$-\frac{1}{4} = a$$

Thus, the equation of the parabola is

$$y^2 = 4\left(-\frac{1}{4}\right)x$$
$$y^2 = -x$$

(9-1)

8. Make a rough sketch of the ellipse, locate the focus and x intercepts, then determine y intercepts using the special triangle relationship.

$$\frac{x^2}{b^2} + \frac{y^2}{a^2} = 1$$
$$b = \frac{6}{2} = 3$$
$$a^2 = 4^2 + 3^2 = 16 + 9 = 25$$
$$a = 5$$
$$\frac{x^2}{9} + \frac{y^2}{25} = 1$$

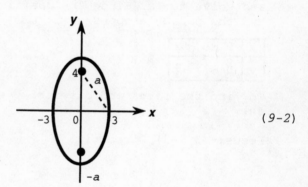

(9-2)

9. Start with $\frac{y^2}{a^2} - \frac{x^2}{b^2} = 1$ and find a and b.

$$b = \frac{8}{2} = 4$$

To find a, sketch the asymptote rectangle, label known parts, and use the Pythagorean Theorem.

$$a^2 = 5^2 - 4^2$$
$$a^2 = 9$$
$$a = 3$$

Thus, the equation is

$$\frac{y^2}{9} - \frac{x^2}{16} = 1$$

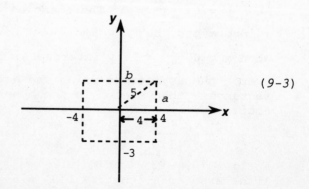

(9-3)

10. $x^2 + 4y^2 = 32$
$\quad\ x + 2y = 0$

We solve for x in terms of y in the second equation, then substitute the expression for x into the first equation.

$$x = -2y$$
$$(-2y)^2 + 4y^2 = 32$$
$$4y^2 + 4y^2 = 32$$
$$8y^2 = 32$$
$$y^2 = 4$$
$$y = \pm 2$$

For $y = 2$ $\qquad$ For $y = -2$
$\quad x = -2(2)$ $\qquad\quad x = -2(-2)$
$\quad\ \ = -4$ $\qquad\qquad\ \ = 4$

Solutions: $(-4, 2)$, $(4, -2)$

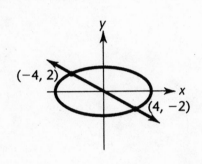

Graphs:
$x^2 + 4y^2 = 32$

$\dfrac{x^2}{32} + \dfrac{y^2}{8} = 1$ Ellipse

$a = \sqrt{32}$, $b = \sqrt{8}$, major axis on the x axis

x intercepts: $\pm\sqrt{32}$

y intercepts: $\pm\sqrt{8}$

$x + 2y = 0$

Straight line through origin, slope $-\dfrac{1}{2}$.

$(9\text{-}2,\ 6\text{-}3)$

11. $16x^2 + 25y^2 = 400$
$\quad\ 16x^2 - 45y = 0$

We eliminate x by multiplying the second equation by -1, then adding the two equations.

$$16x^2 + 25y^2 = 400$$
$$\underline{-16x^2 + 45y\ \ \ = 0}$$
$$25y^2 + 45y\ \ \ = 400$$
$$5y^2 + 9y - 80 = 0$$
$$(5y - 16)(y + 5) = 0$$
$$y = \frac{16}{5} \quad \text{or} \quad y = -5$$

For $y = \dfrac{16}{5}$ $\qquad\quad$ For $y = -5$

$\quad 16x^2 = 45y$ $\qquad\qquad\ 16x^2 = 45(-5)$

$\quad 16x^2 = 45\left(\dfrac{16}{5}\right)$ $\qquad\quad x^2 = -\dfrac{225}{16}$

$\quad\ \ x^2 = 9$ $\qquad\qquad$ No real solution

$\quad\ \ \ x = \pm 3$

Solutions: $\left(3, \dfrac{16}{5}\right)$, $\left(-3, \dfrac{16}{5}\right)$

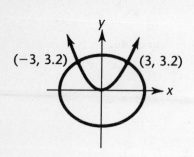

Graphs:

$16x^2 + 25y^2 = 400$

$\dfrac{x^2}{25} + \dfrac{y^2}{16} = 1$ Ellipse

$a = 5$, $b = 4$, major axis on the x axis.

x intercepts: ± 5

y intercepts: ± 4

$16x^2 - 45y = 0$ Parabola, opens up.

(9-1, 9-2, 6-3)

12. $x^2 + y^2 = 10$
$16x^2 + y^2 = 25$

We eliminate y by multiplying the first equation by -1, then adding the two equations.

$$\begin{array}{r} -x^2 - y^2 = -10 \\ \underline{16x^2 + y^2 = 25} \\ 15x^2 \qquad\quad = 15 \end{array}$$

$$x^2 = 1$$
$$x = \pm 1$$

For $x = \pm 1$
$x^2 + y^2 = 10$
$1 + y^2 = 10$
$\qquad y^2 = 9$
$\qquad y = \pm 3$

Solutions: $(1, 3)$, $(1, -3)$, $(-1, 3)$, $(-1, -3)$

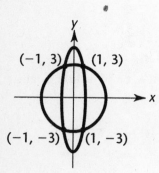

Graphs:

$16x^2 + y^2 = 25$

$\dfrac{x^2}{\frac{25}{16}} + \dfrac{y^2}{25} = 1$ Ellipse

$a = 5$, $b = \dfrac{5}{4}$, major axis on the y axis.

x intercepts: $\pm\dfrac{5}{4}$

y intercepts: ± 5

$x^2 + y^2 = 10$ Circle, radius $\sqrt{10}$
center $(0, 0)$.   (9-2, 6-3)

13. $16x^2 + 4y^2 + 96x - 16y + 96 = 0$
$16x^2 + 96x + 4y^2 - 16y = -96$
$16(x^2 + 6x + ?) + 4(y^2 - 4y + ?) = -96$
$16(x^2 + 6x + 9) + 4(y^2 - 4y + 4) = -96 + 144 + 16$
$16(x + 3)^2 + 4(y - 2)^2 = 64$
$$\dfrac{(x + 3)^2}{4} + \dfrac{(y - 2)^2}{16} = 1$$

This is the equation of an ellipse with center at $(-3, 2)$. The equations of translation are $x' = x + 3$, $y' = y - 2$. Making these substitutions, we obtain

$$\dfrac{x'^2}{4} + \dfrac{y'^2}{16} = 1$$

We graph this in the $x'y'$ system, following the process discussed in Section 9-2.

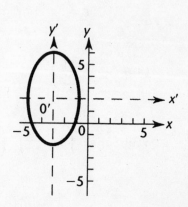

(9-4)

14. $x^2 - 4x - 8y - 20 = 0$

$$x^2 - 4x = 8y + 20$$
$$x^2 - 4x + 4 = 8y + 24$$
$$(x - 2)^2 = 4(2)(y + 3)$$

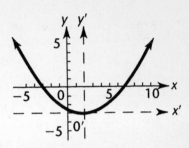

This is the equation of a parabola opening up with vertex at $(h, k) = (2, -3)$. The equations of translation are $x' = x - 2$, $y' = y + 3$. Making these substitutions, we obtain

$$x'^2 = 8y'$$

We graph this in the $x'y'$ system, following the process discussed in Section 9-1.

(9-4)

15.

$$4x^2 - 9y^2 + 24x - 36y - 36 = 0$$
$$4x^2 + 24x - 9y^2 - 36y = 36$$
$$4(x^2 + 6x + ?) - 9(y^2 + 4y + ?) = 36$$
$$4(x^2 + 6x + 9) - 9(y^2 + 4y + 4) = 36 + 36 - 36$$
$$4(x + 3)^2 - 9(y + 2)^2 = 36$$
$$\frac{(x + 3)^2}{9} - \frac{(y + 2)^2}{4} = 1$$

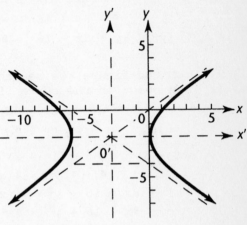

This is the equation of a hyperbola with center at $(-3, -2)$. The equations of translation are $x' = x + 3$, $y' = y + 2$.
Making these substitutions, we obtain

$$\frac{x'^2}{9} - \frac{y'^2}{4} = 1$$

We graph this in the $x'y'$ system, following the process discussed in Section 9-3.

(9-4)

16. Here is a computer-generated graph of $x^2 = y$ and $x^2 = 50y$, $-10 \le x \le 10$, $-10 \le y \le 10$.

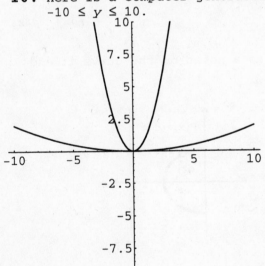

In order to make the graph of $x^2 = y$, in the window $-m \le x \le m$, $-m \le y \le m$, look like the graph at the left of $x^2 = 50y$, we need a magnification by a factor of 50. Then $\left(\dfrac{x'}{50}\right)^2 = \dfrac{y'}{50}$ becomes $x'^2 = 50y$. A magnification by a factor of 50 requires $\dfrac{10}{50} = 0.2 = m$. Here is the required graph of $x^2 = y$ in the window $-0.2 \le x \le 0.2$, $-0.2 \le y \le 0.2$.

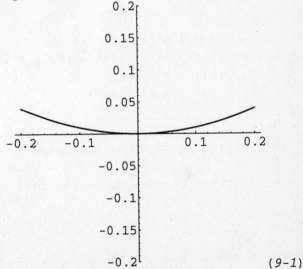

(9-1)

17. From the figure, we see that the point $P(x, y)$ is a point on the curve if and only if

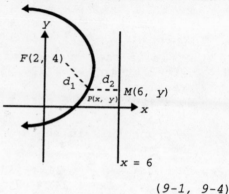

$$d_1 = d_2$$
$$d(F, P) = d(M, P)$$
$$\sqrt{(x - 2)^2 + (y - 4)^2} = \sqrt{(x - 6)^2 + (y - y)^2}$$
$$(x - 2)^2 + (y - 4)^2 = (x - 6)^2$$
$$x^2 - 4x + 4 + (y - 4)^2 = x^2 - 12x + 36$$
$$(y - 4)^2 = -8x + 32$$
$$(y - 4)^2 = -8(x - 4) \text{ or}$$
$$y^2 - 8y + 16 = -8x + 32$$
$$y^2 - 8y + 8x - 16 = 0$$

(9-1, 9-4)

18. From the figure, we see that the point $P(x, y)$ is a point on the curve if and only if

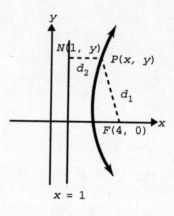

$$d_1 = 2d_2$$
$$d(F, P) = 2d(N, P)$$
$$\sqrt{(x - 4)^2 + (y - 0)^2} = 2\sqrt{(x - 1)^2 + (y - y)^2}$$
$$(x - 4)^2 + y^2 = 4(x - 1)^2$$
$$x^2 - 8x + 16 + y^2 = 4(x^2 - 2x + 1)$$
$$x^2 - 8x + 16 + y^2 = 4x^2 - 8x + 4$$
$$-3x^2 + y^2 = -12$$
$$\frac{x^2}{4} - \frac{y^2}{12} = 1$$

This is the equation of a hyperbola.

(9-3, 9-4)

19. From the figure, we see that the point $P(x, y)$ is a point on the curve if and only if

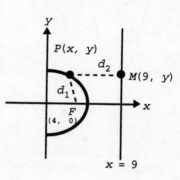

$$d_1 = \frac{2}{3}d_2$$
$$d(F, P) = \frac{2}{3}d(M, P)$$
$$\sqrt{(x - 4)^2 + (y - 0)^2} = \frac{2}{3}\sqrt{(x - 9)^2 + (y - y)^2}$$
$$(x - 4)^2 + y^2 = \frac{4}{9}(x - 9)^2$$
$$9[(x - 4)^2 + y^2] = 4(x - 9)^2$$
$$9(x^2 - 8x + 16 + y^2) = 4(x^2 - 18x + 81)$$
$$9x^2 - 72x + 144 + 9y^2 = 4x^2 - 72x + 324$$
$$5x^2 + 9y^2 = 180$$
$$\frac{x^2}{36} + \frac{y^2}{20} = 1$$

This is the equation of an ellipse.

(9-2, 9-4)

20. First find the coordinates of the foci in the translated system.
$$c'^2 = 4^2 - 2^2 = 12$$
$$c' = \sqrt{12}$$
$$-c' = -\sqrt{12}$$
Thus the coordinates in the translated system are
$$F'(0, -\sqrt{12}) \text{ and } F(0, \sqrt{12})$$

Now use
$$x = x' + h = x' - 3$$
$$y = y' + k = y' + 2$$
to obtain
$$F'(-3, -\sqrt{12} + 2) \text{ and } F(-3, \sqrt{12} + 2)$$
as the coordinates of the foci in the original system. *(9-4)*

21. First find the coordinates of the focus in the translated system. Since $a = 2$ and the parabola opens up, they are $(0, 2)$. Now use
$$x = x' + h = x' + 2 = 0 + 2 = 2$$
$$y = y' + k = y' - 3 = 2 - 3 = -1$$
to obtain $(2, -1)$ as the coordinates of the focus in the original system. *(9-4)*

22. First find the coordinates of the foci in the original system.
$$c'^2 = 3^2 + 2^2 = 13$$
$$c' = \sqrt{13}$$
$$-c' = -\sqrt{13}$$
Thus the coordinates in the translated system are
$$F'(-\sqrt{13}, 0) \text{ and } F(\sqrt{13}, 0)$$
Now use
$$x = x' + h = x' - 3$$
$$y = y' + k = y' - 2$$
to obtain
$$F'(-\sqrt{13} - 3, -2) \text{ and } F(\sqrt{13} - 3, -2) \text{ as the coordinates of the foci in the original system.}$$ *(9-4)*

23. Before we can enter these equations in our graphing utility, we must solve for y:
$$x^2 - 3y^2 + 9x + 7y - 22 = 0 \qquad 4x^2 + 5x + 10y - 53 = 0$$
$$3y^2 - 7y - (x^2 + 9x - 22) = 0 \qquad\qquad 10y = 53 - 5x - 4x^2$$
$$y = \frac{53 - 5x - 4x^2}{10}$$

Applying the quadratic formula yields
$$y = \frac{7 \pm \sqrt{49 + 4(3)(x^2 + 9x - 22)}}{6}$$
$$y = \frac{7 \pm \sqrt{12x^2 + 108x - 215}}{6}$$
Entering these three equations into a graphing utility produces the graph at the right.

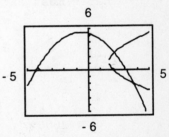

The curves intersect for x between 2 and 2.4 and between 3.6 and 4. Zooming in on these intervals, we obtain the following graphs:

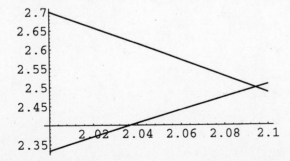 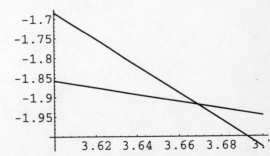

To two decimal places, the coordinates of the points of intersection are $(2.09, 2.50)$ and $(3.67, -1.92)$. *(9-4)*

24. From the figure, we see that the parabola opens up, hence its equation must be of the form $x^2 = 4ay$. Since $(4, 1)$ is on the graph we have

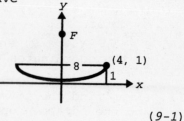

$$4^2 = 4a \cdot 1$$
$$16 = 4a$$
$$a = 4$$

Thus a, the distance of the focus from the vertex, is 4 feet.

$(9-1)$

25. From the figure, we see that the x intercepts must be at $(-5, 0)$ and $(5, 0)$, the foci at $(-4, 0)$ and $(4, 0)$. Hence $a = 5$ and $c = 4$. We can determine the y intercepts using the special triangle relationship

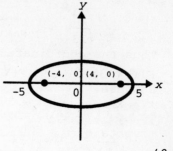

$$5^2 = 4^2 + b^2$$
$$25 = 16 + b^2$$
$$b = 3$$

Hence, the equation of the ellipse is

$$\frac{x^2}{5^2} + \frac{y^2}{3^2} = 1$$

$(9-2)$

26. From the figure, we can see:

$d + a = y_1 =$ the y coordinate of the point on the hyperbola with x coordinate 15. From the equation of the hyperbola $\dfrac{y^2}{40^2} - \dfrac{x^2}{30^2} = 1$, we have $a = 40$, hence $d + 40 = y_1$, or $d = y_1 - 40$. Since the point $(15, y_1)$ is on the hyperbola, its coordinates must satisfy the equation of the hyperbola. Thus,

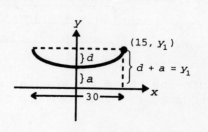

$$\frac{y_1^2}{40^2} - \frac{15^2}{30^2} = 1$$
$$\frac{y_1^2}{40^2} = 1 + \frac{15^2}{30^2}$$
$$\frac{y_1^2}{40^2} = \frac{5}{4}$$
$$y_1^2 = 2000$$
$$y_1 = 44.72$$

depth $= y_1 - 40 = 4.72$ feet.

$(9-3)$

CUMULATIVE REVIEW EXERCISE (Chapters 8 and 9)

1. (A) Since $15 - 20 = 10 - 15 = -5$, this could start an arithmetic sequence.

(B) Since $\frac{25}{5} = \frac{125}{25} = 5$, this could start a geometric sequence.

(C) Since $\frac{25}{5} \neq \frac{50}{25}$ and $25 - 5 \neq 50 - 25$, this could start neither an arithmetic nor a geometric sequence.

(D) Since $\frac{-9}{27} = \frac{3}{-9} = -\frac{1}{3}$, this could start a geometric sequence.

(E) Since $(-6) - (-9) = (-3) - (-6) = 3$, this could start an arithmetic sequence. *(8-3)*

2. $a_n = 2 \cdot 5^n$

(A) $a_1 = 2 \cdot 5^1 = 10$
$a_2 = 2 \cdot 5^2 = 50$
$a_3 = 2 \cdot 5^3 = 250$
$a_4 = 2 \cdot 5^4 = 1,250$

(B) This is a geometric sequence with $r = 5$. Hence,
$a_n = a_1 r^{n-1}$
$a_8 = 10(5)^{8-1}$
$\quad = 781,250$

(C) $S_n = \dfrac{a_1 - ra_n}{1 - r}$

$S_8 = \dfrac{10 - 5(781,250)}{1 - 5}$
$\quad = 976,560$ *(8-3)*

3. $a_n = 3n - 1$

(A) $a_1 = 3 \cdot 1 - 1 = 2$
$a_2 = 3 \cdot 2 - 1 = 5$
$a_3 = 3 \cdot 3 - 1 = 8$
$a_4 = 3 \cdot 4 - 1 = 11$

(B) This is an arithmetic sequence with $d = 3$. Hence,
$a_n = a_1 + (n - 1)d$
$a_8 = 2 + (8 - 1)3$
$\quad = 23$

(C) $S_n = \dfrac{n}{2}(a_1 + a_n)$

$S_8 = \dfrac{8}{2}(2 + 23)$
$\quad = 100$ *(8-3)*

4. $a_1 = 100 \quad a_n = a_{n-1} - 6 \quad n \geq 2$

(A) $a_1 = 100$
$a_2 = a_1 - 6 = 94$
$a_3 = a_2 - 6 = 88$
$a_4 = a_3 - 6 = 82$

(B) This is an arithmetic sequence with $d = -6$. Hence,
$a_n = a_1 + (n - 1)d$
$a_8 = 100 + (8 - 1)(-6)$
$\quad = 58$

(C) $S_n = \dfrac{n}{2}(a_1 + a_n)$

$S_8 = \dfrac{8}{2}(100 + 58)$
$\quad = 632$ *(8-3)*

5. (A) $8! = 8 \cdot 7 \cdot 6 \cdot 5 \cdot 4 \cdot 3 \cdot 2 \cdot 1 = 40,320$.

(B) $\dfrac{32!}{30!} = \dfrac{32 \cdot 31 \cdot 30!}{30!} = 992$

(C) $\dfrac{9!}{3!(9 - 3)!} = \dfrac{9!}{3!6!}$
$= \dfrac{9 \cdot 8 \cdot 7 \cdot 6!}{3 \cdot 2 \cdot 1 \cdot 6!} = 84$ *(8-4)*

6. (A) $\dbinom{7}{2} = \dfrac{7!}{2!(7 - 2)!} = \dfrac{7!}{2!5!} = \dfrac{7 \cdot 6 \cdot 5!}{2 \cdot 1 \cdot 5!} = 21$ (B) $C_{7,2} = \dfrac{7!}{2!(7 - 2)!} = 21$

(C) $P_{7,2} = \dfrac{7!}{(7 - 2)!} = \dfrac{7!}{5!} = \dfrac{7 \cdot 6 \cdot 5!}{5!} = 42$ *(8-4, 8-5)*

7. First, write the equation in standard form by dividing both sides by 900.
$$25x^2 - 36y^2 = 900$$
$$\frac{x^2}{36} - \frac{y^2}{25} = 1$$

In this form the equation is identifiable as that of a hyperbola.

When $x = 0$, $-\frac{y^2}{25} = 1$. There are no y intercepts, but $b = 5$.

When $y = 0$, $\frac{x^2}{36} = 1$. x intercepts: ± 6.

Sketch the asymptotes using the asymptote rectangle, then sketch in the hyperbola.
Foci: $c^2 = 5^2 + 6^2$
$c^2 = 61$
$c = \sqrt{61}$
Foci: $F'(-\sqrt{61}, 0)$, $F(\sqrt{61}, 0)$
Transverse axis length $= 2(6) = 12$
Conjugate axis length $= 2(5) = 10$

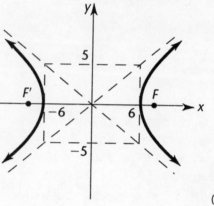

$(9\text{-}3)$

8. First, write the equation in standard form by dividing both sides by 900.
$$25x^2 + 36y^2 = 900$$
$$\frac{x^2}{36} + \frac{y^2}{25} = 1$$

In this form the equation is identifiable as that of an ellipse. Locate the intercepts:

When $y = 0$, $\frac{x^2}{36} = 1$. x intercepts: ± 6.

When $x = 0$, $\frac{y^2}{25} = 1$. y intercepts: ± 5

Thus, $a = 6$, $b = 5$, and the major axis is on the x axis.
Foci: $c^2 = a^2 - b^2$
$c^2 = 36 - 25$
$c^2 = 11$
$c = \sqrt{11}$
Foci: $F'(-\sqrt{11}, 0)$, $F(\sqrt{11}, 0)$
Major axis length $= 2(6) = 12$
Minor axis length $= 2(5) = 10$

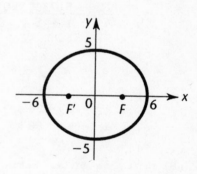

$(9\text{-}2)$

9. $25x^2 - 36y = 0$ is the equation of a parabola. For convenience, we rewrite this as $25x^2 = 36y$. To graph, assign y values that make $25x^2$ a perfect square (y must be positive or zero for x to be real) and solve for x. Since the coefficient of x is positive, a must be positive, and the parabola opens up.

x	0	$\pm\frac{6}{5}$	$\pm\frac{12}{5}$
y	0	1	4

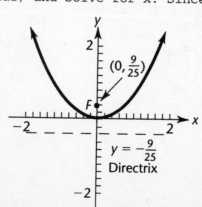

To find the focus and directrix, solve
$$4a = \frac{36}{25}$$
$$a = \frac{9}{25}$$

Focus: $\left(0, \frac{9}{25}\right)$. Directrix: $y = -\frac{9}{25}$

$(9\text{-}1)$

10. (A) The outcomes can be displayed in a tree diagram as follows:

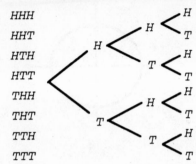

HHH

HHT

HTH

HTT

THH

THT

TTH

TTT

(B) O_1: First flip of the coin
N_1: 2 outcomes
O_2: Second flip of the coin
N_2: 2 outcomes
O_3: Third flip of the coin
N_3: 2 outcomes

Applying the multiplication principle, there are $2 \cdot 2 \cdot 2 = 8$ possible outcomes. *(8-5)*

11. (A) O_1: Place first book N_1: 4 ways
O_2: Place second book N_2: 3 ways
O_3: Place third book N_3: 2 ways
O_4: Place fourth book N_4: 1 way
Applying the multiplication principle, there are $4 \cdot 3 \cdot 2 \cdot 1$ arrangements.

(B) Order is important here. We use permutations to determine the number of arrangements of 4 objects. $P_{4,4} = 4! = 24$. *(8-5)*

12. P_1: $1 = 1(2 \cdot 1 - 1) = 1 \cdot 1 = 1$
P_2: $1 + 5 = 2(2 \cdot 2 - 1)$
$\qquad 6 = 6$
P_3: $1 + 5 + 9 = 3(2 \cdot 3 - 1)$
$\qquad\qquad 15 = 15$ *(8-2)*

13. P_1: $1^2 + 1 + 2$ is divisible by 2
$\qquad 4 = 2 \cdot 2$ true
P_2: $2^2 + 2 + 2$ is divisible by 2
$\qquad 8 = 2 \cdot 4$ true
P_3: $3^2 + 3 + 2$ is divisible by 2
$\qquad 14 = 2 \cdot 7$ true   *(8-2)*

14. P_k: $1 + 5 + 9 + \cdots + (4k - 3) = k(2k - 1)$
P_{k+1}: $1 + 5 + 9 + \cdots + (4k - 3) + (4k + 1) = (k + 1)(2k + 1)$ *(8-2)*

15. P_k: $k^2 + k + 2 = 2r$ for some integer r
P_{k+1}: $(k + 1)^2 + (k + 1) + 2 = 2s$ for some integer s *(8-2)*

16. The parabola is opening either up or down and has an equation of the form $x^2 = 4ay$. Since $(2, -8)$ is on the graph, we have:
$2^2 = 4a(-8)$
$4 = -32a$
$a = -\dfrac{1}{8}$
Thus, the equation of the parabola is
$x^2 = 4\left(-\dfrac{1}{8}\right)y$
$x^2 = -\dfrac{1}{2}y$
$y = -2x^2$ *(9-1)*

17. Make a rough sketch of the ellipse, locate focus and x intercepts, then determine y intercepts using the special triangle relationship.

$$\frac{x^2}{a^2} + \frac{y^2}{b^2} = 1$$

$$a = \frac{10}{2} = 5$$

$$b^2 = a^2 - c^2 = 5^2 - 3^2 = 25 - 9 = 16$$

$$b = 4$$

$$\frac{x^2}{25} + \frac{y^2}{16} = 1$$

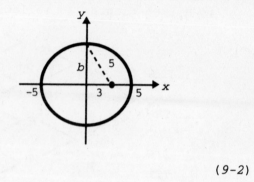

$(9-2)$

18. Start with $\frac{x^2}{a^2} - \frac{y^2}{b^2} = 1$ and find a and b.

$$a = \frac{16}{2} = 8$$

To find b, sketch the asymptote rectangle, label known parts, and use the Pythagorean Theorem.

$$b^2 = (\sqrt{89})^2 - 8^2$$

$$= 89 - 64$$

$$b^2 = 25$$

$$b = 5$$

Thus, the equation is $\frac{x^2}{64} - \frac{y^2}{25} = 1$

$(9-3)$

19. $\sum_{k=1}^{5} k^k = 1^1 + 2^2 + 3^3 + 4^4 + 5^5 = 1 + 4 + 27 + 256 + 3,125 = 3,413$

$(8-1)$

20. $S_6 = \frac{2}{2!} - \frac{2^2}{3!} + \frac{2^3}{4!} - \frac{2^4}{5!} + \frac{2^5}{6!} - \frac{2^6}{7!}$

$k = 1, 2, 3, 4, 5, 6$

Noting that the terms alternate in sign, we can rewrite as follows:

$$S_6 = (-1)^2 \frac{2}{2!} + (-1)^3 \frac{2^2}{3!} + (-1)^4 \frac{2^3}{4!} + (-1)^5 \frac{2^4}{5!} + (-1)^6 \frac{2^5}{6!} + (-1)^7 \frac{2^6}{7!}$$

Clearly, $a_k = (-1)^{k+1} \frac{2^k}{(k+1)!}$

$$S_6 = \sum_{k=1}^{6} (-1)^{k+1} \frac{2^k}{(k+1)!}$$

$(8-1)$

21. $\frac{a_2}{a_1} = \frac{-36}{108} = -\frac{1}{3} = r. \quad |r| < 1.$

Therefore, this infinite geometric series has a sum.

$$S_\infty = \frac{a_1}{1 - r}$$

$$= \frac{108}{1 - (-\frac{1}{3})}$$

$$= 81$$

$(8-3)$

22.

	Case 1	Case 2	Case 3
O_1: Select the first letter	N_1: 6 ways	6 ways	6 ways
O_2: Select the second letter	N_2: 5 ways	6 ways	5 ways (exclude first letter)
O_3: Select the third letter	N_3: 4 ways	6 ways	5 ways (exclude second letter)
O_4: Select the fourth letter	N_4: 3 ways	6 ways	5 ways (exclude third letter)
	$6 \cdot 5 \cdot 4 \cdot 3 = $ 360 words	$6 \cdot 6 \cdot 6 \cdot 6 = $ 1,296 words	$6 \cdot 5 \cdot 5 \cdot 5 = 750$ words

$(8-5)$

23. Here are computer-generated graphs of $\{a_n\}$ and $\{b_n\}$ (the dots suggest a straight line).

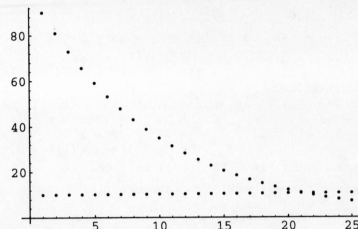

The graph indicates that the least positive integer n such that $a_n < b_n$ is between 21 and 23. From the table display we note

$$n = 21 \qquad a_n = 10.9419 \qquad b_n = 10.63$$
$$n = 22 \qquad a_n = 9.84771 \qquad b_n = 10.66$$

confirming that 22 is the least positive integer n required. *(8-3)*

24. (A) $P_{25,5} = 25 \cdot 24 \cdot 23 \cdot 22 \cdot 21 = 6,375,600$

(B) $C_{25,5} = \dfrac{25!}{5!(25-5)!} = \dfrac{25!}{5!20!} = \dfrac{25 \cdot 24 \cdot 23 \cdot 22 \cdot 21 \cdot 20!}{5 \cdot 4 \cdot 3 \cdot 2 \cdot 1 \cdot 20!} = 53,130$

(C) $\dbinom{25}{5} = \dfrac{25!}{5!(25-5)!} = 53,130$ *(8-4, 8-5)*

25. $\left(a + \dfrac{1}{2}b\right)^6 = \displaystyle\sum_{k=0}^{6} \binom{6}{k} a^{6-k} \left(\dfrac{1}{2}b\right)^k$

$= \dbinom{6}{0}a^6 + \dbinom{6}{1}a^5\left(\dfrac{1}{2}b\right)^1 + \dbinom{6}{2}a^4\left(\dfrac{1}{2}b\right)^2 + \dbinom{6}{3}a^3\left(\dfrac{1}{2}b\right)^3 + \dbinom{6}{4}a^2\left(\dfrac{1}{2}b\right)^4$

$\qquad + \dbinom{6}{5}a\left(\dfrac{1}{2}b\right)^5 + \dbinom{6}{6}\left(\dfrac{1}{2}b\right)^6$

$= a^6 + 3a^5b + \dfrac{15}{4}a^4b^2 + \dfrac{5}{2}a^3b^3 + \dfrac{15}{16}a^2b^4 + \dfrac{3}{16}ab^5 + \dfrac{1}{64}b^6$ *(8-4)*

26. In the expansion of $(a + b)^n$, the exponent of b in the rth term is $r - 1$ and the exponent of a is $n - (r - 1)$. Here, in the first case, $r = 5$, $n = 10$.

Fifth term $= \dbinom{10}{4}(3x)^6(-y)^4$

$\qquad\qquad = 153,090x^6y^4$

In the second case, $r = 8$, n is still 10.

Eighth term $= \dbinom{10}{7}(3x)^3(-y)^7$

$\qquad\qquad\quad = -3,240x^3y^7$ *(8-4)*

27. P_n: $1 + 5 + 9 + \cdots + (4n - 3) = n(2n - 1)$

Part 1: Show that P_1 is true.

P_1: $1 = 1(2 \cdot 1 - 1) = 1$ True

Part 2: Show that if P_k is true, then P_{k+1} is true.

Write out P_k and P_{k+1}.

P_k: $1 + 5 + 9 + \cdots + (4k - 3) = k(2k - 1)$

P_{k+1}: $1 + 5 + 9 + \cdots + (4k - 3) + (4k + 1) = (k + 1)(2k + 1)$

We start with P_k:

$1 + 5 + 9 + \cdots + (4k - 3) = k(2k - 1)$

Adding $4k + 1$ to both sides:

$$
\begin{aligned}
1 + 5 + 9 + \cdots + (4k - 3) + (4k + 1) &= k(2k - 1) + 4k + 1 \\
&= 2k^2 - k + 4k + 1 \\
&= 2k^2 + 3k + 1 \\
&= (k + 1)(2k + 1)
\end{aligned}
$$

We have shown that if P_k is true, then P_{k+1} is true.

Conclusion: P_n is true for all positive integers n. (8-2)

28. P_n: $n^2 + n + 2 = 2p$ for some integer p.

Part 1: Show that P_1 is true.

P_1: $1^2 + 1 + 2 = 4 = 2 \cdot 2$ is true.

Part 2: Show that if P_k is true, then P_{k+1} is true.

Write out P_k and P_{k+1}.

P_k: $k^2 + k + 2 = 2r$ for some integer r

P_{k+1}: $(k + 1)^2 + (k + 1) + 2 = 2s$ for some integer s

We start with P_k:

$k^2 + k + 2 = 2r$ for some integer r

Now,
$$
\begin{aligned}
(k + 1)^2 + (k + 1) + 2 &= k^2 + 2k + 1 + k + 1 + 2 \\
&= k^2 + k + 2 + 2k + 2 \\
&= 2r + 2k + 2 \\
&= 2(r + k + 2)
\end{aligned}
$$

Therefore,

$(k + 1)^2 + (k + 1) + 2 = 2s$ for some integer s $(= r + k + 2)$

$(k + 1)^2 + (k + 1) + 2$ is divisible by 2.

We have shown that if P_k is true, then P_{k+1} is true.

Conclusion: P_n is true for all positive integers n. (8-2)

29. We are to find $51 + 53 + \cdots + 499$. This is the sum of an arithmetic sequence, S_n, with $d = 2$.

First, find n:

$a_n = a_1 + (n - 1)d$

$499 = 51 + (n - 1)2$

$n = 225$

Now, find S_{225}

$S_n = \dfrac{n}{2}(a_1 + a_n)$

$S_{225} = \dfrac{225}{2}(51 + 499)$

$\phantom{S_{225}} = 61,875$ (8-3)

30. $2.\overline{45} = 2.454545\ldots$

$\phantom{2.\overline{45}} = 2 + 0.454545\ldots$

$0.454545\ldots = 0.45 + 0.0045 + 0.000045 + \cdots$

This is an infinite geometric series with $a_1 = 0.45$ and $r = 0.01$. Thus

$S_\infty = \dfrac{a_1}{1 - r} = \dfrac{0.45}{1 - 0.01} = \dfrac{0.45}{0.99} = \dfrac{5}{11}$

Hence, $2.\overline{45} = 2 + \dfrac{5}{11} = \dfrac{27}{11}$

(8-3)

31. Here is a computer-generated graph of the sequence

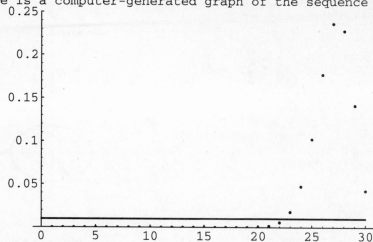

The solid line indicates 0.01. The largest term of the sequence is a_{27}. From the table display,

$$a_{27} = \binom{30}{27}(0.1)^{30-27}(0.9)^{27} = 0.236088$$

There are 8 terms larger than 0.01, as can be seen from the graph or the table display (details omitted).

(8-4)

32. $4x + 4y - y^2 + 8 = 0$

$\qquad 4x + 8 = y^2 - 4y$

$\qquad 4x + 8 + 4 = y^2 - 4y + 4$

$\qquad 4x + 12 = (y - 2)^2$

$\qquad 4(x + 3) = (y - 2)^2$

This is the equation of a parabola opening right with vertex at $(h, k) = (-3, 2)$. The equations of translation are $x' = x + 3$, $y' = y - 2$. Making these substitutions, we obtain

$\qquad 4x' = y'^2$

We graph this in the $x'y'$ system.

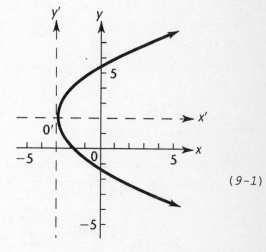

(9-1)

33. $\qquad x^2 + 2x - 4y^2 - 16y + 1 = 0$

$\qquad x^2 + 2x - 4y^2 - 16y = -1$

$x^2 + 2x + ? - 4(y^2 + 4y + ?) = -1$

$x^2 + 2x + 1 - 4(y^2 + 4y + 4) = -1 + 1 - 16$

$\qquad (x + 1)^2 - 4(y + 2)^2 = -16$

$$\frac{(y + 2)^2}{4} - \frac{(x + 1)^2}{16} = 1$$

This is the equation of a hyperbola with center at $(-1, -2)$. The equations of translation are $x' = x + 1$, $y' = y + 2$. Making these substitutions, we obtain

$$\frac{y'^2}{4} - \frac{x'^2}{16} = 1$$

We graph this in the $x'y'$ system.

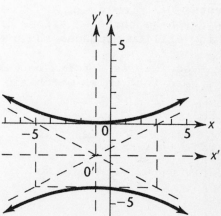

(9-3)

34.
$$4x^2 - 16x + 9y^2 + 54y + 61 = 0$$
$$4x^2 - 16x + 9y^2 + 54y = -61$$
$$4(x^2 - 4x + ?) + 9(y^2 + 6y + ?) = -61$$
$$4(x^2 - 4x + 4) + 9(y^2 + 6y + 9) = -61 + 16 + 81$$
$$4(x - 2)^2 + 9(y + 3)^2 = 36$$
$$\frac{(x - 2)^2}{9} + \frac{(y + 3)^2}{4} = 1$$

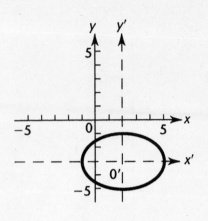

This is the equation of an ellipse with center at (2, -3). The equations of translation are $x' = x - 2$, $y' = y + 3$. Making these substitutions, we obtain
$$\frac{x'^2}{9} + \frac{y'^2}{4} = 1$$
We graph this in the $x'y'$ system.

(9-2)

35.

	Allowing digits to repeat	No digit can be repeated
O_1: Selecting first digit	N_1: 10 ways	N_1: 10 ways
O_2: Selecting second digit	N_2: 10 ways	N_2: 9 ways
$\vdots$	$\vdots$	$\vdots$
O_9: Selecting ninth digit	N_9: 10 ways	N_9: 2 ways

$$10 \cdot 10 \cdot 10 \cdot 10 \cdot 10 \cdot 10 \cdot 10 \cdot 10 \cdot 10 = 10^9 \text{ zip codes}$$

$$P_{10,9} = 10 \cdot 9 \cdot 8 \cdot 7 \cdot 6 \cdot 5 \cdot 4 \cdot 3 \cdot 2$$
$$= 3,628,800 \text{ zip codes}$$

(8-5)

36. Before we can enter these equations in our graphing utility, we must solve for y:

$$5x^2 + 2y^2 - 7x + 8y - 48 = 0 \qquad e^x - e^{-x} - 2y = 0$$
$$2y^2 + 8y + 5x^2 - 7x - 48 = 0 \qquad e^x - e^{-x} = 2y$$
$$y = \frac{e^x - e^{-x}}{2}$$

Applying the quadratic formula yields

$$y = \frac{-8 \pm \sqrt{64 - 4(2)(5x^2 - 7x - 48)}}{4}$$

$$y = \frac{-8 \pm \sqrt{448 + 56x - 40x^2}}{4}$$

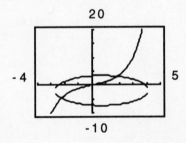

Entering these three equations into a graphing utility produces the graph shown at the right.

The curves intersect for x between -2.4 and -2.2 and between 1.8 and 2. Zooming in on these intervals, we obtain the following graphs:

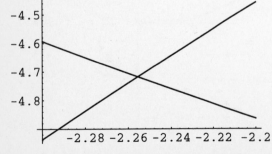

To two decimal places, the coordinates of the points of intersection are (-2.26, -4.72) and (1.85, 3.09).

(9-4)

37. Prove P_n: $\dfrac{1}{1\cdot3} + \dfrac{1}{3\cdot5} + \dfrac{1}{5\cdot7} + \cdots + \dfrac{1}{(2n-1)(2n+1)} = \dfrac{n}{2n+1}$

Part 1: Show that P_1 is true.

P_1: $\dfrac{1}{1\cdot3} = \dfrac{1}{2\cdot1+1}$

$\dfrac{1}{3} = \dfrac{1}{3}$ P_1 is true

Part 2: Show that if P_k is true, then P_{k+1} is true.

Write out P_k and P_{k+1}:

P_k: $\dfrac{1}{1\cdot3} + \dfrac{1}{3\cdot5} + \dfrac{1}{5\cdot7} + \cdots + \dfrac{1}{(2k-1)(2k+1)} = \dfrac{k}{2k+1}$

P_{k+1}: $\dfrac{1}{1\cdot3} + \dfrac{1}{3\cdot5} + \dfrac{1}{5\cdot7} + \cdots + \dfrac{1}{(2k-1)(2k+1)} + \dfrac{1}{(2k+1)(2k+3)} = \dfrac{k+1}{2k+3}$

We start with P_k:

$\dfrac{1}{1\cdot3} + \dfrac{1}{3\cdot5} + \dfrac{1}{5\cdot7} + \cdots + \dfrac{1}{(2k-1)(2k+1)} = \dfrac{k}{2k+1}$

Adding $\dfrac{1}{(2k+1)(2k+3)}$ to both sides, we get

$\dfrac{1}{1\cdot3} + \dfrac{1}{3\cdot5} + \dfrac{1}{5\cdot7} + \cdots + \dfrac{1}{(2k-1)(2k+1)} + \dfrac{1}{(2k+1)(2k+3)}$

$$= \frac{k}{2k+1} + \frac{1}{(2k+1)(2k+3)}$$

$$= \frac{k(2k+3)}{(2k+1)(2k+3)} + \frac{1}{(2k+1)(2k+3)}$$

$$= \frac{k(2k+3)+1}{(2k+1)(2k+3)}$$

$$= \frac{2k^2+3k+1}{(2k+1)(2k+3)}$$

$$= \frac{(2k+1)(k+1)}{(2k+1)(2k+3)}$$

$$= \frac{k+1}{2k+3}$$

We have shown that if P_k is true, then P_{k+1} is true.

Conclusion: P_n is true for positive integers n. $(8-2)$

38. $(x-2i)^6 = [x+(-2i)]^6 = \displaystyle\sum_{k=0}^{6}\binom{6}{k}x^{6-k}(-2i)^k$

$= \binom{6}{0}x^6 + \binom{6}{1}x^5(-2i)^1 + \binom{6}{2}x^4(-2i)^2 + \binom{6}{3}x^3(-2i)^3$

$+ \binom{6}{4}x^2(-2i)^4 + \binom{6}{5}x(-2i)^5 + \binom{6}{6}(-2i)^6$

$= x^6 - 12ix^5 - 60x^4 + 160ix^3 + 240x^2 - 192ix - 64$

$(8-4)$

39. Let $P(x, y)$ be a point on the parabola.

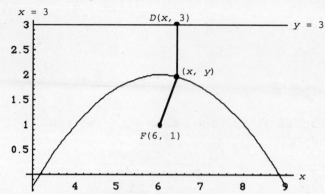

Then by the definition of the parabola, the distance from $P(x, y)$ to the focus $F(6, 1)$ must equal the perpendicular distance from P to the directrix at $D(x, 3)$. Applying the distance formula, we have

$$d(P, F) = d(P, D)$$
$$\sqrt{(x - 6)^2 + (y - 1)^2} = \sqrt{(x - x)^2 + (y - 3)^2}$$
$$(x - 6)^2 + (y - 1)^2 = (x - x)^2 + (y - 3)^2$$
$$x^2 - 12x + 36 + y^2 - 2y + 1 = 0 + y^2 - 6y + 9$$
$$x^2 - 12x + 4y + 28 = 0$$

$(9\text{-}1)$

40. Make a rough sketch of the ellipse, locate focus and x intercepts, then determine y intercepts using the special triangle relationship.

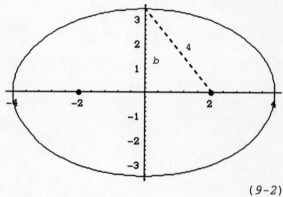

$a = 4$
$b^2 = a^2 - c^2$
$\quad = 4^2 - 2^2$
$\quad = 12$
$\quad b = \sqrt{12} = 2\sqrt{3}$
y intercepts: $\pm 2\sqrt{3}$

$(9\text{-}2)$

41. Sketch the asymptote rectangle, label known parts, and use the Pythagorean Theorem.

$a = 3 \qquad c = 5$
$b^2 = c^2 - a^2$
$\quad = 5^2 - 3^2$
$\quad = 16$
$\quad b = 4$

Thus, the length of the conjugate axis is $2b = 8$.

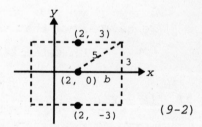

$(9\text{-}2)$

42. We can select any three of the seven points to use as vertices of the triangle. Order is not important.

$$C_{7,3} = \frac{7!}{3!(7 - 3)!} = 35 \text{ triangles}$$

$(8\text{-}5)$

43. P_n: $2^n < n!$ $n \geq 4$ (for n an integer, this is identical with the condition $n > 3$)

Part 1: Show that P_4 is true
P_4: $2^4 < 4!$
$\quad \quad 16 < 24$ True.

Part 2: Show that if P_k is true, then P_{k+1} is true.
Write out P_k and P_{k+1}:

$\quad P_k$: $2^k < k!$

$\quad P_{k+1}$: $2^{k+1} < (k + 1)!$

We start with P_k:

$\quad 2^k < k!$

Multiplying both sides by 2: $2 \cdot 2^k < 2 \cdot k!$

Now, $k > 3$, thus $1 < k$, $2 < k + 1$, hence $2 \cdot k! < (k + 1)k!$

Therefore, $2^{k+1} = 2 \cdot 2^k < 2 \cdot k! < (k + 1)k! = (k + 1)!$

$\qquad\qquad 2^{k+1} < (k + 1)!$

Thus, if P_k is true, then P_{k+1} is true.

Conclusion: P_n is true for all $n \geq 4$. $\hfill (8\text{-}2)$

44. To prove $a_n = b_n$ for all positive integers n, write:

P_n: $a_n = b_n$

Proof: Part 1: Show P_1 is true.

P_1: $a_1 = b_1$. $a_1 = 3$. $b_1 = 2^1 + 1 = 3$.

Thus, $a_1 = b_1$

Part 2: Show that if P_k is true, then P_{k+1} is true.
Write out P_k and P_{k+1}.

$\quad P_k$: $a_k = b_k$

$\quad P_{k+1}$: $a_{k+1} = b_{k+1}$

We start with P_k:

$\quad a_k = b_k$

Now, $a_{k+1} = 2a_{k+1-1} - 1 = 2a_k - 1 = 2b_k - 1 = 2(2^k + 1) - 1$

$\qquad\qquad = 2 \cdot 2^k + 2 - 1 = 2^1 \cdot 2^k + 1 = 2^{k+1} + 1 = b_{k+1}$

Therefore, $a_{k+1} = b_{k+1}$

Thus, if P_k is true, then P_{k+1} is true.

Conclusion: P_n is true for all positive integers n. Hence, $\{a_n\} = \{b_n\}$. $\hfill (8\text{-}2)$

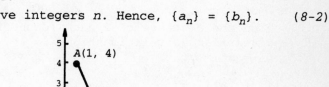

45. Make a rough sketch of the situation.

We are given $d(P, A) = 3d(P, B)$.
Applying the distance formula,
we have

$$\sqrt{(x - 1)^2 + (y - 4)^2} = 3\sqrt{(x - x)^2 + (y - 0)^2}$$
$$(x - 1)^2 + (y - 4)^2 = 9[(x - x)^2 + (y - 0)^2]$$
$$x^2 - 2x + 1 + y^2 - 8y + 16 = 9y^2$$
$$x^2 - 2x - 8y^2 - 8y + 17 = 0$$

is the equation of the curve. Completing the square relative to x and y, we have

$$x^2 - 2x + 1 - 8\left(y^2 + y + \frac{1}{4}\right) - 1 + 2 + 17 = 0$$

$$(x - 1)^2 - 8\left(y + \frac{1}{2}\right)^2 = -18$$

$$\frac{\left(y + \frac{1}{2}\right)^2}{\frac{18}{8}} - \frac{(x - 1)^2}{18} = 1$$

This is the equation of a hyperbola. $\hfill (9\text{-}3)$

46. We are asked for the sum of an infinite geometric series.

$a = \$2,000,000(0.75)$

$r = 0.75 \quad |r| \le 1$, so the series has a sum,

$$S_\infty = \frac{a_1}{1 - r}$$

$$= \frac{\$2,000,000(0.75)}{1 - 0.75}$$

$$= \frac{\$2,000,000(0.75)}{0.25}$$

$$= \$6,000,000$$

(8-3)

47.

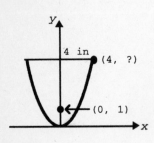

From the figure, we see that the parabola can be positioned so that it is opening up with axis the y axis, hence it has an equation of the form $x^2 = 4ay$. Since the focus is at $(0, a) = (0, 1)$, $a = 1$ and the equation of the parabola is $x^2 = 4y$.

Since the depth represents the y coordinate y_1 of a point on the parabola with $x = 4$, we have

$4^2 = 4y_1$

$y_1 =$ depth $= 4$ in.

(9-1)

48.

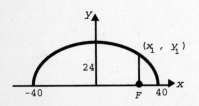

From the figure, we see that the x and y intercepts of the ellipse must be 40 and 24 respectively. Hence the equation of the ellipse must be

$$\frac{x^2}{(40)^2} + \frac{y^2}{(24)^2} = 1$$

or $\frac{x^2}{1,600} + \frac{y^2}{576} = 1$

To find the distance c of each focus from the center of the arch, we use the special triangle relationship. $a = 40$, $b = 24$

$c^2 = a^2 - b^2$

$ = 40^2 - 24^2$

$c^2 = 1024$

$c = 32$ ft.

To find the height of the arch above each focus, we need the y coordinate y_1 of the point P whose x coordinate is 32. Since P is on the ellipse, we have

$$\frac{32^2}{1,600} + \frac{y_1^2}{576} = 1$$

$$\frac{1,024}{1,600} + \frac{y_1^2}{576} = 1$$

$$0.64 + \frac{y_1^2}{576} = 1$$

$$\frac{y_1^2}{576} = 0.36$$

$$y_1^2 = 576(0.36)$$

$$y_1^2 = 207.36$$

$$y_1 = 14.4 \text{ ft.}$$

(9-2)